Beautiful China: 70 Years Since 1949 and 70 People's Views on Eco-civilization Construction

Jiahua Pan · Shiji Gao · Qingrui Li · Jinnan Wang ·
Dekai Wu · Chengliang Huang
Editors

Beautiful China: 70 Years Since 1949 and 70 People's Views on Eco-civilization Construction

CEPG

Springer

Editors
Jiahua Pan
Chinese Academy of Social Sciences
Beijing, China

Shiji Gao
Development Research Center of the State Council
Beijing, China

Qingrui Li
China Ecological Civilization Research and Promotion Association (CECRPA)
Beijing, China

Jinnan Wang
Chinese Academy of Environmental Planning
Beijing, China

Dekai Wu
China Environment Publishing Group Co., Ltd.
Beijing, China

Chengliang Huang
Chinese Academy of Social Sciences
Beijing, China

ISBN 978-981-33-6741-8 ISBN 978-981-33-6742-5 (eBook)
https://doi.org/10.1007/978-981-33-6742-5

Jointly published with China Environment Publishing Group Co., Ltd.
The print edition is not for sale in China (Mainland). Customers from China (Mainland) please order the print book from: China Environment Publishing Group Co., Ltd.

This Springer imprint is published by the registered company Springer Nature Singapore Pte Ltd.
The registered company address is: 152 Beach Road, #21-01/04 Gateway East, Singapore 189721, Singapore

Foreword I

Continuously Promote the Formation of a New Pattern of Modernization and Harmonious Coexistence of Mankind and Nature

This year marks the 70th anniversary of the founding of the People's Republic of China. To celebrate the anniversary, the Chinese Academy of Social Sciences (CASS) Think Tank for Eco-Civilization Studies, Research Institute for Resources and Environment Policies affiliated to the Development Research Center of the State Council, the Chinese Ecological Civilization Research and Promotion Association, the Environmental Planning Institute of the Ministry of Ecology and Environment, and the China Environmental Publishing Group co-sponsored and completed the compilation of *Beautiful China 1949–2019: 70 People's Views on Ecological Civilization*, which I believe is a meaningful thing to do.

Since the 18th National Congress of the Communist Party of China (CPC), the Central Committee of the CPC with Comrade Xi Jinping at its core has attached great importance to the ecological civilization. Comrade Xi Jinping has emphasized the importance of focusing on changes in the main contradictions in the Chinese society in the new era, meeting the Chinese people's ever-growing needs for a beautiful environment, sticking to the people-centered development philosophy, promoting harmonic coexistence between mankind and nature in a modern Chinese society, promoting an ecological civilization, building a community with a shared future for mankind, resolutely winning the battle against environmental pollution, building in an all-round way a great modern socialist country that is prosperous, strong, democratic, culturally advanced, harmonious, and beautiful. He has given a series of important instructions and statements and proposed scientific concepts which formed Xi Jinping Thought on Ecological Civilization. All this has catalyzed a historic change toward sustainability in the Chinese society. Reviewing what China has done in environment conservation in the past 70 years is of great significance to China's effort to open up a new chapter at a new historical juncture in accordance with Xi Jinping Thought on Ecological Civilization, build a modern ecological civilization in which man and nature coexist harmoniously, and realize the Chinese dream of national rejuvenation.

In the early days of the People's Republic of China, there was a lot of work to be done. In the summer of 1950, after the catastrophic floods in Anhui and Henan, Comrade Mao Zedong said, "We must strengthen flood control along the Huai River." After that, China began to carry out water conservancy projects such as flood control, irrigation, river dredging, and construction of canals in a well-planned and step-by-step manner. As a systems thinker and visionary leader who always kept the people's interest in mind, Comrade Mao Zedong believed that the country should take radical measures to eliminate flood risk, prioritize water conservancy, flood control and drought relief, and support flood diversion projects along the Yangtze River, Jingjiang River, Yellow River, and Haihe River. In 1956, Comrade Mao Zedong issued a call for actions to increase the forest area in China. At the same time, he pointed out that afforestation was not something that could be achieved overnight, and it may take a long time to reach the forest growth goal. In June 1972, the Chinese government sent a delegation to attend the first UN Conference on the Human Environment in Stockholm, attracting worldwide attention. In 1973, the State Council convened the first national environmental protection conference in China, and reviewed and approved the Provisions on the Protection and Improvement of the Environment (Draft), listing environmental protection as one of the functions of governments at all levels. This conference is deemed as the first milestone in the People's Republic of China's environment protection history.

After China adopted the reform and opening-up policy, environment protection legalization in the country began to pick up the pace. Comrade Deng Xiaoping clearly emphasized the need to strengthen the rule of law in the socialist society and concentrate efforts to formulate a number of important laws, including the Forest Law, the Grassland Law, the Environmental Protection Law, and other laws on forestry, forestation, and environment protection. In February 1979, the Sixth Session of the Standing Committee of the Fifth National People's Congress passed the Forest Law of the People's Republic of China (Trial Implementation) and set March 12 each year as the Arbor Day; in September, the 11th meeting of the Standing Committee of the Fifth National People's Congress passed China's first basic law on environmental protection—Environmental Protection Law of the People's Republic of China (Trial Implementation). In December 1981, the Fourth Session of the Fifth National People's Congress passed the Resolution on Carrying out the National Voluntary Tree Planting Campaign, which stipulated that planting trees is an obligation of Chinese citizens. In 1983, China held the second National Environmental Protection Conference, officially making environmental protection a basic national policy. After that, China formulated and promulgated more environmental protection laws and regulations, including the Water Pollution Prevention and Control Law, the Air Pollution Prevention and Control Law, and the Water Law. In December 1989, the Environmental Protection Law was formally promulgated and became the basic law of environmental protection in China. By the early 1990s, China had formed a relatively complete environmental protection legal system.

The 1990s saw the rise in global awareness of sustainable development and environmental protection legislation in China entering a new stage. In June 1992, at the UN Conference on Environment and Development held in Rio de Janeiro, the UN adopted the Rio Declaration on Environment and Development, Agenda 21 and other documents and China promised to set out on the path toward sustainable development. In March 1994, China released China's Agenda 21: White Paper on China's Population, Environment and Development in the 21st Century, specifying that changing the development strategy and setting on the path toward sustainable development is the right choice for China when it is seeking ways to accelerate economic growth and solve environmental issues. In 1997, at the 15th National Congress of the Communist Party of China, sustainable development was first written into the report of the conference as a strategic choice of the country. During this period, the Central Committee of the Communist Party of China with Comrade Jiang Zemin as the core focused on maintaining ecological security proposed the strategy of returning farmland to forests and closing mountains for afforestation and issued a call for actions to make the country beautiful again.

The new century has seen China's social development entering a new phase. The Central Committee of the Communist Party of China led by General Secretary Hu Jintao advocated the "scientific development concept," which is a summation of the "comprehensive, coordinated, and sustainable development," emphasized sustainable development and utilization of nature alongside harmonious coexistence between mankind and nature, regarded "harmonious coexistence between mankind and nature" as one of the basic characteristics of a harmonious socialist society, called for correct handling of the relationship between quantity/quality and speed/efficiency, promoted coordinated development of resources, environment, and economy with both short-term and long-term development goals in mind, and placed the goal of transforming into a sustainable society at a prominent place in the country's industrialization and modernization strategy. The proposal of "building an ecological civilization" was put forward for the first time at the 17th National Congress of the Communist Party of China in 2007, marking the formal establishment of the socialist ecological civilization concept, a new development concept of the Communist Party of China in governing and rejuvenating the country. The proposal of the concept was a major step in the advancement of theories, paths, and models of human civilization development.

The theoretical system of socialism with Chinese characteristics has entered a new era. The Central Committee of the Communist Party of China with Comrade Xi Jinping at its core is keeping a close eye on the changes in the main social contradictions in the new era and integrating ecological civilization into the overall plan for promoting all-round economic, political, cultural, social, and ecological progress and the Four-Pronged Comprehensive Strategy. It emphasizes that building an ecological civilization is of fundamental importance to China's sustainable development, continuously enriches the Marxist theory concerning the relationship between man and nature, puts the new development concept in practice, and deepens the reform of the ecological civilization system. The CPC Central Committee's determination to build an ecological civilization and the intensity and effectiveness of the efforts it has made for this purpose are unparalleled, opening up a new chapter in ecological conservation. A large number of environmental protection projects related to people's livelihoods, including air, water, and soil pollution prevention and control projects, have achieved historic results. The quality of natural ecosystems has continued to improve, and the scope and extent of ecological degradation have been reduced. Ecosystem services have been improved. China's efforts to protect and restore environment have been fruitful. The overall environmental conditions are improving. The Chinese people are enjoying benefits resulting from these actions. Urban and rural environments have become more livable, and the Chinese people's lives are getting better. The construction of China's ecological civilization has entered the fast lane.

China is a large country with a population of nearly 1.4 billion. It is particularly important for China to promote sustainable production and lifestyles, accelerate the establishment of an ecological civilization system that respects nature and green development and integrates resource conservation and environmental friendliness into mainstream production activities and lifestyles. At present, we should adhere to high moral standards, emphasize humanity, seek to understand the relationship between man and nature, transform the harmonious symbiosis between man and nature into interdependence and mutual love between people, and transform equality, justice, fraternity, and harmony in human societies into harmony between humans and nature. In fact, environmental degradation caused by human activities affects the health and life quality of human beings. Comrade Xi Jinping pointed out that building an ecological civilization is in everyone's interest, and as part of nature, everyone should participate in environment protection which can truly benefit us instead of taking the role of a bystander, outsider, or critic. It is necessary to create a culture that encourages society-wide participation in environmental protection and governance. We hope that the publication of this book will help further consolidate social consensus on jointly building an ecological civilization and strengthen the determination of the Chinese people to open up a new chapter in the historical journey of building an ecological civilization.

Beijing, China

Zongxing Chen
Vice Chairperson of the 11th National Committee of the Chinese People's Political Consultative Conference;
President, Chinese Ecological Civilization Research and Promotion Association

Foreword II

Build Academic and Discourse Systems of Ecological Civilization in Accordance with Xi Jinping Thought on Ecological Civilization

On May 17, 2016, in his important speech at the symposium on philosophy and social sciences, General Secretary Xi Jinping put forward the strategic task of "accelerating the development of philosophy and social sciences with Chinese characteristics" for the first time, calling for efforts to fully reflect Chinese characteristics in guiding ideology, discipline system, academic system, and discourse system. General Secretary Xi Jinping also expounded the three principles for accelerating the development of philosophy and social sciences with Chinese characteristics: (i) inheritance and nationality; (ii) originality and time relevance; and (iii) systematism and professionalism. He clarified a series of major theoretical and practical issues concerning the development philosophy and social sciences in China. The important speech delivered by Xi on May 17, 2016, is compiled into a programmatic document that shows the advantages of Marxism and guides the long-term development of philosophy and social sciences in the new era.

Ecological civilization is a field of study under philosophy and social sciences. Nowadays, the Chinese people have embarked on a historical journey to realize the great rejuvenation of the Chinese nation. The strategic position of building an ecological civilization is unprecedentedly high. The Communist Party of China (CPC) points out at its 19th National Congress that building an ecological civilization is a strategy crucial for the millennium to come. General Secretary Xi Jinping said at the National Environment Protection Conference that building an ecological civilization is of fundamental importance for the sustainable development of the country. Since the 18th National Congress of the CPC, General Secretary Xi Jinping has issued a series of important speeches, expositions, and instructions on ecological civilization, forming a system of science-based ecological civilization theories. Theoretical researchers in ecological civilization should follow Xi Jinping Thought on Ecological Civilization, pay more attention to the practical problems concerning China's efforts to build an ecological civilization in the new era, implement the people-centered development concept, and continue to enrich and expand Marxist people–nature relationship theory, and continuously advance the ecological civilization.

I. Ecological Research is a Crucial Part of the Philosophy and Social Science System with Chinese Characteristics

Ecology is an integral part of the broad and profound Marxist system of theories and knowledge. General Secretary Xi Jinping pointed this out clearly in his speech at a symposium on philosophy and social sciences, adding that ecology represents "an achievement in adapting Marxism to China's conditions and is the cultural manifestation of such effort" and that it is a major part as well as essential sub-field of "philosophy and social sciences with Chinese characteristics." The human–nature relationship has remained a key topic in the Marxist philosophy. Whether the relationship between human and nature is discordant or harmonious, strained or symbiotic, is a question that humanity needs to deal with. How the question is answered is one of the most important measures to distinguish Marxism from non-Marxism. First, Marxism sees nature as the basis for human survival. Marx points out: "Nature is man's inorganic body—nature, that is, in so far as it is not itself the human body. Man lives on nature—means that nature is his body, with which he must remain in continuous interchange if he is not to die." Second, nature is the mother of all life, and human and nature form a community of life. As long as human beings treat nature well, nature will bestow gifts on humans. Marx believes that the natural world can be divided into "nature-in-itself" and "humanized nature." Social production is the intermediary between human and nature. It not only continuously drives the transformation of nature-in-itself into humanized nature but also effectively establishes a harmonious unity between human and nature. Third, human beings can never triumph over nature, even with the aid of modern science and technology. Industrial civilization emphasizes that nature can be conquered by mankind and regards human beings as the central element of existence on the Earth, the rule maker for the planet, and the masters of the world. The real problem is that, while natural science and technology change social production methods and people's lifestyles, they also bring potential and uncontrollable risks. To some extent, the increase in the vulnerability of modern ecosystems is precisely due to people's eagerness for technological innovation and neglect of potential adverse effects in the long term. Engels points out: "All hitherto existing modes of production have aimed merely at achieving the most immediately and directly useful effect of labor. The further consequences, which only appear later on and become effective through gradual repetition and accumulation, were totally neglected." Fourth, contempt for dialectics cannot go unpunished. Humans must respect and protect nature and conform to the laws of nature. Marx said: "A human plan that is not based on the great laws of nature will only bring disaster." Only by following the laws of nature can humans effectively prevent detours in the development and utilization of nature. Damage caused by humans to nature will eventually and inevitably harm human beings themselves.

"Ecology" is the manifestation of inheritance and nationality of philosophy and social sciences with Chinese characteristics. Cultural soft power should be leveraged to build ecological civilization which is an aspect of the hard power of the country. In his speech at the symposium on philosophy and social sciences, General Secretary Xi Jinping also pointed out that "ecology" is the manifestation of the inheritance and nationality of philosophy and social sciences with Chinese characteristics. Confucianism, Taoism, and Buddhism were the three main philosophies of ancient China. Under their combined effect, the Chinese nation has formed its own unique cultural system, at the core of which were doctrines of *zhong yong*, harmony and tolerance. These doctrines advocate the adoption of an attitude of respect for nature, reflecting the unique values and worldviews of the Chinese people. Without full confidence in the Chinese culture, without a rich and prosperous culture, the Chinese nation will not be able to rejuvenate itself. General Secretary Xi Jinping pointed out that the rich philosophical thoughts, ethical values, humanist ethics, social teachings, and moral virtues that are a crucial part of traditional Chinese culture can provide inspirations for people who want to understand and transform the world, improve state governance and shape morals and values of a society. The wisdom of ancient Chinese people concerning ecology such as the unity of nature and man still enlightens us today. It inspires us to keep improve ourselves and achieve harmonious coexistence with nature. We should be confident in the reconstruction of contemporary ecological civilization theory and practice paradigms with the traditional values of Chinese culture.

We must promote the transformation of human ecology. Ecological research is not included in philosophy or social sciences either in China or other countries. The "ecological" elements in the western philosophy and social science system are scattered in fields such as ecological ethics, sustainable development, ecological economy, green, and low-carbon. Ecological research is still not an independent field of study. In China, ecological research is categorized under environment, natural resources, and ecological protection, and its status as an independent discipline under philosophy and social sciences is often overlooked. Even though there are some ecological studies conducted from perspectives of philosophy and social sciences, they are in the fragmentation pattern of Western Humanities and social sciences. Among the disciplines of higher education and scientific research institutions in China, "ecology" is categorized under natural science instead of humanities and social sciences within the framework of the philosophy and social science system. Ethics can be used as a tool to govern society, and all societies should value the power of ethics. Environmental ethics advocates the adoption of an attitude of respect for life and nature, studies the entities and processes of life and nature, and calls for actions to care for nature, future generations, and all lives. Its emergence reflects the improvement of human beings' moral realm. General Secretary Xi Jinping pointed out that the country should protect the environment like one protects his eyes and treat the environment like one treats his life. This makes us feel that this planet is an ecological home that "fits our natural sensibility." To build an ecological civilization, we need to build a culture that encourages us to continuously deepen our love for nature.

II. Xi Jinping Thought on Ecological Civilization is an Important Part of Xi Jinping Thought on Socialism with Chinese Characteristics for the New Era and a National Strategy and Mission

General Secretary Xi Jinping has always attached great importance to work concerning the ecological environment. Harmonic coexistence of man and nature is an unremitting pursuit of General Secretary Xi Jinping. When working in Zhengding, Xiamen, Ningde, Fujian, Zhejiang, Shanghai, and other cities, General Secretary Xi Jinping prioritized work concerning the ecological environment although at that time economic development was considered the most important task by governments at all levels. General Secretary Xi Jinping has published a lot of documents and put forward a lot of profound, thought-provoking expositions, instructions, proposals, and visions on ecological civilization in a systematic manner. Just as Engels pointed out in *Dialectics of Nature-Introduction*, "It was the greatest progressive revolution that mankind has so far experienced, a time which called for giants and produced giants - giants in power of thought, passion, and character, in universality and learning." Xi Jinping Thought on Ecological Civilization and harmonic coexistence between man and nature are consistent, inspiring, and philosophical.

Since the 18th National Congress of the CPC, guided by Xi Jinping Thought on Ecological Civilization, China has made great strides in promoting the development of ecological civilization. Since the 18th National Congress of the CPC, under leadership of the Party Central Committee with Comrade Xi Jinping at its core, the CPC and all ethnic groups in the country have given full play to the advantages of China's socialist system to concentrate efforts on major issues and leveraged resources accumulated since the adoption of the reform and opening-up policy to advance the development of ecological civilization, solve ecological and environmental problems, and step up pollution prevention. A series of fundamental, pioneering, and long-term initiatives have been launched to strengthen environment protection and push the development of ecological civilization to a new level. "Ecological civilization" is an expression with Chinese characteristics as well as distinctive characteristics of this era. It encapsulates wisdom of the Chinese people. Regarding ecological civilization and the Chinese dream, General Secretary Xi Jinping pointed out: "Promoting the development of ecological civilization and building a beautiful China is an important part of the great rejuvenation of the Chinese nation which is the greatest dream of the Chinese people." General Secretary Xi Jinping has also emphasized that development of ecological civilization is an important part of the Five-Sphere Integrated Plan and the Four-Pronged Comprehensive Strategy. Xi's theory about the relationship between ecological civilization and the Chinese dream emphasizes the strategic position of ecological civilization and answers the question why building an ecological civilization is to build a prosperous, democratic, civilized, harmonious, beautiful modern socialist country; his theory about the relationship between ecological civilization and the Five-Sphere Integrated Plan highlights the strategic position of ecological civilization and explains what socialism is and comprehensive development as an inherent attribute of socialism; Xi's theory about ecological civilization and the Four-Pronged Comprehensive Strategy answers questions concerning how to build ecological civilization and introduces strategic measures, methodology, and practice adopted to promote the development of ecological civilization. It must be noted that building ecological civilization is an important action plan of China's ruling party, the CPC.

III. Community of Shared Future for Mankind as Summation of Xi Jinping Thought on Ecological Civilization in Global Context

The community with a shared future is a vision of global governance proposed by General Secretary Xi Jinping. Xi is a compassionate, global-minded Chinese leader. His philosophy on ecological civilization reflects the Chinese wisdom. Xi calls for global actions to build a global ecological civilization by promoting the development of a community with a shared future and common interests for mankind. China is an increasingly important participant, contributor, and leader in this process.

China is an increasingly important participant in the development of a global ecological civilization. Since the 18th National Congress of the Communist Party of China, the development of China's ecological civilization has become an important driving force for the development of a community with a shared future for mankind. China has actively undertaken its due international responsibilities, made enormous inroads in addressing climate change, participated in international exchanges and cooperation in the fields of ecological civilization, energy and resource security, and disaster relief. China's energy consumption and carbon dioxide emissions per unit of GDP have dropped significantly. China has announced the establishment of an RMB 20 billion South–South cooperation fund to help other developing countries tackle climate change. International cooperation in fields such as clean energy, disaster prevention and relief, environment protection, adaption of agriculture to climate change, and low-carbon smart cities continues to advance. China has also strengthened the protection of wild animals and their habitats and severely cracked down on the illegal trade of wild animals and ivory and other animal products. It attaches great importance to and has made significant inroads in desertification prevention and control, providing a Chinese model for the international community in managing the ecological environment. From General Secretary Xi Jinping's attendance at the Paris Conference on Climate Change where the Paris Agreement was signed to the Katowice Global Climate Change Conference in Poland, from the United Nations' 2030 Agenda for Sustainable Development to the G20 Hangzhou Summit, China has played an important role in advancing the Paris Agreement, helping developing countries tackle climate change, and pursuing the 2030 Agenda and the Sustainable Development Goals (SDGs).

China is an increasingly important contributor to the development of a global ecological civilization. Since the onset of the twenty-first century, environmental problems have arisen on a global scale never seen before and have seriously affected the development of the increasingly interconnected global village. With a population of nearly 1.4 billion, China accounts for one-fifth of the world's population and one-fourth of the land area of Asia. China's solutions to its own environmental problems are a great contribution to the world. For example, General Secretary Xi Jinping has repeatedly emphasized the need to protect the Lancang River, which is the seventh longest river in the world and the third longest river in Asia. It flows into Ho Chi Minh City in Vietnam before entering the South China Sea. This is the first time that the Communist Party of China, as the world's largest ruling party, provided a solution to global environmental issues and a sustainable development paradigm for developing countries.

China is an increasingly important leader in the development of a global ecological civilization. China, striving for national rejuvenation, is leading global ecological civilization actions. The twenty-first century saw human beings entering a new stage of civilization, that is, ecological civilization. This is an inevitable trend, unaffected by human will. Just as General Secretary Xi Jinping pointed out that human beings have experienced hunter-gatherer civilization, agricultural civilization, and industrial civilization. Ecological civilization is a stage of higher level after industrial civilization and is a precondition for the harmonious coexistence between man and nature. Ecological civilization conforms to the law of evolution of human civilization. Based on the law of evolution of human civilization, mankind is increasingly aware that whether it is an industrialized country or a developing country that has not yet completed industrialization, it needs to abandon ethical values, production methods, consumption methods, and institutional mechanisms of industrial civilization and adopt those of ecological civilization. In reality, western developed countries traditionally have a strong bias toward those who are ideologically different from them. The great rejuvenation of the Chinese nation featuring a universally applicable and the most inclusive value system and an international discourse system is to be accepted and recognized by the world and help China take the leadership role in global action to build a community with a shared future for mankind. Ecological civilization undoubtedly has such systems. The concept of ecological civilization and its development path advocated and practiced by China encourage different countries to steer away from traditional industrial civilization and promote the development of ecological civilization.

IV. Promote the Transformation of Human Civilization Paradigm Under the Guidance of Xi Jinping Thought on Ecological Civilization

It should be noted that we lack a complete, science-based theoretical system of ecological civilization, both domestically and internationally, especially a discourse system widely recognized and consciously adopted by countries around the world to effectively respond to foreseeable and unforeseeable difficulties, risks and trials. China is working hard to eliminate the cumulative impacts of human activities on the environment over the past several decades. It has stepped up pollution prevention and control and actively promoted green technologies, industrial practices, green systems, and ecological culture that best fit the community of shared future for mankind. It is striving to integrate its plan to build an ecological civilization with the Chinese dream, i.e., the great rejuvenation of the Chinese nation, with the efforts to build a prosperous, democratic, civilized, harmonious, and beautiful socialist modern country, and with the global action to build a community with a shared future for mankind under the guidance of Xi Jinping Thought on Ecological Civilization. It is important to build academic and discourse systems of ecological civilization for the new era and attach importance to theoretical and strategic perspectives of the systems.

First, we need to consider the development of discipline and academic systems. We should promote the development of disciplines, theories, and concepts of philosophy and social sciences and introduce new concepts, new paradigms, and new expressions concerning socialism with Chinese characteristics that are easy to understand and accept by the international community for best practices and theoretical innovations. It is necessary to focus on issues of concern to the international community, actively participate in the initiation and formulation of international rules, standards, and laws, enhance China's international influence, and increase China's involvement in international rule-making. Despite the overall improvement of the environment in Western Europe, North America, and Japan, which are the first group of countries and regions to achieve industrialization, China now deeply understands the urgency of building a community with a shared future for mankind. It should be noted that industrial civilization will still in its heyday in the foreseeable future, and developed countries in the West will still occupy dominant positions in fields relating to ecological technology, ecological industry, environmental protection, and sustainable development. Figuring out how to build an ecological civilization-related discourse system with Chinese characteristics and how to promote the ecological civilization concept and model in parallel with the global action to build a community with a shared future for mankind is the mission and responsibility of the ecological civilization think tank. It is necessary to strengthen the study of academic, discourse, methodology, industry, and environmental governance systems concerning ecological civilization.

Second, we should strengthen strategic, forward-looking, realistic, and countermeasure research. The Chinese government has set several related goals for the near future. One of the goals is to build a moderately prosperous society in an all-round way by 2022. What should the ecological civilization be like? What are the challenges? Will there be opportunities? Another important goal is the Fourteenth Five-Year Plan for Economic and Social Development (2021–2025). What are the major problems and challenges faced by China in the development of ecological civilization during 2012–2025? It should be noted that the more we have achieved, the more types of challenges and the more difficulties we will encounter. The 19th National Congress of the CPC has established another two important goals. One is to achieve basic modernization in 2035, and the other is to achieve the second centenary goal—the great rejuvenation of the Chinese nation—and build a prosperous, strong, democratic, civilized, beautiful, harmonious modern socialist country by 2050. It is necessary to plan ahead for challenges, develop strategies, and set measurable goals concerning the development of ecological civilization before 2035 and 2050. Some forward-looking and strategic studies should be carried out to identify opportunities and challenges for green development before 2020, 2035, 2050, and other major strategic time points, and some strategic measures should be adopted to seize the opportunities and address the challenges.

Third, we should expand global horizons and jointly build a global ecological civilization. As repeatedly pointed out by General Secretary Xi Jinping, China is an increasingly important participant, contributor, and leader in the development of a global ecological civilization. It can be said that we have academic advantages and skill sets required to participate in the international dialogue on ecological civilization and lead the international discourse. However, we must also realize that, regarding the development of ecological civilization and the international discourse system itself, China's voice in the international arena is still very small. Ecological civilization think tanks should broaden their vision, focus on the goals of the Paris Agreement and the 2030 Agenda for Sustainable Human Development, take the initiative to go global, build an academic platform, continuously strengthen collaborative innovation, tell the international community stories about the development of ecological civilization in China, and play an important role in promoting green development on a global scale.

This year marks the 70th anniversary of the founding of the People's Republic of China. Although we did not use the term ecological civilization in the 1950s and 1960s, the Chinese government has made a series of efforts to explore ways to build an ecological civilization and drive continuous development of the concept of ecological civilization during the 70 years since the founding of People's Republic of China. The decades-long process of trial and error has left the Chinese people a precious legacy. After the founding of the People's Republic of China, Comrade Mao Zedong called for actions to strengthen flood control along the Huai River and the Yellow River, green the motherland within two hundred years to pursue Marxism. China's experience in developing ecological civilization in the past 70 years proves that the Chinese people have the confidence and ability to make their own country beautiful again, and the solutions and experience China provides are useful for the development of global ecological civilization and a community with a shared future for mankind.

We will continue to draw lessons from history and leverage the valuable legacy accumulated by China in the past 70 years to drive the development of ecological civilization in the new era. This time, the Think Tank for Ecological Civilization Studies affiliated to the Chinese Academy of Social Sciences, Research Institute for Resources and Environment Policies affiliated to the Development Research Center of the State Council, the Chinese Ecological Civilization Research and Promotion Association, the Environmental Planning Institute of the Ministry of Ecology and Environment, and the China Environmental Publishing Group published the book *Beautiful China 1949–2019: 70 People's Views on Ecological Civilization* to help people better and more comprehensively understand Xi Jinping Thought on Ecological Civilization, build a society-wide consensus on ecological civilization, implement the sustainable path, and pursue happiness. On behalf of the Chinese Academy of Social Sciences, I would like to express my gratitude to institutions that have participated in the project for their support.

Beijing, China

Fuzhan Xie
President and Secretary of the Leading
Party Members' Group of the Chinese Academy
of Social Sciences

Foreword III

Protect the Environment and Improve People's Livelihood—Celebration of the 70th Anniversary of the Founding of the People's Republic of China

History tells us that thriving ecosystems are a precondition for thriving human civilization, and the decline of ecosystems will lead to the decline of human civilization. Since the founding of the People's Republic of China 70 years ago, the CPC has been working hard to protect the Chinese people's entitlement to the pursuit of happiness, rejuvenate the Chinese nation, and promote environmental protection. After entering the new era, the CPC Central Committee with Comrade Xi Jinping as the core has been working to promote the development of ecological civilization, underline the perseverance in building beautiful China, strengthen environment protection, and significantly bolster the Chinese people's sense of gain, happiness, and security derived from the ecological environment.

I. Promote Environment Protection Practices that Truly Benefit People

Over the past 70 years, the most important responsibility of the CPC has always been to serve and benefit the people. The CPC regards environment protection as a major people's livelihood project. It has made continuous efforts to deepen the understanding of environment protection and drive the development of ecological civilization.

The strategic position of environment protection policies continues to improve. The first national environmental protection conference was held in 1973 when environmental protection was put on the central government's agenda. In the 1980s, environmental protection was made a basic national policy; in the 1990s, the sustainable development strategy was made a national strategy. Since entering the new century, the CPC has vigorously promoted the transformation of the country to a resource-saving and environment-friendly society. In the new era, the development of ecological civilization has been incorporated into the Five-Sphere Integrated Plan for development of socialism with Chinese characteristics. Building a beautiful country has become the goal of the CPC. The development of China's ecological civilization has entered the fast lane.

China has stepped up environmental pollution control. With the advancement of building an ecological civilization, China has continuously increased the intensity of environmental pollution control. In the 1970s, the Guanting Reservoir Pollution Control Project became China's first water pollution control project; in the 1980s, China launched a comprehensive prevention and control of industrial pollution initiative alongside technological transformation; in the 1990s, China launched the "33211" project to improve the environment of key cities, river basins, and territorial waters. After entering the new era, China has issued and implemented three major action plans for air, water, and soil pollution prevention and control. The quality of the ecological environment has continued to improve, and life satisfaction of the Chinese people has continued to rise.

China is making remarkable strides in environment protection. In 1956, China built the first national nature reserve. In 1978, it decided to launch the Three North Shelterbelt Project. In 1981, China initiated a nationwide voluntary tree planting program. Later, China adopted a series of key projects, including natural forest projection and conversion of farmland to forest or grassland, to continuously build the country's ecological security. In the new era, China insists on giving priority to protection and focusing on natural restoration. It has launched forest, river, lake and grassland ecological protection and restoration projects, adopted green practices, drawn up ecological red lines, and strengthened biodiversity protection. At present, 474 national nature reserves have been established nationwide, and the total area of land-based nature reserves of various types has reached more than 1.7 million square kilometers. China is becoming more and more beautiful.

Relevant laws and regulations are improving day by day. In 1978, the state should protect the environment and natural resources and prevent pollution and other public hazards which was written into China's Constitution. In 1979, the Environmental Protection Law of the People's Republic of China (Trial Implementation) was passed at the 11th meeting of the Standing Committee of the Fifth National People's Congress. In 1989, the 11th meeting of the Standing Committee of the Seventh National People's Congress passed the Environmental Protection Law of the People's Republic of China, marking the formation of a legal framework for environmental protection in China. In the new era, China has made remarkable strides in bolstering environmental legislation and law enforcement. It has formulated and revised the Environmental Protection Law, the Environmental Protection Tax Law, the Air, Water, and Soil Pollution Prevention and Control Law, and the Nuclear Safety Law. The Standing Committee of the National People's Congress, the Supreme People's Court, and the Supreme People's Procuratorate have defined environmental crimes, strictly implemented relevant laws in an unprecedented way, and achieved remarkable results, respectively.

Public participation in environment protection is increasing. China has been working to mobilize the whole society for environmental initiatives. Public environmental awareness and public participation in the development of ecological civilization are increasing. In 1985, China launched the June 5 Environment Day Campaign.

The *Communiqué on the Environment in China* was published for the first time in 1990, and China began to release official data on environmental quality in 2007. In the new era, the Chinese government encourages citizens to embrace simple, moderate, green, healthy and low-carbon ways of life and say no to waste, builds an environmental governance system involving the whole society, makes environmentalism part of the mainstream culture, and advocates for respect and a caring for the natural world.

II. Improve the Environment for the Common Good of the People

Over the past 70 years, the CPC has persisted in protecting the ecological environment as one of the ways to improve people's life quality. Especially since the 18th National Congress of the Communist Party of China, Comrade Xi Jinping has put forward a series of new concepts, new ideas, and new strategies concerning the development of ecological civilization which formed Xi Jinping Thought on Ecological Civilization, laying a theoretical foundation for historical changes in China's environment protection policy.

The CPC regards protecting the ecological environment as one of its responsibilities and missions. The ecological environment is a major political issue related to the party's mission and purpose, as well as a major social issue related to people's livelihood. The most fundamental reason for historic changes in China's environmental protection policy during the past 70 years, especially after the Eighteenth National Congress of the CPC, lies in the continuous strengthening of the party's leadership in the development of ecological civilization. The past experience tells us that to build an ecological civilization and protect the ecological environment, it is necessary for the CPC to strengthen party members' consciousness to maintain political integrity, think in big-picture terms, follow the leadership core, and keep in alignment, strengthen the authority and centralized leadership of the CPC Central Committee, and shoulder the political responsibility for the development of ecological civilization. The CPC should faithfully implement the decisions of the CPC Central Committee, assume the same responsibilities as the government for environment protection, and strive to build a politically strong, capable, hardworking, persistent, responsible, and dedicated environment protection team.

Environmental issues are now considered top priority for the Chinese people. Over the past 70 years, the Chinese government has shifted the focus of its work from poverty reduction to environment protection. Environmental quality is becoming an increasingly important indicator of happiness for the Chinese people. In response to the collective rise in consciousness about the health of the environment, the Chinese government should follow the people-centered development path and make ecological and environmental issues a top priority. Air, water and soil pollution and rural environmental problems have greatly affected the Chinese people's life quality. We must start by solving prominent ecological and environmental problems and create a good environment for people to live and work in.

The Chinese government places equal emphasis on development, people's livelihood, and environment protection. Experience in the past 70 years tells us that development is the key to solving all problems in China, and environmental problems must also be solved through development. Economic development must not come at the expense of the environment, and vice versa. Environmental benefits can be transformed into economic benefits. Improving the environment can lead to a rise in productivity. A good environment itself has infinite economic values. It can bring various benefits and drive sustainable economic and social development. To fundamentally solve environmental problems, we must implement new development concepts, promote the development and adoption of environmentally friendly spatial patterns, industrial structures, production methods, and lifestyles, reduce the environmental impact of human activities, and ensure there is enough time for ecosystems to regenerate.

China is mobilizing the whole society for environment protection. The environment is the fairest public good. Building an ecological civilization is a cause that affects and involves all people. Everyone should participate in environment protection which can truly benefit us instead of taking the role of a bystander, outsider, or critic. To mobilize the whole society for environment protection, we should continue to enhance public environmental awareness, set sustainability codes of ethics and conduct, build a society-wide environmental governance system, and encourage people to reduce energy consumption and pollution discharge and be part of a nationwide movement about protecting the environment.

III. Timely Respond to the Collective Rise in Consciousness About the Health of the Environment

According to the report of the 19th National Congress of the CPC, it is necessary to create more material and spiritual wealth to improve the people's life quality and provide more high-quality public goods concerning environment protection in response to the collective rise in consciousness about the health of the environment. At present, environmental quality in China has steadily improved, but may deteriorate in the case of a slight slack in environment protection efforts.

It must be noted that China still faces many environmental issues, including limited environmental capacity, fragile ecosystems, heavy pollution, and low ecological security. The unique geographical environment has exacerbated regional imbalances in environmental quality. For example, bad weather events caused by heavy pollution occur frequently in autumn and winter in northern China; some rivers, lakes, and sea areas are still heavily polluted; the country is still under massive pressure to control soil environmental risks and crack down on illegal transfer and dumping of solid waste and hazardous waste; ecological degradation and biodiversity decline have not been effectively reversed in some regions, and the contradiction between biodiversity conservation and economic development still exists. There are two main reasons for this situation: First, it is closely related to China's national conditions and development stage. China is still undergoing industrialization, urbanization, and agricultural modernization. Heavy industries are still the dominant contributors to GDP, and coal is still the leading fuel for power generation in China. Road transportation is still the primary transportation mode. The level of pollution is still increasing at a fast pace, putting ecosystems under huge pressure. Second, it is related to poor implementation of environmental policies. Due to inadequate understanding of the green development concept and weak capabilities, some local governments still place economic development above environmental protection on their agenda.

To effectively solve these problems, we must follow the guidance of Xi Jinping Thought on Socialism with Chinese Characteristics for a New Era and Xi Jinping Thought on Ecological Civilization, strengthen environment protection, and win the people's trust and bring real benefits to the people by improving environmental quality. We should put new development concepts into practice, and take the high-quality green development path, and prioritize environment protection; emphasize stability, pragmatism, overall planning, and comprehensive implementation of pollution prevention and control policies; follow the rules, scientifically plan relevant actions, adapt to local conditions, and create diverse and symbiotic ecosystems; strengthen the role of the central government's environment protection supervision agency, step up environment protection supervision and poverty alleviation of key areas, pay attention to the reasonable demands of enterprises, and solve environmental issues that the public most worry about.

To celebrate the 70th anniversary of the founding of the People's Republic of China and summarize the practices, great achievements and valuable experience of China in environment protection and development of ecological civilization in the 70 years, the Think Tank for Ecological Civilization Studies affiliated to the Chinese Academy of Social Sciences, Research Institute for Resources and Environment Policies affiliated to the Development Research Center of the State Council, the Chinese Ecological Civilization Research and Promotion Association, the Environmental Planning Institute of the Ministry of Ecology and Environment, and the China Environmental Publishing Group have compiled *Beautiful China 1949–2019: 70 People's Views on Ecological Civilization* which I believe is a meaningful thing to do. On behalf of the Ministry of Ecology and Environment, I would like to express my respect and heartfelt gratitude to people who have cared about, supported, and participated in environmental protection and the development of ecological civilization.

Beijing, China

Ganjie Li
Minister of Ecology and Environment
and Secretary of the Ministry's Leading Party
Members' Group

Foreword IV

Build a Solid Foundation for Ecological Civilization and Take the Green Development Path

Since the 18th National Congress of the Communist Party of China (CPC), the CPC Central Committee with Comrade Xi Jinping as its core has summarized the laws of human civilization development, created a long-term vision for, and taken a holistic approach to the development of ecological civilization in China. In the past six years, more and more Chinese people have realized "lucid waters and lush mountains are invaluable assets." The development of the top-level design and institutional system of ecological civilization has been accelerated. China has stepped up pollution control and environment protection and restoration, made remarkable strides in green development, and significantly improved environmental quality.

In particular, Xi Jinping Thought on Ecological Civilization released at the 2018 National Conference on Eco-environmental Protection is a landmark achievement of the CPC in theoretical innovation. It is a summation of ecological concepts, thoughts, and strategies put forward by General Secretary Xi Jinping since the 18th National Congress of the CPC. It provides fundamental guidelines for the development of ecological civilization and modernization of harmonious coexistence between man and nature in the new era. It is an important part of Xi Jinping Thought on Socialism with Chinese Characteristics for a New Era. Xi Jinping Thought on Ecological Civilization provides guidelines and principles for China's efforts to make the country beautiful again and build an ecological civilization. We should accelerate the reform of the ecological civilization system under the guidance of Xi Jinping Thought on Ecological Civilization, actively drive the development of ecological civilization, promote green production methods and lifestyles, and place equal emphasis on development, people's livelihood, and environment protection.

To take the path of green development and modernization, we must thoroughly understand and consciously follow the guidance of Xi Jinping Thought on Ecological Civilization and follow the Eight Principles, that is, (i) thriving ecosystems are a precondition for thriving human civilization; (ii) we must promote harmonious coexistence between man and nature; (iii) "lucid waters and lush mountains are invaluable assets"; (iv) a good environment is the fairest public good; (v) mountains, rivers, forests, farmlands, lakes, and grasslands are interconnected ecosystems; (vi) we should adopt the strictest institutional and legal systems to protect the environment; (vii) we should mobilize the whole society for the beautiful China campaign; and (viii) as a major global power, China should shoulder corresponding responsibilities for the development of the global ecological civilization.

Taking the path of green development means that we should strengthen pollution prevention and control, faithfully perform tasks relating to the development of ecological civilization, and comprehensively promote sustainable development.

First, we must make environment a top priority and take the path of green development. We should follow the guidance of Xi Jinping Thought on Ecological Civilization, properly handle the relationship between economic development and environmental protection, and establish the concept that protecting the environment means protecting productivity and improving the environment means developing productivity. Green development and ecological civilization development are long-term causes. We should keep in mind that the environment is a key factor that impacts people's life quality and happiness. We should make the environment a top priority, be far-sighted, and plan the future with wisdom. We owe future generations a beautiful world. We must increase the public's knowledge of green development and ecological civilization, promote environmental awareness, create a strong sustainability culture, and effectively raise people's sense of personal responsibility concerning environmental issues.

Second, green development practices should be embedded into every economic and social process. To ensure the sustainable development of the Chinese nation, we must start from the source, promote the development and adoption of environmentally friendly spatial patterns, industrial structures, production methods, and lifestyles, solve resource and environmental constraints, and promote high-quality development. We must incorporate green development concepts and practices such as ecology, low-carbon, energy-saving, and emission reduction into the whole process of development, especially in the development of products, services, and industries, highlight green creativity, green innovation and design, green manufacturing and production, green procurement and logistics, green services and sales, and green consumption and recycling, promote the greening of the entire economy, and ultimately realize the sustainable development of the whole society.

Third, we must accelerate the development of a green modern industrial system. To take the path of green development with the environment as a top priority, it is necessary to build a green industrial system on the basis of "industrialization of environment protection and ecologicalization of industries" and leverage the green industrial system to promote green development with the environment as a top priority. The current focus is on optimizing the economic structure and energy structure, promoting resource conservation and recycling on all fronts, fostering energy-saving and environmental-friendly industries, clean production industries, and clean energy industries, strengthening environmental supervision to force companies to improve management and accelerate innovation, driving industrial transformation, promoting high-quality development, and realizing resource decoupling and impact decoupling.

We should pay attention to key tasks concerning environment-first green development of industries, speed up technological advancement and innovation, focus on the development of strategic emerging industries with high added value, high technology content, strong competitiveness and extendable industrial value chains, and vigorously cultivate new kinetic energy, new business formats, and new economy, and promote green and sustainable economic development by supporting reform and innovation.

Fourth, we must continue to improve the ecological civilization system and the supporting policy system. To advance the development of ecological civilization, in addition to talent, technology, and capital, we also need a complete supporting policy system. The policy system includes not only macro-level principled regulations but also restrictions. We need the strictest system and the strictest set of rules to make sure that we stick to the path of environment-first green development. It is necessary to formulate and implement a comprehensive system of policies and laws that are conducive to the development of a modern economic system in accordance with the principles of ecological civilization and green development, strengthen punishment and accountability, and step up awareness raising. We must create a strong policy system and legal environment for promoting the development of ecological civilization.

Fifth, we must deepen policy and theoretical research to support green development and ecological civilization development. As the world's most populous country and the world's second-largest economy, China is advancing green development and modernization. It is unprecedented in modern history, and the problems and difficulties facing China are also unparalleled. It is urgent to summarize and learn from the past experience in human civilization development and then, on the basis of this, find innovative solutions to form a green development theoretical system that is in line with China's national conditions, highlights Chinese characteristics, and is based on China's experience. We should create a good atmosphere, focus on the requirements of green development, seek solutions from natural sciences, engineering sciences, economic sciences, social sciences, law, and other disciplines, deepen policy and theoretical research on green development and ecological civilization, develop systems and policies that promote green development and ecological civilization development, and deepen relevant theoretical analysis. China's theoretical innovation and successful practices in the field of green development and ecological civilization will make a great contribution to green modernization of other developing countries as well as the development of ecological civilization and sustainable development of the whole world.

China is making a continuous effort to advance ecological civilization development and green development. Over the past 70 years, the country has made great strides in building a prosperous, democratic, civilized, harmonious, and beautiful modern country. Today, it has finally stood at a new starting point and entered a new stage of development. Looking back on the great achievements of China in building ecological civilization in the past 70 years, and reflecting on the new mission of green development in the new era, we will find that only by following the guidance of Xi Jinping Thought on Ecological Civilization, further consolidating the consensus of the whole society, and accelerating the development of ecological civilization and green development can we build a modern society where man and nature coexist harmoniously and advance the beautiful China campaign.

Beijing, China

Jiantang Ma
Secretary of the Leading Party Members' Group of the Development Research Center of the State Council

Preface

To celebrate the 70th anniversary of the founding of the People's Republic of China, summarize the practices, great achievements and valuable experience of China in environment protection and development of ecological civilization in the 70 years, promote Xi Jinping Thought on Ecological Civilization, and drive the development of ecological civilization in China, the Think Tank for Ecological Civilization Studies affiliated to the Chinese Academy of Social Sciences, Research Institute for Resources and Environment Policies affiliated to the Development Research Center of the State Council, the Chinese Ecological Civilization Research and Promotion Association, the Environmental Planning Institute of the Ministry of Ecology and Environment, and the China Environmental Publishing Group have compiled *Beautiful China 1949–2019: 70 People's Views on Ecological Civilization.*

As of the time of publication, 75 manuscripts from 75 people specializing in the development of ecological civilization policies and theories were selected from the recommended articles. Among them are 15 articles by party and government leaders at or above the deputy ministerial level, and 60 articles by experts and scholars. Among the experts and scholars are bureau-level academic officials, academicians, university presidents, professors participating in the Changjiang Scholars Program, and post-80s scholars. The content covers, among others, Xi Jinping Thought on Ecological Civilization, ecological culture, industrial economy, legal system, and ecological security. The editorial board of the book also classified the manuscripts accord to their content. A comprehensive review of all the manuscripts reflects that the authors are committed to promoting continuous innovation and development of ecological civilization theories. The manuscripts were written based on Marxism–Leninism, Mao Zedong Thought, Deng Xiaoping Theory, Three Represents, Scientific Outlook on Development, and Xi Jinping Thought on Socialism with Chinese Characteristics for a New Era, Xi Jinping Thought on Ecological Civilization, the instructions of the National Conference on Eco-environmental Protection, as well as the major documents on China's ecological civilization over the past 70 years. The content of the manuscripts covers the basic course, historical context, laws, experience, lessons, and internal logic, theories, and conceptual innovations concerning the development of ecological civilization in the past 70 years, especially studies and implementation of Xi Jinping Thought on Ecological Civilization. They reflect the unremitting efforts of people from all walks of life to study and explore ways to promote the development of ecological civilization, the Chinese people's support for the CPC in promoting economic development, environmental protection, and harmonious coexistence between man and nature, as well as their sense of responsibility for making the country beautiful again.

This book and related activities have received strong support from the supervisory unit of the organizer and its leaders. They have placed high expectations on the book. Chen Zongxing, Vice Chairman of the 11th National Committee of the Chinese People's Political Consultative Conference and Chairman of the China Association for the Study and Promotion of Ecological Civilization, Xie Fuzhan, Dean and Secretary of the Leading Party Members' Group of the Chinese Academy of Social Sciences, Li Ganjie, Minister of Ecology and Environment and Secretary of the Leading Party Members' Group of the Ministry of Ecology and Environment, and Ma Jiantang, Secretary of the Leading Party Members' Group of the Development Research Center of the State Council wrote prefaces for the book. The editorial board hereby expresses its gratitude to them for accepting our invitation to write a preface to this collection.

Xi Jinping pointed out that "important goals are the coordinates of our work." This project was launched by the organizers to realize these goals, open up horizons, and establish coordinates for China's efforts to build ecological civilization. As Chen Zongxing pointed out, it is something that should be done and must be done well. Over the past 70 years, from the approval of the "32-character environmental protection policy" and the first environmental protection document of the PRC at the first environmental protection conference in Beijing in 1973 to the proposal of the argument that "lucid waters and lush mountains are invaluable assets" by Xi Jinping, the Chinese people have continuously explored new ways to promote economic development and environmental protection, achieve the harmonious coexistence of man and nature, and constantly deepen the understanding of the regularity of ecological civilization development. This journey is full of ups and downs, but great achievements have been made. Since the 18th National Congress of the Communist Party of China, Xi Jinping has stood at the strategic height of upholding and developing socialism with Chinese characteristics and realizing the great Chinese Dream. He has profoundly and systematically answered why and what kind of ecological civilization to build, how to build ecological civilization and other major theoretical and practical issues. Xi Jinping Thought on Ecological Civilization effectively guides the development of ecological civilization in China and promoted historic reforms toward sustainability in the Chinese society. By continuing to study, understand, and practice Xi Jinping Thought on Ecological Civilization while looking back on history, we will be able to more deeply appreciate the great significance of Xi Jinping Thought on Ecological Civilization, keep in mind that thriving ecosystems are a precondition for thriving human civilization, and strive for harmonious coexistence between man and nature. We hope that the publication of this book will help us achieve the above goals. If so, the editorial board will be very pleased. However, despite the best efforts of the editorial board and the authors, the book still has a lot of room for improvement. There might be some mistakes and omissions we fail to notice. Any constructive criticism will be appreciated.

Beijing, China

Jiahua Pan
Shiji Gao
Qingrui Li
Jinnan Wang
Dekai Wu
Chengliang Huang

Contents

Furthering the Reform of the System and the Set of Institutions for Developing an Ecological Civilization

Building a Global Ecological Civilization and an Ecological Security System

China's Efforts to Build an Ecological Civilization over the Past 70 Years

Changes in Environmental Policy in the Past 40 Years of Reform and Opening-Up

Zhenhua Xie

The past 40 years have seen China, with a population of nearly 1.4 billion, grow from a poor country to the second largest economy in the world. China's environmental protection policy has also changed, with emphasis shifting from the control of wastewater, exhaust gas, and solid waste to the development of ecological civilization. China has gradually become a participant, contributor, and leader in global sustainable development. Looking back on China's past environment protection practices, we find that environment protection is closely related to social and economic development. We have different environmental problems at different stages of social and economic development. A country's environment management system and structure depend on its stage of economic development and the needs of its society. China's environment governance system and model have been continuously improved to keep up with the time. In the future, as socialism with Chinese characteristics advances into a new stage, China should step up its efforts to build the ecological civilization, strive for a beautiful China, work with other countries to build a clean and beautiful world, and contribute to global sustainable development.

Z. Xie (✉)
Committee on Population, Resources and Environment, The National Committee of CPPCC, Beijing, China

J. Pan et al. (eds.), *Beautiful China: 70 Years Since 1949 and 70 People's Views on Eco-civilization Construction*, https://doi.org/10.1007/978-981-33-6742-5_1

1 Over the Past 40 Years, Socioeconomic Progress Has Driven the Transformation of China's Environmental Protection Management System, Which in Turn Has Laid an Institutional Cornerstone for the Balance Between Socioeconomic Development and Environmental Protection

Forty-seven years have passed since environment protection was officially put on the Chinese government's agenda in 1972. Over these years, China's environmental protection system would improve significantly every decade or so. The Environmental Protection Leadership Group of the State Council has been renamed the Ministry of Ecology and Environment. The development of the environment protection system in China is in essence a process of adaption to changes of China's economic and social conditions brought by the reform and opening-up. China's environment protection management system and model are constantly adjusted to address new environmental problems.

The years 1972–1988 were the first crucial period in the China's environmental protection history. During this period, China established the Environmental Protection Leadership Group of the State Council and then reorganized it into the State Administration of Environment Protection (directly under the State Council) and upgraded environmental protection into a priority in its national system of macro-management.

Before 1972, although environmental pollution had already become a serious problem in many places, the Chinese government believed that there was no environmental pollution in socialist countries, and industrial pollution was the product of capitalism. It was not until the Chinese government sent a delegation to participate in the United Nations Conference on the Human Environment in Sweden in June 1972 that this mindset began to change. In August 1973, the State Council held the first national environmental protection conference and passed the *Environmental Protection and Improvement Regulations (Interim)*, proposing to plan environment protection in a holistic and rational manner, improve resource allocation efficiency, reduce the negative environmental impact of human activities, and encourage everyone to participate in environment protection, officially putting environment protection on the Chinese government's agenda. In October 1974, the Environmental Protection Leadership Group of the State Council was formally established.

In December 1978, at the Third Plenary Session of the Eleventh CPC Central Committee, it was decided China will launch its reform and opening-up and shift its focus to economic development. After the session, China began to deploy the household contract responsibility system in rural areas, pushed fast forward the economic reform and opening-up, and worked hard for the emancipation of productive forces. In August 1980, Shenzhen Special Economic Zone was established. In 1984, 14 coastal cities opened gates to the outside world. A large number of labor-intensive industries were relocated from Japan, South Korea, Hong Kong and Taiwan to the

Chinese mainland. Government authorities at all levels, rural collectives, and social organizations all focused their efforts on attracting investment and facilitating the development of enterprises. China's environmental protection also began to get on track.

In the meanwhile, the Chinese government became increasingly aware of environmental problems caused by economic development and the transfer of production activities from other countries. In 1979, China promulgated its first environmental protection law—the Environmental Protection Law of the People's Republic of China (Trial). After that, environmental legislation in China gradually got on track. In August 1982, the Standing Committee of the National People's Congress passed the Marine Environment Protection Law. In May 1984 and September 1987 respectively, the Water Pollution Prevention and Control Law and Air Pollution Prevention and Control Law were adopted. In the meantime, China also began to strengthen environmental management and system building. In May 1982, at the 23rd meeting of the Standing Committee of the Fifth National People's Congress, it was decided that the State Development Commission, the State Urban Development Administration, the State Construction Engineering Bureau, the State Bureau of Surveying and Mapping, and the State Council Environmental Protection Leadership Group Office shall merge to form the Ministry of Urban and Rural Development and Environmental Protection and that the Environmental Protection Bureau shall be established under the Ministry of Urban and Rural Development and Environmental Protection. At the second National Environmental Protection Conference held at the end of 1983, then Vice Premier Li Peng declared environmental protection as one of China's basic, long-term national policies. The Environmental Protection Commission was established under the State Council in May 1984, with Li Peng as chair. The office of the Environmental Protection Commission was set up under the Ministry of Urban and Rural Development and Environmental Protection and its responsibilities were performed by the Environmental Protection Bureau on its behalf. In December 1984, the Environmental Protection Bureau of the Ministry of Urban and Rural Development and Environmental Protection was renamed National Environmental Protection Agency. However, it was still affiliated to the Ministry of Urban and Rural Development and Environmental Protection.

In 1988, the Ministry of Urban and Rural Development and Environmental Protection was reorganized into the Ministry of Construction. The National Environmental Protection Agency became a sub-ministerial agency directly affiliated to the State Council. It was the functional department of the State Council responsible for overseeing environmental protection work nationwide. It enjoyed a high level of independence, having its own staffing quotas, funding, and supplies. In the same year, the CPC Central Committee and the State Council launched a pilot civil service reform program at the National Environmental Protection Agency, set up new positions to meet the needs for environmental protection, and recruited from society a large number of environmental protection officials. This program laid the foundation for professional administration of environmental protection.

1989–1998 was the second major phase in China's environmental protection history. During this period, the pressure for environment protection continued to increase and the Chinese government launched the "33211" and "One Control and Two Standards" environmental improvement projects. In 1998, the National Environmental Protection Agency was reorganized into the State Environmental Protection Administration.

After the political turmoil of 1989 in China, the drastic changes in Eastern Europe, and the disintegration of the Soviet Union, Deng Xiaoping recognized the necessity of economic reform and opening-up in speeches during his South China tour in 1992. At that time the Yangtze River Delta was developing rapidly and there was a building frenzy across the country involving nearly 10,000 economic development zones and industrial parks. However, rapid economic development led to a drop in arable land and serious environmental degradation and pollution. There was a folk song at the time, depicting the deterioration of water quality in the Huaihe River. According to the song, water in in the Huaihe River could be used to wash rice and vegetables in the 1950s, irrigate farmland in the 1960s; then the water was seriously polluted in the 1970s and almost all fish and shrimps died in the 1980s. In the 1990s, the water quality further deteriorated, drawing attention from the whole society. In the meantime, environmental issues such as damages to the ecosystem, soil erosion, and desertification became increasingly serious. Sandstorms in Beijing become increasingly fierce, and major ecological disasters such as drying up of a section of the Yellow River and flooding of the Yangtze River occurred frequently.

In order to solve these problems, the Standing Committee of the National People's Congress accelerated environmental protection legislation. In December 1989, the Environmental Protection Law was revised and officially promulgated. In the 1990s, the Air Pollution Prevention and Control Law and the Water Pollution Prevention and Control Law were revised, and the Solid Waste Pollution Prevention and Control Law and the Noise Control Law were formulated and promulgated. China's legal system for environment protection began to take shape. At the same time, the "33211" pollution control project, which was the first large-scale pollution control project in China, was launched. "33211" means three rivers (Huaihe River, Haihe River, and Liaohe River), three lakes (Dianchi Lake, Taihu Lake, and Chaohu Lake); two control areas (sulfur dioxide and acid rain control areas), one city (Beijing) and one sea (the Bohai Sea). The "33211" project started from addressing water pollution in the Huaihe River. According to the requirements of the State Council, all key factories in the Huaihe River Basin must meet the wastewater discharge standards by midnight on December 31, 1997, and all factories that failed to do so would be shut down for rectification. In 1995, then Vice Premier Zou Jiahua and State Councilor Song Jian, on behalf of the State Council, listened to the report on the Chinese government's work concerning environmental protection and set out the goal that by 2000, the country's gross pollutant discharge should not be higher than that of 1995 and all environmental function zones and industrial pollution sources should meet relevant standards (i.e., as set out for the "One Control and Two Standards" project). During this period, the primary method adopted to control gross pollutant discharge was offsetting pollutant discharge from new projects by reducing discharge from other pollution sources in the same area. During the same period, China also began to implement six key ecological projects, including the one that turns farmland back into forests.

Another important event during this period was the United Nations Conference on Environment and Development held in Brazil in 1992. At the conference, the concept of sustainable development was put forth and Agenda 21 was adopted. After participating in the conference, China, as a developing country, compiled its own *Agenda 21: White Paper on China's Population, Environment and Development in the 21st Century* and formulated its own sustainable development goals in 1994. Then, in 1998, the State Environmental Protection Agency was promoted to the ministerial level, the functions of the former Environmental Protection Commission of the State Council and the pollution prevention responsibilities of the Ministry of Electric Power Industry and other relevant authorities were transferred to the upgraded State Environmental Protection Administration.

The third period in China's history of environment protection was 1999–2008. During this period, Chinese environment protection authorities mainly focused on curbing the rapid growth of gross discharge of major pollutants, promoting the development of a circular economy, building a resource-efficient and environment-friendly society and establishing the Ministry of Environmental Protection.

China acceded to the WTO in December 2001, followed by rapid social and economic development and continuous growth of the chemical, energy, steel and other heavy industries. China leapt to the top of the world in terms of production capacity and output of these industries. The consumption of resources and energy rose rapidly, and the gross discharge of major pollutants also increased significantly. Among the main objectives set in the Tenth Five-Year Plan of China, the total sulfur dioxide emission control target was raised. The Chinese government realized the necessity to step up efforts to improve energy efficiency and reduce the gross discharge of pollutants. During 2006–2010, the Chinese government developed national targets for the reduction of major pollutant discharge and energy consumption per unit of GDP and included them in the Eleventh Five-Year Plan. The national targets were then broken down to form targets for each province, autonomous region and municipality.

The number of environmental facilities and power plant desulfurization facilities built during 2006–2010 in China exceeded the total number of such facilities built in the country since founding of the People's Republic of China in 1949. Two policies played a key role in promoting the construction of these facilities. One is the strict implementation of energy conservation and emission reduction targets, which led to the construction of major local environmental projects; the other is the integration of environmental considerations into economic policies, an example of which is the policy concerning electricity price and the desulfurization coal-fired power plants, which led to the construction frenzy of desulfurization projects and prompted China to build the world's largest clean coal power system.

China also made new progress in environmental legislation and law enforcement during this period. To further improve the country's environment, the Chinese government revised the Air Pollution Prevention and Control Law, the Water Pollution Prevention and Control Law, the Solid Waste Pollution Prevention and Control Law, and the Marine Environment Protection Law, and issued the Radioactive Pollution Prevention and Control Law, the Environmental Impact Assessment Law, Cleaner Production Promotion Law, Circular Economy Promotion Law, and other environmental laws and regulations. In terms of institutional arrangements, in order to strengthen environmental law enforcement and local administrative intervention, the State Environmental Protection Administration established six environmental inspection centers in Northeast China, North China, Northwest China, Southwest China, East China, and South China respectively in 2006. In July 2008, the State Environmental Protection Administration was reorganized into the Ministry of Environmental Protection and became a ministerial-level authority under the State Council.

The fourth period in China's history of environment protection was 2009–2018. Promoting the development of the ecological civilization was included in the Five-Sphere Integrated Plan for the development of socialism with Chinese characteristics. The Chinese government made continuous effort to improve environmental quality, promote green development and reduce pollution, and established the Ministry of Ecology and Environment.

Since the 18th National Congress of the CPC, the CPC Central Committee with Comrade Xi Jinping at its core has attached great importance to environmental protection and ecological civilization, incorporated ecological civilization into the Five-Sphere Integrated Plan for the development of socialism with Chinese characteristics, promoted green development and the harmonious coexistence of man and nature, strengthen pollution control, promulgated the *Ten Regulations on Air, Water, and Soil Pollution Control*, developed the *Master Plan for the Ecological Civilization System Reform*, and put into place the central environmental protection inspection system and other important systems. According to the new requirements for the development of the ecological civilization, major amendments have been made to a series of laws including the Environmental Protection Law, the Air Pollution Prevention and Control Law, the Water Pollution Prevention and Control Law, the Solid Waste Pollution Prevention and Control Law, and the Marine Environment Protection Law. The Environmental Protection Law was revised in 2014 to give teeth to the law. The National People's Congress adopted the Soil Pollution Prevention Law in 2018. A relatively complete legal system for environmental protection in China took shape. China started to assume a leading role in the global efforts for the development of the ecological civilization.

In October 2017, at the 19th National Congress of the CPC, it was pointed out that as socialism with Chinese characteristics had entered a new era, the principal contradiction facing Chinese society has evolved, and that what China now faced was the contradiction between unbalanced and inadequate development and the people's ever-growing needs for a better life. After China entered the new era of socialism with Chinese characteristics, the requirements for the environmental protection underwent significant changes. First, the focus of institutional arrangements and governance concepts should shift from growth to environment protection. This means that resource conservation and environment protection authorities must play a more important role. Second, the primary function of the environment protection system should gradually change from decentralized management of resource and environmental factors to a holistic approach to the management of ecosystems. Third, the responsibilities of business owners and environment regulators must be clearly defined and a system of checks and balances should be formed. Fourth, the distribution of power and responsibilities between central and local governments must be clear and reasonable and a relatively independent monitoring, evaluation and supervision system should be put in place.

In March 2018, at the first meeting of the 13th National People's Congress, a plan for institutional reform at the State Council was adopted, and one of the major moves included therein was to establish the Ministry of Ecology and Environment, putting pollution prevention and control functions uniformly under this one authority while also mandating it with such responsibilities as climate change response and marine environment protection, environmental inspection and urban and rural pollutant discharge regulation. This reform achieved the following results: First, the supervision functions in the three major areas of environmental governance, i.e. pollution prevention, ecosystem protection, and nuclear radiation protection, were put together in accordance with the roadmap of major institutional reforms, laying a solid institutional foundation for future attempts to address the problems fragmentation of the structure of government authorities and responsibilities. This is also a major highlight and the most prominent feature of this round of institutional reform. Second, the reform clarified the responsibilities of natural resource owners for development management and the responsibilities of resource regulators for supervision and law enforcement, and to a certain extent realized the separation between implementation and supervision as required in the system design. Third, environmental protection actions were better unified and aligned, resulting in heightened authoritativeness, and ecological protection, urban and rural pollutant discharge control, administrative regulation, law enforcement, and supervision were fully integrated in the overall responsibility structure. Of course, there are still many problems to be solved before we can fully ensure the effectiveness of the reform. Yet still, the reform has laid a solid foundation for the country's effort to build a moderately prosperous society in all respects and achieve the goal of building a beautiful China.

All in all, along with the implementation of the reform and opening-up strategy, the past forty years have seen continuous improvement of China's environment protection system.

2 Over the Past Forty Years, China Focused on the Need to Solve Major Environmental Problems in Different Stages of Development, and Made Continual Efforts to Develop and Improve Environmental Protection Governance Systems and Models

China has adjusted its environmental governance system and adopted corresponding methods to address major environmental problems it faced in different stages of development and achieved great results. There are many lessons to learn, mainly in the following three aspects.

First, we should balance between environmental protection and development and build an ecological and environmental protection system that is suitable for the development stage we are in by taking combining theory and practice.

Environmental issues are in essence development issues. They arise from development and need to be resolved in the process development. We need not only to solve environmental problems step by step, but also to propose innovative development concepts. Since China launched its reform and opening-up, the Chinese government has put forward many new concepts to balance between economic development and environmental protection offering guidance for ecological and environment protection. In the first stage, it put forward the requirements of "three synchronizations" and "three unifications," and the whole society was required to pay attention to environmental issues arising in the process of economic growth and urban and rural development. In the second stage, it focused on the environmental governance of key river basins and regions, adjusted the economic structure, shut down small paper mills and other heavily polluting factories, promoted healthy economic growth, and strove to achieve its environmental protection goals. In the third stage, it put in place a gross pollutant discharge control system, and balanced between environment protection and economic development at that macro level. In the fourth stage, it vigorously promoted the development of the ecological civilization, holding that "lucid waters and lush mountains are invaluable assets," and made the fundamental change to a holistic approach towards economic development and environmental protection. Since the twenty-first century, many new concepts such as circular economy (2003), resource-efficient and environment-friendly society (2004), and low-carbon development (2009) have been proposed. In the report to the 18th National Congress of the CPC, developing ecological civilization is included as part of the Five-Sphere Integrated Plan (2012). These concepts reflect the continuous deepening of the Chinese government's understanding of the relationship between environment and development. The concepts proposed in each development stage are not only the results of research into the primary contradiction and the national conditions of China at that time, but also the products of inheritance and evolution of existing ideas. Environmental issues and socioeconomic development needs drove conceptual and theoretical innovations, and such innovations in turn provided guidance for environment protection practices and an environmental governance system in line with the characteristics of the current development stage.

As early as the early 1980s, the Chinese government made environmental protection a basic national policy, proposed "three synchronizations" and "three unifications" for a good balance between development and environment protection, and began to build an independent and authoritative environment protection institution. However, in the initial stages of development, the Chinese government was not able to distinguish between environmental governance and environmental development in theory and considered development as its top priority. The fact that China was still in the preliminary stage of socialism determined that China should focus on economic growth at the time. Some local governments overemphasized development and did much more for economic growth than for ecological and environmental protection. This was perhaps natural given the stage of economic development the country was in at the time. Then, the Chinese government's approach to environment protection evolved along with changes in China's national conditions. It distinguished clearly between government environmental management and corporate environmental governance, as well as between environmental management and environmental development, and established a national environmental protection authority directly under the State Council. To separate development and environment protection in institutional arrangements, the State Environmental Protection Administration was moved out of the charge of the Ministry of Urban and Rural Development and Environmental Protection whose top responsibility was to promote development.

Since 2000, under the guidance of a series of development concepts, China's environment governance system has significantly expanded its integrated management, planning, and policy coordination functions. This change has been prompted by the increasing public awareness of the massive environmental cost of rapid social and economic development as China adhered to a growth-first approach in its development. Around 2003, the State Environmental Protection Administration published the National Environmental Security Strategy Report, pointing out that in China, the environmental cost of economic development had offset the benefits of economic growth in many regions, and it was meaningful at the time to understand the importance of environmental protection to national security and development. More and more people gradually realized that environmental problems would affect economic growth ultimately, and that it was necessary for China to adopt a more sustainable economic development model and a big-picture decision-making method in order to strengthen environment protection. This realization altered the direction of the reform of China's environmental management authorities. The State Environmental Protection Administration was reorganized into the Ministry of Environmental Protection, a ministerial-level department under the State Council, with the power of deliberation. The government departments responsible for overseeing economic sectors also increased their engagement in environmental protection and played an important role in promoting energy conservation, emission reduction, circular economy, and low-carbon development.

With regard to circular economy, the then President Jiang Zemin attended the second General Assembly of the Global Environment Facility in 2002 and pointed out that only by promoting resource-efficient circular economy and stepping up environment protection efforts could the world achieve the sustainable development goals. In 2008, learning from Germany, Japan and other countries as well as based on the experience of some local pilot projects, China promulgated the Circular Economy Promotion Law, promoting a system of resource utilization where consumption reduction, reuse and recycling of resource factors prevailed and products were managed for their environmental impact throughout their life cycles. Also, energy consumption, carbon emission, resource productivity and other indicators were included in the Twelfth Five-Year Plan. At present, China is working on the problem of urban waste, promoting the recycling of materials from urban mines, setting up special funds for the development of circular economy, launching "waste-free city" pilot projects, and supporting the development of the recycling industry.

With regard to low-carbon development, the international community regards carbon footprint control as a top priority. China must cut its carbon dioxide emissions and transform its development model. In other words, China must adopt a low-carbon development approach in order to achieve sustainable development and build an ecological civilization. In the past few years, the Chinese government set targets for the improvement of energy efficiency, the reduction of carbon intensity, the development of non-fossil energy and the increase of carbon accumulation in forests, promoted the transformation of the country's development model, adjusted industrial and energy structures, and made great strides in energy conservation, emissions reduction and response to climate change. In 2017, China's carbon dioxide emissions per unit of GDP dropped by 46% from the 2005 level, which is equivalent to reducing carbon dioxide emissions by 4.1 billion tons, already exceeding the carbon intensity reduction target that China committed to (by 40–45% by 2020). As the CPC Central Committee became more aware of China's environment issues, the Chinese government launched an institutional reform in 2018, transferring the function of responding to climate change from departments responsible for overseeing economic sectors to the Ministry of Ecology and Environment.

Since the 18th National Congress of the CPC, ecological civilization has become an important development concept advocated by the ruling Party, and the focus of the Chinese government has gradually shifted from economic growth to environment protection, which has promoted environmental awareness and driven the reform of the environmental governance system. In 2015, the CPC Central Committee and the State Council issued the *Opinions on Accelerating the Development of Ecological Civilization*, summarizing China's experience in environment protection and innovations in environmental policies and systems over the past more than 40 years. Environment protection is included in the five-sphere development plan. The main concern of the global community is how to achieve sustainable development in economic, social and environmental dimensions. According to the State Council's institutional reform plan in 2018, the central government established the Ministry of Ecology and Environment in accordance with the principle of one matter being managed by one department and the separation of regulation from ownership. This greatly enhanced the authority of the ministry and strengthened pollution reduction at the sources, process management, punishment of polluters, and environmental accountability. The reform better coordinated environment protection efforts with production, distribution, consumption and other activities, forming a long-term mechanism emphasizing both incentives and restrictions and maximizing economic efficiency of environment protection measures.

Second, we should take comprehensive measures to improve the quality of the ecological environment and continuously improve public awareness of the benefits offered by the ecological environment to human beings.

Since 2016, China has been working to shift the focus of its environmental governance system from gross pollutant discharge control (as required by the Eleventh Five-Year Plan and Twelfth Five-Year Plan) to the improvement of environmental quality. If we put things in historical context, we will find that China has always emphasized the close relations between environment quality and life quality and the clever use of incentives and restrictions to ensure that the environmental quality is up to relevant standards.

When implementing the "33211" environmental pollution control project, the Chinese government adopted a system of "one control and two standards" based on environmental function zoning, with focus on regions, river basins and cities. Environmental function zoning is a method that determines environment quality standards for an area based on its social conditions, functions, natural environment and environmental self-purification capabilities. Companies must meet the emissions targets, and local governments must ensure that the environment quality of environmental function zones meet relevant standards. If companies meet their emissions targets but environmental function zones fail to deliver an environment quality up to their respective standards, measures for total pollutant discharge control would apply. This reflects that there is a connection between environmental quality improvement goals and the measures adopted.

The pollution control projects carried out in the mid-1990s in China played an important role in slowing down environmental deterioration, and the environmental quality of some key river basins did improve significantly. From 2006 to 2015, the Chinese government mainly tried to reduce pollution by setting gross pollutant discharge control targets. At the time, the discharge of major pollutants exceeded the environment's self-purification capacity and the environmental quality was declining. To curb the downward trend of environmental quality by reducing gross pollutant discharge, a pollutant discharge management system, headed by a dedicated institution for total pollutant discharge control, was established. Thus, significant progress was made in gross pollutant discharge control. Some studies based on data from NASA satellites and meteorological vehicles show that China's sulfur dioxide emissions fell by more than 70% from 2005 to 2016.

Since 2016, China has been working on an environmental governance model centered on improving the quality of the ecological environment, and has been advancing with three major action plans for air, water and soil quality improvement. Indicators closely related to people's living standards, such as air quality and water quality indicators, were included in the Thirteenth Five-Year Plan as mandatory targets. In order to significantly improve environmental quality, the Ministry of Environmental Protection established three new divisions for air, soil, and water quality improvement respectively in 2017 to replace the pollution prevention division and the gross pollutant control division, and the ministry's core responsibility became the management of air, soil and water quality with clearly defined targets. In 2018, *The Opinions of the CPC Central Committee and the State Council on Comprehensively Strengthening Environmental Protection and Pollution Prevention and Control* was promulgated, setting out the overall environmental governance goals, which were to improve the overall quality of the ecological environment, significantly reduce the total amount of major pollutant discharge, effectively control environment risks, and build an ecological civilization in line with the goal of completing the building of a moderately prosperous society in all respects by 2020.

Third, we should build an environment governance system that is led by the CPC and the government, integrates the market mechanism and enlists the support of the private sector and the public. Under this system, the CPC and the government should share the same responsibilities for environmental governance and broad social participation should be ensured to maximize the effectiveness of actions.

Before 2012, China focused on strengthening government leadership, corporate participation, and unified supervision of environmental protection, but failed to pay sufficient attention to corporate environmental responsibility and accountability. The emphasis was on government control. The system failed to effectively leverage the market mechanism and public participation lacked effective guidance. The problems of insufficient and excessive participation of the public coexisted. Therefore, in the future, it is necessary to further emphasize the role of the market in resource allocation, put into place effective economic incentives, and gradually form a coordinated governance pattern in which the government, the private sector, and the public cooperate with and supervise one another.

In this pattern, what is the most important is a joint management mechanism that emphasizes comprehensive coordination and promotes the clear definition of roles and responsibilities of various players. Since its establishment, the country's environmental governance authority has always been responsible for coordination for the implementation of environmental protection strategies and various relevant tasks across different authorities. The CPC and the government share responsibilities for environmental governance and the leadership team of the environmental governance agency are responsible for not only overseeing the performance of environmental management-related tasks but also building an integrity culture among CPC members in the agency ("*yi gang liang ze*"). These practices have historical origins. Starting from the time of the Office of the Environmental Protection Leadership Group of the State Council, all national environmental governance institutions primary responsibility of the national adopted this management approach. During 1984–1998, there was the Environmental Protection Commission under the State Council, which was headed by a vice premier of the State Council. During that period, the State Council convened quarterly meetings to review and plan its work concerning environmental protection. As the national environmental protection authority, the commission was responsible for coordinating across all relevant departments to ensure efficient implementation of environmental protection plans. Relevant departments were assigned different responsibilities for environmental protection, and need to report regularly at the meetings of the Environmental Protection Commission. This was where the "*yi gang liang ze*" management approach originated.

In 1998, with the State Environmental Protection Administration established to replace the Environmental Protection Commission of the State Council, the national authority for environmental protection was upgraded to a higher level, but its power for comprehensive coordination was weakened to some extent. Then, putting great emphasis on environmental protection, the CPC Central Committee and the State Council launched a high-level annual event, i.e. the Central Symposium on Population, Resources and Environment Work. The event was chaired by the General Secretary of the CPC Central Committee and participants included ministers, provincial party secretaries and governors, playing very important roles for some time.

We should continue with the environmental governance system that emphasizes both incentives and restrictions and engage local governments in environmental governance. With regard to restrictions, since 2006, the central government has been carrying out supervision on the fulfillment of environmental protection responsibilities across the country primarily through mandatory energy conservation and pollution reduction targets. After the 18th National Congress of the CPC, the central government put forward the requirement that Party and government bodies should share responsibilities for environmental protection and put in place the national environmental protection inspection system to oversee efforts to meet this requirement. With regard to incentives, a typical example is the naming of national model cities for environmental protection and eco-friendly provinces, cities and counties, the former of which was proposed during the Ninth Five-Year Plan period (1996–2000). In 1997, six cities including Zhangjiagang and Dalian were named national model cities for environmental protection. Since then, nearly 100 cities have applied for this title and efforts to win it boosted enthusiasm for environmental protection and improved environmental protection capabilities and management of various cities. It is a cost-efficient way to effectively promote sustainable development of cities. In addition, the Chinese government also named eco-friendly provinces, cities and counties.

Fourth, we should build a complete environmental protection legal system, clarify the environmental responsibilities of the government, enterprises, and the public, and establish a government governance system that is better aligned to China's national conditions.

Over the past 40 years, China successively revised and formulated basic laws such as the Constitution, the General Principles of Civil Law, the Property Law, the Tort Liability Law, and the Criminal Law. Some of these basic laws contain provisions on environmental protection and sustainable resource management. For example, the General Principles of Civil Law and the Property Law contain a wide-ranging set of provisions on the ownership of natural resources and usufruct rights to natural resources, the Tort Liability Law contains a special chapter on tort liability for damage caused by environmental pollution, and the Criminal Law contains a special chapter on environment-related crime.

At the same time, the Clean Production Promotion Law, Circular Economy Promotion Law, Urban and Rural Planning Law and other laws related to the transformation of industrial or energy structure and land planning have been formulated, laying a solid legal foundation for the promotion of green development and reasonable land use planning. China also promulgated more than 10 environment protection and pollution prevention laws, including the Environmental Protection Law, Marine Environment Protection Law, Water Pollution Prevention and Control Law, Air Pollution Prevention and Control Law, Solid Waste Pollution Prevention and Control Law, Noise Control Law, Soil Pollution Prevention and Control Law, Radioactive Pollution Prevention and Control Law, Environmental Impact Assessment Law, Wildlife Protection Law, Water and Soil Conservation Law, Desertification Prevention and Control Law, and Island Protection Law. A relatively complete legal and institutional system for environmental protection thus took shape in China. Relevant administrative systems and technical specifications were established for all major links of environment protection and pollution prevention. With regard to administrative approval of projects, there are the environmental impact assessment system and the "three simultaneous" environmental evaluation system. With regard to supervision and inspection, there are systems and policies concerning pollutant discharge permits, gross pollutant discharge control, on-site inspections, emergency response, pollution remediation within a specified time limit, compulsory shutdown of polluting factories, and other compulsory administrative measures. There are also laws and regulations stipulating administrative, civil or criminal penalties for actions polluting or harming the environment. In the meanwhile, the Chinese government increased the use of environmental-economic measures, raised pollution charges and expanded the scope of pollution charges collection, and promoted the pilot project of environmental insurance and pollution rights trading systems. Social management systems also took environmental management into consideration, and environmental information disclosure and public participation gradually became the norm.

3 On the Basis of the Achievements Already Made in Economic Reform and Opening-Up, China Will Continue to Promote the Development of Ecological Civilization in the Country and Across the World, Strive to Build a Beautiful China, Work with Other Countries for a Clean and Beautiful World, and Contribute to Global Sustainable Development

The Beautiful China campaign was launched at the 19th National Congress of the CPC, depicting the vision of a clean and beautiful world: China should fundamentally improve its environment, basically achieve the goal of a beautiful China by 2035, and into a great modern socialist country that is prosperous, strong, democratic, culturally advanced, harmonious, and beautiful by mid-twenty-first century. China will not only rise as a great country in the world in terms of its industrial, economic, and ecological development, but also make a unique contribution to global development. Chinese ideas, Chinese wisdom and Chinese solutions will provide inspirations for developing countries seeking to embark on the path towards green development and prosperity.

First, China will deepen the reform of the environment governance system in order to meet the requirements of a beautiful China.

The reform of China's environmental governance system will not be accomplished overnight, nor will it be a once-and-for-all solution. China need to further advance functional transformation, improve efficiency, and strengthen mechanism innovation and capacity building. (1) China should further improve its environmental governance system by accelerating functional transformation, clarifying responsibilities, promoting the modernization of the system and building governance capabilities. It should continue to clarify the responsibilities of government authorities, especially in regard to environmental supervision, climate change response, regulation of the nature reserve system, establishment and development of regional and river basin management authorities, and the allocation of power and financial resources between central and local levels. (2) China should accelerate the integration and transformation of relevant functions within environmental governance authorities. With the aim of continuously improving environment quality, efforts should be made to better define and clarify the interrelationships between various systems and highlight core systems. Synergy should be formed between conventional pollution prevention and control and climate change response, and the accountability system should be strengthened for local governments. Also, efforts should be made to improve the ability to respond to institutional changes, step up the efforts to build a nature reserve system with national parks at its core, and promote the reform towards a national park system. (3) China should strengthen mechanism building and innovation, integrate different functions concerning the development of ecological civilization, and improve the efficiency of relevant systems. Establishing a ministry to take charge of environmental governance in the country can solve problems concerning the coordination across different environmental factors, but it is necessary to build an effective mechanism to balance between environment management, natural resource management and economic development. It is recommended that the central government create an institution to oversee the development of ecological civilization, formulate green development strategies, roadmaps, and timetables, and set priorities for the various relevant issues. (4) China should step the effort to build a modern environment governance system. It should move towards a coordinated governance pattern in which the government, enterprises, and the public cooperate with and supervise one another, so that centralized government management of natural resources and independent environmental monitoring can truly be effective. It should continue to improve market-based environmental protection, stimulate the vitality of market entities, and encourage NGOs and the public to participate in environment management and monitoring, (5) China should comprehensively strengthen capacity building in government authorities in charge of natural resources conservation of and environmental governance. It should continuously improve the investigation, monitoring, statistics, and assessment systems, and offer guidance to local government departments to improve their capabilities to complete the increasingly onerous management tasks.

Second, China should have a top-level design, a strategic roadmap, a timetable and an implementation path for the Beautiful China campaign.

China's experience over the past 40 years shows that environmental protection involves complex and ever-changing contexts and relationships, and the situation differs greatly across the country. If actions are not taken with a clear strategic goal in mind and with sound coordination, it will not be possible to form synergy. In the past, the Chinese government clarified its environment-related strategic goals and development directions at each stage of development thanks to seven national environmental protection conferences, five-year environmental protection plans, and a series of action plans, took a step-by-step approach to key issues, reforms and mechanisms, improved policies and systems, and promoted the continuous improvement of the environmental protection system. China should continue on this path if it wants to effectively modernize the environmental governance system and improve environmental governance capabilities.

To achieve the goals of the Beautiful China campaign by 2035 and mid-twenty-first century, we must keep track of the changes in the primary contradiction of society in the new era, follow the guidance of Xi Jinping Thought on Ecological Civilization, and adhere to the basic policy of "conservation, protection, and restoration." We should find an effective way to coordinate between high-quality development and high-level environmental protection, make environmental protection the top priority of the country, promote green development, and continuously increase the supply of high-quality ecological products to meet the people's growing needs for a good ecological environment.

We must think about the future in a visionary manner and systematically formulate strategic goals and targets for ecological protection, environmental governance, resource and energy security, and climate change response, and coordinate control, protection, and governance efforts. We should keep in mind that pollution reduction and ecological restoration will be a tough and protracted battle, clarify the mid-to-long-term goals and targets in the 14th Five-Year Plan and 15th Five-Year Plan and the goals for the year 2035 and 2050, and take a step-by-step approach to significantly improve environmental quality by 2035 and comprehensively improve the development level of ecological civilization by 2050.

Third, we should stick to the path of green development with a focus on China itself while taking a global perspective, actively respond to climate change, contribute to global sustainable development, and work with other countries for a clean and beautiful world.

China's ecological civilization development and environment protection efforts are an important part of the global efforts to promote sustainable development. The achievements made by China in the past 40 years, especially since the 18th National

Congress of the CPC, have demonstrate the country's ability to firmly pursue green development and achieve harmony between man and nature.

China has actively participated in climate change negotiations, pushed forward the conclusion and the follow-up negotiations of the Paris Agreement and the establishment of a fair, reasonable, cooperative and win-win global climate governance system, and called for global actions to build a community with a shared future for mankind. China will continue to work with other countries to strengthen South-South cooperation, advance the implementation of the Green Belt and Road Initiative, enhance cooperation with international organizations and multilateral financial institutions, build consensus for win-win cooperation, and play a more active role in global climate governance.

China will unswervingly fulfill its commitments and make greater contributions to global green and low-carbon development. The relationships between climate, environmental, and economic issues must be clarified, emission reduction and environment pollution control must be considered as part of the big picture of economic and social development, and the whole society must be mobilized to ensure the fulfilment of China's pledge to peak its emissions around 2030 and significantly improve its ecological environment by 2035. We hope other countries would be able to draw inspirations from Chinese ideas, Chinese wisdom and Chinese solutions, and work with China to build a clean and beautiful world.

Over the past forty years, China has achieved unprecedented economic success, but it has also been faced with a series of deep-seated problems. Currently, China is in a crucial stage of ecological civilization development and environmental protection. Under the leadership of the CPC Central Committee, the country has embarked on an arduous journey of building a beautiful country. Although there will be huge challenges, we believe as long as we work hard and persevere, we will eventually achieve the goals of a Beautiful China and witness the spread of ecological civilization across the whole world.

Environmental Legislation in the Past 70 Years

Zhongmei Lv

The year 2019 marks the 70th anniversary of the founding of the People's Republic of China. Over the past 70 years, China has promulgated a full range of environmental laws, developed an environmental legislative system with Chinese characteristics, actively participated in global environmental governance, and led the world onto the path towards environmental rule of law. A review of the 70-year history of building the environmental rule of law and an analysis of the context in which the environmental rule of law has evolved over the years can help us build confidence in the environmental rule of law in the new era and is of profound significance to China's efforts to accelerate the development of the environmental rule of law in the new era, improve the environmental legislative system, and achieve the goals of the Beautiful China campaign.

1 Moving into the Forefront of the World in Environmental Legislation

After the founding of the People's Republic of China in 1949, as socialist economic development progressed, many environmental problems gradually emerged in China and China started to use legal means to protect the environment. Globally, starting from since the 1940s, incidents of pollution-related health hazards were frequently seen, arousing great concern worldwide and calling for environmental legislation in

Z. Lv (✉)
Committee on Social and Legal Affairs, The National Committee of CPPCC, Beijing, China

China Law Society, Beijing, China

J. Pan et al. (eds.), *Beautiful China: 70 Years Since 1949 and 70 People's Views on Eco-civilization Construction*, https://doi.org/10.1007/978-981-33-6742-5_2

the modern sense.[1] In 1972, China sent its first delegation after it claimed its lawful seat in the United Nations to participate in the United Nations Conference on the Human Environment (the Stockholm Conference) and then in 1973, it promulgated its first environmental protection administrative regulation, embarking on the path of dedicated environmental protection legislation around the same time as the rest of the world. Since then, China has actively participated in environmental affairs at the United Nations and offered proposals for global environmental governance. It has also been exploring effective ways to promote environmental rule of law with Chinese characteristics, vigorously supporting environmental legislation and law enforcement, and encouraging public participation. The country has made remarkable strides in this area.

(1) **Environmental legislation in China started in a difficult period (1949–1979)**

After the founding of the People's Republic of China, the Chinese government began promoting economic development in a planned manner. This period saw economic ups and downs as well as the beginning of environmental legislation in China. China's environmental legislation started at roughly the same time as many other countries.[2] First, the Chinese government established state ownership of natural resources and promulgated relevant laws and regulations. The 1954 Constitution stipulates that the ownership of mines, waters, and forests, wasteland and other natural resources that are state-owned vest in the people. From 1951 to 1959, the Chinese government promulgated the *Interim Regulations of the People's Republic of China on the Mining Industry, Regulations on Government Land Acquisitions, Interim Regulations on the Protection of Mineral Resources, Factory Safety and Sanitation Regulations, Interim Guidelines for Soil and Water Conservation of the People's Republic of China, Drinking Water Quality Regulations* and other relevant laws and regulations. Second, the Chinese government enacted administrative regulations on environmental protection. In August 1973, the first national environmental

---

[1]From a global perspective, many countries have had laws concerning nature and the environment since ancient times. For example, in China, the law of the Shang Dynasty stipulated that the hands of anyone who dumped ashes on a street would be chopped off. However, although these laws did have the effect of protecting the environment, they were not specifically designed for this purpose. Therefore, scholars of jurisprudence do not regard them as environmental legislation. A consensus in the academic circle is that the concept of environmental protection in the modern sense was first put forward in the Stockholm Declaration on the Human Environment; "environmental law" in the modern sense is a law aimed at protecting the environment formulated by human beings on the basis of re-understanding the relationship between man and nature. For example, the National Environmental Policy Act of 1969 enacted by the United States is one of such environmental laws.

[2]On June 11, 1972, *People's Daily* published an article (on the 5th page) titled "The head of the Chinese delegation made a speech at the United Nations Conference on the Human Environment and expounded China's propositions on environmental issues," which the first news report on environmental protection in China. On June 16, 1972, *People's Daily* published an article (on the second page) titled "Economic Development and Environmental Protection," arguing that the Chinese government should attach equal importance to economic development and environmental protection. Because of these understandings, China's focus was on the control of industrial waste when it first started promoting environmental legislation.

protection conference passed the *Regulations of the State Council on the Protection and Improvement of the Environment (Draft)*, and *Opinions on Strengthening Environmental Monitoring* and *Interim Regulations on Natural Reserves* were also formulated. Third, environmental protection was incorporated in the Constitution. On March 5, 1978, the First Session of the Fifth National People's Congress approved the revision of Article 11, Paragraph 3 of the Constitution of the People's Republic of China. The revised article stipulates that the state shall protect the environment and natural resources and prevent pollution and other public hazards. Fourth, a comprehensive environmental legislation system began to take shape. In September 1979, the Eleventh Session of the Fifth National People's Congress passed the Environmental Protection Law of the People's Republic of China (tentative), which clarified the objectives, tasks, principles and scope of application of the law, and established the basic principle that whoever causes the pollution shall bear the responsibility for treatment and control, and provided for environmental impact assessment, the "three simultaneous" system, pollution discharge fees, and pollution remediation within a specified time limit. Apparently, China has learned from the painful lesson of developed countries which failed to effectively control pollution during their industrialization process, and attached great importance to establishing a legal system that paid equal attention to both economic development and environmental protection.[3]

(2) **China's environmental legal system gradually took shape alongside the advancement of the economic reform and opening-up (1979–2013)**

After the promulgation of the Environmental Protection Law (Trial), China entered the fast track of environmental legislation. The status of environmental protection as a basic national policy was recognized in the Constitution. China's environmental legal system and environmental governance system began to take shape. Milestones in environmental legislation during this period are as follows: First, the status of environmental protection as a basic national policy was inscribed in the Constitution and a dedicated environmental legislation committee was set up under the country's national legislature. Article 9 of the Constitution, revised in 1982, stipulates that "the state shall promote sustainable use of natural resources and protect endangered animals and plants. No organization or individual may use natural resources without authorization or damage natural resources." Article 26 stipulates that "the state shall protect and improve the living environment and ecological environment, and prevent pollution and other public hazards." In March 1993, it was decided at the First Session of the Eighth National People's Congress to establish the Environmental Protection Committee (later renamed Environment and Natural Resources Protection Committee) under the National People's Congress in the following year to effectively advance the environmental legislation process. Second, an environmental

[3]In December 1978, the Central Committee of the Communist Party of China forwarded *the Summary of the Environmental Protection Work Report of the State Council's Environmental Protection Leading Group (ZF(1978) No. 79)*, which pointed out: "Eliminating pollution and protecting the environment are an important part of China's efforts to promote the development of socialism in the country and achieve the four modernizations...We must keep in mind that environmental pollution is a major obstacle to sustainable development."

legal system roughly took shape with the Environmental Protection Law at the core and pollution prevention and control and natural resources protection laws as the backbone, supplemented by environmental protection regulations promulgated by the relevant authorities. By 2013, China had promulgated four comprehensive environmental protection laws, six environmental pollution prevention laws, 13 natural resources and ecological protection laws, two laws on the promotion of cleaner production and circular economy, and two laws on sustainable use and development of energy. In addition, about 10 civil, criminal, administrative and economic legislations contained stipulations related to environmental protection. The State Council has enacted more than 60 environmental administrative regulations. The relevant departments of the State Council promulgated over 600 environmental administrative regulations and more than 1200 national environmental standards. Third, the environmental governance system and relevant mechanisms gradually took shape. In 1982, it was decided at the 23rd meeting of the Standing Committee of the Fifth National People's Congress to set up the Environmental Protection Bureau under the Ministry of Urban and Rural Development and Environmental Protection. In the three rounds of national institutional reforms, the national authority for environmental governance changed from the National Environmental Protection Bureau to the State Environmental Protection Administration to the Ministry of Environmental Protection. In the meanwhile, eight systems of environmental management (the construction project environmental impact assessment system, the "three simultaneous" system, the pollutant discharge fee management system, the environmental protection target management system, the quantitative assessment system for comprehensive urban environmental improvement, the pollution permits system, centralized pollution control system, and the pollution remediation system) gradually came into being. During this period, the Chinese government also adopted a series of prevention and control measures with a focus on prevention, established the principle that whoever caused the pollution shall bear the responsibility for pollution treatment and control, and rolled out an environmental management enhancement policy. Fourth, dedicated local environmental courts and tribunals were established. In 2007, the Guizhou Provincial Higher People's Court gave approval for the establishment of China's first dedicated environment court, the Qingzhen Town Environmental Protection Tribunal in Guiyang City. As of the end of 2013, more than 170 such environmental courts, tribunals, and collegiate panels had been established across the country, exploring ways and criteria for determining facts and applying laws in environmental cases. Fifth, China began to actively participate in international environmental governance. In 1992, the Chinese government sent a delegation to attend the United Nations Conference on Environment and Development and signed the *Framework Convention on Climate Change* and the *Convention on the Protection of Biological Diversity*. Before the conference, 41 environment ministers from developing countries were invited to Beijing to discuss environmental issues and they together released the *Beijing Declaration*, which clarified the common position and propositions of developing countries on environmental issues. In 1994, China became the first developing country to release a national sustainable development agenda—*China's Agenda 21: White Paper on China's Population, Environment and*

Development in the 21st Century, and sustainable development became part of the country's long-term economic and social development plan for the first time. China also took an active part in international climate change negotiations and related conferences, and signed and approved the Kyoto Protocol. By 2013, China had joined in 37 multilateral environmental treaties, and signed environmental cooperation agreements with Japan, the United States, Mongolia, the DPRK, Canada, India, South Korea, Russia and some other countries.

From a global perspective, the development of environmental legislation during this period was closely related to the evolution of the UN's environmental protection concepts. The following are three of the most important milestones. First, before and after the United Nations Conference on the Human Environment held in Stockholm in 1972, the United Nations adopted special resolutions calling on member states to discuss legal issues concerning environmental protection and developed countries formulated basic environmental protection laws and incorporated environmental protection in their constitutions. Second, before and after the United Nations Conference on Environment and Development held in Rio de Janeiro in 1992, the United Nations proposed the concept of sustainable development. Countries around the world re-examined their environmental laws or formulated new environmental laws in accordance with this concept. Many countries also started compiling environmental codes. Third, before and after the Johannesburg Summit 2002, the United Nations called for actions by all countries to achieve sustainable development. Countries around the world refined and improved their national environmental legislation in accordance with the UN Millennium Development Goals. Based on the environmental legislative process of the Standing Committee of the National People's Congress, it was obvious that China kept in step with the rest of the world in environmental legislation. There are three important periods in environmental legislation in China since 1979. The first period is from 1982 to 1990 which is during the early stage of China's economic reform and opening-up. During this period, the Standing Committee of the National People's Congress enacted ten environmental laws, including the Marine Environment Protection Law (1982), Water Pollution Prevention and Control Law (1984), and Air Pollution Prevention and Control Law (1987). The second period is from 1993 to 2003 during which China started building a socialist market economy. During this period, the Standing Committee of the National People's Congress formulated and revised 14 environmental laws, including the Air Pollution Prevention and Control Law (1995), Solid Waste Pollution Prevention and Control Law (1995), and Water Pollution Prevention and Control Law (revised in 1996). The third period is from 2003 to 2013 during which China began to embark on a path of science-based development. During this period, the Standing Committee of the National People's Congress formulated and revised 11 laws, including the Radioactive Pollution Prevention and Control Law (2003), Solid Waste Pollution Prevention and Control Law (revised in 2004), and Renewable Energy Law (2005). China's environmental legal system thus developed rapidly in sync with the rest of the world in environmental legislation. China is playing an increasingly more important role in global environmental governance.

(3) **China ushered in a new era for environmental rule of law (2013–present)**

Since the 18th National Congress of the CPC, under the guidance of Xi Jinping Thought on Ecological Civilization, the Chinese government has elevated ecological civilization development to the strategic height of state governance, and called on the international community to jointly build a community with a shared future for mankind. The country has ushered in a new era for environmental rule of law. The following events are milestones during this period: First, the term ecological civilization was included in the Constitution. In March 2018, at the first session of the 13th National People's Congress, a constitutional amendment was adopted, providing that "promoting the coordinated development of ecological civilization" is a fundamental task of the country (the preface) and that providing guidance for the country's efforts to build ecological civilization is among the responsibilities of the State Council (Article 89). Second, environmental laws reflected new concepts. In 2014, after deliberation, the eighth meeting of the Standing Committee of the 12th National People's Congress adopted an amendment to the Environmental Protection Law, which defined Environmental Protection Law as a comprehensive law, made environmental protection a top priority of the Chinese government, and established a system of joint governance by multiple parties. In August 2018, the fifth meeting of the Standing Committee of the 13th National People's Congress reviewed and approved the Soil Pollution Prevention and Control Law, which defined the term "risk prevention" and provided for the protection of the health of citizens. Third, the reform of the ecological civilization system developed rapidly. Since 2014, the state has issued more than 40 reform plans involving the development of ecological civilization, including the *Opinions on Accelerating the Development of Ecological Civilization*, and the *Master Plan for the Reform of the Ecological Civilization System*, making comprehensive and systematic arrangements for the development of ecological civilization in all aspects, including overall goals, basic concepts, primary principles, key tasks, and institutional support. The central government established the Ministry of Ecological Environment and the Ministry of Natural Resources in an attempt to further improve relevant management systems and mechanisms. Fourth, remarkable achievements were made in specialized and environmental law enforcement. In July 2014, the Supreme People's Court established an environment tribunal to promote dedicated and specialized judicial practices at the national level. As of December 2018, 1272 specialized environmental courts and tribunals have been set up across the country to hear environment-related civil cases, criminal cases, and public interest cases. A 3 + 2 litigation model that handled both traditional and new litigation types and heard both common and special cases was established. The Supreme People's Procuratorate urged procuratorates to play a more active role in environmental public-interest litigation. In 2018, the Public Interest Litigation Procuratorate of the Supreme People's Procuratorate was established and corresponding institutions were also set up at lower levels to handle environment-related public interest litigations. Fourth, China is becoming an increasingly important participant in, contributor to, and leader for the development of a global ecological civilization. In 2015, Chinese President Xi Jinping attended the Paris Climate Conference and delivered a speech, calling on countries to jointly build a community with a shared future for mankind, deliver environmental justice, promote technological innovation, and embark on a path towards inclusive and sustainable development. China formally submitted its first nationally determined contribution plan to the United Nations as scheduled during the global climate change negotiations, and proactively pushed forward the negotiation process for the Paris Agreement. At the UN Climate Change Conference, China actively promoted negotiations concerning the implementation details of the Paris Agreement. It also joined the "Towards a Global Pact for the Environment" working group, attended relevant meetings, participated in various projects, and made clear its views on relevant issues. China attended many environmental conferences held by the UN, participated in themed discussions and deliberation, and made a continuous effort to promote concepts and practices concerning the development of ecological civilization, strengthen communication with all stakeholders, and assume a more constructive leading role in the development of a global ecological civilization and a community with a shared future for mankind.

2 Promoting the Environmental Rule of Law with Chinese Characteristics

Over the past 70 years, the Chinese government has made continuous efforts to promote environmental rule of law with Chinese characteristics. In May 2016, the United Nations Environment Assembly (UNEA) released *Green is Gold: The Strategy and Actions of China's Ecological Civilization*, pointing out that lucid waters and lush mountains are invaluable assets and that China's ecological civilization strategy offers a sustainable development solution to the international community.

1. **Environmental legislation with Chinese characteristics**

In 1973, the Office of the Preparatory Team for the First Environmental Protection Conference of China led the formulation of *Industrial Waste Control Standards (Trial)*, and promulgated *Regulations on the Protection and Improvement of the Environment (Trial)*. In September 1979, the national legislature adopted the Environmental Protection Law of the People's Republic of China (Trial) at the 11th meeting of the Fifth National People's Congress. In the initial stage, the focus of China's environmental legislation was on the control of industrial waste. At that time, addressing conflicts between different classes was the key task of the Chinese government, but the principles established by the *Regulations on the Protection and Improvement of the Environment (Trial)* and Environmental Protection Law (Trial) were reasonable given the country's national conditions at the time. Environmental legislation at the time reflected the Chinese government's commitment to strengthening pollution prevention. Whether by enacting new laws or revising old laws, the Chinese government has always proposed solutions to problems at the time with China's national conditions in mind. It has achieved three major transitions in environmental legislation in the past 70 years. First, the focus of environmental legislation transitioned from the control of industrial waste to the development of ecological civilization. With the promulgation of the 1978 Constitutional Amendment, the 2018 Constitutional Amendment, Environmental Protection Law (Trial) (1979) and Environmental Protection Law (Amendment) (2014) and the adoption of national strategies concerning environmental protection, the coordination between environmental protection and economic development, scientific outlook on development and green development, the value orientation of the Chinese government from development to environmental protection. Second, the scope of environmental legislation expanded from pollution control to ecological security. After the environmental legal system with the Environmental Protection Law at the core and natural resource conservation and pollution prevention as the main tasks has roughly taken shape, the Chinese government worked to accelerate legislation on ecological security, and paid more attention to the environmental legislation system. It incorporated ecological security into the National Security Law as an important part of the overall national security system, and provided for the "green principle" and added relevant content into the *General Principles of Civil Law*.[4] Third, the overall goal of environmental legislation has transitioned from the control of urban pollution and pollution by enterprises to sustainable development. China's environmental pollution spread from urban to rural areas and environmental problems became more and more serious. In response to this, the overall goal of China's environmental legislation shifted from controlling urban pollution and pollution by enterprises to promoting sustainable development and clarifying the government's responsibilities for the environment. In the Environmental Protection Law revised in 2014, "promoting the development of ecological civilization and sustainable economic and social development" was established as the legislative purpose. The government also established an environmental monitoring system, laying the foundation for the formulation and revision of more specific laws.

2. **An environmental law enforcement system that best fits China's national conditions**

After the State Council set up the Environmental Protection Leadership Group in 1973, China's environmental protection law enforcement system underwent many changes. The national environmental governance authority changed from the Environmental Protection Bureau of the Ministry of Urban and Rural Development to the Ministry of Ecology and Environment and the Ministry of Natural Resources. The national environmental law enforcement authority was upgraded to a higher status, and their functions were strengthened with more diverse legal means. Overall, China embarked on an environmental law enforcement path that best fits its national conditions. First, the law enforcement model shifted from administrative control to governance by multiple parties. From *Environmental Protection Law* to more than 30 specialized environmental legislations, China shifted away from its reliance

[4]In March 2017, *General Principles of Civil Law* was approved at the Fifth Session of the 12th National People's Congress. It stipulates that "All individuals and entities shall use resources in a sustainable manner and protect the environment when interacting with each other" (Chapter I, Article 9), which is called the "green principle." This provision is of great significance for regulating natural resource development and utilization activities and ensuring ecological safety.

on administrative orders, regional law enforcement, and post-event supervision. It clearly defined the rights and responsibilities of the government, enterprises, and individuals in environmental protection and established a modern environmental governance system characterized by joint governance by multiple parties, cross-regional law enforcement, and society-wide participation. On the one hand, governments and environmental protection authorities at all levels were granted powers of supervision and law enforcement. On the other hand, the People's Congress got to supervise local governments; disclosure to the public of environmental protection assessment results became a performance indicator for government officials; a strict administrative accountability system for environmental protection authorities was also established. Second, in the past, the environmental law enforcement authorities only supervised enterprises. Now, their job covers both enterprises and government bodies. In response to such an expansion of the regulation scope of environmental laws from enterprises to governments and from cities to rural areas, the Ministry of Ecology and Environment promulgated over 50 regulations, standards, and guidelines to support the implementation of the Environmental Protection Law (Revision) and clarify the responsibilities of law enforcement authorities and the rights and obligations of persons and entities, improved law enforcement procedures, established law enforcement mechanisms, and increased the transparency of environmental law enforcement. In the meanwhile, the central government rolled out green accounting, environmental auditing, party and government responsibility synchronization and lifelong accountability reform programs, strengthened environmental inspection, and prompted government bodies and enterprises to perform their environmental protection obligations and responsibilities by conducting interviews, approving quota, and fostering accountability. Third, the Chinese government encouraged environmental law enforcement institutions to collaborate. While authorizing environmental and natural resources management institutions to supervise and manage the ecological environment and natural resources, the law also stipulates the functions and powers of the relevant departments, requiring each department to clearly divide labor within its mandatory scope of power, and establish communication, cooperation, and coordination mechanisms to promote interdepartmental collaboration in law enforcement. At present, the Ministry of Ecology and Environment, the Ministry of Natural Resources and relevant departments have established coordination mechanisms for transfer of environmental violation cases, joint release of technical guidelines, pre-notification of public interest litigation with judicial authorities, environmental administrative consultation and litigation and other work. These mechanisms have played an important role in promoting the effective implementation of environmental law.

3. **An environmental justice system that reflects the wisdom of the Chinese people**

With the establishment of the first dedicated environmental court in 2007 and the establishment of the environmental resources tribunal in 2014 by the Supreme People's Court, judicial specialization in environmental law in the country picked up pace. In the past years, under the guidance of Xi Jinping Thought on Ecological

Civilization, the Supreme People's Court issued relevant judicial policies, emphasizing the need to adhere to principles set out in Xi Jinping Thought on Ecological Civilization when hearing environmental cases, the harmonious coexistence of man and nature, respect for the environment's capacity, and the belief that lucid waters and lush mountains are invaluable assets. It attaches importance to the ecological environment, advocates high-quality economic development, protects the basic rights and interests of the people, upholds green justice concepts such as the laws of nature, environmental protection first, and green development, and embraces the path of green justice with Chinese characteristics. First, the Chinese government adopted a bottom-up approach to promoting judicial specialization in environmental law. Marked by the establishment of local environmental courts, judicial specialization in environmental law began from the local level, with intermediate people's courts and lower-level courts playing a major role. These local environmental courts handled many major environmental pollution incidents. Local government officials pay high attention to judicial specialization in environmental law. Since 2010, the Supreme People's Court has issued a number of relevant documents and established environmental courts to promote judicial specialization in environmental law. Second, a set of rules were set out for the handling of environmental lawsuits. China made continuous efforts to advance judicial specialization in environmental law and integrate jurisdiction over environmental civil proceedings and some environmental administrative proceedings. To support the development of the Yangtze River Economic Belt, the Xiongan New Area and the national strategy of integrated development of the Beijing-Tianjin-Hebei region, the Supreme People's Court worked to reform the jurisdiction system for environmental cases in the Yangtze River Basin, the Beijing-Tianjin-Hebei region, and Sanjiangyuan; To meet the needs of judicial authorities and advance the reform of the ecological civilization system, the Supreme People's Court issued judicial interpretations, further refined and improved court procedures and rules for public-interest environmental cases, as well as lawsuits concerning mining rights, marine resources and environment damages. Third, the focus of the environmental case handling mechanism shifted from specialization to standardization. A 3 + 2 litigation model that handles both traditional and new litigation cases and hears both general and special cases was established. Arrangements for procedures of environmental criminal trials reflect the country's commitment to strengthening environmental protection. The rules for judging environmental civil cases emphasize the protection of ecological interests, while those for environmental administrative cases pay more attention to the substantive review of ecological and environmental protection. Clearly, China has been standardizing the handling of traditional and new environmental cases in parallel with advancing the judicial specialization in environmental law.

4. A path with Chinese characteristics to environmental rule of law

In addition to advancing environmental rule of law, China has also been beefing up education on environmental rule of law, and raising environmental awareness of the whole society. Nearly one thousand environmental NGOs and tens of thousands of volunteers have been active in pollution control and ecological protection, initiated public interest environmental legal actions, urged the government to handle environment-related matters according to law, and provided necessary help to the government. Improvement is seen in the public's awareness of green consumption, low-carbon lifestyle, environment issues, and their rights to participate, express opinions, and supervise actions. The country has also been building up the legal basis for public participation in environmental protection and promoting legal education on environmental laws. First, the Chinese government encourages the public to actively participate in environmental protection. The State Council promulgated and revised the *Measures for Public Participation in Environmental Protection*. The document defines the rights of citizens, legal persons and other entities to obtain environmental information and participate in and supervise environmental protection activities, emphasizes lawful, orderly, voluntary, and convenient public participation, sets out the procedural rules for public participation, and encourages the public to actively participate in environmental protection activities. The Supreme People's Court, the Ministry of Ecology and Environment, and the Ministry of Civil Affairs jointly issued documents to protect the rights of NGOs to initiate public interest environmental legal actions. More than 700 NGOs were qualified for initiating environmental public interest legal actions by the Ministry of Civil Affairs. Various non-profit environmental foundations, organizations and volunteer programs are developing rapidly and have actively participated in environmental decision-making, environmental law enforcement, and environmental protection actions. Second, China has been strengthening legal education on environmental law. To respond to the urgent need for professional talent in the initial development period of environmental rule of law, The State Council issued the *Decision on Strengthening Environmental Protection During the Period of National Economic Adjustment* in 1981, which clearly stated that "training of environmental protection talent should be included in the national education plan, environmental science should be included in the curriculum of primary and secondary schools, higher education institutions and secondary vocational schools must offer environmental protection courses for students majoring in science, engineering, agriculture, medicine, economics, and law, and colleges and universities with sufficient resources should offer environmental protection programs." According to this, colleges and universities across the country began to launch environmental and resources protection law programs. So far, 16 doctoral programs and 100 master's degree programs of environmental and resources protection law major are offered to candidates. At the same time, environmental legal education has also been included in the country's law popularization plans, national examinations for law professionals, civil servant training courses, and legal courses for non-law majors in undergraduate programs. The education authority attaches importance to both specialized talent development and universal education in the field of environmental protection. Third, China has been working to raise public awareness of environmental rule of law. Article 6 of the Environmental Protection Law stipulates: "Citizens should raise their environmental awareness, adopt a low-carbon and frugal lifestyle, and actively fulfill their environmental protection obligations." In order to enhance citizens' environmental awareness, publicity activities have been organized in various forms, including community-based training programs, the naming of the person of the year in the field of environmental law on the National Constitution Day, summer camps for volunteers participating in environmental protection actions, Open Day activities at courts, with NPC deputies and CPPCC members in attendance, pollution maps and urban environmental information disclosure indexes released by environmental NGOs, national press conferences held regularly to publish environmental quality information, etc. Creative methods have been coming out one after another for the promotion of environmental laws.

3 Conclusion

In the 70 years since the founding of the People's Republic of China, especially since the reform and opening-up, China has made remarkable strides in the development of environmental rule of law under the guidance of the theory of socialist ecological civilization with Chinese characteristics. Sound and fully science-based legislation guarantees good laws and good governance, strict law enforcement ensures the authority of the laws, a well-designed justice system upholds fairness and justice, and law-abiding citizens play a key role in promoting the development of humanity. The tasks of promoting environmental protection and advancing environmental rule of law are intertwined. With the improvement of the environmental legal system, China's environmental governance system and capabilities have also gradually improved. However, China's ecological and environmental problems are still quite serious. After a period of rapid economic development characterized by high investment, high consumption and high pollution, it is necessary to transition from unsustainable development to green development, and promote low-carbon and environmentally friendly production methods and lifestyles. This is by no means easy to achieve, though, and China needs a step-by-step plan.

Since the 18th National Congress of the CPC, General Secretary Xi Jinping has put forward a series of new concepts, new ideas, and new strategies for the development of ecological civilization, and set out the following six basic principles: (i) We must promote harmonious coexistence of man and nature; (ii) lucid waters and lush mountains are invaluable assets; (iii) a good environment is the fairest public good; (iv) mountains, rivers, forests, farmlands, lakes and grasslands are interconnected ecosystems; (v) we should adopt the strictest institutional and legal systems to protect the environment; and (vi) as a major country in the world, China should shoulder corresponding responsibilities for the development of global ecological civilization. General Secretary Xi Jinping also proposed that the government should accelerate the development of five systems—namely, ecological culture system, ecological economic system, target and responsibility management system, ecological civilization system, and ecological security system. These important instructions provided theoretical guidance for the development of environmental rule of law in China in the new era. Despite the achievements already made in China in advancing environmental rule of law, the country is now standing at a new starting point. It is still far away from achieving the goals of the Beautiful China campaign and needs to continue to promote environmental rule of law.

Environmental Protection in China in the Past 70 Years

Jinnan Wang

The development of the People's Republic of China in the past 70 years is the greatest miracle in the modern world. The past 70 years have seen China, with a population of more than 1.3 billion, grow from a poor country to the second largest economy in the world. The Chinese government has worked to keep its environmental protection policy up to date, with emphasis shifting from the control of wastewater, exhaust gas, and solid waste to the development of ecological civilization. With environmental protection at the center of modernization, it has proposed a grand blueprint for building a beautiful China, and is gradually moving toward the center stage of global environmental governance. China is joining hands with other countries to build a community with a shared future for mankind and a clean and beautiful world. This unprecedented ambition attracted worldwide attention. In particular, China's achievements in environmental protection and ecological civilization development since the 18th National Congress of the CPC received praise and recognition from the Chinese people and the international community.

1 Timeline of Environmental Protection in China

Since the founding of the People's Republic of China, remarkable strides have been made in the country in environmental protection despite some ups and downs, with emphasis shifting from the control of wastewater, exhaust gas, and solid waste to the development of ecological civilization and to building a beautiful China. Awareness of the importance of environmental protection has been raised to an unprecedented

Qin Changbo also contributed to this article.

J. Wang (✉)
Chinese Academy of Engineering, Beijing, China

Chinese Academy of Environmental Planning, Beijing, China

J. Pan et al. (eds.), *Beautiful China: 70 Years Since 1949 and 70 People's Views on Eco-civilization Construction*, https://doi.org/10.1007/978-981-33-6742-5_3

level. The development of China's attitudes and policies towards environmental protection can be roughly divided into five stages.

Stage I (1949–1978): take-off

After the founding of the People's Republic of China, the country desperately wanted to end poverty. It regarded air pollution and land reclamation as a necessary evil in order to promote development. Environmental pollution did not receive much attention from the central government at the time. There was even a big misunderstanding of the relationship between man and nature. Many people believed man could conquer nature in the end. Before 1972, it was generally believed in China that environmental pollution was a problem that only appeared in a capitalist society and China did not have environmental pollution. But in fact, in the early 1970s, environmental pollution in many places across the country was already very serious. Due to a lack of experience, urban infrastructure remained insufficient and extensive growth of resource-based industries prevailed. Low productivity resulted in excessive waste of resources. Environmental pollution and degradation began to pose a serious challenge to the country. Premier Zhou Enlai requested industrial waste management and utilization be incorporated into the national plan for economic development. There were only a few pollution control measures taken during this period, including building vegetation barriers for air pollution abatement and simple sewage treatment and smoke and dust removal facilities for polluting projects. The country did very little to protect the environment at the time.

In June 1972, the Chinese government sent a delegation to participate in the United Nations Conference on Human Environment held in Sweden. In August 1973, the State Council of China held the first National Environmental Protection Conference and adopted the *Environmental Protection and Improvement Regulations (Interim)*, proposing to plan environmental protection in a holistic and reasonable manner, improve resource efficiency, reduce the negative impact of human activities on the environment, and encourage everyone to participate in environmental protection, officially putting environmental protection on the Chinese government's agenda.

After the National Environmental Protection Conference in 1973, all provinces, autonomous regions, and municipalities, as well as relevant departments of the State Council established environmental governance institutions and environmental research and monitoring institutions, and strengthened pollution prevention and control as well as the treatment and utilization of wastewater, exhaust gas, and solid waste. At the same time, the country stepped up pollution control in heavily polluted areas, including the control of water pollution in rivers and lakes such as Guanting Reservoir, Fuchun River, Baiyangdian Lake, Ya'er Lake (in Wuhan), and air pollution in cities such as Beijing, Tianjin, Zibo, Shenyang, Taiyuan, and Lanzhou. The experience accumulated during this period proved to be useful for later river and urban pollution control efforts.

Stage II (1978–1992): environmental protection as a fundamental national policy

In December 1978, at the Third Plenary Session of the Eleventh CPC Central Committee, it was decided that China would launch its reform and opening-up and shift its focus to economic development. Economic reform soon started from the rural areas and then moved on to urban areas. Many labor-intensive industries, including some of the most polluting industries, were relocated from Japan, South Korea, Hong Kong and Taiwan to coastal areas of the Chinese mainland. This led to further environmental degradation and pollution increased with economic growth in many places across the country.

At the Second National Conference on Environmental Protection in 1983, the Chinese government framed environmental protection as a fundamental national strategy and announced its general policy for environmental protection, calling for simultaneous planning for economic development, urban and rural development, and environmental protection, simultaneous implementation of relevant plans, and coordination of solutions to economic, social and environmental problems. Environmental protection thus moved from a marginal position in state governance to a central position. China began to pay more attention to pollution control. The establishment of environmental protection as a basic national strategy laid a solid foundation for the government's efforts to strengthen environmental protection. This was in line with the concept of sustainable development put forward by the international community in the late 1980s. It was also suitable for China's realities.

The environmental policy and legislation system and environmental governance system began to take shape. In 1979, the Environmental Protection Law (Trial), the basic environmental law in China, was officially promulgated. Japan enacted its first basic environmental law in 1967, Sweden in 1969, the United States in 1970, the United Kingdom in 1974, and France in 1976. During this period, China's pollution control system began to take shape. The Chinese government issued eight polices concerning environmental protection targets, quantitative assessment of urban environmental improvement, pollution permits, centralized pollution control, pollution remediation, environmental impact assessment, the "three simultaneous" system, and the collection of pollution charges. It also set up dedicated environmental management institutions. In 1982, the Ministry of Urban and Rural Development and Environmental Protection was established under the central government and there was an environmental protection bureau under the ministry. In the 10 years before that, a temporary agency, the National Environmental Protection Office, performed environmental governance duties. In 1988, the Environmental Protection Bureau was separated from the Ministry of Urban and Rural Development and Environmental Protection, and became a vice-ministerial level agency directly under the State Council under the name "State Environmental Protection Administration." This reform has laid the foundation for dedicated environmental governance in China. Later, the national environmental governance authority was changed to State Environmental Protection Administration, Ministry of Environmental Protection, and Ministry of Ecology and Environment successively.

Stage III (1992–2002): establishment of the sustainable development strategy

During this period, China's cities and industrial parks developed rapidly, resulting in a decline in the total area of arable land and an increase in environmental pollution. The Huaihe River Basin and other areas faced serious environmental problems and sandstorms in Beijing became stronger and more frequent. The Yellow River dried up at some points for the first time in 1998. There was a folk song at the time, depicting the serious problem of environmental pollution in the country. According to the song, water in in the Huaihe River could be used to wash rice and vegetables in the 1950s and irrigate farmland in the 1960s, water quality in the river began to deteriorate in the 1970s, almost all fish and shrimps in the river died in the 1980s, and in the 1990s, the water quality further deteriorated.

During this period, the Chinese government launched the "One Control and Two Standards" program, requesting that by 2000, the country's gross pollutant discharge should be no higher than the 1995 level, and all environmental function zones and industrial pollution sources should meet relevant standards. In order to solve the pollution problem, the state has decided to launch the "33211" pollution control program for three rivers (Huaihe River, Haihe River, and Liaohe River), three lakes (Dianchi Lake, Taihu Lake, and Chaohu Lake), two control areas (sulfur dioxide and acid rain control areas), one city (Beijing) and one sea (the Bohai Sea). The implementation of the program led to significant improvement in environmental quality in key river basins. The country also launched a number of key environmental pollution control projects nationwide to control the total amount of major pollutants, as well as six key environmental remediation projects, including the project to turn farmland back to forests. It also implemented a dedicated action plan for pollution control for the Huaihe River and the Taihu Lake.

The Chinese government also developed a sustainable development strategy during this period. At the United Nations Conference on Environment and Development held in Brazil in 1992, the concept of sustainable development was put forward and *Agenda 21* was adopted. China attended the conference as a developing country and soon compiled *China's Agenda 21: White Paper on China's Population, Environment and Development in the 21st Century* and formulated its own sustainable development goals in 1994. In 1998, the Chinese government promoted the State Environmental Protection Administration from vice-ministerial to the ministerial level and stepped up environment protection efforts.

Stage IV (2002–2012): implementation of the scientific development concept

During this period, China began to emphasize both speed and quality of economic development. Environmental issues were included in the key considerations of the scientific development concept, and the importance of the transformation of economic growth model was also emphasized. The country began to embark on a sustainable development path by building capacities required for sustainable development, improving the environment and life quality of citizens, raising the efficiency of resource utilization, and promoting harmony between human beings and nature.

The primary environmental management method at this stage was controlling the total discharge of pollutants. After China acceded to the WTO, its economy grew sharply and the country leapt to the top of the world in terms of production capacity and the output of the chemical industry and energy, steel and other heavy industries. Resource consumption rose rapidly and the gross discharge of major pollutants also increased significantly. During 2006–2010, the Chinese government stepped up its environmental management efforts and formulated national targets for the reduction of major pollutant discharge and energy consumption per unit of GDP and included them in the Eleventh Five-Year Plan. The national targets were then broken down for each province, autonomous region and municipalities. The Chinese government launched pollution reduction projects, strengthened target-based environmental management, and requested local governments to attach greater emphasis to environmental protection. As a result of these efforts, the deterioration of environmental quality in the country was effectively curbed.

The increasing importance of pollution control and decision-making at a macro level to the effectiveness of environmental protection determined the development of national environmental management policies and the direction of institutional reforms. Environmental problems and economic problems are intertwined. Good economic policies must take environmental issues into consideration. During this period, the government provided financial incentives to support environmental protection initiatives. Financial incentives for desulfurization and denitrification at power plants pushed China forward in the competition for the world's largest clean coal power system. The urban sewage treatment pricing policy provided financial guarantee for the construction and operation of the largest urban sewage treatment system in the world. The number of environmental facilities and power plant desulfurization facilities built during 2006–2010 in China exceeded the total number of such facilities built in the country since 1949. In July 2008, the State Environmental Protection Administration became the Ministry of Environmental Protection, a ministerial-level department under the State Council. Moreover, regional and river-basin specific environmental issues began to receive more attention. In 2006, the State Environmental Protection Administration set up six regional environmental supervision and management institutions respectively for Northeast China, North China, Northwest China, Southwest China, East China, and South China.

Stage V (2012–present): ecological civilization and the Beautiful China campaign

With the development of heavy and chemical industries, China's resource consumption and pollutant discharge reached very high levels, and its air quality continued to deteriorate. Extreme weather occurred in many places, including the Beijing-Tianjin-Hebei region and its surrounding areas, seriously affecting people's daily life. Chinese people became increasingly aware that environmental problems affected not only economic development but also their life. Since the 18th National Congress of the CPC, the CPC Central Committee with Comrade Xi Jinping at its core has attached great importance to ecological civilization development and environmental protection, and incorporated ecological civilization development into the Five-Sphere Integrated Plan and the Four-Pronged Comprehensive Strategy for development. It has promoted sustainable development and a sustainable lifestyle in an all-round manner, promulgated air, water and soil pollution control policies and the *13th Five-Year Plan for Environmental Protection*, set up environmental inspection teams, and strengthened pollution prevention and control, improving environment quality continuously.

At the 19th National Congress of the CPC held in October 2017, General Secretary Xi Jinping announced that socialism with Chinese characteristics had entered a new era and once again emphasized the importance of ecological civilization development to the achievement of the country's sustainable development goals. Tasks such as promoting the development of an ecological civilization and adopting a strict environment protection system were included in the revised Party Constitution. At the first meeting of the 13th National People's Congress, an amendment to the Constitution of the People's Republic of China was adopted and ecological civilization was officially included. In May 2018, Xi Jinping delivered an important speech at the National Environment Protection Conference, systematically expounding his thought on ecological civilization and answering the question of why we should build ecological civilization, what kind of an ecological civilization we should build, and how to build it. This is a milestone in ecological civilization development and environmental protection in China.

2 Outstanding Achievements in Environment Protection

In the past 70 years, China has made great strides in social and economic development, reform and opening-up, and environmental protection. Under the guidance of Xi Jinping Thought on Ecological Civilization, local governments made unprecedented efforts to raise environmental awareness, promote environmental stewardship, and improve environmental protection practices. In China, theoretical understanding and practices concerning the development of an ecological civilization have undergone fundamental changes.

First, the Chinese government established a complete theoretical system for the development of an ecological civilization, i.e. Xi Jinping Thought on Ecological Civilization, which sets out the fundamental principles of ecological civilization development and environmental protection in China. At the 19th National Congress of the CPC, it was made clear that promoting harmony between human and nature was one of the primary tasks of China, and building an ecological civilization is vital to sustain the Chinese nation's development. The meeting also clarified the goals of the ecological civilization system reform and the Beautiful China campaign. The complete theoretical system for the development of ecological civilization, i.e. Xi Jinping Thought on Ecological Civilization established at the National Environment Protection Conference expounds the positioning, basic concepts, essential relationships, political requirements, goals, practices, foundation, and international perspectives of ecological civilization development and clarifies the eight principles of ecological civilization development: (i) thriving ecosystems are a precondition for a thriving human civilization; (ii) we must promote harmony between human and nature; (iii) lucid water and lush mountains are invaluable assets; (iv) a good environment is the fairest public good; (v) mountains, rivers, forests, farmlands, lakes and grasslands are interconnected ecosystems; (vi) we should adopt the strictest institutional and legal systems to protect the environment; (vii) we should mobilize the whole society for the Beautiful China campaign; and (viii) as a major country in the world, China should shoulder its due responsibilities for the development of an ecological civilization in the world.

Second, China is winning the battle against pollution. It has solved many major ecological and environmental problems. In 2018, the percentage of days with good air quality in 338 cities at and above prefecture level was 79.3%, and the average concentration of fine particulate matter (PM 2.5), a major indicator of air quality, was 43 $\mu g/m^3$, a decrease of 24.6% compared with 2015. Up to 36% of the cities met the requirements for air quality. In terms of water environment, the quality of 71.0% of water bodies was at or higher than Grade III, already exceeding the target set for the 13th Five-Year Plan period; the quality of 6.7% of water bodies was lower than Grade V. The country's chemical oxygen demand, ammonia nitrogen emissions, sulfur dioxide emissions, and nitrogen oxide emissions fell by 8.5%, 8.9%, 18.9%, and 13.1% respectively, and the carbon dioxide emissions per unit of GDP fell by 14.2%. The country strengthened coordination between different ecological protection and pollution prevention mechanisms, and comprehensively improved the stability and functions of various ecosystems. It also strengthened the control of non-point source pollution caused by agricultural activities, improved the rural environment, launched large-scale greening projects, and accelerated the comprehensive control of soil erosion and desertification. The country continued to implement the projects to turn farmland to back to forests and grasslands as well as pilot projects of cropland rotation and fallowing, and built national parks, ecological corridors and ecological networks for biodiversity conservation.

Third, ecological and environmental protection has accelerated supply-side structural reforms and effectively promoted high-quality development. China stepped up the effort to cut backward production capacity and overcapacity. As of 2018, the country had cut 155 million tons and 810 million tons of capacity in steel and coal production, respectively. It shut down enterprises illegally discharging pollutants, strengthened the environmental regulation on enterprises and industrial clusters, and took effective measures to prevent bad money from driving out good money. The market environment of industrial products became healthier and better regulated, and the capacity utilization rate of major industries increased significantly. Measures were also taken to optimize the energy and transportation structures of the country. As of 2018, the country had built the world's largest clean coal power sector, and the capacity of coal-fired power plants with ultra-low-emissions in China reached more than 810 million kW, accounting for 80% of the total installed capacity of coal-fired power generation in the country. Clean energy accounted for 22.1% of China's energy consumption, and the share of coal power dropped to 59%. In 2018, China's railway freight transportation increased by 19.9% compared with the 2015 volume, accounting for 8.0% of the total, and the coal and iron ore port in the Beijing-Tianjin-Hebei region reduced the region's reliance on road transportation for coal and iron ore. The country continued to improve its motor vehicle retiring and repairing system, and the production, sales and ownership of new energy vehicles has been on the rise.

Fourth, the top-level design for the reform of the ecological civilization development system has been basically completed, laying a solid foundation for environmental protection in the new era. The Chinese government has issued relevant policy documents such as *Opinions on Accelerating the Development of Ecological Civilization*, *Overall Plan for Ecological Civilization System Reform*, *Measures for the Evaluation and Assessment of Ecological Civilization Development Targets*, and *CPC and Government Leader Accountability for Environmental Damage (Trial)*, and created the top-level framework and development targets for the ecological civilization development system. The Ministry of Ecology and Environment was established during this period and environmental protection gained weight in the central government's decision-making process. A nationwide pollution permit system was put into place. As of September 2019, China had issued more than 70,000 pollution permits to companies engaged in 24 major industries including thermal power and paper making. The reform of the ecological civilization development system progressed smoothly with steady steps and the environmental legislation, governance, and law enforcement mechanisms and capabilities continuously improved.

Fifth, the environmental awareness of Chinese citizens has been significantly improved, and green lifestyles and green consumption have gained popularity. The country took effective measures to raise the environmental awareness and promote a sustainable way of thinking in the whole society. Chinese people's understanding of the relationship between economic development and environmental protection changed profoundly. More and more people now accepts the concept that lucid waters and lush mountains are invaluable assets. There are diverse channels for people to access environmental information. Environmental monitoring authorities are required to publish air quality data (from hourly to annual, from individual cities to regions). Environmental public interest litigation and social supervision mechanisms gradually improved, and green consumption and green lifestyles are gaining popularity.

Sixth, local governments have actively explored typical models of ecological civilization development and environmental protection. Since the 13th Five-Year Plan period, the Ministry of Ecology and Environment has named 95 national demonstration cities and counties (districts) for ecological civilization development, and constructed 29 "Green is Gold" innovation centers. In 2018, the Ministry of Ecology and Environment named 11 cities and 5 special economic zones including Shenzhen "waste-free cities." Various reform measures and innovations have been introduced in Fujian, Jiangxi, and Guizhou ecological civilization pilot zones. Sixteen provinces and more than 1000 cities and counties are striving to become ecological provinces, cities and counties. The 263 Action Plan of Jiangsu and efforts to build Guangdong Green Development Demonstration Zone have achieved remarkable results. The Ten Million Project in Zhejiang and the Saihanba Forest Farm have won Guardian of the Earth award granted by the United Nations Environment Programme. Zhejiang Province launched a beautiful countryside campaign and held up Anji, where a thriving bamboo industry greatly boosted the local economy, as a model for rural sustainable development.

Seventh, China's ecological civilization strategy and actions have contributed to the sustainable development of the world. In 2016, the United Nations Environment Programme released the report *Green is Gold: The Strategy and Actions of China's Ecological Civilization*, pointing out that China's ecological civilization strategy and actions contributed to the sustainable development of the world by not only addressing China's own environmental challenges but also serving as a demonstration for other countries with similar economic, environmental and social challenges. In 2019, the United Nations Environment Programme and UN-Habitat released *A Review of 20 Years' Air Pollution Control in Beijing* and *Making Cities Sustainable Through Rehabilitating Polluted Urban Rivers: Lessons from China and Other Countries*, pointing out that Beijing's experience in controlling air pollution and China's experience in restoring polluted rivers can serve as excellent examples for other countries and cities. China is working with other countries to advance the Paris Agreement, the Green Belt and Road Initiative and other international conventions.

3 Blueprint for Building Beautiful China

At the 19th National Congress of the CPC, the Beautiful China campaign was launched, depicting the vision of a clean and beautiful world. The meeting also set the following goals: China should fundamentally improve the environment, and basically achieve the goal of the "Beautiful China" campaign by 2035, and build a prosperous, democratic, civilized, harmonious, beautiful modern socialist country by the middle of the twenty-first century. China's experience in the past 70 years shows that environmental protection involves complex and changeable contexts and relationships, and the situation across the country is very different. If actions are not coordinated with a clear strategic goal in mind, we will be unable to produce a synergistic effect. To achieve the goals of the Beautiful China campaign by 2035 and the mid-twenty-first century, we must keep up with the change in the primary contradiction faced by the Chinese society in the new era, follow the guidance of Xi Jinping Thought on Ecological Civilization, and adhere to the basic policy of "conservation, protection, and restoration." China should make environmental protection the top priority, promote green development, and continuously increase the supply of high-quality ecological products to meet the people's growing demand for a beautiful ecological environment. It should be visionary and systematically plan key tasks and objectives in relation to ecological protection, environmental governance, risk prevention, and response to climate change; coordinate for high-quality economic development and high-level environmental protection; develop top-level designs, blueprints, schedules and pathways for environmental protection with the main goals and milestones of the 14th and 15th Five-Year Plan periods, as well as the targets for 2035 and the 100th anniversary of the founding of the People's Republic of China in mind, and take a stepwise approach to environmental improvement.

The strategic roadmap for building a beautiful China from the present to the middle of the twenty-first century is aligned with China's economic and social development plan and the plan for solving its ecological and environmental problems. It can be summarized as follows:

By 2025, on the basis of completing the building a moderately prosperous society in all respects, China will consolidate the results of its critical battle against pollution, advance four major structural reforms, continue to improve environment conditions, including air quality, ensure the water quality of all water bodies is better than Grade V, effectively control soil pollution, significantly reduce the total discharge of major pollutants, foster environmental management capability, and make remarkable strides in building a beautiful China.

By 2030, China will comprehensively its improve environmental conditions; water and air quality will basically meet relevant standards; water quality across the country will meet the requirements set for functional zones, the functions of water ecosystems will be basically restored, the quality of soil environments nationwide will be steadily improved, the total amount of pollutants discharged into the environment will be drastically reduced; traditional environmental problems will be effectively dealt with; there will be mechanisms to remove environmental risks to human health; results of the Beautiful China campaign will be seen clearly in some places, and China will have established a virtuous cycle of development and environmental protection.

By 2035, the environmental quality will be radically improved and the goals of the Beautiful China campaign will be basically achieved. First, the spatial distribution, industrial structure, production methods, and lifestyles that are conducive to resource conservation and environmental protection will come into shape. The country will gradually transition to a green, low-carbon and circular economy, and green development methods and lifestyles will become common practice. Second, the carrying capacity of the environment will be greatly improved. The environmental quality nationwide will be up to relevant standards; the air, water and soil quality will steadily improve; environmental risks will be under control in all aspects; the functions of various ecosystems will be restored with steady steps; and China will be more beautiful. Third, the modernization process of China's environmental; governance system and capabilities will be basically completed.

By the mid-twenty-first century, China will become a great modern socialist country that is prosperous, strong, democratic, culturally advanced, harmonious and beautiful; green development approaches and lifestyles will be widespread, and the environment will be significantly improved together with the living standards of people in a modern socialist country; ecosystems will run in a virtuous cycle; nature will regain its tranquility, harmony and beauty; human and nature will be in harmony; the ecological civilization of China will become more mature; the modernization process of the country's environment governance system and capabilities will be completed; and the goals of the Beautiful China campaign will be realized.

China will continue to build on its achievements in the past 70 years, follow the guidelines of Xi Jinping Thought on Ecological Civilization, and strive to achieve the goal of building a beautiful China. This will not only turn China into a leading, modern industrialized and green economy but also enable China to gradually become a center of a world for environmental protection and an important participant, contributor, and leader of global environmental protection initiatives. China will continue to offer valuable ideas and solutions for the advancement of the global ecological civilization, and work with other countries to build a community with a shared future for mankind and a clean and beautiful world.

Development and Effects of Nature Conservation in China in the Past 70 Years

Jixi Gao

In the 70 years since the founding of the People's Republic of China, nature reserves have increased in both quantity and variety and the focus of China's nature conservation efforts has shifted from the construction of protected areas to the development of a regional ecological security pattern. China's nature conservation efforts have been continuously strengthened, leading to remarkable results. In 1956, China built its first nature reserve—Dinghushan Nature Reserve. After more than 60 years, China's nature conservation system has gradually matured. The nature conservation system consists of geological parks, natural and cultural heritage sites, wetland parks, aquatic germplasm reserves, marine protected areas, and specially protected islands. On this basis, China successively added ecological function areas (2008), ecologically fragile areas (2008), key ecological function areas (2011) and other key ecological conservation areas to the system. In 2011, China first put forward the proposal of "delineating the ecological conservation red line." In 2015, China launched a pilot reform of its national park system and started building a nature conservation system in which national parks played a major role. These two major measures further enriched China's nature conservation system and significantly accelerated the formation of a sound national ecological security pattern.

China has made remarkable strides in the following three aspects of nature conservation: (i) improvement of the protected area system; (ii) development of the ecological protection red line system; and (iii) the national park system reform.

Zou Changxin and Xu Mengjia also contributed to this article.

J. Gao (✉)
Satellite Environment Application Center, Ministry of Ecology and Environment, Beijing, China

J. Pan et al. (eds.), *Beautiful China: 70 Years Since 1949 and 70 People's Views on Eco-civilization Construction*, https://doi.org/10.1007/978-981-33-6742-5_4

1 A Protected Area System to Effectively Protect Important Natural Resources and Natural Landscapes

According to statistics, China has more than ten categories of protected areas, including nature reserves, scenic areas, forest parks, world cultural and natural heritage sites, geological parks, wetland parks, public drinking water source areas (PDWSAs), water conservancy scenic areas, enclosed areas where desertification is in progress, special marine reserves (including marine parks), germplasm resource reserves, and national parks (pilot). As of 2018, there were more than 10,000 protected areas, including 3766 at the national level. The total area of the various protected areas on land accounted for about 18% of China's total land area, higher than the world average. Specifically, nature reserves accounted for about 14.8% of the country's land area and more than 80% of the total area of all protected areas. Scenic areas and forest parks accounted for about 3.8% of the total area of protected areas. The proportions of other types of protected areas were relatively small. At the beginning, protected areas were mainly established according to industries and ecological elements, and were guided by Chinese Academy of Sciences and authorities in charge of the management of ecological environment, forests, grassland, agriculture rural areas, natural resources, housing and construction, water conservancy, and oceans. After the institutional reform of 2018, all types of protected areas were put under the unified administration of the Ministry of Natural Resources, but local governments were still responsible for the management of the human resources, property, and assets of protected areas within their respective jurisdictions.

Over the years, protected areas have played an important role in protecting biodiversity, natural landscapes and natural heritage, maintaining national and regional ecological security, and ensuring the sustainable development of China's economy and society.

1. **Nature reserves**

A nature reserve is a protected area of land or water body of importance for rare and endangered flora and fauna, or features of special interest, which is reserved and managed for purposes of conservation. With the rise of the global biodiversity conservation movement and the improvement of environmental awareness, countries around the world attach great importance to nature reserves, taking them as a symbol of an advanced civilization.

In 1956, China established its first nature reserve, Dinghushan National Nature Reserve, in Guangdong. Over the past 60 years, the CPC Central Committee, the State Council, local governments at all levels and relevant departments all attached great importance to the protection of the natural environment and sustainable use of natural resources, and built a large number of nature reserves for this purpose. China's nature reserves have grown in quantity, size, and variety. A complete nature reserve system with a basically reasonable layout is now in place with its functions strengthening with time.

As of the end of 2016, there were 2750 nature reserves in China (excluding Hong Kong, Macao and Taiwan, the same below), and the total area of them registered 147.33 million hectares. The land area of nature reserves was approximately 142.88 million hectares. Nature reserves on land account for 14.88% of China's land area.

Nature reserves are mainly set aside for the protection of natural ecosystems, wildlife and natural heritage. The country has built nature reserves for over 90% of terrestrial natural ecosystems, 89% of national-level key protected wild animal and plant species and most important natural heritage sites. Wild populations of some rare and endangered species in the country are gradually recovering. Giant pandas in the wild totaled more than 1800, making its way from the list of endangered species into the list of vulnerable species; the populations of species such as Amur tiger, Amur leopard, Asian elephant, and crested ibis have increased significantly; elk, once extinct in the wild, was reintroduced through the establishment of nature reserves for elk and its population grew steadily, regarded by the international community as an example of successful biodiversity conservation.

Natural reserves are divided into two levels: national and sub-national. There are 446 national nature reserves, accounting for only 16.2% of the total, covering 96.95 million hectares, 65.8% of the country's total protected areas and 9.97% of the country's land area. The total number of sub-national nature reserves reached 2304, covering an area of 50.39 million hectares. Among them, 870, or 31.6%, are provincial nature reserves, covering an area of 37.56 million hectares, 25.5% of the total area of nature reserves in the country; 414 are municipal ones, 15.1% of the total, covering an area of 4.96 million hectares, or 3.4% of the total area of nature reserves; and 1020, or 37.1% of the total, are county-level nature reserves, covering 7.86 million hectares, or 5.3% of the total.

Since the establishment of the first nature reserve in 1956, China has basically formed a nature reserve system with a basically reasonable layout and a full range of functions as well as a relatively complete system of policies, regulations and standards for nature reserves. It also has a relatively complete nature reserve management system and a research support system. These systems have played a big role in promoting resource protection, research and survey, as well as publicity and education.

In recent years, more and more nature reserves have begun to engage in international exchanges and join relevant international protection networks. As of the end of 2016, 33 nature reserves in China, including Xilingol Reserve in Inner Mongolia and Laotie Mountain National Nature Reserve on Snake Island in Liaoning Province, had been included in the International Man and Biosphere Reserve Network of UNESCO; 46 nature reserves, including Shuangtai River Nature Reserve in Liaoning and Xianghai Nature Reserve in Jilin, had been included in the Ramsar Convention's list of internationally important wetlands; 37 nature reserves, including Wuyishan Nature Reserve in Fujian and Zhangjiajie Nature Reserve in Hunan, have been officially recognized as UNESCO World Heritage sites; and 40 nature reserves, including Wudalianchi Nature Reserve in Heilongjiang and Lushan Nature Reserve in Jiangxi, had been included in the Global Geoparks Network.

2. **Other types of protected areas**

In order to better protect the country's natural resources, biodiversity and ecological environment, in addition to nature reserves and national parks, China has also built a large number of other types of protected areas such as scenic spots, forest parks, geological parks, and wetland parks.

(1) Scenic areas. Scenic areas are places with nice views or of cultural or scientific significance. They are usually places where natural and cultural attractions and beautiful views concentrate. People visit these places to enjoy nice views or conduct scientific and cultural activities. They are natural and cultural heritage sites established by the government according to law, and have functions concerning ecological protection, cultural heritage protection, scientific research, tourism, leisure and regional development as well as ecological, aesthetic, scientific, and cultural values. The scenic areas in China have distinct Chinese characteristics. They have beautiful natural views and rich cultural heritage. They are like museums of natural history and cultural history. They are the embodiment of harmonious coexistence between human and nature. They are the common wealth of the Chinese nation.
In 1982, China formally established its system of scenic areas, with national and provincial scenic areas. As of the end of 2017, the State Council had approved 244 national-level scenic areas in nine batches, covering an area of approximately 100,000 km^2; provincial governments had approved more than 700 provincial scenic areas, covering an area of approximately 90,000 km^2. The two categories together made a total area of about 190,000 km^2. A total of 32 national and eight provincial scenic areas are UNESCO World Heritage sites.
(2) Forest parks. A forest park refers to a large area of forest that has been made into a park for people to enjoy. It has rich biological resources and relatively concentrated natural and cultural attractions. It is an area of forest with multiple functions that are managed for the primary purpose of conservation. It provides people with various forms of tourism services and people can carry out scientific and cultural activities in it. Forest parks are divided into three levels: national, provincial/municipal, and county. In September 1982, China officially approved its first forest park—Zhangjiajie National Forest Park in Hunan. As of the end of 2017, the total number of national forest parks reached 881, and the total area reached 12.7862 million hectares (including planned forest parks), accounting for 1.3% of the country's total land area.
At present, China has created a forest park construction, management and development system that covers geological, water, meteorological, cultural, and other resources. The establishment and development of this system has not only enabled the effective protection of a large number of precious natural and cultural heritage sites in forest areas, but also effectively promoted environmental protection and nature conservation on the whole. It has played an important role in protecting forest scenic resources, promoting the ecological

culture, meeting the needs of the public for a better life, and driving targeted poverty alleviation.

(3) Geological parks. The concept of a geological park dates back to the 1980s. In February 1999, UNESCO launched the Geoparks Program, which gave rise to the term "geopark." Following the principle that geological heritage should be simultaneously developed and protected, geoparks have played a major role in the protection of geological heritage and the environment, popularizing geoscience, providing employment opportunities, advocating science-themed tourism, and improving public scientific literacy. The remarkable benefits brought by geoparks have been universally recognized by local governments and all sectors of society. Geoparks have become one of the best ways to use natural resources in a sustainable way.
China is one of the few countries in the world with a full range of geological heritage sites distributed widely across the country. China's geoparks are divided into international, national, provincial, and county-level geoparks. As of 2017, China had built 239 national geoparks and more than 100 provincial geoparks, 35 of which had been included in the list of UNESCO Global Geoparks. In Taiwan of China, there are also village-level geoparks. Thus, China has basically put in place a sound geopark network featuring all types of geological heritage and covering the entire country.

(4) Wetland parks. A wetland park is an area of wetland turned into a park for wetland ecosystem conservation, demonstration of the functions and sustainable use of the wetland ecosystem, and science popularization and education. Wetland parks have cultural and aesthetic values and are managed for scientific research and public enjoyment. Wetland parks are natural ecosystems adapted to cities. They are public parks with multiple functions. At the same time, they are also open systems that play an important role in maintaining urban ecological balance and promoting sustainable development. They are managed mainly for science education, sustainable use of resources, and public enjoyment.
Since joining the Ramsar Convention on Wetlands of International Importance in 1992, China has been working to strengthen wetland protection and promote sustainable use of wetlands, and giving full play to the functions of wetlands for the benefit of both the present and future generations. In 2003, the State Council approved the Plan for the National Wetland Protection, providing new impetus for the development of wetland parks in China. Xixi Wetland Park formally joined the national wetland park pilot program of the State Forestry Administration in 2005. As of 2017, the total areas of natural wetland parks in China reached 21.85 million hectares, and 706 national wetland park pilot projects had been approved nationwide. Among them, 98 had passed relevant inspections and were formally named national wetland parks and 49 had been recognized as internationally important wetlands.

(5) Drinking water sources. China is unique in being both water-rich and water-poor. China is rich in water resources in terms of the total quantity: The total quantity of freshwater resources in China registers about 2.8 trillion m^3, accounting for 6% of the world total and ranking fourth in the world. However,

China is also listed by the United Nations as one of the most water-scarce countries as its per capita quantity of water resources is only 2200 m^3, merely a quarter of the world average, ranking 121st in the world. Rapid economic development and urbanization have exacerbated the water crisis in Chinese cities. Therefore, the protection of water resources is particularly important.
Public drinking water source areas (PDWSAs) are divided into two categories: surface water protection areas and groundwater protection areas. It is estimated that about three quarters of the world's PDWSAs are surface water protection areas and only one quarter are groundwater protection areas. A 2016 survey showed that more than 2400 surface water protection areas were established at the national level, of which 618 were included in the National List of Important Public Drinking Water Source Areas (2016).

(6) Marine protected areas (including marine parks). A marine protected area is an area of sea that has special geographical features, ecosystems, biological and non-living resources and is managed through effective means for conservation and marine development and utilization. Marine protected areas are scientifically planned, managed in accordance with relevant standards and utilized in a sustainable manner primarily for conservation purposes. Marine protected areas are open for fully science-based and rational development and utilization activities while effectively protecting marine ecosystems and restoring resources, thereby maintaining a good balance between protection of the marine environment and resource utilization.
Since China established its first national marine protected area in 2005, marine protected areas in China have been developing rapidly. At present, a network of various marine protected areas and marine parks, including those dedicated to the protection of special geographical conditions, marine ecosystems, and marine resources, has taken shape. As of 2014, China had 56 national marine protected areas including 30 marine parks, covering a total area of 69,000 km^2.

(7) World heritage sites. A world heritage site is a site on the UNESCO World Heritage List that contains cultural and natural heritage considered to be of outstanding value to humanity. As a well-known ancient civilization, China ratified *The Convention Concerning the Protection of the World Cultural and Natural Heritage* in 1985. As of July 2017, China had 52 sites on the UNESCO World Heritage List, ranking top in the world along with Italy. Among them, 32 sites are cultural sites, 12 are natural heritage sites, 4 are cultural and natural heritage sites, and 4 are cultural landscape sites.
China's World Heritage sites has effectively protected important ecosystems and natural heritage, including habitats and natural ecosystems of endangered and rare species such as giant pandas and Yunnan golden monkeys, *danxia* landforms, karst landforms, granite landforms, sandstone landforms, paleontological fossils, and other geological heritage, beautiful natural landscapes such as mountains, lakes, forests, and unique cultural landscapes such as religion, mountains and rivers, ancient architecture, farming practices. They follow the principles of ecological civilization development (respecting the nature,

following the law of nature, and protecting nature; coordinating between development and environmental protection; lucid waters and lush mountains are invaluable assets; mountains, rivers, forests, farmlands, lakes and grasslands are interconnected ecosystems, etc.) and have played a special role in promoting the development of ecological civilization.

2 Ecological Red Lines to Effectively Maintain Ecological Integrity of China

Entering the twenty-first century, China saw its ecological protection work develop rapidly, and the top-down implementation of the relevant top-level design is progressing in an orderly manner. In July 2008, the Ministry of Environmental Protection and the Chinese Academy of Sciences issued the National Ecological Function Zoning Plan, which set aside 50 important national ecological function zones covering 2.37 million km^2, accounting for 24.8% of the country's total land area. In order to strengthen the protection of ecologically fragile areas, control ecological degradation, restore ecosystem functions, and improve the quality of the ecological environment, in September 2008, the Chinese government released the *General Plan for the Protection of the National Ecologically Fragile Areas*, which clarified the geographical distribution, current status and characteristics of the country's eight ecologically fragile areas and the guiding ideology, principles and tasks of ecological protection. The document provides scientific guidelines for restoring and rebuilding a sound ecological environment in ecologically fragile areas. In June 2011, the *National Plan for Major Function Zones* was released, dividing the country into the following major function zones: (i) According to development approach, there are areas for optimized development, key development areas, areas for development with restrictions, and areas where development is prohibited. (ii) According to development purposes, there are urban areas, main agricultural production areas and key ecological function areas. The National Plan for Major Function Zones systematically and comprehensively outlines the strategic pattern for maintaining ecological security of China. At the core of the pattern are the Qinghai-Tibet Plateau Ecological Barrier, the Loess Plateau-Sichuan-Yunnan Ecological Barrier, the Northeast Forest Belt, the Northern Desertification Prevention Belt, and the Southern Hilly Area Ecological Belt. The pattern is formed in accordance with the principle of respecting nature and following the law of nature.

In 2011, delineation of ecological red lines became an important strategic task of the Chinese government [*Opinions of the State Council on Carrying out Key Environmental Protection Tasks* (GF [2011] No. 35)]. The Chinese government delineated important and key ecological function zones as well as terrestrial and marine ecologically sensitive and vulnerable areas for permanent protection, which reflected China's determination to strengthen ecological protection by compulsory measures at the national level.

On February 7, 2017, the General Office of the Central Committee of the Communist Party of China and the General Office of the State Council issued *Several Opinions on Delineating and Strictly Observing Ecological Red Lines*, which clarified the overall requirements and specific arrangements for the delineation of ecological red lines. Since then, the topics of ecological red line research has expanded from zoning to basic theories, delineation methods, and management measures. Literature on ecological red lines have become more comprehensive, multidimensional and practical. The delineation of ecological red lines is a project led by the state.

Compared with the existing protected areas across the world, the ecological red line system of China consists of three types of red lines: red lines for ecosystem services areas, red lines for disaster mitigation and control areas, and red lines for biodiversity maintenance areas. The system integrates various existing protected areas, and adds areas that are extremely important to ecosystem services and areas with extremely sensitive and fragile ecological environment. The ecological protection zoning system has become more comprehensive and formed a more scientific distribution pattern with more prominent regional functions and more stringent control measures. It is fair to say that the system is a major improvement to the existing international protected area system. The ecological red line system has strengthened environmental protection and ecosystem services, and played an important role in maintaining ecological security and promoting sustainable development of the country. It not only effectively protects biodiversity and important natural landscapes, but also plays an important role in purifying the air and building the capacity of the water environment. Therefore, ecological red lines are considered another lifeline of the country following the red line for cultivated land in China.

At the same time, ecological red lines are also land development control lines, which is of great significance to the improvement of the country's overall land development pattern.

3 A National Park System to Maintain the Harmonious Coexistence Between Mankind and Nature

It is stated in the *Decision of the Central Committee of the Communist Party of China on Major Issues Concerning Comprehensively Deepening Reform* adopted at the Third Plenary Session of the 18th Central Committee of the CPC in November 2013 that the country should build a national park system. Building a nature reserve system with national parks at its core is very important to ecological and environmental protection in China.

The two major functions of national parks are protection of the ecological environment and natural resources and moderate development of tourist resources. National parks provide effective protection on a large scale and ensure that the development and utilization of resources is appropriate and not in conflict with the protection goals. They not only protect the integrity of ecosystems but also provide the public with opportunities and places for tourist activities, scientific research, education, and entertainment. They provide an effective protection and management model with which we can reasonably better balance environment protection with resource development and utilization. This model coordinates between conservation and development. It not only effectively protects the ecological environment and biodiversity, but also greatly promotes the development of local tourism and economy and the sustainable use of resources.

At present, there are nearly 10,000 national parks in nearly 100 different countries around the world. Different countries have different definitions for national parks. In China, a national park refers to a large area of land or ocean designated and managed by the state for protection of typical natural ecosystems and scientific protection and rational use of natural resources. National parks are one of the most important types of protected areas in China. They are areas where development activities are prohibited according to the *National Plan for Major Function Zones*. They are sufficiently protected by the national ecological red line system and the most stringent protection measures.

National parks in China put ecosystem protection first. They protect places that should be protected most and leave precious natural heritage for future generations. They are owned by the country and reflect national interests and the country's image. They are public assets owned by all Chinese people. National parks focus on improving ecosystem services, disseminating environmental knowledge, and providing the public with opportunities to get close to and understand nature.

At present, China has approved 10 national park pilot projects, namely the Sanjiangyuan National Park Pilot Project, the Siberian Tiger and Amur Leopard National Park Pilot Project, the Giant Panda National Park Pilot Project, the Qilian Mountain National Park Pilot Project, the Shennongjia National Park Pilot Project in Hubei, the Wuyi Mountain National Park Pilot Project in Fujian, the Qianjiangyuan National Park Pilot Project in Zhejiang, the Nanshan National Park Pilot Project in Hunan, the Great Wall National Park Pilot Project in Beijing, and the Pudacuo National Park Pilot Project in Yunnan. At present, these pilot projects have produced and put into use various management regulations or methods, and rolled out ecological and environmental protection, environmental accountability, and natural resource audit policies. Aiming primarily to maintain the integrity of ecosystems, national parks cover nature reserves, geological parks, forest parks, scenic areas and other types of protected areas with a certain area. The national park system ensures that a protected area only has one brand and one management agency. Many national park pilot projects, such as the Sanjiangyuan National Park Pilot Project, the Shennongjia National Park Pilot Project, the Wuyi Mountain National Park Pilot Project, the Nanshan National Park Pilot Project, the Qianjiangyuan National Park Pilot Project and the Siberian Tiger and Amur Leopard National Park Pilot Project, have set up administration or management committees to take the place of the original institutions of various protected areas in their respective regions. After such reorganization efforts, previously unconnected protected areas were connected. Currently, these national park administration or management committees are working to ensure a clear-cut division of labor for each unit and employee.

It was pointed out in the report to the 19th National Congress of the CPC that it is necessary to step up the effort to build a natural reserve system with national parks at its core. For some time in the future, relevant authorities will continue to build the protected area system in accordance with relevant requirements in the Five-Sphere Integrated Plan and the Four-Pronged Comprehensive Strategy, focus on improving the effectiveness of management and protection, prevent unreasonable development and utilization, shift the focus from quantity to quality and from extensive growth to intensive growth, and vigorously promote the protection and restoration of natural ecosystems.

Thriving ecosystems are a precondition for a thriving human civilization. To provide Chinese people with more high-quality ecological products, achieve the goals of the Beautiful China campaign, and make the world a better place for future generations, we must give priority to conservation and passive restoration, keep in mind that mountains, rivers, forests, farmlands, lakes and grasslands are interconnected ecosystems, strictly implement the ecological red line system, build a complete protected area system, optimize the distribution of the country's ecological network, ensure that economic development activities are in harmony with the carrying capacity of the environment for development, and maintain national and regional ecological security.

Environmental Protection and Sustainable Development in China: Review and Outlook

Yi Wang

Environmental protection in China formally started 40 years ago in the 1970s. On the one hand, China has made remarkable strides in both development and environmental protection; on the other hand, it is still under tremendous environmental pressure. The country still has a long way to go before it can truly achieve green, low-carbon and sustainable development. Domestic and international environment has changed and China has much more to do to continuously promote the development of an ecological civilization and home and abroad while maintaining high-quality social and economic development.

It is important to be aware that environmental protection has entered a new stage in China. First, as the income of Chinese citizens increases and the Chinese economy transitions to a new normal, environmental awareness grows steadily and a new technological revolution that drives green, smart, and inclusive growth is brewing in China. These tremendous socioeconomic changes have laid the foundation for the transformation of the relationship between the environment and development. Second, after the Five-Sphere Integrated Plan was proposed in the report to the 18th National Congress of the Communist Party of China (CPC), ecological civilization development in China entered the stage of top-level design and systematic advancement. With the introduction and implementation of a series of reform programs and relevant policies, the pattern of green transformation and development started to take shape. In any case, establishing a sound environmental governance system and laying the institutional foundation for China's efforts to promote green transformation and development, solve major resource and environmental problems, and realize good coordination between the government, enterprises, and social forces for governance will be key to the development of ecological civilization and to China's economic and social development for a considerable period of time in the future.

Y. Wang (✉)
Standing Committee of the National People's Congress, Beijing, China

Science and Technology Strategy Consulting Institute, Chinese Academy of Sciences, Beijing, China

J. Pan et al. (eds.), *Beautiful China: 70 Years Since 1949 and 70 People's Views on Eco-civilization Construction*, https://doi.org/10.1007/978-981-33-6742-5_5

1 A Brief Review of Development and Environmental Protection in China

The history of environmental protection in China is a history of balancing between environmental protection and development needs. Based on a comprehensive comparative analysis, I divide China's process of environmental protection and development into four stages: Stage 1 centers around the building environmental awareness, Stage 2 is marked by the adoption of a problem-oriented approach, Stage 3 features the scaling up of environmental governance, and Stage 4 is a stage of systematically promoting environmental protection. In this section, I will list the progress made at each stage, summarize the experience and lessons of late-starting countries in pursuing sustainable development, and give a summary of China's theories and practices relating to the balance between development and environmental protection and its contributions to the global efforts towards environmental sustainability and sustainable development.

1. **Stage 1—building environmental awareness (1973–1978)**

China had its first National Environmental Protection Conference in 1973, marking the official inclusion of environmental protection into the Chinese government's responsibilities. At the conference, the Chinese government promulgated the *Environmental Protection and Improvement Regulations (Interim)*, proposing to plan environmental protection in a holistic and reasonable manner, improve resource efficiency, reduce the negative environmental impact of human activities, and encourage everyone to participate. In 1974, the Environmental Protection Leadership Group of the State Council was established, and environmental protection was officially included in the agenda of the central government. The focus of the group at the time was the treatment and utilization of wastewater, exhaust gas, and solid waste.

In fact, environmental protection had always been a controversial topic in China before this. Although the Great Leap Forward, the Cultural Revolution and the Third Front Movement caused various environmental problems in China, the general understanding among the general public at that time was that environmental problems were unique to capitalist countries and had nothing to do with China as it was a planned economy. In 1972, a Chinese delegation attended the United Nations Conference on the Human Environment. It was only until then that Chinese official began to realize that many environmental issues are global problems and we should not turn a blind eye to them. With the Office of the Environmental Protection Leadership Group of the State Council as the competent authority specifically responsible for environmental protection, China made significant progress in environmental protection.

2. **Stage 2—adoption of a problem-oriented approach (1979–1998)**

China began to step up environmental protection efforts towards the end of 1978, in parallel with the launch of the reform and opening-up. Milestones in this stage included the enactment of the Environmental Protection Law (Trial) in 1979, the inscription of environmental protection as a basic national policy in 1983, and the adoption of sustainable development as a national strategy in 1996.

After the launch of the reform and opening-up, China made a series of major decisions concerning environmental protection, formulated many important policies and laws, and proposed a strategy for simultaneously promoting economic growth, urban and rural development, and environmental protection. China first had a dedicated national environmental governance authority in 1988 and upgraded it once and again after that. In 1998, the Environmental Protection Bureau was reorganized into the State Environmental Protection Administration (a ministerial-level institution) directly under the State Council in charge of environmental protection. The Chinese government also internalized the UN Sustainable Development Goals (SDGs) and formulated *China's Agenda 21*, and then broke down national sustainable development targets by sector.

Due to the strategy of prioritizing economic growth and the limited resources and capabilities in various aspects, China's environmental protection efforts mainly focused on conventional pollution control and pollution prevention in key areas. During this period, the Chinese government launched a project to reduce pollution in "three rivers, three lakes, two control areas, one city and one sea" (i.e. the Haihe River, the Huaihe River, the Liaohe River, the Taihu Lake, the Dianchi Lake, the Chaohu Lake, the sulfur dioxide pollution control area, the acid rain control area, the city of Beijing, and the Bohai Sea) and achieved remarkable results. China's environmental policies at this stage were oriented towards problems and challenges encountered in reality, aiming primarily to reduce pollution and meet environmental targets. However, due to the huge ecological debt accumulated over the past years and the failure to get a grasp on the big picture, the country in general still faced serious environmental problems.

3. **Stage 3—scaling up environmental governance (1999–2012)**

After 1999, China truly entered the stage of large-scale environmental infrastructure construction and ecological restoration. With the significant increase of the country's economic strength, under the dual pressure of the 1998 Yangtze River flood and the massive sandstorm in North China, the central government initiated a large-scale ecological restoration project focusing on protecting natural forests and turning farmland back into forests and grasslands. The project built environmental infrastructure, especially sewage treatment plants, in urban areas to pay the ecological debt, reduce the environmental impact of economic growth, and lay the foundation for overall environmental improvement.

At the same time, after China's accession to the WTO, the rapid growth of manufacturing had brought about a huge increase in resource consumption and pollutant emissions, especially carbon dioxide emissions, in the country. Buoyed by the sheer size of China's economy and population, China saw most of its environmental indicators rose rapidly to the top of the world, which not only attracted worldwide attention but also put increasing pressure on pollutant emission reduction.

Another feature of the environmental policies in this stage was the guiding role of advanced concepts, which was a result of Chinese people's deepened understanding of the relationship between the environment and development, the emergence of new problems, the expansion of areas affected by environmental problems, and the characteristics of China in its current stage of development. After 2002, China put forward a series of new concepts such as scientific outlook on development, resource-efficient and environment-friendly society, circular economy, low-carbon development, and ecological civilization. During 2006–2015, the Chinese government began to implement legally binding energy consumption and emission reduction targets and build a comprehensive indicator system consisting of emission intensity, total emissions, and structural indicators and a corresponding accountability mechanism. It embarked on a path of "proposing new ideas, setting goals, launching pilot programs, and rolling out a replicable model." It also enacted a series of environmental laws, such as the Environmental Impact Assessment Law, the Renewable Energy Law, and the Circular Economy Promotion Law. This was conducive to problem solving and the implementation of relevant plans, but it also made it difficult to respond to specific problems in a systematic, scientific and differentiated manner.

4. **Stage 4—systematically promoting environmental protection (2013–present)**

After the concept of ecological civilization was proposed at the 18th National Congress of the CPC in 2012, China started to promote environmental protection in a systematic manner. Since 2013, the CPC Central Committee and the State Council have issued 40 important policy documents such as the *Decisions of the Central Committee of the Communist Party of China on Several Major Issues of Comprehensively Deepening Reform*, *Opinions of the Central Committee of the Communist Party of China and the State Council on Accelerating the Construction of Ecological Civilization*, and *Overall Plan for the Reform of the Ecological Civilization System*. It also developed a theoretical system for the development of an ecological civilization with Xi Jinping Thought on Ecological Civilization at its core.

Along with the rapid development brought about by the reform and opening-up, China has accumulated more than 40 years of experience in environmental protection. It now has the capabilities required to develop top-level design and realize systematic transformation. The theory of ecological civilization development in the new era emphasizes protecting the environment with sound systems, attaches importance to the modernization of environmental governance systems and capabilities, and points out that the quality of the environment is the key and that "we should prioritize the environment and promote green development." However, from a more comprehensive perspective, since China is still a developing country in the primary stage of socialism, development and environmental protection should still be advanced in good coordination.

After more than five years of hard work, the environmental quality in China has improved at an unprecedented rate and China's environmental protection system has seen dramatic changes. At the same time, China's economic structure changes, the society undergoes transformation, and the situations at home and abroad have seen profound changes, it is now necessary for China to adjust its timeframe, path, and priorities of development. Standing at a new starting line, we must pursue a deeper understanding of the concept of ecological civilization, accurately grasp the basic situation that China is in and the challenges faced, plan for the development of a socialist ecological civilization with Chinese characteristics, actively participate in global environmental governance, and contribute to the global response to climate change and global sustainability.

Over the past 40 years, one of the most important lessons that the Chinese people learned is to make development and environmental protection mutually supportive. Since the launch of the reform and opening-up policy, China has experienced rapid economic growth and continuous social transformations, and gained deeper and deeper understanding of the relationship between development and the environment. On the one hand, due to the constraints posed by the current development stage, comprehensive capabilities, and international division of labor, we have to make appropriate trade-offs between environmental protection and economic growth and choose to prioritize different development goals, still failing to avoid the pollution-first path. On the other hand, with the continuous improvement of the Chinese government's environmental governance capabilities and Chinese people's environmental awareness, environmental protection has climbed up in the list of the Chinese government's policy priorities. It should be pointed out here that China's understanding of the environment and its ecological value also changed greatly with time and it is thus necessary and important that we adjust our environmental protection measures, standards, and policies as we move forward. The ultimate goal is to strike a balance between the environment and development and maintain harmony between human and nature.

We should also see that, although the targets of China's ecological environment governance system are clear, there are still different views on the relationship between the what the reform aims at and what is currently in place, the combination of top-level design and basic-level practices, comprehensive advancement, priorities in system development, goals and objectives, improvement of governance capabilities, the pace of the reform and relevant action plans. We must realize that ecological civilization development and sustainable development are systematic projects which require consensuses and participation of the whole society. We need to comprehensively deepen our understanding, overcome the constraints imposed by vested interests, solve theoretical and practical problems, keep the appropriate pace, promote innovation and systematically advance the development of an ecological civilization.

2 Main Issues and Trends in Ecological Civilization Development

1. Main issues

The task of building an ecological civilization, which is highly relevant to the long-term development of China, remains onerous. Land use planning needs to be further improved. The spatial layout of economic activities is currently not in harmony with the spatial layout of resources. A lot of agricultural and protected land is illegally used for other purposes. The country still lacks a comprehensive vision for future land and sea development. The supply of high-quality ecological goods and services is insufficient, and the country's forest stock per unit area is less than 80% of the global average. The way we use our resources is still unsustainable. The energy consumption per unit of GDP is still about twice the world average, and resource productivity is far below the world average. Economic restructuring and relocation of industries still face a lot of challenges, and technological innovation and industrial competitiveness in the field of sustainable development remain relatively low.

The large gap between reality and expectation concerning environmental protection is a major issue to be tackled in China. First, current environmental protection measures have not yet produced results up to the expectations of the general public. Smog is common in big cities during winter and water pollution remains a prominent problem seriously affecting Chinese people's perception of what efforts have been made. Second, the benefits of some ecological projects, such as projects concerning ecosystem improvement, land use improvement, and efficient use of resources, are not easily felt in people's daily life. Third, the implementation of some environmental policies, such as land trade and tree protection policies, may directly affect the vital interests of some groups. Fourth, there is a gap between people's need for rapid improvement in environmental quality and their willingness to pay for it. The "not-in-my-backyard" (NIMBY) attitude affects the process of economic restructuring and urban transformation.

The eagerness for success during the transition period often leads to low effectiveness of actions. China is shifting its focus from development to environmental protection now. The gradual transformation of ideas, systems, and actions is crucial to solving environmental problems. However, this transition cannot be accomplished overnight. It must go through a long process and be achieved in a systematic manner. We must respect relevant rules and avoid seeking instant benefits. For example, when the theoretical foundation remains weak, we should not over emphasize the role of the environmental property rights market, including natural resource balance sheets, emission rights trading market, water rights market, energy rights market, and energy savings market. Market-based solutions have not really gained much success worldwide due to barriers such as basic conditions, time and space constraints, and inter-system conflicts.

China still has a long way to go before it could build an effective ecological civilization system. Fragmentation is a prominent problem. A long-term mechanism has not yet been established, and we still need to work on coordination. The ecological civilization system is still in its infancy, with many constraints and few incentives. Some policy documents have just been promulgated but not yet implemented. Some systems are difficult to build and are still in the pilot stage. It is difficult to coordinate and balance between multiple stakeholders involved in some systems. Since some systems are based on theoretical assumptions, the high transaction costs and a lack of practical work pose big challenges. In addition, due to shortcomings of the accountability and fault tolerance mechanisms, "perfunctory" or "formalistic" behavior are not uncommon in local authorities.

2. **Trends**

In the short term, the pressure on resources and the environment will likely remain high. The next five to ten years are expected to be critical for the overall improvement of resource efficiency and environment quality in China. Internationally, for cities such as London and Los Angeles, it was 20–30 years before air pollution was effectively curbed. According to our analysis, China's resource- and pollution-intensive industries will reach their peaks or plateaus in the five to ten years. This means that both conventional pollutant discharge and consumption of major raw materials will peak around 2020. The pressure on China's resource reserves and environment is still high. From 2020 to 2030, as China's total population reaches and passes the peak and industrialization and urbanization in the traditional sense will be basically completed, the total fossil energy consumption and carbon emissions of the country will also peak or plateau and then start to decline, and the quality of the environment is expected to improve on the whole.

The new technological revolution is sweeping across different sectors and is characterized by intelligence, integration, environment-friendliness, inclusiveness and interaction. At the same time, as AI penetrates fast into people's daily life, green lifestyles spread relatively slow. It takes time to popularize the concepts of green production and consumption. AI makes progress very rapidly, but green technologies move forward at a much slower pace and require high costs to employ. Green consumption models such as quality-oriented consumption, collaborative consumption, and service-oriented consumption have emerged, but public awareness is not yet high enough. Consumers seem more inclined to adopt green consumption behavior that brings health, safety, and economic benefits. Currently, they are not willing to pay a premium price for green products.

Anti-globalization sentiments are growing and the leading position for global green development is vacant. Anger over the unfairness of globalization has led to the rise of populism and protectionism in various countries and a decline in the international community's interest in the transformation towards green and low-carbon development. Realism gradually gained an upper hand. After the United States announced its withdrawal from the Paris Agreement, the international community has high expectation that China will take over leadership in global climate mitigation. In any case, the uncertainty in China–U.S. economic relations will have a profound impact on the global political and economic order and give rise to concerns over security risks.

3 Theoretical Basis for Building a Comprehensive Ecological Civilization System

Building an ecological civilization system is not only about formulating a new set of rules under the guidance of the environment-first value orientation, but also about the inheritance and development of existing institutional arrangements. To build a good ecological civilization system, we must first identify the reasons behind the failure of a system or management model. In fact, natural ecosystems and their problems are much more complex than we believe. They have multiple dimensions such as time, space, structure, and function. Therefore, it is necessary for us to thoroughly understand the laws of natural evolution and the corresponding management models, so as to find a scientific way to build the system.

1. **Integrity of natural ecosystems and diversity of institutional arrangements**

A natural ecosystem demonstrates spatial integrity after long-term evolution. That is, living organisms and their environment form a whole. Environmental problems caused by social and economic activities of human beings have become local, regional and then global problems. The most prominent ones in China are regional air pollution (smog) and river basin water pollution. However, an environmental problem may affect multiple administrative divisions, making it difficult for the current administration system to solve such problems. From the perspective of structure, elements of the environment (water, soil, air, etc.) are all interconnected, and these elements are often under the jurisdiction of different authorities, causing a lack of integrity and coordination in environment-related issues. From the perspective of functions, an element of the environment often has multiple functions. For example, water is not only fresh water resources, but can also be used for power generation and provide various ecological services. To achieve the optimal allocation of different functions of water at the river basin level, we need regulations to coordinate well between different management institutions. However, under the current system, there are various institutional obstacles to the comprehensive management of resources and environment.

According to traditional economic theories, environmental problems are caused by market failures resulting from externalities, and policy failures may aggravate environmental problems caused by market failures. In order to give full play to the roles of the market, the government and the public, in addition to adopting administrative orders and implementing environmental regulations, we can adopt the following three methods to deal with market failures and policy failures: (i) we can improve natural resources markets where competition for natural resources trigger exclusion, clearly define the property rights of resources and the environment (such as water resources and forest resources) and eliminate market distortions caused by inappropriate policies such as unreasonable government subsidies; (ii) we can create new markets for energy-saving and pollution/emission rights trading and make good use of economic incentives (such as environmental taxation); and (iii) we should not use any market-based approach to tackle environmental problems relating to natural assets which are public goods and environmental resources or ecological services that are difficult to monetize, and it is necessary to improve information quality and governance structure and encourage stakeholders and the public to participate in collective actions based on long-term balanced interests and consensus.

Therefore, in addition to ensuring ecological integrity, we need to make corresponding institutional arrangements for different types of environmental problems.

It is obviously very difficult for a single institution to oversee environmental governance issues in China, a large developing country with a vast territory and complex and diverse environmental problems. However, no matter which reform solution is adopted, it is necessary to establish a framework for coordination within the complex environmental governance system.

2. **Path dependence and cognitive inertia in China's environmental governance system**

China's current environmental governance system, formed after evolution over a long period of time, has the following characteristics: First, the system does not have a comprehensive design due to a lack of proper understanding in the old days. Instead, it has evolved as different problems emerged. Individual policies were issued before there was any overarching legislation. Related management functions also fell under different authorities and became departmental vested interests. Comprehensive management is only expressed in words but not backed by deeds. Second, traditional environmental management arose from a centrally planned economic system. The system fails to reasonably and clearly assign functions and powers and there is significant overlap in responsibilities. At the same time, there is a lack of effective coordination mechanisms and institutional arrangements for solving cross-departmental or cross-regional issues. Third, the system mainly adopts administrative methods, especially administrative approval and emission control yet penalties alone are inadequate to deter potential offenders. Other problems include poor enforcement, lack of effective economic incentives, lack of environmental management targets, and insufficient institutional arrangements for public participation in environmental governance.

Currently, efforts are made to break down the targets of the ecological civilization system reform to share the responsibilities across different levels. Although this is absolutely necessary for the reform, there are also potential concerns. Because stakeholders have different understandings of ecological civilization development, their actions are probably not well-aligned. Insufficient planning and departmental misdirection may result in redundancy and inefficiency, thus affecting reform progress. There are various signs that cognitive inertia remains one of the major obstacles in the work of relevant departments, who, overly eager to secure their current status, fail to think creatively.

At present, different departments have taken different paths in both the setting of and targets and plans and the running of pilot projects. Conflicts will therefore be inevitable. For example, different departments and local authorities have formulated or are working on different indicator systems ecological civilization development that are not compatible with one another. This not only reflects differences in perceptions, but also shows that these government departments have forgotten the purpose of formulating the indicator system in the first place. What the general public wants may be just as simple as seeing blue sky or being able to swim in clear waters. As for ecological civilization pilot projects, in addition to the first batch, China has also launched water, ocean, forest, town, demonstration area, and national ecological civilization pilots and national park system reform pilot projects, but not much has been done to review the results, summarize the experience and learn the lessons.

We should be aware that sufficient emphasis must be laid on top-level design rather than department-led initiatives when promoting the reform and development of an ecological civilization system. Therefore, it is necessary to build a comprehensive ecological civilization governance system, further strengthen overall coordination, timely summarize the problems and lessons of pilot projects, solicit opinions from various stakeholders, and avoid simply breaking down overall reform targets to form departmental targets. We should also prevent departments from using ecological civilization development as an excuse for putting departmental interests first, which goes against the purpose of the ecological civilization system reform.

3. **Basic principles for building an environmental governance system**

As mentioned above, building an ecological civilization system is a long-term task. We should maintain a fully science-based approach and hold firmly onto the strategic focus and priority areas, make a clear timetable and roadmap, improve important existing mechanisms, accelerate the development of relatively mature mechanisms, identify and fully investigate controversies, strengthen overall coordination and hierarchical classification, and promote the development of the ecological civilization system in an orderly manner.

Based on the basic theories of modern ecology, economics and management science, the experience of environmental protection in developed countries, the characteristics of modern environmental governance systems, and the current goals and requirements set by the Chinese government for the modern environmental governance system, the development of the environmental governance system should follow the principles set out below:

(1) We should assign roles and responsibilities and reshape the mindset of governance authorities on the premise of a balance between development and environmental protection. The relationship between development and environmental protection is not only the most important relationship in relation to China's environmental protection, but also a relationship that is very tricky to deal with. In developing countries, neither side should be neglected. It is necessary to set this balance as an important part of the goal of the environmental governance system. The government needs to fully consider the requirements of environmental protection when allocating public management responsibilities, managing budgets and developing the evaluation system for government officials, and establish a comprehensive decision-making mechanism for maintaining balance between development and environmental protection. Similarly, enterprises need to strike a balance between pursuing corporate interests and protecting public environmental interests, and comply with relevant national laws and standards and assume corporate social responsibilities. The public should actively participate in environmental protection activities and adopt sustainable and green lifestyles.
(2) We should reasonably allocate responsibilities and resources based on the principle that the government, enterprises and the public should share responsibilities for the environment. The government, enterprises and the public are the

key stakeholders in a modern environmental governance system. Based on the modern governance theory and international experience, it is necessary to regulate the rights and obligations of key stakeholders in environmental governance in accordance with the law, and to rationally set up the three regulatory mechanisms for government regulation, market allocation and public participation in environmental governance in order to produce synergy. Currently, the Chinese government's environmental governance functions are quite clearly defined. Once the government effectively performs its functions, the next step is to focus on building the power of enterprises, NGOs and individual citizens and letting them fully play their roles by reforming the decision-making process and leveraging the market mechanism.

(3) We should analyze the characteristics and problems of natural ecosystems and how they evolve, and on this basis reform and improve our environmental protection system and mechanisms. The goal of the environmental governance system is to properly handle the relationship between man and nature, and between people. We must respect the characteristics and the pattern of evolution of ecosystems when promoting the development of the environmental protection system and mechanisms. A natural ecosystem demonstrates spatial integrity after long-term evolution. That is, living organisms and their environment form an integral whole. Environmental problems caused by social and economic activities of human beings grow into local, regional and global problems.

Based on this, to create an environmental governance system, it is necessary to consider not only the overall characteristics of the environment, but also the natural, social and regional characteristics of the various elements of an ecosystem (water, soil, air, etc.). We should consider the economic nature and ecological and public good nature of natural resources and ecosystem services. In the meantime, we should also consider both the quantity and quality of environmental resources, as well as the externalities and public good nature of ecological goods and services. Other factors that need to be considered include the scales of time and space that we are dealing with, i.e. whether they are global, regional, or local, as well as what functions they perform, at what time, and how to coordinate for a solution of various existing environmental problems. For example, it will not do to simply put the management of biological resources and mineral resources together. Another example to the point is water management. Water not only serves as fresh water resources, but also can be used for power generation and provide various ecological services. Environmental issues concerning water involve both water quality and spatial distribution. We need to adopt a holistic approach to water resource management. As for waste management, we should consider classification, recycling, and the final disposal of waste in order to gain more with less effort. Therefore, we must, on the basis of fully understanding relevant laws, carefully define the roles and understand the limitations of the government and the market, and design sound environmental protection and governance systems and mechanisms accordingly.

4 Suggestions for Improving the Ecological Civilization System

We should strengthen the CPC's leadership in building an ecological civilization and promoting system and mechanism innovations. To tackle the problem of low enthusiasm in some local authorities, we should enhance monitoring, evaluation, and supervision in the process of the reform to keep track of progress; make good use of data platforms, scientific evaluations, audits, inspections, reviews, accountability systems, damage compensations and the like in coordination; provide more support for specific regions (especially central and western China) where more active efforts are made for innovations, environmental protection tasks are especially onerous, and transition to green development is more difficult. Thus, we will encourage innovations for ecological civilization development at the local level and boost local stakeholder engagement.

We should step up our efforts to build an ecological civilization and promote the rule of law. Efforts should be made to promote the "greening" of the legal system and coordinate for the formulation and revision of relevant laws, regulations and standards in accordance with the Five-Sphere Integrated Plan; step up legislation efforts in key fields such as land resources management, environmental protection, and climate change response, and coordinate for the formulation and revision of laws and regulations on delegated legislation concerning ecological civilization development; study the feasibility of regional and river basin-based legislation; accelerate the improvement of the legislative capabilities of local legislatures, especially legislatures of cities with districts under their jurisdiction who have only recently been granted legislative power; strengthen the enforcement of ecological civilization-related laws and build dedicated trial mechanisms, institutions and procedures in a rational manner.

We should prioritize environmental protection, promote green development, and formulate a plan for the Beautiful China campaign as well as for the transition to a green economy. We should push for the formulation of a plan for the Beautiful China 2050 campaign; define strategies, goals, policies and action plans for green, low-carbon and sustainable development towards 2035 and 2050; build consensuses, pool resources and promote stakeholder participation; forge good political, social and market expectations for green innovation and green transformation; develop industrial and product standards based on targets for energy conservation, environmental protection, and emission reduction so as to promote the green transformation and upgrading of the manufacturing industry; and establish green transformation funds to support to the relocation of industries and enterprises to the central and western regions with relevant technologies and financial resources.

We should clarify the relationship between systems, drive system integration with breakthroughs towards the completion of basic tasks, and advance pilot projects with enhanced efforts, so as to effectively reduce the costs of the ecological civilization system reform. Specifically, we should do more research into the relationship between reform tasks, further clarify the basic functions of each system, and on this basis, incorporate relevant tasks into a comprehensive framework designed with clearly defined priorities. In regard to resources-as-assets management, the reform of the natural resource property rights system should be advanced in good coordination with the clarification of the rights of all stakeholders, and the reform should be planned in a way that stakeholders at different levels and in different categories are dealt with in a differentiated manner. In terms of infrastructure construction for ecological civilization development, progress should be accelerated towards a unified resource and environment information platform and give authorizations so that different authorities can share the information on the platform according to their respective scopes of responsibilities.

We should comprehensively promote the reform of the natural resource property rights system and coordinate between asset management, licensing and administrative supervision. On the basis of the reform of the natural resource property rights system, we should incorporate the reforms of asset management, administrative supervision, natural resource property rights market, pricing, taxation, and licensing system into a well-designed systematic reform framework. Based on the relationship between the property rights system and government regulation, pilot projects should be launched for the natural resource property rights system reform. The reform should be planned in a holistic manner with easy tasks performed before difficult ones with sound connections and steps with limited consensus and high costs should wait till later. In addition, the current environmental problems in China are mostly related to the lack of effective legislation for property rights of natural resources and the transfer of such property rights. Therefore, it is necessary to first establish an effective natural resource property rights system and corresponding markets, and then clarify the rights and responsibilities of the central and local governments and other stakeholders in relation to different types of environmental public goods.

We should improve the reform evaluation and supervision mechanism, build a flexible error correction mechanism, and dynamically adjust reform plans. First, we should build an independent third-party evaluation mechanism. The client should be the reform coordinator instead of the plan maker and third-party evaluation institutions and expert pools should be gradually established. Third-party evaluation reports should all bear signatures, so that the signatory can assume responsibility for the quality of the report. There should also be technical specifications in place for third-party evaluation. Second, we should put in place strict accountability and incentive mechanisms, truly and effectively hold relevant stakeholders accountable for ineffective implementation, and reward those who have made remarkable progress. Third, we should build reform error correction and process management mechanisms, summarize experience and correct errors in each stage in a pragmatical and timely manner. For tasks that are not well implemented, it is necessary to adjust and improve the way things have been done or cancel the task. For practices that deviate from the instructions of the central government, rectifications should be made in time.

We should effectively expand market and public participation mechanisms, and establish a sound ecological civilization governance system. There should be a unified and standardized environmental data platform to promote cross-departmental and cross-regional sharing of environmental information. Efforts should be made to improve the mechanisms for public participation, communication and consultation in environmental impact assessment of projects in the planning and construction phases. Guidance should be provided for the healthy and orderly development of environmental NGOs. The market-based incentive system should be improved and improvement should also be made to the licensing and licensee rights protection systems, and the feedback mechanism for private investment in environmental governance. More financial and taxation incentives should be given and intellectual property rights protection should be strengthened to encourage green innovations.

The Mass Line: An Important Methodology Used by China to Promote the Development of Ecological Civilization in the Past 70 Years

Yunfei Zhang

Since the founding of the People's Republic of China, the Communist Party of China has creatively used the mass line to promote the development of an ecological civilization. It attached great importance to promoting the development of ecological civilization through mass efforts. It not only launched environmental (green) campaigns with Chinese characteristics, but also built an environmental (green) governance system with Chinese characteristics. The mass line provides strong and lasting driving forces for the development of an ecological civilization in China.

1 Innovations in Applying the Mass Line to Ecological Civilization Development

The CPC led the Chinese people through various revolutions, construction processes, and reforms and based on this proposed the mass line, which is a combination of the epistemology of Marxism and Marx's view on the masses: The Party must serve the needs of the masses, rely on the masses for its strength, and draw inspirations from the masses. The mass line is one of the most important tools of the Party. Since the founding of the People's Republic of China, the CPC has used the mass line in a creative way to promote the development of an ecological civilization.

In the early days of the People's Republic of China, the CPC, led by Mao Zedong, used the mass line in environmental governance. In 1952, Zhou Enlai proposed to engage the masses in soil and water conservation. In 1957, Mao Zedong proposed the Patriotic Health Campaign. In 1957, Zhu De proposed to call upon the masses to plant trees across the country. In addition, the water conservancy authorities proposed in the

Y. Zhang (✉)
National Academy of Development and Strategy, Renmin University of China, Beijing, China

School of Marxism Studies, Renmin University of China, Beijing, China

J. Pan et al. (eds.), *Beautiful China: 70 Years Since 1949 and 70 People's Views on Eco-civilization Construction*, https://doi.org/10.1007/978-981-33-6742-5_6

early 1950s that a mass water conservancy campaign should be carried out to control floods. Soil and water conservation, sanitation, afforestation, and water conservancy facility construction are the basic projects for an ecological civilization. Therefore, the above-mentioned mass campaigns are all of great significance to ecological civilization development. On this basis, at the United Nations Conference on Human Environment held in 1972, the Chinese delegation put forward that China's policy for environmental protection was "sound planning, reasonable layout, comprehensive utilization, turning harm into benefits, relying on the masses, engaging everyone, for the benefit of the people." This is how the Party's mass line was applied in environmental protection. In 1973, this policy was officially announced at the first National Environmental Protection Conference and in 1979, it was included in the Environmental Protection Law of the People's Republic of China (Trial). Thus, the mass line became an important feature of China's environment work.

After 1978, on the basis of restoring and developing the Party's mass line, the CPC, led by Deng Xiaoping, initiated the nationwide voluntary tree planting movement. On February 23, 1979, the sixth meeting of the Standing Committee of the Fifth National People's Congress assigned March 12 to be China's Arbor Day. On March 5, 1980, the Central Committee of the Communist Party of China and the State Council issued the *Instructions on Vigorously Promoting Tree Planting and Afforestation*, requesting people in urban and rural areas and from all walks of life be mobilized to plant trees. On March 8, 1981, the *Decision of the Central Committee of the Communist Party of China and the State Council on Several Issues Concerning the Protection of Forests and the Development of Forestry* stated that everyone is responsible for greening the motherland. On December 13, 1981, the Fourth Session of the Fifth National People's Congress adopted the *Resolution on Carrying out the National Voluntary Tree Planting Campaign*, which was a call for "everyone to plant trees every year," citing the ancient Chinese legend of a foolish old man moves mountains with perseverance. On the Arbor Day of 1982, the State Council issued the *Measures for the Implementation of the National Voluntary Tree Planting Campaign*. This is how the mass afforestation movement was carried out and institutionalized.

After 1992, in order to fulfill China's solemn commitment to the international community, proceeding from the country's large population and low per capita resource endowment, the CPC, led by Jiang Zemin, established sustainable development as a major strategy for the country's modernization and began to encourage "public participation." In 1994, *China's Agenda 21* proposed that participation of the public would determine how the sustainable development goals would be achieved. Therefore, sustainable development requires support and participation of the public and social groups. On August 3, 1996, the *Decision of the State Council on Several Issues Concerning Environmental Protection* stated that China should establish a public participation mechanism, give play to the role of social organizations, encourage the public to participate in environmental protection, report and expose cases of violations of environmental protection laws and regulations. Thus, public participation became an important way of applying the mass line to sustainable development in a market economy.

In the process of building a moderately prosperous society in all respects, after 2002, the CPC, led by Hu Jintao, put forward the strategic vision of a scientific outlook on development and building a harmonious society. Sustainable development is part of the scientific outlook on development, and a harmonious society is a society in which man and nature coexist in harmony. It was also proposed that population, resources and environment work involves vital interests of the people, and the fundamental interests of the overwhelming majority of the people must be taken as the starting point and end point. It is necessary to fully mobilize the enthusiasm, initiative and creativity of the people, to meet the needs of the people and promote all-round development of the people, to improve people's quality of life and health conditions, and to create good production conditions and a good life for the people. A good environment creates good conditions for the long-term development of the Chinese nation. This pointed out the direction for the application of the Party's mass line to China's pursuit of sustainable development. On this basis, on October 15, 2007, the 17th National Congress of the CPC creatively put forward the concept, principles and goals of an ecological civilization and pointed out that we must rely on the people and mobilize all factors to build a harmonious socialist society. Thus, in building an ecological civilization, China effectively connected public participation with the mass line.

As China became the world's second largest economy and launched its campaign to complete the building of a moderately prosperous society in all respects, on November 8, 2012, the 18th National Congress of the CPC incorporated ecological civilization as part of the overall deployment for socialism with Chinese characteristics, and required faster advance towards a social management system featuring Party leadership, government responsibility, social collaboration, public participation, and legal support. Since the 18th National Congress of the CPC, the CPC Central Committee with Comrade Xi Jinping at its core has attached great importance to the role of the mass line in building an ecological civilization. At a voluntary tree planting event in Beijing, which Xi had participated in for several consecutive years, he emphasized that efforts must be made to guide the people to actively participate in voluntary tree planting activities and translate the idea of building a beautiful China into the conscious actions of all people. On April 25, 2015, the CPC Central Committee and the State Council issued the *Opinions on Accelerating the Development of an Ecological Civilization*, encouraging the public to actively participate in environmental protection activities. In September 2015, the *Overall Plan for Ecological Civilization System Reform* issued by the Central Committee of the Communist Party of China and the State Council proposed that the public participation mechanism should be improved. On October 18, 2017, the report to the 19th National Congress of the CPC proposed to build an environmental governance system with the government in the leadership position, enterprises as the main players, and social organizations and the public as participants. On May 18, 2018, President Xi Jinping put forward at the National Environment Protection Conference that "since developing an ecological civilization is a cause in which the entire public participates and has a stake, we must translate the idea of building a beautiful China into the conscious actions of all people." This confirms the important position of the mass line in building an ecological civilization.

In short, since the founding of the People's Republic of China, the CPC has creatively used the mass line in building an ecological civilization and established it as an important principle of and the main path towards a socialist ecological civilization in China.

2 The Mass Line in Ecological Civilization Development

Since 1949, under the leadership of the CPC, the Chinese people, who have turned themselves into the masters of the country, have demonstrated high enthusiasm in socialist construction and creatively carried out many ecological civilization development activities with Chinese characteristics.

Resource conservation. China is a country with a large population and a low per capita share of resources. Resource conservation is a must. In the early days of the People's Republic of China, to fight corruption, waste and bureaucracy, the CPC Central Committee issued instructions to launch struggles against the "three evils." Mao Zedong claimed corruption and waste to be the biggest crimes so it was pointed out that the broad masses, including members of democratic parties and people from all walks of life must all be mobilized for the struggles. On January 4, 1952, the Central Committee of the Communist Party of China issued the *Instructions on Mobilizing the Masses to Start Struggles Against the "Three Evils" within a Time Limit*, requiring all institutions to mobilize the masses to fight within a time limit. On this basis, on June 3, 1957, *Instructions of the State Council on Further Promoting the Campaign to Increase Production and Save Resources* stated that in economic construction, we should rely on the people to save natural resources. After 1978, China took resource conservation as a basic national policy, and clearly set building a resource-saving society as an important goal of ecological civilization development. After the 18th National Congress of the CPC, in response to the serious problems of formalism, bureaucracy, hedonism, and extravagance, the Party launched activities to see members command and act on the Party's mass line and issued the eight-point decision on improving work conduct. In this process, the broad masses of the people carried forward the tradition of diligence and thrift of the Chinese nation, carried out a series of resource-saving activities, changed the social atmosphere and saved natural resources.

Patriotic health campaign. Themed on ensuring the health of the people, the Chinese government launched the patriotic health campaign. In 1953, it was proposed to combine health work with mass movement and incorporate the patriotic health campaign centered on the elimination of the "four pests" into socialist construction. As Mao Zedong pointed out, "the elimination of the four pests is a big clean and sanitation campaign... If all the people are mobilized and achievements can be made, I believe people's attitude will change, and the morale of our Chinese nation will be heightened." Through hard work, a preliminary victory was achieved in eliminating the "four pests." At the same time, Mao Zedong called upon the people eradicate schistosomiasis. In March 1951, Mao Zedong sent schistosomiasis control personnel to Yujiang County, Jiangxi Province, to do some investigation, and for the first time confirmed the county as a place with high schistosomiasis prevalence. In April 1953, Mao Zedong sent doctors to the county for important experimental research. On September 27, according to an investigation report from members of democratic parties, Mao Zedong raised the issue of schistosomiasis prevention and control, and appointed Xi Zhongxun, then Secretary-General of the State Council, for the task. In 1956, Mao Zedong issued a call to "mobilize the whole Party and all people to eliminate schistosomiasis." In the same year, Mao Zedong instructed relevant authorities to send two expert teams to the county to inspect schistosomiasis prevention and control. A mass movement was thus initiated in Yujiang County to eradicate schistosomiasis, which succeeded in 1958. For this achievement, Mao Zedong praised the creativity of the people with the famous poem line "the six hundred million Chinese people are all great people like Shun and Yao." After 1978, the patriotic health campaign became routine. In 2002, under the wise leadership of the CPC and thanks to persistence in the mass line, China triumphed over SARS. In sum, the patriotic health campaign ensured a clean environment, better health of the people, and a higher capacity for China to pursue sustainable development.

Environmental protection. Faced with the environmental pollution problem resulted from a lack of experience, Chinese people spontaneously launched many environmental protection campaigns. A good example is the one led by Li Shuangliang, a retiree of Taiyuan Iron and Steel Company. Since the establishment of the company in 1974, a slag yard gradually formed on the premise gathering a total of 10 million m^3 of slag, which was not only a waste of resources, but also a source of heavy pollution in Taiyuan, seriously affecting the production and life of local people. Due to a shortage of time and funding, company authorities and experts found no feasible solution to this problem. After he retired in 1983, Li Shuangliang took the initiative to work on this asking for no input from the authorities. After more than 10 years of hard work of Li and his peer slag yard employees, up to 1.309 million tons of scrap iron and steel were reclaimed and used to produce all kinds of products, creating an economic value of RMB 330 million. On this basis, they built 2500 m of protection slopes around the yard, 20 m wide at the bottom and 13 m high, and planted more than 70,000 trees, making the yard a beautiful garden. Li Shuangliang spirit was thus formed, as a symbol of trust in the masses, relying on them, and having faith in employees for corporate operation. In 1988, Li Shuangliang won the Global 500 gold medal. In addition, as environmental protection became a basic national policy of China, many people took active actions to report and fight against environmental pollution, and carry out waste classification and recycling campaigns, making great contribution to environmental protection in China.

Soil conservation. Soil erosion is a serious problem affecting people's lives and production. There was a large-scale mass soil and water conservation campaign in

China. On December 26, 1952, Zhou Enlai pointed out that "soil and water conservation is a long-term and comprehensive task requiring mass efforts. It must be done in a way suitable for the actual production needs and the mass must be mobilized and organized for long-term efforts before the expected results can be attained." In November 1955, Mao Zedong announced it necessary to solicit advanced experience from local people to make a good plan for soil and water conservation. In 1959, the Water and Soil Conservation Committee of the State Council proposed that China must do a good job in water and soil conservation by relying on the its people and while striving to develop production. Based on the above, the mass soil and water conservation movement was launched with strong momentum. Changting County in Fujian Province is a place of serious red soil erosion in southern China and soil and water conservation experimental zone was established here in December 1949. Yet by 1983, only very simple and basic work for soil erosion control had been done and the problem remained serious. According to remote sensing data of a 1985 survey, soil erosion was still seen on 1.462 million mu of land in the county, accounting for nearly one third of the county's total area. In 1999 and 2001, Comrade Xi Jinping, then governor of Fujian Province, made two trips to Changting to inspect and guide soil erosion control work. Then, in late 2011 and early 2012, Xi Jinping, as President of China, gave two more important instructions on the soil erosion control in Changting. As of the end of 2012, soil erosion was put under control on 1,281,900 mu of land, 87.7% of the total affected area. Specifically, thorough treatment was completed on 1.0108 million mu of land while only 451,200 mu still suffered from soil erosion, making a success rate up to 69%. It can be seen that adhering to the mass line and giving play to the enthusiasm and creativity of the people in soil erosion control is lies at the core of Changting's experience. The mass soil and water conservation campaign built up China's capacity for sustainable development.

Afforestation. After the founding of the People's Republic of China, in order to green the motherland, a massive afforestation campaign was launched. On September 30, 1953, the State Council's *Instructions on Mobilizing the Masses to Carry out Afforestation and Reforestation Work* stated that only by initiating and continuing with a mass afforestation campaign can we bring about a fundamental change in China's basic situation of a low total area of forests; and only if we can gradually reduce natural disasters, increase agricultural production, add to the income of the people in mountainous areas, and accumulate more timber resources can we offer support to the country's economic development. In the 70 years since the founding of the People's Republic of China, a large number of role models have come out in China's afforestation efforts. For example, in Youyu County, Shanxi Province, forest covered only less than 0.3% of the total land area before 1949. Now, 70 years of hard work by the local people under the leadership of the county Party committee and the country government have pushed up the forest coverage rate of the county to 54%, turning mountains green and waters clear. According to available statistics, in the past 70 years, local farmers in Youyu have delivered more than 200 million working days of volunteer service. President Xi Jinping mentioned Youyu spirit for five times. On September 28, 2012, Xi instructed that Youyu is a representation of spirit of serving the people wholeheartedly, facing difficulties squarely, and working hard for long-term benefits. Nationwide, since 1978, in response to the call of the Party and the state, the Chinese people have actively participated in afforestation campaigns. As of 2018, a total of 15.5 billion people had participated in tree planting campaigns across the country, planting 70.5 billion trees (including other environmental services converted into quantity of trees planted). In particular, the Three North Shelterbelt is a miracle on earth.

Water resource conservation. After the founding of the People's Republic of China, facing the coexistence of severe floods and droughts, Mao Zedong regarded water conservancy as a major issue concerning the vital interests of the Chinese people, and demanded trust in and mobilization and organization of the masses for the construction of water conservancy projects. From June to July 1950, a severe flood hit the Huaihe River Basin. At the end of August, with instructions from Mao Zedong, a dedicated national conference was held in Beijing for the management of the Huaihe River. After the meeting, the State Council issued the *Decision on Huaihe River Management* and in November, Huaihe River Management Committee was established. Hundreds of thousands of workers soon gathered on construction sites along the Huaihe River. In May 1951, Mao Zedong made the call that we must make sure the Huaihe River is fully treated. Ten large reservoirs, many more medium-sized ones, and hundreds of small ones were soon built with the simplest tools and massive manpower and four flood storage projects and 18 flood discharge areas were put into use. By the end of the 1970s, although there had been many floods, no major disaster hit. In 1952 and 1963 respectively, Mao Zedong call for treatment work for the Haihe River and the Yellow River. In 1958, Mao Zedong delivered in volunteer service at the construction site of Shisanling Reservoir in Beijing. Before 1949, China had only 23 large and medium-sized reservoirs, while from 1949 to 1976, more than 85,000 reservoirs of all types were built nationwide. In terms of drought control, the Hongqi Canal in Linxian County, Henan Province, is great example. Linxian County, located at the junction of Henan, Shanxi and Hebei provinces, used to suffer a lot from severe droughts and water shortage. There had been efforts to get things better but the results were not satisfactory. Then, in 1959, the county decided to dig through the Taihang Mountains to bring in water from the Zhuozhang River. The project began in 1960 and was completed in 1969, pushing up the total area of irrigated farmland by 600,000 mu. More than 70,000 people participated in the construction work, and 81 gave their lives. During the construction period, Mao Zedong repeatedly emphasized that water conservancy project must take irrigation, flood control, and soil and water conservation all into consideration. After 1978, water conservancy work in China mainly took the form of specialized construction projects. From 1978 to 2017, the number of reservoirs nationwide increased from 75,669 to 98,795, with a total capacity of 996.3 billion m^3 including 639.3 billion m^3 of new capacity. The total length of dikes nationwide increased from 130,000 to 306,000 km, more than seven times the periphery of the Earth. Thus, China greatly enhanced its capacities for flood and drought control, as well as its capabilities for sustainable development.

Disaster prevention and mitigation. There are many types of disasters in China, with wide distribution, high frequency, and heavy losses caused. China is actually one the hardest hit by severe disasters in the world. After the founding of the People's Republic of China, under the leadership of the Party, the Chinese people have launched various disaster prevention and mitigation campaigns. The fight against the 1998 flood is a typical example. In 1998, abnormal climate brought floods to the Yangtze River, Nenjiang River and Songhua River. To fight against the extraordinary natural disaster, the People's Liberation Army and the Armed Police Force of China dispatched over 360,000 people, and local Party committees and government organizations mobilized more than 8 million people. Adding to these were many more supportive forces and the total number of direct participants in the fight amounted to over a hundred million. Many more extended care and support through different channels. As a result of united efforts, we finally won this battle against floods. Another example is the relief work after the 2008 Wenchuan Earthquake. After hearing about the devastating earthquake in Wenchuan, Sichuan, 13 peasants from Tangshan City, Hebei Province, formed a volunteer team and traveled day and night over thousands of miles to the disaster area to participate relief work headed by the People's Liberation Army, together with local government officials and other volunteers. Up to 25 survivors were rescued from under the rubbles, and more than 60 dead bodies were found. Realizing that many students were temporarily out of school in the disaster-hit area, they sent 246 students from the disaster area to Yutian County, Tangshan City, to attend Yinhe Middle School for a year before they returned to their rebuilt homes in Sichuan. Some of these 13 farmers are survivors of the 1976 Tangshan earthquake and their descendants. What they did is a great example of the Chinese tradition of rewarding every bit of help one receives and helping hands from all walks of life to those in need. They contributed to the reconstruction of the disaster-stricken areas and to China's capacity for sustainable development. As president Xi Jinping pointed out, "we must adhere to the mass viewpoint and mass line, and broaden the channels for the masses to effectively participate in public security governance." In this sense, mass disaster prevention and mitigation campaigns are also green campaigns with Chinese characteristics.

In short, since the founding of the People's Republic of China, the Chinese people have actively responded to the calls of the Party and the government, delivered volunteer services in all kinds of construction projects with a daring spirit and the signature selflessness of Lei Feng for a more environment-friendly China. This is a strong driving force for ecological civilization development in China.

3 Valuable Experience in Using the Mass Line in Ecological Civilization Development

In the process of implementing the Party's mass line in ecological civilization development, it is necessary to mobilize people across society to pool forces for the ultimate success in this grand campaign. Here is some valuable experience in the green campaign and the pursuit of green governance with Chinese characteristics.

We should promote the development of an ecological civilization for the benefits of the people. Western environmental movements are often based on a nature-centered, as opposed to human-centered system of values. Western ecological governance policy makers often regard creating higher surplus value as the primary goal, ignoring the needs and interests of the working class. Different from these, China's model of ecological civilization development is designed based on the CPC's mass line, with the interest of the people in mind. President Xi Jinping once pointed out that the purpose of both economic growth and environmental protection is to improve people's living standards. We should create more material and spiritual wealth to improve the people's life quality and provide more high-quality public goods concerning the environment in response to the collective rise in people's awareness of the environment and health. In response to the Chinese people's growing environmental awareness, we should create more environmental and public health benefits for the people by focusing on solving outstanding environmental problems, speeding up the improvement of the quality of the environment, providing more high-quality ecological products and striving to achieve social fairness and justice. For example, we have launched many mass environmental campaigns to control air, water, and soil pollution in urban and rural areas, aiming to improve the environmental quality as well as the life of citizens. Obviously, the purpose of building an ecological civilization is to meet the people's growing demand for a better environment. To achieve this, we must adhere to the Marxist political stance and serve the people wholeheartedly in the process of promoting the development of ecological civilization.

We should increase public participation in the development of an ecological civilization. Top-level design and adherence to professional paths are of course important to the development of an ecological civilization, but what's more important is to engage the masses. In accordance with the Party's mass line, we have always regarded the masses as the main stakeholder in the development of an ecological civilization. The CPC Central Committee and the State Council have pointed out that "we must continue to carry out the mass Beautiful China campaign. This campaign mobilizes the public to participate in environmental protection and aims to benefit the whole country. We must strengthen the publicity and education relating to ecological civilization development, establish a system of values and codes of conduct for this purpose, and encourage the masses to take the initiative to participate in the Beautiful China campaign." Specifically, the government has insisted on relying on the people, effectively achieved scientific, democratic, legal, and ecological decision-making, and accelerated the country's transformation towards sustainable development. For example, in 2007, Xiamen City in Fujian Province decided to launch the PX (p-Xylene) project but many Xiamen citizens expressed their concerns about the project's negative impact on the environment and health hazards in a rational way. In the face of opposition, the Party and government authorities of Fujian Province and Xiamen City decisively terminated the project, providing a model for democratic decision-making on ecological issues. In the social sphere, we have gradually improved our understanding of the role of mass movements and NGOs in the public domain, and regarded them as an important means for individuals to play a role in a market economy. For example, in 2005, the relevant authorities decided to lay waterproof and leak-proof membranes at the bottom of the Yuanmingyuan Lake. Considering that this practice may have a negative impact on the environment of the site, some NGOs raised objections. The national environmental authority thus held public hearings to gather opinions from the general public and in the end, the project was terminated. This is how the government has been raising public environmental awareness and promoting public participation in environmental protection. We also request companies to strengthen environmental management, promote dedication to environmental conservation and adopt workplace practices for higher sustainability. Taiyuan Iron and Steel Group, which successfully solved the problem of its slag yard, and the Saihanba Forest Farm, which showcased strong determination to control desertification, are great examples in this regard. We have also been actively encouraging people to adopt green lifestyles as a way to support the country's pursuit of higher sustainability. For example, we have selected environmental ambassadors, called upon people to go car-free for one day, set air conditioners at 26 °C, eat up what's on their plates, and launched "earth station" and other mass environmental campaigns. These have not only effectively raised people's environmental awareness, but also promoted the development of an ecological civilization.

The achievements of ecological civilization development are to be shared by the people. Since the founding of the People's Republic of China, we have creatively employed approaches such as people-centered development, inclusive development, and common prosperity to promote ecological civilization development, and build an environmental benefit sharing system. "Inclusive development means the benefits of development on economic, political, cultural, social, and ecological fronts reach all and people's legitimate rights and interests in all aspects are fully protected." First of all, the prerequisite for sharing environmental benefits is shared ownership of environmental resources. China's Constitution stipulates that all natural resources and urban land are owned by the state, that is, by the people; natural resources, land in rural areas and city suburbs, homesteads, and mountains that are collectively owned according to the law are collectively owned. In this way, an economic system is established to ensure the sharing of environmental benefits. Second, we use the mass line to ensure the sharing of environmental benefits. The masses are not only the main players in the development of an ecological civilization, but also the main actors in environmental management. We guide people to effectively protect the environment and enjoy environmental benefits. The state not only supervises people's fulfillment of their obligation to protect the environment, but also effectively protects people's right to enjoy environmental benefits. Last but not least, our goal is to ensure the sharing of environmental benefits among the people. For example, it was proposed in the 13th Five-Year Plan that residents in impoverished areas where hydropower and mineral resources development projects are launched should be compensated with ownership of equity in the projects. The state should improve the mechanism for the sharing of the benefits from resource development, especially in impoverished areas. In terms of environment management, we have introduced a horizontal environmental protection compensation policy. On the whole, the CPC regards a good environment as the fairest and most inclusive public good.

Of course, the effectiveness of ecological civilization development must also be evaluated by the masses. In the meantime, while implementing the mass line in the development of an ecological civilization, it is necessary to adhere to Party leadership, regard the people as masters of the country, and promote the rule of law. This can prevent the mass line from turning into blind mass movements and populism.

In short, the mass line has been an important tool used by China to promote the development of an ecological civilization in the past 70 years.

Evolution of Socialist Ecological Civilization from Ideology to a Form of Society

Chengliang Huang

1 Achievements and Ecological Problems of an Industrial Civilization

The emergence of an industrial civilization is the historical roots of all modern conflicts. On the one hand, as Friedrich Engels said, it not only promotes the progress of material civilization, but also is conducive to the development of spiritual civilization. On the other hand, while advancing the capitalist civilization, the industrial civilization overly exploited natural resources, showing a strong anthropocentric tendency, which ultimately would lead to a deterioration of the balance between man and nature. The anthropocentric tendency of an industrial civilization emphasizes the conquest of nature by humans. This tendency allows humans to set rules for the world with peace of mind, and use such rules to develop and utilize the natural world, treating nature as a source of wealth instead of a friend.

In the 20–30 years around the middle of the twentieth century, the industrial civilization, which was "overly intoxicated by the victory of humans over the natural world," suffered "retaliation from the natural world." Around the world, shocking events of pollution frequently occurred, causing deaths and harm to humans. Some experts listed out eight most deadly pollution disasters in history. For example, there is the historic smog in Los Angeles from May to October 1943, a result of auto exhaust, which caused many LA residents to suffer from eye redness, laryngitis, and worsened respiratory conditions, killing more than 400 people over 65 years old. The Donora smog in the United States, which lasted from October 26 to 30, 1948, was caused by emissions of sulfur dioxide and pollutants, sickening 42% of local

The original article, with some minor differences from this one, first appeared in *Guizhou Social Sciences* (eighth issue of 2015).

C. Huang (✉)
Research Insitute for Eco-civilization,
Chinese Academy of Social Sciences, Beijing, China

J. Pan et al. (eds.), *Beautiful China: 70 Years Since 1949 and 70 People's Views on Eco-civilization Construction*, https://doi.org/10.1007/978-981-33-6742-5_7

residents and killing 17 within four days. The London smog from December 5 to 8, 1952, caused more than 4000 deaths.

The occurrence of the above-mentioned environmental disasters aroused wide attention in the United States and some other countries, and set off an environmental movement in the West. Politicians and intellectuals in the United States employed various means to promote environmental protection. For example, in 1962, American biologist Rachel Carson published the book *Silent Spring*, disclosing that the use of pesticides had a bad effect on the country's environment; in 1963, U.S. President John F. Kennedy wrote an introduction for *The Quiet Crisis*, a book calling for environmental protection; American historian Lynn White published *The Historical Roots of Our Ecological Crisis* at an academic conference, which led to an increase in the American religious circle's attention to environmental issues; American politician Gai Rod Nelson delivered speeches at major colleges and universities in the United States to inform American college students of the fact that the environment had deteriorated in the country and encourage them to engage in environmental protection; and U.S. newspapers and magazines published articles related to environmental issues. From the late 1960s to the 1970s, thanks to the joint efforts of people from all walks of life in the United States, the country saw the founding of a large number of environmental organizations related to land use and animal protection, and California and other states that faced with serious environmental problems issued laws related to the use of automobiles and industrial fuels. On April 22, 1970, 20 million people attended events across the United States to celebrate the first Earth Day. The U.S. Congress adjourned for the day. No vehicles were allowed on Fifth Avenue, Manhattan, the busiest street in New York City. Hundreds of thousands of people took to the street to demand a clean environment. Two years later, the United Nations convened the Conference on the Human Environment and adopted the *Declaration of the United Nations Conference on the Human Environment*, calling for prudent decisions to ensure environmental sustainability for future generations.

Since then, modern Western philosophers have emphasized that it is necessary to oppose anthropocentrism and promote pan-humanism. Biocentrism and ecocentrism are two representative ecological political philosophies. Biocentrism holds that all living things are morally considerable. It was first proposed by Albert Schweitzer in his book *Civilization and Ethics*. Schweitzer believes that ethics must be extended to all animals and plants, and human beings should cherish all lives and hold them equal to their own. Theoretically speaking, treating all lives with respect and awe is consistent with the Buddhist belief that all lives are equal and that it is wrong to kill. Biocentrism recognizes the intrinsic value of all lives and establishes the same ethical relationships between them as those between human beings. In fact, Schweitzer was indeed profoundly influenced by Buddhism from the East. He is also the author of the book *Indian Thought and Its Development*. He believes in the Buddhist teaching that human beings should not injure, abuse, oppress, enslave, insult, torment, torture, or kill any creature or living being. In response to the tragic human struggles caused by the two world wars and the unprecedented destruction of the ecological environment, Schweitzer proposed the concept of reverence for life and extended the principle of love to animals. This is a huge step forward in the history of ethics. Ecocentrism, as its name suggests, is an ethical philosophy that extends the scope of ethics from humans to ecosystems. Prominent scholars supporting this position include Aldo Leopold and Holmes Rolston. Aldo Leopold puts forward the idea of land ethics and the argument that human beings should see land as a community to which they belong. He reminds people to pay attention to elements in the land community that lack commercial value. With "capital" at the core, in the early and middle stages of the industrial era when natural resources were plundered and used without any caution and the linear economic model from raw materials to waste was adopted to promote economic growth, Leopold's view was obviously prophetic. It reminds human beings to pay attention to the long-term interests of mankind in the integrity of and balance with the land community. Holmes Rolston was an influential philosopher in contemporary Western environmental ethics, and the founder of the international academic journal *Environmental Ethics*. He had deep influence on Mr. Yu Mouchang, a Chinese scholar in environmental ethics and former president of the Chinese Society for Environmental Ethics. Holmes Rolston proposed the theory of natural value in his books such as *Philosophy Gone Wild* and *Conserving Natural Value*, which emphasized the intrinsic value and system value of the natural world and recognized the instrumental value of non-human nature as a means to human well-being. He also proposed the diversity of natural values, pointing out that maintaining and promoting the integrity and stability of the ecosystem is the responsibility of human beings. However, it is obvious that the popularity of anthropocentrism confined to the inherent pattern of the industrial civilization has contributed to the creation of unprecedented material wealth, but has brought the Earth's ecosystems on which mankind depends to survive to the brink of a collapse; Pan-humanism gives meaning to all lives and affirms the value of nature, but it may push humanity, who "knows the difference between nature and man," back into the jungle. It is meaningless and impossible to talk about the value of nature without giving due consideration to mankind.

2 Logic Behind and the CPC's Creative Contribution to Human Civilization

The prominent conflicts between man and nature also attracted the attention of Marxist theorists. In the 1970s, a combination of the Western environmental movement and socialist thoughts gave rise to eco-socialism, one of the most important ten Marxist ideologies. Eco-socialism believes that capitalism is the root cause of the global ecological crisis because (i) capitalism regards the pursuit of profit as the ultimate goal and nature as the source of wealth; (ii) capitalism pursues capital expansion, and in order to expand production and minimize production costs, capitalist countries had been trying to exploit natural resources in the third world to the greatest extent, which caused an ecological crisis in underdeveloped countries. Therefore, eco-socialism believes that the essence of environmental problems is social inequality and that the solution to environmental problems lies in getting rid of the capitalist system. Eco-socialism argues that only socialism can truly solve the problem of social inequality and thereby eradicating environmental inequality.

Eco-socialism regards socialism as the fundamental solution to environmental problems because it believes that the essence of socialism is fairness. However, in reality, China, deeply influenced by capitalist countries, bases its economy largely on the traditional industrial civilization model, in which humans exploit the natural world without a limit, and it is difficult to balance between human and nature and solve problems such as resource shortage, ecological degradation, and environmental pollution. As Chinese Xi Jinping said, "now the total population of developed countries is less than 1.3 billion. When the modernization of China add 1.3 billion people to that, the figure will be more than doubled. Using natural resources and exploiting the natural world in the old way is definitely unsustainable!"

Based on the judgment that threats to the security of human civilization in the future are likely to come from environmental disasters and energy wars caused within the industrial civilization that has lasted more than 200 years and given that China's industrialization occurred after the industrialization of Western developed countries had already brought about global energy and environmental crises, if socialist China wants to resolve the conflicts between man and nature, it must work out a new civilization model that maximizes energy efficiency and is worthy of emulation to avoid further disasters for human civilization. For now, as China's economic and social development enters the new normal, China needs to adopt production, growth and development models that are fundamentally different from the traditional extensive industrialization model. As President Xi Jinping pointed out, "if we still adopt the extensive growth model, even if the goal of doubling GDP is achieved, what kind of pollution disasters will we face? The environment may not be able to handle them at all." As for cultural development in the new normal, China needs to reflect on and adjust its ethics, moral principles, and behavior in the ecological sphere, and promote sustainable, healthy, and green consumption patterns and lifestyles. The civilization thus formed is an ecological civilization, which was mentioned in the report to the national congress of the Communist Party of China, the largest political party in the world.

In October 2007, the term "ecological civilization" was mentioned in the report to the 17th National Congress of the CPC, which was a first, making historical contribution to the advancement of human civilization. Following the emphasis on "material civilization" and "spiritual civilization" from the 12th to the 15th National Congress of the CPC, and the "political civilization" added at the 16th National Congress of the CPC, "ecological civilization" now joined the cohort. The significance of the ecological civilization theory lies in the fact that the it summarizes and unifies the dialectical relationship between man and nature. As Karl Marx once wrote, "we know only a single science, the science of history. One can look at history from two sides and divide it into the history of nature and the history of men. The two sides are, however, inseparable; the history of nature and the history of men are dependent on each other so long as men exist." If history belongs to the past according to traditional thinking, the ecological civilization reflects the current state of China and our beautiful vision for the future. As Vladimir Lenin pointed out, "a revision of the 'form' of Engels' materialism, a revision of his natural-philosophical propositions is not only not 'revisionism,' in the accepted meaning of the term, but, on the contrary, is demanded by Marxism." Ecological civilization reflects the relationships between man and nature, society and nature, man and man, and society and society across the entire history of mankind.

3 Evolution of the Logic and Socialist Ecological Civilization in the New Era

The mentioning of ecological civilization in the report to the National Congress of the CPC delivered by General Secretary Hu Jintao was groundbreaking. Since the 18th National Congress of the CPC, General Secretary Xi Jinping has made a series of important expositions on ecological civilization to profoundly, systematically and comprehensively answer major theoretical and practical questions encountered by China in ecological civilization development, marking the evolution of socialist ecological civilization from an ideology to a form of society.

1. **Ecological civilization is a very important part of the Five-Sphere Integrated Plan for development of socialism with Chinese characteristics**

The 18th National Congress of the CPC focused on integrating the overall basis of China being in the preliminary stage of socialism, and the overarching goals of socialist modernization and the great rejuvenation of the Chinese nation, and developed the four-sphere integrated plan for of building socialism with Chinese characteristics (comprising the economic, political, cultural and social spheres) to the Five-Sphere Integrated Plan by adding ecological civilization development to it. Xi Jinping once pointed out: "The 18th National Congress of the Communist Party of China incorporated the construction of ecological civilization into the Five-Sphere Integrated Plan for development of socialism with Chinese characteristic, and proposed to vigorously promote the construction of ecological civilization, strive to build a beautiful China, and realize the sustainable development of the country. This demonstrates the deepening of our understanding of the evolutionary laws of socialism with Chinese characteristics and our firm determination to accelerate the construction of ecological civilization."

2. **Ecological civilization is an important part of the Chinese Dream of a great rejuvenation of the Chinese nation**

The Chinese Dream is to realize the great rejuvenation of the Chinese nation, which includes economic prosperity, national rejuvenation, happiness of the people and social harmony. It is the Chinese people's vision and expectation for the country's future. General Secretary Xi Jinping pointed out: "Promoting the development of ecological civilization and building a beautiful China is an important part of the great rejuvenation of the Chinese nation which is the greatest dream of the Chinese people." Under the leadership of the CPC, Chinese people from all ethnic groups across the country have been working together towards the Chinese Dream and an ecological civilization. Fundamentally, to enhance happiness of the whole society, it is necessary to consider both material and non-material factors. As Xi Jinping pointed out, "money has an important impact on most people's lives, but a good environment is also an important part of a happy life, and money cannot buy it. Even if you have a lot of money, you won't be happy if the quality of the air and drinking water is poor." China is still at a critical stage of industrialization. Under the leadership of the CPC, Chinese people are moving towards a socialist ecological civilization. This is not only a historical opportunity but also an irreversible trend of the times.

3. **Ecological civilization development has been or is being integrated into all aspects and the whole process of economic, political, cultural, and social development of China**

China is integrating the its pursuit of an ecological civilization into economic development. Xi Jinping pointed out: "We want not only clean water and green mountains, but also mountains of gold and silver. It is better to have clean water and green mountains than mountains of gold and silver. Green is gold. We must not promote economic development at a price of environmental degradation." In modern society, green science and technologies such as ecological technology, recycling technology, system management science, complex systems engineering, clean energy and environmental protection technology are advancing and have become a decisive factor in productivity improvement and the transformation of production methods, which is a direct driving force for the rise of the green industry. Now, the green industry and green economy have become an important part of the new normal of China's economic development and presented an important opportunity for China to grow into a major economic power.

China is integrating ecological civilization building into its political development. Xi Jinping pointed out: "We must completely change our concept - that is, we should no longer talk about development in terms of GDP growth rate. We must put the environment in a prominent position in the evaluation system of economic and social development." "The most important thing is to improve the evaluation system of economic and social development, and to incorporate indicators such as resource consumption, environmental damage, and ecological benefits that reflect the status of ecological civilization construction into the evaluation system." From an institutional point of view, we must establish a sound resource and environmental management system, move faster towards the establishment of a land development and protection system, a system for paid use of resources, an ecological compensation system, an environmental accountability system, and an environmental damage compensation system. These systems can reflect market supply and demand, the scarcity of resources, the value of the environment, and intergenerational fairness.

China is integrating ecological civilization building into its cultural development, Xi Jinping pointed out that the excellent traditional Chinese culture, such as the ideas of "unification of nature and man (*tian ren he yi*)" and "following the laws of nature (*dao fa zi ran*)" offers inspirations on how to solve the problems faced by human beings in contemporary times. Mr. Ji Xianlin once explained *tian ren he yi* in this way: "*Tian* means nature; *ren* means human; *he yi* means mutual understanding and friendship." *Tian ren he yi* not only is an important proposition of Confucianism made two thousand years ago, but also part of the mainstream Chinese philosophy over all the years. It has an important impact on the unique worldview, value system, mindset, and problem-solving methods of the Chinese people. *Dao fa zi ran* lays the foundation for the idea of an ecological civilization which emphasizes harmony between man and nature, and is highly consistent with modern environmentalism. To build an ecological civilization, we must protect and creatively use our cultural heritage, and answer the two essential questions posed by historians and philosophers: Where do we come from? Where are we going?

Ecological civilization is integrated into China's social development. Xi Jinping pointed out: "We must pay more attention to the construction of ecological civilization. This is also what the public want us to do." "The people are highly concerned about environmental issues. It can be said that the importance of the environment to the people's happiness will inevitably continue to grow." In recent years, pollution issues, such as smog, drinking water safety, and high heavy metal content in the soil, have attracted more and more attention. We must focus on strengthening the prevention and control of air pollution, fine particulate matter (PM2.5) pollution, heavy metal pollution and soil pollution in key industries and regions. It is necessary to give full play to the political advantages of the socialist system, and strive to build a public participation system in which the CPC and the government assume the leadership role and the whole society works together to improve the environmental quality. We should continuously step up the implementation of the mass line tactics in the development of an ecological civilization, improve people's understanding of the concept and encourage everyone to actively participate in environmental protection.

4. **Building a global ecological civilization has become a common cause of the international community**

According to Marx and Engels, the historical trend of global ecological development reveals that to solve ecological problems, we must promote the internationalization of environmental governance—that is, we should see environmental governance in a global and historical perspective. The conflicts between countries over response to climate change directly affect not only the modernization of developing countries, including the great rejuvenation of the Chinese nation, but also developed countries' pursuit of a redistribution of global environmental benefits. Xi Jinping had deep thoughts about the relationship between China's development and the development of the world. He once said: "China will continue to undertake its due international obligations, deepen international exchanges and cooperation in the field of ecological civilization, promote the sharing of environmental benefits, and work together with other countries to build a beautiful world." In the current era, in response to major global challenges, China has taken up environmental commitments commensurate with its level of development by incorporating climate change response into its economic and social development plans and vigorously promoting green growth. In November 2014, President Xi Jinping and US President Barack Obama announced in Beijing that China and the U.S. reached an agreement on climate change response. The United States pledged to cut its greenhouse gas emissions to 26–28% below the 2005 level by 2025, which refreshed its previous commitment reducing it to 17% below the 2005 level by 2020. China officially pledged for the first time that it aimed to hit peak emissions by 2030 and that the share of non-fossil fuel in its total energy demand will increase to 20% by 2030.

Dialectics of nature originates from practice and is tested in practice at all times. It is not rigid dogma or empty preaching, but a guide to action. It requires people to broaden their horizons and enliven their thoughts, not to simply stick to rules. As the outpost and guard for the natural sciences, it must constantly absorb nutrients from natural sciences and keep up with their development. The task of building a global ecological civilization ultimately concerns fairness between people, how contemporary people treat the legacy of their ancestors, the gap in development between East and West and North and South, and the relationship between us and our descendants. However, for each and every generation, the key to building an ecological civilization is to figure out how to achieve harmony between man and nature. Marx writes: "... communism, as fully developed naturalism, equals humanism, and as fully developed humanism equals naturalism; it is the genuine resolution of the conflict between man and nature and between man and man – the true resolution of the strife between existence and essence, between objectification and self-confirmation, between freedom and necessity, between the individual and the species." From ideology to a form of society, socialist ecological civilization has not only realized the sublimation of civilization ideologies in the new era, but also laid a foundation in the ecological front for the realization of the Chinese Dream (i.e., the great rejuvenation of the Chinese nation) and a cornerstone of social equality for the country's move towards communism.

Environmental Planning in China (1949–Present)

Jun Wan

Environmental planning and environmental protection started at the same time and have been developing simultaneously in China. In August 1973, the Chinese government held its first national environmental protection conference and issued the *Environmental Protection and Improvement Regulations (Interim)*, proposing to plan environmental protection in a holistic and reasonable manner, improve resource efficiency, reduce negative environmental impact of human activities, and encourage everyone to participate in environmental protection efforts. Planning environmental protection in a holistic manner is at the core of the policy, which establishes the fundamental and leading position of planning in various environmental management systems. Since the first national environmental protection plan, nine five-year national environmental protection plans have been prepared and implemented thought the specific names changed over time. The status of the plans was raised from internal plans to departmental plans and then to national plans approved and issued by the State Council. An environmental planning system with Chinese characteristics has been created and played an important role in the country's environmental protection endeavor.

1 Simultaneous Development of Environmental Planning and Environmental Protection

The Fifth Five-Year Environmental Protection Plan (1976–1980) is China's first five-year environmental protection plan. In 1975, the Environmental Protection Leading Group of the State Council issued *Circular on the Formulation of the Ten-Year Environmental Protection Plan and the Fifth Five-Year Environmental Protection Plan*

J. Wan (✉)
Chinese Academy of Environmental Planning, Beijing, China

J. Pan et al. (eds.), *Beautiful China: 70 Years Since 1949 and 70 People's Views on Eco-civilization Construction*, https://doi.org/10.1007/978-981-33-6742-5_8

(1976–1980), proposing that large and medium-sized industrial and mining enterprises with serious pollution problems should strengthen the prevention and control of industrial waste in accordance with national standards. At the first National Environmental Protection Conference, the Chinese government designated 18 key cities for environmental protection, including Beijing, Shanghai, and Tianjin. In these cities, industrial and domestic sewage must be treated in accordance with national standards. The country stepped up pollution control in the Yellow River, Huaihe River, Songhua River, Lijiang River, Baiyangdian Lake, Guanting Reservoir, Bohai Sea and other water systems and major ports and water quality in these places was significantly improved. *The Fifth Five-Year Environmental Protection Plan* started the chapter of environmental planning in China. It proposed that environmental protection should be incorporated into the national economic and social development plan. It also, for the first time, proposed measures for the treatment of wastewater, exhaust gas, and solid waste, marking the beginning of the era of point-source industrial pollution treatment in the country.

The Six Five-Year Plan period saw environmental protection included in the national development plan for the first time as a separate chapter. *The Sixth Five-Year Plan for National Economic and Social Development (1981–1985)* is the first of its kind to contain an independent chapter on environmental protection and it was approved on December 10, 1982, at the Fifth Session of the Fifth National People's Congress. The plan set out specific targets and measures for preventing pollution, treating pollution from old enterprises, improving industrial waste treatment capability, strengthening environmental protection planning, environmental monitoring, environmental research, environmental legislation and environmental law enforcement. It also proposed that construction projects must be screened for environmental impact and the "three simultaneous" system must be put in place for every construction project. The plan also allocated RMB 17.88 billion from the fund for fixed asset investment to support employee housing, urban construction, and environmental protection. The importance of environmental protection was fully affirmed in the plan. It marked the shift of the focus of China's environmental management efforts from industrial waste treatment to pollution prevention, laid a solid foundation for the establishment of an environmental impact assessment system and the "three simultaneous" system, and promoted the standardization and institutionalization of environment impact assessment.

The Seventh Five-Year Plan period welcomed China's first standalone national environmental protection plan issued by the Chinese government. In April 1987, the National Planning Commission and the Environmental Protection Commission of the State Council issued the *National Environmental Protection Plan for the Seventh Five-Year Plan Period (1986–1990)*, which became the first standalone five-year environmental protection plan issued by the Chinese government. The plan highlighted the comprehensive improvement of the urban environment and the prevention and control of industrial pollution, and emphasized that regions and industries should set out their own environmental protection goals. It also emphasized the constraint of environmental capacity and total discharge-based pollution control. It proposed that "industrial activities in areas with high population density and industrial concentration should be gradually relocated to areas with large environmental capacity." The Seventh Five-Year Plan proposed that "economic development, urban development, and environmental protection should be planned and promoted simultaneously to maximize economic, social and environmental benefits" and that a prevention-centered comprehensive environmental planning and protection method should be adopted. These measures served as the strategic guidelines for China's environmental protection work. The provisions on the environmental responsibilities of polluters and developers in the plan was a key principle for environmental protection in China which had far-reaching impact and was incorporated into the Environmental Protection Law of China.

During the Eighth Five-Year Plan period, environmental targets were incorporated into the national five-year plan and decomposed into targets for provinces, autonomous regions and municipalities. In July 1992, the National Environmental Protection Agency issued the *National Environmental Protection Ten-Year Plan and the Eighth Five-Year Plan Outline* (hereinafter referred to as the *Eighth Five-Year Plan Outline*), proposing that the focus of pollution prevention and control should gradually shift from concentration control to total discharge control, from end-of-pipe treatment to whole-process prevention. It also put forward the goal of total industrial dust emission control and proposed to carry out total pollution control in key industrial pollution source areas, river basins, and sea areas. The plan also proposed to focus on environmental protection and coordinated economic and social development, strengthen environmental management and promote scientific and technological progress. The main environmental targets in the *Eighth Five-Year Plan Outline* were integrated into the *Eighth Five-Year Plan for National Economic and Social Development* after balancing various considerations. In the meanwhile, the main targets were decomposed into targets for provinces, autonomous regions, and municipalities, which were separately listed in the plan and assigned to relevant local governments for implementation. *The Eighth Five-Year Plan* also contained a national technical outline for environmental planning to support the establishment of a macro-level environmental planning system for the purpose of promoting the sustained and coordinated development of the economy and the environment and formulating source-specific pollution control targets. The focus of environmental protection planning thus shifted from pollution control planning to pollution prevention planning. Environmental protection was integrated into the national economic and social development plan, marking the inclusion of environmental protection in the country's development decision-making.

The Ninth Five-Year Plan period saw China's first environmental protection five-year plan approved and issued by the State Council. In July 1996, the State Council reviewed and approved the *Ninth Five-Year Plan for National Environmental Protection and 2010 Targets*. As the first national environmental protection plan approved and issued by the State Council, it clarified the sustainable development strategy. Sustainable development and environmental targets were also clearly stated in the national economic and social development plan. The plan also proposed the "One

Control and Two Standards"[1] system, the *Major Pollutants Total Emission Control Plan*, the *Plan for China's Trans-Century Green Project*, and the policy of equal emphasis on pollution prevention and environmental protection. During 1996–2000, the country made remarkable strides in environmental legislation in accordance with *The Ninth Five-Year Plan for Environmental Protection*. The Chinese government also revised the Air Pollution Prevention and Control Law, Water Pollution Prevention and Control Law, and Marine Environment Protection Law, and promulgated a plethora of environmental protection laws and regulations, including Noise Pollution Prevention and Control Law, Implementation Rules of the Water Pollution Prevention and Control Law, and Regulations on the Environmental Management of Construction Projects. The amendment to the Criminal Law added provisions on the crime of causing damage to the environment and the crime of dereliction of environmental supervision duty, which was an important step in environmental legislation.

The 10th Five-Year Environmental Protection Plan (2001–2005) highlighted environmental protection in key areas, total pollution control, and major environmental protection projects. In December 2001, with the approval of the State Council, the State Environmental Protection Administration, the State Planning Commission, the State Economic and Trade Commission, and the Ministry of Finance jointly issued the *10th Five-Year National Environmental Protection Plan* (hereinafter referred to as the *10th Five-Year Plan*), proposing to continue to focus on pollution prevention and control in key areas identified during 1996–2000, including "Three Rivers, Three Lakes, and Two Control Areas," as well as Beijing, and the Bohai Sea, and to step up water pollution control in the Three Gorges Reservoir and along the South-to-North Water Transfer Project. The government also compiled and implemented the *Plan for the Decomposition of National Total Pollution Control Targets for Major Pollutants during 2001–2005*, which set out and broke down the control targets for six major pollutants, designated water environment functional zones across the country and completed ecological functional zoning for 31 provinces, autonomous regions and municipalities. During 2001–2005, the People's Congress of Guangdong Province approved and implemented the *Pearl River Delta Regional Environmental Protection Plan* and the *Integrated Environmental Protection Plan of Guangdong Province*, which marked a big step forward in China's efforts for major regional environmental protection plans.

The 11th Five-Year Plan period saw emphasis laid on implementation for the first time. On November 22, 2007, the State Council issued the *11th Five-Year National Environmental Protection Plan*, which was the first five-year plan for environmental protection issued by the State Council. The plan set the target of reducing sulfur dioxide emissions and chemical oxygen demand by 10% from the 2005 levels by 2010. For the first time, the environmental targets were included as the binding

[1]"One Control and Two Standards": By the end of 2000, all provinces, autonomous regions, and municipalities must control the discharge of major pollutants in their jurisdictions within the total emission targets set by the state, and industrial pollution sources must meet the pollutant discharge standards set by the central or local government. The air and surface water must meet the environmental quality standards set by the country for this functional zone.

targets in the *11th Five-Year Plan for National Economic and Social Development. The 11th Five-Year National Environmental Protection Plan* also set out the evaluation and assessment requirements for the implementation of environmental protection plans. National environmental quality targets were broken down to for each province, autonomous regions and municipalities, and environmental protection tasks were broken down assigned to specific departments. The central government also increased investment in environmental protection, which greatly improved the status and execution of the environmental protection plan. *The 11th Five-Year National Environmental Protection Plan* proposed many key projects such as the national survey of pollution sources. It also proposed to strengthen the management capability of the environmental protection system, and prioritize the improvement of the environmental protection system. The plan contained content related to climate change. It attached importance to the development of environmental protection systems and long-term mechanisms, emphasized active participation in global environmental governance and demonstration of China's commitment to environmental sustainability. The Chinese government listed environmental protection as a binding indicator for the transformation of economic growth model, indicating the integration of environmental protection into the country's economic development.

The 12th Five-Year Environmental Protection Plan (2011–2015) outlined goals and measures for total pollution control, environmental quality improvement, environmental risk prevention, and the promotion of environmental public services. On December 15, 2011, the State Council formally issued the *12th Five-Year National Environmental Protection Plan.* The emission control targets proposed in the plan for chemical oxygen demand, sulfur dioxide, ammonia nitrogen, and nitrogen oxides were included in the *12th Five-Year Plan Outline* as binding indicators. The plan also proposed to build pollution reduction, environmental quality improvement, environmental risk prevention and environmental service equity systems. It was proposed that the control targets for total discharge of major pollutants should be used as the basis for decision-making concerning regional and industrial development. Regional environmental governance authorities should apply limits and restrictions when issuing pollution permits to local governments that had not completed emission reduction tasks. The plan stepped up regional total pollution control requirements, especially requirements for the total discharge of major heavy metal pollutants in key areas for comprehensive prevention and control. During 2011–2015, the country stepped up its efforts to make annual plans, mid-term evaluation and final assessment systems for five-year plan implementation, and achieved remarkable results in addressing environmental problems and reducing heavy metal pollution in accordance with the plan.

The 13th Five-Year Ecological and Environmental Protection Plan (2016–2020) focused on improving environmental quality and integrating environmental protection and environmental governance. On November 24, 2016, the State Council issued the *13th Five-Year Plan for National Ecological and Environmental Protection* under the guidance of the new development philosophy emphasizing innovative, coordinated, green and open development for all. The plan proposed to promote the implementation of the Five-Sphere Integrated Plan, put forward the overarching goal of achieving overall improvement the country's environmental quality by 2020, and listed seven major tasks including strengthening air, water and soil pollution prevention and control. The plan highlighted the systematic connections between environmental quality improvement, total emission reduction, environmental protection, and environmental risk prevention and control; regarded environmental quality improvement as the overall goal and evaluation criterion, broke down goals and tasks to regions, river basins, cities, and control units, compiled a list of environmental tasks, and refined environment management. The title of the plan was in itself a reflection of the integration of environmental protection and ecological protection. The plan prioritized green development and reform, significantly strengthened the link between green development and ecological and environmental protection, and insisted on tackling ecological and environmental problems from the source. The plan also played an important role in promoting the establishment of a central environmental inspection system to supervise the performance of the main responsibilities by local ecological and environmental authorities.

2 Fundamental Guiding Role of Environmental Planning

The evolution of environmental plans in China is a process in which the country's emphasis on environmental protection continuously increased. The Chinese government no longer only pays lip service to environmental plans. The national environmental plans reflected the will of the state and had a huge impact on people's lives. Environmental planning played an increasingly important role in environmental protection.

First, a national environmental plan is a blueprint that sets out environmental protection goals and tasks based on goals of national development strategies, designed to transform beautiful visions concerning the environment into reality. The social and economic development of China has always been strongly plan-based and goal-oriented and the environmental plans also take into consideration the strategic goals of social and economic development. Thus, a goal-oriented planning system is fully in place in China. Since the adoption of the reform and opening-up strategy, China has proposed to build a moderately prosperous society by 2000 as part of its "three steps" strategy. In addition to ecological and environmental issues, the five-year ecological and environmental protection plans during 1986–2000 also focused on designing environmental goals and tasks to support the country's efforts to fulfill this goal. Entering the twenty-first century, at the 16th National Congress of the CPC, the Party proposed the strategic goal of building a moderately prosperous society in all respects by 2020, so the five-year ecological and environmental protection plans during 2006–2020 focused on improving the country's environment and reducing major pollutant discharge by 2020 and strengthening environmental protection in all fields to ensure that the beautiful national development blueprint can be gradually transformed into reality.

During 1986–1995, China still lagged behind many countries in urbanization and industrialization and environmental pollution in the country was mainly from point sources. The focus of China's environmental protection efforts then was on the treatment of industrial wastewater, waste gas and residue, and the environmental plans mainly focused on industrial pollution control. During 1996–2000, China began its transformation towards sustainable development and set relevant targets in the Tenth Five-Year Plan. During 2006–2010, the reduction of total discharge of major pollutants became binding targets for development evaluation. The country began to focus on solving outstanding environmental problems, reflecting the strategic position of environmental protection in China. At the 18th National Congress of the CPC, the Chinese government incorporated the building of an ecological civilization into the Five-Sphere Integrated Plan for development of socialism with Chinese characteristics. General Secretary Xi Jinping said: "Luci water and lush mountains are valuable assets." Environmental authorities at all levels followed Xi's instruction in environmental planning during 2011–2020.

Therefore, the goals, tasks, and policies of relevant plans reflected what China has been through in regard to environmental protection, the outstanding problems to be solved, and the main measures to be taken at this stage. China has gradually built a national-provincial-city planning system and a comprehensive planning-special planning system to ensure the fulfilment of its development goals and environmental goals at all levels and in all fields.

Second, national environmental plans establish the fundamentals of environmental protection in the country, which is the foundation for systematical environmental protection.

The Chinese government has been working to keep its environmental protection policies up to date, with emphasis shifting from conquering the nature to the control of wastewater, exhaust gas, and solid waste, the development of ecological civilization, the Five-Sphere Integrated Plan for modernization, the beautiful China campaign and cooperation with other countries to build a clean and beautiful world. The dialectical relationship between development and environmental protection has been continuously adjusted and balanced. The focus of planning is continuously adjusted based on the most serious environmental issues and the planning authority's understanding of environmental protection at the time, and the purpose is to establish the fundamentals of environmental protection appropriately.

Five-year comprehensive plans for environmental protection have played an important role in guiding environmental protection efforts. A national environmental protection plan makes national-level arrangements for ecological and environmental protection, and the binding targets set out in the plan constitute mandatory requirements that must be adhered to. The target system includes various elements and key tasks in each field and is designed to ensure that all tasks will be completed in time. Priority is given to important tasks, and continuity and coherence of are maintained over the long run. The Chinese government has been strengthening the supervision and evaluation of the implementation of environmental protection plans.

Local governments are encouraged to maintain a strong commitment to environmental sustainability, and to avoid major adjustments in their environmental protection priorities due to external fluctuations, especially changes in economic, policy and even meteorological conditions, so that the progress of long-term, fundamental tasks will not be affected. Major measures such as energy conservation and emission reduction during 2006–2010 and the current pollution prevention and control campaigns are all designed in accordance with the target and task system established in five-year comprehensive plans.

As the country strengthens assessment and inspection, it is especially necessary to avoid fragmentation of work and a passive work attitude. We must comprehensively consider fundamental, long-term, structural and layout-related environmental issues in the administrative area when making environmental plans and build consensuses.

Third, an effective platform for dialogue and exchange of ideas has been built in the process of preparing and implementing environmental protection plans and has played an irreplaceable role in building environmental consensuses.

Environmental protection involves a wide range of areas. The roles and responsibilities of multiple departments need to be coordinated for the preparation and implementation of environmental plans, and consistency should be ensured between the goals and tasks at the national, provincial, and city levels. Actions should be taken to respond to the demand of the government, enterprises, and the public. For a long time, through the establishment of a multi-departmental, multi-level environmental planning mechanism, China has built channels for communication and dialogue between governments at all levels. The scope of planning is not just limited to mechanisms, systems, and division of responsibilities. The key is to ensure all work activities are appropriately coordinated in the planning stage.

For 2011–2020, the *13th Five-Year Environmental Protection Plan* was formulated. The previous boundaries for planning were broken and the scope of planning was expanded to cover, among other things, ecological protection, environmental governance, environment restoration and environmental monitoring, laying the foundation for a new round of institutional reform. At the same time, the central government leveraged the environmental planning and evaluation system to propel local governments to establish a dialogue platform or mechanism to promote environmental protection in accordance with environmental plans. For example, in order to achieve the goals of the 13th Five-Year Plan, Jiangsu Province launched the *263 Special Action Plan* and established a special working mechanism participated in by multiple departments, which greatly promoted the resolution of prominent environmental problems.

As China's socialist modernization has entered a new era, the society presented diverse demands for ecological products and services and social groups from different regions, industries, and classes have different environmental requirements. It is, therefore, necessary to make good use of the environment planning platform, strengthen dialogue and communication, broadly build consensuses among all stakeholders, and ensure that environmental protection activities can be carried out in an orderly manner in accordance with the national development roadmap.

3 Compilation and Implementation of Environment Plans to Support the Beautiful China Campaign

At the 19th National Congress of the CPC, the Chinese government outlined a roadmap for socialist modernization with Chinese characteristics in the new era and launched the Beautiful China campaign, which provided guidance for environmental protection in the new era. When designing environmental plans, we must follow Xi Jinping Thought on Ecological Civilization, focus on improving the quality of the environment, systematically develop a roadmap for environmental protection, and strive to give play to the fundamental guiding role of environmental plans on the grand journey of building a beautiful China.

First, we should follow the guidance of the environmental protection strategy, systematically develop a roadmap for environmental protection to support the grand goal of building a beautiful China. Now standing at a new historical starting point, we are striving to build a moderately prosperous society in all respects and a beautiful China. The world is undergoing major changes unseen in a century. In such a context, when making environmental plans, we must have an objective understanding of the stage of development that China is currently in and the historical process of environmental governance in the country. We must seek truth from facts and act within our capabilities. In addition to improving the overall quality of the environment, we should work to fundamentally improve the environment, and systematically develop a roadmap for the strategic goals and different stages of environmental protection in different regions and fields.

Second, we should think systematically, promote high-quality development and high-level environmental protection simultaneously, and build a beautiful country in which men coexist with nature in harmony. Environmental protection should be considered in the context of overall social and economic development and the grand picture of the historical process of the great rejuvenation of the Chinese nation. We should prioritize environmental protection, promote green development, respect nature, follow its laws, and protect it. Improvement of the quality of the environment is a key aspect of high-quality development. To improve the quality and efficiency of development, we must improve the quality of the environment. Green development is regarded as a fundamental solution to various ecological and environmental problems. To achieve the simultaneous advancement of high-quality development and high-level environmental protection, we must strive to form sound spatial patterns and industrial structures, and promote production methods and lifestyles that are conducive to resource conservation and environmental protection.

Third, we should strengthen operation and implementation, build an efficient plan preparation and implementation system, and make further efforts for scientific planning, targeted governance, systematic management, and comprehensive implementation. In response to the country's need to modernize its governance systems and capabilities in the new era, we should deepen the research on planning theories, methods and techniques, separate comprehensive planning from special planning, improve the national-province-city-county planning and implementation system and the compilation-implementation-evaluation- supervision system for plans, adopt a differentiated approach to environmental governance, develop effective reform plans, supervise plan implementation, improve relevant systems and policies in practice, and develop a sound roadmap for the Beautiful China campaign.

Environmental Protection Attempts in the Early Days of the People's Republic of China

Lianhui Zhang

The convening of the first National Environmental Protection Conference in August 1973 marked the beginning of modern environmental governance in China. Almost all retrospective studies on environmental protection in China regard this conference as the starting point of modern environmental governance in the country, while, due to the lack of information and other reasons, environmental protection activities prior to this conference are rarely discussed. Even if there is such discussion, the information given is usually quite vague. However, in fact, without the long-term efforts made by the Chinese government in environmental protection before the conference, it would be difficult for the country to kick-start the environmental efforts in the modern sense smoothly in 1973. This article aims to deepen people's understanding of China's environmental protection efforts before the first National Environmental Protection Conference. We have found and examined sufficient historical records on the environmental protection efforts of the Chinese government during that period, and laid out the characteristics of different stages and the connections of these efforts with environmental protection activities after the second National Environmental Protection Conference.

1 Infancy Stage of Environmental Protection in China (1949–1957)

After the founding of the People's Republic of China, ecological and environmental problems began to appear with the advancement of industrialization and economic growth. At the time, China's economy was just on the right track and the industrialization process just started. Environmental pollution and damage only occurred in some areas and were not very serious. Therefore, Chinese people's environmental

L. Zhang (✉)
School of Economics, Zhongnan University of Economics and Law, Wuhan, China

J. Pan et al. (eds.), *Beautiful China: 70 Years Since 1949 and 70 People's Views on Eco-civilization Construction*, https://doi.org/10.1007/978-981-33-6742-5_9

awareness was very low at the time. However, in order to deal with environmental problems, relevant authorities issued some policy documents and regulations with environmental protection functions, and some cities also took measures to protect the environment.

In terms of industrial pollution prevention and control, in order to deal with the sanitation issues involved in planning, site selection, design, and industrial waste treatment concerning new industrial projects in cities, the Ministry of Health established a sanitation supervision office in 1953 to oversee sanitation activities under the guidance of Soviet consultants. This office should be considered the first environmental protection agency established by the central government of China. The proposal of "preventive sanitation supervision" can also be regarded as the ideological origin of pollution reduction at the sources emphasized in subsequent pollution control policies. From May to June 1954, the Ministry of Health held the first National Industrial Sanitation Conference, and formulated an industrial sanitation policy, proposing to strengthen leadership, promote steady progress in the development of a sanitation system that was designed to support production, rely on the masses, and prioritize pollution reduction at the sources. Due to the need for uniform sanitation standards and regulations as a basis for sanitation supervision, the State Council adopted the *Factory Safety and Sanitation Regulations* in 1956, and the Ministry of Health and the State Construction Commission jointly issued the *Interim Sanitation Standards for Industrial Enterprises* and *Joint Directive on Sanitation Supervision in Urban Planning and Development*. These three documents offered guidance for the country's efforts to prevent pollution, ensure drinking water safety and plan urban development reasonably. In June 1957, the Third and Fourth Offices of the State Council issued the *Circular on the Management of Toxic Wastewater and Emissions from Industrial and Mining Enterprises*, which clearly stated that attention should be paid to the prevention and control of industrial pollution. The circular was considered a substantial environmental policy document. In order to better prevent and control industrial pollution, the Chinese government established the policy of "comprehensive utilization of industrial waste" in 1956. This was a fundamental policy concerning the treatment of industrial pollution for more than ten years. Under the guidance of these documents, polluting industrial enterprises, especially the 156 large and medium-sized projects that had been intensively constructed during that period, adopted certain prevention and control measures, such as sewage purification and installation of dust removal equipment. These measures reduced pollution in the country to a certain extent. In urban development, more attention was paid to the distribution of pollution sources. Polluting enterprises were built in industrial areas far away from downtown areas, and forest isolation belts were built between urban areas and industrial areas to protect urban residents from the hazardous impact of industrial waste. For example, the *Wuhan City Master Plan* formulated in 1954 took into account environmental considerations when planning for urban development. The Wuhan Iron and Steel Corporation (WISCO) and the Wuhan Roulian Plant, for which construction began during 1953–57, were located on the north and south banks of the lower reaches of the Yangtze River in Wuhan, 20 km from Wuchang and Hankou. Large-scale machinery manufacturers such as Wuhan

Heavy Machine Tool Plant, Wuhan Boiler Plant, and Wuhan Turbine Engine Plant were also built in a newly planned suburban industrial area. Considering industrial pollution problems in the factory area, the logistics and dormitory area of WISCO was located five kilometers away from the factory site, with a green isolation belt between the two areas. During the period, in order to collect first-hand data on environmental pollution, the health departments of a few cities such as Beijing and Chongqing carried out surveys on pollution sources and environmental conditions. For example, in 1954, 1955, and 1956, Chongqing conducted a survey to collect data on water quality, pollution and self-purification capacity of the Chongqing section of the Yangtze River and the Jialing River, as well as the pollution of the two rivers caused by industrial waste. It also collected data on dust, toxic gases, production noise and industrial wastewater. We could say that Chongqing was a pioneer city in environmental monitoring in China. In terms of environmental management, the "three simultaneous" system on environmental protection in construction projects (i.e., environmental protection facilities and the main part of the construction project should be designed, constructed, and put into use at the same time) proposed after the first National Environmental Protection Conference in 1973 began to come into shape. In July 1957, the Nanjing Municipal People's Committee issued a circular requesting that new factories and mines should consider measures to treat wastewater and waste gas in the design stage, and send design drawings and related materials to the relevant authority for approval. It can be said that the idea of the "three simultaneous" system originated from this requirement, but unfortunately this policy document was shelved. Environmental monitoring is an important means to understand environmental conditions in a timely manner and support informed decision-making concerning pollution prevention and control. In the 1950s, Shanghai, Zibo and other cities began to carry out various environmental monitoring activities, but the results were less than satisfactory. However, it was the beginning of environmental monitoring work in China.

To prevent urban environmental pollution, some central cities made great efforts to dredge rivers and relocate polluting enterprises. For example, from 1949 to 1952, Beijing repaired more than 220 km of old sewers and removed 160,000 m^3 of silt from old ditches, including the famous Longxu Ditch. In September 1950, the Beijing Municipal People's Government decided to move all factories using or producing flammable and explosive materials to the southern suburbs or other areas. In 1951, it decided to relocate the timber factory near Yongdingmen and the leather factory in the downtown area to the southern suburbs, and relocate factories using or producing inflammable and explosive materials to Shazikou, Dahongmen, and Tiejiangying areas in the southern suburbs. From 1949 to 1953, Fuzhou dredged 22 rivers, dug out a total of 360,000 m^3 of sludge, and cleared 572 ditches. The Patriotic Hygiene Campaign launched in 1952 played a huge role in improving the urban environment. In terms of the prevention and control of urban pollution, in addition to wastewater and waste gas management, cities such as Beijing, Nanjing, and Qiqihar issued policies and regulations specifically for the prevention and control of noise pollution. For example, in October 1953, the Beijing Municipal People's Government issued the *Circular on Reducing Urban Noise Pollution*; in July 1954, the it issued one more circular requesting to continue to reduce urban noise pollution; in May 1955, the Beijing Municipal People's Committee issued the *Regulations on Reducing Urban Noise*. Nanjing City promulgated the *Regulations on Reducing Urban Noise* in May 1956. In June 1956, Qiqihar City issued the *Circular on Reducing Urban Noise*.

In terms of natural resources and ecological protection, the Ministry of Forestry and Reclamation issued the *Interim Regulations for the Protection of Forests (Draft)* in February 1951. In December 1952, at the 163rd political affairs meeting, the State Council adopted the *Instructions of the State Council on Mobilizing the Masses to Participate in Drought Prevention and Relief Campaigns and Soil Conservation Efforts*. In July 1953, at the 185th meeting the State Council issued *Instructions on Mobilizing the Masses to Participate in Afforestation and Forest Protection*. In March 1955, the CPC Central Committee forwarded Wang Huayun's *Summary Report on Further Promoting Soil Conservation*. In 1956, the State Council issued the *Interim Regulations on the Protection of Mineral Resources*. In 1957, the State Council issued the *Interim Soil Conservation Plan Outline of the People's Republic of China*. In 1956, China established the first comprehensive nature reserve—Dinghushan Nature Reserve. Regulatory systems concerning natural resources and ecological protection began to take shape. It is worth mentioning that China made remarkable achievements in afforestation and forest conservation during this period. From 1950 to 1957, forest area in China increased by 235.964 million mu; from 1950 to 1952, 62.1 million mu of hills and mountains were closed for afforestation; and in 1956, more than 58.35 million mu of hills and mountains were closed for afforestation. Afforestation is an important means to improve the environment and prevent soil erosion.

Obviously, during this period, the Chinese government made a lot of efforts in environmental protection. However, for various reasons, the results were less than satisfactory. At the beginning of the country's industrialization process, the national government drew upon the experience of the Soviet Union and attached importance to pollution issues and actively adopted measures to reduce pollution. For example, the *Interim Sanitary Standards for the Design of Industrial Enterprises* promulgated by the Chinese government in 1956 was formulated in imitation of the *Sanitary Standards for Design Industrial Enterprises* promulgated by the Soviet Union in November 1954. China's environmental protection efforts in the 1950s were largely based on the experience of the Soviet Union. Naturally, the Chinese government's understanding of environmental protection at that time was also influenced by the Soviet Union. For example, at that time, China did not have the concept of environmental protection like the Soviet Union did. Environmental protection measures, especially pollution prevention and control measures, were mostly implemented under the guidance of the concept of "environmental sanitation." The agencies implementing environmental protection measures were mainly health departments. It was not until the first National Environmental Protection Conference that Chinese people's environmental awareness began to rise. After 1957, China began to break away from the Soviet model and go its own way, and China's environmental protection also embarked on an independent track.

2 Twists and Turns in China's Environmental Protection (1958–1969)

The Great Leap Forward that began in 1958 caused huge environmental pollution and damage within a short period of time, but efforts were still made to protect the environment despite all challenges. There were certain achievements in environmental protection during the period of national economic adjustment, but then serious setbacks occurred due to the Cultural Revolution.

1. **Serious environmental problems caused by the Great Leap Forward and the environmental protection efforts of the central government**

During the Great Leap Forward, China's environment deteriorated significantly for the first time since 1949. To promote industrialization of the country, in the second half of 1958 alone, tens of millions of laborers across the country were pulled from agriculture into the manufacturing sector. They were encouraged to set up backyard steel furnaces, where citizens could turn scrap metal into usable steel, and worked in small manufacturing units. More than 600,000 backyard steel furnaces, 59,000 small furnaces, 4000 small power stations, 9000 small cement plants, and 80,000 farm tool repair plants were built across the country. The number of industrial enterprises soared from 170,000 in 1957 to more than 600,000 in 1959. The number of small polluting enterprises with backward technologies increased rapidly, and the seriously polluting heavy industries and chemical industry were the pillars of China's economy. At the same time, regulations promoting environmental protection in the fields of industrialization and urban development were criticized and rejected. Under the situation of chaotic management and the absence of pollution control measures, the problem of industrial pollution became increasingly serious in the country. In the agricultural sector, a one-sided policy of "grain is key" was implemented. During that period, Chinese people were overly eager to catch up with industrialized countries soon. There were many slogans that declared war on nature. This led to a significant reduction in area of forests, grasslands, and lakes, and the country's environment significantly deteriorated. Environmental deterioration caused by the Great Leap Forward quickly aroused wide attention.

In response to the environmental problems that occurred at that time, the central government adopted a series of countermeasures, mainly to (i) prevent and control industrial pollution and (ii) stop deforestation and restore the normal order of the forestry economy. The prevention and control of industrial pollution were still guided by the comprehensive utilization policy. From 1958 to 1966, the relevant departments of the central government successively issued *The Official Letter on Wastewater Management of Small Petroleum Plants (May 1958), The Joint Circular on Enterprises' Responsibility of Urging Their Industrial and Mining Subsidiaries to Implement the State Council's Instructions on Disposal of Toxic Wastewater (September 1959), Report on the Hazards, Treatment and Utilization of Industrial Wastewater (March 1960), Circular on Strengthening the Management of Domestic Wastewater and Industrial Waste Produced by Factories and Mines (April 1962), Circular on Strengthening the Treatment, Utilization and Management of Industrial Waste (June 1964)* and other documents. It also held a series of meetings, including the National Meeting on the Reuse of Wastewater for Irrigation (October 1958), National Conference on Industrial Wastewater Treatment and Comprehensive Utilization (November–December 1959), Second Urban Development Meeting (September–October 1963), Urban Development Meeting (July 1965), and New Factory and Mine Sanitation Meetings (March 1966), repeatedly calling for actions to strengthen the management of urban and industrial waste. At the National Conference on Industrial Wastewater Treatment and Comprehensive Utilization, the Chinese government demanded that the treatment of industrial wastewater be taken as an integral part of the design of new factories. For the first time, the idea of "three simultaneous" was proposed at a central meeting or in a central government's document. The *Interim Sanitary Standards for Design of Industrial Enterprises* was revised and renamed as *Sanitary Standards for Design of Industrial Enterprises*, and the revised document was promulgated in April 1963. During this period, the focus of industrial and urban waste treatment was wastewater treatment, and the primary method was reusing wastewater for irrigation. The reuse of wastewater for irrigation followed the two steps of treatment and reuse, and complied with the comprehensive utilization policy. In the meantime, some industrial enterprises were shut down, suspended, merged, or relocated; the chaotic industrial layout was also corrected to some extent. In early 1960, the State Council approved the *Interim Regulations on Management of Radioactive Materials*, which included provisions on the prevention of radioactive pollution.

From 1960 to 1963, to restore the normal order of the forestry economy, the Chinese government issued a series of policy documents, including *Urgent Instructions on the Current Policy Issues of Rural Communes* (November 1960), *Regulations on Resolute Correction of Adjustment Errors and Compensation* (June 1961), *Regulations on the Determination of Forest Rights, Protection of Forests and Forestry Development (Draft)* (June 1961), *Amendment to Rural Commune Regulations* (September 1962), and *Forest Protection Regulations* (May 1963). In 1964, the Ministry of Forestry stressed that in forest management importance should be attached to planting, protection, harvesting and the species structure. The above regulations and policies effectively restored the normal order of the forestry economy and played an important role in making up for the environmental damage caused by deforestation. In addition, to promote natural resources conservation and environmental protection, the State Council issued the *Instructions on Proactive Protection and Reasonable Utilization of Wild Animal Resources* (1962) and *Mineral Resources Protection Regulations* (1963), and established a number of nature reserves with comprehensive conservation functions before 1965, further improving the system of natural resources conservation and environmental protection.

2. **Environmental protection efforts of local governments from 1958 to 1969**

During this period, local governments actively responded to the call of the central government and played a more active role in environmental protection, especially in the prevention and control of industrial pollution. This also become an important feature of China's environmental governance at this stage. Local governments mainly made efforts in the following four aspects:

(1) Establishment of environmental protection agencies. The establishment of environmental protection agencies laid the organizational foundation for environmental protection. At this stage, a few provincial-level administrative units were established in Beijing, Tianjin, Shanghai, Heilongjiang and Xinjiang, as well as cities with a relatively high level of industrial activities such as Anshan, Wuhan, Harbin, Nanjing, Nanchang, Qiqihar, Baoding, Qingdao, and Jilin for environmental protection, mostly industrial waste management offices. However, as far as I know, the central government had not established any similar agency before 1969. Compared with local sanitation supervision agencies, these agencies had clearly specified environmental protection duties and were more similar to modern environmental protection agencies. The emergence of this type of government entities reflected the transition from environmental sanitation to environmental protection. Most of these environmental protection agencies were established during the period of national economic adjustment between the end of the Great Leap Forward and the beginning of the Cultural Revolution.

(2) Promulgation of pollution prevention policies and regulations. The formulation of specialized environmental protection regulations and policies laid the institutional foundation for environmental protection. In the process of industrial pollution prevention and control, the governments of Jiangxi, Heilongjiang, Wuhan, and Harbin issued documents and regulations according to local conditions. For example, from 1960 to 1965, Harbin successively promulgated eight policies and regulations emphasizing the management of industrial waste and domestic sewage. In terms of content, these policies and regulations were more detailed and regionally specific than those issued by the central government. They were also issued during the adjustment period of China's economy.

(3) Environmental surveys. Environmental surveys constituted the basis for objective assessment of environmental conditions and evidence-based policy-making concerning the environment. During this period, the central government did not directly organize environmental surveys, but Beijing, Qinghai, Heilongjiang, Chongqing, Anshan, Baoding, Wuhan, Foshan, Jilin, Nanhai and other provinces, cities and counties carried out environmental surveys which primarily focused on industrial pollution. These surveys were basically conducted by health and epidemic prevention agencies during the period of national economic adjustment. On the one hand, it shows that that the dominant environmental concept at the time was still environmental sanitation. On the other hand, it also shows that even if some provinces and cities did establish environmental protection agencies such as waste management offices, such agencies still lacked the administrative power and capabilities necessary to organize environmental protection efforts.

(4) Adoption of environmental management measures. During this period, at the level of local environmental governance, the "three simultaneous" idea was

> more clearly expressed. As mentioned above, in 1957, a circular issued by Nanjing City already proposed the preliminary idea of the "three simultaneous" system. On this basis, in December 1965, the Municipal Planning Commission, the Urban Construction Department, and the Health Department of Nanjing City jointly reported to the Nanjing Municipal People's Committee: "The waste treatment facilities of new, expansion and rebuilding projects should be designed and constructed simultaneously with the main part of the projects. The design documents shall be submitted to the waste management department, which, together with health, public security and labor departments will give opinions and sign the design documents. Urban development, design, and construction departments should supervise this process." Obviously, the logic behind the "three simultaneous" system was already very clear. However, due to various reasons, this policy was shelved.

A few cities began to step up environmental monitoring. An environmental monitoring network began to take shape. For example, in 1963, Qiqihar City's Waste Management Office, together with the Sanitation and Epidemic Prevention Station, the Planning Division of the Urban Construction Bureau and other departments and agencies of the city government, began to test water quality of drainage projects of newly built, rebuilt and expanded factories and enterprise and strictly control the use of various seepage wells. On this basis, in 1967, the Qiqihar Municipal Sewage Treatment Office began to build a wastewater monitoring network, which consisted of the downtown wastewater monitoring area, the southwest industrial wastewater monitoring, and Fularji industrial wastewater monitoring area, with a total of 10 points for regular inspection and monitoring.

3 Awakening of Chinese people's Environmental Awareness and Accelerated Development of Environmental Protection Administration (1970–July 1973)

Although the ultra-leftist ideology at the time denied that China had environmental problems, the increasingly serious environmental problems in reality and the health hazards and economic losses caused by these problems had attracted the attention of then Premier Zhou Enlai. On June 26, 1970, when Zhou Enlai met with the representative of the Military Management Commission of the Ministry of Health, he pointed out: "The health system must focus on health concerns, especially health hazards of air and water pollution." Concerning industrial pollution problems in the United States, Japan and other countries, he pointed out: "Chairman Mao emphasized preventing pollution, including air and water pollution, at its source. We must make use of waste gas and wastewater. Capitalist countries did not solve the problem. We socialist countries must solve the pollution problem." Later, Zhou Enlai repeatedly emphasized the need to pay attention to environmental issues. It is estimated that, from 1970 to 1973 (before the first National Environmental Protection Conference),

Zhou Enlai made 29 speeches on environmental protection. It is precisely because of the repeated mentioning by Zhou Enlai that people began to pay more attention to environmental protection. Zhou Enlai presided over the economic reform after 1971. The stabilization of domestic political and economic situation created favorable conditions for people to pay more attention to environmental issues and adopt environmental protection measures.

In the 1960s, environmental disasters occurring frequently in western industrialized countries gripped the entire international community. For this reason, the United Nations decided to hold the first human environment conference in Stockholm in June 1972 and sent an invitation to China. In those days, under leadership of Zhou Enlai, the Chinese government overcame various challenges and sent a delegation composed of representatives from the State Planning Commission, the Ministry of Fuel and Chemical Technology, the Ministry of Health and the Ministry of Foreign Affairs to attend the conference. This conference let many Chinese people, at least the Chinese participants, realize that China also had environmental problems. The Chinese delegation came to the conclusion that "the environmental problems in Chinese cities are no less serious than those in Western countries, and ecological problems in China are far more serious than those in Western countries." From ignorance of environmental problems and refusing to recognize environmental problems to recognition of the seriousness of environmental problems is a major change in understanding. The report after the delegation returned to China prompted Zhou Enlai to immediately convene a national conference to study environmental issues. On January 8, 1973, the State Planning Commission submitted the *Request for Convening a National Environmental Protection Conference* to the State Council, suggesting that the National Environmental Protection Conference be held on August 5 of that year. Subsequently, a conference preparation team was established. After approval by the central government, the State Planning Commission issued an advance notice on January 20, notifying relevant departments and agencies that the first National Environmental Protection Conference would be hold in early August. After the notice was issued, all central departments and agencies and governments of provinces, cities and districts began to prepare materials required by the conference. After sufficient preparations, China's first National Environmental Protection Conference was held in Beijing in August 1973. In the process, Chinese people's awareness of environmental protection began to awaken.

During this period, due to Zhou Enlai's repeated mentioning and the awakening of Chinese people's awareness of environmental protection, and in order to support the convening of the first national environmental protection conference, central and local governments began to step up environmental protection efforts. During this period, the focus of China's environmental protection efforts was still on the treatment and comprehensive utilization of industrial waste. Especially worth mentioning are following achievements:

(1) The Chinese government proposed the first environmental protection policy since the founding the PRC. In 1971, to better guide environmental protection efforts, Zhou Enlai proposed to plan environment protection in a holistic,

reasonable manner, improve resource efficiency, reduce negative environmental impact of human activities, and encourage everyone to participate. At the first UN Conference on the Human Environment, the Chinese delegation officially announced to the world that China was stepping up environmental protection in accordance with this guideline. Since then, various local governments began to emphasize the implementation of this guideline in environmental protection. After discussion at the first National Environmental Protection Conference, this guideline was formally established as a national environmental protection policy.

(2) A large number of environmental protection agencies were established during this period. In 1971, under the supervision of Zhou Enlai, the State Planning Commission established the Waste Reuse Leading Group. This was the first official environmental protection agency established by the central government. In January 1973, the Office in Charge of the Preparation of the Environmental Protection Leading Group of the State Council was established. Urged by the central government, many local governments set up environmental protection agencies. Beijing, Gansu, Hubei, Guangdong, Guizhou, Hebei, Henan, Liaoning, Yunnan, Zhejiang, Hunan, Shandong, Jilin, Ningxia, Inner Mongolia and other provincial-level administrative units established new environmental protection agencies or reorganized original environmental protection agencies such as the industrial waste management office. Tianjin, Shanghai, Heilongjiang, and Xinjiang already established environmental protection agencies in the 1960s. There were 19 provinces, autonomous regions, and municipalities that had established environmental protection agencies prior to the first National Environmental Protection Conference. In addition, Changchun, Chengdu, Dalian, Guiyang, Nanjing, Wuhan, Zhengzhou, Chongqing, Xiangfan, Yichang and other central cities had also set up or restored such institutions. In June 1972, the country's first trans-provincial environmental protection agency—the Guanting Reservoir Water Conservation Leading Group—was established. Subsequently, a number of other environmental protection leading groups were established to protect the Yellow River Basin, Huai River Basin, Yangtze River Basin, Songhua River Basin, Pearl River Basin, and Taihu Lake Watershed. As of July 1973, a network of environmental protection organizations covering at least three levels of administrative units (i.e., central, provincial, and prefecture levels) had roughly taken shape. The establishment of the above-mentioned environmental protection agencies created favorable organizational conditions for the smooth start of modern environmental protection work in the future. At the same time, it also established the basic pattern of combining regional governance and river basin governance in China's environmental protection. Of course, these environmental protection agencies were not in superior-subordinate relationships. This was an important organizational feature of environmental governance before the institutionalization.

(3) Meetings were held to study the issue of pollution control. In response to the increasingly serious environmental pollution, the central government and some

local governments held several meetings to study the issue of pollution control. At the National Planning Conference in 1971 and 1972, the Chinese government proposed to step up industrial pollution prevention and control. From December 16, 1970 to February 19, 1971, the National Planning Meeting for the year of 1971 convened by the State Council specifically discussed the management of industrial waste and safe recycling of hazardous industrial waste. New projects must take proper measures to safely recycle and dispose of industrial waste. Projects under construction should also take proper measures. At this meeting, Zhou Enlai emphasized the necessity to recycle industrial waste. At the 1972 National Planning Conference held by the State Council from December 16, 1971 to February 12, 1972, the Chinese government proposed that waste recycling should be put on the agenda and effective measures should be taken to avoid unnecessary waste and eliminate public health hazards. In addition, the central government also held a series of conferences to study the issue of pollution control. In April 1972, the State Construction Commission and the State Planning Commission held a chimney dust removal meeting in Shanghai. In December 1972, the State Construction Commission entrusted the Institute of Building Research with job of organizing the Meeting on the Treatment and Reuse of Wastewater Containing Phenol and Phenolic Derivatives. From February to March 1973, the Ministry of Light Industries also successively convened the Symposium on Treatment of Wastewater of Printing and Dyeing Industries and the Symposium on Treatment of Wastewater of the Leather Industry. Propelled by the central government, many provincial and municipal governments also held meetings to convey the message of the above-mentioned meetings. These environmental protection meetings played an important role in forcing Chinese people to recognize the seriousness of environmental pollution and the importance of environmental protection, and in promoting the implementation of environmental protection measures.

(4) Both the central government and local governments carried out extensive pollution investigations. During this period, both the central government and local governments carried out extensive pollution investigations, reflecting an increase in the Chinese government's attention to environmental issues. At the level of the central government, in order to collect data on industrial pollution in the country, the Military Management Commission of the Ministry of Health issued the 4th Five-Year Plan for National Health Service (1971–1975) (Draft) on February 7, 1971 according to Zhou Enlai's instructions, requiring that all local governments investigate air and water pollution caused by industrial waste in the region within one or two years and collect data on harmful substances, pollution levels, etc., and that the health sector should assist the industrial sector in treatment and recycling of industrial waste. For the first time, the central government decided to carry out nationwide industrial pollution surveys. On February 12, the Main Points of Health Work Plan in 1971 (Draft) issued by the Military Management Commission of the Ministry of Health reiterated the above requirements. On April 27, the Military Management Commission of the Ministry of Health issued the *Notice on Investigation*

of Air and Water Pollution Caused by Industrial Waste, requesting surveys be carried out to find out the impact of industrial waste on rivers, the air, and water sources in the region as soon as possible. A national industrial pollution survey was officially launched. In December 1971, to support the national industrial pollution survey, the Military Management Commission of the Ministry of Health organized a seminar to facilitate exchange of experience in industrial pollution investigation. At the seminar, participants exchanged their experience in industrial pollution investigation, industrial pollution control and prevention, revision of sanitary standards, sampling and analysis, discussed the main points of the plan for industrial pollution control and prevention in 1972 (including the issue of safe reuse of wastewater for irrigation) and prepared materials for the first UN Conference on the Human Environment. Before the first National Environmental Protection Conference, the Chinese government gained preliminary knowledge of pollution in the Bohai Sea, the East China Sea, the Yangtze River, the Yellow River, the Songhua River, the Fuchun River, the Taihu Lake, and the Pearl River, groundwater pollution in some cities, air pollution in Chengdu, Jilin and other cities, as well as the hazards of industrial waste and pesticide residues through a recent survey. Environmental surveys provide important factual materials and enable people to have a more objective and clear understanding of environmental pollution and the urgency of environmental protection.

(5) Environmental management measures were further improved. During this period, the idea of the three-simultaneous system was clearly expressed for the first time in the document of the central government. In June 1972, the *Report of the State Planning Commission and the State Construction Commission on Pollution in the Guanting Reservoir and Solutions* approved by the State Council proposed that "waste treatment and reuse facilities should be designed, constructed and put into use at the same time as the main part of a factory construction project." At the same time, Beijing and Yunnan Province formulated regulations for three-simultaneous systems of construction projects. The three-simultaneous system gradually become an important environmental management tool. Another environmental management policy at the time was treatment of major pollution point sources within a specified time limit. In 1972, Beijing received complaints from local residents about acid and benzene-containing waste gas from 11 factories located in densely populated areas, including the Hepingli Chemical Factory and Beijing Lead Wire Factory. These factories were ordered to solve pollution problems caused by them within a specified time limit. On May 7, 1973, the Hubei Provincial Reform Committee forwarded the report on the pollution of Wuchang East Lake and treatment opinions, requesting the extinguishing agent factory affiliated to Wuhan University, Wuhan Second Pharmaceutical Factory, Qingshan Thermal Power Plant, Wuhan Heavy Machinery Factory, Wuhan Heavy Instrument Factory, Wuhan Thermometer Factory, Hubei University of Medicine, and the six nursing homes for patients with tuberculosis in Huangjiawan solve the pollution problems caused by them within a specified time limit. Otherwise,

they would be relocated or shut down. Beijing, Guangdong, Heilongjiang, Hubei, Yunnan, Shandong, Wuhan, Harbin, Qiqihar and other provinces and cities issued more specific and detailed industrial pollution control policies and regulations; Beijing, Anhui, Yunnan, Nanjing, and Qiqihar set up special funds for pollution control; Wuhan formulated an industrial waste management plan. In summary, from 1970 to before the first National Environmental Protection Conference, under the leadership of the central government, China stepped up environmental governance, and Chinese people's environmental awareness began to awaken. This created favorable conditions on organizational, ideological and experience accumulation fronts for the convening of the first national environmental protection conference in August 1973 and the smooth development of environmental protection work following the conference. It can be said that Zhou Enlai, the first UN Conference on Human Environment, early environmental protection efforts, and the national economic reform had paved the way for the smooth convening of the first National Environmental Protection Conference in 1973.

4 Conclusion

(1) Before the first National Environmental Protection Conference was held, the new Chinese government had made efforts to step up environmental protection. Even though these efforts were not properly planned and the results were less than satisfactory, these efforts at least can prove that environmental protection had not been lacking in China before the first National Environmental Protection Conference. Of course, most Chinese people's environmental awareness was low for most of this period, the Chinese government's environmental protection efforts were guided by the concept of "environmental sanitation" which is different from environmental protection in the modern sense.

(2) The environmental protection policies before and after the first National Environmental Protection Conference are mostly in the same vein. The "32-character policy" for environmental protection approved at the first National Environmental Protection Conference had already been proposed and put into practice before the meeting. Environmental protection agencies established by the central government and local governments after the first National Environmental Protection Conference were modeled after industrial waste management offices and other environmental protection agencies built before the meeting. The focus of the environmental protection efforts of the Chinese government before 1973 and during the period from 1973 to 1978 was the treatment and reuse of industrial waste. The concept of preventing pollution at its source was put forward in the 1950s. The three-simultaneous system and the policy of rectification within a specified time limit established after the first National Environmental Protection Conference had been proposed and implemented prior to the conference. The environmental governance pattern

combining regional governance and river basin governance had roughly taken shape before 1973. The Chinese government had also made efforts in other major aspects of environmental protection, such as environmental monitoring, environmental status investigation, and environmental legislation, before the first National Environmental Protection Conference. In fact, China's environmental protection efforts in the modern sense started on the basis of its previous environmental protection efforts.

(3) From the perspective of environmental protection initiation mechanism, the leadership of the central government (i.e., top-down approach) has always been a major feature of China's environmental protection efforts, which is different from the bottom-up approach in industrialized countries such as Japan and the United States. It is mainly because China implemented a planned economic system at that time, and under the planned economic system, the formulation and implementation of important decisions mainly relied on the central government. At the same time, under the planned economy system, the opinions of important national leaders often had a huge important on the country's policy-making. For example, Zhou Enlai's vision of environmental protection was a key factor in the rise of modern environmentalism in the country. Therefore, Zhou Enlai has been heralded as the founder of environmental movement in the PRC.

Efforts, Experience in and Inspiration of Environmental Protection in Rural Areas Over 70 Years Since the Founding of the People's Republic of China

Yanqiang Du

"Grass grows vibrantly by the pool and birds sing happily on the willow tree." The verse vividly depicts a scene commonly seen in the beautiful and livable countryside. Rural areas assume multiple functions such as providing agricultural products, offering ecological services and carrying forward folk customs. They are also the places where people can enjoy the natural landscape and recall their love of nature. In other words, the beautiful countryside is the source of quality agricultural products, the backyard and garden of cities, and the spiritual home of people who are away from their hometown.

In 1949, China's agricultural output totaled RMB 32.6 billion, and its population was 546 million, around 90% of whom lived in rural areas. Grain yield was 113.18 million tons, or 208.9 kg per capita. In 2017, China's agricultural output reached RMB 11.46 trillion, and the population grew to 1.39 billion, 58.2% of whom lived in urban areas. Grain yield went up to 617.93 million tons, or 445 kg per capita. The amount of fertilizers used for farming grew from 78,000 tons in 1952 to 58.59 million tons in 2017. The amount of pesticides used increased from 73.3 tons in 1990 to 178.3 tons in 2015. Such statistics show that China's agricultural production expanded exponentially over the past 70 years, while overuse of pesticides and fertilizers were also clearly seen. China is a major agricultural economy. Over 90% of its territory is rural areas. Under the urban-rural dual structure, how should we proceed with environmental protection in rural areas?

General Secretary Xi Jinping stressed efforts to improve rural living environment, build a beautiful countryside and retain beautiful idyllic scenes for the people on many occasions.

The CPC included rural revitalization in its national strategies at its 19th National Congress. The *Proposals of the CPC Central Committee and the State Council on Implementing Rural Revitalization Initiative* (draft) released in January 2018 made it clear that rural revitalization is a consensus and joint action of the whole Party.

Y. Du (✉)
Nanjing Agricultural University, Nanjing, China

J. Pan et al. (eds.), *Beautiful China: 70 Years Since 1949 and 70 People's Views on Eco-civilization Construction*, https://doi.org/10.1007/978-981-33-6742-5_10

The general offices of the CPC Central Committee and the State Council introduced a three-year action plan for improving living environments in rural areas in February 2018, stressing faster moves to shore up weak links regarding rural environments. Simply put, environmental protection in rural areas has drawn unprecedented attention. It is a major task and will remain so for a long period of time to come.

This paper reviews the evolution of environmental protection policies in rural areas over the past 70 years since the People's Republic of China (PRC) was founded in 1949. After attempts at different environmental issues and changes of policy focuses at different stages, major good practices are identified as well as difficulties that remain. Presenting experience of the UK, the US and Japan in environmental protection in rural areas, the paper put forth several issues that China must properly handle in environmental protection in rural areas in the new era.

1 Changes and Characteristics of China's Environmental Protection Policies for Rural Areas

Stage I (1949–1976): No policy. The stage can be further divided into three periods. The first period lasted from 1949 when the People's Republic of China was founded to 1957 when the first Five-Year Plan period ended. The focus of the nation was to recover national economy and build up pillar industries. There were no clear environmental protection policies or goals. The second period lasted from 1958 to 1965. Major events such as the Great Leap Forward and the irrational pursuit of borrowed boom in the iron and steel industry brought severe damage to the environment. Agricultural output slumped. The third period started in 1966 and ended in 1976. It was a period when leftist ideas led to serious consequences in society. The national economy was on the brink of a collapse, and environmental pollution and deterioration was very serious. For example, under wrong policies, a large number of factories that involved hazardous substances were set up deep in mountainous areas. With a sole focus on agricultural development, forests, grasslands and lakes were destroyed and turned into farmland, leading to environmental deterioration in rural areas. In 1973, guiding principles and policies for environmental protection was proposed on the national level with corresponding institutions established, but the focused was on industries and urban areas. Generally speaking, the country paid more attention to agricultural development and grain yields, while it failed to take good care of the environment in rural areas. Moreover, overuse of fertilizers and pesticides and excessive land reclamation brought negative impacts on the rural environment.

Stage II (1977–1994): Inception. The stage ran from the fifth to the eighth Five-Year Plan period. Bourgeoning township businesses in the early days of the reform and opening up were accompanied by serious industrial pollutions. In the meantime, the implementation of the production responsibility system boosted rural and agricultural development. For example, the total agricultural output of 1980 reached RMB 218.7 billion, and the grain yield was 320 million tons. In 1990, the figures were RMB

403.7 billion and 435 million tons, respectively. However, the heavy use of fertilizers and pesticides caused widespread non-point source pollutions. For example, organochlorine pesticides seriously affected vegetable and grain production. In 1983, the State Council decided to totally ban HCH and DDT. The *Decision on Strengthening Environmental Protection* released in 1984 and the *Proposals for Implementing Eco-Agriculture and Strengthening Environmental Protection in Agricultural Production* issued the 1985 urged efforts to promote ecological agriculture. The Environmental Protection Law that took effect in 1989 stipulates that governments at various levels shall strengthen environmental protection in agriculture and use fertilizers, pesticides and plant growth hormones in a rational manner. In general, environmental protection in China made progress at this stage, but the focus was still mainly on controlling industrial pollution and curbing the fast deterioration of the urban environment. For rural areas, there were no specific or targeted actions or policies to protect the environment. An undeniable fact is that some environmental management systems that was taking shape at the stage provided the basis and organizational support for pollution control in rural areas.

Stage III (1995–2001): Explorations. Rural environment was mentioned in the 1995 communique for China's environmental conditions for the first time. According to the communique, environmental pollution was expanding from urban areas to rural areas at a fast pace. A preliminary survey showed that two thirds of farmland plots over 10 million hectares in area were polluted. The 1999 communique for China's environmental conditions said the environmental quality in some places and urban areas was slightly improved, but the rural environment got worse. At the stage, agricultural and rural environmental pollution was characterized by the coexistence of point and non-point pollution, domestic and agricultural pollution, and local and migrated pollution. Under such a circumstance, the 9th Five-Year Plan explicitly required that population growth be contained and arable land and the environment be protected for sustainable development of agriculture and the rural economy. In 1998, the National Environmental Protection Agency established a division for environmental protection in rural areas. In 1999, the National Environmental Protection Agency introduced *Several Guidelines on Strengthening Environmental Protection in Rural Areas*, the first policy document specifically for environmental protection in rural areas. *The Emission Standards for Pollutants from Livestock Breeding Industry* were issued in 2001. In general, environmental protection in rural areas at this stage focused on improving water quality and toilets, and preventing pollutions from the livestock breeding industry. Policy guidelines took shape, but the overall objective was to prevent and control agricultural pollution.

Stage IV (2002–2012): Acceleration. At this stage, fertilizers, pesticides and livestock breeding were the main sources of pollution in rural areas. In 2002, the amount of pesticides used logged 131.1 tons, while 443 kg of fertilizers were used for every hectare of land on average (the safety cap in developed countries was 225 kg per hectare). According to the first general survey on pollutant sources in China released in 2007 showed that COD and ammonia nitrogen produced in the livestock breeding industry accounted for 95.8% and 78.1% of the total agricultural pollutant discharges respectively. The figures were 41.9% and 41.5% nationwide, respectively. In 2012,

180.6 tons of pesticides, and 528 kg of fertilizers were used per hectare. *The 10th Five-Year Plan for Environmental Protection* made containing agricultural point and non-point pollution and domestic pollution in rural areas and improving rural environment major tasks in environmental protection in rural areas. The Fifth Plenary Session of the 16th CPC Central Committee put forth for the first time the building of socialist new countryside. In 2008, the Ministry of Environmental Protection was established. A special fund for environmental protection in rural areas was also set up. A total of RMB 500 million was invested for this purpose and 2012, it increased to RMB 5.5 billion. In 2010, trial guidelines for treatment in contiguous polluted rural areas were rolled out. In general, since the 16th National Congress of the CPC, the CPC Central Committee and the State Council have been demanding efforts to foster and implement sustainable development and paying increasingly more attention to environmental protection in rural areas. In particular, a variety of legislations, policies and standards for environmental protection in rural areas were formed at the stage. Moreover, pollution treatment expanded from singe areas and demonstration centers to various sectors and contiguous regions.

Stage V (2012–present): Deepening. Since the 18th National Congress of the CPC, building an ecological civilization has been incorporated into all aspects of China's social and economic development. Environmental protection in rural areas is inherently a part of efforts to build an ecological civilization. In the 2013 No. 1 document, China's central authorities pledged to advancing the development of an ecological civilization in rural areas and building a beautiful countryside. In the same year, the Ministry of Agriculture issued a document on a campaign to build a beautiful countryside. The Environmental Protection Law revised in 2014 contain provisions on monitoring and warning against agricultural pollution sources, comprehensively improving the rural environment, preventing and treating agricultural non-point pollution, and allocating fund for environmental protection in rural areas. These provisions laid a solid foundation for advancing environmental protection in rural areas. In the same year, the State Council released *Guidelines on Improving the Living Environment in Rural Areas*. The No. 1 document of the central government in 2015 made it clear that agricultural environment and the living environment in rural areas should be improved. In April 2015, the Ministry of Agriculture released *Guidelines on Fighting Agricultural Point and Non-Point Pollution*, setting a target of reducing agricultural water use and water pollution. Specifically, the total amount of water used for irrigation was to be kept at 372 billion cubic meters, the quality of irrigation water should be up to standards, the use of fertilizers and pesticides was to be reduced, and agricultural waste was to be disposed of in environment-friendly ways. In November 2015, the Ministry of Housing and Urban-Rural Development and other authorities released *Guidelines on Advancing of Rural Waste*. In 2017, the Ministry of Environmental Protection and the Ministry of Finance jointly issued the *13th Five Year Plan for Comprehensive Improvement of Rural Environment*. In 2018, the CPC Central Committee and the State Council released a three-year action plan for improving the living environment in rural areas. In general, this stage is a period when environmental protection in agriculture and rural areas was comprehensively deepened. Instead of the fragmented and separated policies in the past, pollution prevention and control was effectively combined with agriculture and farmers. From the top-level design for comprehensive improvement of rural environment to supporting policies in specific areas such as sources of drinking water, domestic garbage and sewage, pollution from the livestock breeding sector and the use of fertilizers and pesticides, green development and building a beautiful and livable countryside, became the major goals of environmental protection in rural areas.

2 Institutions for Environmental Protection in Rural Areas and Relevant Difficulties

It can be seen from the evolution of policies that the central government pays high attention to environmental protection in rural areas and that notable results have been achieved over the past years. However, environmental protection in rural areas remains an arduous task. Across the country, 40% of villages have no garbage treating facilities, and 78% do nothing to treat wastewater. To further explore paths to build institutions for pollution control in rural areas, this paper does analysis from the angles of national policy, departmental regulations and technical standards.

In regard to national policies, the core of institutions for environmental governance in rural areas is reward-based environmental governance. It is about incentives. The central authorities allocate special funds as rewards to encourage local governments to work on rural environment protection and increase input in this field. The following-up policy on improving the living environment in rural areas is also at the core, but it does not take the place of reward-based governance. The policy of improving the living environment in rural areas is more inclusive and brings more benefits to the people. Instead of focusing only on environmental issues, environmental protection in rural areas also cover public facilities, living conditions and sanitary conditions. However, early work mainly focused on waste treatment in rural areas. In 2018, the CPC Central Committee and the State Council released a plan for rural revitalization. In a systematic approach, efforts were made to build rural areas with thriving businesses, pleasant living environments, social etiquette and civility, effective governance, and prosperity. The plan provides an effective institutional support to push green agricultural development and continuously improve living environments in rural areas.

As for departmental regulations, environmental protection in rural areas involves departments in charge of agriculture, environmental protection, water resources, housing and urban-rural planning, land and resources and health. Every department issues regulations for a specific area within their scope of responsibilities, resulting in fragmentation and a waste of resources. Two environmental governance policies had comparatively wider coverage. The first is about contiguous polluted rural areas, rolled out by the former Ministry of Environmental Protection, and the other is the beautiful countryside initiative launched by the former Ministry of Agriculture. The former aimed to advance the policy of reward-based environmental governance, and prevent fragmented efforts for pollution control. The latter covered all parts of the agricultural industry, farmers' ways of production and life, and the environment in rural areas. In 2018, China reorganized some of its ministries, establishing the Ministry of Agriculture and Rural Affairs, putting together the agriculture- and rural-affairs-related functions previously scattered in multiple authorities. The function of environmental supervision in rural areas was transferred to the new Ministry of Ecology and Environment. Such changes helped improve cross-departmental coordination and the efficiency of pollution control in rural areas.

In terms of technical standards, standards for pollution emissions from the livestock breeding sector, rural domestic waste control and rural drinking water sources have been put in place. However, these standards are not specific enough for China, a country with great differences in resource endowment and economic development across its vast rural areas. Taking rural domestic wastewater as an example. There are no specific provisions defining the scope of application of the standard, and the control indicators selected and the emission caps set were not quite appropriate. Many provincial governments design standards and manage rural wastewater based on the emission standard for urban sewage treatment. Some even adopted the one-size-fits-all approach in livestock breeding and agricultural production, banning breeding and farming as a mandatory measure. Ecological protection was confronted with farmers' fights for livelihoods. Local conditions, economic development, population conditions and people's demands for the improvement of living environments were not considered as they should be in environmental protection.

Furthermore, environmental protection in rural areas mainly depended on administrative forces or special campaigns such as that to improve the conditions in contiguous polluted rural areas and build a beautiful countryside. Under the current political system in which local governments face pressure from superior authorities to finish various tasks, and with a prevailing decentralized fiscal system, the way of environmental protection in rural areas is an appropriate and wise choice. On the one hand, environmental protection in rural areas is a public good. The state government transfers resources to fund local projects with the aim of spurring the economy. In the meantime, the practice ensures effective input in or full coverage of public services in the environmental sector in rural areas. It also stimulates local governments' enthusiasm. On the other hand, local governments face heavy debts. Without specific projects, they will not be able to acquire special funds for environmental protection in rural areas and then use the fund to fill the gap in funding for infrastructure construction projects. Moreover, most villages and townships have very limited fiscal revenues at their disposal, and are unable to afford construction and maintenance for environmental protection projects. They need specific projects to get urgently needed financial support. There are also flaws to the practice of protecting rural environment through specific projects, though. First, the current governance capacity cannot meet the demand. In other words, getting projects approved and reaching project acceptance standards become the primary goal of governments at the basic level, and long-term operation and maintenance of the projects (whether environmental governance is sustainable and whether the cost matches the returns) become secondary or not a goal at all. In reality, many domestic wastewater treatment facilities went idle after being used occasionally during inspections. Some local residents dubbed them "environmental protection garbage." Second, by taking all responsibilities, the governments tend to ignore social resources and the collective strengths of rural residents in protecting the environment. Also, current fiscal expenditure cannot fund environmental protection efforts in over 600,000 villages across the country. If each village needs RMB 1.1 million to treat pollution, the construction expenses will be as high as RMB 660 billion, excluding expenses for operations and maintenance. In comparison, the average annual input from the central authorities are estimated at RMB 4.4 billion. For RMB 660 billion to be provided, 150 years will be needed. Third, environmental protection projects and resources allocated for the projects come from different departments. For example, the housing and urban-rural planning authority allocates funds for treating domestic solid waste and wastewater, the agricultural authority for treating pollution from the livestock breeding and biogas, the health authority for improving water facilities and toilets, and the water resources authority for improving drinking water and dredging rivers. Limited funding for environmental protection in rural areas is fragmented and scattered, resulting in low efficiency in resource allocation. This is incompatible with the nature of environmental protection as a systematic project.

3 Practices and Experience of Foreign Countries in Rural Environment Governance

In the process of urbanization, industrialization and modernization, various countries faced the common problem of a withering countryside. The evolution of rural areas in developed countries shows that a declining share of agriculture in GDP or a diminishing rural population do not necessarily mean a decline of the countryside, though. Rather, the countryside of many countries presents beautiful landscape and an enchanting and serene lifestyle. Particularly in the post-industrialization era, rural development and rural environmental protection show huge potential.

According to statistics by the World Bank, there were 3.34 billion people living in rural areas across the world by the end of 2016, accounting for 46.1% of the world population (in 2007, urban residents outnumbered rural residents for the first time in world history). Of the 3.34 billion rural residents, 580 million were from China. By 2035 when China basically realizes modernization, there will still be 400 million people living in rural areas in the country. The shares of rural population in Singapore, Japan, the UK, the US and China were 0.1, 17, 18 and 44% in the world total. As agricultural population moves rapidly to non-agricultural sectors, each country has to experience a process of rural construction, agricultural development and farmer migration. Many developed countries and regions have completed the process or dealt with the pains coming along with it, and restored beautiful countryside scenes.

The UK was one of the first countries in the world to start urbanization. It was also the country witnessing the earliest and largest population mobility from rural areas to urban areas, as well as the fastest decline of rural population. The share of agriculture fell sharply in its GDP. In 1801, the share was 32.5% and in 1901, it was 6.1%. However, there was no drastic decrease in the share of agricultural output after 1901. In 1951, the share was 4.7%. Up to 77% of the land in the UK is in rural areas (the median of Europe is 40%). The country became aware of the consequences of a declining countryside and rural underdevelopment early enough, and took measures to reverse the trend. It integrated agricultural development and rural environmental governance into urban development. With a radical approach, it contained agricultural pollution at the source. It replaced labor force with capital to support cultural and ecological protection in rural areas, and offered tax incentives and government subsidies among other types of financial support to control point and non-point pollution. In the meantime, the country improved its legal system for agricultural pollution control. For example, the *Control of Pollution Act* criminalized discharge of pollutants into water. The UK also values effective communication between the government and the public and information transparency.

The United States was founded on agriculture. In 1787, the country had 90% of its population working in agriculture. The urban population did not outnumber the rural population until 1920. In the early twentieth century, population mobility facilitated inclusive development of agriculture, but the environment was the cost. Farmland diminished, heavy use of pesticides and fertilizers caused the hardening of the soil and the loss of organic matter, and water pollution was serious. Across the country, 40% of the total area of 47 major rivers, lakes and pools were contaminated. In response, the government made laws and policies to protect the environment. For example, in 1936, it introduced the first legislation on point and non-point source pollution control. The country also widely applied advanced technologies in agriculture. Agricultural mechanization, precise fertilizing and GPS among other efforts enabled the country to reverse the trend of increasing environmental pollution. Every farmer can manage at least 3000 mu of farmland. In general, the U.S. employed a systematic approach to environmental protection in rural areas. It used various projects for the end and comprehensively used technological and economic means to solve different kinds of pollutions and meet the demands of various farms. It also solved problems relating to technology adaptability and long-term mechanism. The country encouraged farmers to use environmentally friendly technological alternatives of their own accord, and also encouraged social organizations to contribute to environmental protection in rural areas.

Japan began to experience hollowing of the countryside in the 1960s. The ripple effect diminished agricultural population and pushed population aging in rural areas. At that time, agricultural non-point source pollution was extremely serious, and highly polluting industries including steel making and mining moved to rural areas in clusters, causing pollutions to farmland, water sources and foods. To handle the serious pollutions in rural areas, Japan introduced laws and regulations for rural development on the whole. It also increased input in environmental improvement and fostered environmental activist groups. Laws and regulations for environmental protection and preventing rural areas from declining include the Village Vitalization Act and the Special Law for Sparsely Populated Areas and Rural Areas. With the law supporting specific non-profit activities, the government guided urban non-governmental organizations to promote environmental protection in rural areas. In the countryside development movement, Japan highlighted planning, and integrated agricultural development and environmental protection. The government, businesses and consumers worked jointly for the operation, services and supervision of environmental protection projects. In particular, the country strengthened organizational efforts at basic-level agricultural associations. Other efforts included increasing financial input, bringing in technologies and fostering people's environmental awareness.

4 Four Major Issues that Should Be Well Handled in Environmental Protection in Rural Areas in the Future

Environmental protection in rural areas aims to improve people's well-being. It is also an important task of the rural revitalization campaign. Protecting the environment in rural areas is also what needs to be done to complete the building of a moderately prosperous society in all respects. Based on problems concerning environmental protection in rural areas and drawing experience from foreign countries, the author, speaking from his findings from field trips, believes that in environmental protection in rural areas, the following issues need to be properly handled.

First, we should try to strike a between environmental improvement and rural revitalization. Pollution control is the cornerstone of building a beautiful and livable countryside. Rural revitalization is the safeguard for sustainable investment in environmental improvement. More specifically, after figuring out what rural areas have and pooling resources for agricultural and rural development, the plan for environmental protection should be in tandem with local plans for rural revitalization. In particular, it should be aligned with future population size and distribution, the conditions of ecological elements such as farmland, water and forests, and the functional layout of villages in order to avoid redundant construction and mismatch of resources. For example, road pavement in rural areas will drastically increase the cost of sewage network construction, a large quantity of resources was invested in sparsely populated remote areas without cost-benefit analysis. Meanwhile, the implementation of the rural revitalization initiative should be spearheaded by green development, and wisdom should be drawn from rural traditions. Highly profitable and high-quality industries should be employed to spur growth of the rural collective economy. Fostering thriving industries should be incorporated into the broader process of improving the living environment in rural areas. Environmental issues in rural areas should be examined from a development-oriented perspective. The ultimate goal is to bring both wealth and a good environment to the people.

Second, government investment and public participation should be balanced. A good environment is a typical public good. Governments at various levels, particularly township Party committees and governments, should take the lead. However, this does not mean that the governmental should take all responsibilities or that social forces and local residents should be excluded from decision-making, project funding, operation and maintenance. More specifically, considerations should be given to both infrastructure construction and follow-up capability building for operating and maintaining the projects through appropriate methods and rational investment. Codes of conduct should be improved amid efforts to promote self-education and awareness raising among farmers in regard to environmental protection. On this basis, efforts should be made to encourage basic-level Party organizations, capable brains and investors to make their contribution. Farmers' wills should be respected, and communities should play their due role in resource allocation. Capable brains in farming and organic agricultural production are to offer their opinions on prioritizing pollution control.

Third, both facility construction and long-term maintenance should be highlighted. This is key to successful pollution control in rural areas. Targeting the problem of the lack of maintenance after environmental protection facilities are put in place, long-term mechanisms should be created. Specifically, based on financial capabilities of the village and villagers' willingness, an appropriate method for funding, operating and maintaining the projects should be determined. For example, the cost of pollution can be assumed by all stakeholders, which include the central government, the local government, the village and residents. Suitable work plans and technical patterns should be determined based on local geographical conditions, traditions, economic conditions and farmers' expectations. Due considerations should be given to rural landscape protection and pollution control. Projects should never be built only for vanity or a good result in government performance evaluation. It is inappropriate to simply copy the approach used in urban environmental protection when it comes to environmental governance in rural areas. Bad choices may include building man-made wetlands and planting trees by rivers and lakes to help treat domestic wastewater. Kitchen waste and compostable wastes can be recycled into compost. More affordable and easy-to-learn ways can be proposed to ensure the long-term operation of environmental protection facilities.

Fourth, both end-of-the-pipe pollution and pollution at the source should be sufficiently treated. In governing the environment in rural areas, the big picture of agricultural and rural development should be considered. To fundamentally improve rural environment, efforts should be made to advance agricultural mechanization, improve people's sanitation habits, and improve public facilities and services at the township and village levels. End-of-the-pipe pollution treatment with suitable technologies and projects for pollution control at the sources in production and everyday life scenarios should be combined. Specifically, support should be offered to translate the ecological advantages of rural areas into strengths in economic development and form a mutually beneficial pattern for the environment and the economy. Measures for this purpose include using local special industries and agro-tourism among other strengths to improve infrastructure for environmental protection in rural areas. In this way, pollution control and integrated industrial development can reinforce each other. In less developed rural areas, environmental protection projects can be aligned with medium- and long-term rural development plans, based on the restoration capabilities of semi-natural ecosystems and integrating seed funds for progress in regard to agriculture, farmers and the countryside. The core lies in exploring affordable, easy-to-operate and effective ways to advance rural environmental governance.

Learning and Implementing Xi Jinping Thought on Ecological Civilization

Some Gains from Learning Xi Jinping Thought on Ecological Civilization

Geping Qu

Since the 18th National Congress of the Communist Party of China (CPC) incorporated the development of ecological civilization into the Five-Sphere Integrated Plan, I have been pleased with the decision of the CPC Central Committee on significantly improving the status and role of environmental protection. This is exactly the great situation that comrades in environmental protection departments have been looking forward to for many years. The 19th National Congress of the CPC has set a beautiful China as an essential goal for building a strong modern socialist country. From the perspective of the Five-Sphere Integrated Plan covering material, political, cultural, social, and ecological domains, that also means a modern socialist power is not only "prosperous, democratic, civilized and harmonious" but also "beautiful," highlighting the holistic and synergistic nature of development. The decision has also made the socialist modernization with Chinese characteristics more distinct and the features of a new era more prominent. At the First Session of the 13th National People's Congress, ecological civilization was written into the Constitution. It was also decided to establish the Ministry of Ecology and Environment, indicating that we have elevated the ecological civilization to the national will through the Constitution and provided organizational guarantee for its development. Since 2016, the central environmental protection inspection team has been inspecting the environmental work across China with unprecedented and productive practices and efforts.

These events have pleasantly surprised and heartened me. It is visible to me that China has not only found a Chinese-style way to protect the environment but also effectively combined environmental protection with economic, social and cultural development, as well as material, political, intellectual and social fields.

It is a revolutionary change in China's environmental protection from the traditional one based on pollution control to the one focusing on the development of ecological civilization. The significance of this change cannot be overestimated.

G. Qu (✉)
China Environmental Protection Foundation, Beijing, China

J. Pan et al. (eds.), *Beautiful China: 70 Years Since 1949 and 70 People's Views on Eco-civilization Construction*, https://doi.org/10.1007/978-981-33-6742-5_11

What I am concerned about is how to realize this transformation and how to do it well.

1 Moving from a New Starting Point to a New Stage in the Continuous Self-reform

Regarding environmental protection, China has markedly different characteristics from those of developed countries. In the industrialization, developed countries neglected environmental protection, leading to pollution. Later, they had to deal with the pollution at a high cost. Premier Zhou Enlai had been far-sighted, pointing out time and again that China, a developing country, should learn from lessons of developed countries and start protecting the environment before pollution become serious, so as to avoid following their old path of "pollution before treatment." However, people lacked the experience of pollution in developed countries, especially in the chaos of the "cultural revolution." It was difficult to accept the idea of environmental protection. Some believed that pollution was a product of capitalism, and did not exist in socialist China; some argued what was the most important and urgent was to thrive the economy; and others held that "treatment after pollution" was a rule and China was no exception. Despite these interruptions, Premier Zhou worked to promote environmental protection and establish the "National Environmental Protection Office" under the Environmental Protection Leading Group of the State Council. I was then deputy director of the Office. That was a temporary agency under the leadership of Premier Zhou and the State Council, so it was quite influential. At that time, environmental protection departments were weak and short-staffed. Although they did their utmost to promote environmental protection, they were unable to prevent environmental pollution and ecological damage. They followed the path of treating pollution and restoring environment as they went along, raising the public awareness of environmental protection in this process. For a long time, environmental regulators had been passive in their work. That is why I have the sentiment that Sisyphus repeatedly pushes a rock up the slope. In fact, what we are pushing is not the cold boulder of unchanging size. The environmental protection we have promoted is a daunting task. Even when we are frustrated, such an effort, not the boulder, is moving from a new starting point to a new stage. With deep insight, the director of an environmental protection agency in a provincial capital city said: "Environmental protection, a magical field always changing and reforming itself, has been advancing slowly, difficult but charming."

Sisyphus was punished by the gods and had to "engage in futile and hopeless work." But it reveals a higher fidelity that lifts up the boulder and denies the gods. "It drives the gods out of this world. It makes of fate a human matter, which must be settled among men." From this point of view, the spirit of fearlessness and hard struggle expressed in the myths of ancient China, such as the foolish old man Yu Gong removing the mountains, the bird Jingwei trying to fill the sea, the goddess

Nuwa mending the sky, and the archer Hou Yi shooting the sun, can better reflect that of environmentalists in China.

2 The Overall Environmental Protection Has Been Encouraging in New China Over the Past Seven Decades

In the four decades since the first *Environmental Protection Law* was enacted in 1979, environmental protection has changed remarkably. Only after exploring the way forward at first, the "three major policies" (giving priority to prevention while combining prevention with treatment, who pollutes and who treats, and strengthening environmental management) and the "eight systems" (environmental impact assessment, the "three-simultaneousness" system, the charge system for sewage discharging, the target responsibility system for environmental protection, the quantitative assessment system for comprehensive urban environmental improvement, the sewage-discharging permit system, the system for pollution control within a time limit, and the centralized pollution control system) were finally formulated and implemented throughout the country, so that environmental protection, especially environmental management, can be conducted in a rule-based manner. Now we must push environmental protection to a broader domain with ecological civilization as the main body. Party and state leaders have set the direction for the development of ecological civilization, empowering us with the confidence in our path, theory, system and culture.

Around 1980, environmental protection bureaus were set up across China, but their staff members knew nothing about environmental protection. Some studying chemistry or specializing in chemical engineering, if transferred to such bureaus, would be technical backbones. Nowadays, professionals and those holding master's and doctor's degrees make up a big proportion in the environmental protection team, greatly enhancing the quality of the entire team.

In the 1980s and 1990s, the important periods for spreading knowledge about environmental protection, almost everyone could not tell the difference between environmental protection and sanitation, perceiving it as cleaning up streets. People knew, of course, little about the discipline of environmental protection. Only a few technicians in this sector had an idea about the term TSP (total suspended particulates), but now almost everyone knows about PM 2.5 (fine particulates) and wears a mask in smoggy weather.

In 1987, the Environmental Protection Committee of the State Council held a national conference on air pollution prevention and control in Taiyuan, Shanxi Province, where pollution was much serious. According to a comrade from the environmental protection bureau, a biking to work would take him more than half an hour, and, without glasses, dust always got into his eyes. When he wore glasses, his eye parts were like the eyes of a panda. He could write words on his dark pants covered with dust. At that time, there was also a popular saying in Lanzhou to describe the local air quality: “You cannot tell the sun from the moon, the day is almost as dark as the night, and the nostrils are as black as chimney walls.” The pollution is hard for people to imagine nowadays.

While the current quality of the environment leaves a lot to be desired, tremendous changes have been made in the environmental quality across China. In today’s Taiyuan, even if you wear a shirt for three to five days, the collar is not as dirty as it was when you wore it for an hour. This accomplishment is truly gratifying to an old environmentalist like me.

According to the 19th National Congress of the CPC, we will build a moderately prosperous society in all respects by 2020. In this decisive period, it is necessary to fight the battle against air, water and soil pollution. Fighting the battle against pollution lays a solid foundation for meeting the goal of building a moderately prosperous society in all respects, achieving socialist modernization by 2035, fundamentally improving the ecological environment and developing a beautiful China. With the fundamental improvement of the environment, it is tempting and feasible to build China into a prosperous, democratic, civilized, harmonious and beautiful modern socialist power by 2050.

Pollution prevention and control, an arduous task, must be refined to fight seven milestone battles, including keeping our skies blue, pollution control of diesel trucks, treatment of dark and fetid bodies of water in urban areas, comprehensive improvement of the Bohai region, protection and restoration of Yangtze River ecosystems, safeguarding of water sources, and handling of agricultural and rural pollution. The battles are aimed at the chronic intractable problems that we have been addressing for years. The arrangements that are now in place are much better and more scientifically sound than before. Keeping our skies blue, for example, aims to take Beijing-Tianjin-Hebei and surrounding areas, the Yangtze River Delta region and the Fenwei Plain as key areas to further reduce PM 2.5 concentrations during the autumn, winter and heating periods by strengthening joint pollution prevention and control, and intensifying efforts to adjust and optimize industrial structure, energy mix, transport structure and composition of land use. As a newly-added key region, the Fenwei Plain, a general term for the Fen River Plain and Wei River Plain and their tablelands and terraces in the Yellow River Basin, includes some areas of Shanxi, Shaanxi and Henan provinces. It is a new battleground for environmental protection, deepening our understanding of cross-regional environmental issues. The Party committees and governments of the three provinces have worked together with the Ministry of Ecology and Environment to take comprehensive measures for joint pollution prevention and control to deal with the challenging problem. What they are doing embodies China’s positive and determined approach to environmental protection, signaling that we have conditions and ability to tackle chronic problems. That is our strength in winning the battle against pollution.

There is only one year left until 2020. Facing a tight schedule, heavy burdens, and tremendous pressure, we can only redouble our efforts to win the seven battles in just one year.

3 Reflections on Environmental Work

1. **Learning and practicing Xi Jinping Thought on Ecological Civilization**

The environmental work in the charge of the Ministry of Ecology and Environment is vital to how China pursues its development path and the living quality of the people. Therefore, it is necessary to further raise the awareness of environmental issues. The most fundamental point is to study and practice Xi Jinping Thought on Ecological Civilization. It is a powerful ideological weapon and fundamental guideline to ensure the development of ecological civilization by the Party and the country, indicating that our Party's philosophy and way of governance have entered a new theoretical and practical realm. The establishment of Xi Jinping Thought on Ecological Civilization as the guideline unveils a fundamental and historic change in the status of ecological civilization in the cause of socialist construction. We must consciously take an ecologically civilized approach to lead overall economic and social development. Never again must we regard ecological civilization and environmental work as just one task for one department.

2. **Moving faster to build systems for an ecological civilization**

It is also a huge systematic program that requires joint efforts of the entire society. The program includes a cultural system in which environmental concepts and values are norms, an economic system aimed at integrating industries and ecosystems, a system of responsibilities and objectives centered on improving environmental quality, a system of institutions backed by modernized governance systems and capacity, and a safety system that emphasizes the maintenance of well-functioning ecosystems and the effective prevention and control of environmental risks. Twenty years ago, China's Agenda 21 introduced the concept of the environment as a source of "natural capital." The environment is generally considered to be a natural resource. When we develop natural resources including their various elements, do we consider that natural resources are also important natural assets? In this way, there emerge three concepts including natural resources, natural assets and natural capital. The industrialization of ecology is made possible only if their relations and transformation conditions are identified. If we can use the concepts of resources, assets and capital to understand the current treatment of air, water and soil pollution, we can heighten our awareness of "lucid waters and lush mountains are invaluable assets" and step up efforts to deepen the environmental work.

3. **Transforming the economic development mode as the top priority in building a system of ecological economy**

As we all know, ecological problems arise in economic development and are closely related to the way the economy develops. Back in early 1980s, Xu Dixin, a famous Chinese economist, proposed to rebalance the relations between economic development and ecological environment and to study ecological economics. In 1984, I attended the inaugural meeting of the China Ecological Economics Society and served as the vice president. From then to now, the ecological economics and ecological economy in China have developed significantly. We should make good use of rich research results and practical experience to conscientiously summarize the practice of ecological economy in China. Also, great importance should be attached to "ecological industrialization" and "industrial ecologization," so as to effectively transform ecological factors into production factors and ecological wealth into material wealth. We should lay a more solid foundation for economic development by utilizing the advantages of ecological resources and ecosystems, and truly integrate the transition into the development of an ecological economy, so that they can become a strong material basis for building an ecologically civilized society and produce fruitful results.

4. **Building a global community with a shared future**

Since the 18th National Congress of the CPC, the idea of a "global community with a shared future" has become a global concept that General Secretary Xi Jinping, from a global perspective, has been promoting a broader vision in the governance of China. At the 19th National Congress of the CPC, General Secretary Xi called on China to be "an important participant, contributor, and torchbearer in the global endeavor for ecological civilization."

Since attending the United Nations Conference on the Human Environment in 1972, I have partaken in many international meetings and deepened my understanding of the concept of "we have only one earth." All the undertakings that China has done to protect the earth's environment have been a driving force behind China's participation in international cooperation.

Xi Jinping Thought on Ecological Civilization has raised the environmental improvement to an unprecedented level, opened up a new theoretical and practical domain of Marxist man-nature relations, established a sound worldview, values, practice theory and methodology for the ecological civilization, a brand-new form of civilization for human society, and aimed at transforming the ecosystems shared by all mankind into a global community with a shared future. As the second largest economy, China has a direct impact on the competition among developed countries regarding the global living environment and redistribution of ecological capital in realizing the great rejuvenation of the Chinese nation. The ecological environment has increasingly become a comprehensive issue inseparable from political affairs, the economy, livelihood projects, international governance and global games. Therefore, while participating in major international actions to address climate change, we should also promote green development and growth at home, make our due contribution to building a community with a shared future for all mankind through multilateral cooperation mechanisms such as the Belt and Road Initiative. We should also secure the new international right of discourse and demonstrate a responsible major country.

5. **The entire society must undertake the task of developing an ecological civilization**

If pollution control and ecological restoration are still at a certain distance away from each citizen, then the ecological civilization, which embodies higher quality development and living standard, is closely related to the well-being of everyone. All the people should step out of the agricultural and industrial civilizations and move towards the ecological civilization. We need to raise our ecological awareness to promote profound changes in our ideas and concepts, improve our ecological literacy and behavioral norms, rebalance the relations between humans and nature, the environment and themselves, and increase our engagement in the environmental protection. The most practical action is to conserve water, electricity and energy as well as report pollution but never to drop litter.

Continuously Advancing China's Development of an Ecological Civilization to a New Stage

Cungen Chen

On May 18, 2018, the CPC Central Committee held the National Conference on Eco-environmental Protection, the highlight of which was the illumination of Xi Jinping Thought on Ecological Civilization. General Secretary Xi Jinping delivered the important speech entitled *Pushing China's Development of an Ecological Civilization to a New Stage* at the conference, putting forward the general guideline and strategy for promoting ecological civilization and building a beautiful China in the new era.

1 The Profound and Extensive Essence of Xi Jinping Thought on Ecological Civilization

Xi Jinping Thought on Ecological Civilization, the result of long-term in-depth thinking and rich practice by General Secretary Xi Jinping, has carried forward the millennium-old outstanding ecological culture of the Chinese nation, inherited and developed Marxism, and embodied the socialist concept of ecological civilization. As an integral part of Xi Jinping Thought on Socialism with Chinese Characteristics for a New Era, it is profound and extensive, with a solid theoretical foundation, a rich cultural background and a wealth of practical experience, which are highlighted in the following aspects.

1. **Promoting harmony between man and nature is the basic guideline for achieving sustainable economic and social development**

The history of human civilization is a course along which human beings understand, adapt to, utilize, interact with, and transform nature and promote the development of

C. Chen (✉)
Work Committee of the Central Government Departments under the CPC Central Committee, Beijing, China

J. Pan et al. (eds.), *Beautiful China: 70 Years Since 1949 and 70 People's Views on Eco-civilization Construction*, https://doi.org/10.1007/978-981-33-6742-5_12

human society. Marx believed that humans cannot survive without nature, the means of life and production come from nature, nature provides the objective conditions for human existence, humans produce, live and reproduce in their interaction with nature, and nature would be a gift for human beings if they treat nature well. "If man has conquered the forces of nature by science and creative genius, then the forces have taken their revenge on man," said Engels, "Do not revel too much in our victories over nature. For every such victory, nature has taken revenge on us." The dialectical relationship between man and nature is the eternal theme of human development. In the recorded history of human civilization, the decline of ancient Egyptian and Babylonian civilizations, the disappearance of the Loulan civilization, the evolution of China's Loess Plateau and Hexi Corridor from the ancient lush forests and abundant waters and grasslands to the current ravines and *gobi* deserts are all lessons learned from the degradation of the environment caused by man's excessive exploitation of nature and the resulting civilizational crisis. General Secretary Xi Jinping's assertion that man and nature form a biotic community with a shared future reveals in a nutshell the true meaning of the relationship between man and nature. He pointed out that, in order to achieve sustainable development, we must respect, adapt to and protect nature, and "protect the ecological environment like we protect our own eyes and treat it like we treat our own life." From the perspective of historical development, he focuses on prosperity of our country, rejuvenation of our nation and the well-being of our people, making his assertion that "a civilization may thrive if its natural surroundings thrive, and will suffer if its natural surroundings suffer" and that "building an ecological civilization is vital for sustaining the development of the Chinese nation."

2. **Maintaining a good ecological environment is a basic condition for achieving well-being for all**

Since the reform and opening up, China has seen rapid economic and social development, which has greatly raised the people's standard of living and enriched their intellectual life and will soon successfully accomplish the building of a moderately prosperous society in all respects. However, environmental problems have become an obstacle to the building of a well-off society in all respects, as evidenced by heavy pollution, growing heaps of urban waste, drinking water pollution, and excessive heavy metal content in soil. They have led to strong public response and great social concern. The people's expectations for clean water, fresh air, safe food and a beautiful environment are growing stronger. The quality of the environment has a direct impact on the people's well-being and is becoming increasingly prominent. General Secretary Xi Jinping has proactively responded to the concerns and aspirations of the people, clearly stating that "the environment is people's livelihood, green hills are beautiful, and blue skies are happiness," that a good environment is the most benefit to people's livelihood, and that it is the Party's ruling mission and goal to promote the development of an ecological civilization and provide the people with more high-quality products and a beautiful environment. He has also demanded that building an ecological civilization be vigorously promoted, more quality products from nature be provided, and the people's growing need for a beautiful environment be continuously met.

3. **Protecting lucid waters and lush mountains and obtaining material assets are the basic way to achieve green development**

In response to the widespread problem of the dichotomy between ecological conservation and economic development, General Secretary Xi Jinping repeatedly stressed on different occasions that "we want both lucid waters and lush mountains as well as mountains of gold and silver." We prefer the former to the latter, and "lucid waters and lush mountains are as valuable as gold and silver." The former represents the natural environment and ecological benefits, while the latter the socio-economic value and economic benefits. We need to be committed to socio-economic development as well as environmental protection, pursuing the harmony between man and nature and the balance between socio-economic development and ecological conservation. When there is an imbalance between them, we prefer the environmental protection. Nevertheless, talents and ideas hold the key to balancing lucid waters and lush mountains with assets. General Secretary Xi Jinping pointed out that "we need to foster our respect for the value of nature and natural capital. Natural ecosystems have value. The protection of nature is a process of increasing the value of nature and the value of natural capital." Therefore, while strictly protecting the environment, we should make good use of lucid waters and lush mountains, and let them continue to yield mountains of gold and silver by vigorously developing ecological industries.

4. **Making the overall planning for protection and management of mountains, rivers, forests, farmlands, lakes and grasslands is the basic way to achieve coordinated development**

Mountains, rivers, forests, farmlands, lakes and grasslands are integral natural ecosystems that are interconnected and inseparable. Lush trees and well-structured forests can effectively trap precipitation, conserve water sources, consolidate soil fertility, prevent flooding and soil erosion, slowly transport constant water into rivers and lakes, irrigate crops and replenish soil nutrients, and create and maintain a comfortable and beautiful environment for human habitation. In the large ecosystem of mountains, rivers, forests, farmlands, lakes and grasslands, people, as a part of it, are the ultimate beneficiaries and the most active and influential factors. The huge ecosystem maintains a dynamic balance through material and energy flows. Once one factor changes, the original balance will be broken, and a new balance will be reached through dynamic and successional changes. Historically, numerous examples of serious system imbalance and major natural disasters ultimately caused damage on humans themselves due to the greedy and unreasonable exploitation of nature. The Italians in the Alps, for example, cut down and exhausted seemingly unrelated fir forests, destroying local alpine livestock farming and causing flooding. In China's early days, disasters caused by excessive logging of forests, deforestation for reclamation, reclamation of farmland from lakes, indiscriminate reclamation and grazing, and dredging of rivers abounded, leading to huge and painful losses and trauma to economic and social development and to the people's production and life. General Secretary Xi Jinping stressed that "we must take all of the ways that we impact nature into account from a broad and long-term perspective, and avoid being penny-wise and pound-foolish, biting off more than we can chew or emphasizing one thing while neglecting another, for this would ultimately cause long-term environmental damage on a system-wide scale."

5. **Reinforcing rigid restraints on the red-line legal system is the basic guarantee for fighting the battle against pollution**

For a long time, in order to speed up economic development and increase GDP in some places, local governments make up plans, launch projects, and advance rashly, with little regard for the impact on ecosystems and the environment. Even some industrial, mining and chemical enterprises, where money is paramount, save money as much as possible in a bid to gain maximum benefits, and do not take measures to clean up pollutant emissions from production, or reduce emissions only to go through the formality. General Secretary Xi called for the resolute eradication of this practice of pollution before treatment and not treating at all, and for the resolute rejection of this path of seeking short-term gain at the expense of future generations. He pointed out that only the strictest system and the strictest rule of law can provide a reliable guarantee for the development of ecological civilization. Therefore, it is necessary to establish an effective legal system for ecological civilization that constrains exploitation and promotes green, cyclic and low-carbon development by following the idea of strict prevention at the pollution source, strict control of the pollution process and severe punishment of consequences. We must improve the complete system of ecological civilization with clear property rights, pluralistic participation, and equal emphasis on incentives and constraints, so as to provide legal and institutional safeguards for building an ecological civilization. All local governments should earnestly implement the requirements of General Secretary Xi Jinping. They should be determined and courageous to rise to the challenge, regulate matters according to the law, use the system to regulate people, strictly investigate violations of the law, make the law and the system a rigid constraint in the fight against pollution, integrate economic and social development with ecological civilization, and comprehensively promote green development.

6. **Promoting co-governance among countries as a basic strategy for achieving global ecological civilization**

Since industrialization, Western developed countries have followed the path of "pollution before treatment." Greenhouse gas emissions, rising global temperatures, melting Arctic icebergs, frequent extreme weather events and many other irreversible ecological problems on a global scale have brought immeasurable disasters and losses to humans and posed hidden development hazards that are difficult to predict and prevent. General Secretary Xi Jinping stressed that humans constitute a community with a shared future and building a green home is the common dream of mankind. No country can be immune to the global challenges posed by the ecological crisis. Therefore, all countries should be encouraged to join hands to take care of the earth and protect our home. Promoting the construction of the Belt and Road Initiative is an effective way to achieve the global governance led by China. Through active engagement in global environmental governance, China will continue to increase its voice and influence in international affairs, lead countries and regions along the Belt and Road in establishing the concept of ecological civilization that respects nature and green development, guide the formation of an international order and international cooperation that protects the environment and promotes sustainable development, and allow the concept and practice of ecological civilization to benefit the people of all countries along the routes.

2 Remarkable Progress in Building an Ecological Civilization

In recent years, under the leadership of General Secretary Xi Jinping, mechanisms and institutions for ecological civilization have been gradually established, related systems have been continuously improved, and ecological conservation and environmental governance have been advanced steadily, yielding remarkable results.

1. **Continuous improvement of institutional mechanism and institutional design**

    (1) Establishing the guiding ideology of building an ecological civilization. Building an ecological civilization has been incorporated into the Five-Sphere Integrated Plan and the Four-Pronged Comprehensive Strategy. It has been written into the newly revised Constitution of the Communist Party of China as an important element. It has become the unified will and goal of the entire Party and the people of China. This is also a fundamental, global and historic strategic deployment by the CPC Central Committee for ecological civilization and socio-economic development at this critical moment of comprehensive success in building a moderately prosperous society.

    (2) Putting in place the institutional mechanism for building an ecological civilization. The new round of institutional reform has resulted in the unified management of natural resource assets, the use of national land and space, and ecological protection and restoration. The reform has integrated the supervision and management of forests, grasslands and wetlands, accelerated the construction of national parks and other reserves, combined the responsibilities of departments involved in the supervision of environmental protection, and realized an effective mechanism for the "integration of multiple regulations" on the use of land, the unified supervision of pollution emissions, and the unified coordination of ecological protection and restoration, thus providing an institutional mechanism for the development of an ecological civilization.

    (3) Improving the top-level institutional design for developing an ecological civilization. The Chinese government has promulgated such major

guiding policies as the *Opinions on Accelerating the Development of an Ecological Civilization* and the *General Plan for the Reform of the Ecological Civilization System*. By doing so, it has refined many laws and regulations regarding the development of an ecological civilization, and established a number of systems for the evaluation of the objectives in this ecological campaign, the off-office auditing of natural resource assets, and accountability for damage on the environment. It has also smoothly implemented such policies and systems as the reform of green finance, the compilation of balance sheets for natural resources, environmental protection taxes and compensation for ecological conservation. The government has launched a comprehensive national experiment in the reform of the ecological civilization system, further perfecting the institutional system for building an ecological civilization.

(4) Building an ecological civilization as a priority. Party committees and governments at all levels have made the development of an ecological civilization an important task in the new era. All departments, institutions and social organizations have proactively made planning and taken action. Building an ecological civilization has become the consensus of society as a whole. People across the country have united in their determination to develop an ecological civilization, build a beautiful China and realize the Chinese dream of the great rejuvenation of the Chinese nation.

2. **Significant achievements in ecological conservation and environmental governance**

(1) Identifying the bottom line and objectives of ecological conservation. All parts of the country have conscientiously drawn the red line for ecological conservation; and they have identified goals, tasks, timetable and road map for the development of ecological civilization, giving a boost to green development in an orderly manner.

(2) Rapid development of greening and beautification throughout the country. From 2012 to 2016, 11.6 billion trees were planted, over 500 million *mu* of land area were afforested, forest coverage reached 22.98%, forest area amounted to 3.2 billion mu, forest growing stock was 16.5 billion m^3, 300-plus cities carried out the construction of national forest cities, and there were 2750 nature reserves nationwide. All the achievements have made China one with the largest increase in forest resources across the world in the same period.

(3) Significant results made in air pollution control. Through vigorous remediation, we have banned and closed down high-energy-consumption and high-polluting enterprises, and vigorously promoted pollution prevention and control, leading to a reduction in the average concentration of respirable particulates (PM 10) in 338 cities at the prefectural level and above nationwide. In the Beijing-Tianjin-Hebei region, the Yangtze River Delta and the Pearl River Delta, the average concentration of PM 2.5 has dropped significantly. In Beijing, for example, the average concentration

of PM 2.5 declined from 89.5 mcg/m^3 in 2013 to 51 mcg/m^3 in 2018, with more blue skies and fewer people wearing masks for outdoor activities than previous years.

(4) The pollution of surface water has been reduced. The proportion of water body failing to meet Grade V standard decreased. In 2018, the proportion of state-controlled sections of Grade I–III surface water bodies increased to 71.0%, while that of water body failing to reach Grade V standard dropped to 6.7%.

3 Building an Ecological Civilization Still Faces Severe Challenges

Over the past 40 years of reform and opening up, China has made tremendous achievements in its economic development that have attracted worldwide attention, with the economic aggregate ranking second in the world. Also, with the rapid development of the economy, many ecological problems have been accumulated. In some places, environmental pollution, ecological imbalance, soil erosion and frequent disasters, a cause for concern, have become an outstanding weak area for developing an ecological civilization and realize the great rejuvenation of the Chinese nation, as well as a worry in governance and a pain for people's livelihood.

"The southeastern part of the Hu Huanyong Line covering 43% of China's land area is rich in water and heat, but it carries about 94% of the country's population and 96% of national GDP, which puts enormous pressure on population, development and environment. The northwestern part of the Line making up 57% of the country's land area has harsh natural conditions and fragile ecosystems, so it supports only about 6% of the population and lags behind in economic development. In the southeastern part, care should be taken to ensure that no major ecological risk arises and to vigorously promote the harmonious coexistence between man and nature. In the northwestern part, scientific restoration and comprehensive governance of ecosystems should be stepped up to promote their gradual recovery and enhance their functions, providing more space and greater environmental capacity for human, economic and social development.

With regard to agriculture, China is still based on traditional farming. Its development mode is still to transform nature and take resources from nature, with the pollution of fertilizers, pesticides and irrigation becoming increasingly prominent. Transforming traditional agricultural thinking and farming methods is still a long way to go. The pollution from industrial waste, exhaust gas and sewage is also shocking in vigorously pushing forward the development of industrialization. Currently, the global economy is under increasing downward pressure, international trade frictions are growing, factors of instability and uncertainty are becoming more prominent, global challenges and unexpected problems are unpredictable, and the pressure on the external environment for development is increasing sharply. China has also entered the new normal of readjustment from high-speed growth to high-quality development. In advancing the development of an ecological civilization, there are still a lot of hurdles to cross and a host of stubborn maladies to cure. Arguably, the situation is still grim due to multiple pressures.

Now, the primary contradiction in Chinese society has transformed into that between imbalanced and inadequate development and the people's growing needs for a better life, and the people's demands for a beautiful environment have already become an important aspect of this challenge. General Secretary Xi Jinping demanded that we must step up our efforts to build an ecological civilization and solve environmental problems with the determination to make painful changes, the courage to stand and fight, and the will to commit every last ounce of our effort. It is an unavoidable and urgent responsibility to strengthen the development of an ecological civilization which should be pushed forward with vigorous efforts.

4 Some Ideas for Promoting Ecological Civilization

The development of an ecological civilization involves a wide range of areas and a lot of efforts, relying on scientific and technological progress, an advanced and developed economy, a sound legal system, coordinated social development and the prosperity of the people. People and ideas hold the key to achieving these conditions. We need to start with management of people and promote harmony between man and nature. The following aspects should be strengthened to actively stimulate and bring into play the initiative of people in developing an ecological civilization.

1. **Fostering ethics of ecological civilization in ideological concepts**

It should at least include the "four virtues": the virtue of respecting nature, the political virtue of green development, the social morality of protecting the environment and the moral character of frugality and abstinence from extravagance. The virtue of respecting nature means that reverence for nature comes from your inner mind, you should act according to the laws of nature, and actively pursue the harmonious coexistence between man and nature. Regarding the political virtue of green development, the development for future generations and for long-term peace and security aims to reach a balance between immediate and long-term interests. Both material assets and beautiful environments are emphasized equally in a sound manner. We must put an end to striving for resources for personal gains at a high cost of the environment. The social morality of protecting the environment requires that everyone is responsible for protecting the environment and working for clean production, low-carbon travel and green living. We should foster our moral character of frugality and abstinence from extravagance to live a hardworking, thrifty, simple, civilized and healthy way of life, cherish every drop of water, every grain of rice, every piece of green space, and every degree of electricity, and refrain from wastefulness, extravagance and squandering of resources.

2. **Letting the concept of mountains, rivers, forests, farmlands, lakes and grasslands together forming a biotic community take root in the people's hearts**

General Secretary Xi Jinping pointed out that just as humans need the fields for food, the fields need the rivers, the rivers need the mountains, the mountains need the soils, and the soils need the forests and grasslands, with mountains, rivers, forests, farmlands, lakes and grasslands forming a biotic community. They are a unified natural system, and an organic chain of interdependence and close links. In terms of policy formulation, planning and project implementation, we must take a holistic approach to the overall situation from the perspective of system engineering, comprehensively consider the biological and ecological characteristics of regional mountains, rivers, forests, farmlands, lakes and grasslands, as well as their dynamic laws of succession, and make planning for pollution prevention and control and ecological protection in all natural ecosystems. We need to seek the best combination of systemic governance and protection, and conduct departmental collaboration, overall consideration, integrated policy, multiple measures, and collaborative promotion. Instead of each department taking charge of its own duty, we must seek collaboration among multiple departments to implement the integrated ecological conservation and restoration of mountains, rivers, forests, farmlands, lakes and grasslands.

3. **Strengthening technological innovation to increase the driving force for development**

Science and technology are the first productive forces. Many thorny challenges that must be overcome in the development of an ecological civilization are technical problems. Therefore, we should strengthen scientific and technological innovation, improve the quality of scientific and technological supplies, take technological innovation as a tool to dismantle the "fence wall" that hinders developing an ecological civilization, open the "last kilometer," promote the precise connection of technological innovation with ecological civilization, and let science and technology play their due role in the development of an ecological civilization. Therefore, for one thing, we should strengthen the supply-side innovation of science and technology, step up the research of new energy, new materials, new processes and new modes, lay emphasis on low carbon, environmental protection, green and efficiency, and explore new ways for resource recycling and conservation. For another, it is necessary to take the improvement in the technological content of production as an important driving force for economic transformation and green development, and to work for the popularization and application of existing research results so that science and technology can really be turned into productive forces. In order for technological innovation, we should put talent training and optimization mechanism in a prominent position, cultivate talent through institutional innovation, boost technical R&D through mechanism optimization, and provide an inexhaustible source of power for the scientific and technological revolution with the supply-side reform as the main line.

4. **Vigorously boosting the development of green industries**

China boasts 4.7 billion mu of woodland, 6 billion *mu* of grasslands, 800 million mu of wetlands, 3.9 billion mu of desert, 87,000 species, and other natural resources. Any sector can produce huge economic value and social benefits. For example, bamboo can produce shoots, and be made into furniture, bamboo charcoal, medicinal materials, textiles, chemical raw materials and other products. In particular, the recently hotly-discussed bamboo winding composite material has been made using natural properties of high axial tensile strength. The material can be utilized in urban construction, municipal utilities, transport, water conservancy and other fields, replacing a large number of metal, plastic, cement and other traditional materials. This will greatly reduce energy consumption and emissions. China enjoys 1392 5A and 4A tourist scenic spots, most of which are located in mountainous, central and western regions, making ecotourism and green industries a shortcut to advance regional green development. Similar situations can be found across China. The key to green development is the way of thinking. We should firmly implement the five major development concepts of "innovation, coordination, green, openness and sharing," vigorously facilitate the integration of industry and agriculture, develop agriculture with industrial concepts, equip agriculture with advanced industrial technologies, and enhance agricultural modernization. We should double efforts to integrate urban and rural areas, introduce urban development concepts into the countryside, guide the development of rural green industries with urban management models, and give a boost to rural development. We should also intensify efforts to integrate primary, secondary and tertiary industries, develop industrial chains that are ecologically green, low-carbon, eco-friendly and recyclable, maximize resources utilization, reduce energy consumption and pollution, heighten the quality of economic and social development, and form a green model of production, life and consumption.

5. **Putting strict institutional constraints on environmental protection**

The system holds the key to governing the country by the rule by law. The outstanding problems in China's environmental protection are incomplete systems, lax law enforcement, inadequate enforcement, and ineffective punishment. To protect the environment and build an ecological civilization, we must first improve the system and implement responsibilities. We need to establish a comprehensive assessment mechanism for promoting green development, develop a balance sheet of natural resources, implement a compensation system for ecological damage, strictly assess the accountability mechanism, and implement a responsibility system for key officials for the development of an ecological civilization. Those who have caused serious consequences to the environment must be held accountable. We should implement the lifelong accountability system, forming a high-pressure situation for the protection of the ecological environment.

6. **Raising the awareness of society as a whole in building an ecological civilization**

Everyone has a stake in building an ecological civilization, so we must strengthen the education of the entire population in this regard. The publicity for ecological civilization should be introduced into schools and families, so that teachers will become role models for students, parents their children, adult minors, and farmers moving to the cities. We must guide society as a whole to develop the ideological concepts and conduct codes of an ecological civilization that respects, adapts to and protects nature, raise the ecological awareness of the entire population, standardize ecological behaviors that are embodied by the consistency between words and deeds, and foster cultural and living habits. Only in this way can we lay a solid and reliable ideological foundation for, reach a highly-unified consensus on, and actively and extensively engage in the development of an ecological civilization.

A Systematic Understanding of the Highlights, Difficulties and Key Points of Ecological Civilization Development

Jun Li

China has entered a critical period of ecological civilization development under pressures from multiple aspects. A careful review of China's efforts in ecological civilization development in recent years and an in-depth analysis of achievements, problems, and key points are of great significance for accelerating development in this regard and building a beautiful China. This paper proposes three suggestions concerning the highlights, difficulties, and key points of China's ecological civilization development efforts.

1 Highlights

The Central Committee of the Communist Party of China (CPC) with Comrade Xi Jinping at its core has set ecological civilization development as a great cause concerning the welfare of its people, the future of the Chinese nation, the fulfillment of the Two Centenary goals, and the Chinese Dream of national rejuvenation since the 18th National Congress of the CPC. A package of major policies has been adopted to initiate radical and comprehensive changes in regard to an ecological civilization. We have made a number of unprecedented achievements in the following five aspects.

1. **Unprecedented attention: ecological civilization as part of the Five-Sphere Integrated Plan**

With an eye on the bigger picture, the CPC Central Committee has included ecological civilization as one of the major tasks for building socialism with Chinese characteristics, and made the historic leap from "two civilizations" to Three-Sphere Integrated Plan, Four-Sphere Integrated Plan, and now Five-Sphere Integrated Plan. Meanwhile,

J. Li (✉)
Hainan Provincial Committee of CPC, Haikou, China

J. Pan et al. (eds.), *Beautiful China: 70 Years Since 1949 and 70 People's Views on Eco-civilization Construction*, https://doi.org/10.1007/978-981-33-6742-5_13

ecological civilization has been inscribed in the Party Constitution and the Constitution of the People's Republic of China, turning it into the Party's proposition, a national will and a task for all Chinese. Giving ecological civilization unprecedented historical significance means that the Party has brought its governance philosophy and methods to the next level. General Secretary Xi Jinping clearly stated that it is necessary for us to deeply understand the significance for incorporating ecological civilization development into socialism with Chinese characteristics. He also repeatedly emphasized the need to incorporate ecological civilization development into all aspects of work and put ecological progress in juxtaposition with our pursuit of economic, political, cultural, and social progress. These highlights the dominant role of ecological civilization in the national strategy, enriched the connotation of building a moderately prosperous society in all respects, and made the cause of socialism with Chinese characteristics more attractive. Moreover, ecological civilization development does not only involve resource conservation and environmental protection, but also requires fundamental changes to production methods and lifestyles. Taking the lead and in the meantime serving as a coordinator and carrier in China's all-round endeavor for development, ecological civilization development is indispensable for, rather than against our pursuits in other aspects.

2. **Scientific top-level design: unprecedented reform of the ecological civilization system**

The CPC Central Committee adopts a problem-oriented and target-based approach to the reform of the systems that has been severely restricting and hindering the development of an ecological civilization over years. The following approaches have been used. First, the theoretical structure with multiple pillars has taken shape. The central government successively issued more than 40 reform plans, basically forming a system for ecological civilization development based on the *Opinions on Accelerating the Advancement of Ecological Civilization* and the *Integrated Plan for Promoting Ecological Progress* and the eight modules of property rights of natural resource assets, land development and protection, spatial planning system, resource management and conservation, paid use and compensation of resources, environmental governance system, market system, performance assessment and accountability. An ecological environmental protection system for whole-process governance has been formed to strengthen pollution reduction at the sources, process management, punishment of polluters, and environmental accountability. Second, significant progress has been made in the management system reform. The decentralized ecological and environmental protection responsibilities was reorganized in the reform to clearly define responsible institutions and put the responsibilities for supervision, administration and law enforcement of ecological and environmental affairs in urban and rural areas in the charge of dedicated authorities. The Ministry of Ecology and Environment was established to push forward the reform of vertical oversight and inspection system in institutions below the provincial level and the reform of comprehensive administrative law enforcement regarding ecological and environmental protection. Much work has been done to remove the shortcomings of the existing systems and mechanisms, and accelerate the building of an ecological environment management system that covers all regions and full processes in all directions. Third, responsibilities have been clearly defined with an accountability system put in place. We have further clarified the environment quality accountability system for local governments, and changed an environmental governance system to uphold government leadership, enterprises' participation as main players, and public participation, with the relevant Party committee at the helm. It is emphasized that the Party committee and the government should share responsibilities for environmental protection, enterprises should ensure fulfillment of their environmental responsibilities, the general public should be guided with active efforts to make their contributions, and legal means should also be used to raise awareness of ecological and environmental protection.

3. **Unprecedented measures: strong determination to solve all problems in ecological environment governance**

In the face of serious environmental problems, the CPC Central Committee has adhered to the people-centered development philosophy and taken a series of radical, groundbreaking and long-term measures to resolutely combat pollution. On the one hand, China has been addressing the thorniest problems in the fight against pollution. It has implemented thoroughly the three major action plans for the prevention and control of air, water and soil pollution, and focused on the seven major landmark battles against pollution (i.e. protecting blue sky, reducing exhaust emissions by diesel trucks, protecting drinking water sources, cleaning up black and malodorous water bodies, Yangtze River environment protection and restoration, comprehensive protection of the Bohai Sea, and contamination control in agriculture and rural areas), with emphasis on solving the outstanding problems in the three aspects and the biggest concerns of the general public. As a result, the quality of the ecological environment has improved markedly. In 2018, the proportion of days with good air quality and with severe pollution in 338 cities at and above the prefecture level increased by 1.3% and fell by 0.3% year-on-year respectively. PM 2.5 concentration in those cities decreased by 9.3% year-on-year, with a cumulative decrease of 22% since the start of the 13th Five-Year Plan period. The peak concentration, pollution intensity, duration and scope of influence of heavily polluted weather were significantly reduced. Water quality improved greatly as remarkable results were achieved in the campaign to clean up black and malodorous water bodies in cities, with 1009 of the 1062 water bodies in 36 cities cleaned up completely or basically. In addition, the security of drinking water in rural areas was effectively guaranteed. Nationwide, 106 million mu of land was covered with trees or other types of vegetation, pushing the forest coverage rate up to 22% from the 16.6% of the beginning of the twenty-first century. As the ecological environment continued to improve, people's sense of happiness and fulfillment grew constantly. On the other hand, a "zero tolerance" approach was adopted to investigate and deal with illegal acts that damage the ecological environment. The strictest ever *Environmental Protection Law*, was amended and put into effect, under which a well-defined environmental monitoring system was established to crack down on environmental violations with measures such as daily fines and a "dual penalty system" for both the person and the company responsible for pollution.

A sound system of ecological and environmental inspections was also established at both central and provincial levels with good results achieved through continuously reinforcing the responsibility for ecological and environmental protection for Party committees and governments at all levels.

4. **Unprecedented efforts to transform the development model: achievements in pursuing green development**

China has adopted in all aspects a new development philosophy emphasizing long-term gains. Under the theme of building a green, low-carbon, and sustainable economy, China has been improving its industrial structure constantly by cultivating new industries, transforming and upgrading traditional industries, phasing out backward capacity and promoting energy conservation and emissions reduction. In this way, we have achieved economic growth, and reduced energy and resource consumption and environmental pollution. Currently, the proportion of heavy and chemical industries in the country is decreasing, while that of high-tech industries and equipment manufacturing continues to rise. Eco-agriculture and new industries are flourishing, and the value added of the services sector has contributed about 60% of the economic growth. The environmental protection industry has been rapidly developing and expanding. In 2018, China witnessed an increase of 43% year-on-year in its ecological and environmental protection and environmental governance industry, and it is expected that the annual operating revenues of the industries will exceed RMB 1.5 trillion and reach RMB 2 trillion by 2020. The formation of a modern industrial system of environmental services has been accelerated with third-party governance and environmental stewardship services as its core. As its energy consumption structure underwent positive changes, China has become the world's largest country in the use of new and renewable energy, with the total emissions of major pollutants, energy consumption and water consumption per unit of GDP continuing to decline. Green development has become an essential force driving China's quality development.

5. **Unprecedented international cooperation for an ecological civilization: build a community with a shared green future**

China has actively undertaken its obligations as a major country and become an important participant, contributor and leader in the construction of a global ecological civilization through in-depth participation in global environmental governance, carrying out international exchanges and cooperation, and advocating the building a beautiful Earth. First, China's green development strategy has been widely recognized. In 2013, the Governing Council of the United Nations Environment Programme (UNEP) decided to promote the concept of ecological civilization proposed by China; and in 2016, the United Nations released *Green Is Gold: The Strategy and Actions of China's Ecological Civilization* to introduce China's experience. Second, China has redoubled its efforts to advance the green development strategy. It has set a great example by releasing *China's National Plan on Implementation of the 2030 Agenda for Sustainable Development*, implementing *China's*

National Plan on Climate Change (2014–2010), and promoting the *Paris Agreement*. It also launched the Belt and Road Initiative International Green Development Coalition, calling for global efforts to lay solid foundation for an ecological civilization, and pursue green development. Third, China has made greater contributions to green development. At present, China has emerged as the biggest contributor to the protection of the ozone layer as it phased out more than 50% of the total discharge of ozone-depleting substances by developing countries. It has actively launched global cooperation and established the China South-South Climate Cooperation Fund to support developing countries in dealing with climate change. Besides, China has shared its experience and technologies widely to help developing countries effectively improve their ecological environment.

2 Difficulties

On the whole, China's ecological civilization development has been moving forward steadily with a positive trend. However, the poor environmental foundation, a late start in overall environmental protection, and the unbalanced regional and industrial development, among other factors, resulted in outstanding challenges and problems that still need to be tackled. With a view to help deal with such challenges, we hereby point out the following five difficulties.

1. **Difficulty in matching words with actions**

In recent years, the concept of ecological civilization has been increasingly recognized, which has led to a general consensus on the importance of building an ecological civilization. Nevertheless, the mismatch between words and actions has been constantly observed since to achieve conspicuous results in this regard, deep-rooted problems must be addressed, especially problems related to vested interests and existing conducts. For example, some government officials paid only lip service to ecological civilization, not making sufficient efforts in reality; some leaders dared not hold accountable those who have been gaining profits from polluting the environment; and some local poor-performing local government officials acted against the natural law to build ecological civilization in undue haste, which would definitely bring harms rather than benefits. In addition, some enterprises and individuals still went their own way to make profits at the cost of the environment. For example, among the typical cases reported by the central environmental protection inspection work groups, some enterprises were two-faced, making but not truly implementing rectification plans when they were required to redress what they had been doing. Besides, the public also rarely took the initiative to participate in environmental endeavors despite their attention to environmental issues.

2. **Difficulty in balancing between development and environmental protection**

On the one hand, some government leaders lacked a clear understanding of the relationship between development and environmental protection and tended to take them as separate issues. For example, some valued only development, pursuing economic growth through impractical projects at the cost of the environment; some focused only environmental protection, taking it as the excuse for inaction when it came to the pursuit of development; and some could not come up with innovative strategies to promote green development, or failed to balance between development and protection. On the other hand, it takes time for China to complete the transformation in its development model. China is still in the process of industrialization, and has not thoroughly changed its industrial structure with high energy-consumption, high-emissions, and high-pollution industries still as pillars. It also has not got rid of the negative impacts of what has been done previous, or fully solved the environmental problems caused by a unidirectional, linear and extensive economic model of "resources—products—pollution." Meanwhile, China is still relatively underdevelopment in the R&D and application of technologies related to ecological civilization development, making it difficult for it to see great achievements in both development and environmental protection.

3. **Difficulty in forming synergy between partial and systematic protection**

An ecosystem is a unified and interdependent natural system. Ecological civilization development, especially environmental protection and governance, must be promoted in a holistic and systematic way which is line with the special characteristics and inherent laws of an ecosystem. All elements have to be taken into account to carry out overall protection, systematic restoration and comprehensive management to enhance the capacity of the ecosystem. At present, we are deepening the reform of the management system for an ecological civilization, which cannot be accomplished at one stroke. First, departmental functions need to be better integrated and further improved. After the new round of institutional reform, the Ministry of Ecology and Environment was established, taking charge of all aspects of environmental protection by bringing together functions that were previously scattered among various departments, solving the problems of overlapping functions, contradictory actions, and poor coordination. However, these problems are not yet fully solved, especially when it comes to comprehensive administrative law enforcement where further efforts are needed for better integration and coordination. Second, it will take some time for the reform towards a vertical management system to take effect as the reform, targeted at monitoring, supervision and law enforcement by environmental protection authorities at and below the provincial level is currently being rolled out nationwide. All in all, explorations are still needed to find solutions that really work in practice to such problems as intervention from local governments in environmental law enforcement and failure to fulfill environmental protection obligations. Third, institutions and mechanisms for cross-regional and cross-basin ecological management need to be further improved. As some water bodies span multiple provinces, autonomous regions and municipalities, ecological and environmental management for them requires effective coordination between different administrative divisions. However, due to the different stages of economic and social development, local governments tend to make different choices between environmental protection and economic development. Therefore, it is often difficult to form consensuses and achieve effective coordination, especially in the management of environmental elements such as water and air, as relevant responsibilities are hard to define.

4. **Difficulty in matching needs with input**

As the major contradiction facing the Chinese society has changed, more people now regard a good ecological environment as a major factor in their happiness. However, on the one hand, China has a long way to go to restore its ecosystems and lags behind many other countries in terms of its construction of environmental protection infrastructure for a variety of reasons. On the other hand, investment in environmental protection generally does not generate as good results as expected by the public. One reason for this is a lack of incentives for social capital join the efforts, which may be the result of an incomplete market mechanism, imbalance between input and output in the environmental protection industry, and the fact that the environment is easy to destroy but difficult to restore. According to statistics, in 2016, investment in environmental protection accounted for only 1.3% of China's GDP, of which direct investment in pollution control facilities accounted for 0.6% of GDP and 0.73% of total investment in fixed assets; and the central and local governments spent 1.2% of their budgets in this regard. In comparison, in developed countries, investment in environmental protection accounted for 6–8% of GDP at the peak, and the average level is 2–3%. China's total investment in environmental protection is therefore seriously inadequate, and falls far behind what the country intends to do according to its advanced philosophy regarding an ecological civilization.

5. **Difficulty in handling historical problems with current policies**

A raft of deeply rooted problems formed due to various reasons over time hinder China's progress in building an ecological civilization. Some projects or enterprises have failed to keep up with new plans and policies for ecological and environmental protection amid rapid economic and social development, industrial restructuring and city upgrading. For instance, some enterprises engaged in the chemical industry, smelting, power generation, mining and other pollution-intensive production activities are located at places that used to be city suburbs and had to be shut down or relocated as fast urbanization brought residential communities all around them. This involve some projects that were once strongly advocated by the government but gradually came under more restrictions and even total prohibition as relevant standards and requirements got increasingly stringent for ecological and environmental protection. For example, land reclamation shifted from an initiative backed by the government to an area subject strict control, because blind reclamation and inappropriate use of reclaimed land were observed with a lack of restrictions. Some projects were built without sound laws, regulations, institutional arrangements or plans in place. One conspicuous problem, for example, is how to deal with some emergency facilities, rural housing, schools, and hospitals that were built without a clear plan. These problems are unlikely to be solved soon as heavy vested interest may be involved and there is not yet clear laws, regulations and policies to follow in this regard.

3 Key Points

The building of an ecological civilization is essential for not only harmony between man and nature, but the sustainable development of the Chinese nation. Therefore, unswerving efforts are needed all along the way. Currently, China's efforts to build an ecological civilization are in a crucial phase in which more high-quality ecological goods need to be provided to meet the growing demands of the people for a sound and clean environment. Yet this is also a phase full of opportunities as conditions are ripe and all abilities and capacities needed are in place to resolve prominent environmental problems. Here are five key points that require targeted efforts if we are to overcome all challenges and move forward towards an ecological civilization in China.

1. **Understanding and implementing Xi Jinping Thought on Ecological Civilization**

Xi Jinping Thought on Ecological Civilization, based on China's actual conditions for ecological civilization development, thoroughly answers the major theoretical and practical questions of why we need to build an ecological civilization, what exactly to build and how to do it; and proposes six principles for this endeavor in the new era: the harmonious coexistence between humans and nature must be upheld, our natural environment is precious, there is no welfare more universally beneficial than a sound natural environment, our mountains, rivers, forests, farmlands, lakes and grasslands together form a biotic community, the strictest regulations and laws must be applied in protecting the environment, and joint efforts must be made in building a global ecological civilization. Xi Jinping Thought on Ecological Civilization embodies traditional wisdom of Chinese civilization regarding the ecological environment, develops the Marxist concept of ecological conservation, and serves as a basic guideline and ultimate criterion for the building of an ecological civilization in the new era. We must further learn and understand it and put it truly into practice. We must further deepen our learning through fully and accurately grasping the essence, principles, objectives, priorities and scientific methods of it contains; embrace it from the bottom of our hearts; and completely change our mindset and approach to keep up with the times. We must do more to implement Xi Jinping Thought on Ecological Civilization with strong determination, make unremitting efforts, and avoid lip service, inaction, or practicing formalities for formalities' sake.

2. **Pushing forward the reform of institutions and mechanisms for an ecological civilization**

The key to overcoming difficulties in ecological civilization development lies in continuously deepening relevant reform efforts. To promote the reform of institutions and mechanisms for an ecological civilization, we should focus on solving problems encountered, streamline procedures in a reasonable and effective manner, give priority to more urgent problems, and make substantive breakthroughs, so a number of less difficult problems can be solved naturally as more difficult once are properly addressed. Hence, it is imperative to continuously improve the market mechanism and focus on building a government-led, enterprise-based governance system for an ecological civilization, since the ecological civilization development and the provision of ecological products are not only for public interest and universal benefit, but also for profits in the market. Being "government-led" means that the government should fulfill its responsibilities for establishing rules and regulations, and formulating plans and policies, in order to regulate and guide the society and the market. However, it does not mean that the government should take on all duties, nor should it take administrative orders as the only way of doing its job. Being "enterprise-based" require us to tap into the commercial value of all kinds of ecological goods based on the concept that lucid waters and lush mountains are invaluable assets, and make full use of market tools to attract more market entities to join the effort. More specifically, this means that we need to improve the pricing mechanism for resources and the environment, and the long-term mechanism for concrete gains to act as the guide based on reward and punishment rules, compensation for environmental damages, and various other specific measures. For example, Inner Mongolia tried an approach that combines government guidance, enterprise investment, and public participation with the support of technological innovations such as growing liquorice for soil improvement, in its efforts to combat further desertification in the Kubuqi Desert, delivering good results mainly in four aspects: Deserts turned green, the liquorice business thrived, soil was improved, and local people were helped out of poverty. This pattern that benefits all is worthy of replication as it solves the problem of high investment and low output caused by excessive government engagement in environmental governance and the sole emphasis on restrictions rather than offering incentives.

3. **Learning from advanced international practices**

Learning is a shortcut to success. Objective rules in ecological civilization development make some stages difficult, if not impossible to stride over. As developed countries encountered ecological and environmental problems earlier, they have gained much experience in such areas as ecological protection legislation, environmental-economic policies, green, low-carbon and sustainable economic development, awareness raising for ecological and environmental protection, and the upgrading of the industrial structure. On the one hand, we must leverage our latecomer advantage to draw lessons from what they have gone through, avoiding pitfalls as much as possible, especially that of "pollution first and treatment later." On the other hand, we can learn advanced concepts, technologies and management models from them to promote ecological progress and green development from a higher starting point. For instance, in 1872, the United States passed a law to establish the world's first national park, the Yellowstone, which, hosting the world's most complete temperate ecosystem in the northern hemisphere, and was listed as a world natural heritage site in 1978. Building national parks has become a widespread convention for the conservation of nature and culture, producing a series of advanced conservation ideas and patterns. In 2015, the Chinese government drew on foreign experience and launched a pilot national park system in nine provinces to gather administration responsibilities from multiple departments for a raft of nature reserves, scenic areas, natural heritage sites, forest parks and geoparks. Besides, urban construction waste, a prominent problem in China today, is deemed as "urban treasure" Germany, Japan, the United States, and many other countries. In those countries, waste treatment and recycling, which is at the core of sustainable economic development, has been developed into a new industry, which not only saves land and reduces environmental pollution at a low cost, but also creates new growth drivers and creates a win–win situation for both the economy and the environment. There is clearly much we can learn in our efforts to build an ecological civilization.

4. **Pursuing law-based ecological civilization development**

As General Secretary Xi Jinping stressed, only the tightest regulations under the strictest legal system can provide a reliable guarantee for ecological civilization development. This calls for the joint efforts of the legislature, law enforcement agencies, the judiciary and the public to fully practice the rule of law in building an ecological civilization. As China's legal system for ecological civilization development is still on the path of further improvement, we should now lay emphasis on law enforcement. We should change the situation of easy and low-cost violation that resulted from a lack of heavy-enough penalties by improving our systems for credibility assessment based on environmental protection performance and for administrative law enforcement, and impose heavier punishments. In addition, we should deepen the reform of approval procedures for environmental and resource projects, improve the system for environment-related public interest litigation, strengthen synergy between administrative law enforcement and judicial procedures in regard to the ecological environment, and improve the new mechanism for environmental protection featuring connection, cooperation and mutual supervision between the environmental protection authorities, the police, procuratorates and courts, in a bid to effectively enhance the rule of law.

5. **Giving full play to the guiding role of target-based ecological civilization assessment**

The performance of Party committees, governments and officials at all levels is evaluated largely according to what superiors say about them. We should make full use of the assessment system as the central government has issued the *Measures for Target-Based Evaluation and Assessment of Ecological Development*, the *Index System for Green Development,* and the *Target-Based Assessment System of Ecological Civilization Development*, which provide the baseline. Firstly, we should actively guide officials at all levels to form the correct view on political achievements that economic development and ecological progress both represent political achievements and the latter is of greater historic significance since it benefits the people and the country. We should raise people's awareness of ecological civilization development, urge local Party committees and governments to adhere to the principle of sound integration and coordination between development and environmental protection, and resolutely curb the "GDP impulse" which would lead to development at the cost of the environment. We should unswervingly pursue "green GDP," and thoroughly abandon the traditional production and consumption mode featuring large-scale production, consumption and emissions. Secondly, we should explore different ways of performance assessment. We should set reasonable assessment indicators based on the functions, socioeconomic conditions and natural resources endowment of each place, and set weights for indicators to reflect the new development concept. Currently, in Fujian, Hainan and some other provinces, GDP-based assessment has been abolished for some cities and counties, with a differentiated assessment system taking its place, which is a useful exploration. Thirdly, we should make better use of the assessment results. We will combine the assessment results with the appointment and dismissal of officials, transfer payment, and ecological compensation fund. We will resolutely hold accountable officials who cause serious consequences due to actions for merely formalities' sake, ignorance of environment protection, or blind decision-making with a veto. Fourthly, we should explore a society-wide comprehensive evaluation system for the coordinated development of the economy, society and the ecological environment. We will strengthen publicity and education on ecological civilization development, guide the public to form correct development concepts and values, and accelerate changes towards lifestyles and consumption patterns that are economical, green, low-carbon, eco-friendly and healthy, so that everyone will become a protector, builder and beneficiary of a better ecological environment.

Theoretical Reflection on Xi Jinping Thought on Ecological Civilization

Yong Ren

The National Conference on Eco-environmental Protection held in 2018 officially established Xi Jinping Though on Ecological Civilization as a guiding principle. The scientific foundation of the thought is laid as eight basic principles: First, civilization may thrive if its natural surroundings thrive, and will suffer if its natural surroundings suffer. Second, harmonious coexistence between humans and nature must be upheld. Third, our natural environment is precious. Fourth, there is no welfare more universally beneficial than a sound natural environment. Fifth, our mountains, rivers, forests, farmlands, lakes and grasslands together form a biotic community. Sixth, the strictest regulations and laws must be applied in protecting the environment. Seventh, it is necessary to translate the idea of building a beautiful China into the conscious actions of all people. Eighth, joint efforts must be made in building a global ecological civilization. Xi Jinping Though on Ecological Civilization upholds the general principles of historical materialism and dialectical materialism, and takes roots in China's national conditions, and the fine traditional Chinese culture. It is a new understanding, new development and new practices of the laws of the development of human society, the building of socialism and the governing of the Party in a new era for building a modern socialist county after four decades of reform and opening-up of China. It is a new contribution to the theory of socialism with Chinese characteristics, and is of particularly importance. It is also of global significance as it represents the latest development of Marxism in a time when human society has experienced industrial civilization for two centuries and entered the new era of sustainable development.

Xi Jinping Thought on Ecological Civilization originates from theories and practices in the following three aspects. First, it inherits and develops Marxist ideas on the relationship between man and nature. It summarizes and theorizes the approaches to dealing with the relationship between man and nature in the history of human society as well as that of China's development. According to this thought, civilization may

Y. Ren (✉)
Environmental Development Center, Ministry of Ecology and Environment, Beijing, China

J. Pan et al. (eds.), *Beautiful China: 70 Years Since 1949 and 70 People's Views on Eco-civilization Construction*, https://doi.org/10.1007/978-981-33-6742-5_14

thrive if its natural surroundings thrive, and harmonious coexistence between man and nature must be upheld. These are the basic principles and strategies in advancing the building of an ecological civilization. Second, it originates from the ideas and practices of deepening reforms on all fronts and governing the county in accordance with the law. "The strictest regulations and laws must be applied in protecting the environment." This provides a mechanism for and a path to building an ecological civilization. In other words, an ecological civilization can only be built by improving the system and capacity for ecological and environmental governance. Third, it is firmly based on related findings of natural sciences and social sciences. The logic of an ecological civilization begins with environmental problems brought about by the industrial civilization and its relationship with economic, political, cultural and social development. The eight principles listed out above represent an insightful understanding of the relationship between the environment and economy, society, global governance and politics as well as of theories of ecological and environmental sciences. The principles provide methods, targets and institutional support for building an ecological civilization.

A major aspect of the significance of Xi Jinping Thought on Ecological Civilization lies in that it makes a strategic arrangement for ecological civilization development as part of the great cause of socialism with Chinese characteristics. With such strategic arrangement, historic, upending and fundamental changes are taking place in the way we protection our environment.

1 The Principle of Lucid Waters and Lush Mountains Being Invaluable Assets and the Laws of Environmental Economics

The type, extent and influence of environmental problems are closely associated with industrialization and urbanization. Since the launch of the reform and opening up, China's economic development or industrialization process has gone through five stages: the stage from 1978 to 1980 when agriculture was the focus; the stage from the mid-1980s to the mid-1990s when the light industry was the focus; the stage in the mid-1990s when the economy began to be driven by the heavy industry; the stage beginning in the late 1990s when the heavy industry dominated the economy; and the current stage of an economic new normal. In line with the characteristics of different stages, the conflict between the environment and economic development began to drastically deteriorate in the 1990s. Before the early and middle 1990s, China's environment saw local deterioration, but was still sound on the whole. Entering the stage when high-energy-consuming and high-pollutant-discharging heavy industry dominated economic activities, China faced an environmental situation featuring local improvement and general deterioration, posing increasing pressure.

China's understanding of the relationship between the environment and economic development evolved with the conflicts between the environment and the economy.

It also deepened as time went by. At 15th National Congress of the CPC held in 1997, the Party made the first statement on the relationship between the environment and the economy, saying that population growth and economic development posed huge pressure on the environment and resources. It was further stressed at the 16th National Congress of the CPC that the conflicts between ecological environment, natural resources and economic and social development were becoming increasingly prominent. Then at the 17th National Congress of the CPC, it was made clear that the environmental cost of economic growth was too huge. It was stressed at the 18th National Congress of the CPC that unbalanced, poorly coordinated and unsustainable development remained a big problem.

It was point out at the 2014 Central conference on Economic Work that the carrying capacity of resources and the environment had reached or was approaching the ceiling. The 5th Plenary Session of the 18th CPC Central Committee stated that environmental pollution, particularly air, water and soil pollution had emerged as a prominent weak point that threatened the completion of the building of a moderately prosperous society in all respects. At the National Conference on Eco-environmental Protection held in 2018, it was stated that the environment "is stable and goes on an upward trajectory, but the achievements of environment protection are not stable." This statement has two-found meanings. First, after arduous efforts, the overall situation of environmental deterioration was contained. Second, in an era of economic new normal featuring gear shifting and restructuring, the impact of economic and social activities on resources and the environment was going down. However, China was still facing great pressure, high risks and grave problems in the field of environmental protection. China was in a critical period when it was confronted with overlapping pressures from environmental protection and various other aspects. So, the CPC made it clear at the 19th National Congress that "we have a long way to go in protecting the environment."

In sync with such an understanding, China's strategic arrangement for dealing with the relationship between the environment and economic development started and was unfolded gradually in the 1990s. The process was basically parallel with the global efforts to balance between the environment and development. After the 1992 United Nations Conference on Environment and Development, China took the lead in 1994 in releasing its own Agenda 21, which put forth the idea of changing traditional development path and embracing sustainable development for the first time. In 1996, the 9th Five-Year Plan and the 2010 visions were formulated, putting into action the strategy for invigorating China through science and education and the sustainable development strategy. China also vowed to pursue transformations in its socialist market economy and its growth path.

In general, before the Scientific Outlook on Development was fully implemented, China's understanding of laws in environmental economics as well as its strategic arrangements and practices for balanced development of the economy and the environment was at a stage featuring a focus on economic growth and neglect of environmental protection. In other words, wealth outweighed lucid waters and lush mountains. For example, though Article 4 of the 1989 Environmental Protection Law requires coordination between environmental protection and economic and social development, in practice, environmental protection lagged far behind economic development. In 2005, the State Council made a decision on implementing the Scientific Outlook on Development and strengthening environmental protection. For the first time, the document said economic and social development should be in good coordination with resources and the environment. The revised Environmental Protection Law of 2014 adopted this principle and included putting economic and social development in harmony with environmental protection in Article 4.

In the early twenty-first century, Anji, Zhangjiagang and some other places in coastal provinces of Zhejiang, Fujian and Jiangsu in east China emerged as good models for coordinated development of the environment and the economy. They stood out in preserving lucid waters and lush mountains while pursuing and maintaining prosperity. When the result of environmental protection was translated into tangible economic and social benefits, protecting the environment equalled protecting the productive forces, and lucid waters and lush mountains were indeed recognized as invaluable assets.

After the CPC pledged to build an ecological civilization at its 17th National Congress, particularly after the 18th National Congress, China's understanding of the laws for balance between the environment and the economy and its strategic arrangements and practices for balanced development underwent systematic changes marked by the Five-Sphere Integrated Plan, the Four-Pronged Comprehensive Strategy and the green development concept. When the economy entered a new normal, and shifted from high-speed growth to high-quality development, the relationship between the environment and the economy changed holistically and fundamentally. The environment was regarded as precious resources and valuable natural assets. Like land and technology, it is a factor of production that has a bearing on high-quality development. In the meantime, a sound environment is the result of high-quality development and a gauge for high-quality development. A sound environment and high quality development are what should be pursued. The two aspects reinforce each other. This is the theoretical meaning of the principle that lucid waters and lush mountains are invaluable assets.

Therefore, this principle put forth by Xi Jinping reveals the relationship between the environment and the economy as well as the law governing this relationship. A vivid condensation of environmental economics, it provides guidance for balanced development of the environment and the economy. The principle is thus an important part of Xi Jinping Thought on Ecological Civilization.

Xi expounded on the dialectical relationship between the environment and economic development during his inspection tour to Hainan in 2013. The effectiveness of environmental protection efforts, at its very bottom, depends on the economic structure and the approach to economic development. We should not deplete natural resources and use up all that the environment can offer to pursue economic development. Protecting the environment does not mean sacrificing economic growth.

Xi said in 2014 that it was the intrinsic requirement of sustainable development to properly handle the relationship between environmental protection and development or the relationship between lucid waters and lush mountains and material wealth. This is also an overarching principle in advancing modernization. In 2017, Xi further noted that the practice of seeking economic growth at the cost of the environment should be abandoned, and that efforts should be made to ensure that a sound environment becomes the source of better life for the people, support for sound and sustainable development of the economy and society and the focus of efforts to polish China's national image. In response to misunderstandings of the principle as was shown in practice, Xi pointed out during the "two sessions" of 2019 that it was necessary to maintain the strong determination to strengthen ecological development. Fundamentally, protecting the environment and developing the economy are mutually reinforcing. When meeting with difficulties in economic development, [local governments] should not seek to blindly launch new projects with the aim of trading the environment for economic growth, nor stepping over the ecological red lines. In the process of gear-shifting from high-speed economic growth to high-quality development, pollution control is a key point. We must press ahead despite all difficulties. We must maintain our focus on protecting the environment.

In regard to the economic theory regarding the relationship between a good environment and material wealth, or the relationship between environmental protection and productivity, Xi made an analogy during the "two sessions" of 2014: Why are lucid waters and lush mountains invaluable assets? Fish seeks waters with aquatic plants to live, and birds choose to perch on fine woods. If conditions are good, who wouldn't want to invest, pursue a career, live and travel in a place with lucid waters and lush mountains? In this sense, lucid waters and lush mountains are natural assets as well as social and economic assets. In 2015, Xi stressed at the Central Conference for Poverty Reduction and Development that it is necessary to make better use of the land, labour force, assets and natural scenery in poor areas through reform and innovation, and turn resources into assets, money into equity, farmers into shareholders and lucid waters and lush mountains into sources of income for the poor population.

Regarding how to put the principle of lucid waters and lush mountains being invaluable assets into action, Xi stressed, "We must pursue development that is innovative, coordinated, green, open and shared, and accelerate the formation of spatial patterns, industrial structures, modes of production and lifestyles conducive to resource conservation and environmental protection. We must give the environment the time and space that it needs to rest and recuperate." "We must establish an economic system aimed at integrating industries and ecosystems; and promote green development across the board." Green development is essential for establishing a modern and high-quality economic system, and is key to resolving pollution problems. It is imperative to adjust the economic and energy structures, improve the distribution of spaces available for development, adjust the distribution of industries within regions and river basins, grow industries that promote energy conservation, environmental protection, clean production and clean energy, facilitate all-round conservation and recycling of resources, ensure that the systems of production and everyday life overlap in their circular use of resources, advocate a green and low-carbon lifestyle of moderation and frugality, and oppose extravagance and unjustifiable consumption.

2 The Principle of Sound Natural Environments Being the Most Universally Beneficial Aspect of Public Well-Being, and all People in Action to Build a Beautiful China, and Relevant Laws in Environmental Sociology

The relationship between the environment and society and laws thereof are about the impact of the environment on people's production and life, people's understanding of, attitude towards, and behavior regarding the environment, as well as laws reflected in the process.

In China, the relationship between the environment and society roughly evolved through three stages. Before the 1980s, vegetation deterioration, soil erosion and desertification had serious impact in some local regions, but environmental pollution was not serious on the whole. In general, people's top concern at the time was food. They were indifferent to the environment. Some even had extreme thought that socialist planned economy would not generate pollution to the environment in the early 1970s.

In the mid- and late 1990s, public awareness of environmental protection gradually improved and the public got more actively involved in environmental conservation as people's living standards were higher and public education programs were carried out. Also, with the overall environmental condition deteriorating quickly, pollution incidents and environmental emergencies in some places seriously affected people's production and everyday life, and some people realized the necessary of safeguarding their environmental rights.

In the twenty-first century, particularly over the past 10 years, the relationship between the environment and society showed three distinctive characteristics. First, as China began to build a moderately prosperous society in all respects, the public grew increasingly aware of environmental protection and regarded a sound environment part of a quality life. Shifting their focus from securing access to food and shelter and survival to calling for a sound environment is a general law governing the relationship between the environment and society. Second, apart from local water and soil pollution, air pollution, mainly manifested by smog exerted serious impact on people's production and everyday life nationwide. The public was thus highly dissatisfied with the environmental quality and the government's efforts to protect the environment. In the meantime, marked by the water pollution incident of the Songhua River in 2005, China had entered a stage of frequent environmental emergencies. The public was more willing to safeguard their environmental rights and many started to take concrete actions. Third, with the PX project incident in Xiamen in 2007 as a landmark event, the public, dissatisfied with the situation of environmental pollution, became likely to react radically to construction projects that may generate pollution. In other words, the Not in My Backyard phenomenon emerged. Environmental problems generated serious influences on social and economic order, and posed greater social risks.

China's economic and social development underwent extensive adjustment in an era of social media, featuring overlapping conflicts, and conspicuous information cascade phenomena. In this era, it is inexpensive to mobilize mass actions and such action can easily expert huge influence. This means that the relationship between the environment and society became a sensitive topic. It is generally stable, but it is very necessary to guard against grey rhino incidents. Of course, if public enthusiasm for and participation in environmental campaigns can be well harnessed, the structure and effect of social and environmental governance can be improved across the board.

Time is the matrix of thoughts. Xi, with a forward-looking mindset, acutely grasped the pulse of the times and revealed the underling laws governing the relationship between the environment and social development.

On the economic, social and political nature of environmental issues, Xi said in 2013 that China had entered to a stage when environmental problems happened frequently after more than 30 years of rapid development, regarding environmental protection as a major economic issue as well as a major social and political issue. What a situation would it be when the economy grew, but people were shy of happiness or even very dissatisfied? So, we cannot regard ecological progress, environmental conservation and green and low-carbon lifestyle simply as an economic issue. It is about politics. On an inspection tour to Hainan in 2013, Xi said a sound natural environment is the fairest public good and the most universally beneficial welfare. In 2017, he said, "if the economy grew, but the ecosystem was destroyed, the environment degraded, and people lived in smog, had no access to safe food, clean water, fresh air, or liveable environment, that would not be a moderately prosperous society, or modern society people aspired for. At the National Conference on Eco-environmental Protection, Xi further stressed that "the environment is a major political issue which bears upon the mission and purpose of the CPC as well as a major social issue which bears upon public well-being."

On the relationship between the environment and public well-being, Xi said in 2013 that the people are highly attentive to environmental issues. It is safe to say that the environment will have an increasingly important position in the measurement for a happy life. As the economy and society progress and people's livelihoods are improved, environmental issues would be more likely to spark public dissatisfaction or even public incidents when they were not properly handled. In 2016, he said frequently happening environmental pollutions had become problems that generate great public anxiety. "We must put every effort into finding solutions to them." The most direct and vivid thesis on the relationship between the environment and public well-being came from his remarks in 2015 when he joined lawmakers from Jiangxi for discussions during the third session of the 12th National People's Congress. "The environment is crucial to the well-being of the public, since blue skies and verdant surroundings allow us to feel the beauty and joy of life. We shall protect ecosystems as preciously as we protect our eyes, and cherish them as dearly as we cherish our lives."

On the basis of Xi's theses on the environment and society, the CPC included a beautiful environment in the principal contraction facing Chinese society in the new era for socialism with Chinese characteristics at its 19th National Congress. In the report to the congress, the CPC stressed, "The modernization that we pursue is one characterized by harmonious coexistence between man and nature. In addition to creating more material and cultural wealth to meet people's ever-growing needs for a better life, we need also to provide more quality ecological goods to meet people's ever-growing demands for a beautiful environment." At the National Conference on Eco-environmental Protection, Xi offered an approach to ensuring public welfare regarding the environment. "We must ensure that the environment benefits the people, stressing the resolution of prominent environmental problems that impact public health, and constantly meet people's ever-growing demands for a beautiful environment."

Xi's thesis on the relationship between the environment and public well-being takes its root in a people-centered approach. The people-centered approach is a core of Xi Jinping Thought on Socialism with Chinese Characteristics for a New Era. It carries forward and develops Marxist historical materialism and sums up the CPC's experiences in leading revolution, construction, reform and development. It is also the calling of the times for CPC members to remain true to their original aspiration and mission. At the beginning when Xi took office, he told reporters from home and abroad, "The people's wish for a happy life is our goal." Just because Xi puts people on his mind and acts on the people-oriented philosophy, he is capable of making insightful observations on the laws regarding the relationship between the environment and social development, guiding efforts in promoting ecological progress and conserving the ecosystems.

Based on scientific grasp of environment and social development laws, Xi put forth a slew of innovative mechanisms for improving environmental and social governance. He said at the 41st group study session of the political bureau of the 18th CPC Central Committee that a green action system featuring joint participation of the government, businesses and the general public should be established. At the National Conference on Eco-environmental Protection, he stressed, "Since developing an ecological civilization is a cause in which the entire public participates and has a stake, we must translate the idea of building a beautiful China into the conscious actions of all people. Every person is a protector, contributor, and beneficiary of the environment. No one can remain aloof, choosing to critique from the sidelines rather than participate."

Guarding against and forestalling social risks on the environmental front is also an important task in environmental and social governance. At the National Conference on Eco-environmental Protection, Xi said, "we must effectively forestall environmental risks. Environmental security is a key component of national security, and an important safeguard for the continuation of healthy socioeconomic development. We must make management of environmental risks a regular practice, carefully constructing a multi-levelled environmental risk prevention system that monitors these risks at every stage. We must conduct strict prevention and control of key environmental risks, such as those stemming from garbage incineration and p-xylene, forestall and defuse these problems related to the so-called "not in my back yard" effect, prevent this effect from cropping up in environmental protection projects, and focus on raising our capacity to deal with environmental emergencies."

3 Biotic Community and Ecological Systems

An ecological system is a unit formed by biocenosis and the inorganic environment. The system being a unit is determined by the spatial structure, material and energy exchanges of various elements in the system. In the end, a special functional relationship is formed. Therefore, it is necessary to take a holistic approach to environmental protection so as to achieve the best results.

According to the ecological systems theory, water, air, soil, living things and other elements in the environment form a unit featuring universal connection. Comprehensive and systematic methods should be employed to manage the ecological system. As the system has multiple functions including providing products, adjusting the environment and offering aesthetic value, a multi-target management approach should be employed. Therefore, ecological system management is about applying general rules and laws of the ecological system and employing comprehensive tactics to solve problems. Compared with traditional management methods, ecological system management aims to maintain a sound ecological system. It encompasses the management of resources and the environment, pollution control and ecological conservation. It focuses on preserving the integrity of the ecosystem and the functions of the system. To achieve the core goal of maximizing public interests, multi-target management measures are employed. Joint governance of the environment is achieved through unified decision making, planning and actions, as well as cross-department and inter-regional cooperation.

Xi put forth the principle that our mountains, rivers, forests, farmlands, lakes and grasslands together form a biotic community at the Third Plenary Session of the 18th CPC Central Committee. At the Fifth Meeting of the CPC Central Committee Leading Group for Financial and Economic Affairs held in 2014, Xi said, "our mountains, rivers, forests, farmlands, lakes and grasslands together form a biotic community. Just as humans need the fields for food, the fields need the rivers, the rivers need the mountains, the mountains need the soils, and the soils need the forests and grassland. It is a cycle that never ends."

As for how to put the theory into practice, Xi put forth a series of guiding principles and institutions for taming the water, restoring the ecosystem, and managing the territorial space and cities.

Xi said during the sixth group study session of the Political Bureau of the 18th CPC Central Committee that the territorial space is where an ecological civilization is developed. It is necessary to plan well the development of territorial space. "Maintaining a balance between population, resources and the environment, and promoting economic, social and ecological efficiency, we will develop an overall plan for developing our land, and allot space to production, to daily life and to ecological development as appropriate, in order to leave more space for nature's self-restoration, more fertile farmland for agriculture and blue sky, lush mountains and lucid waters for future generations." In 2014, Xi stressed at the Fifth Meeting of the CPC Central Committee Leading Group for Financial and Economic Affairs that a holistic approach should be adopted to water regulation. Water management should not focus only on waters, but take into consideration various elements in the natural system. A systematic approach should be employed. The ecological system is an organic being. Water management should be well coordinated with efforts to manage the mountains, forests and fields. At the meeting, Xi also said that the practice of placing emphasis on skyscrapers above the ground while neglecting underground pipeline network and vegetation should be stopped in urban planning and construction. [Governments] should consciously reduce the intensity of urban development to retain and restore a certain proportion of ecological space, and build sponge homes and sponge cities. At the Central Conference on City Work held in 2015, Xi stressed efforts to integrate urban planning to allot appropriate space for production, daily lives and the ecosystem to make cities liveable. In ancient China, cities were regarded as places where people thrived. In urban development, the intrinsic connection between spaces for production, daily life and the ecosystem should be fully understood to make the space for production highly intensive and efficient, that for daily life comfortable, and that for the ecosystem beautiful. As for development of the rural areas, Xi said during an inspection tour in Yunnan in 2015 that in building new socialist countryside, a path suitable to the specific conditions of rural areas should be followed. It is necessary to follow the underlying laws governing rural development, reflect the characteristics of the countryside to the fullest extent, reserve the rural culture and beautiful scenery, and enable the people to recall their rural roots with fond memories.

At the National Conference on Eco-environmental Protection, Xi stressed, "We must make plans that take all factors into consideration and simultaneously implement multiple comprehensive measures to ensure that our efforts to build an ecological civilization permeate all fields, regions and processes."

4 Joint Efforts to Build a Global Ecological Civilization and Global Environmental Governance

The international environment and development cause began at the United Nations Conference on the Human Environment held in 1972. It was advanced with landmark events such as the UN Conference on Environment and Development held in Rio de Janeiro, Brazil, in 1992, the World Summit on Sustainable Development held in Johannesburg, South Africa, in 2002, and the UN Conference on Sustainable Development, or Rio +20, that took place in Rio de Janeiro in 2012. The significance of the international environment and development cause lies in that environmental issues have been included in the agenda for global cooperation, which is leading to solutions to environmental problems and that countries around the world have reached a consensus on pursuing sustainable development. Global environmental governance is about a cooperation system the international community built to advance the cause. It includes the formulation and fulfillment of various environmental conventions and mechanisms as well as institutions and non-governmental organizations supporting international cooperation on the environment front.

China is a major contributor to the building and progress of the international environment and development cause. It is also an active participant in global environmental governance. China made substantial contributions and reaped benefits as well. Since the introduction of the reform and opening up policy in 1978, through bilateral and multilateral mechanisms, China has benefited a lot as it learned from the policies regarding environment management and sustainable development, borrowed experience, attracted investment, brought in technologies and have its people trained from the outside world. It played an important role in advancing China's environmental protection and sustainable development, promoting related legislation and the formation of management policies, and enhancing technical capabilities.

As China emerged as a leading country in terms of economic and comprehensive strengths, it has moved to a position like never before in the international community. China's position and role in the international environment and development cause and global environmental governance system also changed fundamentally. The change means that starting from active participation, China is now leading the cause; starting from aligning itself with international standards, it is now contributing institutional innovations; starting from observing the rules, it is now upholding and making the rules; and starting from bringing in various factors, it is now going out with what it can offer.

At such a critical and historic moment, Xi, a man with a global vision and a broad mind, put forth the idea of building a community with a shared future for mankind. Jointly working towards a global ecological civilization is a concrete embodiment of the idea. As early as in 2013, Xi noted that environmental protection, response to climate change and energy security are common challenges facing the world. China is ready to continue assuming its international obligations, conduct exchanges and cooperation with other countries in building an ecological civilization, and make sure that the benefits of such endeavors are shared by all. China is also willing to join hands with the world to build a beautiful planet. At the 70th session of the UN General Assembly in 2015, Xi said, "to build a sound environment is vital for mankind's future. All members of the international community should work together to build a sound global eco-environment." These are not just empty worlds, but what China has done. In 2015, without China's leading contributions, the Paris Agreement would not be reached. Over the last five years since the Belt and Road Initiative was proposed and implemented, green development has become a basic principle for related practice and lead to substantial progress in many areas.

In modernizing a country with a large population of over 10 billion at such a high speed within such a short period of time, China faced greater pressure from natural resources and the environment than any other country. In building an ecological civilization, China needs cooperation with the international community. Meanwhile, its efforts to advance ecological progress are a demonstration that may help with global sustainable development. At the National Conference on Eco-environmental Protection, Xi stressed, "joint efforts must be made in building a global ecological civilization. We must be deeply involved in global environmental governance, create solutions for environmental protection and sustainable development around world, and take an active role in international cooperation on climate change."

5 Xi Jinping Thought on Ecological Civilization and the Overall Plan for Developing a Socialist Ecological Civilization

The CPC pledged at its 18th National Congress to strive to usher in a new era of socialist ecological progress. Under the guidance of Xi Jinping Though on Ecological Civilization, the building of a socialist ecological civilization was gradually incorporated into the path, theory, system, culture and strategic goals and tasks of socialism with Chinese characteristics. The foundation of this is that people's need for a beautiful environment has become one important aspect of the major contradiction facing Chinese society.

Regarding the path of socialism with Chinese characteristics, it was pointed out at the 18th National Congress of the CPC that we must, under the leadership of the Communist Party of China and basing ourselves on China's realities, take economic development as the central task and adhere to the Four Cardinal Principles and the policy of reform and opening up. We must release and develop the productive forces, develop the socialist market economy, socialist democracy, and advanced socialist culture and a harmonious socialist society and promote socialist ecological progress. We must promote well-rounded development of the person, achieve prosperity for all over time, and make China a modern socialist country that is prosperous, strong, democratic, culturally advanced and harmonious. The Five-Sphere Integrated Plan is a structural arrangement in line with the path of socialism with Chinese characteristics.

In terms of socialism with Chinese characteristics, the Third Plenary Session of the 18th CPC Central Committee included mechanisms for ecological civilization into the system of institutions for socialism with Chinese characteristics, putting it together with institutions for the economy, politics, culture and society. It was made clear that an ecological civilization is a component of China's governance system and part of the Four-Pronged Comprehensive Strategy. Unprecedented progress and achievements had been made in deepening comprehensive reform and advancing law-based governance.

In terms of the system of theories of socialism with Chinese characteristics, Xi Jinping Thought on Ecological Civilization, as an important constituent of Xi Jinping Thought on Socialism with Chinese Characteristics for a New Era, is also a guiding principle for building a socialist country with Chinese characteristics.

On socialist culture with Chinese characteristics, the CPC said at its 19th National Congress, "socialist culture with Chinese characteristics is derived from China's fine traditional culture, which was born of the Chinese civilization and nurtured over more than 5000 years; it has grown out of the revolutionary and advanced socialist culture that developed over the course of the Chinese people's revolution, construction and reform under Party leadership; and it is rooted in the great practice of socialism with Chinese characteristics. At the sixth group study session of the Political Bureau of the 18th CPC Central Committee, Xi noted that Chinese civilization gave birth to extensive wisdom on ecology over more than 5000 years. From the philosophical idea of achieving harmony between man and nature and that the law of the Dao is its being what it is, to classic poem lines which loosely translated as "don't hit birds during spring, as their nestlings are waiting for them," and maxims urging people to stay frugal, Chinese wisdom still provides guidance and enlightenment today. Obviously, insightful ancient Chinese observations on the relationship between man and nature are parts of socialist culture with Chinese characteristics, and fostering a socialist ecological culture with Chinese characteristics is of fundamental importance in building an ecological civilization. Therefore, at the National Conference on Eco-environmental Protection, Xi stressed, "We must act quickly to establish a cultural system in which environmental concepts and values are norms." The system is also the first one of the ecologically focused systems. In this way, an ecological culture will play a foundational role in changing people's mindset and behavior for the building of an ecological civilization.

About the relationship between ecological progress and socialism with Chinese characteristics, Xi made a generalization when joining lawmakers from the Inner Mongolia Autonomous Region for discussions during the "two sessions" of 2019: since its 18th National Congress of the CPC, the Party's theories on ecological civilization have been developing and improving. Ecological progress is in the first aspect of the Five-Sphere Integrated Plan. Among basic policies for adhering to and developing socialism with Chinese characteristics in a new era, one is ensuring harmony between man and nature. Green development is one aspect of the new development philosophy. Pollution prevention and control is one of the three most urgent tasks. All these show the Party's grasp of the laws governing ecological progress, the position of an ecological civilization in China's and the CPC's cause in the new era, and the arrangement and requirement of the Party for the development of an ecological civilization.

Putting Xi Jinping Thought on Ecological Civilization into Action and Promoting Green Development

Shiji Gao

1 Distinctive Features: For the People, Science-Based, Practice-Oriented, for the World and of the Times

A distinctive feature of Xi Jinping Thought on Ecological Civilization is that it serves the people. "The people-oriented approach to development is a fundamental principle of Marxist political economics." "The original aspiration and mission of Chinese Communists is to seek happiness for the Chinese people and rejuvenation for the Chinese nation." This is both the starting point and the ultimate goal of Xi Jinping Thought on Socialism with Chinese Characteristics for a New Era. The aim of building an ecological civilization is to improve people's well-being and meet people's increasing needs for a better life. Xi Jinping said, "A good environment is the fairest public goods and most inclusive benefits for the people. Wealth is important for people's survival, but lush mountains and lucid waters are also an important part of a happy life and cannot be replaced by money. When you had money, but the air and drinking water were tainted, how could life be happy?" As China's per capita GDP exceeded $9000 and its Engle coefficient (the proportion of food expenses) continued to decline, people are asking for an ever better environment. "A good environment is the fairest public goods." In simple words, Xi expounds on what people's well-being is at the current stage of development, and provides an insightful explanation of the people-oriented approach to development. China's central authorities respond to people's call, and ensure that a sound environment benefits the people. Making environmental protection a top priority, China rolled out a slew of action plans to prevent and control air, water and soil pollution, and address prominent problems

The article is based on "Building Ecological Civilization, Advancing Green Development" by Gao Shiji and Li Zuojun. The original article was published in the ninth issue of Social Sciences in China in 2019.

S. Gao (✉)
Research Institute of Resources and Environment Policies, Development Research Center of the State Council, Beijing, China

J. Pan et al. (eds.), *Beautiful China: 70 Years Since 1949 and 70 People's Views on Eco-civilization Construction*, https://doi.org/10.1007/978-981-33-6742-5_15

endangering people's health as a response to the most urgent concerns regarding people's life.

The feature of serving the people is also reflected in Xi's other expositions on the concept of ecological civilization, including the calls for public participation and joint governance. To best serve the interests of the people, it is necessary to mobilize all to engage in the endeavor, translate the initiative of building a beautiful China into conscious actions of all people, and encourage everybody to do their bit for the environment. In this way, the problem of environmental pollution can be addressed at the lowest social and economic cost, ecological progress can be made in a sustainable manner, and harmony between man and nature can be achieved. It is a responsible action for the people, for the survival of the Chinese nation, and for the sustainable development of the world to practice a green and low-carbon lifestyle and eschew extravagancy and unnecessary consumption.

Xi Jinping Thought on Ecological Civilization is fully based on science. An embodiment of dialectics, it builds on and further enriches Marxist historical materialism. According to historical materialism, material forces of production are the basis of all social activities, and the laws of nature must be observed in tapping into the potential of forces of production. Development should never be achieved at the expense of the environment. With the development history of human civilization in mind, Xi stressed efforts to advance ecological progress and noted that protecting and improving the ecological environment equalled developing productivity.

As he pointed out, "man and nature share a common future. We, as human beings, must respect nature, follow its ways and protect it. Only by observing the laws of nature, can mankind avoid costly blunders in its exploitation."

Xi Jinping Thought on Ecological Civilization is scientific in that it represents a correct understanding of the law governing the cycle of matter in nature, and of the shared future of man and nature. In 2013, using a holistic and creative approach, Xi proposed the idea of "a shared future of our mountains, waters, forests, farmlands and lakes," and noted the importance of coordinated and concerted action to protect and restore them. Xi Jinping Thought on Ecological Civilization respects the integrity and underlying laws of the ecosystem. It enriches and develops Marxist views on nature, system and ecology. While reflecting the conventional wisdom of the Chinese nation in dealing with the relationships between nature and man, it arrives at conclusions that echo the latest results of modern ecological studies and research of the relationships between economic activities and the environment.

The science-based nature of Xi Jinping Thought on Ecological Civilization can also be seen in its emphasis on technological innovation. Green development is a requirement of an ecological civilization. Green development of industries and the development of green industries represent the future of technological and industrial reforms and have a promising prospect. There is strong evidence that technological innovations and progress are indispensable for the advancement of a civilization. To solve environmental problems and promote the building of an ecological civilization, we need scientific discoveries, technological inventions and product innovations. We must leverage technological innovations to solve problems in green development and promote a new development pattern featuring harmony between man and nature.

The science-based feature of Xi's thought is also shown in its respect for the economic imperatives of green development and in its examination of what lessons we can draw from the old development path of American and European countries. "The problem of the environment is essentially about our approach to development." It is important that we adopt a holistic approach to conservation, prevent problems at the source, impose strict oversight on the process, and avoid the old path of polluting first and fixing it later. China's modernization is a great process never seen in history before.

Xi Jinping Thought on Ecological Civilization is practice-oriented. It takes root in Chinese people's practices of building a modern country, and is committed to solving practical problems in development. It fully reflects the unity of the practical, revolutionary and scientific nature of Marxism, and is the latest achievement in adapting Marxism to the Chinese context. When heading Fujian and Zhejiang, Xi stressed resource conservation and environmental protection, and called for efforts to advance green development for sustainability and harmony between man and nature. This is the source of the principle of lucid waters and lush mountains being invaluable assets and the practice-oriented feature of Xi Jinping Thought on Ecological Civilization. Since the 18th National Congress of the CPC, with Xi at the helm, China has made coordination in environmental protection one of the three focuses of the plan for coordinated development in the Beijing-Tianjin-Hebei region; promoted well-coordinated environmental conservation and avoided excessive development in the Yangtze River Economic Belt; and highlighted green development in advancing the Belt and Road Initiative. These practices show that Xi Jinping Thought on Ecological Civilization is not only of generality, but also of direct practicality in shaping the world and advancing China's modernization drive. In comparison, mainstream Western economics or non-mainstream leftist economics either focus on symptoms rather than fundamental reasons, or is no more than scholasticism or Eco-utopia. They regard their principles as post-modern pursuits, and fail to value the feasibility of sustainable development.

Xi Jinping Thought on Ecological Civilization is for the world. In its unprecedented modernization drive, China chose to make all-out efforts to pursue ecological progress rather than the path of development first and clean up later as did Western developed countries and some developing countries. What China has done offers experience and gives a possible option to other developing countries and makes major contributions to global sustainable development.

China is implementing the *National Plan on Climate Change (2014–2020)*, and was among the first to release the *National Plan for the Implementation of the 2030 Agenda for Sustainable Development*. Fulfilling its obligations as a major developing country, China played a decisive role in enabling a resolution at the Paris UN Climate Change Conference, and won global acclamation. At the 27th session of the UN Environment Programme (UNEP) Governing Council held in February 2013, ecological civilization put forth by China was included in the conference resolution. At the second session of the UN Environment Assembly held in May 2016, China's former ministry of environmental protection and the UNEP jointly released *Green is Gold: The Strategy and Actions of China's Ecological Civilization*. China also worked to include ecological civilization and green finance into the G20 leaders' communique in 2016. Under the guidance of Xi Jinping Thought on Ecological Civilization, China has already become a major participant, contributor and champion in the movement to build a global ecological civilization. China is ready to work in synergy with people from all over the world for an open, inclusive, clean and beautiful world that enjoys lasting peace, universal security and common prosperity, and for a community with a shared future for mankind in a new historical epoch.

Xi Jinping Thought on Ecological Civilization is of the times. Drawing inspirations from the traditional wisdom of ancient China, Xi Jinping Thought on Ecological Civilization reflects the latest requirements for harmonious coexistence of man and nature in an era of industrialization and IT application. It is of the times. What it shows is that after people have access to decent food, clothing, housing and transportation, they are working for harmonious coexistence with nature to ensure the environment sustains eternity of the human race. With environmental protection movements sweeping across the globe in the past 50 years, and the global quest for peace and development after the Cold War ended, environmental protection and response to climate change have become an equally urgent international issue to poverty reduction and eradication. Development is still the basic route to modernization. Green development is the way to modernization and an ecological civilization. Xi Jinping Thought on Ecological Civilization offers a scientific answer to tough questions raised by followers of eco-centralism and anthropocentricism. With a definite political stance and value orientation of socialist ecological civilization, the thought draws a clear distinction between socialist ecological civilization and green capitalism. He said that a new global vision is necessary to achieve sustainable development. When the old path leads nowhere, innovation is the way out. We should follow the law of the Dao and achieve harmony between man and nature and find a way to sustain the development of the Chinese nation. Building an ecological civilization is vital for sustaining the development of the Chinese nation.

2 China's Principles, Institutional Support and Governance System for Green Development

Xi Jinping Thought on Ecological Civilization is mainly represented by the insightful observation that lucid waters and lush mountains are invaluable assets. The idea not only highlights the multi-faceted values of mountains, rivers, forests, farmlands, lakes, and grasslands among other natural resources with regard to human wealth and public welfare, but also points out the ultimate goal of building an ecological civilization. The immediate economic value of a sound environment lies in the financial returns generated by tourism, green agriculture and forestry as their commercial value can be measured directly. However, more important is the huge social value of a sound environment. Enjoying lucid waters, green mountains, blue sky and safe food, people have both their physical and mental health secured. In this way, individual's well-being is improved and public welfare is enhanced. In the dialectical relationship between the duality of labor and the duality of goods in a market economy, lucid waters and lush mountains stand out with their social value of being the fairest public good. The dissimilated nature as a result of industrial civilization can be reshaped to the most valuable social wealth through green development. Xi's idea affects our view on the value of natural resources and the environment, and also affects our understanding and measurement of national wealth.

To better understand the value of a sound environment and find a way to sustainable development, international organizations and academics have been looking for effective methods to measure the value of natural resources and the environment. The World Bank and some academics have put forth many such methods. The carrying capacity of the ecosystem does not remain unchanged. It relies on technological advancement, people's preferences, production structure, and consumption pattern. It is also affected by human factors and ecological interactions. The carrying capacity of the ecosystem, or ecosystem resilience, of a certain area can be improved by increasing energy efficiency, fostering a habit of energy conservation, controlling pollutant discharges, promoting recycling, and restoring the ecological environment. Human development involves a process of managing and improving the wealth portfolio. Sustainable development means that the per capital comprehensive wealth can sustain as the population grows and technology advances. According to the calculation of Kenneth Joseph Arrow, a winner of Nobel Prize in Economics, as China invested constantly in human capital and alternative physical capital, and controlled environmental damage to a certain degree (development within the carrying capacity of the environment), the country is absolutely capable of realizing sustainable development, or maintaining a constant increase of per capita comprehensive wealth within the carrying capacity of the environment. Measuring the value of the ecology is conducive to judging whether an economy could sustain. At the micro level, calculating the value of natural resources and the environment, or the stock and changes of ecological assets will help determine the value and price of ecological services and products. In turn, ecological products or services can be compensated through direct trading or other indirect ways. The ecosystem has its values. Preserving nature is increasing the ecological value and assets. It is protecting and developing the productive forces and deserves to be compensated financially.

"The people's wish for a happy life is our goal." Xi said, "To do this, we should first of all improve the evaluation norms for economic and social development to include resource consumption, environmental damage, ecological benefits and other indicators that can be used to assess ecological improvement, and use them to direct and shape our ecological work." The new principles regarding natural resources and wealth developed by Xi as he criticizes the practice of using GDP as the only indicator to evaluate government performance will lead China to a socialist ecological civilization via a path that is different from that taken by Western countries towards industrialization.

Xi pays special attention to institutional building in pursuing green development and ecological progress. He stressed that institutional building should be the focus of efforts to modernize China's capacity and system for governance on the ecological front. "We must keep the ecological red lines in mind. Our ecological problems are so serious that we have to take the strictest measures." "We must have the strictest possible institutions and legislation in place in order to guarantee ecological progress." Under the guidance of Xi Jinping Thought on Ecological Civilization, the system for ecological civilization is made complete with supporting mechanisms, a raft of reform measures have been implemented, and green development has solid institutional support.

The *Integrated Reform Plan for Promoting Ecological Civilization* sets a principle for structural reform in the ecological domain. "We must stick to the right direction of reform, improve market mechanisms, give full play to the leading and supervising roles of the government, spur businesses' enthusiasm and self-discipline, and mobilize social organizations and the public to participate in and supervise the process." The principle is different from that on the economic front as it stresses the government's leading and supervising roles in building an ecological civilization. This shows an important feature of China's path towards innovation. By the principle, the *Integrated Reform Plan for Promoting Ecological Civilization* sets the goal for structural reform in the ecological domain. It is to establish a complete system for ecological civilization development, which is made up of eight specific mechanisms: mechanisms for natural resource property rights management, territorial space development, spatial planning, total amount management and all-round conservation of resources, paid resources consumption and ecological compensation, environmental governance, market system for environmental governance and ecological conservation, and performance evaluation and accountability in the ecological domain. The system features clearly-defined property rights, extensive participation from various sectors, and equal emphasis on incentives and restraints. It is a way to modernize China's capacity and system for governance in regard to the development of an ecological civilization.

Xi Jinping Thought on Ecological Civilization goes beyond the dichotomous approach to the relationship between the market and the government, thus breaking the dilemma resulted therefrom. China better addressed the "tragedy of the commons" than some other developing countries based on private ownership. It is because it appropriately handled the relationship between the government and the market that China performed more effectively in planning, implementing and constructing major ecological conservation projects than most developing and developed countries. China's socialist system is likely to show its strengths more and more in promoting global green development and building an ecological civilization.

The leadership of the CPC is the most important feature and the greatest strength of socialism with Chinese characteristics. Xi values the CPC's leading role in pushing forward green development. The Party charts the course, crafts overall plans, and designs policies and tasks to effect the structural reform towards an ecological civilization. Xi Jinping Thought on Ecological Civilization offers a set of effective methods for China's green development. In the meantime, from enshrining ecological civilization in the Constitution to improving legislation, and intensifying law enforcement and judicial services, China is promoting the rule of law in the ecological domain and combining top-down and bottom-up supervision models. This is in line with the goal of rule-based governance and the modernization of China's capacity and system for governance. Xi stressed that an accountability system holding both the Party and government chiefs responsible should be put in place. The central authorities introduced a system of accountability for environmental damage, and established the central environmental protection inspection system as effective mechanisms to enforce policies. From the perspectives of politics and administration, the top-down accountability system represents a distinctive feature of China's system. The state also encourages the public to participate in building an ecological civilization, and invites them to supervise pollutant discharge and law enforcement. The central authorities also offer timely encouragement and promotion to local governments' innovative approaches to pollution control and environmental protection. By strengthening top-level design, spurring local innovation and mobilizing public participation, China has formed strong synergy in advancing ecological progress. The unique strength of China's system as is shown in the ecological domain is forming an effective comprehensive system to treat serious environmental pollution. Xi made a visionary explanation of the system for ecological civilization development from five perspectives and called for efforts to win the protracted war against pollution. "We must move faster to build systems for an ecological civilization. We must act quickly to establish the following ecologically focused systems: a cultural system in which environmental concepts and values are norms, an economic system aimed at integrating industries and ecosystems, a system of responsibilities and objectives centered on improving environmental quality, a system of institutions backed by modernized governance systems and capacity, and a safety system that emphasizes the maintenance of well-functioning ecosystems and the effective prevention and control of environmental risks. By moving faster to build systems for an ecological civilization, we will guarantee that the goals of the Beautiful China Initiative are essentially met. By the middle of the century, we will complete all-round improvements in the material, political, intellectual, social and ecological domains. At that time, environmentally friendly ways of living and developing will be fully formed, human and nature will coexist in harmony, modernization of our national governance system and capacity in the environmental field will be fully realized, and our efforts to build a beautiful China will be successful."

3 Green Development is Essential for the Building of a High-Quality Modern Economic System

At the National Conference on Eco-environmental Protection, Xi said, "we must promote green development across the board. Green development is essential for establishing a modern and high-quality economic system and the key to resolving pollution problems. We must adjust the economic and energy structures, optimize the distribution of territorial space available for development, adjust the distribution of industries within regions and river basins, foster growing industries that promote energy conservation and environmental protection, clean production and clean energy, promote all-round conservation and recycling of resources, ensure that the systems of production and everyday life overlap in their circular use of resources, and advocate a green and low-carbon lifestyle of moderation and frugality and oppose extravagance and unjustifiable consumption. Green development is an integral component of efforts to achieve economic transformation and establish a modern and high-quality economic system. Without the system for green development, there would be no modern economic system for high-quality development. Green development should be the hallmark of a modern and high-quality economic system in a moderately prosperous society in all respects.

Guided by the new development concept that features innovative, coordinated, green and open development for the benefit of all, green development, the latest achievement of socialist political economics with Chinese characteristics, is an important constituent of Xi Jinping Thought on Socialist Economics with Chinese Characteristics for a New Era. Xi stressed, "a modernized economy is an organic whole in which every link, every level and every area of social and economic activities interact and connect with each other." It is made up of an innovation-driven industrial system that promotes coordinated development, a single open market system featuring orderly competition, an income distribution system reflecting efficiency and promoting fairness, a system for balanced development between urban and rural areas and between regions, a green development system that is energy-efficient and environmentally friendly, an opening-up system that is diversified, balanced, secure and efficient, and an economic system in which both the market and the government play their due roles. The systems form a unified whole and should be built and advanced with concerted efforts. The system for green development included in the modern economic system put forth by Xi is the embodiment of Xi Jinping Thought on Ecological Civilization in the economic field. In implementing the systems, a people-oriented approach should be followed, the Party should lead the economic work and ecological endeavor, coordinated efforts should be made to advance the Five-Sphere Integrated Plan and the Four-Pronged Comprehensive Strategy, and the relationship between the government and the market should be properly handled. To build a modernized economy, complete the transformation from high-speed growth to high-quality development, and initiate changes in regard to efficiency, quality and driving forces, China should follow the green development approach, improve resource efficiency, protect the ecological environment, provide more quality ecological products, and meet people's need for material and cultural life and for a beautiful environment.

From a global perspective, green development, which is also referred to as sustainable development, green growth, green economy, or inclusive green development, has become an important part for the change of development paradigms across the world, and is an important direction of innovation on development theories. Despite differences in meanings, main contents and evaluation systems among other aspects, the theories and policy frameworks all aim to seek a path to green development featuring thriving businesses, good life and a healthy environment. Xi said, "in building a modernized economy, we can draw experience from developed countries, but should be our own system and based on the realities in China."

To accelerate green development and build a modernized economy, it is imperative for China to implement the *Integrated Reform Plan for Promoting Ecological Civilization* and address institutional barriers in line with the requirements for deepening overall reform. Under the guidance of Xi Jinping Thought on Ecological Civilization, a systematic evaluation on the reforms in the ecological domain should be made to identify existing problems in the reforms and create an enabling social and economic environment for green development. First, systems for natural resource property right management, natural resource assets management and natural resource property trading should be established and improved to underpin green development. Second, the reform of the energy system supporting green development should be advanced. Reforms in key areas including oil, natural gas and power supply should be accelerated. A pricing system reflecting the real cost of the energy should be established. An energy management system where the government can better play its role should be put in place. An energy supervision system ensuring fair competition and energy security should be built. A modern energy system that is green, low-carbon, secure and efficient should be put fully in place sooner. Third, the relationship between the central authorities and local governments should be clearly defined. Key links in the vertical management system for environmental supervision should be the focus of the reform. By the principles of a modernized oversight system in a modernized market economy, a modernized oversight system that is open, transparent, professional, highly efficient, and accountable should be put in place. The oversight system for the environment at both the central and local levels should be refined. Forth, the overall framework for green development evaluation should be refined, and a performance evaluation and accountability system in the ecological domain should be put in place. In line with the *Evaluation Methods for Ecological Civilization*, the content of the evaluation should be improved, a complete, systematic and procedure-based evaluation process should be put in place and the result of the evaluation should play its due role. Fifth, the rule of law should be strengthened to ensure green development. Legislation for key areas of green development should be accelerated. The natural resource law, the environment law, the energy law for transforming to green and low-carbon development, among other laws, as well as supporting laws and regulations should all be formulated and improved. The reform of the judicial system in the ecological domain should be deepened. Faster moves should be made to enhance law enforcement capacity building for green development.

Institutions and policies provide perpetual impetus for green development and ecological civilization development. Apart from accelerating institutional building, we should formulate a raft of supporting economic policies, and build a complete policy system to support businesses in acquiring technologies for green development, the government in managing and supervising green development, and the general public in participating in the building of an ecological civilization. This is of great significance to China's green development.

Building a green production system is at the core of an economic system aimed at balancing industrial development and ecological conservation. To this end, it is necessary to encourage technological innovations and new business models related to resources and the environment, and build industrial clusters and business ecosystems to support green development. For developing countries, the formulation and implementation of effective industrial policies led by the government is very important for the upgrading and transformation of traditional industries. Industrial policies should work in tandem with other related policies. In a world where green development is widely encouraged, there is little controversy over supportive policies for green industries. From developed countries to developing countries, there are policies to support renewable energy development, the transformation to clean and low-carbon energy, and social and economic transformation.

China's practice in implementing supportive policies for green development will bring the world benefits in two aspects. First, it will lower the cost of turning industries green, and improve the well-being of the Chinese population, which is 19% of the world population. Second, it will generate huge technological spillover and benefit consumers. China will provide the most cost-effective green products for the world and play a demonstration role in global green development. The significance of China's successful industrial policies has been verified in the ICT sector. In the clean energy sector, the policies have also generated preliminary results. China now leads the world in new energy development. It is the largest manufacturer of wind power equipment and solar photovoltaic components, and is the country with the largest installed capacities of wind power and solar power. China is leading global transformation toward clean energy consumption. There is room for improvement when it comes to China's supportive policies for green development. Efforts should focus on making better use of market mechanisms, combining incentives and restraint measures, forestalling and defusing production overcapacity in new energy sector, fostering faster technological progress, enabling technological breakthroughs and innovation on business models to sustain green development, and making greater contributions to global green development and ecological progress.

Connotations of Xi Jinping Thought on Ecological Civilization

Jianjun Zhao

1 The Formation of Xi Jinping Thought on Ecological Civilization

A retrospect on Xi Jinping's growth and career experiences as well as some research on his speeches about ecological progress lead to the following timeline of the formation of Xi Jinping Thought on Ecological Civilization:

1. **The rudimentary stage**

From 1969 to 1975, Xi Jinping spent seven years of his youth life in Liangjiahe Village, Yanchuan County of northwest China's Shaanxi Province. To improve the fragile environment and people's living conditions in Liangjiahe, Xi went to Mianyang, Sichuan Province, at his own expense and learned how to build bio-gas digesters. It was a time when Chinese people just began to make their attempts at building bio-gas digesters. Xi led villagers in Liangjiahe to build dozens of such digesters. Thus the local people no longer had to rely on kerosene lamps for lighting, or on firewood for cooking. More importantly, this brought the idea of recycling to the village. The bio-gas digesters alleviated the pressure on the local environment and protected the fragile ecosystem. In a sense, the Loess Plateau is the birth place of Xi Jinping Thought on Ecological Civilization.

In Zhengding, Xi put forth an idea of developing "extensive agriculture" based on local realities. He called on local people to change their mindset about farming, improve the agricultural technology, build a more reasonable economic structure for better balanced development, and use land in more efficient ways. With Xi at the helm, Zhengding County formulated a master plan for economic, technological and social development in 1985. The document highlighted environmental protection, pollution control, rational use of resources and ecological balance as important tasks in the modernization drive and urgent needs of production and life. According to the

J. Zhao (✉)
Philosophy Department, Party School of the Central Committee of the CPC, Beijing, China

J. Pan et al. (eds.), *Beautiful China: 70 Years Since 1949 and 70 People's Views on Eco-civilization Construction*, https://doi.org/10.1007/978-981-33-6742-5_16

master plan, from 1986 to 1990, an ecological agricultural system featuring mutually reinforcing and coordinated development of farming, forestry and herding should be preliminarily set up; and from 1991 to 2000, there should be a better ecosystem, a sound economic system, an energy system that meet the needs, a flexible information system, an advanced labor system and a strong technological system. With plain language, such goals represent new a development vision and specific ways to achieve it, expressing the idea of promoting harmony between man and nature.

In Ningde, Xi pondered on how to vitalize the underdeveloped eastern Fujian Province and help local people shake off poverty. Xi was well aware of the local conditions: Agriculture was the mainstay, while coastal and mountainous areas had vastly different economic structures. Based on this, Xi called for all-round development of farming, forestry, herding, fishing, and agricultural product processing based on specific conditions. He made forestry development as a strategic issue, saying that when the mountains got green, eastern Fujian would get rich. It represents the insightful understanding that forestry was the key to poverty alleviation in eastern Fujian. Xi also attached importance to technology for better use of resources. He advocated driving agricultural development with technologies and depending on the technology to develop and use resources that could not be used before, or to extend the use of existing resources. Such an approach to resource development is based on a virtuous cycle and dynamic balance of the natural system. It emphasizes respecting and complying with natural laws and realizing sustainable development featuring a good balance between human production and consumption and the natural ecosystem.

2. **The exploration stage**

After Xi was transferred to Zhejiang Province, his ideas on an ecological civilization were further developed. During this stage, Xi's ideas on ecological development were extended to cover the ocean ecosystem. Based on the Green Zhejiang initiative proposed by the CPC Zhejiang Provincial Committee, Xi further expounded on the relationship between ecological development and civilization development. At this stage, Xi Jinping Thought on Ecological Civilization became a relatively complete system. The focus shifted to interactions between ecological conservation and economic development. He put forth strategies for ecological advancement suitable for local conditions. The thought acquired philosophical significance as it evolved from an approach targeted at individual or special cases to a more general principle. When working as Party Secretary of Zhejiang, Xi said, "We pursue harmony between man and nature. In simple words, we want lucid waters and lush mountains, as well as material wealth." In an op-ed titled "Lucid Water and Lush Mountains Are Invaluable Assets" published in *Zhejiang Daily*, Xi said, "When ecological advantages are translated into strengths in ecological agriculture, eco-friendly industry and eco-tourism, lucid waters and lush mountains will become invaluable assets."

3. **The formation stage**

The Beautiful China campaign was proposed for the first time at the 18th National Congress of the CPC. Ecological civilization was included in the Five-Sphere Integrated Plan. In his speeches on building a beautiful China, ecological culture was gradually upgraded to ecological civilization. Against the backdrop of building a beautiful China, Xi said ecological civilization plays an important role in building a Beautiful China and realizing the Chinese Dream. At this stage, Xi began to think about the ecosystem in a holistic and systematic manner from the perspective of top-level design. From law-based handling of ecological issues and legal support for an ecological civilization as was decided at the Fourth Plenary Session of the 18th CPC Central Committee, it can be seen that Xi's thought on law-based ecological civilization development had moved to the lever of superstructure. At this stage, Xi also proposed ideas about international cooperation in ecological civilization building and drawing ecological red lines, among other topics. Such eco-philosophical ideas echoed China's practices in building a beautiful China. Ideas of an ecological civilization was upgraded from simple ideas to the superstructure level and to philosophical guidelines from the grand perspective of state governance and civilization progress.

It was pointed out at the 19th National Congress of the CPC that building an ecological civilization is vital to sustain the Chinese nation's development. This shows the important position of ecological civilization development in China's socialist modernization drive. As socialism with Chinese characteristics entered a new era, Xi stressed in the report to the 19th National Congress of the CPC that efforts should be made to speed up reform of the system for developing an ecological civilization and building a beautiful China, promote green development, solve prominent environmental problems, strengthen the protection of ecosystems, and reform the environmental regulation system. The report to the 19th National Congress of the CPC offered insightful and specific answers to the question of how to build an ecological civilization in the new era. Xi's ideas on ecological civilization laid a solid foundation for the building of a socialist ecological civilization in the new era and opened a new chapter in China's endeavor in this regard.

2 The Principle of Lucid Waters and Lush Mountains Being Invaluable Assets is an Important Constituent of Xi Jinping Thought on Ecological Civilization

The principle of lucid waters and lush mountains being invaluable assets is an important constituent of Xi Jinping Thought on Ecological Civilization. This indicates the importance of pursuing harmony between man and nature and offers a way to realize the value of both man and nature.

1. **Presenting and developing the principle**

In 2005, Xi Jinping, then Party Secretary of Zhejiang Province, put forth the idea that lucid waters and lush mountains are invaluable assets during an inspection tour to Yucun Village in Tianhuangping Township of Anji County. In 2006, Xi further expounded the idea. The understanding of the relationship between a good natural environment, i.e. lucid waters and lush mountains, and material wealth goes through three stages. At the first stage, lucid waters and lush mountains are exploited for material wealth. No or little consideration is given to the carrying capacity of the environment. At the second stage, both material wealth and lucid waters and lush mountains are valued, but the conflict between economic development and resource shortage and environmental deterioration start to play out. People begin to realize that the environment is the foundation of their survival and development and that it is necessary to keep the mountains green so that we don't have to worry about firewood. At the third stage, people realize that lucid waters and lush mountains could generate a constant stream of material wealth, and that are themselves invaluable assets. The evergreens planted are money spinners. Ecological strengths can be turned into economic benefits. The two become a harmonious unity.

In a speech delivered at Nazarbayev University in Kazakhstan on September 7, 2013, Xi explained the dialectical relationship between material wealth and the environment. He said, "We need lucid waters and lush mountains. We also need material wealth. If we have to choose between the two, it will be the environment. And lucid waters and lush mountains are invaluable assets." In simple words, Xi fully explained the meaning of the principle. First, the two form a dialectical unity. He said economic development and environmental protection are both important for sustainable development. They are inseparable and mutually reinforcing. When economic development contradicts environmental protection, the environment should be always put first. The environment can never be sacrificed for material wealth, even if it means some losses of economic benefits. At a meeting of the Political Bureau of the CPC Central Committee held on March 24, 2015, the *Guidelines on Accelerating Ecological Civilization* was introduced, and the principle that lucid waters and lush mountains are invaluable assets was included in an official document of the central authorities and became a guiding principle for building an ecological civilization. On October 18, 2017, the principle was included in the report to the 19th National Congress of the CPC and became one of the basic national policies as China proceeded with socialism with Chinese characteristics in a new era.

2. **The scientific connotation of the principle**

First, lucid waters and lush mountains are examples of natural assets. In a broad sense, lucid waters and lush mountains refer to natural resources offered by the environment, including water, land, forests, air, fossil energy, and various ecosystems comprised of basic elements. Ecological resources have ecological significance in the first place, which means that natural resources can provide ecological products and services. For example, they can help adjust the climate, protect the soil and conserve biodiversity. In addition, natural resources should be of economic value, which means the development and exploitation of natural resources can support human production and consumption. Therefore, protecting lucid waters and lush mountains is how we make it possible to keep and use their economic values.

Second, lucid waters and lush mountains are where competitiveness lies. A sound environment could bring material wealth. The environment is not only the foundation for people's survival and economic development, but also an important part of a region's competitiveness. A sound environment is what supports the development of a city. It is also a draw to attract investment and talent from elsewhere. More and more highly competitive regions turn to build their advantages by improving the environment in the local area. As China pushes its ecological civilization drive forward, such practices will become mainstream. Keeping lucid waters and lush mountains is a must for boosting the economy.

Third, lucid waters and lush mountains are the foundation of industrial development. China devotes great energy to boosting green development and ecological industries, for which lucid waters and lush mountains serve as the foundation. Eco-agriculture, eco-friendly industry and eco-tourism all depend heavily on natural resources. A sound environment is the foundation and core for developing new ecological industries. Each place should explore ways to strike a balance between economic development and environmental protection based on local conditions. Protecting the ecological environment offers great support to the development of productive forces. The goal is to secure both a sound environment and economic prosperity.

Fourth, lucid waters and lush mountains are the source of happiness. What is a beautiful China? Everyone may have a different understanding, but blue sky, green land and clean water are probably widely recognized elements of a beautiful China. The boom in tourism among Chinese people, on the one hand, reflects people's pursuit of a colorful cultural life, and on the other hand, shows their dissatisfaction with urban life. They aspire to escape the hustle and bustle of the concrete jungle in big cities, and seek asylum in nature. However, what China faces is complex ecological conditions and problems. Thus, the principle that lucid waters and lush mountains are invaluable assets should be incorporated into our overarching development strategy. An ecological civilization should be built with all-round efforts so as to improve the environment, raise the quality of life, and enable every person to live in a world where the sky is blue, the land is green, the air is fresh, the food is safe, the water is clean and the scenery is beautiful.

3. The theoretical value of the principle

First, the principle reflects a vision of green development. Economic development and environmental protection may be contradictory, but are also interdependent. They form a unity of opposites. Conciliation can only be achieved through green development which translates lucid waters and lush mountains into material wealth. If this approach is not adopted, development will only be temporary as it will drain the resources sooner or later. In the meantime, to translate lucid waters and lush mountains into material wealth, a sound environment is necessary. Therefore, green development is a prerequisite and a necessity if we are to meet people's need for progress. The principle that lucid waters and lush mountains are invaluable assets reflects rational thinking on the relationship between rapid development and sustainability. To strike a balance and build an ecologically advanced society featuring harmony between man and nature, the green development path should be followed.

Second, the principle is about realizing the values of both man and nature. Man and nature are equally valuable in the development of human society. Man used to see only their own value and made the mistake of human-centralism. Or they focused on only environmental protection and ignored development after discovering the value of the environment, thus missing golden opportunities for social progress. Neither of them is the right path for harmonious or coordinated development of human society and the natural environment. Therefore, the only way to realize the values of both man and nature is to give equal emphasis to the two sides. Guided by the vision of green development, the principle that lucid waters and lush mountains are invaluable assets reflects the state that our society is currently in. Lucid waters and lush mountains are representatives of nature, and by saying that they are invaluable assets, we highlight their value for development. The vision of green development does not only focus on the value of nature, but also takes into consideration all the ecological value that is contained in where technological progress is heading and which production methods we choose adopt. When trying to push the human civilization forward, we should pay attention to ecological indicators when so as to keep track of the impact of technological advances on productivity. Rather than seeing the environment a warehouse of an endless stream of resources, we should regard it as an integral part of our pursuit of development and an important constituent of our future wealth. By recognizing lucid waters and lush mountains as invaluable assets, we reshape the value system of human society, achieve harmony between the value of man and the value of nature, and then realize the value of both man and nature through green development.

3 Harmonious Coexistence Between Man and Nature is the Essence of Xi Jinping Thought on Ecological Civilization

Harmonious coexistence between man and nature is a new requirement based on China's ecological conditions, social development and people's demand in the new era. The pursuit of "harmony between man and nature," a hallmark of socialist harmonious society, proposed in the early twenty-first century, means developed industries, an affluent life and a sound natural environments. The concept of "harmonious coexistence between man and nature" is the upgraded version of pursuing harmony between man and nature. The former indicates that man and nature are interdependent, interactive and form a biotic community. In the new era, the following four principles should be followed in pursuing harmonious coexistence of man and nature.

1. **The unity between fostering ideas and improving institutions**

Both ideas and institutions are important in achieving harmonious coexistence between man and nature. Ideas guide actions. Institutions regulate actions. Both should play their due role. In the report to the 19th CPC National Congress, Xi said, "to ensure harmony between man and nature... we must realize that lucid waters and lush mountains are invaluable assets and act on this understanding." As early as in 2005, Xi proposed the scientific thesis of lucid waters and lush mountains being invaluable assets. The thesis highlights the value of nature. Natural ecology has values. Preserving nature is improving and accruing natural assets, which is protecting and developing productive forces. Notably, the value of nature is not produced because of human activities, but is showed through human activities. Different activities only change the way the nature's value is reflected. In the industrial civilization, people chose extensive development and developed at the coast of resources and the environment due to lack of knowledge and immature technology among other social reasons. People at the time crudely exploited the environment for material wealth. The old saying of relying on what people get in nature for a living reflects such idea. To act on the principle of lucid waters and lush mountains being invaluable assets, we must change our mindset and develop the technology. On this basis, we must fundamentally change the way people treat nature. Rather than selling minerals, we can turn the beautiful scenery to a source income. We can preserve the environment and make the environment benefit us financially in a sustainable manner.

New ideas cannot be fostered overnight. To make ecological preservation hold great appeal among the public, continuous efforts should be made. At present, it is imperative to speed up ecological building. Institutions should play a supportive role at the critical moment. The report to the 18th CPC National Congress said environmental protection must rely on institutions. The third plenary session of the 18th CPC Central Committee stressed that a systematic and complete system should be put in place for building an ecological civilization. The report to the 19th CPC National Congress elaborated efforts to deepen the system for building an ecological civilization to realize a beautiful China initiative. The strictest possible system for ecological protection should be established and implemented, and the supporting mechanisms for the system for building an ecological civilization should be put in place so that strong institutional safeguard can be provided to ensure the harmony between man and nature.

2. **Unity between economic development and ecological protection**

To deliver on the two centenary goals, realize the Chinese Dream of national rejuvenation, and steadily improve people's lives, we must continue to pursue development as the Party's top priority in governance; we must unleash and development productive forces. China has far overtaken Japan in terms of GDP and consolidated its position as the second largest economy. Contributing more than 30% of the world's economy, China is the first engine for global economic growth. The reason lies in the fact that China has been making development a top priority. As socialism with Chinese characteristics has entered a new era, the principal contradiction facing Chinese society has evolved. What we now face is the contradiction between unbalanced and inadequate development and the people's ever-growing needs for a better life. Overall, China's social productivity has substantially improved and the country has emerged as a global leader on many fronts. A more prominent problem is imbalance and inadequacy in development, which constrains efforts to meet people's ever-growing needs for a better life. The principal social contradiction has evolved, but fundamentally, what we face is the contradiction between people's needs and social production. The key to dissolve the contradiction is development, development that highlights quality and efficiency, development that pursues harmonious coexistence between man and nature.

Striking a balance between economic development and environmental protection is actually an aspect of pursuing harmony between man and nature. Economic development is about human society's economic development. The harmony between man and nature cannot be about one individual's harmony with nature, but about the entire human society's harmony with nature. This is what Mar stressed. The relationship between man and nature is a relationship between human society and nature, rather than the relationship between individual and the outside world. In a sense, striking a balance between economic development and environmental protection is what we should do in pursuing harmonious coexistence between man and nature. The former is the prerequisite of the latter. To say striking a balance between economic development and environmental protection: first, it means that only by coordinating economic development with environmental protection, can the economy grow in a stable and sustainable manner. Economic development decides people's living standards, but environmental protection decides whether people can survive. Therefore, the latter is a fundamental issue. Economic development must be subject to and rely on ecological protection. Second, it means that the relationship between economy and the environment should be well handled in economic development. By changing economic growth model, optimizing economic structure, transforming driving forces for economic growth, and building a modern economic system, coordination between economic development and ecological preservation can be realized.

3. **Unity between green production and green lifestyle**

The mode of production and people's lifestyle are also important standards in measuring harmony between man and nature. The mode of production has twofold implications – material mode of production and social mode of production. A green mode of production is no exception. In the material sense, compared with traditional mode of production, green production introduces the concepts of recycling and low-carbon development into the entire process of material production and the life cycle of products, aiming to minimize environmental impact and energy consumption throughout the life cycle. In social terms, green production expands simple social relationships and connects people with nature. Instead of taking only human production development per se into account, green production also values the balance in development between man and nature. Green production is a natural, harmonious and sustainable mode of production. It is an active mode of production and one that benefits the people.

A green lifestyle refers to efforts of advocating green products, participating in green volunteer services, and guiding the public to foster ideas of green, low-carbon and sharing lifestyles and environmental protection. People can develop a green lifestyle in terms of consumption, transportation and their way of life. By building green homes, green campuses and green communities, the entire society will develop a natural and healthy lifestyle. While people fully enjoy the convenience and comfort brought by green development, harmonious coexistence between man and nature can be achieved. Green production and green lifestyle are closely connected and mutually reinforcing. Green production is the prerequisite and foundation of a green life. Only when green products are available can people consume them and travel in a green manner. Only when more quality ecological products are provided, can people's ever-increasing demands for a beautiful environment be satisfied. The demand of a green life is a reflection of green production. Demands formed in everyday life can guide adjustments in production, boost green industries and provide continuous driving forces for green production.

4. Unity of current development and a sustainable future

It was stressed in the report to the 19th National Congress of the CPC that "what we are doing today to build an ecological civilization will benefit generations to come. Building an ecological civilization is vital to sustain the Chinese nation's development." Xi called on us to have a strong commitment to a socialist ecological civilization and work to develop a new model of modernization with humans developing in harmony with nature. "We must do our generations' share to protect the environment," he said. To develop in harmony with nature, we must balance current development and future sustainable development. Development cannot be restrained to the present, but should sustain the entire Chinese nation. Sustainable development is actually development for generation by generation. People of the current generation cannot grow at the expense of future generations. The current development should not squeeze the resources and space of generations to come. In retrospect, the Mesopotamian civilization, the ancient Babylonian civilization and the ancient Egyptian civilization were all buried in history and failed to sustain because they all failed to balance the relationship between man and nature. Harmonious coexistence between man and nature is an important guiding principle for sustainable development. Here, "man" doesn't mean people in a country or people of the current time, but humans generation by generation. This requires compliance with the principle of sustainable development in our current pursuit of development. The relationship between man and nature should be balanced. Man should strive to achieve harmonious coexistence with nature for generations. In this way, sustainability for human society and nature can be realized.

Major Development of Productivity Theory

Zujiao Li

Since the 18th National Congress of the CPC, General Secretary Xi Jinping has stressed on multiple occasions that "we must properly handle the relation between economic development and environmental protection, and uphold the philosophy that protecting and improving the environment is protecting and increasing productivity." This statement, representing a new requirement on China's sustainable economic and social development on a comprehensive and strategic level, is an action guide for the whole Party and the whole country to build a beautiful China and realize the Chinese Dream of the great national rejuvenations. By revealing the relation between ecological environment and productivity from the ideological and theoretical perspective, it marks a major development of the productivity theory.

General Secretary Xi Jinping's remark that "protecting and improving the environment is protecting and increasing productivity" is a major development of the productivity theory. First of all, the idea that "environmental productivity is also a kind of productivity" it contains inherits and develops Marx's view that "natural productivity is also a kind of productivity."

As we all know, the earth surface, where mankind carries out all activities, consists of two highly connected worlds—the human world comprising the social sphere, techno-sphere and noosphere, and the natural world comprising the lithosphere, atmosphere, hydrosphere and biosphere. According to the Marxist productivity theory, social material production and natural material production take place simultaneously in these two worlds; correspondingly, there is the social productivity and natural productivity driving each of the two material production processes. In

This article was originally collected in the book *Building Ecological Civilization, Building Beautiful China—Studying and Implementing General Secretary Xi Jinping Strategic Thought on Ecological Civilization* edited by the former National Forestry Administration and published by China Forestry Press. Later it was published on the *Green China* magazine with revisions by the author himself.

Z. Li (✉)
Economic Development Research Center of Former National Forestry Administration, Beijing, China

J. Pan et al. (eds.), *Beautiful China: 70 Years Since 1949 and 70 People's Views on Eco-civilization Construction*, https://doi.org/10.1007/978-981-33-6742-5_17

other words, productivity also consists of two parts—social productivity and natural productivity. As Marx pointed out, on all levels of social development, social material production not only includes man's productive activities, but also the productivity of nature itself (*Collected Works of Marx and Engels*, Vol. 26, published by the People's Publishing House in 1972, Page 500). In short, natural productivity is also a kind of productivity.

Marx, the founder of Marxist theories, had full scientific basis to regard natural productivity as a kind of productivity. Social material production and re-production both stem from the combination of social material productivity and natural material productivity, without the latter there is no social production or reproduction to speak of. The process of social material production and reproduction also includes the process of natural material production and reproduction. Even when man isn't laboring, the physical, chemical, and biological processes of natural material production are still at play. Other than the social material production process, material products generated from natural material production have the same value as man-made products in meeting human needs.

It is a pity that this important Marxist view didn't receive enough attention for a long time. The traditional mainstream economics only admits the value of products deriving from human labor, but not the value of ecological products and services provided by the naturel world, or ecosystem, to sustain human life and production. It only admits the social material production and productivity, not the natural material production and productivity. Influenced by this economics theory, man, on the one hand, regards the various natural resources we draw from earth's ecosystem as priceless and infinite. On the other hand, they think the natural environment' capability of self-adjustment and self-purification is unlimited and its capability and capacity of bearing and accommodating the wastes generated from human life and production is unlimited. Therefore, man has continuously increased its plundering of natural resources and has discharged wastes into the environment without restraint, which has led to a series of environmental problems such as resource depletion, environmental pollution and ecological degradation, even global environmental crisis.

It was amid the aggravating global environmental crisis that General Secretary Xi Jinping has timely put forth the scientific view of "protecting and improving the environment is protecting and increasing productivity." It has not only brought new hope of protecting and improving the ecological environment and curbing global environmental crisis while promoting economic and social development, but also opened our eyes to the Marxist truth that "natural productivity is also a kind of productivity" by revealing that "environmental productivity is also a kind of productivity."

From the perspective of modern ecology, what Marx called the "productivity of nature itself" means the "productivity of natural ecosystem," or "ecological productivity" in short—this is a common sense today with the maturity of ecological theories. The term "ecosystem" or "ecology" didn't appear in Marx's works because the science of ecology wasn't invented back then. The "ecological environment" mentioned by General Secretary Xi Jinping, the combination of ecology and environment, encompasses all external conditions on the earth surface that man relies on for survival, life and production, including all kinds of ecosystems and their environments and resources. This is the ecological definition of "natural ecosystem." It is clear that the idea "environmental productivity is also a kind of productivity" contained in General Secretary Xi's remarks is equal to "ecological productivity is also a kind of productivity," which is essentially the same as Marx's "natural productivity is also a kind of productivity" in scientific definition. However, Xi's profound thought proposed today when mankind is facing a global environmental crisis has more contemporary imprints than Marx's view put forth hundreds of years ago, and is therefore of greater realistic significance. That is why we say Xi's scientific statement has inherited and developed the Marxist view that "natural productivity is also a kind of productivity."

That General Secretary Xi Jinping's remark that "protecting and improving the environment is protecting and increasing productivity" is a major development of productivity theory is also reflected in its implication that "the capability of protecting and improving the ecological environment is also a kind of productivity," which deepens and enriches the connotations of "productivity".

For many years, the mainstream economic academia has defined productivity as "the human power of understanding, utilizing and reforming nature and obtaining materials from nature" or "the human power of utilizing and reforming nature." Therefore, productivity is often generalized as the relation between man and nature, whereby the human power of reforming and conquering nature is emphasized. That's how "productivity" has been understood since the Industrial Revolution, or during the entire industrial civilization.

Based on this understanding, man has always placed itself outside or above the natural world, taken nature as a kind of alien force, and regarded productivity development as a one-way process of man claiming resources from nature. While doing that, it has turned a blind eye to the significance of nature providing man with various products and services, to the importance of man protecting and improving such natural functions, and to the bearing capacity of the natural ecosystem. The result is that while man has made tremendous economic and social achievements and significantly improved their material and cultural life, the natural ecosystem has suffered from ever more serious damages. Various kinds of mineral, water and land resources are depleted, the area of forests and wetlands is shrinking sharply, water and soil loss is serious, land desertification and stony desertification have aggravated, air, water and soil pollution is grave, biodiversity is diminishing, acid rain is causing more damages, the ozone layer is destroyed and depleted, and global greenhouse gas emission is increasing... All these have led to a slew of worsening environmental issues such as global warming, which have greatly affected the sustainable socioeconomic development and the sustainable improvement of our material and cultural life, even affected human health directly. Facts have proven that the traditional definition of "productivity" is neither feasible nor sustainable.

From a philosophical point of view, the traditional definition of "productivity" does not conform to the law of unity of opposites advocated by Marxist material dialectics for several reasons.

First, since productivity is the relation between man and nature, it does not just include the opposition between the two sides, but also their unity. In other words, while man is utilizing and reforming nature, they are also protecting and improving it, and come to harmony with it through constant adjustment, so that nature becomes more suitable for human survival and development by better meeting their production and living needs.

Second, productivity, as a human initiative, is realized through the material and energy exchanges between man and nature, and the precondition and foundation for such exchanges is the structural integrity and functional sustainability of natural ecosystem. When man is utilizing and reforming nature, they should also respect, adapt to and protect nature—only by doing that can they obtain the necessary and sustainable materials, energies and natural environment for their survival and development.

Using the law of unity of opposites of Marxist material dialectics, we can come to a comprehensive, complete and scientific definition of productivity—it is the human power of utilizing, reforming, protecting and improving nature and obtaining material resources from it in a sustainable way. In other words, the power of utilizing and reforming nature as well as the power of protecting and improving nature are both productivity.

It must be pointed out that given the ecological perception that "the natural world refers to natural ecosystem" and General Secretary Xi's remark that the ecological environment is natural ecosystem, the idea that "the power of protecting and improving nature is also a kind of productivity" is totally consistent with the important view contained in General Secretary Xi's remarks, namely "the power of protecting and improving the ecological environment is also a kind of productivity."

This shows that compared with the traditional concept of "productivity", the important view by General Secretary Xi Jinping has indeed injected it with new and scientific connotations, which is why we say the General Secretary's important remarks have enriched and developed the concept.

It's worth mentioning that "the power of protecting and improving the ecological environment is also a kind of productivity" contained in Xi's remarks also largely reflects man's awakening to the global environmental crisis it is faced with and its attitude to correctly cope with it. As stated in the *Declaration of United Nations Conference on Human Environment* (the Stockholm Declaration), "A point has been reached in history when we must shape our actions throughout the world with a more prudent care for their environmental consequences. Through ignorance or indifference we can do massive and irreversible harm to the earthly environment on which our life and well-being depend." "The protection and improvement of the human environment is a major issue which affects the well-being of peoples and economic development throughout the world; it is the urgent desire of the peoples of the whole world and the duty of all Governments."

This further tells us that General Secretary Xi's remarks are not only of great importance in guiding our economic and social development, but are also a major contribution to all mankind following the general trend of "protecting and improving the human environment. We have every reason to believe that as man is faced with a global environmental crisis today, the significant theoretical and practical significance of this important viewpoint for promoting economic and social sustainability will gradually come to the fore.

That General Secretary Xi Jinping's remark that "protecting and improving the ecological environment is protecting and increasing productivity" is a major development of the productivity theory is also reflected in the implication that "the ecological environment is an important factor of productivity," which enhances and elevates people's understanding of the relation between ecological environment and productivity.

"The human power of reforming natural objects into material resources meeting their needs through production includes laborers with certain knowledge, experience and skills, means of labor, mainly means of production, and objects of labor. Laborers are the principal active factor." This is the typical description of productivity by mainstream economics. For a long time, people have made many meaningful explorations on the definition of productivity, and have put forth the scientific assertion that "science and technology are primary productivity," which offers us a new angle to find out the scientific definition of productivity. However, the relation between ecological environment and productivity is hardly ever mentioned. This is exactly why we usually neglect environmental factors when developing productivity. Some places even pursue economic development through unrestrained resource consumption, environmental pollution, and ecological damage, which has resulted in increasingly prominent issues on energy, resources, and the ecological environment. As General Secretary Xi Jinping said, "we've owed too much debts on the environmental front. If we don't strengthen this work now, we'll have to pay a much higher price in the future."

Environmental factors have a direct bearing on the sustainable economic and social development of China, but they have been seriously ignored. It is amid such a grave situation that General Secretary Xi made the scientific assertion that "protecting and improving the ecological environment is protecting and improving productivity," which reveals the true relation between the ecological environment and productivity. The implication it contains—ecological environment is an important factor of productivity—stresses that every aspect of the ecological environment is closely connected with productivity.

First, a sound ecological environment is the precondition for the survival and life of laborers, the primary factor of productivity. This is because, according to the principles of human ecology, the existence and life of man relies on the sustained functioning of natural ecosystem, which ensures the supply of energies and materials. As Marx said, "The first premise of all human existence... is that men must be in a position to live in order to be able to 'make history'. But life involves before everything else eating and drinking, a habitation, clothing and many other things." The "eating and drinking, a habitation, clothing and many other things" all come from a sound ecological environment. In General Secretary Xi's words, "a sound ecological environment is the most equitable public product and the most inclusive wellbeing," which applies the same to everyone. Without a sound ecological environment, say, without fresh air, clean drinking water and healthy food, the existence and life of laborers would be at risk, not to mention developing social productivity.

Second, the environmental state, which is the object of human labor, is a determinant factor for productivity development. This is because all material resources, be it land, mineral, water, or biological resources represented by the forest, come from the natural ecosystem, or ecological environment. The environmental state directly affects and determines the development of productivity. To be more specific, the global resource aggregate and the total bearing capacity of global ecosystem determine the limit on global productivity development; the resource aggregate and ecosystem's bearing capacity in different regions determine the scale and speed of regional productivity development; the ecological space and the types and quantities of resources in different regions determine the structure and layout of regional productivity development. This can go on.

At last, the environmental state is also an important factor that affects the functioning of the means of labor, mainly the instruments of labor. This is because the means of labor can only exert their due effects in suitable conditions, the absence of which would lead to their malfunction. Some abnormal meteorological disasters that we are familiar with can cause the malfunction of certain means of labor in certain regions. For instance, the frost in some southern provinces in early 2008 paralyzed the traffic because highways were frozen and cars couldn't move, and it seriously damaged the electricity and communications lines, because of which electric automotive couldn't work, plants stopped running, and civil power use and some communication services were cut off. In 2009, fifteen provinces and cities in northern China, the Huanghuai region, northwest and Jianghuai region were hit by a serious drought that lasted more than three months, not only endangering the winter wheat and domestic animals, causing trouble in farmers' lives and production, but also creating an all-round emergency of industrial production and domestic water use in cities.

All in all, General Secretary Xi's remark that "protecting and improving the environment is protecting and increasing productivity" is a major breakthrough on the productivity theory of mainstream economics on three levels. First, it contains the idea of "environmental productivity is also a kind of productivity," which inherits and develops the Marxist view of "natural productivity is also a kind of productivity"; second, it contains the idea of "the power of protecting and improving the ecological environment is also a kind of productivity," which deepens and enriches the connotations of "productivity"; third, it contains the idea of "ecological environment is an important factor of productivity," which elevates our understanding of the relation between ecological environment and productivity. Xi's scientific assertion is of immense significance in guiding us to correctly handle the relation between economic development and environmental protection and should be carefully studied, deeply understood and rigorously implemented.

Xi Jinping Thought on Ecological Civilization: Essence, Characteristics and Significance

Hai Yu

The National Conference on Eco-environmental Protection held in May 18, 2018 was a milestone in China's efforts of developing ecological civilization and protecting the ecological environment. A signature outcome and the biggest innovation of the conference was that the Xi Jinping Thought on Ecological Civilization was officially proposed and established. General Secretary Xi put forth this thought bearing in mind the fine traditional culture of the Chinese nation, the trends of the time and the world, the people's wishes and expectations, the Five-Sphere Integrated Plan, and the Four-Pronged Comprehensive Strategy. Standing at the historical and strategic height of adhering to and developing socialism with Chinese characteristics and realizing the Chinese Dream of great national rejuvenation, General Secretary Xi answered the following major theoretical and practical questions: why we should build an ecological civilization, what kind of ecological civilization to build, and how we build it. The answers, which constitute the scientific and systematic Xi Jinping Thought on Ecological Civilization, reflect the historical mission, governing philosophy and responsibilities of the CPC, and make a key component of the Xi Jinping Thought on Socialism with Chinese Characteristics for a New Era. They represent the collective wisdom of the CPC on major theoretical and practical issues concerning ecological progress and environmental protection, and guide us to make historic achievements and transformations in those fields. Accurately grasping and applying the essence, characteristics and significance of the Xi Jinping Thought on Ecological Civilization is of great importance in our endeavors to win the tough battle of pollution prevention and control and build a beautiful China.

H. Yu (✉)
Environmental Strategy Department, Policy Research Center for Environment and Economy, Ministry of Ecology and Environment, Beijing, China

J. Pan et al. (eds.), *Beautiful China: 70 Years Since 1949 and 70 People's Views on Eco-civilization Construction*, https://doi.org/10.1007/978-981-33-6742-5_18

1 Essence

The Xi Jinping Thought on Ecological Civilization covers a wide range of aspects of ecological progress, including historical positioning, basic concept, core relations, political requirements, targets and direction, practical methods, fundamental guarantee and international vision. It reflects the scientific perception of man-nature relation, the profound insight into the laws governing human civilization development, the nature, and economic and social development, and the clear vision of the lasting development of the Chinese nation. The thought provides the direction, guideline and motivation of building a beautiful China, underpins the establishment of the "Four Confidences," and offers the Chinese vision, solution and contribution to the world's sustainable development. The rich, systematic and profound Xi Jinping Thought on Ecological Civilization can be understood in eight aspects.

1. **Civilization prospers when ecology prospers**

This fully reflects the incisive historical outlook of Xi Jinping Thought on Ecological Civilization. General Secretary Xi stressed that building an ecological civilization is vital to sustain the Chinese nation's development, and our current endeavors for that purpose will benefit generations to come. Promoting ecological progress concerns people's wellbeing and China's future, and is the fundamental guarantee for realizing the Chinese nation's lasting development and great rejuvenation. The ecological environment is the most basic condition of human existence and the essential cornerstone for sustainable development. In the history of both Chinese and world civilization, changes in ecological environment directly affect the ups and downs of civilizations. We need to adhere to the basic state policy of resource conservation and environmental protection, treat the ecological environment as preciously as we treat our lives, and keep walking along the development path of advanced production, affluent life, and sound ecology, so as to lay a solid foundation for lasting national development and leave a beautiful home of blue sky, green land and clear water to our offspring. This incisive historical outlook stems from the insight to the historical lesson that civilization prospers/declines when ecology prospers/declines. It is imbued with both traditional Chinese philosophy and the philosophical ideas of Marxist historical and dialectical materialism. Never in history and nowhere in the world has human civilization ever deviated from this universal law of social development.

2. **Harmony between man and nature**

This fully reflects the scientific natural outlook of Xi Jinping Thought on Ecological Civilization. General Secretary Xi stressed in the report of the 19th National Congress of the CPC that "We should ensure harmony between man and nature. Building an ecological civilization is vital to sustain the Chinese nation's development." He also emphasized, "Humanity and nature are a community of life. There is no substitute for ecological environment. When we use it, we take it for granted. But once damaged, it will struggle to recover." The National Conference on Eco-environmental Protection even elevated the development of an ecological civilization to be the "fundamental program." Man lives on earth, relies on nature for existence and development, and creates and develops human civilization through the interactions and mutual effects with nature. In this process, the man-nature relation has transformed from man relying on nature to utilizing it and then to harmonious co-existence with it. Nowadays the human society has gradually come to such a consensus—man lives thanks to nature; man and nature are in a symbiotic relation; and damages to nature will eventually hurt man itself. Man must respect, accommodate and protect nature, otherwise nature will take revenge—this is an objective law that no one can resist.

3. **Lucid waters and lush mountains are invaluable assets**

This fully reflects the green development outlook of Xi Jinping Thought on Ecological Civilization. General Secretary Xi has repeatedly underscored how important it is to "correctly handle the relation between economic development and environmental protection, keep in mind that protecting and improving the ecological environment is protecting and developing productivity, and more consciously promote green, circular and low-carbon development without sacrificing the environment for short-term economic growth". Lucid waters and lush mountains are invaluable assets. This assertion reveals the essential relation between development and protection, and points out the methodology and new path for achieving inherent unity, mutual promotion and symbiotic coordination between them. The adherence to green development is a profound revolution of the development outlook and values. Lucid waters and lush mountains are not only natural and ecological wealth, but also social and economic wealth. Protecting the ecological environment means protecting nature's value and adding value to nature's capital, and protecting the potential of socioeconomic development. If we can translate environmental advantages into economic and social advantages, lucid waters and lush mountains will be an endless source of wealth. We need to adopt and implement new development concepts, properly handle the relation between development and protection, and foster the green development pattern and lifestyle, in the endeavor to push forward socioeconomic development and environmental protection in parallel.

4. **A sound ecology is the most inclusive wellbeing**

This fully reflects the basic livelihood outlook of Xi Jinping Thought on Ecological Civilization. General Secretary Xi said, a good environment is part of the public's wellbeing, and green mountains and blue skies bring delight and happiness to the people; protecting the ecological environment is a major political issue related to the CPC's mission and purpose and a major social issue concerning people's livelihood; and a good ecological environment is the most inclusive form of public wellbeing. He also urged us to pursue ecological benefits for the people with the priority to solving prominent ecological problems. These significant remarks have enhanced our realization of the importance of ecological civilization, laid down the key principles that must be observed when promoting ecological civilization in the new era, and illustrated the people-centered development philosophy. They are a vital guide on our work to advance ecological civilization and continuously meet people's ever-growing need for a beautiful ecological environment.

5. **Mountains, waters, forests, farmlands, lakes and grasslands are a community of life**

This fully reflects the overall system outlook of Xi Jinping Thought on Ecological Civilization. An ecosystem is an integrated natural system of interdependent natural elements and circular ecological chains. This community of life is the basic material foundation for human existence and development. Viewing mountains, waters, forests, farmlands, lakes and grasslands as a community of life, we must seek a new way of governance from a systematic and broader perspective. Instead of taking a fragmented and palliative approach that only treats the symptoms, we, in light of the ecosystem's integrity, systematicness and internal laws, must make an overall plan that takes all relevant factors into consideration, and adopt multiple measures simultaneously to advance environmental protection in all respects. The natural and ecological factors, either on or at the foot of mountains, above or underground, in land, rivers or oceans, or upstream or downstream of river basins, should be considered comprehensively for general management and control, protection and governance. This will improve the ecosystem's circular capability, maintain ecological balance and lead to the best effect of systematic governance.

6. **Protect the ecological environment with the strictest regulations and laws**

This fully reflects the strict rule of law contained in Xi Jinping Thought on Ecological Civilization. During the 6th group study of the Political Bureau of the CPC Central Committee in May 2013, General Secretary Xi pointed out that only the strictest regulations and laws can provide reliable guarantee for ecological civilization. An accountability system has to be established to hold accountable, for life, those whose blind decisions in disregard of ecological environment have caused severe consequences. The "most rigorous" ecological "rule of law" put forth by General Secretary Xi demonstrates the CPC Central Committee's resolute attitude and holds the key to the development of ecological civilization. To implement the strictest environmental protection system, the principles of "prevention at source, process control, damage compensation and accountability" have to be followed. Moreover, several major systems must be established and perfected, including the red line safeguarding resource and ecological environment, property right of natural resource assets and management of their use, ecological protection and compensation, and performance assessment and accountability.

7. **Mobilize the whole society to build a beautiful China**

This fully reflects the whole-of-people approach of Xi Jinping Thought on Ecological Civilization. General Secretary Xi stressed that the development of ecological civilization is closely related with everyone and everyone should practice and promote it. A beautiful ecological environment is shared by the whole society, which in turn should jointly build, protect and govern it. Everyone is a protector, builder and beneficiary of the ecological environment, and no one is a bystander, an outsider or a critic. We should talk less, do more and get involved. Building ecological civilization and a beautiful China is a long-term and formidable systematic program that calls for the concerted efforts of all Chinese. Everyone should take part actively. Environmental protection is the main approach and fundamental measure of building an ecological civilization, and the united action of all people is the foundation and guarantee for them both. We need to take faster steps to create a social action system where all residents take part in environmental protection, everyone fulfills the responsibility for building an ecological civilization, cherishes resources and protects the environment. Only by doing that can we turn our vision of a beautiful China and man-nature harmony into reality.

8. **Jointly promote a global ecological civilization**

This fully reflects the vision of a win–win world upheld by Xi Jinping Thought on Ecological Civilization. General Secretary Xi emphasized that the mankind is a community of a shared future, building a green home is our common dream, and protecting the ecological environment, dealing with climate change and ensuring energy and resource security is the common challenge faced by the whole world. No country can keep aloof because a global challenge calls for global efforts of all countries. The international community should join hands in creating an ecological system that respects nature and advocates green development, building a beautiful world, and safeguarding the planet that we all call home. As a responsible major country, China has become an important participant, contributor and leader in the global efforts toward ecological civilization. Building an ecological civilization is an effective endeavor by China, the largest developing country, for sustainable development, and the Chinese philosophy, solution and contribution for global environmental governance.

2 Characteristics

The Xi Jinping Thought on Ecological Civilization is the outcome of General Secretary Xi's personal thoughts, scientific perceptions and concrete practices combined with the great undertaking of socialist development with Chinese characteristics through mutual complementarity and enhancement. Backed by a deep theoretical basis and practical support, it has evolved from a primitive idea into a systematic thought. It is the fundamental philosophy and action guide on China's ecological civilization campaign and environmental protection because of its profound intrinsic characteristics that differentiate it from other or previous concepts and that inject it with strong vitality. We must adhere to the Xi Jinping Thought on Ecological Civilization as our guideline for a long time.

1. **Theoretical innovation**

The Xi Jinping Thought on Ecological Civilization, which has been developed based on the summary of historical experience and lessons drawn from human civilization development and the profound thinking of its significance, reflects the insight to the new situation, new goals, new contradictions, new issues, and new requirements of

China's environmental protection in the new era. Following the traditional Chinese philosophy of "man and nature in one" and "following nature's course", the thought dialectically develops and sublimates the traditional theory of environmental protection, inherits and deepens the theory of world sustainable development, and enriches and perfects the theory of socialism with Chinese characteristics. It is the latest theoretical achievement of Marxist ecology and development outlook. For instance, General Secretary Xi said "lucid waters and lush mountains are invaluable assets," which clearly indicates that lucid waters and lush mountains and valuable assets are not in opposition against each other, but are instead in a relation of dialectical unity, mutual complementarity and win–win interaction. This abandons the traditional, rigid mindset that ecological environment and economic development are in conflict, and points out the direction, road and approach of achieving the coordinated progress in ecological priority and green development.

2. **Scientific assertion**

The Xi Jinping Thought on Ecological Civilization has taken form based on Comrade Xi Jinping's long years of working experience and profound thinking. It represents the latest understanding of the laws governing human civilization development, nature, economic and social development, and environmental protection in the new era, and is a scientific truth able to withstand the test of practice, history and the people. For instance, General Secretary Xi's remark that "civilization prospers when ecology prospers" reflects a scientific perception of the laws governing the historical evolution of man-nature relation. He also said "the quality of ecological environment has improved amid stability but the achievements are not solid enough," and China is in a critical window period of tackling tough difficulties in seeking ecological progress. This is a clear-minded perception and scientific conclusion of the foundation, situation, challenges and opportunities regarding the current ecological progress and environmental protection in China. It draws out the coordinates for us to make decisions and implement policies in a scientific manner to win the critical battle of pollution prevention and control.

3. **Comprehensive guidance**

The Xi Jinping Thought on Ecological Civilization is a systemic and comprehensive judgment of ecological progress and environmental protection from the perspective of promoting the Five-Sphere Integrated Plan and the Four-Pronged Comprehensive Strategy in a holistic manner. It sets out the guidelines, world outlook and methodology for ecological advancement and environmental protection, not the local or partial ideas and policies. For example, the profound outlook on history, scientific outlook on nature, green development philosophy, whole-of-people approach and the vision of a win–win world contained in the thought demonstrate the different dimensions and perspectives it encompasses. Moreover, guided by the Xi Jinping Thought on Socialism with Chinese Characteristics for a New Era, the thought also reflects the comprehensive thinking of the basic strategy and action outline for the development of ecological civilization and environmental protection, and the all-round effects and role of this undertaking toward the goal of building a moderately prosperous society in all respects and a socialist modern power. In particular, the statement that "mountains, waters, forests, farmlands, lakes and grasslands are a community of life" vividly presents the systematic outlook and methodology of preserving natural resources and ecological environment.

4. **Far-reaching influence**

Ecological progress and environmental protection is a painstaking task that requires sustained efforts and involves numerous difficulties. General Secretary Xi Jinping said, building an ecological civilization is a fundamental and millennial program vital to sustaining the Chinese nation's development, and it will benefit generations to come. Rooted in history, taking the present situation into consideration and eyeing the whole world and long-term development, the Xi Jinping Thought on Ecological Civilization is not only the fundamental philosophy that guides us to win the tough battle of pollution prevention and control, but also a beacon that leads us to build a beautiful China and a socialist modern power and search for a path of global ecological civilization. It is a glorious thought of historic significance and deep-going and far-reaching influence.

5. **Effective practices**

Practice is the only standard to test truth. The Xi Jinping Thought on Ecological Civilization has become the guiding thought that is observed and upheld by the whole Party, whole country and whole society because past practices under its guidance have yielded remarkable achievements. Therefore, it is a truth that stems from practices, guides practices and is tested by practices. Since the 18th National Congress of the CPC, China, guided by this thought, has carried out a series of fundamental and innovative work aimed at long-term results, which have led to historical and comprehensive changes in our environmental protection. A new vista of a beautiful China is gradually unfolding. The past five years have seen China making unprecedented efforts on the development of ecological civilization and environmental protection, as reflected in the depth of perception, intensity of pollution control, frequency of new regulations, rigor of enforcement and inspection, and speed of environmental improvement. Going forward, the Xi Jinping Thought on Ecological Civilization will continue to guide us to advance our ecological civilization to new heights.

6. **Systematic contents**

The Xi Jinping Thought on Ecological Civilization is also characterized by its airtight system and its ability to answer such major theoretical and practical questions as why we build an ecological civilization, what kind of ecological civilization to build, and how to build it. For example, why do we build an ecological civilization? The reason is that civilization prospers when ecology prospers, developing ecological civilization is vital to sustaining the Chinese nation's development, a sound ecological environment is an inclusive wellbeing, and we want to ensure lasting peace, stability and happiness for the country, the nation and the people. What kind of ecological civilization do we build? We need to build a strong modern power of man-nature harmony, preserve a sound environment and ecosystem while making consistent economic achievements, and leave both the present and future generations a beautiful home of blue sky, green land and clean water. Meanwhile, no country can really stay detached and intact in the face of global challenges such as environmental crisis. The international community should join hands in seeking a road to global ecological civilization. How to build an ecological civilization? We must keep in mind that lucid waters and lush mountains are invaluable assets, explore a new path to advance ecological priority and green development in parallel, and balance the relation between development and protection, so that the sound ecological environment will generate more green wealth. We must protect the ecological environment with the strictest laws and regulations, intensify front-end prevention, process control and back-end punishment, and place equal emphasis on formulation and execution, punishment and incentive, so as to make the regulations real deterrents and provide the institutional guarantee for ecological civilization and environmental protection. We must regard mountains, waters, forests, farmlands, lakes and grasslands as a community of life, respect the ecosystem's integrity and internal laws, carry out comprehensive and systematic protection, recovery and governance, and take a holistic view of all natural and ecological elements to reinforce ecosystem's circulation mechanism and maintain ecological balance. We must follow the mass line of ecological civilization, and create a social system where all residents take part in environmental protection, everyone fulfills the responsibility for building an ecological civilization, cherishes resources and protects the environment, so as to build a beautiful China and realize man-nature harmony.

3 Significance

The Xi Jinping Thought on Ecological Civilization, featuring clear theme, rigorous logic, complete system and rich connotations, provides us with the ideological ammunition, direction, basis, guidance and motivation to promote ecological progress and build a beautiful China at the new historical starting point. The though demonstrates the historical mission, governing philosophy and responsibility of the CPC. It is of theoretical, historical, realistic and global significance in our endeavors to secure the decisive victory in building a moderately prosperous society in all respects, a beautiful China, and a strong modern power.

1. **Theoretical significance of Xi Jinping Thought on Ecological Civilization—enriching and developing the Marxist view on man and nature**

With the deep insight, acute judgment and strategic resolve of a Marxist politician and theorist, General Secretary Xi Jinping has put forth the thought on socialist ecological civilization with Chinese characteristics. The thought inherits the theory, spirit and character of the Marxist thought on man-nature relation, and accurately grasps the new situation, new contradictions, new features, and new issues concerning the man-nature relation in China in the new era. Fully absorbing the current situations, it lays down a series of new concepts, new thoughts and new strategies that must be observed for China to achieve modernization featuring man-nature harmony in the new era, thus pointing out the direction for us to blaze a path of civilization development characterized by advanced production, affluent life and sound ecology. The thought is teeming with Marxist truths. Stemming from and guiding our practices, the Xi Jinping Thought on Ecological Civilization provides the fundamental philosophy for advancing ecological civilization and environmental development in the new era. It opens the new horizon of the Marxist thought on man-nature relation by uplifting it to a new height in the twenty-first century, and makes decisive and historical contribution to its enrichment and development. It is a new achievement in combining the basic Marxist principles with China's actual conditions.

2. **Historical significance of Xi Jinping Thought on Ecological Civilization—being an organic and important part of the Xi Jinping Thought on Socialism with Chinese Characteristics for a New Era**

The Xi Jinping Thought on Ecological Civilization demonstrates the General Secretary's immense political courage and strong sense of responsibility. On the strategic height of upholding and developing socialism with Chinese characteristics and realizing the Chinese Dream of great national rejuvenation, he, following the general trend and commanding the overall situation, integrates ecological civilization with socialism with Chinese characteristics, socialist modernization and great rejuvenation of the Chines nation. The thought reveals the essence, characteristics, laws and paths of building an ecological civilization, blazes a new realm of its theories, thoughts, and practices, sheds light on the participation, leadership and contribution concerning global ecological civilization. The thought is a scientific action guide and powerful spiritual drive for ecological civilization and environmental protection in the new era. The Xi Jinping Thought on Ecological Civilization represents the crystallization of the practices and collective wisdom of the Party and the people on ecological civilization and environmental protection, and reflects the socialist outlook on ecological civilization. It is the action guide for the whole Party and the whole people to achieve the dream of a beautiful China, and an indispensable component of the Xi Jinping Thought on Socialism with Chinese Characteristics for a New Era. It lays the ideological foundation and forges a spiritual synergy for us to win the tough battle of pollution prevention and control, advance ecological progress, build a beautiful China and lead the global ecological efforts. We must uphold and develop this guiding thought consistently.

3. **Realistic significance of Xi Jinping Thought on Ecological Civilization—being the fundamental philosophy and action guide for our endeavors in developing ecological civilization and protecting the ecological environment**

On the height of world outlook and methodology, General Secretary Xi Jinping has expounded on the connotations and strategies of socialist ecological civilization for a new era and supported it with the "eight-point adherence" theory. Being of great political, strategic and guiding importance, this thought is the banner and beacon that guides our efforts to build a strong modern power of man-nature harmony and realize the dream of a beautiful China. We must firmly uphold the "Four Consciousness", truly apprehend the essence and rich connotations of this thought, and consistently apply it in our thinking and practices, so as to break new ground in the cause of ecological civilization and environmental protection.

4. **Global significance of Xi Jinping Thought on Ecological Civilization—contributing Chinese thought, plan and values for world sustainable development**

The Xi Jinping Thought on Ecological Civilization accurately grasps and comprehensively addresses a wide range of issues related with resources and the environment on the new height of human civilization progress. It calls for efforts on all fronts including economy, politics, culture, society and environment, and aims at harmony between man and nature, environment and economy, and man and society on a higher level. The thought not only provides the scientific concept and methodological guidance for achieving the Chinese nation's sustainable development, but also makes huge contributions to the theories and practices of world sustainable development. Committed to a global ecological civilization through concerted efforts, and taking into account both domestic and international situations, the Xi Jinping Thought on Ecological Civilization not only strives to create a sound productive and living environment for the Chinese people, but also contributes to global ecological security by promoting a clean and beautiful world. Guided by this thought, China has become an important participant, contributor and leader in advancing ecological civilization worldwide. This thought and the China-advocated ecological civilization are consistent with the United Nation's goals, and many of China's valuable ideas and best practices have received close attention and positive comments from the international community, becoming a global narrative and shared value.

Deeply Studying and Apprehending the Practical Character and Requirements of Xi Jinping Thought on Ecological Civilization

Kanping Hu

Practice is the source of knowledge. It drives cognitive development by providing necessary conditions and hones the subject's cognitive ability. Practice is the standard for truth and the fundamental purpose of cognition as well. The essence of Marxist epistemology is that practice comes first, everything should be based on reality, and practice plays a decisive role in cognition.

At the National Conference on Eco-environmental Protection held in May 2018, General XiJinping summarized China's historical achievements and changes in the field of ecological civilization and environmental protection since the 18th National Congress of the CPC. He also made a scientific analysis of the current tasks and challenges, elaborated on the significance and principles of intensifying ecological efforts in the new era, and made detailed deployments for strengthening environmental protection and fighting the tough battle of pollution prevention and control. A landmark outcome of this conference was the establishment of the Xi Jinping Thought on Ecological Civilization, which commands the strategic height of adhering to and developing socialism with Chinese characteristics and realizing the Chinese Dream of the great rejuvenation of the Chinese nation. Rooted in traditional Chinese culture, following the trend of the time and meeting the wishes of the people, the thought has been a powerful guideline on our way to making historical ecological achievements. The Xi Jinping Thought on Ecological Civilization answers several major theoretical and practical questions, such as why we must build an ecological civilization, what kind of ecological civilization to build and how to build it. This is a momentous theoretical and practical innovation by the CPC and the "fundamental philosophy guiding our ecological practices in the new era".[1] This thought that "stems from

[1] *Opinions of the State Council of the CPC Central Committee on Comprehensively Strengthening Eco-environmental Protection and Resolutely Implementing Pollution Prevention and Control* released on June 16, 2018.

K. Hu (✉)
Department of Research and Communication, China Ecological Civilization Research and Promotion Association, Beijing, China

J. Pan et al. (eds.), *Beautiful China: 70 Years Since 1949 and 70 People's Views on Eco-civilization Construction*, https://doi.org/10.1007/978-981-33-6742-5_19

practice and has been tested by practice"[2] has a distinct practical character and is oriented by problems, by the time and by the people. Deeply studying and apprehending its practical character and requirements is of great importance for advancing green development on all fronts and building a beautiful China at a faster pace.

1 Xi Jinping Thought on Ecological Civilization: Stemming from Practice, Tested by Practice

The Xi Jinping Thought on Socialism with Chinese Characteristics is extracted and elevated from practical experience, and has played a critical role in promoting our practices. On the one hand, this thought is the fundamental guide for us to achieve historical transformation as it, with its theoretical and practical significance and strategic foresight, has led the Party and the nation to create a new situation in their undertakings across the board. On the other hand, the practical value of this thought is of far-reaching influence and will be fully manifested in future practices. It is stated in the *Constitution of the Communist Party of China* that the Xi Jinping Thought on Socialism with Chinese Characteristics for a New Era is "a crystallization of the practical experience and collective wisdom of the Party and the people." As an important component of this thought, the Xi Jinping Thought on Ecological Civilization has also taken root, germinated, matured and borne fruits through the Party's and the people's practices of ecological civilization. It is also derived from and tested by practice.

From a practical point of view, a main source of the Xi Jinping Thought on Ecological Civilization is the thoughts and reflections drawn from the governing experience. Comrade Xi Jinping has always attached great importance to studying and resolving practical issues in ecological civilization development. Whether working in local areas or in the central government, he has proposed a string of new assertions with strategic vision and theoretical innovation. These new assertions, which are down-to-earth and widely supported by the people, have all stemmed from the explorations and practices of ecological endeavors. They are the crystallization of those practices and contribute typical concepts to the Xi Jinping Thought on Ecological Civilization.

Following a geographical trajectory of Xi Jinping's work experience from local areas to the central government, researchers have divided the formation of the Xi Jinping Thought on Ecological Civilization roughly into the following stages. The first stage was when he worked in Liangjiahe, Yan'an. In the 1960s and 1970s, Xi was Secretary of the Party Branch of Liangjiahe in northern Shaanxi's Yan'an, where he began to think about the ecological evolution of the loess plateau. He led the local people to improve the ecology by building dams and farmlands, develop

[2]Guiding Committee for the Compilation and Review of Materials for National Training of Government Officials. *Promoting Ecological Civilization, Building a Beautiful China*. Beijing: People's Publishing House, Party Building Books Publishing House, 2019.

production, and build the first methane tank in the province using straws and livestock and poultry feces. This is his earliest practices in environmental protection. The second stage was when he worked in Zhengding, Hebei Province. In consideration of the local reality, Xi Jinping put forward the idea of "big agriculture" in a bid to realize the balance between agriculture and ecology, coordinate ecological and economic targets, and create a sound and circular productive structure. When leading the compilation of the *Plan for Economic, Technological and Social Development of Zhengding*, he stressed that "protecting the environment, eliminating pollution and maintaining eco-balance is an important task of modernization and an urgent demand of people's life and production," and insisted on including in the plan that "we'd rather give up economic benefits than cause pollution; we will prohibit pollution transfer to the countryside." These plain words presented new development concepts as well as practical approaches, and conveyed the important thought of harmony between man and nature. The third stage was when he worked in Fujian. Comrade Xi Jinping said east Fujian should "make full use of its mountains and sea" and urged it to coordinate social, economic and ecological benefits. He initiated the strategic idea of "urban ecological progress" and ecological civilization, and called for efforts to develop the local economy and boost related industries by leveraging its environmental advantages, so as to drive economic growth and liberate and develop local productivity. When serving as the governor of Fujian Province, Comrade Xi personally headed up the provincial leading group of ecological civilization, launched the most massive eco-protection program in the provincial history, and proposed with great vision to build Fujian into an ecological province. The fourth stage was when he worked in Zhejiang. After setting the goal of building a "green Zhejiang," he further elaborated on the relation between ecological progress and civilization development. He repeatedly demanded intensifying the assessment of performance in ecological protection and preservation and establishing related systems, suggested "exploring a performance assessment system that encourages circular economic development and corresponding incentive and restrictive mechanisms", so as to "gradually steer circular economic development onto the law-based track". Xi's article "Lucid Waters and Lush Mountains Are Invaluable Assets" published on *Zhejiang Daily* made it clear that "if we can translate environmental advantages into eco-economic advantages, such as eco-agriculture, eco-industry and eco-tourism, lucid waters and lush mountains will become invaluable assets." The fifth stage is the construction of beautiful China. In this period, Xi Jinping begins to comprehensively contemplate the ecosystem as a systematic entirety from the perspective of top-level design, under which a raft of views containing ecological philosophy and the practices of building a beautiful China verify, complement and promote each other. Amid the overall vision of state governance and the systematic thinking of civilization development, the Xi Jinping Thought on Ecological Civilization has been enriched and sublimated.

The most profound practical foundation for the Xi Jinping Thought on Ecological Civilization comes from the theoretical results formed during the great practice of promoting ecological progress since the 18th CPC National Congress. While China has worked world-shocking wonders in economic and social development, it has also encountered a lot of eco-environmental problems, and people are making growing demands on clean air, safe food and a beautiful environment. The 18th CPC National Congress proposed to work hard to build a beautiful country, achieve lasting development of the Chinese nation, give high priority to ecological progress and incorporate it into all aspects and the whole process of advancing economic, political, cultural, and social progress. Moreover, "the CPC leads the people to develop socialist ecological civilization" is written into the Constitution of the CPC. The 3rd plenary session of the 18th CPC Central Committee proposed to accelerate the establishment of a systematic and complete eco-civilization system, the 4th plenary session demanded protecting the eco-environment with rigorous laws and regulations, and the 5th plenary session made green development a new development concept. The report of the 19th CPC National Congress vowed to develop China into a great modern socialist country that is prosperous, strong, democratic, culturally advanced, harmonious, and beautiful. It recognized "ensuring harmony between man and nature" as one of the basic approaches to upholding and developing socialism with Chinese characteristics in the new era, and underscored the goal of "providing more quality ecological goods to meet people's ever-growing demands for a beautiful environment." On March 11, 2018, the 1st session of the 13th National People's Congress passed the *Amendments to the Constitution of the People's Republic of China*, whereby ecological civilization was officially included in this fundamental law, achieving a high level of unity among the CPC's assertion, the nation's will and the people's wishes. As an array of major strategic measures, guidelines and policies have been issued and implemented, the whole Party and the whole nation have become more conscientious and proactive in carrying out the green development concept. The solid steps that have been taken in developing eco-civilization have yielded remarkable results. The Xi Jinping Thought on Ecological Civilization reflects the CPC Central Committee's deepened understanding of the laws governing economic and social development as well as nature. As the latest achievement in the Party's perception of the laws governing ecological progress and socialist modernization, the thought has effected profound changes in the Party's concepts and approaches concerning both development and governance. The Party and the nation have adopted exploratory measures to promote eco-civilization, made some progress and accumulated some experience, all of which have laid an important and solid foundation for the creation and development of the Xi Jinping Thought on Ecological Civilization.

2 Xi Jinping Thought on Ecological Civilization Has a Distinct Practical Character

Xi Jinping pointed out in the report of the 19th CPC National Congress that "the era is the mother of thought; practice is the fount of theory." This important judgment conveys the Marxist view on practice, reflects the Party's valuable experience in advancing theoretical innovations over the years, and reveals the laws governing such innovations.

Practice orientation is the distinct character of Marxism and the reason why it is scientific and vigorous. Marxism, a scientific system with a practical character, stems from and evolves through practice, aims to change the world, and is continuously supplemented, corrected and perfected by new practices, thus becoming a theory with practice as its essence. Theory relates to practice, practice endorses and enriches theory, and theory guides practice—this is the key feature of Marxist epistemology and materialistic dialectics.

The Xi Jinping Thought on Ecological Civilization takes practice as the standard for truth. It evolves along with the time and practice, keeps deepening itself, summarizing experience and making theoretical innovations, insists on the dialectical unity between theoretical guidance and practical explorations, and realizes the benign interaction between theoretical and practical innovations. It is amidst such unity and interaction that the thought has emerged and evolved with a distinct practical character.

The thought upholds practice as the top standard. Practice determines perception—being its source as well as destination. Xun Zi said "knowing something is not as good as practicing it," Lu You believed "one must practice something to know it thoroughly," and Wang Yangming accentuated the "unity of knowledge and practice"—all these remarks by ancient sages highlight the relation between perception and practice. Theory would become rigid doctrines once it is divorced from practice, thus losing its vigor and vitality. Practice would become purposeless and aimless if not guided by correct theories. During the formation and development of the Xi Jinping Thought on Ecological Civilization, it has consistently highlighted practical rationality, emphasized concrete results, and adhered to the practical standard. It aims to effectively transform the world based on the correct perception of it, and upholds the view and basic requirement of "practice first" advocated by Marxist methodology.

The thought sticks to problem orientation. Comrade Xi Jinping said, "Problem is the starting point and motivation of innovation," and "theoretical innovation can only start with problems. In a sense, the process of theoretical innovation is the process of discovering, screening, studying and solving problems." The Xi Jinping Thought on Ecological Civilization has answered on a deep level several major theoretical and practical questions, namely why we should build eco-civilization, what kind of eco-civilization to build and how to build it. It makes a point of studying and resolving the major issues emerging in the course of reform and development, pressing subjects we face nowadays, and hotspot issues people are generally concerned with. The thought is highly targeted and instructive.

The thought values investigation and research. Conducting in-depth investigation and analysis of objective realities is the groundwork for succeeding in all work. Comrade Xi Jinping's important statements about ecological progress and eco-environmental protection are premised on a thorough investigation of the whole facts of matters, an accurate grip over their essence and laws, and an in-depth study of the thoughts and measures for resolving problems. Insights have to be extracted from the facts obtained through investigations in order to promote theoretical innovation.

The thought takes the people as the subjects of practice. Ecological civilization is a cause jointly undertaken by the whole people. The Xi Jinping Thought on Ecological Civilization takes the people as the subjects of practice, thinks and apprehends issues from the people's stance, and reflects the interests and will of the most majority of people. Taking people's aspiration for a beautiful life as the starting point and the goal, it gathers their collective wisdom and taps and fully absorbs the people's new creations and new experience.

The thought insists on plain and honest expressions. Lenin once wrote a simple equation that "Maximal Marxism = Maximal popularity and simplicity." Comrade Xi Jinping is good at extracting important conclusions from specific facts and experience, and his important statements on ecological progress are all colloquial and vivid. "Protecting and improving the eco-environment is protecting and developing productivity," "clean waters and green mountains are invaluable assets," "a sound eco-environment is the most equitable public product and the most inclusive wellbeing for the people," "we must protect the eco-environment as preciously as we protect our eyes and treat it as dearly as we treat our lives," and "a good environment is part of the public's wellbeing, green mountains and blue skies bring delight and happiness to the people"...these viewpoints, important as they are, are all conveyed in simple, direct and colloquial languages so that they are easily understood and applied by the people.

3 Deeply Grasping and Comprehensively Implementing the Practical Requirements of Xi Jinping Thought on Ecological Civilization

To study and practice the Xi Jinping Thought on Ecological Civilization, it is necessary to deeply understand its practical character and requirements and combine what we learn with what we think and do. We must conscientiously arm our minds with this thought and use it to guide our practice and work, and combine the thought with actual conditions and apply it for tangible effects when promoting innovations in ecological work.

Developing eco-civilization is a long-term, complicated, arduous and systemic campaign. At the National Conference on Eco-environmental Protection, General Secretary Xi Jinping called for accelerating the establishment of an eco-civilization system. This includes the ecological culture system with ecological values as the norm, ecological economy system featuring industry-ecology integration, target and responsibility system centered on improving the eco-environmental quality, ecological institution guaranteed by the modernization of governance system and capability, and ecological security system focused on sound ecosystem circulation and effective prevention and control of environmental risks. He, by combining theories with practices, made overall arrangements and specific requirements on building eco-civilization.

Ecological progress has to be promoted comprehensively, but it should also be implemented step by step with priorities according to the different conditions in different periods. In my opinion, to comprehensively implement the Xi Jinping Thought on Ecological Civilization, we must give priority to the following practical requirements.

First, we should intensify the Party's leadership and take up the political responsibility for advancing ecological progress. The Party leads our work on all fronts nationwide. The Party's leadership is the fundamental political guarantee for reinforcing ecological advancements and eco-environmental protection. The Xi Jinping Thought on Ecological Civilization profoundly embodies the CPC's initial aspiration and mission of striving for happiness for the Chinese people and rejuvenation of the Chinese nation. Since the 18th CPC National Congress, our ecological efforts have seen historical transformation and made unprecedented achievements, which should essentially be attributed to the firm leadership of the CPC Central Committee with Xi Jinping at the core. To put the Xi Jinping Thought on Ecological Civilization into practice for tangible effects, the key lies in enhancing the Party's self-development, reinforcing its leadership, and resolutely shouldering the political responsibility for eco-civilization development. This, first of all, accurately reveals the foundation and source of power of the CPC from a historical perspective, and it also, in view of China's current development stage and the people's demand for a beautiful environment in the new era, showcases the CPC's commitment to its purpose and meets the people's expectations for a beautiful life. CPC committees and governments of all levels should strictly implement the requirement of "equal responsibility in Party and government positions," and urge the leading officials to earnestly fulfill their responsibility for ecological progress. The overall design, organization and leadership of ecological work must be strengthened, major issues should be addressed in a coordinated way, and efforts should be made to guide, push and supervise how the local areas and departments have carried out the decisions, deployments, and policy measures made by the CPC Central Committee and the State Council.

We need to take resolute steps to win the tough battle of pollution prevention and control to lay a solid foundation for building the eco-civilization. With the sustained and rapid development of Chinese economy and the considerable improvement of people's lives, the people have an ever-growing demand for ecological products such as clean air and water, safe food and a good environment. To win this battle, we must concentrate on tackling major eco-environmental problems, particularly the improvement of eco-environmental quality and the handling of prominent issues that have caused serious public complaints, such as the pollution of air, water and soil. Aiming at obvious improvement of air quality, we should step up joint prevention and control, basically eliminate heavily polluted days, and bring back blue sky, white clouds and starry nights for the people. The action plan for water pollution prevention and control will be implemented strictly to ensure safe drinking water, basically remove fetid water bodies in cities, and bring back green-banked clear water with fish swimming inside. The action plan for soil pollution prevention and control will be implemented comprehensively to ensure a safe living environment for the public, with the emphasis on key areas, industries and pollutants and on the control and recovery of soil pollution. The rural environment rectification campaign will be implemented continuously to create a beautiful countryside and retain the pastoral landscape of singing birds and fragrant flowers for rural residents.

We must further reform the eco-environmental supervision system and protect the eco-environment with the most rigorous laws and regulations. General Secretary Xi Jinping said, "Only by implementing the most rigorous laws and regulations can we provide reliable guarantee for ecological advancement." Since the 18th CPC National Congress, the CPC Central Committee with Xi Jinping at the core has accelerated the reform of eco-civilization system on all fronts, which comprises such aspects as property right of natural resource assets, spatial development and protection, spatial planning, resource aggregate management and all-round conservation, paid use of resources and ecological compensation, environmental governance, eco-protection market, and ecological work assessment and accountability. During the organizational reform of the Party and the nation in 2018, the State Council set up the Ministry of Natural Resources and the Ministry of Ecology and Environment. This has effectively rectified the chronic multi-point leadership and governance over China's eco-environmental work, clarified the duties of stakeholders and supervisors, and straightened out the administrative system and mechanism for eco-civilization, turning a new page in our efforts for ecological progress. At present, most of the outstanding environmental problems in China result from incomplete systems, the lack of appropriate rules and enforceable laws, inadequate implementation, and ineffective punishment. We must speed up institutional innovation, and establish a complete eco-civilization system featuring clear-cut property right, multi-party engagement, and equal emphasis on incentive and restriction. We should also improve a law-based system characterized by strict front-end prevention, process control and back-end punishment, so as to remove the institutional and mechanism obstacles that impede ecological progress. The top priority now is to earnestly implement the reform measures already adopted to promote eco-civilization, work out new reform plans in a timely manner, and push for the modernization of our ecological governance system and capability, including perfecting the natural resource asset management system and reforming the eco-environmental monitoring system. Regulation should implemented strictly as an impregnable restraint and deterrent. Anyone who has crossed the line on eco-environmental protection should be punished. This has drawn a clear and definite bottom line that should never be crossed.

We must lead our ecological work with innovation and give full play to science and technology as the primary productive force. Comrade Xi Jinping stressed that "green, circular and low-carbon development is the direction of technological and industrial revolution in modern times and is the most promising sector. China has great potential in this sector and can create many new economic growth points." Ecological progress not only needs theoretical and institutional innovation, but also, and more importantly, needs to be powered by green technologies. Green tech innovation is vital for pollution control, energy mix optimization and ecological recovery because it provide inexhaustible driving force. We need to foster a market-oriented green tech innovation system, develop green finance centered on the directions and priorities of strategic emerging industries, and expand the industries of energy conservation and environmental protection, clean production, and clean energies. The concentration of innovative resources, which mainly refer to innovative institutions, talents and capital, should be taken as a key and strategic measure toward that end.

We must promote green development across the board and shift to a green development, productive and living mode more quickly. Efforts should be made to speed up the establishment of the legal system and policy orientation in favor of green production and consumption, and to set up and improve the economic system of green, low-carbon and circular development. We need to foster a market-oriented green tech innovation system, develop green finance, and expand the industries of energy conservation and environmental protection, clean production, and clean energies. We need to advance the revolution of energy production and consumption, and build a clean, low-carbon, secure and efficient energy system. We should also promote the all-round conservation and circular utilization of resources, launch the national water conservation program, reduce energy and material consumption, and realize the circular connection between the production system and living system.

Comrade Xi Jinping stressed, "Developing eco-civilization is closely related with everyone, and everyone should practice and promote it." In developing eco-civilization, no one is a bystander or an outsider. We need to strengthen the publicity and education of eco-civilization, enhance the public's environmental awareness, and create a public action system to urge the formation of a civilized and healthy lifestyle and consumption mode that is moderate, green and low-carbon. We should oppose luxury, waste and irrational consumption, carry out a series of initiatives such as thrifty government organs, green families, schools and communities, and green travel. These efforts are intended to foster a favorable social atmosphere joined and contributed by the whole society and to make the "beautiful China" campaign a conscientious action of all Chinese.

Governance Dimension of Xi Jinping Thought on Ecological Civilization

Zhen Lin

It is the CPC's unwavering aspiration and mission to build China into a great modern socialist country that is prosperous, strong, democratic, culturally advanced, harmonious, and beautiful by the middle of the twenty-first century. Modernizing the systems and capabilities of ecological governance is the only way and necessary guarantee for building a beautiful China. Since the People's Republic of China was founded 70 years ago, China has explored many innovative experiences in ecological governance, blazed a trail of ecological progress with Chinese characteristics, developed the Xi Jinping Thought on Ecological Civilization, and made historic achievements in eco-environmental protection. The formation of the Xi Jinping Thought on Ecological Civilization is attributed to the General Secretary's experience in eco-environmental construction and protection and the lessons he has drawn from it, and also to his systematic thinking on eco-civilization and sustainable development. On the dimension of modernizing the ecological governance system and capability, this paper reviews Xi Jinping's green governance experience when he was working as the head of local governments, elaborates on the connotations of the eco-governance system, and offers suggestions on enhancing the governing capability.

1 Xi Jinping's Experience in Local Green Governance

1. **Liangjiahe: the first methane tank in northern Shaanxi**

In January 1969, the 15-year-old Xi Jinping left Beijing to the loess plateau in northern Shaanxi that was known for its harsh natural environment. From 1969 to 1975, Xi stayed in Liangjiahe for seven years, a period that changed his life. Xi once recalled with deep feelings that "when I came to the loess at the age of 15,

Z. Lin (✉)
School of Humanities and Social Sciences, Beijing Forestry University, Beijing, China

J. Pan et al. (eds.), *Beautiful China: 70 Years Since 1949 and 70 People's Views on Eco-civilization Construction*, https://doi.org/10.1007/978-981-33-6742-5_20

I felt bewildered and confused; but when I left there at the age of 22, I had a firm goal in life and was confident to achieve it. As a servant of the people, I had my root in the plateau because it nurtured my conviction to do something concrete for the people. Wherever I go, I'm always the son of that loess land." Wu Hui, a villager at Liangjiahe, recalled that "1972 and 1973 marked a turning point in Jinping's life… I remember him talking to me, 'why cannot I take root in Liangjiahe? Why cannot I stay here and do something good for the people? I will go my own way, do my own things!' Jinping's world outlook and values already took shape back then."

During that period, a milestone event took place in the environmental protection history of China and the world. In June 1972, China assigned a delegation to participate in the first UN Conference on Human Environment held in Stockholm, the capital of Sweden. In August 1973, the first National Conference on Eco-environmental Protection was held in Beijing, lifting the curtain on our environmental cause. The conference determined the principles for environmental protection, namely "overall planning, rational layout, comprehensive utilization, turning hazards to treasures, relying on and mobilizing the people, protecting the environment, and benefiting the people." Xi Jinping, who was in the habit of reading books and newspapers, knew about these major events and was more or less influenced by them.

In January 1974, a piece of news about methane development in Sichuan published by the *People's Daily* drew Xi Jinping's attention. Immediately after the Spring Festival, he borrowed money and went to Sichuan to learn about the methane technology. When he came back to Yanchuan, he, in consideration of local climate and other conditions, organized research, experiments and construction, and succeeded in building Shaanxi's first methane tank in Liangjiahe. After the first methane bulb was lit up, he led the villagers to build tens of methane tanks, which basically met the cooking and lighting demand, and he himself became a well-known "specialist" in methane tank construction. Motivated by him, the Yanchuan County quickly had a surge of building methane tanks that was known as an "energy revolution." By September 30, 1975, the county had more than 3200 methane tanks distributed in all 15 communes, and all the 47 production brigades used methane.

2. Zhengding: we'd rather give up economic benefits than cause pollution

In April 1982, Xi Jinping was transferred to Zhengding County of Hebei Province and became secretary of the county's CPC Committee in July 1983. In the early 1980s, China's environmental protection entered a new stage. In March 1982, the Ministry of Urban–Rural Construction and Environmental Protection was formed, under which there was the Environmental Protection Bureau. On December 31, 1983, the second National Conference on Eco-environmental Protection was held, which made environmental protection a basic state policy. At that conference, Yu Guangyuan, vice president of the Chinese Academy of Social Sciences, put forth the vision of "big cruciform agriculture" that was similar to the integration of the primary, secondary and tertiary industries today, as well as the idea of building ecological counties and provinces. Having inspected and studied Zhengding's conditions deeply, Xi Jinping proposed the new strategy of "accelerating 'two transformations' by leveraging on cities and introducing talents." He invited Yu Guangyuan and other experts as advisors, envisioned "big agriculture" based on local realities, and adopted the "semi-suburban" economic mode. He led the county to establish a reasonable and balanced economic structure and a production structure of sound circulation, raise land utilization rate, and strive at agro-ecological balance and the unity of ecological and economic targets. As a result, Zhengding's history as a high-yield poor county came to an end. Considering that the development of township enterprises and commodity economy may cause environmental pollution, Xi Jinping presided over the formulation of the *Master Plan for Economic, Technological and Social Development of Zhengding County* in 1985, which set the basic goals in environmental protection by the end of the twentieth century—prohibiting destruction of natural environment, preventing new pollution and addressing existing pollution sources. The *Master Plan* specifically stressed that "we'd rather give up economic benefits than cause pollution; we will prohibit pollution transfer to the countryside." The *Master Plan* also prioritized forestry development as a key to developing eco-agriculture and maintaining eco-balance, and laid down concrete measures for economizing on water and protecting underground water resources, conducting energy research, and promoting new energies.

3. **Xiamen: protecting natural resources, building a beautiful city**

From June 1985 to June 1988, Xi Jinping was successively the member of the standing committee of CPC Xiamen Committee and deputy mayor and executive deputy mayor of the city. As the leader, trailblazer and builder of the newly established Xiamen Special Economic Zone, Xi paid close attention to the top-level design for its development, and presided over the formulation of the *Economic and Social Development Strategy of Xiamen 1985–2000*, which pointed out the development direction for the city. In the meantime, he attached great importance to the protection of eco-environment and historical culture during urban construction. He took a clear-cut stance on the overexploitation of natural resources then. "Can we pursue construction at the price of local destruction? I think the answer is clear. Xiamen should not pursue development at such a price." "Destruction because of ignorance is no longer the main problem. The current tendency is destruction during construction, which is not necessarily caused by uneducated people, but such a behavior reflects a kind of ignorance, or irresponsibility." He suggested feasible measurers from an overall perspective. "The general principle is that maximal eco-environmental protection should be implemented on the island, and the use of resources outside the island and in the suburb should also be managed, planned and approved more strictly…We used to say that when we live by mountains and waters, we should make use of them as much as possible, but destruction of them must be prohibited resolutely. This is the duty of governments on all levels." Regarding the excessive deforestation on the Wanshi Mountain in Gulangyu, Xi Jinping said many trees on Gulangyu were introduced by overseas Chinese, including a lot of rarities, and called for detailed work to protect these ancient and rare trees. He also instructed on the overall planning of the Wanshi Mountain area. "There are very few places where the natural landscape and cultural landscape exist in harmony. It is necessary to view Gulangyu as a national treasure and plan its construction and protection with that in mind." Regarding the water pollution of Yundang Lake, Xi Jinping chaired a special meeting on its overall governance on March 30, 1988, which kicked off a massive and tough campaign to address the environmental pollution in Xiamen. The meeting decided to form a comprehensive governance mechanism, set up a leading group of Yundang Lake governance comprising relevant departments and experts, and creatively put forth the principles of "law-based governance, pollution interception, de-sludging and bank construction, introducing fresh water, and beautifying the environment." Xi Jinping's providence and perseverance laid a solid foundation for building Xiamen into an international modern garden city that boasts blooming economy, civilized society, rational layout and beautiful environment.

4. **Fujian: from poverty alleviation through ecological development to building an ecological province**

From 1988 to 1990, Xi Jinping was secretary of the CPC Ningde Committee. To change the backward outlook of east Fujian, he proposed to fully utilize the local natural resources, saying that "the forest is also a reservoir, a coffer and a grain warehouse." Seeing that mountains and forests held the key to east Fujian's economic development, he took forestry development as of strategic importance for reviving the region, while stressing that "this kind of development is not one-dimensional, but comprehensive. We cannot simply pursue economic benefits, but aim at coordinated social, economic and ecological benefits." When working as secretary of the CPC Fuzhou Committee, Xi proposed and presided over the formulation of the *20-year Strategic Plan for Economic and Social Development of Fuzhou*, which set out the strategic goals, steps, deployments and priorities of the city's socioeconomic development for three years, eight years and 20 years, known as the "3820" program. While promoting the leap-frog and unconventional progress of Fuzhou's economy featuring high speed and high benefits, Xi underscored the importance of "urban ecological progress" and vowed to build Fuzhou into an "open coastal city that is clean, beautiful, comfortable and quiet with its eco-environment basically restored in sound circulation."

When working as Deputy Secretary of the CPC Fujian Committee and Governor of Fujian Province, Xi Jinping actively promoted the building of an ecological province. He attached great importance to the overall protection of the Minjiang River basin, and called for greater efforts to protect vegetation and build ecological forests in the upper reaches of the river. He pledged to fundamentally address the pollution of the Xihu Lake to create a beautiful and comfortable living environment for the people. He inquired about and inspected the governance of Mulan Stream multiple times and rooted up the flooding problem there. He also instructed on the control of water and soil erosion in Changting and made "Changting experience" a national paradigm for addressing such problems.

In 1999, Hainan Province took the initiative to decide to build an ecological province. In 2000, Xi Jinping, the then Governor of Fujian Province, also put forth the strategic vision of building an ecological province, and personally headed up the leading group for that purpose and guided the formulation of the *Outline of the Master Plan for Building an Ecological Province in Fujian*. This forward-looking plan maintained the clean waters and green mountains of Fujian and left the most valuable resources and most competitive advantages for the province to become the first national experimental zone for developing ecological civilization.

5. **Zhejiang: promoting ecological progress, building a green province**

In 2002, Xi Jinping was transferred from Fujian to Zhejiang as secretary of the CPC Zhejiang Provincial Committee. Zhejiang's economy was already one of the best in the country then, so how to resolve chronic difficulties and push economic and social development to a new level was what was constantly on his mind. On December 18, Xi Jinping, not even in office for a month yet, presided over the 2nd plenary (extended) session of the 11th CPC Zhejiang Committee, at which he proposed to "build a 'green Zhejiang' at a faster pace with the construction of an ecological province as an important carrier and entry point, and strive to coordinate population, resources and environment." In 2003, Zhejiang officially launched the ecological province initiative, which Xi Jinping demanded be carried on year after year, term after term, in the endeavor to succeed in making Zhejiang an ecological province with economic prosperity, natural beauty and culturally advanced society. He published an article titled "Civilization Prospers when Ecology Prospers—Promoting Ecological Progress, Building a Green Zhejiang" on the 13th issue of *Qiu Shi* magazine that year, in which he made the famous assertion that "civilization prospers when ecology prospers, civilization declines when ecology declines." Xi regarded promoting ecological progress and developing green Zhejiang as a massive systematic program concerning the overall situation, and a program with popular support that will benefit generations to come. At the 4th plenary (extended) session of the 11th CPC Zhejiang Committee held in July, Xi Jinping made the significant decision and deployments to further exert "advantages in eight aspects" and implement "measures on eight fronts," including further exerting Zhejiang's ecological advantages and building an ecological and green province. Since then, the "Double Eight Strategy" has become the roadmap for Zhejiang's comprehensive and deepened reform.

During his term in Zhejiang, Xi Jinping's thinking of eco-civilization was deeper and more multi-faceted. First, the theory of "lucid waters and lush mountains are invaluable assets" was formed in this period. When he inspected Yucun Village in Tianhuangping Town of Anji County on August 15, 2005, Xi said "ecological resources are the most valuable resources, and lucid waters and lush mountains are invaluable assets. We should never promote economic growth by sacrificing the environment. We cannot have it both ways, so we must know what to give up and what to adhere to, and pursue the development path of man-nature harmony." Later he elaborated multiple times on the relation between "mountains and waters" and "invaluable assets," recognizing their conflicts and dialectical unity. In practice, the understanding of this relation went through three stages from excessive exploitation to protection and eventually to win–win development. Second, environmental governance efforts were coordinated between urban and rural areas. Based on in-depth inspections, Xi Jinping found that "indoor modernization, outdoor mess" and "garbage and sewage everywhere" was still a common phenomenon in the countryside. Therefore, he personally directed the rural environmental governance program across Zhejiang Province and followed up on its implementation by holding an annual on-site meeting to give instructions and make case studies. Today this demonstration program has been promoted nationwide as a paradigm. Third, the systematic outlook on building ecological province was established. Xi Jinping urged the establishment of five main systems underpinning the construction of an ecological province, namely an eco-economic system centered on circular economy, a natural resource guarantee system that is sustainable, an eco-environmental system with beautiful landscape, a demographic ecosystem consistent with the bearing capacity of resources and environment, and a scientific and efficient supportive system. Besides, Xi also paid attention to coordinated land and sea governance, eco-friendly water governance, and institutional and cultural development.

2 Governance Dimension of Xi Jinping Thought on Ecological Civilization

The National Conference on Eco-environmental Protection held on May 18, 2018 marked the establishment of the Xi Jinping Thought on Ecological Civilization. At the conference, General Secretary Xi demanded all parties to abide by the following six principles when promoting ecological progress in the new era. (1) Uphold harmonious co-existence between man and nature, and prioritize conservation, protection and the restoration of nature. Protect the environment as we protect our eyes, cherish it as we treat our own lives, so that the beautiful nature and ecology will last forever in serenity and harmony. (2) Lucid waters and lush mountains are invaluable assets. Implement the concept of innovative, coordinated, green, open and shared development, and work faster to develop spatial layouts, industrial structures, and way of production and way of life that help conserve resources and protect the environment, so as to leave time and space for natural and ecological restoration. (3) A sound eco-environment is the most inclusive form of public welfare. We should protect ecology for people's benefits, work hard to address outstanding environmental problems detrimental to public health, and keep meeting people's growing needs for a beautiful environment. (4) Mountains, rivers, forests, farmland, lakes and grasslands are a community of life. An overall plan must be made that takes all relevant factors into consideration, and multiple measures should be adopted simultaneously to advance eco-environmental progress in all respects. (5) Eco-environment should be protected with the most rigorous laws and regulations. Institutional innovation should be accelerated and implementation strengthened to make regulations the real deterrent. (6) Committed to jointly promoting a global eco-civilization, China should get heavily involved in global environmental governance, help form global solutions to eco-environmental protection and sustainable development, and play a constructive role in international cooperation on climate change. General Secretary Xi made a special point of speeding up the creation of the eco-civilization system and establishing and improving the ecological culture system with ecological values as the norm; the ecological economy system featuring industry-ecology integration; the target and responsibility system centered on improving the eco-environmental quality; the ecological institution guaranteed by the modernization of governance system and capability; and the ecological security system focused on sound ecosystem circulation and effective prevention and control of environmental risks.

The 3rd plenary session of the 18th CPC National Congress made improving and developing the socialist system with Chinese characteristics and the modernization of national governance system and capability the general goal of comprehensively deepening the reform. The national governance system refers to the system of people governing the country under the Party's leadership, which is a complete set of closely linked and coordinated systems, covering the institutions, mechanisms, laws and regulations concerning economy, politics, culture, society, eco-civilization, Party building and many other aspects. The national governance capability refers to the nation's capability of applying various systems to administrate all kinds of social affairs. General Secretary Xi Jinping pointed out that since the reform and opening, the Party has realized and underscored the fundamental and comprehensive nature of the leadership and organizational systems as well as the importance of keeping them stable and consistent. The national governance system and capability, which reflect a nation's institution and its ability to carry it out, complement each other. The modernization of this governance system and capability is a massive process. It is set to make the socialist system with Chinese characteristics more mature and consolidated, so as to put in place a whole set of more advanced, stable and effective systems to advance the Party's and the nation's causes and promote the people's welfare, social harmony and stability, and the nation's lasting peace. It is also aimed to make the administration of all matters of the Party, the state and the society more institutionalized, standardized and procedure-based, so as to continuously enhance the ability to effectively run the state through the socialist system with Chinese characteristics.

The modernization of governance system and capability in the ecological sector is an important component of the overall modernization of state governance system and capability. Since the 18th CPC National Congress, General Secretary Xi Jinping has personally directed and promoted the system development and institutional reform of eco-civilization, which has generated historical achievements and brought about unprecedented changes in our eco-environmental outlook.

There are different angles to comprehend the ecological governance system. The governing subject generally includes the state (government), market (enterprises) and society (the public). In China, eco-civilization is carried out under the centralized and unified leadership of the Party, with the Party committees and governments on various levels playing a leading role, and different departments performing their duties in decision making, execution and supervision respectively. Enterprises should take conscientious actions and exercise self-discipline while giving full play to the market mechanism and sci-tech innovation, and contribute to energy conservation and emission reduction. Social organizations and individuals should be firmly aware of conservation and eco-environmental protection, practice a green lifestyle and actively play their supervisory role, so as to create a social atmosphere where eco-civilization is advocated by all, in everything, at all times. As far as the object of governance is concerned, ecology is a unified natural system, and mountains, rivers, forests, farmlands, lakes and grasslands are a community of life. Following the inherent ecological laws, we must set about ecological progress in all respects by making an overall plan taking into consideration all relevant factors and adopting multiple and coordinated measures.

On the vertical coordinate, the eco-governance system consists of six layers—global, regional (international), national, regional (provincial), local and grassroots (community), with the establishment of national eco-civilization system in the center. On the lateral coordinate, China, following the thought of "strict front-end prevention, process control and back-end punishment," establishes a complete eco-civilization system featuring clear ownership of property right, multiple participants and equal emphasis on incentive and restraint, as well as a legal system that effectively restricts development behaviors and promotes green, circular and low-carbon development, in a bid to ensure ecological progress with the most rigorous laws and regulations. The *Master Plan for Institutional Reform for Ecological Civilization* issued in 2015 pledged to shape a complete eco-civilization institution comprised of eight systems by 2020, covering property right of natural resource assets, spatial development and protection, spatial planning, resource aggregate management and all-round conservation, paid use of resources and ecological compensation, environmental governance, eco-protection market, and ecological work assessment and accountability.

To modernize our ecological governance capability, the CPC committees and governments on all levels should comprehensively strengthen their capability of systematic, cooperative, scientific, law-based and sustained governance, focused on the five aspects of top-level design, overall arrangement and coordination, tackling key difficulties, institutional development, and the cultivation of eco-culture.

To strengthen the capability of systematic eco-governance, we have to recognize that ecological progress is a systematic project whereby any detail relates to the whole picture, so the top-level design has to be scientifically planned to advance various tasks in parallel and integrate diverse regulations. We should recognize that mountains, rivers, forests, farmlands, lakes and grasslands are a community of life, in which the lifeline of humans rests with farmlands, that of farmlands with water, that of water with mountains, that of mountains with earth, and that of earth with forests and grasslands. Ecological governance calls for holistic, coordinated and balanced measures, which requires us to employ economic, administrative, engineering, technological, legal and publicity approaches comprehensively instead of making one-sided efforts. We need to keep in mind that man has only one earth and all countries exist in a community with a shared future. As such, we should make overall considerations for domestic and foreign factors and actively lead the global ecological efforts.

To strengthen the capability of cooperative eco-governance, we have to insist that a beautiful eco-environment is jointly built, protected, governed and shared by all. Within the government, different departments, levels and regions should cooperate more closely to exert the positive energy of solidarity and synergy and avoid buck-passing and struggle for power and interests. On the international level, different countries and international organizations should communicate more, share experience and jointly deal with challenges in such areas as climate change, ecological restoration, energy conservation and emission reduction, and biodiversity preservation. They should make unremitting efforts to blaze a trail of civilization development featured by advanced production, affluent life and sound ecology, and make joint contributions to achieving the UN's 2030 sustainable development goals.

To strengthen the capability of scientific eco-governance, we have to abide by ecological principles as well as natural and social development laws, respect facts, make scientific plans and take coordinated measures in view of local conditions, so as to build an ecosystem where all living things co-exist in harmony. We must uphold the green, low-carbon and sustainable development philosophy, respect, accommodate and protect nature, and break a new horizon in the modernization drive featured by the harmonious development between man and nature. Efforts should be made to improve the layout of land and space development, strictly control development intensity and hold the eco-protection red line, so as to come at a situation of intensive and efficient productive space, moderate and hospitable living space, and beautiful and comfortable ecological space.

To strengthen the capability of law-based eco-governance, we have to establish the systems, laws, standards and procedures for ecological work, keep improving the legal system and intensify the execution of laws and regulations, so as to make them the impregnable restraint and deterrent. Steps should be taken to impose on producers more legal responsibilities for environmental protection and considerably increase the cost of law violation. A target and responsibility system centered on improving eco-environmental quality and a scientific and reasonable performance assessment system should be established to truly and strictly hold accountable those responsible for eco-environmental damages, for life.

To strengthen the capability of sustained eco-governance, we have to recognize that developing eco-civilization is a long and hard journey that entails a fighting spirit and perseverance year after year, generation after generation. Only by doing that will we achieve the grand goal of building a great modern socialist country that is prosperous, strong, democratic, culturally advanced, harmonious and beautiful by the middle of the twenty-first century.

Theoretical Character and Practical Guidance of Xi Jinping Thought on Ecological Civilization

Maoxing Huang

Ecological civilization is the imperative need for man-nature harmony, and building ecological civilization is vital to sustaining the Chinese nation's development. At the National Conference on Eco-environmental Protection held in May 2018, General Secretary Xi Jinping gave an important speech from the perspective of the people's wellbeing, the nation's future, and the general undertakings of the Party and the country. He comprehensively summarized China's historical achievements and changes in ecological progress and eco-environmental protection since the 18th CPC National Congress, profoundly explained the significance of intensifying ecological efforts, and put forth the important principles, requirements, paths and measures that have to be followed in this cause. His remarks have presented the fundamental philosophy of strengthening socialist ecological progress, and have led the Chinese nation forward on the great expedition of revival aimed at a modern socialist power. General Secretary Xi's profound historical vision, broad world outlook, and insightful summary and thinking of the experience and lessons of human civilization development have greatly enriched and developed the Marxist ecological outlook. They are important contents of the Xi Jinping Thought on Ecological Civilization and lead the way in advancing eco-civilization in the new era.

Ye Qi contributes to this article.

M. Huang (✉)
School of Economics, Fujian Normal University, Fuzhou, China

J. Pan et al. (eds.), *Beautiful China: 70 Years Since 1949 and 70 People's Views on Eco-civilization Construction*, https://doi.org/10.1007/978-981-33-6742-5_21

1 Major Theoretical Innovations of Xi Jinping Thought on Ecological Civilization

General Secretary Xi Jinping inherits the Marxist ecological thought with innovations and summarizes and sublimates the man-nature relation that's at the core of ecological advancement. Taking "harmonious co-existence between man and nature" as a key component of building socialism with Chinese characteristics in the new era, he has systematically answered why we should develop eco-civilization, what kind of eco-civilization to build and how to build it, and developed a multi-dimensional ideological system of eco-civilization.

1. **Eco-civilization carries the historic mission of the great national rejuvenation**

A great rejuvenation is the greatest dream of the Chinese nation in modern times, but the path to rejuvenation is strewn with difficulties and challenges, including protecting the eco-environment, dealing with climate change, and ensuring energy and resource security. At the opening ceremony of the Eco Forum Global Annual Conference Guiyang 2013, General Secretary Xi Jinping emphasized that walking into the new era of eco-civilization and building a beautiful China are the important content of realizing the Chinese Dream of the great rejuvenation of the Chinese nation. He urged embedding the promotion of ecological progress in economic, political, cultural and social advancement across the board. When chairing the group study of the political bureau of CPC Central Committee and during inspection tours, General Secretary Xi repeatedly called on officials to have a clear idea of the urgent and arduous task of eco-environmental protection and the importance and necessity of intensifying ecological efforts. He asked them to resolutely address environmental pollution with a keen sense of responsibility. When instructing ecological work at the end of 2016, General Secretary Xi stressed again that ecological progress is a vital part of the Five-Sphere Integrated Plan and the Four-Pronged Comprehensive Strategy. The report of the 19th CPC National Congress uplifted eco-civilization to be a millennial program concerning the Chinese nation's lasting development and added "beautiful" as a new target in building a great modern socialist country. It's clear that the General Secretary has long taken eco-civilization as an important part of the socialist cause with Chinese characteristics and a critical choice regarding China's future. This is a cardinal development and contribution to the theoretical system of socialism with Chinese characteristics.

2. **Eco-civilization consolidates the foundation for achieving national prosperity and strength and people's wellbeing**

National prosperity and strength and people's wellbeing are the foundation for social stability and lasting peace and security, and this foundation has to be consolidated and guaranteed with highly advanced productivity and harmonious and stable social relations. General Secretary Xi Jinping, breaking the traditional mindset that productivity depends on the input of factors and technological innovation, put forth the creative assertion that "protecting and improving the eco-environment is protecting and developing productivity," which is a major innovation of the productivity theory. He also generalized the dialectical unity between economic development and environmental protection. "Lucid waters and lush mountains are invaluable assets," "a sound eco-environment is the most equitable public product and most inclusive wellbeing of the people"… these statements make it clear that environmental protection doesn't necessarily come at the expense of economic development. Ecological progress can benefit the mankind by generating immense ecological, economic and social benefits. Elevating eco-civilization to a political height, General Secretary Xi Jinping said "developing eco-civilization meets the public expectation and improves livelihood," and "the people look forward to the accelerated improvement of eco-environmental quality. We should actively respond to what they think and expect, strongly promote ecological progress, and provide more quality ecological products to keep meeting their ever-growing needs for a beautiful environment." He creatively regarded ecological progress as an important way to resolve the principal contradiction in the current Chinese society and to lead the country and the people to strength and affluence. This is a vivid illustration of his thought on green state governance and ecological outlook on livelihood.

3. **Eco-civilization expands the strategic horizon of eco-environmental protection**

The eco-environment is a compound system comprising man, nature and society, in which any change relates to the whole system. Therefore, we have to protect the eco-environment at root instead of only addressing the symptoms. General Secretary Xi Jinping pointed out that environmental governance is a systematic program, so we have to carry out the key tasks of eco-civilization development following a corresponding thought, namely ecological progress should be embedded in all aspects and the whole process of economic, political, cultural and social advancement. Eco-civilization is the summation of the material, cultural and institutional attainments that man has achieved in civilization development. As an advanced state of civilization, eco-civilization is the consummation and crystallization through the evolution of primitive, agricultural and industrial civilization. General Secretary Xi Jinping also put forth the innovative thought of "community of life" on the eco-environmental front. "The lifeline of humans rests with farmlands, that of farmlands with water, that of water with mountains, that of mountains with earth, and that of earth with forests and grasslands"—the word "lifeline" highlights the interdependence among the factors. The factor of "grassland" was added at the 19th CPC National Congress, thus giving rise to the concept that "mountains, rivers, forests, lakes, farmlands and grasslands are a community of life," which drives home that all factors and links of the ecosystem are indispensable for sustaining life. In our efforts of holistic eco-governance and joint maintenance of eco-balance, such a systematic and comprehensive view breaks away from the traditional narrow perception of eco-environment, and blazes a new horizon and space for us to understand eco-environmental protection and ecological progress. It is good for creating a new situation of our modernization drive featuring harmony between man and nature.

4. **Eco-civilization demonstrates the unremitting human pursuit for sustainable development**

General Secretary Xi Jinping views the issue of eco-civilization from the development perspective. He has summarized the historical experience and lessons gained through human civilization development, and proposed new thoughts for the future. He has deeply realized that "the eco-environmental contradiction in China hasn't appeared overnight, but has been accumulated over a long period." In the face of severe eco-environmental problems, he asked, "the Chinese civilization has lasted for more than 5000 years. Can we sustain it for another 5000 years and beyond?" Eyeing the continuation of human civilization, General Secretary Xi said "What we are doing today on eco-environmental protection will benefit generations to come. It is a long-term campaign that requires unremitting efforts," and "the Party members should have the mind and will" to promote eco-civilization. He has revealed with vision that the eco-civilization campaign is not a phased task, but an eternal mission that concerns and accompanies mankind's sustainable development. With the continuous advancement of our ecological initiative, General Secretary Xi has deepened his understanding of how to succeed in it. The 18th CPC National Congress included ecological progress as part of the Five-Sphere Integrated Plan for the first time, and the 3rd and 4th plenary session of the 18th CPC Central Committee elevated it to the institutional level, vowing to "deepen the reform of eco-civilization institution, quickly set up its pillars, and put ecological work on the institutional and law-based track." Later the 5th plenary session of the 18th CPC Central Committee proposed green development, and the 19th CPC National Congress urged the accelerated reform of eco-civilization institution. The constantly enhanced perception of ecological progress and the innovation of this thought are leading the Chinese people toward the new era of socialist eco-civilization.

2 Practical Guidance of Xi Jinping Thought on Ecological Civilization

As China's ecological campaign has come to a critical window period of tackling tough problems, the CPC Central Committee and the State Council, under the guidance of the Xi Jinping Thought on Ecological Civilization, have made a series of major decisions and deployments on ecological work since the 18th CPC National Congress. The top-level design, institutional framework and policy system for ecological progress have been established for the present and beyond, and strong and effective environmental governance initiatives have been carried out with clear priorities, in a bid to systematically advance the eco-civilization institutions. Leading the whole nation in a new and great cause of building a sound eco-environment, the Party and the country have gradually fostered a social movement of "protecting the eco-environment as we protect our eyes, and cherish it as we cherish our lives."

1. **Green development is the new guideline that leads China's ecological progress**

Green development is one of China's five development concepts. At the National Conference on Eco-environmental Protection, General Secretary Xi Jinping described green development as the imperative need for building a high-quality modern economic system and the fundamental measure to address pollution. It is an important path for the Chinese economy to switch to the stage of high-quality development. Given the fact that Chinese economy has entered a new normal, we, on the front of production, have implemented the innovation-driven development strategy, supply-side structural reform and tighter control of excess production capacity. We have also promoted tech innovation, optimized industrial structure, and developed green industry and circular economy, in order to establish a resource-saving and environment-friendly industrial system and a green productive mode. On the front of living, China advocates a consumption concept characterized by civility, thrift, green and low carbon. It urges the people to shift their lifestyle and consumption mode by promoting green travel and garbage sorting, and objects luxury and waste, trying to shape a green lifestyle across the society. Green development has been embedded in every aspect of our economic and social development and become the new path of ecological progress.

2. **Ecological restoration and pollution governance mark a new journey toward a beautiful China**

Eco-civilization development should be people-centric. A beautiful environment of "blue sky, green land and clear water" is the most direct feeling of the people and the goal of the beautiful China initiative. To restore clean water and green mountains, China has put the eco-civilization thought into practice and set about ecological restoration and environmental pollution control by applying targeted measures and solutions. Priority is given to the solution of outstanding eco-environmental problems to achieve the most inclusive public wellbeing. Determined to win the tough battle of pollution prevention and control, China has implemented action plans respectively targeting air, water and soil pollution. It has issued the *Action Plan for Air Pollution Prevention and Control* to address pollution of PM 2.5 and PM 10; the *Action Plan for Water Pollution Prevention and Control* to ensure safe drinking water; and the *Action Plan for Soil Pollution Prevention and Control* that targets rural land pollution to ensure the quality of agricultural products and the security of living environment. China has comprehensively deepened the forestry reform to increase forest coverage rate and carbon sink. By actively adjusting the energy mix, China now has the world's largest installed capacity of hydropower, wind power and solar power, with the ratio of clean energy consumption being on the rise, and it has effectively controlled greenhouse gas emission. Moreover, the governments of provinces, cities, districts and lower-level localities across the country have all rolled out targeted eco-environmental actions, including preventing and curbing desertification and protecting biodiversity. A better eco-environment has substantively improved the people's living quality and sense of gain.

3. **Institutional reform for eco-civilization is the new drive of green development and ecological benefits**

The 3rd plenary session of the 18th CPC Central Committee made accelerating the establishment of eco-civilization institution a major strategic deployment to deep the reform on all front. In 2015, the CPC Central Committee and the State Council printed and distributed the *Master Plan for Institutional Reform for Ecological Civilization*, which determined the top-level design for the reform. The 19th CPC National Congress made further deployments to accelerate the reform. Reform is a vital source of power for our eco-civilization development. Since the 18th CPC National Congress, China has launched a string of reform explorations, such as implementing the functional zoning strategy at a faster pace, bettering the layout of territorial space development, and implementing regional policies of classified management. The *Plan for Reforming State-owned Forestry Farms* and *Guiding Opinions on Reforming State-owned Forestry Farms* have been issued to strongly push the reform of state-owned farms, and the plan for piloting the national park system has been released and great efforts made to launch pilot projects, among others. In August 2016, the CPC Central Committee decided to list Fujian, Jiangxi and Guizhou as the first national experimental zones for eco-civilization. In the past two years, the experimental zones have rolled out a series of reform trials in such areas as establishing the eco-civilization responsibility system, improving the territorial development protection system, intensifying ecological supervision and realizing the value of ecological products, with the focus on innovation and provision of systems and exploration for patterns. A batch of typical cases have emerged and imitable and promotable experience has been generated. Practices have proven that the experiment on eco-civilization reform can give the nature protectors a keener sense of gain and the people more ecological benefits.

4. **Environmental protection responsibility and awareness become the new clarion of joint ecological governance of the whole people**

To make leading officials at all levels more aware of and responsible for protecting and developing the eco-environment, China has taken faster steps of institutional innovation and execution, determined to protect the eco-environment with the most rigorous regulations and laws. The 3rd and 4th plenary session of the 18th CPC Central Committee both proposed to put in place a lifetime accountability system for eco-environmental damages. The *Methods of Holding Party and Government Leaders Accountable for Eco-environmental Damages (Trial)* was also issued to impose environmental restrictions on all leading officials by underscoring matched authority and duty, equal responsibility in the Party and government, and accountability for both behaviors and consequences. China has improved the economic and social development assessment and evaluation system to gradually abandon the traditional GDP-oriented assessment and highlight the ecological orientation by incorporating indicators that reflect the status of ecological progress, such as resource consumption, environmental damages and ecological benefits. The central government has intensified environmental inspection of local areas. In 2017, a nationwide environmental protection inspection was completed to spur local governments to earnestly perform their environmental duties. Moreover, the National Conference on Eco-environmental Protection proposed to establish and improve more quickly an ecological culture system with ecological values as the norm and form a soft restriction system for eco-protection across the society. These strict inspection and accountability systems and the guidance of eco-cultural values have, both on the hard and soft side, reinforced the leading officials,' enterprises' and social entities' awareness of their ecological responsibilities, prompting them to shift from passive participation to profound understanding and conscientious action.

3 Global Contributions of Xi Jinping Thought on Ecological Civilization

The Xi Jinping Thought on Ecological Civilization is an incisive and concise summary of the laws governing the man-nature relation in the new era. This open and inclusive thought condenses the "Chinese experience" gained through long years of ecological efforts, not only drawing out the blueprint and strategic guidance for building a beautiful China, but also contributing the "Chinese plan" to world sustainable development. It embodies China's mission of and responsibility for maintaining global ecological security and facilitates the building of a community with a shared future for mankind.

1. **Xi Jinping thought on ecological civilization has awakened the global ecological awareness**

The Xi Jinping Thought on Ecological Civilization breaks the bondage of material interests, and systematically analyzes the origin of eco-environmental problems from the perspective of human values as well as the logical system of their mutual effects. Being theoretically advanced, convincing and instructive on actions, it has awakened the global awareness of eco-environmental protection and guided all people in the world to act conscientiously and correspondingly. The concept of eco-civilization drew global attention immediately after it was proposed. In February 2012, the 27th Session of the United Nations Environment Programme (UNEP) Governing Council passed the draft decision to promote this China-proposed concept, kicking off its global implementation. In May 2016, the UNEP released a reported titled *Green Is Gold: The Strategies and Actions of China's Ecological*, which indicated that China's eco-civilization endeavours had been recognized and endorsed in the international community and provided experience and references for other countries to deal with economic, environmental and social challenges.

2. **Xi Jinping thought on ecological civilization has enriched the global environmental governance system**

As the world's second largest economy and a main greenhouse gas emitter, China has taken an active part in global cooperation on environmental governance in the spirit of building the "community with a shared future for mankind." It has not only served as a bridge connecting developed and developing countries, but also worked hard to balance the interests of major countries, including carrying out climate diplomacy with major countries and striving to promote and lead the creation of a fair and reasonable global climate governance system aimed at win–win cooperation. China has played a vital role in final signing of the *Paris Agreement*. At the G20 Hangzhou Summit, China and the US took the lead in ratifying the document, giving a strong boost to its official signing. When the US President Donald Trump suddenly announced to withdraw from the *Paris Agreement* on June 1, 2017, China expressed its firm position to resolutely fulfill the document. It has also strengthened the South-South cooperation on climate change, and set up the RMB 20-billion China South-South Climate Cooperation Fund to support other developing countries in this field and consolidate solidarity among them. China has pushed for the perfection of international environmental governance system and rules. It signed and ratified a number of important environmental treaties, including the *Paris Agreement*, the *Minamata Convention on Mercury*, *Montreal Protocol on Substances that Deplete the Ozone Layer*, and the *Nagoya Protocol on Access to Genetic Resources and the Fair and Equitable Sharing of Benefits Arising from their Utilization*, setting a good example in the world.

3. **Xi Jinping thought on ecological civilization offers the Chinese plan for ecological progress in other countries**

China takes it as its duty to safeguard global eco-security. As the largest developing country in the world, it has made the solemn commitment to making special contributions to the solution of global ecological issues. The Xi Jinping Thought on Ecological Civilization, which is highly open and inclusive, is continuously delivering the Chinese plan worldwide. At the Belt and Road Forum (BRF) for International Cooperation in 2017, General Secretary Xi Jinping pledged that "we will pursue the new vision of green development and a way of life and work that is green, low-carbon, circular and sustainable. Efforts should be made to strengthen cooperation in ecological and environmental protection and build a sound ecosystem so as to realize the goals set by the 2030 Agenda for Sustainable Development." China has also provided ecological assistance to other countries, including those along the Belt and Road, and issued the *Guiding Opinions on Promoting the Construction of a Green Belt and Road*. It is committed to creating the cooperation and exchange system, supportive and service platform, and industrial and technological cooperation base for eco-environmental protection to jointly prevent environmental risks. China is working with Belt and Road countries to establish an interconnected environmental impact assessment mechanism, whereby they will share green benefits through industrial output, investment and construction, and technological cooperation. Embedding eco-civilization and green development in the green Belt and Road Initiative not only demonstrates China's fulfillment of responsibility as a regional paradigm in the field, but also highlights the sacred mission of building the Belt and Road into a community of shared interests, responsibilities and future.

On the Value Implications of the Outlook on Sustainable Development of Xi Jinping Thought on Ecological Civilization

Shinan Fang

From the strategic perspective of sustainable development of the Chinese nation and mankind, General Secretary Xi Jinping has systematically expounded on the value and vision of stepping up the building of an ecological civilization to promote harmony between human and nature and achieve sustainable development. This not only fully demonstrates the courage and grit to deal with various types of environmental risks and crises such as global shortage of resources and energy, climate change, natural disasters, desertification, and nuclear proliferation, but also calls for solid action to protect people's ecological rights and meet their expectations for a better life and ultimately build a beautiful China and a beautiful world. Xi's discourses in this regard form the outlook on sustainable development, which is part of his thought on ecological civilization. From an axiological perspective, this new outlook on development is marked by a rich set of values, effective coordination between diversified values, and a turnaround in the way the values are realized.

1 The Outlook on Sustainable Development, as Part of Xi Jinping Thought on Ecological Civilization, Sheds New Light on Development and Has Profound Value Implications

As development is for the purpose of realizing values, sustainable development is to realize sustainable value through development amid harmony between human and nature. The outlook on sustainable development contains well-thought-out views on what is the value of development, why we need to pursue the value of development, and how we can actualize it. It is essentially a new vision that represents fundamental

S. Fang (✉)
Soochow University, Suzhou, China

J. Pan et al. (eds.), *Beautiful China: 70 Years Since 1949 and 70 People's Views on Eco-civilization Construction*, https://doi.org/10.1007/978-981-33-6742-5_22

changes in the way of value realization, from the traditional industrial civilization to ecological civilization for a new era, from every single individual to a community with a shared future for mankind, from the sole pursuit of economic value to the organic combination of political, cultural, social, and ecological value, and from the pursuit of ethnic, regional, and national value to the shared value of mankind. A thorough understanding of these fundamental changes helps us get insight into the outlook on sustainable development.

Since the 18th CPC National Congress, Xi Jinping has delivered a series of important speeches on issues concerning the overall, long-term development of mankind, such as what sustainable development is, why it is emphasized, and how to promote sustainable development. These result in the outlook on sustainable development, a new vision that fundamentally changes the way values are realized and which closely links healthy development for a new era with sustainable development for generations.

As a philosophical concept, value reflects the degree of alignment in the objective, mutually generative, and reciprocal subject-object relations in cognitive and practical activities. It reflects the functions and meanings of the subject and object in the objective relations. Sustainable development is a process in which a full set of diverse, systematic, and long lasting values are pursued and realized. The 18th CPC National Congress looked into the way of value realization by integrating economic, political, cultural, social, and ecological values and incorporated development of an ecological civilization into the five-sphere integrated plan for building socialism with Chinese characteristics, further highlighting the importance of ecological value and development of an ecological civilization. It called on the whole Party and people across the country to accomplish the significant task of "building a beautiful China and realizing the sustainable development of the Chinese nation" In the report delivered to the 19th CPC National Congress on October 18, 2017, Xi Jinping elaborated on the sustainable development of the Chinese nation and proposed the main path of realizing value through sustainable development, that is, "ensuring harmony between human and nature." As he pointed out, "Building an ecological civilization is vital to sustain the Chinese nation's development. We must realize that lucid waters and lush mountains are invaluable assets and act on this understanding, implement our fundamental national policy of conserving resources and protecting the environment, and cherish the environment as we cherish our own lives. We will adopt a holistic approach to conserving our mountains, rivers, forests, farmlands, lakes, and grasslands, implement the strictest possible systems for environmental protection, and develop eco-friendly growth models and ways of life. We must pursue a model of sustainable development featuring increased production, higher living standards, and healthy ecosystems. We must continue the Beautiful China initiative to create good working and living environments for our people and play our part in ensuring global ecological security." At the National Working Conference on Environmental Protection from May 18 to 19, 2018, Xi stressed from a holistic and strategic perspective that "building an ecological civilization is vital for sustaining the development of the Chinese nation." During his inspection of Xiongan New Area on January 16, 2019, he emphasized the importance of developing a well-devised plan for the sustainable development of the new area. "The highest standard of our future life is a pristine environment with blue skies, clear water, and lush trees. Xiongan New Area has a solid foundation and is stepping up efforts in this regard, set to be more livable in the future. By providing a pristine environment, it can realize value and build a good image, making itself a more appealing place," he said.

2 The Outlook on Sustainable Development is a New Vision that Enables Value Realization While Taking into Account Diverse Demands

As a new vision that propels the fundamental changes in the way to realize value, the outlook on sustainable development is established on the basic laws governing the dialectical movement of natural organisms and social organisms that are composed of various value relations. It involves a variety of values and seeks sustainable unity of ecological value and human value, ecological justice and social justice, economic wellbeing and ecological wellbeing, as well as ecological rights and economic rights.

At the core of sustainable development is the relation between human and nature. Nature is the precondition and basis for the survival and development of humans, who depend on the natural environment and create value for both themselves and nature through practical activities to transform the natural environment. It can be said that human and nature can achieve sustainable development through the materialization of their reciprocal relations. Therefore, the sustainable realization of ecological value and human value is a kind of value relation that must be taken into account when it comes to sustainable development. A mutually enhancing relationship between human value and ecological value is both the goal and a measure of sustainable development. People are the most important productive force, a type of value that creates value, and the source of value that promotes the generation and development of all types of value. Hence, development must be people-oriented; that is to say, it must center on the realization of human value. However, human value cannot be created and realized independently; instead, it is realized in the interaction with nature. In that process, due consideration must be given to the value of the natural environment on which humans depend for their survival and development. If the value of the natural environment is underestimated, overlooked, or ignored, human value will become something of an abstraction and nothingness; conversely, if ecological value is overemphasized while human value is ignored, humans will lose their status as value creator for sustainable development, and the ultimate significance of value acquisition will thus fall apart.

Xi Jinping noted, "A civilization may thrive if its natural surroundings thrive, and will suffer if its natural surroundings suffer." This speaks to the intrinsic connection between the natural environment and civilization and between ecological value and human value. When a civilization and its natural surroundings thrive, both human value and ecological value are realized; when a civilization suffers because of its worsening natural surroundings, both human value and ecological value are lost. For this reason, Xi calls on everyone to "protect our natural environment in the same way we would protect our own wellbeing."

The outlook on sustainable development seeks sustainable unity of ecological justice and social justice. To pursue and achieve justice is the primary value of the socialist system. Justice, as a goal of human society, is a state of relative fairness achieved by coordinating all sorts of interest relations through regulations and ethics. Just as people have diverse relations and attributes, justice also enjoys variety, and it takes the form of ecological justice in the domain of human-nature relations. To put it another way, ecological justice is in essence a manifestation of social justice in the field of human-nature relations. It refers to a relatively reasonable and dynamic balance in the use of natural resources, as mirrored in people's equal access to the public good of the natural environment and environmental public services. As Xi points out, there is no public good more equitable and no welfare more universally beneficial than a sound natural environment. This is a profound representation of the philosophy of ecological justice.

Ecological justice manifests itself not only in relations within a generation, a region, or a nation but also in intergenerational relations and international relations. Xi has reiterated that a sound natural environment, as the most equitable public good and most universally beneficial welfare, can be achieved through high-quality development of an ecological civilization.

The outlook on sustainable development seeks sustainable unity of ecological wellbeing and economic wellbeing. People's wellbeing is not restricted to the economic sphere. While economic wellbeing is an essential component of people's wellbeing, a person cannot be merely regarded as homo economicus (or economic man), and their diverse demands cannot be simply generalized as economic demands; improvement in economic wellbeing, though, does not equate with the resolution of the issue of wellbeing. Human beings are products of nature and interact with nature; they are ecological men with qualities conferred by nature and ever-growing needs for a sound natural environment. Therefore, it is necessary to combine ecological wellbeing and economic wellbeing, that is, to pursue economic development while protecting the environment and to attain moderate prosperity while improving people's wellbeing. Xi has on many occasions stressed that the environment is crucial to people's wellbeing, since blue skies and verdant surroundings allow us to feel the beauty and joy of life. Frequent environmental issues have posed an obstacle to economic development and social progress and become a great concern to the public. Therefore, to boost ecological wellbeing, the environmental quality must be improved to meet people's expectations, which is a significant task that must be taken seriously and implemented down to earth. As Xi says, we should make sound natural environment a driver for the improvement of people's lives, the underpinning of the sustainable and healthy development of the economy and society, and a window to demonstrate China's good image. This highlights the value of sustainable development.

3 The Outlook on Sustainable Development is a New Vision that Aims to Realize Value Through Effective Coordination of Diverse and Complex Value Relations

The outlook on sustainable development is a new vision that reflects the fundamental changes in the way of value realization and aims to achieve balanced, sustainable value by coordinating diverse and complex value relations. There are always diverse and complex value relations in the social organism composed of human, nature, and society. Effective coordination of such relations is the prerequisite for promoting sustainable development. Historically, the rise or fall of a civilization is related to how its people perceive and handle environmental issues. Lessons from history, coupled with current environmental problems, demonstrate that whether diverse and complex value relations influencing the development of human society can be well coordinated has a direct bearing on the achievement of sustainable development.

As a new vision with value objectives, demands, and vision, the outlook on sustainable development reflects the unity of opposites between the value of present and future generations and the sustainable use and realization of ecological value. It holds fast to the unity of opposites between national and regional value and global and holistic value and represents the balance between short-term and long-term objectives of the building of an ecological civilization. Diachronic coordination and coexistence of diverse and complex valuation relations as displayed in the outlook on sustainable development reflects the practical logic that development of an ecological civilization can push various value relations to reach a balance through the dialectical process.

The outlook on sustainable development is the result of profound reflection on the way of value realization and coordination of value relations. It came into being and has evolved with a strong sense of crisis about the serious impacts of environmental destruction on the survival and development of present and future generations. It draws upon the concept of sustainable development put forth by the International Union for Conservation of Nature, the United Nations Environment Programme, and the World Wildlife Fund in their report on the conservation of nature in the early 1980s, inherits and carries forward the ideas of Marxism on the relation between human and nature, and was born amid China's battles against air, water, and soil pollution. It is intended to meet people's basic needs for fresh air, clean water, and safe food and respond to the unprecedented tension between human and nature and the accompanying conflicts between human value and ecological value, economic value and natural value, immediate value and long-term value, regional value and national value, as well as national value and global value. It is a new point of view on ecological progress based on comprehensive reflection on the relation between human and nature and its development trends from a holistic perspective.

The outlook on sustainable development of Xi Jinping Thought on Ecological Civilization is an organic system encompassing the central subject, goal, value, basis, conditions, guarantees, and basic requirements of sustainable development, among others. The relations between these basic elements of sustainable development are diverse and complex. Only when the diverse and complex value relations are well coordinated can the goal of sustainable development be achieved. In terms of the relationship between ecology and civilization, the outlook on sustainable development believes that at the core of sustainable development is people, including both present and future generations and both the Chinese people and all others on this planet. To build an ecological civilization is essentially to make all people eco-conscious and realize human value. The goal of sustainable development is to meet the ever-evolving needs of present and future generations for a better life, continuously protect their ecological rights, which is an integral part of human rights, achieve ecological justice featuring equal access to ecological goods and services, and promote free and well-rounded development of every single person. The path towards sustainable development is to achieve balanced and coordinated development in tandem with economic growth, improvement of the environment, and social progress across the board. Moreover, the ultimate purpose of sustainable development is sustainable high-quality development characterized by impressive economic growth and a sound natural environment.

On the Dissemination Channels of Xi Jinping Thought on Ecological Civilization

Chun Shi

At the National Conference on Eco-environmental Protection that opened on May 18, 2018, General Secretary Xi Jinping delivered a speech titled Pushing China's Development of an Ecological Civilization to a New Stage, which was published on Issue No. 3 of *Qiu Shi* journal on February 1, 2019. This important speech embodies Xi Jining Thought on Ecological Civilization and serves as a fundamental guideline for the building of an ecological civilization in the new era.

Based on dialectical materialism and historical materialism of Marxism, Xi Jinping Thought on Ecological Civilization provides a systematic answer to a number of fundamental theoretical and practical issues such as why to build an ecological civilization, what kind of ecological civilization to be developed, and how to do so. Xi Jinping Thought on Ecological Civilization inherits and carries forward Marxism and is content-rich, extensive, and profound, taking the Marxist theory of human-nature relations to a new level. Rooted in Chinese civilization, it carries through the Chinese ecological culture and will lead the Chinese nation towards sustainable development. Therefore, it conveys far-reaching significance.

Xi Jinping Thought on Ecological Civilization is an essential part of his proposition on Socialism with Chinese Characteristics for a New Era. It holds great realistic significance for building socialism with Chinese characteristics, especially an ecological civilization in the new era.

At present and for some time to come, we should vigorously promote Xi Jinping Thought on Ecological Civilization and advance the building of a beautiful China, so that the thought can take root and translate into the conscious action of officials and the masses.

There are mainly five channels to disseminate and promote Xi Jinping Thought on Ecological Civilization.

C. Shi (✉)
Fuyang Municipal Bureau of Ecology and Environment, Fuyang, Anhui Province, China

J. Pan et al. (eds.), *Beautiful China: 70 Years Since 1949 and 70 People's Views on Eco-civilization Construction*, https://doi.org/10.1007/978-981-33-6742-5_23

1 Promoting Xi Jinping Thought on Ecological Civilization Through Party Schools at Various Levels

Party schools at all levels can serve as a main platform for promoting Xi Jinping Thought on Ecological Civilization. Party schools provide education and training for Party members. They are tasked with improving students' abilities to observe and solve problems with the Marxist stance, viewpoint, and methodology through well-planned training and enhancing their political and cultural quality and Party spirit in the new circumstances, thus enabling them to be vanguards and role models.

Party schools should leverage their strengths to promote Xi Jinping Thought on Ecological Civilization, so that every Party member can get hold of its profound connotations. Facilitators should give a comprehensive account of the thought, with a focus on Xi's important discourses on ecological progress and the concept that lucid waters and lush mountains are invaluable assets, among others.

First, ecological progress should be promoted in combination with reports since the 17th CPC National Congress. The report to the 17th CPC National Congress, which took place from October 15 to 21, 2007, for the first time proposed "conservation culture." It vowed to "promote a conservation culture," regarding it as one of the new requirements for attaining the goal of building a moderately prosperous society in all respects. As stated in the report, "We will promote a conservation culture by basically forming an energy- and resource-efficient and environment-friendly structure of industries, pattern of growth and mode of consumption. We will have a large-scale circular economy and considerably increase the proportion of renewable energy sources in total energy consumption. The discharge of major pollutants will be brought under effective control and the ecological and environmental quality will improve notably. Awareness of conservation will be firmly established in the whole of society."

The 18th CPC National Congress that opened on November 8, 2012 further enriched the theory of "ecological progress." The report to the 18th CPC National Congress incorporated ecological progress in the Five-Sphere Integrated Plan, saying that "We must give high priority to making ecological progress and incorporate it into all aspects and the whole process of advancing economic, political, cultural, and social progress, work hard to build a beautiful country, and achieve lasting and sustainable development of the Chinese nation." It not only elevated ecological progress to the same height as cultural and ethical progress and material progress but also put forward several new concepts such as ecological value and new objectives. The report represents a major theoretical breakthrough and practical innovation and holds great realistic significance and profound historical significance.

The 19th CPC National Congress from October 18 to 24, 2017 further elaborated on the concept of "ecological civilization." Based on the ecological progress theories and practices since the 18th CPC National Congress, the report to the 19th CPC National Congress put forward a series of new ideas, requirements, objectives, and plans for development of an ecological civilization and environmental protection. It contained a number of important propositions such as "building an ecological civilization is vital to sustain the Chinese nation's development" and "man and nature form a community of life" and proposed that "We must realize that lucid waters and lush mountains are invaluable assets and act on this understanding, and implement our fundamental national policy of conserving resources and protecting the environment." In terms of the new requirements, it states, "In addition to creating more material and cultural wealth to meet people's ever-increasing needs for a better life, we need also to provide more quality ecological goods to meet people's ever-growing demands for a beautiful environment." It also set a new goal to build a beautiful China by 2035 and laid out new plans for promoting green development, solving prominent environmental problems, intensifying the protection of ecosystems, and reforming the environmental regulation system.

Second, it is important to give an in-depth interpretation of Xi Jinping Thought on Ecological Civilization, so that officials and the masses can recognize the importance of stepping up development of an ecological civilization. The National Conference on Eco-environmental Protection, which opened on May 18, 2018, for the first time systematically summarized and interpreted Xi Jinping Thought on Ecological Civilization, another thought that has been expounded on and proclaimed at a national work conference following Xi Jinping Thought on Socialism with Chinese Characteristics for a New Era and Xi Jinping Thought on Strengthening the Military.

As General Secretary Xi Jinping says, building an ecological civilization is vital for sustaining the development of the Chinese nation. The Chinese people have always revered and loved nature, and China's 5,000-year-long civilization embodies a rich ecological cultural component. These concepts all stress the importance of uniting heaven, earth, and man, following the rules of nature, and using what nature has to offer with patience and restraint, and show that our ancestors well understood the need to properly handle the relationship between man and nature. A civilization may thrive if its natural surroundings thrive, and will suffer if its natural surroundings suffer.

Officials and the masses should be informed that the reason why General Secretary Xi Jinping has repeatedly emphasized the importance of taking environmental issues seriously and handling them properly is that China's environmental capacity is limited, our ecosystems are vulnerable, and we have still not achieved a fundamental reversal of environmental conditions that cause heavy pollution, significant damage, and high risk. Meanwhile, our unique geographical surroundings have exacerbated interregional imbalances. Our efforts to build an ecological civilization are now in a crucial phase in which we must carry forward despite heavy strain and immense pressure, a decisive stage in which we will supply more high-quality ecological goods to meet the growing demands of the people for a pristine environment, and also a period of opportunity in which we have the conditions and abilities necessary to resolve prominent environmental issues.

General Secretary Xi Jinping points out that achieving moderate prosperity throughout society by 2020 is our Party's solemn commitment to the people. Party committees and governments at all levels must take the initiative to incorporate construction of an ecological civilization into plans for economic and social development. We must ensure that Party committees provide leadership, governments give direction, enterprises bear responsibility, and members of the public can participate, and firmly reject the old path of polluting first and fixing it later or any growth model that would damage or destroy the environment.

Finally, it is important to explicate the top-level design for development of an ecological civilization in the new era General Secretary Xi Jinping spearheaded with worst-case scenario, strategic, systematic, and dialectical thinking and with the rule of law in mind. The principles put forth by Xi on stepping up development of an ecological civilization should also be explained clearly.

The first principle is that harmony between man and nature must be upheld. The philosophy of harmony between man and nature has prevailed in China for thousands of years. As stressed by General Secretary Xi Jinping, throughout the process of development, we must prioritize environmental conservation and protection and put the restoration of nature first. We cannot think about what we can take from nature without considering how we can give back to it; we cannot think about development while ignoring our responsibility to protect the environment; and we cannot think about how we can use nature without looking at how he can restore it. We should protect our natural environment in the same way we would protect our own wellbeing, focusing on taking more positive measures that will provide solid foundations and long-term benefits, concrete steps that will help protect and restore the environment, and effective actions that will clean up our surroundings and reveal their natural beauty. This will allow the people to be close to mountains and waters and recall their rural roots with fond memories, ensure that pristine natural vistas are never too far away, and preserve the serenity, harmony, and beauty of nature. General Secretary Xi Jinping has demonstrated to us a charming blueprint of a beautiful China.

The second principle is that our natural environment is precious. General Secretary Xi Jinping proposed that lucid waters and lush mountains are invaluable assets, which is an important concept for development and a major principle for advancing modernization. The phrase expounds the relationship between economic development and environmental protection and shows us the truth that by protecting the environment we are protecting productive forces, and that by improving the environment we are developing productive forces. It therefore points out a new path for achieving coordination between development and conservation.

The third principle is that there is no welfare more universally beneficial than a sound natural environment. The environment is crucial to the wellbeing of the public, since blue skies and verdant surroundings allow us to feel the beauty and joy of life. This has been highlighted by General Secretary Xi Jinping. Just as economic development is for the wellbeing of the people, so is environmental protection. In addition to creating more material and cultural wealth to meet people's growing needs for a better life, we need also to provide more quality ecological goods to meet people's growing demands for a beautiful environment.

The fourth principle is that our mountains, rivers, forests, fields, lakes and grasslands together form a biotic community. Mountains, rivers, forests, fields, lakes, and grasslands are all an integral part of the whole ecosystem and together form a biotic community. Just as humans need the fields for food, the fields need the rivers, the rivers need the mountains, the mountains need the soils, and the soils need the forests and grasslands. This biotic community is the material basis for humanity's subsistence and development. We must take all of the ways that we impact nature into account from a broad and long-term perspective, and avoid being penny-wise and pound-foolish, biting off more than we can chew or emphasizing one thing while neglecting another, for this would ultimately cause long-term environmental damage on a system-wide scale.

The fifth principle is that the strictest regulations and laws must be applied in protecting the environment. As emphasized by General Secretary Xi Jinping, environmental protection must have legal and regulatory backing. We must accelerate innovation of regulations, ensuring that they offer more, have adequate peripheral support, and that they are rigorously enforced. By doing so, we will turn our regulations into rigid and inviolable constraints. We must use our regulations to check the use of power, protect blue skies and increase green coverage, and ensure accountability, thus guaranteeing that the CPC Central Committee's decisions and arrangements on building an ecological civilization are firmly implemented and produce real results.

The sixth principle is that joint efforts must be made in building a global ecological civilization. Ecological conservation and progress cross borders and require concerted efforts from the international community. General Secretary Xi Jinping has stressed that development of an ecological civilization concerns the future of humanity, and building a green living environment is humanity's common aspiration. The whole world must work together to protect the environment and respond to climate change. No country can ignore these issues, and no country will fare well by going it alone. China has already become a major participant, contributor, and champion in the movement to build a global ecological civilization, advocating the cultivation of ecosystems in which Mother Nature and green development come first, and promoting joint efforts toward making the world cleaner and more beautiful.

2 Promoting Xi Jinping Thought on Ecological Civilization at Schools

Xi Jinping Thought on Ecological Civilization should reach out to schools and children. Kindergartens and elementary schools can use vivid images to illustrate concepts such as "lucid waters and lush mountains are invaluable assets" according to the characteristics of different age groups. Middle schools are advised to systematically impart Xi Jinping Thought on Ecological Civilization and expound on it with specific examples. They may provide training for its own faculty in this regard and invite experts to give lectures so that students can get a big picture of Xi Jinping Thought on Ecological Civilization.

It is proposed colleges and universities combine promotion of Xi Jinping Thought on Ecological Civilization with real-life cases and guide students to take field trips to Anji County of Zhejiang and other areas to better understand the profound connotation of the concept that "lucid waters and lush mountains are invaluable assets" and delve into the realistic and historical significance of Xi Jinping Thought on Ecological Civilization. College students who have a good grasp of the thought can be selected to promote it in communities, the countryside, and mountainous areas during vacations or practical training periods.

In keeping with the times, Xi Jinping Thought on Ecological Civilization can be made compulsory and incorporated into textbooks for elementary schools, junior and senior middle schools, as well as colleges and universities. At kindergartens and elementary schools, images that illustrate Xi Jinping Thought on Ecological Civilization can be created for teaching in light of the characteristics of young students.

3 Promoting Xi Jinping Thought on Ecological Civilization at Communities

A community is a group of people living in the same area. A city consists of numerous communities, which are like the "cells." Community offices and subdistrict committees can publicize Xi Jinping Thought on Ecological Civilization by organizing training programs, putting up posters and slogans, systemically introducing the thought on bulletin boards, and so forth.

4 Promoting Xi Jinping Thought on Ecological Civilization in Rural Areas

Currently, a majority of Chinese people live in rural areas. It therefore carries great significance to promote Xi Jinping Thought on Ecological Civilization in rural areas, which should also be an important battlefield. Party members and officials at township levels should strengthen publicity to give rural people an idea of Xi Jinping Thought on Ecological Civilization. According to the characteristics of rural areas, it can be embedded into opera, skit, allegro, and other art forms favored by rural people. In addition, training programs can be organized to introduce typical examples such as Anji County of Zhejiang to give rural people a better understanding of the thought.

5 Environmental Protection Departments Should Deepen Study of Xi Jinping Thought on Ecological Civilization and Act on Its Guidance

Xi Jinping Thought on Ecological Civilization is an important value and methodology and serves as a fundamental guideline for environmental protection work. Environmental protection departments, which plays an essential role in environmental protection, should put the study and implementation of Xi Jinping Thought on Ecological Civilization high on their agenda.

Environmental protection departments should have a good grasp on the rich connotation and tenet of Xi Jinping Thought on Ecological Civilization as well as the underlying Marxist stance, viewpoint, and methodology and take the thought as their guide to action. In-depth study and implementation of Xi Jinping Thought on Ecological Civilization should be a long-term political task of environmental protection departments at all levels. They should be fully aware that Xi Jinping Thought on Ecological Civilization is a well-structured, rigorous logic system and get to grips with the significance and role of development of an ecological civilization in achieving the sustainable development of the Chinese nation, fulfilling the mission of the Party, and improving people's livelihoods. They should realize that Xi Jinping Thought on Ecological Civilization is an integrated, interconnected theoretical system and get a thorough understanding of its ideological core, fundamental requirements, values, and institutional guarantees. They should understand that Xi Jinping Thought on Ecological Civilization is an open, dynamic theoretical system and continuously draw wisdom and nourishment from Xi's latest discourses on ecological progress and help promote the enrichment and involvement of thought in practice. They should understand that Xi Jinping Thought on Ecological Civilization is a scientific system that serves as a work guide and have a good grasp of its principles, practical requirements, and scientific methods. In addition, they should get to the bottom of the concept that lucid waters and lush mountains are invaluable assets and fathom how to put it into practice, properly balance the relation between economic development and environmental protection, and promote high-quality economic development and high-level environmental protection in a coordinated way. Environmental protection departments should continuously advance development of an ecological civilization and translate the requirements of Xi Jinping Thought on Ecological Civilization into the specific practical tasks with Chinese characteristics. Following the guidance of Xi Jinping Thought on Ecological Civilization, they should strive to blaze a trail of high-quality development that gives priority to ecological progress and green development, so that the blueprint for a beautiful China with azure sky, green land, and clear water can become a reality.

Green Development and Ecological Industry System

Work Towards a Green Economy in the 14th Five-Year Plan Period

Xinli Zheng

Green transformation is an inherent requirement of developing a resource-conserving, environmentally friendly economy. By the end of the 14th Five-Year Plan period, namely 2022, China's per capita GDP is expected to reach USD 12,600, bringing China into the ranks of high-income countries, which will be an epoch-making event in the history of human society. At present, the population of high-income economies only makes up 15.7% of the world population, and China's accession will increase the proportion by 18.5–34.2%. 2021 will mark the 100th anniversary of the CPC. The CPC is shaping up to turn a poor and backward China into a high-income country over a century, which will be the greatest contribution it will make to human society.

With a population of 1.4 billion, China will not follow the American mode of high resource consumption but should draw upon the resource-conserving mode of Northern European countries. We will convert waste into resources and develop clean and renewable energy to reduce GHG emissions. By doing so, the modern lifestyles of Chinese people will not bring great pressure on global resources. Facts have proved that it is totally viable to shift towards green development and promote resource-conserving consumption.

China's green transformation is also part of the global imitative to respond to climate change. The UN 2030 Agenda for Sustainable Development sets the goal of achieving net zero carbon emissions by 2050, which hits home the main problem in global development and also dovetails with China's demand for sustainable development. While striving to build a socialist modern country, China is resolute in battling pollution and takes development of an ecological civilization as an unshakable responsibility for achieving the UN's Sustainable Development Goals. Currently, China is faced with two major tasks. The first is to tackle pollution brought about by rapid industrialization in the last 40 years, win the battles against air, water, and soil pollution, and achieve a peak in carbon dioxide emissions

X. Zheng (✉)
Policy Research Office of the CPC Central Committee, Beijing, China

J. Pan et al. (eds.), *Beautiful China: 70 Years Since 1949 and 70 People's Views on Eco-civilization Construction*, https://doi.org/10.1007/978-981-33-6742-5_24

before 2030. The second is to develop a green economic system and put in place market mechanisms and policy systems conducive to the development of the environmental protection industry, including improving environmental legislation and law enforcement systems, refining the property rights system and use control system for natural resource assets, implementing the system of paid use for natural resources and ecological compensation system, and reforming environmental protection and management system. In line with principles of "paid use of resources" and "polluter pays," resource taxes will be gradually extended to all kinds of natural resources and spaces. A market for environment protection will be developed, where a trading system for pollution discharge, carbon emissions, and water rights will be established, and a market mechanism will be put in place to attract private capital investments into environmental protection. The system of third-party environmental rectification will be implemented. In addition, China will create a market-based system for green technology innovation, develop green finance, and spur the development of energy-saving and environmental protection industries as well as clean production and clean energy industries, so as to gradually develop the environmental protection sector into a pillar industry. According to estimates, the output value of China's green economy totaled about RMB 6 trillion in 2018 and is expected to reach RMB 12 trillion by 2025, accounting for approximately 8% of the country's GDP, and that rate will surpass 10% by 2035.

The 14th Five-Year Plan should combine green transformation with urbanization. With rapid advances in urbanization, China's urbanization rate will grow from 59.6% in 2018 to around 80% by 2035, indicating that 280 million people will move from the countryside to cities in the next seventeen years. During the 14th Five-Year Plan period, more than 13 million rural people are expected to cities every year. The largest scale of urbanization ever known in human history is not only changing the mode of production and lifestyle of millions of Chinese people but also having significant impact on the global economy. According to the requirements for integrated urban–rural development, we should continuously optimize the urban structure and layout, stick to the development path of green, resource-conserving, and smart cities, and make urbanization a powerful driving force of economic growth. We will advance the development of three city clusters including the Beijing-Tianjin-Hebei region, the Yangtze River Delta, and the Guangdong-Hong Kong-Macao Greater Bay Area and make them three engines of national economic development. We will boost the development of secondary city clusters centred on provincial capitals, enabling them to fuel the growth of the provincial economy. We will highlight the construction of signature towns centred on county seats in a bid to vitalize the rural areas. In addition, urbanization will prompt the development of the green construction market. According to the *13th Five-Year Plan for the Development of Building Energy Efficiency and Green Building* issued by the Ministry of Housing and Urban–Rural Development, the proportion of green buildings in new buildings in urban areas will increase from 20% in 2015 to 50% in 2020, and energy-saving residential buildings will account for more than 60% of existing urban residential buildings by then.

China will learn green and sustainable development technology and experience from developed counties. The EU's total carbon emissions peaked in the 1980s. In particular, Northern European countries have been leading the world in sustainable development and have embarked on a green and sustainable development path where environmental improvement and economic and social development go hand in hand. On January 15, 2019, five countries in Northern Europe issued a joint statement, vowing to go carbon neutral, that is, achieve net zero carbon emissions, by 2050. They also planned to call on countries around the world to work together to keep the global temperature rise within 1.5 °C by the end of the twenty-first century at the UN General Assembly to be held in the United States in September 2019. The experience of the Nordic countries can be summarized as the following four aspects. First, the government plays a leading role, the legislature enacts laws and rules, and administrative departments put forth specific sustainable development goals, plans, and budget and invite social organizations to conduct third-party assessment. Second, enterprises are actively engaged and set an example for the world in terms of innovation technology and operation models. For example, the Northern European smart grid covers six countries, allowing enterprises on the supply side and users on the demand side to have a free choice and enabling spot and futures trading simultaneously. This effectively avoids water, wind, or solar curtailment. Third, social organizations serve as a bridge between the government and enterprises. For example, the education sector helps children develop environmental awareness and sustainable innovation abilities. Fourth, the function of the income distribution system and the role of the media in guiding public opinion are brought into full play, and the spirit of social contract has taken root, and the social credit system is in place. It can be said that Northern Europe puts a premium on environmental protection. As the saying goes, "Jade can be polished by stones from other hills." We should draw on their experience and strive to make China more beautiful and livable.

It should be noted that China is still facing a grave situation of environmental pollution. We should pool resources to tackle four outstanding problems during the 14th Five-Year Plan period.

(1) Unswervingly tackle smog to return the blue sky to the people

Northern regions dependent on heavy industries are often shrouded in smog, which is caused by high concentrations of particulate matter, posing a threat to public health. This goes against the purpose of economic development that is, improving people's livelihoods so that they can live a longer and healthier life. Therefore, we must put people first, enhance our sense of responsibility, and go all out to prevent and control smog and honour our solemn commitment to keeping the air clean for people.

(2) Take solid action to curb water pollution to provide clean drinking water for the people

Water pollution includes pollution of rivers, lakes, ponds, coastal areas, and groundwater. It can pose a high risk to people's health and has become a pressing issue that must be resolved firmly. The key lies in putting all available resources in the initiative and strengthening supervision over pollution sources. Especially, river basin water pollution across different regions requires concerted efforts of local governments in the upper and lower reaches. In addition, clean-up of black, malodorous water bodies should be a focal point and be completed as scheduled.

(3) Properly handle waste and turn waste into resources

In China, most of household waste is dumped in landfill sites. There are currently more than 6000 compliant landfill sites and 10,000 noncompliant ones across the country, which occupy a mass of land and pose safety risks. The annual amount of construction waste is 5 billion tons in addition to over 20 million tons of existing construction waste, which takes up more than 2 million *mu* (1333.33 km^2) of land. The amount of solid waste such as industrial waste, concrete slugs, coal gangue, tailings, and desulfurized gypsum is also whopping. Household waste and construction waste can both be converted into valuable resources after proper classification and treatment. In some European cities, waste treatment plants are beautifully situated in the city proper and serve as a venue of environmental education for young people. There are already specialized enterprises in China which dig out buried waste for decontamination. Local governments should take classification and disposal of waste, including rural waste, as an important task and incorporate it in their 14th Five-Year Plan.

(4) Address nonpoint source pollution of land and produce organic food

Because of excessive use of chemical fertilizers and pesticides, nonpoint source pollution of land is serious in China. To deal with the problems requires collaboration between agricultural departments and environmental protection departments as well as concerted efforts of the whole country. We must press ahead with the daunting task since it concerns the health of not only our own but also later generations.

China's Philosophy on Development of an Ecological Civilization

Weimin Yang

1 Spatial Development

Development encompasses economic development, human development, and sustainable development, which have different meanings and goals. At the policy level, they should be put together in a specific space to be meaningful; to look at them separately and abstractly will not make much sense.

Beijing has enjoyed remarkable economic development and social progress. It provides high-quality public services and is home to China's best universities and hospitals. However, the city has been grappling with poor air quality over the years, and blue skies with white clouds are a common wish of people living in Beijing. How to achieve sustainable development poses a great challenge to Beijing.

Shenzhen has overtaken Hong Kong to rank 5th among Asian cities by GDP and is reputed for innovation in the world. However, more than 80% of the city's population lives in "urban villages" where the living environments are not satisfactory. For every single individual, without a stable and good living environment, development and wellbeing would be merely an empty talk. Therefore, human development is a major issue for Shenzhen.

Sanjiangyuan, or the Three-River-Source, is situated on the Qinghai-Tibet Plateau, which is known as "the Third Pole of the Earth." It is an important ecological shelter and water conservation area in China and even Asia and also an area sensitive to climate change. Economic activity there will not only have great impact on China's natural environment but also on global climate change. Therefore, China has set up the Three-River-Source National Park in a bid to protect the region to the greatest extent. However, the development of people living in Sanjiangyuan is an important issue. The solution is to encourage local residents, mainly Tibetans, to be ecological

W. Yang (✉)
Committee on Economic Affairs, The National Committee of CPPCC, Beijing, China

General Office of the Central Leading Group for Financial and Economic Affairs, Beijing, China

J. Pan et al. (eds.), *Beautiful China: 70 Years Since 1949 and 70 People's Views on Eco-civilization Construction*, https://doi.org/10.1007/978-981-33-6742-5_25

conservators of the national park, who can get paid for their work to take care of and protect natural resources.

The above three examples show that economic development, human development, and sustainable development should be discussed in a specific space, where the three should be well-balanced. Spatial development is to strike a balance between the three in a specific space; otherwise, people's living quality will be detracted.

Spatial development is development under the vision of ecological civilization. Ecological civilization is a new state of civilization and social pattern following industrial civilization. Although development of an ecological civilization is a long haul, we must stick to this path and do our part. This is what our generation and every one of us should do.

2 Spatial Equilibrium

Spatial equilibrium is closely related to spatial development and is a theoretical basis of spatial development. One requirement for development of an ecological civilization is coordinated development of the population, economy, and environment.

For China, there is still a long way to go from spatial equilibrium. The three city clusters, namely the Beijing-Tianjin-Hebei region, the Yangtze River Delta, and the Pearl River Delta, constitute 41% of China's GDP but only 23% of its population, while less developed regions such as Central China and West China only make up a small part of the country's GDP compared to their large proportion. The gaps in economic development and population between different regions result in disparities in living standards.

A major reason is that restrictions on urban household registration have not yet been lifted across the country, especially in megacities. The household registration system reform is now under way. If the free flow of people is truly achieved, more people in less developed regions may move to developed regions, leading to a basic balance in population and economic size and therefore spatial equilibrium between economic development and human development.

The Gini coefficient is an important indicator that measures economic differences between regions and between urban and rural areas. As we see it, using the Gini coefficient to measure differences in levels of personal income makes great sense, but using it to analyze gaps between regions may have great problems. This is because the Gini coefficient does not take into account the likes of the 4000-m altitude difference between the Qinghai-Tibet Plateau and the Yangtze River Delta and the temperature difference between Heilongjiang and Sanya where temperature can fall below minus 30 °C and rise above 30 °C, respectively.

Theory can be abstract, but economic policy cannot. If economic policy encourages economic development of all regions without taking into account the altitude and temperature differences, many problems, such as ecological and environmental damages, will ensue.

China's geographic and geomorphic conditions are extremely complicated. Blind construction in ecologically fragile areas such as desertified areas can cause damages and even disasters to the ecosystem and environment. Efforts have been made to put an end to such behavior.

In addition to air pollution, Beijing also faces serious water shortage. The city used to depend on groundwater extraction for water supply, which may result in surface subsidence and collapse of buildings and infrastructure. Today, the South-to-North Water Diversion Project diverts water from the Hanjiang River to Beijing. However, if the city's population keeps growing, the project may be unable to provide sufficient water for its use. In view of this, the central government decided to relieve Beijing of functions nonessential to its role as the capital. As people move out of Beijing with those nonessential functions, which will ease the city's water supply pressure, making it possible to achieve coordinated development of the population, economy, and environment.

Coordinated development of the population, economy, and environment involves different aspects. For example, cities may be under great pressure in terms of transportation infrastructure and energy supply. One important reason for the so-called "big city malaises" is excessive population and economic agglomerations, namely imbalances between the population, economy, and environment.

3 Carrying Capacity

The evolution of human civilization is a process in which mankind deals with two basic relations. One is the relation between people, and the other is the relation between human and nature. If the relations between people is not handled properly, it will lead to the collapse of a civilization or a country; the same is true for the relation between human and nature.

The carrying capacity of the environment refers to the economic and population size that the water, soil, and other resources in a specific space can sustain under the premise of keeping nature intact.

As the saying goes, "Each place has its own way of supporting its own inhabitants." Many areas in China, however, have difficulty in supporting local people, let alone enable them to lead an affluent life.

While Europe is mostly plain, China's terrain can be divided into three tiers from west to east. The first is the Qinghai-Tibet Plateau with an average altitude of over 4500 m; the second is the desert areas of Xinjiang, the Inner Mongolia Plateau, and the Yunnan-Guizhou Plateau, with an average altitude of 1000 to 2000 m; the third mainly covers plains with hills and low mountains in between, whose altitude is generally less than 500 m. The regions in the first and second tiers have quite weak carrying capacity, unable to sustain large populations, while the plains, which cover a relatively small area, have stronger carrying capacity.

The Heihe-Tengchong Line, which extends from Heihe in Heilongjiang to Tengchong in Yunnan, was drawn by Chinese economic geographer Hu Huanyong in the

1930s. As he discovered, since the Tang dynasty (618–907 A.D.), regions on the northeast of this line made up 40% of China's total land area but supported 94% of its population. The fifth population census in 2000 and sixth census in 2010 found that this pattern had remained basically unchanged.

China's total area of plains is about the same as that of the U.S. and Europe, but the average area of plains in China is just 860 square meters. This highlights plains as a valuable asset for farming, urban development, and infrastructure. Many people have only a general idea that China has a vast territory without knowing the fact that the land suitable for construction, development, and living is quite limited. A large population, relatively little land available, and limited spaces constitute the basic reality of China.

4 Ecological Goods

There is only agricultural produce and a small quantity of handcrafted products in an agricultural society, while industrial goods and services are dominant in an industrial society. Today, we have an adequate supply of agricultural produce and industrial products and services but fall short in supply of ecological goods.

What are ecological goods? In our opinion, they include fresh air, clean water, a comfortable natural environment, pleasant climate, and more. A large part of ecological goods are created by nature and have existed and been used by mankind all the time. These products of nature seem in inexhaustible supply and free of charge. This is why the concept of ecological goods is often neglected.

Over the past forty-plus years, China's capacity to supply agricultural produce, industrial goods, and services has improved by leaps and bounds; in contrast, its supply capacity for ecological goods has weakened.

Like agricultural produce, ecological goods also need to be produced on "cultivated land," namely ecological spaces such as forests, wetlands, lakes, and oceans.

Clean water sources are a type of ecological goods. China is suffering from a severe water shortage, with the total amount reaching around 50 billion cubic meters a year. The primary reason is that lakes and rivers have become less and smaller.

The area of Poyang Lake, the largest freshwater lake in China, has contracted from around 5100 km^2 in 1954 to over 3000 km^2 today.

It can be seen that the "basins" that hold water are shrinking, including not only lakes and rivers but also groundwater sources, forests, and wetlands. The reason for water shortages is not that there is less rainwater but the "basins" have become small or even disappeared.

In the era of ecological civilization, ecological goods should be regarded as products, which certainly have their worth. General Secretary Xi Jinping stresses that "lucid waters and lush mountains are invaluable assets." This means that lucid waters and lush mountains can translate into wealth through suitable approaches. There are multiple approaches to turning ecological goods into assets, such as purchase of ecological goods by the central government, ecological value exchange between different regions, trading of water rights, emission rights, and carbon emission quotas, premiums of ecological goods, and charges for tourism products.

5 Core Functions

The natural resources and resource endowments of different regions within a country can be disparate, so their functions should be differentiated as well. China is now implementing the system of functional zoning, where territorial spaces are classified into four types, namely development zones to be optimized, key development zones, limited development zones, and prohibited development zones.

The former two types will eventually develop into metropolitan areas like Beijing or Tokyo, and the latter two will ultimately become major agricultural production areas and important ecological areas. China is now implementing the Grain for Green Project in important ecological areas to turn marginal farmland to forests and grasslands in an effort to reduce production spaces while increasing ecological spaces.

Up to now, China has approved five major national parks. The first is the Three-River-Source National Park, which covers an area of tens of thousands of square kilometres and was established on the initiative of General Secretary Xi Jinping. The second is the Northeast China Tiger and Leopard National Park. Northeast China used to be home to a large number of tigers, of which some ran away to Russia; with the establishment of the national park, they are returning to their hometown. The third is the Giant Panda National Park. Giant pandas are suffering from habitat fragmentation and food chain destruction due to human activity. In response, China has decided to set up a giant panda reserve with an area of 27,000 km^2 and designate it as a national park. The fourth is the Qilian Mountain National Park. The Qilian Mountain is an important ecological barrier and water conservation area in West China and one of the most biologically diverse areas in the country. The Qilian Mountain National Park will cover a planned area of more than 50,000 km^2.

The system of functional zoning has established differentiated policies for different regions. Development zones to be optimized such as Beijing have seen their carrying capacity weakening, so development must be optimized there. Key development zones like the Chengdu-Chongqing region have relatively strong carrying capacity and are expected to accommodate more businesses and populations for intensive development and higher efficiency. Limited development zones cannot become congruous urban areas, but small-scale development is allowed in certain areas. In prohibited development zones such as the Three-River-Source National Park and Northeast China Tiger and Leopard National Park, development activities in any form shall be prohibited except for particular reasons.

6 Spatial Patterns

Spatial pattern refers to the spatial distribution and layout of ecological or economic factors. China has established three major strategic patterns, which will therefore form three spatial patterns.

The first is the urbanization strategic pattern encompassing "two horizontal lines and three vertical lines." The two horizontal lines refer to the Lanzhou-Lianyungang Railway section of the new Asia-Europe land bridge and the Yangtze River, and the three vertical lines refer to the coastal, Beijing-Guangzhou, and Baotou-Kunming corridors. The urbanization strategic pattern intends to boost the development of more than 20 key city clusters along these lines and enable them to lead and drive regional development.

The second is the agricultural strategic pattern composed of "seven areas and twenty-three belts." The seven areas include the Northeast Plain, the Huanghe-Huaihe-Haihe Plain, the Yangtze River basin, the Fenhe-Weihe River Plain, the Hetao Irrigation District, and South China and Gansu-Xinjiang agricultural production areas. The seven areas and twenty-three belts are positioned as major agricultural production areas in China, which should not pursue industrialization and urbanization but ensure food security for 1.4 billion Chinese people.

The third is the ecological security strategic pattern composed of "two barriers and three belts." Specifically, they refer to the Qinghai-Tibet Plateau ecological barrier, the Loess Plateau—Sichuan-Yunnan ecological barrier, northeast forest belt, northern sandbreak belt, and southern hilly and mountainous areas.

The Qinghai-Tibet Plateau ecological barrier has great impact on not only the entire natural environment of China but also climate change in East Asia, South Asia, and the world at large. Therefore, it is a conservation area of significance. The Loess Plateau—Sichuan-Yunnan ecological barrier, situated at the intersection of the second tier and third tier of China's terrains, suffers the most serious water loss and soil erosion in the country, which may pose severe risks to ecological security if not handled effectively. In addition, the three belts are also of great importance. For example, the northeast foreign belt has a direct bearing on the ecological security of the entire Northeast Plain.

In short, the "two horizontal lines and three vertical lines" will accommodate a majority of China's population and economic output, the "seven areas and twenty-three belts" will ensure food security for Chinese people, and the "two barriers and three belts" will ensure China's ecological security.

7 Spatial Structure

China's territorial spaces can be divided into the following four categories according to their purposes.

The first is urban space, which includes urban construction space such as that within the Fifth Ring Road of Beijing. Urban space also includes industrial and mining space outside urban residential areas, such as coal mines.

The second is agricultural space, which consists of agricultural production space and rural residential space.

The third is ecological space, which includes not only green ecological space such as natural grasslands, woodlands, wetlands, reservoirs, rivers, and lakes but also non-green space such as heathland, sandy land, saline-alkali land, and plateau deserts. Although the latter may be of little use for humans at the moment, they are an integral part of nature and need to be well protected.

The fourth include miscellaneous space, such as transportation infrastructure space for railways, roads, civil airports, ports and docks, and pipelines, space for water conservancy facilities, and special space for national defence, religious, and other purposes.

There are multiple imbalances in China's spatial structure.

First, compared with the vast production space, ecological space is disproportionately small, unable to meet the needs of 1.4 billion Chinese people for ecological goods. For instance, the forest coverage in Japan is around 70%, while it is only slightly higher than 20% in China.

Second, there is much more industrial space than urban residential space. Currently, industrial and mining space takes up a total of about 60,000 km^2, including 10,000 km^2 of development zones, which are mainly designed for industrial development despite the limited residential space. Industrial and mining space is too large compared to the current industrial value added. According to our estimates, the combined industrial space of Shanghai, Wuxi, and Suzhou exceeds the total of Japan. Therefore, we should take solid steps to cut down on industrial space.

Lastly, rural residential space is enormous. The total area of homestead land in China has reached 170 million *mu* (113,333.33 km^2). As lots of rural people have moved to cities, the area of idle homestead land has exceeded 30 million *mu* (20,000 km^2), which is equivalent to the entire urban residential space. Home prices are rather high in Chinese cities. For example, Shenzhen has a whopping price-to-income ratio of about 25, meaning that on average, it takes twenty-five years of household income to purchase an apartment in the city. Among the many reasons for the hike in home prices is insufficient supply of urban residential land.

To sum up, China's spatial structure needs to be optimized.

8 Development Intensity

Development intensity refers to the proportion of construction space in the total area in a specific space.

The overall development intensity in China is 4%. Given that 60% of China's land area is not suitable for development, this rate is not as low as it seems. In particular, the development intensity of plains is too high.

High development intensity does not guarantee satisfactory development effects. For years to come, we should pay special attention to development gains per unit of land area rather than how much land is developed.

According to Professor Zhou Muzhi, China has overemphasized the pressure of population size and density on the urban environment and infrastructure in urban governance while neglecting that a dense population is an important basis of a city's development vitality. In fact, a high population density is important to economic and social development, and its negative effects on the urban environment are not as great as people worry about.

Urban administrators can draw inspiration from the studies on urban spatial development in China by Professor Zhou Muzhi and Cloud River Urban Research Institute headed by him. The basic reality of a large population, relatively little land available, and limited spaces determine that China must put into good use of every inch of our territory and take the path of intensive spatial development.

9 Spatial Planning

Spatial planning provides guidelines for spatial development. It specifies the criteria for various development and construction activities and constraints on development behavior.

In China, spatial planning involves the state, provincial, and municipal/county levels, where planning at a lower level should conform to and serve that at a higher level. Unified land classification standards should be adopted in planning at different levels; according to functional zoning, land is divided into urban space, agricultural space, and ecological space, with dedicated space reserved for infrastructure. The boundaries of urban construction areas, industrial zones, and rural settlements should be clarified, and the boundaries for the protection of farmland, forests, grasslands, rivers, lakes, and wetlands should be delineated as well. In addition, the leading role and authority of plans should be enhanced.

It is easier said than done. No matter how difficult it is, we will remain resolute in accomplishing what we have envisioned.

10 Spatial Governance

China is now deepening reform on all fronts.

The purpose of reform is to modernize the country's governance, of which spatial governance is an essential part. Spatial governance refers to efforts to achieve a balance between the economy, population, resources, and environment in a specific space.

Spatial governance requires legal backing. We should draw up the law on territorial space development and protection, revise the Land Administration Law, Grassland

Law, and Forestry Law, enact the law on nature reserves or national parks, and improve policies regarding finance, investment land, population, the environment, and performance appraisal.

Local governments should play an active role in spatial governance. For example, in order for county-based governance, greater powers should be delegated to county-level governments.

Green Transformation in China: Progress and Outlook

Yiming Wang

Since reform and opening-up started, China has created a miracle by launching the largest ever industrialization drive in human history. This, however, has brought unprecedented environmental pressure, making it necessary for China to embark on a path of green development distinct from traditional industrialization. Through the past forty years of unremitting exploration, China has made great strides in green transformation but still faced various challenges. For years to come, China will remain at a turning point of the Environmental Kuznets Curve and must stick to the path of green development and strive to build a resource-conserving and environmentally friendly green development system and a new modernization pattern featuring harmony between human and nature.

1 Main Progress and Achievements in China's Green Transformation

Green transformation refers to a shift from a traditional development model characterized by high resource consumption, high emissions, and environmental damage to a green development model where economic growth goes hand in hand with resource conservation, emission reduction, and environmental improvement. Green transformation is not designed to fix the traditional industrialization model but represents a revolutionary change in the way of development.

Since the start of reform and opening-up, China's economy has achieved exceptional development, which, however, has led to a drastic increase in resource and energy consumption and pollutant emissions. In 2018, China's total energy consumption reached 4.64 billion tons of standard coal equivalent, 7.7 times that in 1980. From 1990 to 2017, energy consumption by China's industrial sector grew by about

Y. Wang (✉)
Development Research Center of the State Council, Beijing, China

J. Pan et al. (eds.), *Beautiful China: 70 Years Since 1949 and 70 People's Views on Eco-civilization Construction*, https://doi.org/10.1007/978-981-33-6742-5_26

4.4 times.[1] With the rapid industrial growth, discharges of major pollutants have increased substantially. In 2017, about 3.32 billion tons of industrial solid waste was generated, an increase of 5.7 times over 1990. This pattern of economic growth at the expense of resources, energy, and environmental quality has in fact undermined development quality and efficiency, bringing enormous pressure to the environment. In the process of economic development and reform and opening-up, China has adopted the basic national policy of conserving resources and protecting the environment, implemented the sustainable development strategy, and blazed a trail for green transformation while promoting the transformation of the economic growth model. Since the 18th CPC National Congress, China has incorporated development of an ecological civilization into the Five-Sphere Integrated Plan, established the new concept of green development, and sped up top-down design and institutional building in an endeavour to make historic breakthroughs in green transformation.

1. **Significant results have been achieved in the improving of energy efficiency and adjustment of the energy mix**

China's coal-dominated resource endowments fully justify why it is important to adjust the energy mix and build a modern energy system in green transformation. Since the 17th CPC National Congress, China has made energy consumption per unit of GDP a mandatory indicator, incorporated it into three five-year plans in a row, and taken solid measures to reduce energy consumption in key fields such as industry, construction, and transportation. As a result, energy efficiency has improved greatly, and energy consumption per unit of GDP has been on the decline, falling by 41.5% from 2005 to 2018, or an annual average of 4.0%. Progress has been made in the adjustment of the energy mix, with the proportion of coal in primary energy consumption declining from 67.4% in 2012 to 59.0% in 2018. The use of clean and low-carbon energy has picked up pace. In particular, substantive steps have been taken to promote clean and high-efficiency coal use. Upgrading for ultra-low emissions has been completed for coal-fired units with a total capacity of over 700 million kWh, hitting the target for 2020 ahead of schedule. All new coal-fired units meet ultra-low emission standard, bringing China to the forefront of the world by emission control indicators of coal-fired units. The use of clean energy has seen a significant increase. China has the largest installed capacity of hydropower, wind power, and solar power in the world; the share of non-fossil energy and natural gas in total energy consumption has grown to 14.3% and 7.8%, respectively, and the proportion of electricity in final energy consumption has increased to 25.5%. In short, China has become a world leader in the use of non-fossil energy.

2. **Great progress has been made in the adjustment of industrial structure and development of green industries**

Adjusting the industrial structure and improving industry chains are an important measure of green transformation. Structural adjustment is estimated to contribute

[1]The industrial sector consumed 675.78 million tons of standard coal equivalent in 1990 and 2,944.88 million tons in 2017.

more than 50% to the reduction in carbon emissions. Since reform and opening-up started, China has pressed ahead with structural adjustment while ramping up the industrialization drive. In 2012, the service sector for the first time overtook the industrial sector to be the largest sector in the national economy and the biggest engine of economic growth. The sectoral composition of the economy changed from 27.7:47.7:24.6 in 1978 to 7.2:40.7:52.2 in 2018. Compared with the early days of reform and opening-up, the share of the industrial sector dropped by 7%, while that of the service sector rose by 27.6%. Within the industrial sector, vigorous efforts have been made to phase out backward and excessive capacity in coal, steel, cement, flat glass, electrolytic aluminum, and other industries, speed up upgrading and green transformation of traditional industries, and introduce new technologies, processes, and equipment to reduce energy consumption and emissions. For example, the outdated technology of self-baking anodes for aluminum electrolysis has been eliminated, and the NSP cement technology has become the mainstream technology in cement production. Meanwhile, China is working hard to foster strategic emerging industries such as new energy, energy conservation and environmental protection, next-generation information technology, biology, new materials, and new energy vehicles, develop green services, promote energy performance contracting and water conservation contracting, and build a green industrial system. In recent years, China has vigorously promoted intelligent manufacturing. As a result, "Internet Plus" manufacturing models have cropped up, and industrial Internet has been widely used in such industries as petroleum, petrochemicals, steel, home appliances, apparel, machinery, and energy, giving a strong boost to the green transformation of manufacturing.

3. **Resource conservation and recycling has been pushed forward**

Resource conservation and recycling is an important part of green transformation. In 2002, China promulgated the *Law on the Promotion of Clean Production*, its first law on circular economy, marking the shift from end-point pollution control to whole process control. Since then, China has ramped up green, circular, and low-carbon development, strengthened the management of obligatory targets, imposed controls on the total amount and intensity of energy and water consumption, construction land, and so forth, and raised the standards for the conservation of energy, water, land, and materials and for green mines. It has doubled down on energy conservation and emission reduction in key industries, enterprises, and projects, promoted circular production at enterprises, circular industrial portfolios, and circular transformation of industrial parks, and propelled the shift from the traditional linear growth model of "resources—products—waste" to a sustainable development model where materials flow in a closed loop. The near-zero carbon emission zone demonstration project has been launched, and effective control of carbon emissions has been implemented proactively. A series of resource conservation campaigns have been carried out, such as the energy-saving and low-carbon campaign of 10,000 enterprises, the green building campaign, the low-carbon transportation campaign of 1,000 vehicle, ship, road, and port enterprises, the energy-saving public institution demonstration program, and the circular economy models demonstration and promotion program. These campaigns intend to promote green and low-carbon use and recycling of resources, opening up new space for green development.

4. **Green technology innovation and construction of the standard system have been strengthened**

Technological innovation can bolster high-efficiency and intensive use of energy and resources and is a key measure to promote green transformation. Since reform and opening-up started, China has implemented the strategy of reinvigorating the country through science and education and substantially increased investment in science and technology. In 2014, China overtook Japan and the EU to become the world's second largest economy by R&D investment, with its R&D spending making up nearly one-fourth of the global total. In 2018, China spent nearly RMB 2 trillion in R&D, which accounted for 2.19% of its GDP, exceeding the average of 2.1% of fifteen EU countries. Investment in green technology R&D has also increased significantly. From 1990 to 2014, the number of China's environment-related patents surged by 60 times, compared to only three times in OECD countries; the growth rate of green technology patent applications in the last ten years, especially since 2005, has exceeded that all technology patent applications. In the meantime, China has reduced the transfer cost and boosted the demonstration and application of green technology and promoted the use of new technologies for energy conservation, emission reduction, and comprehensive use of resources. Therefore, green technology innovation has become a powerful driving force of green development. China has promoted the unification of green development standards, underlined the role of science and technology in standard setting, and dynamically raised industry-specific green standards. For instance, the actual energy consumption and emission standard for large coal-fired units in China has been one of the most stringent in the world. Green product and service labelling or certification systems have been established based on international experience, including the standards and labels for production equipment and consumer goods similar to the Energy Star program of the U.S. and the Blue Angel ecolabel of Germany. In a nutshell, the development and diffusion of green technology has provided strategic support for China's green transformation.

5. **Green consumption and lifestyles have gained popularity**

Green consumption plays a vital role in promoting green production and green development. In recent years, China has vigorously encouraged consumers to buy green and energy-saving products such as high-efficiency light-emitting devices and water-saving products such as water-saving faucets, toilets, and washing machines. It has increased efforts in promoting new energy vehicles and stepped up the construction of electric-vehicle charging infrastructure. From 2012 to 2016, total government procurement of China's energy-saving/water-saving products amounted to RMB 746 billion. The number of green consumers on Alibaba's retail platforms surged by 14 times from 2012 to 2015, accounting for 16% of active users. According to estimates, high-efficiency and energy-saving air conditioners, refrigerators, washing machines, flat-panel TVs, and water heaters sold in China in 2017 alone could save roughly 10 billion kWh of electricity, equivalent to a reduction of 6.5 million tons of carbon dioxide, 14,000 tons of sulfur dioxide, 14,000 tons of nitrogen oxides, and 11,000 tons of particulate matter. Green home, school, community, shopping mall, and restaurant campaigns have been carried out across the country; green lifestyles have been advocated, calling on the public to save water and electricity, for example, by setting air conditioners to reasonable temperatures; vigorous efforts have been made in developing public transport and encouraging cycling, walking, and other green mobility options; household waste classification systems have been established in an effort to encourage public participation in waste classification and recycling. Green lifestyles give an impetus to the supply of green products and services and the green transformation of modes of production.

6. **Green finance and green service markets have started to flourish**

To develop green finance and green service markets is a necessary requirement of green transformation. Green finance can funnel funds towards the fields in favor of conservation and efficient use of resources, environmental improvement, and response to climate change, and guide enterprises to produce green and environmentally friendly products. In August 2016, seven central government departments including the People's Bank of China jointly issued the *Guidelines for Establishing the Green Financial System*, calling on the financial industry to provide financial services in various fields such as environmental protection and energy conservation, clean energy, green transportation, and green buildings. For example, green credit, in combination with national energy conservation and emission reduction funds and circular economy funds, should give priority to green development projects. In 2017, the executive meeting of the State Council decided to set up green finance reform and innovation pilot zones in five provincial-level regions including Zhejiang, Jiangxi, Guangdong, Guizhou, and Xinjiang to support the development of green finance. Meanwhile, China is actively promoting the energy rights and carbon trading systems. It will establish a national carbon trading market on the basis of pilot programs, draw up corresponding regulatory rules, and put in place a carbon trading market regulatory system; promote the pollutant emission permit system, and expand the pilot program on paid use and trading of pollutant emission rights to lay the groundwork for further development of carbon trading. The development of green finance and the establishment of carbon trading and emission rights trading markets have provided financing channels and market-oriented tools for green transformation.

7. **Unprecedented efforts have been made in pollution prevention and control**

Increased efforts in pollution prevention and control are vital for promoting green transformation. Since the *Plan for Aggregate Emissions Control* was developed for the first time during the 9th Five-Year Plan period, aggregate emissions control has been incorporated in China's five-year plans and become an obligatory target since the 11th Five-Year Plan. The *Outline of the 11th Five-Year Plan* set the obligatory target of reducing chemical oxygen demand (COD) and sulfur dioxide emissions by

10%. The *Outline of the 12th Five-Year Plan* put forth the obligatory targets of cutting COD and ammonia nitrogen, sulfur dioxide, and nitrogen oxide emissions by 8, 10, 8, and 10%, respectively. Since the beginning of the 13th Five-Year Plan period, China has enacted the new Environmental Protection Law, which is considered "the most stringent in history" and implemented the *Action Plan for Air Pollution Prevention and Control*, the *Action Plan for Water Pollution Prevention and Control*, and the *Action Plan for Soil Pollution Prevention and Control*. As a result, the emissions of some major pollutants have hit their peaks. In 2018, 338 cities in China saw 79.3% of days with good air quality, up 2.6% from 2015; the number of days with heavy pollution accounted for 2.2%, down 1.0%. GHG emissions have decreased dramatically. In 2018, carbon dioxide emissions per unit of GDP fell by 45.8% over 2005, meaning that the target of a 40% to 45% reduction by 2020 was met ahead of schedule. Surface water quality has improved on the whole. In 2018, out of 1,935 river cross sections in the country, 71.0% met grade I-III water quality standards, up 3.2% compared with 2016, while the proportion of those with quality below grade V was 6.7%, down 1.9%. Increased efforts in tackling pollution have improved the environmental quality, paving the way for green transformation.

8. **Reform of the green development system has been accelerated**

The key to promoting green transformation lies in systems and institutions. Since the 18th CPC National Congress, China has rolled out environmental policies with unprecedented frequency and carried out the strictest supervision and law enforcement ever. It successively issued the *Guidelines on Accelerating Ecological Civilization* and the *Integrated Reform Plan for Promoting Ecological Civilization* and formulated more than 40 reform plans regarding ecological progress, which constitute the basis of the ecological civilization system. Smooth progress has been made in all reform tasks. Specifically, China has advanced the reform of the property rights system for natural resource assets, improved the territorial space development and protection system, launched the pilot programs for the spatial planning system reform, strengthened the aggregate resource management and comprehensive conservation systems, pushed forward the system of paid use for natural resources and ecological compensation system, increased efforts in the environmental governance system reform, and basically established the ecological progress performance evaluation system and accountability systems. In particular, the national environmental protection inspection system has been in place. According to the five-step approach of supervision, assignment, tour of inspection, interview, and special inspection, four batches of central environmental protection inspections have been carried out, covering all thirty-one provincial-level regions. Special inspections in key regions, fields, and industries have been carried out. Supervision and law enforcement has been tightened; the policy of "co-responsibility of the Party and governmental officials" and the policy of "one post with dual responsibilities" have been effectively implemented; the accountability system has been strengthened, violations have been handled severely, and a large number of outstanding environmental problems have been resolved. To sum up, the system reform aimed at promoting green development has provided effective institutional support for green transformation.

2 Advantages and Challenges for Promoting Green Transformation

Green development is a long-term, complicated, and uphill task that involves economic, industrial, and technological progress and institutional changes. China has institutional advantages, the latecomer advantage, and the advantage as an enormous economy in promoting green transformation, with large space for green technological transformation of traditional industries. However, various constraints such as "squeezed growth" in industrialization, deficient resource endowments and energy mix, and peaking major pollutant emissions will pose great challenges to China's green transformation.

1. Advantages

The first is China's institutional advantages as a socialist country. Green transformation requires the market to play a pivotal role, of which the premise is that the government effectively fulfills its public functions. China has incorporated ecological progress into its Five-Sphere Integrated Plan, established the new concept of green development, and advanced the construction of a resource-conserving and environmentally friendly green development system. These efforts, together with the government' effective decision-making system and strong execution and society-wide consensus on green development, give China unique institutional advantages in promoting green transformation.

The second is the latecomer's advantage in green development. China's industrialization and urbanization started relatively late and are still at a low level. The new demand for industrial capacity and urban infrastructure can be satisfied by developing green capacity and green infrastructure, thus avoiding the lock-in effect of industrialization and urbanization and bringing enormous green benefits. For example, China is undergoing a sustained urbanization process. In 2018, the urbanization rate was close to 60% and is expected to exceed 70% in the future, meaning that 150 million rural people will flow to cities. This will lend momentum to economic growth and generate great green benefits from the circumvention of the lock-in effect.

The third is the advantage as a major economy. In 2018, China's GDP reached USD 13.6 trillion, accounting for 66% of that of the U.S. and remaining the second largest in the world. The tremendous economic size provides a solid basis for developing green technology and shaping complete industry chains, since the initial costs of green technology R&D can be shared and start-ups can be incubated within the domestic market. What's more, China has an enormous, full-fledged manufacturing system where some industries and technologies are leading the world and other are catching up with developed countries. This is conducive to forming complete industry chains by exploiting industrial and technological strengths in different fields.

The fourth advantage is the broad space for technological transformation of traditional industries. Traditional industries still make up a large part of the economy, with huge space for efficiency improvement. For instance, clean use of traditional energy sources has a broad prospect, and China ranks the first in the world by the share of energy technology R&D spending in GDP and has made great strides in improving the efficiency of coal-fired power plants.

Lastly, China is leading the world in new energy development. China has an abundance of wind energy, solar energy, shale gas, and biogas resources. Since 2005, the capacity of wind turbines has doubled almost every year and is still growing at a high rate. China is also the largest manufacturer of solar PV panels and has the largest solar PV power generation capacity in the world, making it possible for China to reduce its dependence on fossil fuels and improve its energy mix.

2. **Challenges**

The first challenge comes from the centralized discharges of pollutants brought by the industrialization. Compared with industrialized countries, China's industrialization process is marked by "squeezed growth," meaning that a broad range of industrial products have seen an explosive growth and the scale production reached unprecedented levels in a short period of time. As China is becoming a "world factory," massive volumes of pollutants, greenhouse gases, and waste are generated, putting enormous pressure on the environment and resulting in hefty treatment costs. In recent years, China has advanced the supply-side structural reform and stepped up efforts to phase out high-energy-consuming and high-polluting backward capacity and cut excess capacity. However, that will inevitably bring impact on employment, leading to a substantial increase in public expenditures for social security, reliefs, compensations, and placement costs for the unemployed. Additionally, financial institutions have to pay bad debt expenses.

Second, there are challenges in the adjustment of the energy mix and improvement of energy efficiency. Given China's resource endowments, the dominance of coal in the energy mix is unlikely to change in a short period of time. The imbalanced distribution of energy producers and consumers poses a barrier to the development of clean energy in terms of accommodation and transmission channels and cultivation of domestic markets. Therefore, hydropower, wind, and solar curtailment occurs more often than not. Meanwhile, China still lags behind in terms of energy resource use efficiency. According to OECD estimates, in 2016, China's energy output rate was equivalent to 84% of that of the U.S., 57% of that of Germany, and 59% of that of the Japan. In China, one ton of CARBON DIOXIDE can bring in economic benefits of USD2, 150, roughly half of the USD4, 240 of OECD countries. In the past twenty-plus years, green total factor productivity (also known as environmentally-adjusted total factor productivity, which refers to economic growth not explained by human capital, productive capital, and natural capital inputs) has contributed less than 30% to China's economic growth, while the rate has reached 60% in OECD countries. This implies that the dependence of China's economic growth on human capital, productive capital, and natural capital inputs is significantly higher than that of OECD countries.

Third, it is an arduous task to adjust the economic structure and upgrade industrial value chains. Over forty years of reform and opening-up, China's economic structure has seen great changes. Specifically, the share of the service sector rose from 24.6% in 1978 to 52.2% in 2018, which, however, is still lower than that in other emerging market economies such as Brazil, Russia, India, Indonesia, and South Africa. In contrast, the share of the industrial sector has long remained above 40%. Despite the downward trend since 2011, it was still as high as 40.7% in 2018, much higher than that in other emerging market economies. Following Germany, Japan, and the U.S., China became the world's largest manufacturer in 2010. According to data from the National Bureau of Statistics of China, in 2017, China produced about half of the world's crude steel (832 million tons), coal (3.52 billion tons), and cement (2.34 billion tons) and nearly a quarter of the world's automobiles (29.02 million). In 2018, the added value of China's manufacturing accounted for over a quarter of the global total. In recent years, China has accelerated the upgrade of the manufacturing value chain, with its exports from low-end products to middle and high-end products. However, there are still major gaps between China and industrialized countries in basic manufacturing capabilities, such as R&D of core technologies, process innovation, and digital management. In other words, industrial transformation and upgrading will remain a long haul.

Fourth, China is under great pressure to tackle environmental issues. Although the discharges of some pollutants have showed a downward trend in recent years, the amount of discharges has remained enormous, placing huge pressure on the environment. According to statistics, if China's urbanization rate rises by 1%, domestic sewage will increase by 1.15 billion tons, household waste by 12 million tons, construction land by 1000 square kilometres, domestic water consumption by 1.2 billion tons, and energy consumption by 80 million tons of stand coal equivalent. Overall, China's air quality has improved, but the situation remains severe. Since 2013, the annual average concentration of PM 2.5 in Chinese cities has been on the decline, but seasonal fluctuations are notable. The water quality has improved on the whole. While the water quality of main river basins has turned for the better, eutrophication is a major problem affecting the water quality of lakes, and groundwater pollution is also a great concern. By and large, the soil quality is not optimistic, as reflected in serious soil pollution in some regions and acute soil problems in abandoned industrial areas and mining lands.

3 Outlook for China's Green Transformation

For years to come, China will push forward its industrialization and urbanization endeavors and remain at a turning point of the Environmental Kuznets Curve. Pollutant emissions, carbon dioxide emissions, and energy and resource consumption will see a turning point one after another, but China will still face huge resource and environmental pressure. The 14th Five-Year Plan period will be a critical period for green transformation. China must word hard to further reduce energy intensity and carbon intensity, improve the use efficiency of resources, solidify the decoupling of emissions of major pollutants and economic growth, and speed up the development of a resource-conserving and environmentally friendly green development system, thus blazing a new trail of green transformation with Chinese characteristics.

1. **Resource and environmental pressure will reach a peak in the coming years**

Energy demand is expected to peak between 2030 and 2040, while fossil fuel energy consumption and carbon emissions around 2030. China's peak energy demand is estimated at 6–8 billion tons of standard coal equivalent, and per capital peak energy consumption may account for 32–42% of that in the U.S. and 65–84% of that in Japan. In terms of energy mix, China's coal consumption has reached a plateau since 2014, but coal will remain a major source of energy for a long time and still account for more than 50% of China's total primary energy consumption after 2030. Oil demand will see a slow increase and is expected to exceed 8 million tons in 2030. Natural gas consumption will grow steadily at an annual average rate of about 5%, and its share in total primary energy consumption will rise from 7.8% in 2018 to 15% in 2030. Demand for wind energy, solar energy, and other renewable energy sources will increase substantially, and non-fossil energy consumption is expected to surpass that of oil by 2025.

Discharges of major pollutants have reached a turning point. Sulfur dioxide and nitrogen oxide emissions have successively peaked and entered a downward trend, while volatile organic compounds and ammonia emissions are expected to reach their peaks before 2020. The combined emissions of major air pollutants may peak around 2020. Due to the influence of nonpoint source pollution from agriculture, emissions of water pollutants may peak between 2020 and 2025 and then stay more-or-less flat.

It is still uncertain when environmental quality will meet the standard in every aspect. The overall air quality has improved. It is expected that by 2025, 50% of cities will meet the air quality standard, the proportion of days with good air quality in 338 cities at and above the prefecture level will further increase, and the annual average concentration of PM 2.5 will continue to decline, while ozone may become a new major pollutant. There is still uncertainty about the improvement of the overall water quality given the considerable differences in the progress in the water quality of main rivers, lakes, groundwater, and oceans. Compared with air and water quality, to fundamentally improve the soil quality is an even more arduous task.

2. **Greater efforts should be made in green transformation during the 14th five-year plan period**

Following China's success in building a moderately prosperous society in all respects, the 14th Five-Year Plan period will mark the beginning of a new journey of comprehensively building a modern socialist country. During the 14th Five-Year Plan period, we must green transformation, step up efforts to build a resource-conserving and environmentally friendly green development system and a new modernization pattern featuring harmony between human and nature.

First, we will foster a green industry development system. We will promote the transformation and upgrading of green industries, develop new business models of the green economy, boost green technology innovation, and cultivate new entities for the green economy. We will develop green industries such as energy conservation and environmental protection industries, clean production, and clean energy, green agriculture including ecological circular agriculture, and green service industries such as energy performance contracting, water-saving management contracting, and green technology services, while encouraging technological transformation for green production and energy conservation and emission reduction. We will push for transformation in energy production and consumption and build a clean, low-carbon, safe, and efficient energy system. We will upgrade the entire industry chain for key products where conditions permit, strengthen collaboration between relevant entities in technical cooperation, production and use, service, and other areas, and strive to build a green industry development system.

Second, we will put in place a green scientific and technological innovation system. We will increase inputs in the R&D of green technology and fortify connectivity between all parts of the innovation chain; strengthen basic research and carry out research on pollution and its impacts to provide a scientific basis for improving the technical route of environmental management. We will increase investment in the R&D of general green technology, especially green processes and equipment, and beef up the capacity to provide equipment for green technological transformation of enterprises. We will promote collaboration between enterprises, universities, research institutes, and end-users, encourage large enterprises and research institutes to establish technological innovation alliances, and speed up the industrialization of green technology. We will reinforce the protection of intellectual property rights of green technology, increase the enthusiasm of enterprises for green technology and business model innovation, give full play to the decisive role of the market in green technology innovation, route selection, and allocation of innovation resources, and develop a market-oriented green technology innovation system.

Third, we will refine the resource conservation and recycling system. We will promote enterprises to shift towards circular production and implement whole life-cycle management; advocate the rules of 3R (reducing, reusing, and recycling) and develop a circular enterprise evaluation system for key industries; carry out green transformation of industrial parks, push for circular industry chains, high-efficiency resource use, and centralized pollutant treatment, optimize the industry chains of the circular economy, and raise the level of industry linkage and recycling. We will improve the resource recycling system, promote the extended producer responsibility system, put in place the recycled products and materials promotion and use system, and fine-tune the system restricting the use of disposal consumer items. We will improve the circular economy assessment system, tighten circular economy standards and certification, and push forward the implementation of the green credit management system.

Fourth, we will put in place the spatial development and protection system. We will advance the implementation of the system of functional zoning, establish and enforce a unified spatial planning system as soon as possible, improve the spatial governance system, and stick to a single blueprint until the end. We will carry out the campaign on comprehensive management of territorial spaces covering urban areas, rural areas, ecological functional zones, mineral resources development zones, and coastal zones, establish the nature reserve system centred on national parks, and build national ecological security shields.

Fifth, we will develop the green development market service system. We will further boost green finance and experiment with green finance tools such as green loans, green bonds, green insurance, green funds, and green certificate trading. We will improve the carbon trading market, including the information disclosure system, and increase the transparency of allocation of carbon emission quotas; carry out pilot programs with diversified products and ways of trading, draw up relevant regulatory rules, establish accounting rules for carbon assets and carbon trading, and strengthen supervision over carbon trading with overseas entities. We will promote the pollutant discharge rights trading system, extend the pilot programs on paid use and trading of pollutant discharge rights to more regions where conditions permit, and improve the mechanism that allows enterprises to benefit more from emission reduction through pollutant discharge rights trading. In addition, we will promote pollutant discharge rights trading across different administrative regions in key river basins and key regions of the air pollution prevention and control initiative.

Sixth, we will promote green lifestyles and consumption. We will advocate simple, moderate, green, and low-carbon lifestyles, encourage the use of green products, and provide more effective incentives for green mobility. We will establish dedicated circulation channels for green products, support the establishment of green wholesale markets, green malls, energy-saving supermarkets, water-saving supermarkets, and so forth, and encourage marketplaces, malls, supermarkets, souvenir shops, and other businesses to set up dedicated zones for green products. We will promote green consumption through the "Internet Plus" initiative, encourage ecommerce enterprises to sell green products and services independently or in partnership with brick-and-mortar enterprises, and boost online trading of green products and second-hand items to meet the diverse needs of different groups for green products.

Finally, we will improve the green development regulatory system. We will reform the evaluation system and put in place relevant institutions as soon as possible. Specifically, we will deepen the environmental regulation system reform, improve environmental laws and regulations, optimize the regulatory structure, and enhance professional regulatory capabilities to make environmental regulation more effective. We will reform the resource and environmental management system and cultivate an environmental governance system where the government takes a leadership role, enterprises play a major role, and social organizations and the public participate.

The 14th Five-Year Plan period will be a critical period for China to shift towards green development. Only by accelerating green transformation can we lay the groundwork for making a fundamental improvement in the environment and basically attaining the goal of building a beautiful China by 2035 when socialist modernization is basically realized. In addition, China's exploration in green transformation will provide a new option for developing countries to adopt the green development model and contribute China's wisdom and solution to the building of a sustainable community with a shared future for mankind.

Oil-Yielding Peony Industry in China: A Major Move in Building an Ecological Civilization

Yucai Li

Peonies are woody plants native to China. Hailed as the "king of flowers" for their beauty and fragrance, they have been dearly loved by people since ancient times. Peony growing began in the Jin dynasty, gained popularity in the Sui dynasty, and flourished in the Tang and Song dynasties. Peonies have been used for over 2000 years and cultivated for over 1600 years in China. They were first mentioned as medicinal plants in *The Divine Farmer's Material Medical Classic* (Shen Nong Ben Cao Jing), the earliest Chinese herbal textbook compiled in the first or second century, and started to be recognized as ornamental plants in the Northern and Southern Dynasties. In recent years, as the potential of peonies in oil production has been gradually uncovered, "oil-yielding peony" has become an important category in addition to "medicinal peony" and "ornamental peony."

As evidenced by years of research and practice, developing the oil-yielding peony industry is of great significance to improve the environment, reduce China's dependence on exports for edible oil, and ensure food security and people's health. It is an important measure to help get rid of poverty, improve people's lives, promote healthy and sustainable development, and achieve common prosperity and represents a major practice to "make industries eco-friendly and commercialize natural resources," speed up the construction of the "five systems for ecological progress," and advance development of an ecological civilization.

Y. Li (✉)
The State Forestry Administration, Beijing, China

J. Pan et al. (eds.), *Beautiful China: 70 Years Since 1949 and 70 People's Views on Eco-civilization Construction*, https://doi.org/10.1007/978-981-33-6742-5_27

1 Overview of Oil-Yielding Peony

Oil-yielding peony refers to a group of peonies whose seeds can be used to produce edible oil. The way of reproduction determines their oil-yielding potential. Currently, the main oil-yielding peony varieties being promoted in China are "Fengdan" and "Ziban."

As a perennial woody plant native to China, oil-yielding peony is resistant to drought, barrenness, low temperature, and salt-alkali and thrives in partial shade. It can grow in more than twenty provincial-level regions in China, ranging from Heilongjiang and Jilin in the north to the northern parts of Guangdong and Guangxi in the south and from Yunnan, Xinjiang and Tibet in the west to the eastern coastal regions. According to statistics, the total area of regions suitable for planting oil-yielding peony is 4.2 million square kilometres, accounting for 43.75% of China's total land area.

It generally takes three to four years for oil-yielding peony seedlings to produce seeds, and the seed quantity grows year by year. The plants enter the productive period in their seventh or eighth year, which can last thirty to fifty years. Into the productive period, oil-yielding peony can yield around 200 kg of seeds per *mu* (666.67 m^2) in the Yangtze River basin, 300 kg in the Yellow River basin, and 150 kg in Northwest China. For improved varieties grown, managed, and harvested with advanced techniques, the per *mu* yield can reach 400–500 kg.

All parts of oil-yielding peony are of value. The seeds can be pressed to produce high-end edible oils; the petals can be extracted for essential oils which are used in cosmetics; the stamens can be used to make tea, which is good for the urinary system, especially the prostate; the seed coats can be used to extract flavonoids and peony proanthocyanidins, which help improve blood circulation, lower cholesterol, and scavenge free radicals and have antioxidant effects; the fruit pods can be extracted for peony polysaccharides, which can enhance phagocytes and body immunity; the seed meal can be extracted for polysaccharide gums, which have anti-inflammatory and anti-oxidant effects; the leftovers of seeds, fruit pods, and seed coats can be made into nutrition powder or nanophase wood flour, a new type of energy-saving and environmentally-friendly material. So far, hundreds of products such as food, healthcare products, and daily chemical products have been developed from oil-yielding peony, and some of them have gained popularity among consumers.

2 Emergence of the Oil-Yielding Peony Industry

As medical expert Li Shizhen of the Ming dynasty points out in the Compendium of Material Medical, though peony bears seeds, its seedlings grow out of roots. This shows that even more than 500 years ago, people already knew that some peony varieties could produce seeds. Later, some peony growers found that peony seeds contained "oil." In the 1950s and 1960s when edible oils were in short supply, people tried to extract oil from peony seeds. Because of the constraints of the processing techniques, however, such peony oil had poor taste and caused no physical discomfort. From the early 1990s, China started to import large quantities of edible oils and oilseeds; as the demand for edible oils was basically satisfied, the development of oil-yielding peony came to a halt. In the late 1990s, peonies were included in the investigation of woody oil resources. In 2000, based on experiments, testing, and analysis, it was found in Heze City of Shandong Province that apart from seeds that can be extracted for oil, the seed coats, petals, stamens, and seed meal of peonies can be made into various by-products; such findings were reported to the then State Forestry Administration (now State Forestry and Grassland Administration). Under the guidance of the authority, the city carried out small-scale experiments in this regard, which turned out a success. In March 2011, based on a lot of specific materials and safety reports, the then Ministry of Health (now National Health Commission) approved peony seed oil to be a "new resource food." In August 2011 when I worked as Deputy Secretary of the Party Leadership Group and Deputy Minister of the State Forestry Administration, I submitted an investigation report on peony seed oil to the leaders of in charge of agriculture of the State Council, who gave instructions on August 26 that we should get more information and carry out pilot programs. With support from the State Council, we conducted experiments on peony growing and comprehensive use of peony seeds in Heze of Shandong and Luoyang of Henan. The oil-yielding peony industry in China has since come into being.

On March 18, 2013, I submitted a report on the oil-yielding peony industry to General Secretary Xi Jinping, Premier Li Keqiang, and then Vice Premier Wang Yang. All of them made comments on the report, reflecting high attention to the development of the industry from Party and state leaders. Since then, the oil-yielding peony industry has developed rapidly and maintained strong momentum. On the afternoon of November 26, 2013, General Secretary Xi Jinping paid to a visit to Yao and Shun Peony Industry Park in Heze. He was amazed when learning that peonies can be used for not only ornamental and medical purposes but also can be turned into peony seed oil, tea, essential oils, food, and supplements. During the meeting with local officials in charge, he said, "We learned about the development of the peony industry and how it has benefited farmers at Yao and Shun Peony Industry Park and came to know that peonies are not only of value for appreciation and tourism but also can be processed into value-added products. This has expanded our knowledge." The next day, the State Council convened relevant departments directly under the CPC Central Committee to hold a coordination meeting on the development of the oil-yielding peony industry. On December 26, 2014, the General Office of the State Council issued the *Opinions on Accelerating the Development of the Woody Oil Industry* (Guo Ban Fa [2014] No. 68), giving prominence to oil-yielding peony. According to the document, by 2020, 800 key counties growing tea oil camellia, walnut, peony, and other woody oil plants will be built; a number of demonstration bases featuring standard, intensive, large-scale, and industrialized operations will be established; the growing area of woody oil plants will increase from the current 120 to 200 million *mu*, and the annual output of edible woody oil will reach approximately 1.5 million tons.

On December 16, 2015, the eighteenth day after the conclusion of the Central Conference on Poverty Alleviation and Development, I submitted a report explaining the role of the oil-yielding peony industry in targeted poverty alleviation to General Secretary Xi Jinping, Premier Li Keqiang, and then Vice Premier Wang Yang. General Secretary Xi Jinping marked up the report after reading it, and the other two leaders gave instructions, requiring relevant departments to put forth practical proposals. Later, the Ministry of Finance and the State Forestry Administration handed in a joint proposal to the leadership. Then Vice Premier Wang Yang made comments on the proposal, and General Secretary Xi Jinping and Premier Li Keqiang ticked it off. The proposal calls for support for the development of the woody oil industry including peony seed oil by developing relevant plans, carrying out pilot programs in some regions, giving preference in the grain-for-green program, supporting basic research, and revising fund management measures. In January 2019, twenty-one academicians of the Chinese Academy of Sciences and the Chinese Academy of Engineering jointly made suggestions to General Secretary Xi Jinping that the oil-yielding peony industry should be boosted on a national scale. By the end of 2018, the growing area of oil-yielding peony in China reached nearly 10 million *mu*, and there were more than 500 peony seed oil growing and processing enterprises above certain size and over 300,000 people working in the industry chain from seedling growing, planting, and management to processing and sales. The oil-yielding peony industry has started to flourish in more regions where conditions are suitable, the nursery and growing area has increased substantially, and the processing industries of peony seed oil and other relevant products have developed rapidly.

3 Significance of Developing the Oil-Yielding Peony Industry

1. **Growing oil-yielding peony has great ecological value and can help improve the environment**

Since the founding of the People's Republic of China, the country has carried out numerous environmental protection activities such as large-scale forestation, water conservancy, soil and water conservation, and prevention and control of desertification and pollution and built a large number of ecological conservation and environmental protection facilities. These efforts have played a significant role in mitigating natural disasters, guaranteeing rapid and sustained economic growth, and protecting the security of people's lives and properties. For various natural and human reasons, however, there is still not much to rejoice about the environment; for example, soil erosion and desertification remain severe, and the pressure to cut emissions is rising. General Secretary Xi Jinping has reiterated that we should "protect our natural environment in the same way we would protect our own wellbeing" and that by protecting the environment we are protecting productive forces, and by improving the environment we are developing productive forces.

China is one of the countries that suffer the most from soil erosion. According to the results of the first water conservancy census, 2,949,100 km^2 of land was blighted by soil erosion, accounting for 30.72% of the whole territory. Worsening soil erosion has led to sedimentation in rivers, lake, and reservoirs and aggravated flooding and water shortage, causing huge damage to the economy and people's lives. Therefore, piles of materials and funds, together with manpower, have to be invested in the prevention and control of floods and droughts as well as relief and reconstruction every year. oil-yielding peony is a perennial shrub whose planting density (colonization) is about 2000 per *mu*, and there is no need for rotation within thirty to fifty years, effectively preventing soil erosion that is caused by plough if traditional food crops were planted. Moreover, peony has a well-developed root system. Monitoring data shows that growing oil-yielding peony can reduce soil erosion by about 0.8 m^3 per *mu* of land a year compared with barren hills or wasteland, testifying to its role in soil and water conservation.

China is one of the countries most affected by desertification. Decertified areas accounts for about one third of the whole territory. Desertification causes direct economic losses of nearly RMB 100 billion every area. The area of wind eroded decertified land makes up over 60% of total decertified areas, where the climate is harsh, arid, and cold and the land is barren. Oil-yielding peony is resistant to drought, barrenness, low temperature, and salt-alkali and is one of the plants that can effectively help prevent desertification. The peony variety "Ziban" in Lanzhou of Gansu grows in cold, dry, and barren hills at an altitude of above 2000 m and can bear flowers and seeds even with rainfall of only 300 mm. According to research, growing oil-yielding peony in sandy areas (with a coverage of over 60%) can reduce wind speeds by 22.7% and wind erosion by 50%, making it a favoured shrub for future wind prevention and sand fixation projects.

Carbon emissions are a focus of debate in the world and an important bargaining chip in diplomatic negotiations. China is the world's largest emitter of carbon dioxide, accounting for over one fifth of global carbon emissions. Therefore, it is under enormous international pressure. As a responsible major country, China has pledged to reduce its carbon intensity by 40–45% by 2020, compared with 2005 levels. As trees grow, they absorb carbon dioxide in the air, binding and storing the carbon while releasing fresh oxygen. Research shows that to grow by just a cubic meter of stem wood, trees can absorb 1.83 tons of carbon dioxide and release 1.62 tons of oxygen. In recent years, with the opening of the carbon trading market, carbon sink trading has picked up and gained wide attention. According to estimates, one *mu* of oil-yielding peony can bind and store an average 1.7 tons of carbon dioxide during its lifetime, equivalent to carbon sink trading of about USD 6.32. Therefore, oil-yielding peony has great potential in carbon fixation and oxygen production as well as carbon sink trading.

2. **Growing oil-yielding peony can bring high economic benefits and contribute to targeted poverty alleviation and development of an ecological civilization**

Land is the most important means of production for farmers, and the primary sector centred on crop farming is where poverty alleviation efforts pay off the most. Growing plants that adapt to the natural conditions of poverty-stricken areas, align with their productivity and technological levels, and deliver high economic and ecological benefits is of great significance to advance the development-oriented poverty alleviation endeavour, implement the targeted poverty alleviation strategy, and realize the goal of building a moderately prosperous society in all respects by 2020.

First, compared with conventional crops, oil-yielding peony has higher economic value and can increase the income of farmers and help poverty-stricken areas get out of poverty. To increase the income of farmers is a major task of the poverty alleviation initiative and holds key to the building of a moderately prosperous society in all respects. China is a big agricultural country with a large population. Due to the longstanding low prices of agricultural products and poor economic benefits, farmers' income and rural development have been greatly restricted. Mountains, sandy land, and species resources, especially woody oil plants, are advantageous resources in many poverty-stricken areas, presenting great potential for wealth and prosperity. Growing oil-yielding peony can increase farmers' annual income by RMB 4000 per *mu*, providing an efficient and targeted solution for poverty alleviation. Therefore, it is instrumental in improving the lives of people in old revolutionary areas, areas with large ethnic minority populations, border areas, and poverty-stricken areas, addressing the most immediate and realistic issue of poverty that the people are most concerned about, and speeding up the economic and social development of poverty-stricken areas. Furthermore, it can help promote ethnic unity, maintain social stability, and ultimately build a moderately prosperous society in all respects.

Second, with high resistance to stress, oil-yielding peony can grow in a wide range of conditions, suitable for large-scale planting in poverty-stricken areas. Most of poverty-stricken areas are mountainous, hilly, or plateau areas, where cultivated land is not so fertile and production and living conditions are harsh due to frequent droughts and floods and serious desertification and soil erosion. Because of the unfavourable climate and geographical conditions, agricultural outputs are generally low and unstable in those regions. Oil-yielding peony can be planted on a large scale in poverty-stricken areas without affecting crop production, meeting the requirement of "observing the red line for ecological conservation and ensuring no decrease in the area of cultivated land" in the Guidelines of the CPC Central Committee and the State Council on Accelerating Ecological Civilization.

Finally, oil-yielding peony is easy to manage, which is in alignment with productivity and technological levels in rural areas. Nowadays, a large number of young and middle-aged rural people go to the cities to work as wage earners, while the elderly, women, and children left behind, who are incapable of hard manual labour, have to "sow the seeds, spread the fertilizer, and leave it to luck." This to a large extent affects their levels of income. As a perennial shrub, oil-yielding peony only requires simple management practices such as weeding and fertilization and does not need rotation for thirty to fifty years. Therefore, growing oil-yielding peony can save labour, time, and cost and befits the actual situation of rural areas.

3. **Oil-yielding peony features high output and oil yields and can provide safe and healthy edible oils to help ensure national food security**

In 2018, China consumed a total of 38.496 million tons of edible oils, of which only 11.928 million tons were extracted from home-grown oilseed, indicating a gap of 26.568 million tons. In the same year, China imported 8.087 million tons of edible vegetable oils and 94.489 million tons of edible oilseeds, spending over USD 50 billion in foreign currency. Its edible oil self-sufficiency rate dropped from 60% in 2000 to 31% in 2018, well below the national crop security line, and the rate tends to decline further.

General Secretary Xi Jinping has reiterated that we should always have control over our own food supply and be self-sufficient in food supply. The woody oil industry is a traditional industry in China and an important source of healthy and high-quality edible vegetable oils. Compared with other oil plants, oil-yielding peony features high output and oil yields and can turn out 40 kg of high-end edible oils per *mu* of land. Moreover, it thrives in partial shade and can be interplanted with other woody oil plants such as shiny-leaved yellowhorn, purpleblow maple, and tea oil camellia. To fill China's edible oil gap of 26.568 million tons in 2018, we only need to interplant oil-yielding peony with shiny-leaved yellowhorn on 379 million *mu* of land (an average of 30 kg of edible oil per *mu*).

4. **Peony seed oil is nutritious and can help improve people's health**

As General Secretary Xi Jinping pointed out in the report to the 19th CPC National Congress, a healthy population is a key mark of a prosperous nation and a strong country, and we should initiate a food safety strategy to ensure that people have peace of mind about what they're putting on their plates. People now have higher requirements on the quality of edible oils in the hope of living a long, healthy life. Currently, nutritionists at home and abroad believe that the content of unsaturated fatty acids in edible oils, especially alpha-linolenic acid (ALA), is one of the determinants for the quality of edible oils.

The content of unsaturated fatty acids in peony seed oil exceeds 92%, where area percentage of ALA is above 43%. ALA is an important component of human brain cells and tissue cells and an indispensable polyunsaturated fatty acid that cannot be synthesized or replaced within the human body, hence known as "blood nutrient," "vitamin F," and "plant DHA." The World Health Organization and the Food and Agriculture Organization of the United Nations jointly issued a statement in 1993 to promote the adequate intake of ALA worldwide.

In 2014, Food and Chemical Toxicology (FCT), an internationally renowned American academic journal, published a paper titled *α-Linolenic Acid: Nutraceutical, Pharmacological and Toxicological Evaluation*, which was indexed in the Science Citation Index (SCI). As the paper points out, ALA is an essential fatty acid needed for human health and can be acquired via regular dietary intake of foods that contain ALA or dietary supplementation of foods high in ALA. ALA has been reported to have cardiovascular-protective, anti-cancer, neuro-protective, anti-osteoporotic, anti-inflammatory, and antioxidative effects and is safe for use as a food ingredient, based on the limited data currently available.

According to Kuang Tingyun, academician of the Chinese Academy of Sciences, fatty acids in peony seed oil have small molecules and are easy to absorb, and the content of unsaturated fatty acids makes up over 90% of the total, which is significantly higher than that in olive oil, soybean oil, rapeseed oil, and peanut oil. In particular, peony seed oil contains much higher ALA content than other commonly used vegetable oils, and the ratio of linoleic acid to ALA is below 0.6, making it a healthy edible oil with great potential. Researchers at the School of Pharmacy of Yantai University find in their experiments that moderate intake of peony seed oil can protect liver cells from chemical damage while inducing Phase II enzyme activity and reducing free radicals. In addition, researchers at Anhui University of Chinese Medicine find that peony seed oil can ameliorate hyperlipidemia and hyperglycaemia and to some extent improve glucose tolerance.

5. **Oil-yielding peony has high ornamental value and can help meet people's ever-growing spiritual and cultural needs**

In his report to the 19th CPC National Congress, General Secretary Xi Jinping vowed to build China into a great modern socialist country that is prosperous, strong, democratic, culturally advanced, harmonious, and beautiful by the middle of the century. Peonies look both magnificent and graceful. They are known as the "king of flowers" for its "national beauty and heavenly fragrance" and a top hopeful of China's national flower. In a sense, peonies are an indispensable element in the building of a beautiful China. When peonies are in full bloom every year, hundreds of thousands of people swarm to Dianjiang of Chongqing, Tongling of Anhui, Wanhua Hill in Yan'an of Shaanxi, or other places known for peonies. What's more, peonies have become a calling card of tourism for Heze and Luoyang. The 36th China (Luoyang) Peony Culture Festival, which kicked off on April 5, 2018, attracted a total of 26.4731 million Chinese and foreign visitors, bringing in revenue of RMB 24.196 billion. The 27th Heze International Peony Festival, which opened on April 12, 2018, received a total of 9.6363 million visitors from home and abroad, bringing in revenue of RMB 6.385 billion.

The report to the 19th CPC National Congress makes it clear that the principal contradiction facing Chinese society in the new era is that between unbalanced and inadequate development and the people's ever-growing needs for a better life. In addition to bringing economic benefits, oil-yielding peony can please both the eye and the mind and therefore increase people's sense of happiness.

6. **Oil-yielding peony conveys profound cultural significance and can help build up cultural confidence of Chinese nation**

In China, peonies have long been an important part of traditional culture as a symbol of good fortune, wealth, and prosperity. *The Classic of Poetry*, the oldest existing collection of Chinese poetry compiled over 3000 years ago, contains lines that depict someone giving peonies to their beloved to show their affection. After Yang Guang, Emperor Yang of Sui, succeeded to the throne in 604 A.D., he issued an imperial order proclaiming peonies as the national flower, the first national flower in human history. Peonies were considered the national flower in the Tang, Song, and Ming dynasties and explicitly designated as the national flower in the Qing dynasty.

Fast forward to modern times, all generations of Party and state leaders have showed appreciation for peonies. In 1939, Mao Zedong, Zhou Enlai, and other leaders went to Wanhua Hill at the foot of Baota Mountain in Yan'an to admire the beauty of peonies. In 1959, Zhou paid a visit to Luoyang together with foreign guests and said to them, "peonies are the national flower of China. They are gorgeous and elegant, symbolizing prosperity and happiness." In 1994, the National People's Congress instructed then Vice Chairman of the Standing Committee of the National People's Congress Chen Muhua and Minister of Agriculture He Kang to lead a national flower selection, which was organized by China Flower Association. Based on the opinions solicited from all sides, peonies topped the national flower shortlist with 58.06% of votes. *The Commemorative Album of the Selected National Flowers of Some Participating Countries* published during the 1999 Kunming International Horticultural Exposition features peonies with a caption "National flower of China—Peonies (TBD)"; despite the word "TBD," this fully reflects the unrivalled status of peonies as the national flower of China.

The giant panda, also native to China, has become a symbol and a goodwill ambassador of the country. The cherry blossom, or sakura, is one of Japan's national flowers and holds a special place in Japanese culture. According to Japanese records, the cherry blossom originated in the Himalayas in China's territory and later found its way to Japan. At the grand gathering celebrating the 95th anniversary of the founding of the Communist Party of China, General Secretary Xi Jinping stressed that we should "remember where we started in the first place and keep moving forward." This means that we should stick to the Four-sphere Confidence, namely confidence in the path, theory, system and culture of socialism with Chinese characteristics. Xi also highlights that cultural confidence is a kind of more fundamental, extensive, and profound confidence. Vigorously developing the oil-yielding peony industry throughout the country to meet our people's needs for healthy and high-quality foods and gain a foothold in the international market is an important way to boost our cultural confidence and carry forward the traditional peony culture.

Embrace the Philosophy of Ecological Civilization and Promote Socioeconomic Development in Line with Resources and Environment Carrying Capacity

Tan Zhu

For a long time, China's socioeconomic development has largely followed the model during the traditional industrial civilization. Due to the inadequate consideration for resources and environmental factors, the scarcity of natural resources and the limit of environmental capacity have not been well reflected in the allocation of various factors and resources. As a result, the economic development has been at the cost of resources and environment to a certain extent. From a comprehensive and strategic perspective, the 18th CPC National Congress gave high priority to make ecological progress and incorporate it into all aspects and the whole process of advancing economic, political, cultural, and social progress, as well as the Five-Sphere Integrated Plan. According to the philosophy of ecological civilization to respect, accommodate to and protect nature, we need to develop a regional socioeconomic development model based on the carrying capacity of resources and environment. This has become an important topic under the "new normal" in China and a major task to boost ecological progress.

1 The Significance of Ecological Civilization Philosophy for Socioeconomic Development

The ecological progress constitutes a vital solution for China to address the problems arising from rapid industrialization and urbanization, such as environmental pollution, ecological imbalance and resource scarcity, in a bid to achieve sustainable development. From primitive civilization to industrial civilization, the relationship between man and nature has shifted from human dependence on nature, fear of nature to triumph over nature. During the period of industrial civilization in particular, the

T. Zhu (✉)
China Renewable Resources Research Center, Nankai University, Tianjin, China

J. Pan et al. (eds.), *Beautiful China: 70 Years Since 1949 and 70 People's Views on Eco-civilization Construction*, https://doi.org/10.1007/978-981-33-6742-5_28

mankind created unprecedented material wealth and scientific and cultural achievements. With the rapidly increasing capability of the mankind to transform nature, the contradiction between man and nature has intensified accordingly. In this process, human beings have carried out unprecedented and large-scale development of the natural environment, leading to the depletion of natural resources and the deterioration of the environment. In the meanwhile, the survival and development of human beings themselves have been increasingly constrained.

To get rid of the current resource and environmental crisis, we must abandon the "predatory" development model during the industrial civilization, but pursue instead a sustainable development model featuring the "harmonious development between man and nature." In the pursuit of socioeconomic development of yore, we paid more attention to the coordinated relationship between economic development and environmental protection, with the latter giving way to the former from time to time. Such choice had been inevitable, given the historical background and the national conditions characterized by a huge population, weak economy, unbalanced development, as well as the major task of improving people's living standards. After 40 years of reform and opening up, China has made outstanding achievements in economic and social development, while the problem of excessively exploiting the ecological environment has become increasingly prominent. It has become more and more challenging for the ecological environment to support such a large economy to grow rapidly in the traditional and extensive model. The resource supply is facing severe challenges, the environmental quality of some regions has fallen short of the standards for a long time, and regional ecological functions have weakened. In line with the philosophy of ecological civilization, we need to follow the objective law about the nature's limited carrying capacity, making it urgent to solve the problem of overloading resources and environment. Against such backdrop, it has become an important issue for China to develop a socioeconomic development model based on the carrying capacity of resources and environment in the new era of reform and development.

2 The Carrying Capacity of Resources and Environment Serves as the Fundamental Principle and Important Basis for the Development of Ecological Civilization

In a bid to fully implement the philosophy of ecological civilization and in line with the Five-Sphere Integrated Plan, the carrying capacity of resource and environment should be regarded as an important component in the top-level design and incorporated into all aspects and the whole process of advancing economic, political, cultural, and social progress, so as to realize the harmonious development of man and nature. The Third Plenary Session of the 18th CPC Central Committee adopted the "Decision of the Central Committee of the Communist Party of China on Some Major Issues Concerning Comprehensively Continuing the Reform." It highlighted to develop a system for building an ecological civilization, made a strategic plan for the establishment of an early-warning mechanism in relation to resources and environment carrying capacity, and required to implement restrictive measures for regions where water and land resources, the environment and oceanic resources have been excessively exploited. The newly revised Law on Environmental Protection provides a legal basis for the establishment of a regional socioeconomic development model based on the carrying capacity of resource and environment. It stipulates that "The people's governments at and above the provincial level shall organize the relevant departments or commission specialized institutions to survey and assess the environmental condition, and establish a monitoring and early warning mechanism for environmental and resource carrying capacity." In accordance with the *Guidelines of the CPC Central Committee and the State Council on Accelerating Ecological Progress*, we should "think about worst-case scenarios, define and enforce strict controls including caps on resource consumption, minimum standards for environmental quality and red lines for environmental protection, and conduct various development activities within the carrying capacity of resource and environment."

The carrying capacity of resources and environment describes the capacity of the sub-ecosystem to sustain the human socioeconomic subsystem in the regional composite ecosystem under the conditions of ensuring the rational development and utilization of resources and the health of the ecological environment within a certain time and space. Resource and environment constitute the foundation of socioeconomic development. The carrying capacity of the resource and environment refers to, under the goal of sustainable development based on harmony between man and nature, the capability of the resource and environment as the carriers to sustain human economic and social activities. The carrying capacity is subject to the influence of human production and life activities. The carrying capacity of resources and environment is defined with the aim of not degrading the environmental quality for the well-being of mankind and on the basis of the environmental capacity, so as to ensure that the negative environmental impacts that may arise in the process of socioeconomic development are constrained within the limit of the carrying capacity of ecological environment. The carrying capacity of resources and environment can be defined from the following two perspectives. One is the carrying capacity of the ecological environment with environmental quality and ecological functions at the core, including water and atmospheric environment. The other is the carrying capacity of resources based on important resource elements, including water, land, energy and marine resources. Meanwhile, the carrying capacity of resources and that of the environment check and interact each other. Different quantities and ways of resource utilization and human activities exert varied impacts on environmental quality.

3 Major Problems Facing Socioeconomic Development from the Perspective of Resource and Environmental Carrying Capacity

China has long regarded economic development as an important part of national development and even created "invaluable assets" at the sacrifice of "lucid waters and lush mountains." Along the journey, zero or little consideration for the carrying capacity of resources and environment has led to the increasingly prominent contradiction between economic growth and resource scarcity and environmental deterioration. This should be mainly attributed to the inadequate consideration for the constraints of key resource elements and ecological environment on development. As a result, the top-level design for regional development failed to make reasonable plans from the overall and long-term strategic perspective, going against the carrying capacity of resources and environment. An important basis for the relationship between lucid waters and lush mountains and invaluable assets lies in the carrying capacity of resources and environment. Establish a socioeconomic development model based on the carrying capacity of resources and the environment, in an effort to limit the development's impact on resources and environment within the carrying capacity of the nature. This manifests the fundamental role of ecological environment for socioeconomic development and helps to translate the scientific development philosophy that "lucid waters and lush mountains are invaluable assets" into practice. Specifically, we should make efforts to address the following issues.

1. **The decoupling of the socioeconomic development plan from the environmental protection plan**

Most of the existing plans relating to socioeconomic development, particularly the master plan for economic and social development, urban master plan, overall plan for land use, industrial development plan, energy plan, transportation plan, and other plans that have a greater impact on environmental quality, fail to integrate with environmental protection plans. As a result, it's difficult to consider the environmental carrying capacity of the region from a comprehensive and long-term perspective during the development, construction and the implementation of major projects in the region, leading to excessive resource utilization and pollutant discharge in the process. Due to the long-term cumulative effect, the environmental quality fell short of standards, which directly affects human health. Due to the absence of the analysis about coordination and consistency between environmental protection plans and other development plans, the end-point management has failed to transit to prevention and control from the source. The pollutant emission reduction targets in most environmental protection plans haven't truly reflected the fundamental role of resources and environment carrying capacity. The regional environmental protection plans haven't been well coordinated with local environmental capacity.

2. **Failure to clarify the relationship between the increment and the stock of environmental pollution**

Increment and stock are important issues in environmental governance. Currently, the increment of environmental pollutants continues to go upward while the stock hasn't been reduced significantly in many areas. In consequence, continuously accumulating pollutants have depleted the environmental capacity and the environmental quality in some areas have long failed to meet the standards. Without clarifying the relationship between the increment and the stock of environmental pollution, it will be impossible for the areas plagued by the excessive exploitation of resources and environment to meet the environmental quality standards. The relationship between increment and stock is an important scientific issue, featuring a complex dialectical relationship that changes with the actual situations. It is necessary to conduct in-depth research from the perspective of regional environmental capacity, in a bid to understand and grasp the relationship between the two in a comprehensive, systematic and overall manner. At present, our understanding of this issue is still inadequate and the corresponding evaluation methods and management systems are still not sound.

3. **Inadequate systems**

Firstly, the imperfect property rights system of natural resources has led to the predatory exploitation of resources, the damage of the ecological environment and the degradation or even loss of ecological functions. The scarce natural resources and the limited environmental capacity have not been reflected in the top-level design and the management mechanism of the decision-making departments, which hinders the effective allocation of resources. In addition, as the responsibilities of ecological and environmental protection are not clearly defined, different departments and entities have excessively exploited natural resources and ecological environment to pursue economic growth. Secondly, regional differences have not been reflected. At present, regions vary greatly in terms of stage and level of economic development. Given the differences in regional resource and environmental conditions, regions vary significantly in the environmental capacity and resource carrying capacity. However, the major environmental protection systems in relation to pollutant emission reduction and caps on total consumption have failed to fully reflect the regional disparities. The mandatory control indicators for different pollutants have not been adapted to the local resource and environmental carrying capacity. As a result, in some areas where the environmental capacity has exceeded the standard, the environmental quality has continued to deteriorate.

4 Countermeasures and Suggestions

1. **The top-level design should take the carrying capacity of resources and environment into account as a basic factor**

In accordance with the requirements for the harmonious development between man and nature, we should incorporate the philosophy of ecological civilization into all aspects and the whole process of socioeconomic development and create a sound top-level design for socioeconomic development. We should make efforts to control pollution from the source, prevent damage to the environment and ecology, realize the harmonious development between man and nature, establish a bottom-line thinking, and limit various development and construction activities and development plans within the scope of resource and environmental carrying capacity. We should regard the resources and environment carrying capacity as the basic constraint and ponder over the overall issues in an all-round manner while highlighting priorities and main tasks. We should explore to integrate various planning into a single master plan, in a bid to take resources and environmental factors into full account in the decision-making process. In this way, we could address the problem of insufficient consideration for the ecological environment in the process of economic development from the root, thereby solving problems relating to the scale, structure and layout of resources and the environment. We should incorporate the constraints on resource and environmental carrying capacity, protection of ecological space, and cultivated land protection into the urban socioeconomic development plans. The government shall coordinate various departments to formulate the overall plan for urban development, giving consideration to optimize the allocation of resources such as environment, natural resources, economy, and land within the spatial and ecological scope of the administrative area. At the same time, it's necessary to establish a coordination mechanism among departments in charge of the urban planning and construction to ensure the implementation of the plan.

2. **With a view to solving practical problems, keep improving the understanding and application capabilities of science concerning the carrying capacity of resources and environment**

Human's understanding of nature has kept improving over time. However, since human's understanding of the objective laws of nature is an ongoing process that can never be fulfilled, there's no way for man to completely understand, control and conquer nature. Amid the increasingly complex environmental situations, we can continuously improve our scientific understanding of the objective laws of nature in the process of solving various existing problems, while strengthening the supporting and leading role of science and technology to promote continuous improvement of environmental quality. The carrying capacity of resources and environment is an important scientific issue arising from discussions about the long-term coexistence of man and nature, and an important acting point for promoting harmonious coexistence between man and nature. Therefore, we must develop a scientific attitude and carefully study the scientific issues underneath. On this basis, we should continuously improve our understanding of the scientific connotations of the resources and environment carrying capacity, while regarding it as the fundamental principle and important basis for implementing the philosophy of ecological civilization. We should rely on a systematic and perfect institutional system, make the carrying capacity of regional resources and environment an important consideration and constraint factor, and incorporate it into the overall picture of economic and social development.

3. **Further improve relevant systems**

Delineate and strictly observe the red lines for ecological protection. We should regard the red lines for ecological protection as the "bottom line" that must be adhered to in the process of socioeconomic development, so as to ensure that the carrying capacity of resources and environment is effectively put into practice as a binding condition for economic, political, cultural, and social development. The red line for ecological protection describes the minimum space range and the maximum or minimum quantity subject to rigorous protection in order to enhance ecological functions, improve environmental quality, and promote efficient utilization of resources. It consists of red lines respectively for ecological functions, environmental quality and resource utilization. We should study and introduce the technical process and methods for the delineation of red lines for ecological protection as soon as possible. In combination with regional ecological function requirements, environmental quality standards, pollutant discharge standards, and regional environmental management requirements, we should delineate different types of ecological protection red lines and implement strict protection, which shall serve as the bottom line that must be observed under the constraints of resource and environment carrying capacity.

Establish a monitoring and early warning mechanism for the carrying capacity of resources and environment. On the basis of environmental capacity and resource endowments, with the aim toward up-to-standard environmental quality, sustainable utilization of resources, and protection of ecological functions, we should ensure that early warnings are issued when negative environmental impacts of the socioeconomic development may exceed the threshold of environmental carrying capacity. Corresponding to the "pressure—state—response" process of human economic and social activities on the resource and environment, we should establish different types of evaluation systems for the resource and environmental carrying capacity, so as to evaluate its trend of changes and issue early warnings about the overloading of resources and environment based on key threshold values like the upper limit of resource and environmental constraints or the rational scale of population and economic development. Through the mutual influence and interaction between different activities, grasp the real reason for the overload of resources and environment. We should establish a monitoring and accounting system for the carrying capacity of resources and environment, strengthen the standardization of its monitoring and evaluation, and carry out scientific evaluations and demonstrations of regional carrying capacity.

Establish the system of natural resource property rights and the compensation system for the use of resources. We should carry out unified registration for natural ecological spaces and gradually develop a natural resource property rights system with the property rights, rights and responsibilities clearly defined and characterized by effective oversight. We should accelerate pricing reform for natural resources and their products to give full expression to their market supply and demand, the extent of resource scarcity, ecological and environmental damage costs and restoration benefits. In this way, the economical, repeated, multiple and recycled utilization of resources could form an endogenous driving force in economic benefits and provide resource guarantee for sustainable economic and social development.

To sum up, we should adhere to the philosophy of ecological civilization, promote economic and social development based on the carrying capacity of resources and environment, and limit the impact of the development on resources and the environment within the carrying capacity of the nature. It highlights the fundamental role of the ecological environment for economic and social development. Currently, it's a crucial time for the preparation of the "13th Five-Year Plan" both at national and local levels. It is advisable to incorporate the regional resource and environmental carrying capacity into the plans as an important consideration and restrictive factor, dovetail pollutant emissions with regional environmental capacity, and closely associate regional development, urban construction and industrial development with regional resource supply capacity, environmental quality and ecological functions, so as to promote the overall improvement of ecological environmental quality and lay the foundation for building a beautiful China and realizing the Chinese dream of the great rejuvenation of the Chinese nation.

The Strategy to Promote the "Win–Win" Development of Economy and Environment in the Transition Period

Bo Zhang

When we speak of eco-environment protection in general, there would be no objection to it. When it comes to the details, however, environmental protection will stir the vested interests of various parties and opinions vary greatly and even contradict each other from different perspectives. Therefore, the environmental protection actually concerns the relationship between the parts and the whole as well as between the immediate and the long-term benefits. It is not only an issue relating to people's livelihood and economy, but may even develop into a major political problem.

Comrade Xi Jinping has always attached great importance to the ecological progress. Since the 18th CPC National Congress, the CPC Central Committee with Comrade Xi Jinping at the core has put forward a series of new ideas, viewpoints and requirements for ecological progress from the comprehensive and strategic perspectives and developed Xi Jinping Thought on Ecological Civilization. They have provided targeted and precise guidance for the correct understanding of the dialectical relationship between economic growth and environmental protection. Under the guidance of Xi Jinping Thought on Ecological Civilization, this paper conducts a preliminary discussion on the strategy to promote win–win development of economy and environment based on the case study of Shandong Province that forced the transformation and upgrading of traditional industries with environmental standards.

B. Zhang (✉)
Department of Water Ecology and Environment, Ministry of Ecology and Environment, Beijing, China

J. Pan et al. (eds.), *Beautiful China: 70 Years Since 1949 and 70 People's Views on Eco-civilization Construction*, https://doi.org/10.1007/978-981-33-6742-5_29

1 In-Depth Study and Understanding of Xi Jinping Thought on Ecological Civilization

1. **Development of ecological civilization is an inevitable choice for the sustainable development of the Chinese nation**

First, human development activities must respect and protect nature; otherwise they will be retaliated by nature. A civilization may thrive if its natural surroundings thrive, and will suffer if its natural surroundings suffer. This is an irrefutable law.

Second, the old path of the US and Europe will lead to a dead end for us to build a modern country. During a visit to Guangdong Province in December 2012, General Secretary Xi Jinping pointed out that "China's modernization is unique, unprecedented, and all-time great. The total population of developed countries in the world is now less than 1.3 billion, which could be doubled by the modernization of China that's home to a population of 1.3 billion. However, sustainable development will only be empty talk if we follow the old path that relies on resource consumption and environmental pollution."

Third, we owe a huge debt to the ecological environment. If we do not step up efforts from now on, we will have to pay a much heavier cost in the future. On this issue, we have no other choice.

Fourth, the development of ecological civilization concerns the welfare of the people and the future of the nation. The 18th CPC National Congress incorporated the ecological progress into the Five-Sphere Integrated Plan for building socialism with Chinese characteristics. It made clear to vigorously promote ecological progress, build a beautiful China, and realize the sustainable development of the Chinese nation. This indicates a more profound understanding of the laws of socialism with Chinese characteristics and demonstrates our firm will and determination to enhance the development of ecological civilization.

2. **Incorporate the ecological progress into the Five-Sphere Integrated Plan**

Ecological progress requires far more than the efforts of a few departments relating to environmental protection. We must give full play to the political strengths of the socialist system with Chinese characteristics, give more prominence to the development of ecological civilization, and incorporate it into all aspects and the whole process of economic, political, cultural, and social development. In terms of economic development, we should combine the ecological progress with the efforts to adjust the economic structure and shift the growth model. In terms of political development, we should pursue ecological progress while leveraging the advantages of the socialist political system with Chinese characteristics and ensuring that both party committees and governments are held accountable and that officials take responsibility for the protection of ecological environment.In terms of cultural development, ecological progress should be advanced along with advocating green lifestyle and promoting ecological culture. In terms of social development, we should mobilize public participation and national action, while striving to form a large pattern of developing ecological civilization featuring the leading role of the Party and government, the coordination among departments and concerted efforts of the whole society. Such concept and method based on integration are testimony to the Chinese wisdom and Chinese approach for building an ecological civilization. Under the guidance of the strategy, China's development of ecological civilization will yield great results based on the efforts to make unified planning with due consideration for all concerned and overcome difficulties, setting an example of "win–win" development of environment and economy for other developing countries.

3. **Correctly understand and handle the dialectical relationship between economic development and environmental protection**

In 2013, Comrade Xi Jinping put forward the famous statement that "We want both lucid waters and lush mountains, but we also want mountains of gold and silver. We prefer the former to the latter. Besides, lucid waters and lush mountains are indeed mountains of gold and silver."

The three sentences of the above statement are progressive in logic and must be understood as a whole. The first sentence, "we want both lucid waters and lush mountains, but we also want mountains of gold and silver," indicates that our development goal is a win–win situation for both economy and environment. Development is the top priority and economic development is the central task, but what we are pursuing is a win–win situation, instead of developing one at the sacrifice of the other. The second sentence, "we prefer the former to the latter," tells us to stick to the bottom-line thinking in the face of sharp and even irreconcilable contradictions between economic growth and environmental protection. That is to put public health and ecological safety in a more prominent position, with economic growth giving way to environmental protection. The third sentence, "lucid waters and lush mountains are indeed mountains of gold and silver," means that as long as we stick to the path, the environmental protection will force the economic transformation and development, and thereby establishing the core competitiveness of enterprises, industries and even regional development, which could then transform into a higher-quality economic results.

4. **Protect the ecological environment with the strictest system and the most stringent rule of law**

The density of population and enterprises in eastern and central China is very high. Compared with developed countries such as the United States and those in Europe, the same amount of efforts in environmental protection won't deliver the same results. This is an important part of the national condition. The conscious implementation of a strict environmental management system is required by China's basic national conditions, not subject to human will. Therefore, the eco-environment protection must rely on the system and the rule of law. Only by implementing the strictest system and the most stringent rule of law can we provide a reliable guarantee for the development of ecological civilization.

At the sixth group study session of the Political Bureau of the 18th CPC Central Committee in May 2013, Comrade Xi Jinping pointed out that "We must establish the accountability system. What I'm talking about here is the system to hold the leading cadres accountable. We must hold to account on a permanent basis those who cause serious consequences through blind decisions that disregard the environment. When dealing with actions that damage the environment, we must not be soft or make exceptions. Otherwise it will become a mere formality. Some officials leave an environmental mess behind, but they still get promoted without being held accountable. We cannot let it happen. Organizational departments, comprehensive economic departments, statistical departments, and supervision departments must fully implement the accountability system. The concept of red lines for ecological protection must be firmly established. We are now facing quite serious problems in ecological environment, calling for the most draconian measures. Regarding the eco-environment protection, anyone that oversteps the prescribed limit will be punished."

5. **Protect the ecological environment with the focus on prominent problems that directly affect the people**

To protect the ecological environment, we must adhere to the people-centred thinking and start from the environmental problems that are most concerned by the masses, such as heavy air pollution, drinking water safety, black and odorous water bodies. We should think big and act small to resolve the eco-environmental problems on a larger scale. This embodies a working method as well as the people-centred governance of the CPC.

At the 14th meeting of the CPC Central Leading Group for Financial and Economic Affairs in December 2016, Comrade Xi Jinping said, "What're the concerns of the people? They are concerned about the food safety, heating in winter, smog, waste incineration without damaging health, elderly care services, affordable housing, among others. They are more concerned about these issues than the economic growth rate. If we meet the growth target but make no progress in resolving the outstanding issues that are common concern of the people, the people won't agree with us even if we announce to have achieved the goal of building a moderately prosperous society in all respects."

During a panel discussion with deputies to the 3rd Session of the 12th National People's Congress from Jiangxi Province in March 2015, Comrade Xi Jinping further pointed out that "The environment is crucial to the wellbeing of the public, since blue skies and verdant surroundings allow us to feel the beauty and joy of life. We should protect the environment like we protect our eyes and treat the environment the way we treat our lives."

6. **The development of ecological civilization must adopt systematic governance**

Comrade Xi Jinping has put forward profound viewpoints in this regard, with a series of important statements. The systematic governance has been stressed from spatial perspective on the one hand and from working measures on the other hand.

In November 2013, while making explanatory notes for Decision of the Central Committee of the Communist Party of China on Some Major Issues Concerning Comprehensively Deepening the Reform, Comrade Xi Jinping pointed out that, "Our mountains, waters, forests, farmlands and lakes form a community of life. The lifeline of the people comes from the farmland, that of the farmland comes from the water, that of the water comes from the mountain, that of the mountain comes from the earth, and that of the earth comes from the tree. To control the exploitation of natural resources and restore the ecosystem, we must follow the laws of Nature. If people only tend to their own responsibilities, for example, growing trees, regulating rivers or protecting farmlands in isolation, they are prone to gaining in one area and losing in another, which eventually leads to systemic destruction of the ecology."

At the 5th meeting of the CPC Central Leading Group for Financial and Economic Affairs in March 2014, Comrade Xi Jinping went further to note that "We must systematically consider the means of taxation and price. We should distinguish between producers and consumers, drinking water and sewage, surface water and groundwater, water use in urban and rural areas, water use for industrial and agricultural purposes, in an effort to study and put forward a package of solutions for water resource tax, raw water fee, tap water fee, and sewage treatment fee. Based on the actual situation, it shall be implemented level by level and step by step."

Judging from the current development of the ecological civilization in China, the unsystematic problem is still quite prominent, which is a key explanation for the difficulty to form a long-term mechanism in many tasks. We must follow the guidance of General Secretary Xi Jinping's important thoughts in this regard and make painful efforts to solve the problem.

7. **Environmental quality can only get better, not worse**

For a large country in a transitional period like China, the development of ecological civilization must adhere to the bottom-line thinking and in the meanwhile, proceed from the basic national conditions to determine the goals and measures in a realistic manner. At the 5th meeting of the CPC Central Leading Group for Financial and Economic Affairs in March 2014, Comrade Xi Jinping said that "The environmental problems in China have been accumulated over a long time. It didn't happen in one day. However, we will put an end to the environmental deterioration. The CPC members should have such vision and determination."

Towards the end of the inspection tour in Beijing in February 2014, Comrade Xi Jinping pointed out that "The primary task of tackling smog pollution and improving air quality is to control PM 2.5. It's a bit earlier for China to meet the international standards in terms of PM 2.5, since it's difficult for the development stage of China. However, we should notice that the issue has raised the concern of the vast number of leaders and the public, as well as the international community. We must address the problem. We should always make sure to respond to public demands."

When talking about environmental issues, Comrade Xi Jinping often mentioned "to get only better, not worse." It indicates the bottom-line thinking of "not getting worse," reflects the scientific attitude of seeking truth from facts based on the national conditions, and embodies the profound meaning of actively and steadily promoting the development of ecological civilization. This constitutes an important part of the Xi Jinping Thought on Ecological Civilization and we should attach high importance to its study.

2 Shandong Province's Practice of Promoting Transformation and Upgrading of Traditional Industries with Environmental Standards

1. Background

In Shandong Province that's suffering from an acute shortage of water resources, the paper industry plays a major part in water consumption and contamination. In 2002, the industry's fresh water consumption and chemical oxygen demand (COD) emissions exceeded 50% of the total in the province, while contributing merely 3% to the industrial value-added output. Heavy contamination and high water consumption constitute the bottleneck that hampers the sustainable development of the paper industry. Since the mid-1990s, Shandong Province has actively explored solutions to the problem of heavy contamination in the paper industry. Between 1996 and 2002, administrative measures were adopted to shut down 472 small-scale and heavily-contaminating straw-pulp paper production lines, which achieved initial success in curbing the trend of aggravating pollution. However, the administrative measures alone have gradually exposed the problems such as lack of predictability, easy rebound, and high social costs. According to statistics, the shutdown of 41 straw-pulp production lines between 2001 and 2002 laid waste to fixed assets worth RMB 1 billion and left tens of thousands of employees laid off, placing great pressure on social stability. How to improve environmental quality while ensuring economic growth and social stability has become a tricky problem facing the industrial transformation and regional development.

At the beginning of the twenty-first century, Shandong Province set the goal of building a moderately prosperous society ahead of schedule by 2012. Back then, all water in Shandong Province was contaminated. Even pollution-resistant carp and crucian cannot survive in the river, raising concerns about the safety of the people on both banks of the river. Under such circumstances, the environmental protection department proposed that "the stable growth of common fishes" should be a prerequisite for completing the building of a moderately prosperous society. The proposal was accepted by the CPC Shandong Provincial Committee and the Shandong Provincial People's Government and it's planned to achieve the goal in 2010.

At the time, Shandong boasted the second largest economy in China, with traditional industries accounting for a large proportion. To achieve the goal of water environment governance for "stable growth of common fishes," the primary task was to solve the problem of structural pollution in traditional industries like papermaking. However, there are national standards for the pollutant discharge of almost every polluting industry. According to the standards, the polluting industry can "legally" discharge more pollutants than other industries without being held accountable. For example, the national standard for the paper industry stipulates that the COD emission limit is 450 mg/L, higher than the COD concentration in the untreated urban sewage. The drainage volume of papermaking enterprises is generally as high as tens of thousands of tons, even hundreds of thousands of tons, leading to the catastrophe that "a paper mill contaminates a river." The people on both sides and along the lower reaches of the river were suffering from pollution. The environmental problems gave rise to intense social conflicts. Meanwhile, the paper industry itself was plagued by environmental problems as well, struggling to make headway. With the support of the provincial government, the Shandong provincial environmental protection department proposed to formulate and gradually tighten local environmental standards in phases, in a bid to gradually abolish the pollution discharge privileges of heavily-polluting industries within eight years. They developed their own approach of water environment governance to fulfill the goal of "livable water environment for fishes in the rivers."

2. **Process**

In 2003, Shandong Province started from the paper industry that caused the heaviest contamination and launched the first local environmental standard in China—Shandong Provincial Standard for the Discharge of Wastewater Pollutants for Paper Industry. Since then, Shandong Province has embarked on a new path of promoting industrial transformation and restructuring through environmental standards. The standard is implemented in four stages. In the first stage, starting from May 1, 2003, the COD concentration in the wastewater from straw-pulp papermaking shall not exceed the standard limit of 420 mg/L, which is slightly stricter than the original national standard of 450 mg/L. It sent a clear signal to the industry that "standards are about to be tightened." In the second stage, starting from January 1, 2007, the standards were substantially tightened, setting the upper limit of the COD concentration in wastewater from straw-pulp papermaking to be 300 mg/L. Companies with larger scale and more advanced technology and equipment are capable of meeting the discharge standards, while those incapable companies took the initiative to adjust structure. In the third stage, four comprehensive basin-based discharge standards covering the entire territory of Shandong Province were issued, including the Comprehensive Discharge Standard of Water Pollutants along the Route of South-North Water Diversion Project in Shandong Province. The first phase of the basin-based standards is connected with the third phase of water pollutant discharge standards for the paper industry. In this way, the industrial emission standards are docked with comprehensive basin-based discharge standards. The COD concentration of wastewater discharged from straw-pulp papermaking is subject to the standard limit of 150–300 mg/L. In the fourth stage, starting from January 1, 2010, all enterprises within the province's river basins were subject to the unified pollutant discharge standards. The standard limit for the COD shall be 60 mg/L concentration in key protected areas and 100 mg/L in general protected areas, which is 4–7 times stricter than the original national standard. The discharge limit of enterprises is no longer determined by industry, but instead, by the location of the enterprise in the river basin. In essence, it put an end to the "discharge privilege" enjoyed by the heavily-contaminating industries.

3. Results

The implementation of an open and transparent standard system that's foreseeable for the enterprises announced the progress of eliminating backward productivity in the form of digital laws and regulations. It has significantly boosted the technical advancement in environmental protection and promoted the development of advanced productivity in the paper industry. Major paper-making enterprises and relevant scientific research institutions in Shandong Province have invested heavily in convening domestic and foreign experts to tackle key sci-tech problems and remove technical obstacles in the pulping process and advanced wastewater treatment and reuse, leading to the substantial improvement in technology. The major sci-tech project "Research on Advanced Treatment and Reuse Technology of Wastewater from Papermaking" launched by Shandong Academy of Environmental Sciences has achieved an important breakthrough that has been put into application in seven provinces and municipalities. A number of technologies developed by Huatai Group, including "Research and Application of Optimal Control System for Pulp Making and Alkali Recovery Process," won the second-class National Award for Sci-Tech Progress. Shandong Tranlin Group boasts over 200 patents in environmental protection and circular economy, in addition to its technique of "triple uses of grass" that realized the comprehensive utilization of all wheat straw. In July 2011, the Ministry of Environmental Protection went above the policies for grass-pulp paper industry and approved the project for comprehensive utilization of 1.5 million tons of straw per year by Tranlin Group, indicating a major breakthrough in the transformation and development of the paper industry in Shandong Province. In 2014, the State of Virginia in the United States invited wheat-straw papermaking companies to invest and build factories there, offering an award of $ 5 million to companies. As such, the wheat-straw pulping and paper industry in Shandong Province, once labelled as "backward," went through transformation and upgrading under the pressure of local environmental standards, overcome the environmental problems and became the "white swan" in the eyes of Americans.

Since 2003, Shandong has not taken administrative measures to shut down any paper-making enterprises above designated size, and the government has not granted subsidies or provided support for laid-off workers. Under the guidance of local standards, enterprises have taken a path of transformation and restructuring. The advanced enterprises directly aimed at the standards set for eight years later, invested heavily to tackle key technical problems, and played a leading role in the transformation and development of the entire industry. Some enterprises, aware that it is impossible to meet the environmental standards set for eight years later, made use of the transition period defined by the government to gradually change the structure of raw materials and products, and finally "changed their way of life." Some enterprises took the option of mergers and acquisitions with advanced companies. A decade later, the absolute number of wheat-straw pulping and papermaking enterprises in the province has been cut by more than 90% to a dozen from 220 in 2002, but the scale, profits and taxes of the industry have increased significantly. In 2013, the provincial output of machine-made paper and paperboard reached 20.53 million tons, more than doubled than 2002; profits and taxes increased by nearly four times to hit RMB 21.11 billion, ranking first in China; and COD emissions were reduced to 24000 tons, down by 88.2%. At present, all paper-making enterprises in the province have set up "bioindicator pool" at the sewage outlets and met the pollution control standard to only discharge water that's qualified for "stable growth of common fishes." The production mode of the paper industry in Shandong Province has undergone a fundamental change, about five years ahead of the industry in other provinces. Today, the papermaking enterprises funded by national capital that are active in the stock market are mostly in Shandong Province.

Following the approach of tightening standards in a progressive manner, Shandong Province has successively issued 38 local environmental standards and put in place a local environmental standard system covering the entire province, yielding significant results in the prevention and control of water pollution in the river basin. Since 2003, amid medium and high-speed economic growth, Shandong has achieved continuous improvement of water environment quality for 15 consecutive years and successfully established a long-term mechanism for the improvement of water environment. In 2010, all 59 key rivers in Shandong Province have seen restoration of fish growth. In the national evaluations of pollution control in key river basins, Shandong Province won the top place among Huaihe river basins for 9 times and among Haihe river basins for 7 times. Shandong has fundamentally changed its economic development mode at the cost of water environment.

3 Revelations

1. There's no backward industry, but merely backward concepts, standards, technology and management

We should not easily label an industry as "backward" or specify the threshold of designate size below which the companies shall be shut down. Why not promote industrial restructuring by means of standards for energy conservation and environmental protection, quality and safety as well as law-based administration? Despite the size, all companies that could live up to the standards should be allowed to operate normally. Likewise, the companies falling short of the standards won't be allowed for operation no matter how large they are. It's up to the market to decide whether a company can survive. The author doesn't approve of the so-called practice of "cutting one's arm to save one's body." After all, there are only two arms in one body. As a developing region, it is probably not advisable to finish off all backward productive forces with a single blow. According to Confucius, "execute people without having given them instruction is called cruelty." If the government follows a laissez-faire policy and even protect the backward companies in normal times, but adopts draconian measures to "cut one's arm to save one's body" and forcefully shut down companies. It would be nothing less than "tyranny."

Regarding the backward production mode lagging behind the times, it's not advisable for the developing regions to adopt a laissez-faire policy and protect the backward companies, or finish off the backward companies in a single blow. Instead, it's advisable to proceed from the actual situations, create conditions for transformation, improvement and enhancement, and gradually transform it into an advanced and applicable production mode. The practice in Shandong Province shows that as long as we change our concepts and formulate scientific and reasonable guiding standards, we can effectively promote backward industries to optimize their industrial structure, improve production processes, and overcome obstacles of resource and environment. In this way, the backward industries in the traditional sense can rise from the ashes and embark on a path of high-quality development.

2. **Scientific and strict environmental protection measures will not hinder economic development, but play an important role in promoting high-quality development during the transition period**

Shandong Province has formulated local environmental standards for the paper industry that are stricter than the national standards, arousing strong repercussions inside and outside the province. There were two typical views in opposition to the practice. One is that Shandong Province is placing itself under greater pressure by applying much more rigorous standards than the national standards. It will inevitably lead to the failure of the paper industry of Shandong Province in market competition. The other holds that the economic and technical feasibility of the industry should be a top consideration in the formulation of standards. The survey at the time suggested that 85% of papermaking companies could not meet the third-stage standard, making the standards meaningless in practice. These two views may sound quite reasonable, but it's exactly such ostensibly correct views that constitute the ideological barrier that we need to break through in the industrial transformation and restructuring. It is wrong to regard the economic and technical feasibility of corporate pollution control as a prerequisite for decision-making, in disregard of the necessity of safeguarding public health and ecological safety, the inevitable law of economic and social development, and the role of environmental protection to force the optimization of polluting industries.

3. **A win–win situation for economy and environment requires a strategy that integrates the necessity, predictability, guidance and compulsion**

In practice, promoting a win–win situation for economy and environment is often caught in some "dilemmas" and has to go through a painful process to rise from the ashes. Based on lessons drawn from the practice in Shandong Province, the author believes that to alleviate the pain and form a long-term and win–win mechanism, it is necessary to set a reasonable transition period and adopt a strategy that integrates the necessity, predictability, guidance and compulsion. In terms of necessity, it means to develop unified thinking about the necessity to safeguard public health and ecological safety and the inevitable laws of socioeconomic development, and thereby creating a synergy to "prioritize lucid waters and lush mountains over invaluable assets." In terms of predictability, it means to coordinate socioeconomic development and environmental protection, clarify the goal several years in advance, and set a reasonable transition period. In terms of guidance, it means to decompose the goal and the transition period into several stages, formulate, implement and tighten legally binding policies and measures in a progressive manner, and guide enterprises, industries and even regions for restructuring, transformation and upgrading. As for compulsion, it means to resolutely observe the authority of laws and regulations and implement the policies and measures in accordance with the law. The strategy that integrates the necessity, predictability, guidance and compulsion will make it easier to address some tricky problems.

Xi Jinping thought on ecological civilization embodies a theoretical sublimation of a series of new concepts, ideas and strategies put forward by Comrade Xi Jinping on the ecological progress since the 18th CPC National Congress. It is an integral part of Xi Jinping Thought on Socialism with Chinese Characteristics for a New Era. It provides fundamental principles for the building of a beautiful China. As long as we follow the guidance of Xi Jinping thought ecological civilization, seek truth from facts, and make overall plans to force the transformation of development mode with environmental protection and enhance environmental protection with high-quality development, we will surely achieve a win–win situation for both economy and environment on a larger scale, thereby contributing more to the great rejuvenation of the Chinese nation and the global ecological progress.

Green Development of Transportation in China for 70 Years and Its Prospects

Qingrui Li

Since the founding of the People's Republic of China in 1949, under the guidance of the great banner of socialism with Chinese characteristics and the outstanding leadership of the CPC, the transportation industry went through a bottleneck period in pursuing development, a period of initially easing the bottleneck, before "basically meeting" the needs of socioeconomic development. China has been rapid narrowing its transportation gap with world-class countries, even overtaking these countries in certain areas. A modern transportation system is taking shape in China. The level of green development of transportation has steadily improved, along with the depth and breadth, providing a strong guarantee for improving people's living standards and building a beautiful China.

1 The 70-Year History of Green Development of Transportation

Over the past 70 years, China has achieved leapfrog development in transportation infrastructure and its transportation support capabilities have been significantly enhanced. China topped the world in terms of the operating length of high-speed railways, expressways and urban rails and the number of 10,000-ton berths in ports. China is on the front rank in the number of airports and the pipeline mileage. China ranks the first in the world by railway passenger turnover and freight volume as well as road cargo transportation volume and turnover. China ranks the second in the world in terms of the civil aviation passenger turnover and cargo and mail turnover. China has ranked first in the world for more than a decade in terms of the port cargo and container throughput. As of the end of 2018, the national railway operational length reached 131,000 km (including high-speed railway operational length of over

Q. Li (✉)
China Ecological Civilization Research and Promotion Association, Beijing, China

J. Pan et al. (eds.), *Beautiful China: 70 Years Since 1949 and 70 People's Views on Eco-civilization Construction*, https://doi.org/10.1007/978-981-33-6742-5_30

29,000 km), the total highway mileage reached 4,846,500 km (including expressway of 142,600 km), and the navigable mileage of inland waterways reached 127,100 km (including graded waterway of 66,400 km). In addition, there were 23,919 production terminal berths in ports (including 2444 berths of 10,000 tons and above) and 235 civil aviation airports (including 95 general airports capable of receiving more than 1 million passengers).

Transportation is a basic, pioneering and strategic industry in the national economy. It's also a focus of national efforts in eco-environment protection, energy conservation, emission reduction, and response to climate change. At the early stage, the construction of transportation infrastructure consumed a large amount of land, energy and resources, exerting a certain impact on the ecological environment. Pollutants and greenhouse gas emissions generated during transportation have gradually come to the fore. The history of green development of transportation could be divided into three stages respectively marked by the establishment of the Environmental Protection Leading Group by the Ministry of Transportation for the first time in 1974, the 14th CPC National Congress, and the 18th CPC National Congress. The green development has went through the stages from the prevention and control of water pollution from ships and ports, the highway ecological protection as an important part of highway construction, to the promotion of ecological civilization in transportation covering all sectors, regions and processes in an all-round manner under the guidance of concepts about ecological progress as a part of the Five-Sphere Integrated Plan and green development among the five concepts of development.

1. **Actively explore and initiate the prevention and control of water transport pollution (1949—the establishment of Environmental Protection Group by Ministry of Transportation in 1974—prior to the 14th CPC National Congress in 1992)**

During this period, against the backdrop of underdeveloped highway and waterway transportation infrastructure, backward transportation equipment and poor transportation guarantee capacity, China made pioneering explorations to develop the transportation industry, covering such fields as infrastructure, system and institutions, transportation services, legal system building, and opening up. At the first National Conference on Eco-environmental Protection in 1973, the transportation industry was made one of the earliest industries in China to engage in environmental protection. In 1974, the Ministry of Transport established the Environmental Protection Leading Group and set up an office for environmental protection accordingly. In 1987, the Ministry of Transport promulgated the Administrative Measures for Environmental Protection in Transportation Construction Projects (Trial). It marked the official beginning of developing the legal system for environmental protection in the transportation industry. In 1989, the Ministry of Transport adjusted the Environmental Protection Leading Group to the Environmental Protection Committee, consolidating the leadership and organization of environmental protection in the industry.

The prevention and control of pollution from ships and ports was the key task during this period. Active exploration was carried out to reflect the concept of environmental protection at the planning level, improve the ability to tackle oil spill risks, prevent and control the water pollution from ships, and evaluate the environmental impact of water transport construction projects. Since the 1970s, China has vigorously advanced work in the marine oil spill emergency responses and the prevention and control of pollution from ships. In 1983, China signed the "The International Convention for the Prevention of Pollution from Ships, 1973, as Modified by the Protocol of 1978 Relating Thereto" adopted by the International Maritime Organization. At the same year, China issued the Discharge Standard of Water Pollutants from Ships (GB 3552–83). In 1981, in line with the environmental impact assessment system for construction projects stipulated in the Environmental Protection Law, the transportation industry initiated the environmental impact assessment in water transport construction projects. The Ministry of Transport successively introduced the Specifications for Port Environmental Impact Assessment and the Specifications for Port Environmental Protection Design. In 1985, the Ministry of Transport organized and carried out the work of forecasting the pollution of the offshore environment at ports and from ships. At the end of the 1980s, the Ministry of Transport took the lead to specifically make environmental protection demonstration an integral part of the port planning and management, introducing the new idea of emphasizing environmental protection from the planning stage.

2. **Keep pace with the times and make efforts to protect highway ecology (the 14th CPC National Congress in 1992 – prior to the 18th CPC National Congress in 2012)**

During this period, the 14th CPC National Congress launched the reform to establish the socialist market economy system. The 16th CPC National Congress put forward the goal of building a moderately prosperous society in all respects. The 17th CPC National Congress made it clear that "we must give prominence to building a resource-conserving, environment-friendly society in our strategy for industrialization and modernization." In the transportation industry, the rapid development of highway network, waterway network and expressways was promoted in an all-round way. The number of specialized deep-water terminal berths increased rapidly. The scale of railway, aviation, and post and telecommunications networks kept expanding. All these led to a fundamental change to the backward situation of transportation infrastructure in China. Under the guidance of the Scientific Outlook on Development, the transportation industry has actively explored a path of scientific development. National plans were formulated to develop a resource-conserving and environment-friendly transportation industry, including National Planning for Expressway Network, National Planning for the Layout of Coastal Ports, National Planning for the Layout of Highway Transport Hub, and National Planning for Inland Waterway and Ports. In 1993, the Ministry of Transport released the Administrative Regulations on Environmental Protection in the Transportation Industry, clarifying the responsibilities for industrial organizations and the requirements on pollution prevention and control, pollution source treatment, scientific research and design, and education. In 2003, the Ministry of Transport promulgated the Administrative Measures for the Environmental Protection in Transportation Construction Projects, clarifying the requirements on environmental protection in projects from the perspectives of environmental impact assessment procedures and environmental protection facilities. In 2006, the Ministry of Transport issued the Guiding Opinions on the Development of Economical Transportation, proposing to establish the concept of economical transportation. In 2009, the Ministry of Transport introduced the Policy of Developing Resource-conserving and Environment-friendly Highway and Waterway Transportation, providing guidance for the highway and waterway transportation industry to adjust the industrial structure and transform the development mode.

China has made steady headways in energy conservation and environmental protection in the transportation industry, with enhanced efforts in the prevention and control of pollution from water transport and gradual improvement in highway ecological protection. This became a hotspot issue during the period. The highway industry has further promoted green development, adhered to develop typical demonstration projects, and strengthened the development and application of new technologies, materials, processes, and products, laying a solid foundation for the green development of highways in a new era. In 2003, under the guidance of the sustainable development concept, China started the construction of Chuanzhu Temple—Jiuzhaigou Highway in Sichuan Province, China's first demonstration highway in harmony with the natural environment. In 2004, the National Highway Survey and Design Working Conference summarized and promoted the successful experience of the Demonstration Project Chuanzhu Temple—Jiuzhaigou Highway, introducing the new approach for highway survey and design, featuring the adherence to six principles and the establishment of six new concepts. It means to adhere to the people-centred principle while establishing the safety-first concept, adhere to the principle of harmony between man and nature while establishing the concept of respecting nature and protecting environment, adhere to the principal of sustainable development while establishing the energy-conserving concept, adhere to the quality-first principle while establishing the concept about public satisfaction, adhere to the principle about rational selection of technical indicators while establishing the concept of design and creation, and adhere to the principle of systematic thinking while establishing the concept of life cycle cost.) Since 2007, relying on major construction projects such as national expressways, the Ministry of Transport has successively coordinated to implement a number of sci-tech demonstration projects in the expressway construction in mountainous areas, eco-environmental protection and transportation safety. In 2009, in an effort to fully implement the concept of modern project management in highway construction, the Ministry of Transport put forward the five-pronged management requirements (i.e., people-oriented development concept, professional project management, standardized project construction, IT-based management methods, and refined daily management). A three-year campaign of standardized construction was carried out nationwide, which elevated the management of highway construction to a new level. In 2011, the Ministry of Transport successively launched a number of thematic projects on green and low-carbon circular highways.

3. **Comprehensively advance green development and ecological progress (since the 18th CPC National Congress in 2012)**

Since the 18th CPC National Congress, the CPC Central Committee with Comrade Xi Jinping at its core has attached great importance to the development of ecological civilization and focused on promoting eco-friendly growth models and way of life. The 19th CPC National Congress made it clear that building an ecological civilization is vital to sustain the Chinese nation's development, specified to develop eco-friendly growth models and ways of life, and made plans for building China's strength in transportation, green travel, and pollution prevention and control. At the National Conference on Eco-environmental Protection in 2018, General Secretary Xi Jinping pointed out that building an ecological civilization is vital for sustaining the development of the Chinese nation. Ecological civilization has become the fundamental criterion for the sustainable and healthy development of the entire society and various industries.

During this period, the transportation industry thoroughly implemented the new concepts, ideas and strategies on ecological progress that were put forward by the CPC Central Committee with Comrade Xi Jinping as the core. Efforts were made to accelerate the development of integrated, smart, green, and safe transportation, in a bid to promote the transportation development in a scientific and all-round manner. In this way, China achieved positive results in green transportation. The Ministry of Transport has successively formulated and issued a series of policy documents to promote ecological progress in transportation, making it clear to follow the strategy of building a great nation in transportation, pursue in-depth supply-side structural reform as the main task, focus on the implementation of seven major projects, and accelerate to put in place the three systems. Efforts were made to boost the transformation of green transportation from passive adaptation to pioneering example, from pilot projects to comprehensive development, and from government-driven to universal governance involving everyone, in a bid to promote eco-friendly growth models and ways of life and provide solid support and strong guarantee for building a beautiful China for building a beautiful China, enhancing the people's well-being and satisfying the aspirations of the people to live a better life.

2 China's Achievements in the Ecological Progress in Transportation

Over the past 70 years since the founding of the New China, the green development of transportation infrastructure has yielded positive results, the environmental protection in the transportation process has gradually came to the fore, and the ecological progress in the transportation industry has been vigorously promoted.

1. **Continuous improvement of the top-level design**

The Ministry of Transport has continuously improved the environmental protection system, established standards, implemented norms, and carried out systematic planning. It successively issued the *Guiding Opinions on Accelerating the Development of Green, Circular and Low-Carbon Transportation, Master Plan for National Highway and Waterway Transportation Environmental Monitoring Network, Special Action Plan for the Prevention and Control of Pollution from Ships and Ports, Implementation Plan for Control Zones of Pollutant Discharge from Ships in Pearl River Delta, Yangtze River Delta, and Bohai Rim (Beijing-Tianjin-Hebei)*, the *13th Five-Year Plan for Energy Conservation and Environmental Protection in Transportation*, the *Implementation Plan for Promoting the Ecological Progress in Transportation, Opinions on Comprehensively Deepening the Development of Green Transportation*, and the *Opinions on Comprehensively Strengthening the Environmental Protection and the Pollution Prevention and Control*. All these serve as a baton for the green development of transportation industry.

2. **Accelerated restructuring of the transportation industry**

In terms of regional transportation, the Ministry of Transport took the lead in formulating the *Three-Year Action Plan for Advancing the Adjustment of Transport Structure (2018–2020)* to promote the shift from highway to railway and waterway transportation of bulk cargo. According to statistics, in 2018, the national railway freight volume reached 4.025 billion tons, up by 9.1% year-on-year or 337 million tons. The railway freight volume increased by 10.8% in the Beijing-Tianjin-Hebei and surrounding areas, remained the same as the previous year in the Yangtze River Delta region, and rose by 11.4% in the Fen-Wei Plain. The waterway freight volume was 6.99 billion tons, up by 4.7% year-on-year. The highway transportation volume of bulk cargo at coastal ports decreased by about 100 million tons. The railway-and-waterway coordinated transportation volume of containers at key ports increased by 25.5% year-on-year. In terms of urban transportation, efforts were made to vigorously promote the development of "transit metropolis" and urban green freight transportation, prioritize the development of subways, buses, and shared bicycles, and encourage the public to opt for green commuting.

3. **Vigorously promotion of the demonstration projects on green transportation infrastructure**

First, with the special funds for energy conservation and emission reduction, 62 regional thematic demonstration projects on green, circular and low-carbon transportation have been carried out. Second, 63 pilot projects on environmental protection in transportation were implemented, covering transportation environment monitoring network, ecological construction and protection of major transportation infrastructure, among others. Third, 33 typical demonstration projects on green highway were launched in three batches to build tourism roads. Fourth, efforts were made to promote the construction of green ports and green waterways.

4. **The significantly accelerating process of cleaning transportation equipment**

First, vigorous efforts were made to control the pollution from diesel trucks. Joint efforts were made to formulate the *Action Plan for Prevention and Control of Pollution from Diesel Trucks*, put the old implementation plan out of action, promote the establishment of a system for the emission inspection and mandatory maintenance of motor vehicles, and accelerate the application of new and clean energy. The total number of new energy vehicles in China has exceeded 380,000. Second, the prevention and control of pollution from ships and ports were carried out. The emission control areas for ships were established and expanded. It's estimated that by 2020, the emission of sulfur oxides from ships in the control areas will drop by more than 65% compared with 2015. It's encouraged to eliminate the old inland waterway ships. More than 600 old seagoing vessels were dismantled nationwide and 43,800 inland waterway ships were dismantled and rebuilt. Efforts were made to vigorously promote the use of shore power by anchored ships. More than 2400 sets of shore power facilities have been built in coastal and inland rivers across the country, covering more than 3200 berths. In addition, efforts were made to strengthen the capacity to receive and dispose pollutant from ships. Most port cities have established the joint management system for the transfer of pollutants from ships.

3 Promote the Green Development of Transportation at a New Historical Starting Point

In 2018, the *Decisions of the Central Committee of the CPC and the State Council on Comprehensively Enhancing Eco-environmental Protection to Completely Win the Battle against Pollution* introduced strategic plans for pollution prevention and control, in a bid to win the battle to keep our skies blue, our waters clear, and our land pollution-free. It was followed by the promulgation of a slew of special policies closely related to the transportation industry, including the *Three-year Action Plan for Promoting the Adjustment of Transportation Structure (2018–2020)*, *The Action Plan to Win the Battle against Pollution from Diesel Trucks*, and *The Green Travel Action Plan (2019–2022)*. They imposed specific requirements for the green development of the transportation industry, restructuring, changes in growth model, and pollution prevention and control. According to the relevant studies by the Chinese Academy of Engineering, to achieve the fundamental improvement in the quality of ecological environment by 2035, the total emissions of various sectors of the national economy, including the transportation industry, need to be cut by more than two-thirds. Both international experience and domestic practice have shown that with the rapid development of the national economy and the economic restructuring, transformation and upgrading, emissions from the transportation industry in the post-industrial period will continue to grow rapidly, accounting for a rising proportion of the total emissions. The analysis results of PM 2.5 pollution sources in China show that transportation emissions constitute the primary source of pollution in many large and medium-sized cities such as Beijing, Shanghai, Shenzhen, and Hangzhou, and the second or third largest source of pollution in most large and medium-sized cities. It can be seen that with the in-depth development of ecological civilization in China, various requirements for environmental governance are bound to become more stringent. The transportation industry has gradually become a key area in the development of ecological civilization and the pollution prevention and control.

For a period of time in future, China will maintain a relatively high growth rate in transportation infrastructure, equipment, and service demand, while the contradiction between the arduous task of transportation development and the increasingly rigid resource and environmental constraints will become more prominent. To build China into a world leader in transportation, we must increase the capabilities for sustainable development, attach importance to the environmental impact, and assume the responsibility of a major country. The green development of transportation is an essential requirement for satisfying the people's aspirations for a better life, the inevitable trend of high-quality development of transportation, as well as the inherent requirement and important battlefield of building China into a leader in transportation. Green development will also be the basic requirement for transportation infrastructure, equipment and organizations in the future.

According to the overall planning of the Ministry of Transport, by 2020, a green transportation system that's scientifically-designed, eco-friendly, clean, low-carbon, intensive, and efficient will be initially established and the task of transportation pollution prevention and control will be fully completed. By 2035, a new pattern of transportation development that matches the carrying capacity of resources and environment and is coordinated with production, life and ecology will be put in place. The green development of transportation will generally meet the requirements of building China into a leader in transportation.

To promote the green development of transportation, we should stick to the goal of improving the green travel ratio and green logistics indicators, focus on the tasks of comprehensively building a green transportation service system, a green transportation infrastructure system, a clean and low-carbon transportation equipment system, a green transportation governance system, and a green transportation research and innovation system, and pursue the strategy of building China into a leader in transportation. To this purpose, we should focus on the following tasks.

1. **Accelerate the establishment of a green transportation service system**

Firstly, build a green travel system by implementing the *Green Travel Action Plan (2019–2022)*, optimizing the inter-city passenger transport structure, carrying out urban green travel campaigns, innovating the supply of green travel products, and continuing to implement the strategy of prioritizing public transportation. Secondly, establish a green logistics system by implementing the *Three-Year Action Plan for Promoting the Adjustment of Transportation Structure (2018–2020)*, improving the cargo transportation structure, developing efficient urban distribution models, promoting efficient transportation organization models, and strengthening the integrated development of technology and logistics.

2. **Continue to build green transportation infrastructure**

Firstly, make overall planning for the layout of transportation infrastructure. Study and formulate the plan for comprehensive three-dimensional transportation network. Promote the integrated development of railways, highways, waterways, civil aviation, and postal services, based on the national space planning for urban, agricultural and ecological spaces, the redlines for protecting the ecosystems, the permanent basic cropland, and the boundaries for urban development. Secondly, strengthen the intensive use of transportation resources. Promote the intensive use of channel shorelines and land resources, the comprehensive and recycling of resources, and the innovation in energy supply for transportation infrastructure. Thirdly, focus on ecological protection of infrastructure. Promote the development of green infrastructure, implement greening of traffic corridors, and carry out ecological restoration of traffic infrastructure.

3. **Continue to promote clean and low-carbon transportation equipment**

Firstly, promote the professional and standardized transportation equipment. Improve the energy efficiency of transportation equipment and apply advanced and high-performance transportation equipment. Secondly, promote the application of new- and clean-energy vehicles. Thirdly, continue to strengthen the prevention and control of pollution from vehicles and ships. We will implement the *Action Plan to Win the Battle against Pollution from Diesel Trucks*, eliminate and update old diesel trucks with high emission, and comprehensively promote the construction of emission control areas for ships.

4. **Build a modern green transportation governance system**

Firstly, establish a green transportation system. Set up a green transportation planning policy system, improve the green transportation standard system, and strengthen the green transportation data capacity building. Secondly, innovate investment and financing strategies in green transportation. Enhance the guiding role of financial funds at all levels, develop economic policies for the green development of transportation, and accelerate the exploration and application of green transportation market mechanisms. Thirdly, step up implementation, supervision and assessment.

5. **Improve the green transportation research and innovation system**

Firstly, improve the incentive mechanism for green transportation research. Secondly, strengthen the R&D and application of green transportation technology. Thirdly, focus on exchanges and cooperation in green transportation in an all-round way.

The green development of transportation is a profound revolution that requires the wisdom and strength from all sectors of society. On the journey towards green transportation, let us join hands to carry out reform and innovation, work hard, overcome difficulties, and make sustained efforts. Let's work intelligently and diligently to create a beautiful home with blue skies and lucid waters and make greater contributions to promote green development of transportation, build China into a leader in transportation, facilitate ecological progress, and build a beautiful China.

IT-Based Reform and Innovative Development of Ecological Environment

Shaomin Zhang

The 13th Five-Year Plan period (2016–2020) is the decisive stage for China in the building of a moderately prosperous society in an all-round way and the initial stage for new breakthroughs in the transformation of information and communication technology. In the meanwhile, information technology represents new productivity and development goal and has become a pioneering force in leading innovation and driving transformation. At the 19th General Assembly of the Members of the Chinese Academy of Sciences and the 14th General Assembly of the Members of the Chinese Academy of Engineering, General Secretary Xi Jinping gave the instruction to seize the opportunity presented by the integrated development of digitalization, networking, and smart technologies, while leveraging the IT and smart technologies to foster new growth drivers.

In order to win the fight against pollution and protect the ecological environment, the IT application in ecological environment has been brought to the forefront. At the National Conference on Eco-environmental Protection, Minister Li Ganjie of Ecology and Environment emphasized the need to expand the IT application in ecological environment. Information technologies such as big data, Internet Plus, and artificial intelligence are growing to be vital means to modernize the system and capacity for eco-environment governance. In recent years, a host of major IT-based projects have been implemented, including the National Environmental Information and Statistical Capacity Building Project and the Eco-Environment Big Data Project. Consequently, the eco-environment business network has been set up and put into use, the eco-environment monitoring network has made significant headway, the eco-environment information and resource center went online, the IT application in eco-environmental business was further expanded, the eco-environmental IT application standard system has taken shape, and the IT application has played an active role in supporting eco-environmental management.

S. Zhang (✉)
Information Center of Ministry of Ecology and Environment, Beijing, China

J. Pan et al. (eds.), *Beautiful China: 70 Years Since 1949 and 70 People's Views on Eco-civilization Construction*, https://doi.org/10.1007/978-981-33-6742-5_31

1 Overview of the IT-Based Development in Eco-Environment

In China, the IT application in environment started during the 7th Five-Year Plan period and enjoyed a rapid growth during the 9th and 10th Five-Year Plan periods. By virtue of the World Bank loan projects and the Japanese government's grant to build environmental information system in 100 cities in China, we established environmental information organs at ministerial, provincial and municipal levels. They have made notable achievements serving environmental management, basic capabilities in IT application, and organizational team building.

During the 11th and 12th Five-Year Plan periods, a series of major projects were implemented to promote IT application in environment, including the project to enhance the automatic monitoring capacity in state-controlled key pollution sources, the national environmental information and statistical capacity building project, the first national survey of pollution sources, and the first national survey of soil pollution. A "three-tier and four-level" environmental business network covering environmental management agencies at ministerial, provincial, municipal, and county levels, along with the monitoring system for environmental quality and state-controlled pollution sources have been basically completed. The capabilities for the collection and transmission of environmental data have been greatly improved.

Since the beginning of the 13th Five-Year Plan period, under the guidance of Xi Jinping's strategic thinking on building China into a cyberpower and against the backdrop of the rapid development of information technology, the Ministry of Ecology and Environment has successively issued and implemented the *Overall Plan for the Development of Big Data on Ecology and Environment*, the *Implementation Plan for Integration and Sharing of Administrative Information of Ministry of Ecology and Environment*, and a series of other important plans. The monitoring system integrating the use of satellite, aerial and ground stations has been continuously improved. Active efforts have been made to promote the disclosure of administrative information and the development of service platforms, while exploring the innovative application of big data on ecology and environment. In this way, the environmental IT application has made positive progress. In April 2018, in his speech at the national conference on the work of cybersecurity and informatization, General Secretary Xi Jinping pointed out to "practice the new concepts of development and accelerate the IT application." In July 2018, Minister Li Ganjie of Ecology and Environment inspected the Information Center and presided over a symposium on the IT application in ecology and environment. With the focus on winning the battle against pollution, a comprehensive plan for IT-based reform and innovative development was made, launching the IT-based development in ecology and environment to the fast lane.

2 IT-Based Innovations and Achievements in Ecology and Environment

1. Innovation in management model

In response to the current situation and tasks of environmental protection, the Ministry of Ecology and Environment regards reform and innovation as the core means to promote the rapid development of IT application. Since 2018, the *Eco-environmental IT Application Plan (2018–2020)* and the *Guiding Opinions on Strengthening the Cybersecurity and IT Application in Ecology and Environment* have been promulgated to push ahead with the implementation of unified planning, standards, construction, and operation and maintenance, along with the centralized data, funds, personnel, technology, and management. In addition, vigorous innovations have been seen in institutional and team building, IT-based fund management, centralized data sharing, and IT-based leadership mechanisms.

(1) Steadily carry out institutional and team reform

The Ministry of Ecology and Environment set up the Leading Group for Cyberspace Affairs, the Leading Group Office (Cyberspace Affairs Office) and the Information Center, in a bid to enhance the IT-based departments and functions. In this way, the centralized and unified leadership and management of cyberspace affairs has been consolidated significantly. Meanwhile, the IT-based dual management has been carried out within the system of the Ministry of Ecology and Environment. The dual management integrates the information resources of the ministry's subordinate units, including infrastructure, information systems, data resources, and talents, forms a untied and integrated entity to jointly engage in IT application planning and standards, budget declaration, construction and development, system operation and maintenance, data sharing, and cybersecurity, and push ahead with major tasks in IT application deployed by the General Office of the CPC Central Committee, the General Office of the State Office and the Leading Party Group of the Ministry of Ecology and Environment. On such basis, gather the IT strength of the national ecology and environment system through "collaboration," mobilize the forces of the whole society through "cooperation," and create a national "coalition" of IT application in ecology and environment that makes concerted efforts towards the same goal.

(2) Vigorously promote innovation in fund management

Implement centralized management of the funds for IT application. Set up special accounts for the budget of IT application funds. In line with the plan for IT application and under the overall framework, the Office for Cyber Affairs shall be in charge of the review and approval of IT application projects. Strict measures will be adopted to ensure the unified review and the unified account management of IT application funds, in a bid to give full play to the role of "centralized funds."

(3) Continue with data integration and sharing

All of the in-use information system data is incorporated into the eco-environment information resource center. Make concerted efforts to build basic database and data-sharing service platform in the eco-environment business management, so as to provide "one-stop" service for data sharing. Gradually develop a data sharing system with other ministries, commissions and local departments. Share data across departments and various levels through environmental information resource center and data sharing service platforms. Explore the establishment of a model for comprehensive data utilization that's open for all to contribute and gain benefits, so as to continuously improve the productivity of data products.

(4) Actively promote innovation in leadership mechanism

Establish an innovative leadership mechanism for IT application. The minister serves as the head of the Leading Group for Cyber Affairs, responsible to regularly listen to reports on the work of IT application and make plans for major tasks. Other departmental leaders convene meetings on the work of IT application on a monthly basis to supervise and push ahead with the work progress and help resolve difficulties. The Working Rules of the Ministry of Ecology and Environment has made it clear that the main responsible comrades of the Information Center shall attend ministerial meetings as non-voting delegates, so that the Information Center can timely understand the decision and plans of the ministry's party group and provide targeted support services.

2. **Achievements**

Thanks to the above measures for reform and innovation, the IT application in ecology and environment has achieved rapid development. A system incorporating "a network, a Cloud, a database, a map and a portal" has been initially put in place that facilitated the precise management of eco-environmental protection, improved the scientific decision-making for pollution prevention and control, and provided the public with more convenient government services.

(1) Initial establishment of the "five-in-one system"

First, "a network" has been put in place. As of the end of 2018, all units under the Ministry of Ecology and Environment and the eco-environment institutions of all provinces (autonomous regions, municipalities directly under the Central Government, Xinjiang Production and Construction Corps), prefectures, and districts and counties have been incorporated into the eco-environmental business network. As such, an all-encompassing and exhaustive network has been established, providing strong support for major tasks like the data collection for the second survey of pollution sources, the transmission of the automatic monitoring data on pollution sources, and the submission of the environmental statistics during the"13th Five-Year Plan" period.

Second, the development of "a Cloud" has been accelerated. In accordance with the State Council's requirements for the integration and sharing of the government information system, a three-in-one approach of "uploading data into Cloud, database and portal services" has been adopted to integrate the information system on the Cloud. Persistent efforts have been made to strengthen the basic capabilities in the Cloud development in ecology and environment, ensure the Cloud deployment of information system in a convenient and quick way, and provide guarantee for stable and safe operation.

Third, concerted efforts were made to establish "a database." A unified database on stationary sources have been set up, containing basic information about more than 1.8 million companies, such as production and governance facilities, activities, administrative approvals, pollution discharge information, violations of law, and credits.

Fourth, "a map" has been drawn meticulously, delivering a slew of eco-environmental spatial data sets on, covering the environmental quality, pollution sources and natural ecology. The information platform for the map has been established, providing standardized spatial data services for the tasks of defining "Three Lines and One List" (ecological protection red lines, environmental quality bottom lines, resource utilization upper limits, and environmental access negative list), environmental law enforcement and others.

Fifth, all-out efforts have been made to create "a portal." In a bid to continuously promote the development of a comprehensive business portal of the Ministry of Ecology and Environment, all the information systems in use have been incorporated into it, while login, users and authentication for important systems have been unified. The comprehensive business portal has become the unified entrance to the Ministry's administrative affairs, applications and resource.

(2) The scientific decision-making for pollution prevention and control has been improved.

The system for assisting decision-making to strengthen supervision of special actions, the App for sewage outlet information collection, and other information systems have been established. The apps for emergency response to heavy air pollution, agricultural and rural eco-environment management, soil environmental information management, and others have been developed. New technologies such as the Internet of Things and satellite remote sensing, have been applied to promote the monitoring system integrating the use of satellite, aerial and ground stations, providing data services and support for the "7 + 4" special campaigns for pollution prevention and control.

(3) Precise management in environmental protection has made progress.

The information systems for the data collection and management in the second national survey of the pollution sources, the national environmental remote sensing application platform, biodiversity survey, and others have been established. The big data service platform for the environmental protection along the Belt and Road and the data sharing platform for the "Three Lines and One List" along the Yangtze River Economic Belt have been initially set up. The big data has been innovatively applied to environmental impact assessment, pollution discharge permits, and the hotline "12,369" for the public complaint and oversight. The hotspot grid technology was applied to the refined management of atmospheric environment in "2 + 26" cities in the Beijing-Tianjin-Hebei region and the surrounding areas.

(4) More convenient government services were provided for the public.

The Ministry of Ecology and Environment has redesigned and upgraded its official website, featuring the barrier-free access. It constitutes a notable achievement in building a service-oriented government. All portal sites of the subordinate units have been incorporated into the site group, characterized by unified technical specifications, unified technical platform, unified operation and maintenance, and unified security protection, which improved website service capabilities, management levels and overall effectiveness. All non-confidential matters requiring administrative examination and approval shall be handled on the platform of "Internet Plus Government Services," which has effectively promoted the reform of "one website, one portal and one-time visit."

3 Prospects for the Development of IT Application in Ecology and Environment

Despite the rapid development and outstanding progress, the IT application in ecology and environment still faces a number of difficulties, such as the poor coordination of the application system, limited capabilities in data mining and analysis, small talent pool, and a large gap between investment and actual demand. In March 2019, Minister Li Ganjie presided over the first plenary meeting of the Leading Group for Cybersecurity and IT Application of the year, charting the course, setting the goals and assigning tasks for the IT application in ecology and environment at the next stage.

1. **Approaches of development**

Under the guidance of Xi Jinping Thought on Ecological Civilization and the principles established at the National Conference on Eco-environmental Protection, we will fully implement the requirements for unified planning, standards, construction, and operation and maintenance, along with the centralized data, funds, personnel, technology, and management. We will follow a "three-step" approach of consolidating the foundation by closing the loopholes, enhancing the achievements by making breakthroughs at key points, and pushing forward reforms as a whole, in a bid to create the "overall environment" for promoting IT application. We will speed up the construction of "big platforms, big data and big systems" in an effort to enhance the "big security" of IT application, consolidate the achievements of reform and innovation, and further unleash the vitality of reform, stimulate innovation, and promote the leapfrog development of IT application in ecology and environment.

2. **Phased goals**

The Ministry of Ecology and Environment will fulfill the goal of pioneering and leading IT application by making full use of the leading role of innovation, giving prominence to industrial characteristics, maintaining competitive edges, promoting coordinated development, and following the "three-step" approach. In this way, we will provide strong support for winning the battle against pollution.

3. **Key tasks**

We will focus on the following five tasks.

First, create the "overall environment" for IT application. Deepen the IT-based institutional reform to comprehensively promote unified and centralized management. Rely on the dual management and other innovative mechanisms to integrate resources inside and outside the system. Coordinate the management of funds for IT application and bring into play the benefits of capital accumulation. Step up efforts in building a team for cyber affairs and IT application and foster a strong environmental protection corps to provide strategic support. Strengthen publicity and guidance for IT application and create a good atmosphere to accomplish the goals.

Second, build an IT-based "big platform" Make coordinated efforts to promote the building, management and services of a Cloud platform for ecology and environment, strengthen the integration and coordination of resources of the computer rooms, network, computing, storage, security, operation and maintenance within the Ministry, and build a "big platform" for information infrastructure.

Third, develop "big data" for IT application. Strengthen data resource planning, continue to promote the concentration of data resources, gradually build basic databases on environmental quality, pollution sources, and ecological protection, and accelerate the construction of a national database of stationary pollution sources. Step up efforts in efficient development, integration and in-depth utilization of data resources to facilitate the optimization of business process and innovation of management model.

Fourth, develop a "big system" for IT application. Continue to push ahead with system integration. Promote the establishment of a large-scale system for interconnected and coordinated business. With the focus on pollution prevention and control as scheduled, we will improve the capability in comprehensive decision-making. Efforts will be made to build an integrated online government service platform, promote the use of Internet App, and improve the capability for business management and convenient information services.

Fifth, provide guarantee for the "big security" of IT application. We will establish a correct concept of cyber security and strictly implement the main responsibilities and supervision for cyber security. We will carry out the central government's requirements for independent and controllable "cyber security risks" and accelerate the development of safe and reliable information equipment. In addition, we will strengthen security protection measures, improve the capabilities to evaluate security situation and risk prevention, and ensure that the network security is generally under control.

To sum up, in the next step, we aim to achieve the leapfrog development of IT application in ecology and environment, enter the period of smart technology, create an overall environment for the implementation of "unified planning, standards, construction, and operation and maintenance, along with the centralized data, funds, personnel, technology, and management" and build a unified big platform for the "eco-environment cloud" infrastructure featuring stable operation and strong support. We will put in place comprehensive, centralized and efficient big data on ecology and environment, along with a coordinated and distinctive big system for key businesses. We will comprehensively strengthen the security of independent and controllable information network in active defense, fulfill major tasks of IT application, significantly improve the scale and capability of the talent teams for IT application, and make sure to meet the goal of playing a pioneering and leading role in IT application.

Towards Green Development: From an Option to an Objective

Shunze Wu

Green is the color of nature and represents the foundations of a better life for the people. Over the past 70 years since the founding of New China, we have gradually deepened our understanding of green development and improved relevant practices accordingly. After going through a process from limited to profound understanding and from conscious to spontaneous practices, China has embarked on a path of green development with Chinese characteristics. This is an inevitable trend as well as a subjective choice that demonstrates the profound changes in the concept and way of development and the CPC governance. A review of China's journey to explore green development step by step has great theoretical significance and practical value for promoting ecological progress and building a beautiful China.

1 China's Green Development Is Out of Both Historical Necessity and Subjective Choice

From the emergence of the idea, the sublimation of the concept, the establishment of relevant systems, to the practices, the path of green development has been determined by the actual conditions in China, as well as a result of the CPC's active explorations for new ways of development and satisfying the people's aspirations for a better life.

Wang Yong and Liu Yue contributed to this paper.

S. Wu (✉)
Policy Research Center for Environment and Economy, Ministry of Ecology and Environment, Beijing, China

J. Pan et al. (eds.), *Beautiful China: 70 Years Since 1949 and 70 People's Views on Eco-civilization Construction*, https://doi.org/10.1007/978-981-33-6742-5_32

1. **Green development is an inevitable choice for development based on China's resource and environmental endowments**

Due to the limited resources per capita and the relatively fragile ecology, China must take the path of conservation and intensive utilization of resources and environmental protection. According to the national plan for developing functional zones, mountainous areas and plateau make up about 60% of land space in China. Apart from the arable land and existing construction land that must be protected, the land suitable for industrialization, urbanization and other construction in the future covers only 280,000 km^2, accounting for about 3% of the total land area in China. The per capita water in China is only 28% of the world's per capita share, in addition to the problem of unmatched distribution of water resources with land resources and economic development. The ecologically fragile areas above the moderate level account for 55% of the land area in China, of which extremely vulnerable areas make up 9.7%, severely vulnerable areas account for 19.8% and moderately vulnerable areas account for 25.5%. Large-scale and intensive industrialization and urbanization can only be carried out in limited areas suitable for development.

The Hu-Line characteristics in population geology made it a must for China to fully respect the laws of nature and take a development path of maintaining balance between population, industries, resources and the environment. The Hu Line extends from Aihui Country in Heilongjiang Province to Tengchong in Yunnan Province, dividing China into two parts: southeast and northwest. Covering 36% of China's territory, the southeastern China is home to 96% of the national population. In contrast, the northwestern China accounts for about 64% of the national territory, but with merely 4% of the population. According to the Sixth National Population Census in 2010, the proportion of the population in the eastern China decreased by only 2.3% in more than 70 years. The existence and enduring stability of the Hu Line indicates that the overall population distribution in China, featuring dense population in the east and sparse population in the west, has been constrained by resource endowments and natural laws, which in turn affects China's industrial layout and utilization of resources and environment. It can also be said that the "Hu Line" has locked China's economic and geographical pattern, and it would be against the laws of nature to deny it.

To resolve the actual contradiction between economic development and eco-environment, China must take the path of promoting protection and development in a coordinated way. Since the reform and opening up, China's economy has achieved leapfrog development. It took less than half a century for China to accomplish industrialization that consumed two or three centuries in developed countries, creating the "Chinese miracle" that has drawn worldwide attention. However, such "compression of time and space" in economic growth has resulted in corresponding issues in resources and environment. China's total energy consumption and carbon dioxide emissions rose from 990 million tons and 2.29 billion tons in 1990 to 4.26 billion tons and 9.93 billion tons in 2014, registering a three-fold increase. China's energy consumption per unit of GDP is nearly 2.5 times as high as the global average and twice the average in the Asia-Pacific region. The economic structure characterized by heavy industry and extensive development model have brought about fundamental obstacles, making it impossible for the environment and resources to support the sustainable development of the economy. Therefore, we must explore a green path with equal stress on development and protection.

2. **Green development is a scientific choice made by China in the process of deepening understanding of the laws of socioeconomic development**

At the UN Conference on the Human Environment in Stockholm, Sweden in 1972, Premier Zhou Enlai first brought the issue of pollution to the attention. He emphasized that environmental problems should not be regarded as trivial matters and called for immediate actions. Under the instructions of Premier Zhou Enlai, China sent a delegation to the Conference on the Human Environment, which marks an important turning point for China to take the initiative to opt for the path of green development. In 1995, the Fifth Plenary Session of the 14th CPC Central Committee decided to bring about "two fundamental changes." One is to transform from a traditional planned economic system to a socialist market economic system, and the other is to change the mode of economic growth from extensive to intensive. This reflects the inherent requirements of economic growth in the new stage and an objective choice for achieving sustainable development. Since the second half of 2002, China has entered a new round of expanding heavy chemical industry, followed by heavy investment across the country in high energy-consuming and high-emission projects such as steel, cement, chemical, and coal power. This has put tremendous pressure on the resources and environment, leading to the growing resource-related and environmental constraints. In October 2003, the Third Plenary Session of the 16th CPC Central Committee put forward the scientific outlook on development that requires "putting people first, aiming at comprehensive, coordinated and sustainable development, and promoting well-rounded development of the economy, society and people." The Fifth Plenary Session of the 16th CPC Central Committee proposed to "accelerate the development of a resource-conserving and environment-friendly society."

In this way, green development has become the will and the governing concept of the Party and the state. Following the 2008 financial crisis, revolutions in new energy and "Internet Plus" broke out around the world, demonstrating the reflection on the entire industrial civilization. The report at the 18th CPC National Congress first made the "building of a beautiful China" a grand goal of ecological progress. It gave prominence to the development of ecological civilization, incorporating it into all aspects and the whole process of advancing economic, political, cultural, and social progress, as well as the Five-Sphere Integrated Plan for the socialist cause with Chinese characteristics. For the first time, the CPC Constitution amended and adopted at the 18th CPC National Congress stipulated that "the Communist Party of China shall lead the people in building a socialist ecological civilization" and expounded on the statement, consolidating the strategic significance of the ecological progress. In October 2015, the Fifth Plenary Session of the 18th CPC Central Committee introduced the philosophy of innovative, coordinated, green, open and shared development, making green development a basic concept for socioeconomic development during the 13th Five-Year Plan period and beyond.

3. **Green development is an active choice of the CPC to meet people's aspirations for a better life in the process of deepening its understanding of the laws concerning the governance**

The process to improve people's livelihoods could be divided into three stages—problem-solving, improvement and development. It not only requires economic growth, but more importantly, it means satisfying people's growing needs for a stronger sense of happiness with the increasing households' income. Blue skies, lucid water and lush mountains play an important role for people's sense of happiness. The key to governance lies in the peace of the people, to which end we must first deliver benefits for the people. The most inclusive benefit for the livelihood of the people lies in the green development, for which the key is the organic integration of green development and people's livelihood and welfare. It's an objective choice for the CPC to firmly establish the philosophy of green governance in response to the new situations, problems and contradictions facing its governance in the new era. The benign combination of economic development, eco-environment and people's needs in life is an important criterion for the good governance of a country.

2 The Ideological Change Constitutes the Foundation for China's Transformation to Green Development

In the profound reform towards green development, the ideological change is not only a reason behind China's option for the path of green development, but also the endogenous driving force that promotes China to embark on the path. The transformation of the outlook on development and on people's livelihood is the fundamental manifestation that China has taken the path of green development, demonstrating the integration of socioeconomic development and the well-rounded development of the people.

1. **The philosophy that "lucid waters and lush mountains are invaluable assets" provides guidance for green development in terms of the outlook on development**

The outlook on development embodies the needs of socioeconomic development in a certain period at the ideological level, representing the overall and systematic view of a country on its definition and approach to development during the period. Promoting green development is a profound change in the development outlook. General Secretary Xi Jinping's statement that "lucid waters and lush mountains are invaluable assets" clarified the unity of economic development and environmental protection. It is an important manifestation of the change in development outlook. General

Secretary Xi Jinping put forward to "properly handle the relationship between environmental protection and economic development, that is, the relationship between lucid waters and lush mountains and invaluable assets." In a nutshell, "lucid waters and lush mountains are invaluable assets." Epistemologically, it means the integration of economy and environment. In practice, it is to develop eco-friendly growth models and ways of life. In essence, it is to achieve the internal unity of development and protection that reinforces each other. In terms of the goal, it means fulfilling the goal of enriching the country and benefiting the people through green development. The statement that "lucid waters and lush mountains are invaluable assets" fundamentally changed the traditional understanding about zero or low value of ecology and environment, redefined the value and updated the understanding. It broke the constraints in the thinking that simply put development at the opposite side of protection, while clarifying the methodology for realizing the internal unity and harmonious co-existence of development and protection that reinforces each other.

2. **The statement that "a sound ecological environment is the most inclusive welfare for the people" clarifies the fundamental purpose of green development**

A sound ecological environment and natural resources serve as the conditions and foundation for the free and comprehensive development of human beings. The statement that "a good ecological environment is the most inclusive welfare for the people" shows the basic view about the people's livelihood under the framework of the Xi Jinping Thought on Ecological Civilization. Promoting green development and the dramatic increase in natural capital means to develop a sound ecological environment as a growth driver for the people's living standards. In essence, it embodies the people-oriented concept, conducive to achieving the harmony between man and nature, between people, and between man and society. We should promote green development and ecological protection in the interests of the people, focus on solving the outstanding environmental problems doing harm to the health of the people, and consistently meet the people's growing demands for a beautiful environment. In this way, we will deliver benefits to the entire society and every member of the society, which is an important manifestation of an all-round moderately prosperous life shared by all the people.

3 Green Targets and Evaluation Criteria Are the Baton for Navigating China's Green Development

Eco-environmental problems are fundamentally a problem of development mode. The main culprit is the resource consumption and pollutant discharge in the socioeconomic activities of the mankind exceeded the carrying capacity of the earth. Therefore, in essence, the first thing to do to promote green development is to correct the practice of seeking temporary economic growth at the cost of the environment and promote the transformation towards green and intensive mode for socioeconomic development.

1. **Gradually establish the goals of green development in the Five-Year Plan**

The "Five-Year" Plan is an important way for China to promote sustainable and healthy socioeconomic development. With the main indicators for socioeconomic development as its core content, it reflects the basic ideas and charts the course for the development in the upcoming five years. Starting from the 10th Five-Year Plan, China began to incorporate resource-related and environmental indicators into the planning goals. The 10th Five-Year Plan proposed to "attach importance to issues concerning population, resource and eco-environment" and included the reduction of the total discharge of major pollutants as an indicator for planning goal. The 11th Five-Year plan regarded energy conservation and emission reduction as the main indicators of economic and social development. For the first time, the development goals were divided into anticipated and obligatory categories. Regarding the resource and environment, there were six obligatory goals and two anticipated goals. The introduction of obligatory indicators was an important innovation in national governance. In the 12th Five-Year Plan, the proportion of obligatory indicators relating to resource and environment increased to 33.3%, compared with 27.2% in the 11th Five-Year Plan. It clarified the incentive and constraint mechanism for green development. In the 13th Five-Year Plan, the indicators were enriched to cover air quality and surface water quality. The number of the resource-related and environmental indicators rose from eight to ten, all obligatory. Therefore, green development constituted the key note in all sectors and links of socioeconomic development during the 13th Fiver-Year Plan period.

2. **Gradually establish a target responsibility system in the performance evaluation of Party and government officials to promote green development**

The target responsibility system plays a pivotal role in the decision-making and implementation of the Five-Year Plan. In the process of implementing the 11th Five-Year Plan, the State Council decomposed the main objectives and tasks in line with the division of responsibilities and assigned them to each region. In addition, the obligatory indicators were incorporated into the criteria for comprehensive evaluation of socioeconomic development and performance assessment of each region and department. In 2006, the Organization Department of the CPC Central Committee issued the *Trial Measures for the Comprehensive Evaluation of Local Party and Government Leaders and Officials in Line with the Requirements of Scientific Outlook on Development*, including obligatory indicators for the performance evaluation of local leaders. This was of great significance for mobilizing officials at all levels to better realize the goals set in the 11th Five-Year Plan. In the field of ecology and environment, since the introduction of the target responsibility system for environmental protection in 1996, particularly for energy conservation and emission reduction, China has made remarkable achievements. In the 11th Five-Year Plan, the aggregate emissions control has become the most important part of the central government's environmental performance assessment of provinces and large state-owned enterprises. It's also the most important content of environmental governance in each province, acting as the baton for environmental protection.

After the 18th CPC National Congress, the General Office of the CPC Central Committee and the General Office of the State Council have issued the *Measures for the Accountability of Party and Government Leaders for Ecological and Environmental Damage (Trial)*. At the Fifth Plenary Session of the 18th CPC Central Committee, it was made clear to "establish a space governance system consisting of the space planning, utilization control, the audit of outgoing officials' natural resource asset management, and differentiated performance assessment, based on the units of city- and county-level administrative regions." In this way, the connotations of green development have been further enriched. At the end of 2013, the Organization Department of the CPC Central Committee issued the *Notice on Improving the Performance Evaluation of Local Party and Government Leaders and Officials*, requesting to increase the weight of indicators relating to resource consumption and environmental protection. In 2016, the General Office of the CPC Central Committee and the General Office of the State Council promulgated the *Plan for the Pilot Program to Audit Outgoing Officials' Natural Resource Asset Management*. This was an important step in the evaluation of green development and provided guidance for the leaders to pursue green development. In the same year, the General Office of the CPC Central Committee and the General Office of the State Council promulgated the *Measures for Assessing Progress made in Ecological Improvement* and made it clear that the Green Development Index System and the Ecological Progress Assessment Goals System shall respectively serve as the basis for the annual and the five-year evaluation. As such, the performance in green development has officially become an important measure for the evaluation, reward, punishment, appointment and removal of provincial Party and government leaders. In July 2017, the General Office of the CPC Central Committee and the General Office of the State Council issued a circular on the eco-environmental problems in the nature reserve of Qilian Mountain in Gansu Province, for which dozens of leading cadres, including three provincial and ministerial leaders, were held accountable. Since then, some local leaders began to realize the importance of the requirements that both party committees and governments are held accountable, and that officials take responsibility for the protection of ecology and environment. The first round of the central inspection on ecological and environmental protection that started at the end of 2015 realized the full coverage of all 31 provinces (autonomous regions and municipalities) in less than two years. It has therefore become a practical and effective tool for pushing local CPC committees and governments as well as their relevant departments to fulfill their environmental protection obligations.

3. Gradually establish the basic principles of prioritizing ecological protection and green development in major regional development strategies

During the 10th Five-Year Plan period, it's proposed to develop ecological demonstration areas, eco-counties, eco-cities, and eco-provinces. During the 11th Five-Year

Plan period, it's put forward to build a resource-conserving and environment-friendly society. In this way, the connotations of green development have been continuously enriched. However, ecological protection had not yet become the bottom line for economic and social development. After the 18th CPC National Congress, the principle of giving priority to ecological protection has been established in major regional development strategies such as the Yangtze Economic Belt Initiative, Beijing-Tianjin-Hebei Integration Initiative, and Xiong'an New Area Initiative. The Yangtze River Economic Belt Initiative, with ecological conservation and green development at the core, constitutes an integral part of the CPC Central Committee's new ideas, thinking and strategies for national governance. In January 2016, at a symposium on promoting the development of the Yangtze River Economic Belt held in Chongqing, General Secretary Xi Jinping pointed out that to promote the development of the Yangtze River Economic Belt, we must take the long-term interests of the Chinese nation into consideration and take the path of ecological conservation and green development. In April 2018, at the symposium on further promoting of the development of the Yangtze Economic Belt, General Secretary Xi Jinping emphasized again to properly handle the relationship between environmental protection and economic development, explore new ways to coordinately promote ecological conservation and green development. To explore new ways for prioritizing ecological conservation and green development in the Yangtze River Economic Belt, the key is to properly handle the relationship between lucid waters and lush mountains and invaluable assets. On March 5th, 2019, General Secretary Xi Jinping pointed out that it is necessary to explore new ways of high-quality development that prioritizes ecological conservation and green development. Prioritizing ecological conservation as the principle and prerequisite of a major regional development strategy fully reflects the integration of prioritized ecological conservation with decisions for socioeconomic development.

4 The Spatial Planning System Based on Main Functions Is an Important Means to Reshape the Spatial Pattern of Green Development

Green space is a prerequisite and an important dimension for green development. To achieve green development, it is necessary to determine the pace and areas of development based on the main functions of different types of areas, while giving full play to the important roles of space planning and space utilization control in promoting balanced development of space.

1. **Development direction based on the main functions of the region**

If the master plan for national economic and social development could be likened to the commander of the national development planning system, then the plan for developing functional zones would be the general of the national land and space planning. Based on the resource and environment carrying capacity, existing development density and development potential of each region, the plan for developing functional zones strategically divides the national land and space into four categories of functional zones where development must be optimized, prioritized, restricted, or forbidden. This is a major measure to achieve the goal of ecological progress and coordinated regional development.

The basic idea of developing functional zones was first put forward in the proposal for the 11th Five-Year Plan. The proposal for the 12th Five-Year Plan elevated the plan for developing functional zones to a strategy. In 2013, the *Decision of the Central Committee of the Communist Party of China on Some Major Issues Concerning Comprehensively Continuing the Reform* made it clear to "unwaveringly implement the system of functional zones." In the Outline of the 13th Five-Year Plan, the system of functional zones constituted a specific strategic measure underpinning the concepts of coordinated and green development, as well as an important means to promote the coordinated and green regional development. The idea of developing functional zones has gradually developed from a plan, a strategy, a system, to an integral part of the national plan for socioeconomic development.

Following the release of the *National Plan for Functional Zones* in 2010, the provincial plans for functional zones were issued and implemented successively. The plan for functional zones established the system of "nine policies + one system," referring to the fiscal, investment, industrial, land, agricultural, population, ethnic, environmental, and climate change policies, plus the performance evaluation system for the functional areas. In 2017, the *Opinions of the CPC Central Committee and the State Council on Improving the Strategy and System of Functional Zones* was issued. It proposed that by 2020, the county-scale geographical space in consistence with the functional definition will be basically delineated, the strategic layout of functional zones covering land and sea will be accurately implemented, and the spatial planning system that integrates various planning into a single master plan will be established and improved.

2. **Promote the implementation of the strategic layout of functional zones based on the national space planning for "three spaces and three redlines"**

The key to perfecting the strategy and system of functional zones lies in the implementation of the national and provincial strategic layout for functional zones at the city and county levels, on the basis of strictly sticking to the plan for functional zones. The three spaces refer to the urban, agricultural and ecological spaces, while the three redlines are for protecting the ecosystems, the permanent basic cropland, and the boundaries for urban development. The former lays stress on the definition of dominant functions, while the latter focused more on the rigid control of boundaries. The report to the 18th CPC National Congress put forward to ensure that "the space for production is used intensively and efficiently, that the living space is livable and proper in size, and that the ecological space is unspoiled and beautiful," charting the course for optimizing the optimization of three spaces. In the report to the 19th CPC National Congress, it's specified to "complete work on drawing redlines for protecting the ecosystems, designating permanent basic cropland, and delineating boundaries for urban development." The development from "three spaces" to "three spaces and three redlines" has promoted the shift of the core of regulating the use of all territorial space from protecting the single element of land resources to the full range of mountains, water, forests, fields, lakes and grasses. The national planning for "three spaces and three redlines" has become a basic system for coordinating the scientific protection and rational utilization of natural resources. In May 2019, the CPC Central Committee and the State Council issued the *Opinions on Establishing a Geographical Space Planning System and Supervising its Implementation*. It proposed that by 2035, China will comprehensively improve the modernization of the system and capability for regulating the geographical space and basically form a safe, harmonious, competitive and sustainable territorial space layout featuring intensive and efficient use of the space for production, livable and proper-sized living space and unspoiled and beautiful ecological space.

5 Exploring the Path of Integrating Industries and Ecosystems Is a Fundamental Strategy for Green Development to Transform the Growth Model

Industry provides the basic support for green development. At the National Conference on Eco-environmental Protection held in May 2018, General Secretary Xi Jinping pointed out to step up efforts to establish "a sound economic system aimed at integrating industries and ecosystems," setting the two basic directions for China to explore green economic development.

1. **With the focus on energy conservation and emission reduction, adhere to the path of green, low-carbon and circular development of eco-friendly industries**

To promote eco-friendly industries means to incorporate the nature into the closed loop of the economic system. It is a system that generates economic value through economic activities featuring the most intensive utilization of the minimum amount of resources and the minimum- or zero-harm burden on the environment. As early as the 9th Five-Year Plan period, China proposed to shift from the extensive to the intensive economic growth model, offering no specific and well-developed approaches until the end of the 10th Five-Year Plan period. In response to the huge consumption of energy resources and the deterioration of the ecology and environment, the 11th Five-Year Plan officially introduced important measures, such as strengthening energy conservation and emission reduction and developing a circular economy. In 2005, the State Council issued the *Opinions on Accelerating the Development of Circular Economy*, designating the development of circular economy as an important guiding principle for the preparation of relevant plans. It marked China's official launch of the development concept of circular economy. Subsequently, with the development models of circular economy in Japan and Germany as the benchmark reference line, the National Development and Reform Commission and the Ministry of Environmental Protection conducted centralized pilot projects in the form of "circular industrial parks," In 2008, the State Council promulgated the Circular Economy Promotion Law, with a view to promoting the development of a circular economy from micro-, meso- and macro- dimensions. The 12th Five-Year Plan proposed that by 2015, the national industrial system shall achieve a comprehensive solid waste utilization rate of 72%, while more than 50% of national parks and more than 30% of provincial parks shall implement the main goal of recycling transformation. The 13th Five-Year Plan stipulated to establish a more detailed evaluation framework for circular economy and refine the circular economy evaluation indicators. In 2009, at the United Nations Climate Change Conference in Copenhagen, Denmark, then-Premier Wen Jiabao pledged to the world the goal of reducing greenhouse gas emissions. The low-carbon economy with the basic goal of reducing carbon emissions has gradually become another important means for China to transform to the green development of economy.

Another way to promote the eco-friendly industries lies in the eco-friendly restructuring of industries. The 2008 financial crisis reshuffled the economic sectors and brought green economy to the attention of many countries. The industrial restructuring triggered by the financial crisis has accelerated the innovation of clean energy and clean production technology, and thereby opening up new areas of growth. Countries all over the world began to develop new energy and other strategic emerging industries to underpin a new round of economic growth. China designated seven strategic emerging industries, namely energy conservation and environmental protection, a new generation of information technology, biology, new energy, new energy vehicles, high-end equipment manufacturing, and new materials, as well as 23 key areas. Eco-friendliness is the common direction for the development of these industries and areas. In 2012, the report to the 18th CPC National Congress specified "green, circular and low-carbon development" as an important aspect of ecological progress. In 2015, the State Council issued the Opinions on Accelerating the Ecological Progress, positioning green development alongside new industrialization, IT application, urbanization, and agricultural modernization, as a major direction for economic and social development, while clarifying the industrial carrier of green development. The report to the 19th CPC National Congress emphasized to promote green development and establish a sound economic system for green, low-carbon and circular development. This is not only a continuation of China's exploration for the transformation of economic growth model, but also an important foothold for adhering to green development and building a beautiful China in the new era.

2. **Start by realizing the value of ecological goods and explore the mechanisms and paths to transform lucid waters and lush mountains into invaluable assets**

To promote the industrialized ecology means managing eco-environmental resources as a special capital based on theories on ecological services and public products, in a bid to maintain and appreciate the value, while transforming ecological services from free resources to valuable commodities or services in a way for socialized production and market operation. In 2005, *Zhejiang Daily* ran a commentary by Comrade Xi Jinping, titled Lucid Waters and Lush Mountains Are Invaluable Assets, in the column of "Zhejiang, China: A New Vision for Development." Xi wrote that "if we could transform the co-environmental advantages into competitive edge in ecological economy, such as eco-agriculture, eco-industry and eco-tourism, we can turn lucid waters and lush mountains into invaluable assets." In October 2017, the report to the 19th CPC National Congress specified that "we must realize that lucid waters and lush mountains are invaluable assets and act on this understanding," and the newly revised Constitution of the Communist Party of China provides that "the Communist Party of China shall fully understand that lucid waters and lush mountains are invaluable assets."

The industrialized ecology constitutes the industrial pillar for transforming lucid waters and lush mountains into invaluable assets, embodying the concepts about the natural capital and the integration of economic development with nature conservation. The key to ecological industrialization is to explore the path to realize the value of ecological products. At a seminar on boosting the development of the Yangtze Economic Belt in 2018, Xi Jinping emphasized to actively explore and promote the path of transforming lucid waters and lush mountains into invaluable assets, carry out pilot programs to realize the values of ecological products in selected regions, and blaze a government-led, market-oriented operation and sustainable trail for realizing the values of ecological products. The National Development and Reform Commission and the former Ministry of Environmental Protection have respectively advanced the implementation of pilot projects for realizing the values of ecological products and the development of practical innovation bases to put into action the concept that "lucid waters and lush mountains are invaluable assets," in an effort to translate the concept of industrialized ecology into practice. The concept that "lucid waters and lush mountains are invaluable assets" and the practices in this regard have vigorously promoted the green development that integrates economic development with environmental protection, while significantly boosting our confidence in pursuing a model of sustainable development featuring increased production, higher living standards, and healthy ecosystems.

6 The Continuous Improvement of the Eco-Environment Quality Is the Goal and a Concrete Manifestation of Green Development

The excellent eco-environment quality gives prominence to the color of green development. By means of campaigns for prevention and control of environmental pollution and for ecological protection and restoration, we will strengthen areas of weakness in ecology and environment, so as to consolidate the ecological foundation for green development.

1. **Continuously adjust the approaches for eco-environmental governance in response to the eco-environmental problems**

Since 1972 when China sent a delegation to the United Nations Conference on the Human Environment, China has made active explorations and tried to avoid the old path of pollution before protection in developed countries. During the period from the Third Plenary Session of the 11th CPC Central Committee to 1992, environmental protection had developed into a basic national policy and a part of the national plan for economic and social development. In addition, environmental management systems and environmental laws and regulations had taken shape and China has got environmental protection on track. Since the launch of the 9th Five-Year Plan, the environmental protection in China had been characterized by the large-scale ecological construction and environmental pollution treatment. During the period from the 9th Five-Year Plan to the 13th Five-Year Plan, China has gradually deepened the understanding of its eco-environmental issues and improved its approaches for eco-environment governance. Regarding its understanding of environmental issues, the 9th and 10th Five-Year Plans aroused the concerns about prominent eco-environment problems and deteriorating environment. The 11th Five-Year Plan began to pay attention to the problem of excessive consumption of energy resources. During the period of the 12th and 13th Five-Year Plan, resource-related and environmental problems have become a major constraint on economic development. Meanwhile, the eco-environment governance has evolved from partial to systematic governance. During the period of 9th and 10th Five-Year Plans, the focus of environmental treatment was the problems relating to industrial point sources and in major cities. During the period of 11th and 12th Five-Year Plans, with the focus on major basins, regions and cities, the focus of ecological development has shifted to in-advance conservation and the whole process of resource utilization. During the 13th Five-Year Plan period, ecological conservation developed into a part of systemic governance, environmental governance across all sectors of the society and full-process resource management.

2. **Accelerate to improve the weak links of eco-environmental quality to build a moderately prosperous society in all respects**

Since the 18th CPC National Congress, China has resolutely declared war on pollution. Following the promulgation of the Decisions of the Central Committee of the CPC and the State Council on Comprehensively Enhancing Eco-environmental Protection to Completely Win the Battle against Pollution, the action plans for the prevention and control of air, water and soil pollution have been issued and implemented. As such, an all-out fight has been launched to keep our skies blue, our waters clear, and our land pollution-free.

All localities have taken multiple measures to prevent and control air pollution. From 2013 to 2018, the average concentration of inhalable particulate matter (PM 10) in 338 cities at prefecture-level and above dropped by 26.8%, and the average concentration of PM 2.5 in Beijing dropped by 43% from 89.5 $\mu g/m^3$ to 51 Mg/m^3. After the Action Plan for Water Pollution Prevention and Control was issued, all localities have vigorously promoted the investigation and remediation of environmental problems relating to centralized drinking water sources, with the focus on the treatment of black and odorous water bodies in key river basins and cities. In 2018, the proportion of sections with good surface water (Class I to III) across the country rose to 71% from 66% in 2015, while the proportion sections with inferior Class V surface water fell to 6.7% from 9.7% in 2015. In the meanwhile, soil pollution prevention and control have been proceeding in an orderly manner. A total of 214 pilot projects of soil pollution control and remediation technologies have been implemented nationwide. A resolute ban on foreign waste has been imposed and the total volume of solid waste imports in 2018 decreased by 46.5% year on year.

With the accelerating progress in pollution treatment, increasing efforts have been made in ecosystem conservation and restoration. A total of 2,750 nature reserves at various levels have been built across the country, including 474 national nature reserves. Various types of protected land areas cover more than 1.7 million km^2, which effectively protected the natural ecosystem and a majority of key wild animals and plant species. Moreover, the populations of rare and endangered species such as giant panda are gradually restored. The national forest coverage rate has increased from 16.6% at the beginning of the twenty-first century to about 22%.

7 Conclusion

The basic logic behind China's choice to embark on the path of green development could be analyzed from the above six aspects. The choice has been made to satisfy the needs of socioeconomic development and demonstrates the active explorations by the CPC based on the changing social contradictions. China has experienced a journey from the emerging idea of green development to the practice of embarking on the road of green development, from sustainable to scientific development, and then to the development of ecological civilization in the new era. Thanks to the efforts and explorations made by sever generations, the geographical space, industrial mix, way of production and way of life in the interest of conserving resources and protecting the environment have taken shape, while future for the green development have been increasingly promising. At present, despite notable achievements in green development, we still need to step up efforts for in-depth development on a larger scale. The socialist path of green development with Chinese characteristics in the new era indicates an all-round reform and a comprehensive transformation of economic and social development. Only by sticking to the path of green development can we achieve the goal of building a beautiful China.

Follow Xi Jinping Thought on Ecological Civilization in Forestry Development

Lizhe An

As General Secretary Xi Jinping said on many occasions, "the rise and fall of a civilization is closely tied to its relationship with nature." Xi Jinping Thought on Ecological Civilization was adopted as a guide for action at the National Conference on Eco-environmental Protection on May 18, 2018. At the conference, Xi reiterated: "Our mountains, rivers, forests, farmlands, lakes and grasslands together form a biotic community. An ecosystem is an integrated natural system of interdependent and closely related ecological chains. The lifeline of humans rests with farmlands, that of farmlands with water, that of water with mountains, that of mountains with earth, and that of earth with forests and grasslands. This community of life is the material basis for the survival and development of humanity." Both history and reality have shown time and again that forestry is an integral part of an ecological civilization. The development of forestry is crucial to building a beautiful China. It is fair to say that forestry is vital to ecological civilization development. To pursue ecological civilization development in the new era, we should follow Xi Jinping Thought on Ecological Civilization and make continuous efforts to promote forestry development, making our contributions to ecological and environmental conservation and the building of a beautiful China.

1 Forestry as an Integral Part of Human Civilization

Forests are an important part of ecosystems. "Forests are among the world's most productive land ecosystems and are essential to life on Earth," says the United Nations Strategic Plan for Forests 2017–2030. Home to about 80% of terrestrial species, forests are the best repositories of resources, genes, water, carbon and energy. They absorb carbon and produce oxygen, hold water, regulate climate, and provide shelter

L. An (✉)
Beijing Forestry University, Beijing, China

J. Pan et al. (eds.), *Beautiful China: 70 Years Since 1949 and 70 People's Views on Eco-civilization Construction*, https://doi.org/10.1007/978-981-33-6742-5_33

from wind and hold sand in place. Forests also help reduce dust, pollution, and noise, and purify the air. They play a positive role in mitigating climate change and conserving biodiversity, and help prevent land degradation and desertification. Forests also reduce floods, landslides, droughts, dust storms and other natural disasters. Forests are cradles of the human civilization. They are a green gift from nature which has helped mankind survive and thrive.

Human civilization develops through interactions with nature. Primitive people respected, feared, and worship nature. Agricultural civilizations are dependent on nature and follow its laws. Industrial civilizations strive to conquer and transform nature. Now we are in transition to an ecological civilization characterized by harmony between human and nature. The relationship between mankind and forests is an important aspect and a key indicator of the relationship between humans and nature.

In primitive societies, early humans lived in forests. They fed on fruits and animals in the forests, used leaves and animal hides to make clothes, built shelters in trees and used wildfires they collected from forests. Forests form the basis for the development of human civilization in every aspect, and provide basic tools that have facilitated human evolution. Some researchers believe that humans used wood tools for a very long time before the stone age. Instead of submitting to nature, humans have acquired a good understanding of nature, followed its laws and made good use of nature with our intelligence superior to that of other animals. From this perspective, forestry is an important part, and a distinctive feature, of primitive civilization. It is the product of mankind making proper use of natural resources, or to be specific, forests, and reflects the interdependence of humans and nature.

As man developed better life skills, the human society transitioned from hunting and gathering to food cultivation, and entered the era of agricultural civilization. In this stage, forests served as the material foundation for agriculture: In addition to providing the necessities of life and production, forests could also be turned into farmland, which was the foundation of an agricultural civilization. Then, as the population grew, more forests became farmland and tree logs were used as building materials and firewood, resulting in changes in the relationship between mankind and forests. While human activities caused decrease of forest areas, human beings realized the significance of forests and forestry to development. As ancient Chinese politician and philosopher Guan Zhong suggested in his writings *Li Zheng (On Overseeing the Government)*, water conservation and afforestation are crucial to the prosperity of a nation. In *Qing Zhong I (Economic Policies I)*, Guan Zhong claimed that it was the responsibility of state leaders to conserve forests, marshes and grasslands. He also emphasized in *Ba Guan (Eight Observations)* the importance of regulating the use of forests and grasslands, which was elaborated later by Mencius and Xunzi. In *Liang Hui Wang I* of *Mengzi*, Mencius claimed that forests would thrive if people only fell trees at a proper time. Xunzi also said in *Wangzhi* of *Xunzi* that axes must not enter the forest when the plants and trees are flourishing, lest their lives be cut short. China established a special warden system at a very early time to oversee the mountains, forests, rivers and marshes and ensuring proper use of forests.

Wardens were responsible for protecting the natural environment by restricting access to certain areas and enforcing relevant prohibitions.

The Industrial Revolution marks the beginning of the transition to a modern industrial civilization. Forests, an enormous natural treasure, continued to provide mankind with basic raw material and fuel. The intentional massive use of forests was the prelude to mankind's transition from underdevelopment to development. The development of forest industry has produced the capital and raw material required for development of industries, thus boosting industrialization.

Forests have played a vital role in all stages of the human civilization, be it primitive civilization, agricultural civilization or industrial civilization, and have made a great contribution to the development of mankind and the human civilization. Nonetheless, we must be aware that the development of human civilization and forests is accompanied by conflicts. The development of agriculture and industry has caused pressure and impact on forests to varying degrees. As Karl Marx pointed out over a century ago, "the development of civilization and industry in general has always shown itself so active in the destruction of forests." According to the UN Environment Programme, the earth has lost half of its forests, for which human activities are the main reason. While industrialization has boosted productivity and globalization, it has also caused overexploitation of forest resources and greater damage to ecosystems. Statistics show that over ten million hectares of forest is destroyed every year.

Deforestation results in various problems, such as severer drought, flood and soil erosion, biodiversity loss, desertification, etc. posing more and more daunting challenges for the human world. These ecological crises indicate that there are unresolvable contradictions and deficiencies in the development model of mankind in previous eras, especially during the industrial era. This has incurred careful thoughts and deep explorations in human society. As it was said, human civilization started when people cut down a tree for the first time, and it would come to an end when the last tree was felled.

It is against such a backdrop that the concept of ecological civilization emerged as a new form of civilization. It is not a complete negation of the industrial civilization, but transcends the latter while maintaining its strengths. The core of an ecological civilization is harmony between man and nature. It encompasses all previous forms of civilization and seeks improvement from a higher perspective. It balances between economic, social and ecological benefits so as to promote green and sustainable socioeconomic development. It is the path that we must take for further development of the human civilization.

Forests, as the mainstay of land ecosystems, have played an irreplaceable role in all stages of the human civilization. Forests cover nearly 400 million hectares of land on our planet, accounting for 30% of the total. More than one fifth of the world's population (about 1.6 billion people) depend on forests for livelihoods. Therefore, forestry is an important part of an ecological civilization. Forestry finds its foundation in forest resources and constitutes a major front in the endeavor to build an ecological civilization. Forestry plays a key role in ecological development and conservation as it is crucial for the conservation of the ecosystem.

It has been a shared mission and common aspiration of all forestry professionals to create sound ecosystems with clear waters, green mountains and well-conserved biodiversity. We face new tasks and will assume more responsibilities in building an ecological civilization.

2 Accurate Understanding of the Role of Forestry in Building an Ecological Civilization in the New Era

At the 19th National Congress of the CPC, it was announced that socialism with Chinese characteristics had entered a new era. This new era will be an era of building on past successes to further advance our cause, and of continuing in a new historical context to strive for the success of socialism with Chinese characteristics. It will be an era of pursuing ecological progress and building a beautiful China to achieve harmony between mankind and nature following the guidance of Xi Jinping Thought on Ecological Civilization. After the 18th National Congress of the CPC, based on traditional Chinese culture, and in view of the trend of the time and the wishes of the people, General Secretary Xi Jinping clarified major issues in theory and practice relating to the concept and significance of ecological civilization and how to build an ecological civilization from the strategic perspective of upholding and developing socialism with Chinese characteristics and realizing the Chinese dream of national rejuvenation. These remarks constitute Xi Jinping Thought on Ecological Civilization, which has played an important guiding role in China's achievements in the development of an ecological civilization and eco-environmental protection. Xi Jinping values the role of forestry in ecological progress. At a voluntary tree-planting event in Beijing on April 2, 2013, he pointed out that forests are the mainstay and an important resource for the land ecosystem, and an important ecological support for the survival and development of mankind. "It is hard to imagine what would happen to the earth and human beings without forests," he said. Xi Jinping called for persistent efforts to plant trees, saying that it would create a better environment for building a moderately prosperous society in all respects and realizing the Chinese dream of national rejuvenation. He also made remarks on forestry at tree-planting activities in the next few years. In 2014, Xi Jinping stressed that forestry was a fundamental issue concerning the sustainable socioeconomic development. In 2015, he pointed out that planting trees was an important way to build a beautiful environment and an endeavor that will benefit the most people. "China has made an important contribution to mankind in afforestation," said Xi Jinping. In 2016, he emphasized that development of forestry was an important part of building a moderately prosperous society in all respects and an important measure to build an ecological civilization. He said, "It's the common aspiration of mankind to build a green home." In 2017, he acknowledged China's achievements in afforestation over the years, and called for more efforts to improve the environment for the building of an ecological civilization. In 2018, Xi Jinping pointed out that in afforestation, we must attach importance to both quantity and quality, stick to a scientific approach, and make proper plans in light of local conditions; we must make unremitting efforts to pursue green development in a scientific, eco-friendly and thrifty manner, enlarge the area and improve the quality of forests, and improve the quality and stability of ecosystems. In 2019, Xi Jinping reiterated that China still faced serious ecological problems and that the lack of forests and vegetation, and a fragile ecological environment remained problems that need tremendous efforts to tackle. He gave the instruction that we must carry forward the good tradition of caring for, planting and protecting trees, and called for public participation nationwide to promote massive afforestation and achieve tangible results.

Forestry refers to activities to cultivate, protect and use forest resources by means of technologies and management skills. Making full use of forest's functions and utilizing forest resources in a sustainable way, forestry is a fundamental industry for public benefit which facilitates the coordinated development of population, economy, society, environment and resources and contributes to building a clean, beautiful and green environment. In China, the conservation and governance of wetlands, deserts and biodiversity are also considered part of forestry.

In the institutional reform of 2018, the National Forestry and Grassland Administration was established under the Ministry of Natural Resources, taking over all the responsibilities from the former State Forestry Administration, the grassland management responsibility from the Ministry of Agriculture and the management of nature reserves, scenic areas, natural heritage sites, and geological parks from the Ministry of Land and Resources, Ministry of Housing and Urban-Rural Development, Ministry of Water Resources, Ministry of Agriculture, and State Oceanic Administration. The reform significantly expanded the mandate of forestry authorities, putting in their charge forests, grasslands, wetlands, deserts, terrestrial wildlife, national parks and other nature reserves, as well as afforestation and ecological progress, including national voluntary tree planting and greening campaigns in urban and rural areas. As we can see, forestry in the new era covers most terrestrial ecosystems and a few offshore wetland ecosystems and marine and island ecosystems. It plays an indispensable and irreplaceable role in sustainable supply of natural resources and high-quality ecological products, building a beautiful environment in urban and rural areas, and ensuring ecological security.

3 Measures to Promote Forestry Development in the New Era

The report to the 19th National Congress of the CPC specified goals for the three stages on our way to a modern socialist China, i.e. completing the building of a moderately prosperous society in all respects by 2020, basically realizing socialist modernization by 2035, and becoming a great modern socialist country that is prosperous, strong, democratic, culturally advanced, harmonious, and beautiful by 2050.

To achieve these goals, the National Afforestation Commission and the National Forestry and Grassland Administration released the *Opinions on Actively Advancing Massive Afforestation*, setting the following targets:

- By 2020, the forest coverage rate in China should be up to 23.04%, that of rural areas should be up to 30%, the vegetation cover of grasslands should be 56%, achieving a better overall environment and establishing shields for national ecological security.
- By 2035, a framework should have been established for national ecological security the environment should see radical improvement, and the goals of the Beautiful China campaign should have been attained.
- By 2050, ecological civilization development should reach a new high, China should have become a country with advanced forestry and harmony between man and nature.

The goals for forestry development in the new era have been specified. What we need to do now is to mobilize all resources and follow Xi Jinping Thought on Ecological Civilization to strive towards such goals by taking the following measures:

First, we need to strengthen the security system of the forestry sector. We must strictly observe the ecological conservation red lines and combat any violations. We need to advance the greening campaign and intensify our efforts for ecological remediation. We must give better play to the roles of forests in climate change response and improving people's living environment, and reinforce our ecological security shields. We should enhance the quality of afforestation to promote the transformation of man-made forests to natural forests so as to improve the environment comprehensively.

Second, we should build a forest-centered culture. Based on forest resources, we should conduct nature-themed education campaigns and create cultural products reflecting the values of environmental protection to foster a socialist ecological civilization where people respect nature, follows its rules and protects it. We need to advance the development of forest cities, model bases of ecological culture and education bases of ecological civilization. Forestry universities should play an active role in publicizing ecological culture and training environmental protection professionals. We need to encourage simple, moderate, green, and low-carbon ways of life, and take part in the development of eco-friendly institutions, households, schools, communities, etc.

Third, we should improve the economic system of the forestry sector. Forestry is a major source of ecological products. Forest resources utilization must be carefully measured in order to ensure sustainability. We need to press ahead with poverty alleviation through ecological conservation, promote rural development through ecological efforts, and step up the efforts to build beautiful countryside. We must promote the integrated development of the primary, secondary, and tertiary industries in the forestry sector and explore and promote green, circular, and low-carbon development so as to achieve sustainable forestry development. Ecological tourism and forest-centered healthcare industries shall be promoted in a proactive and steady manner, and efforts should be made to explore ways to turn a beautiful environment to economic benefits.

Fourth, we need to improve science and technology relating to forestry. We need to build our innovation capacity in an all-round manner and strive to take the lead in tree planting and grass breeding, ecological remediation and conservation, cultivation and sustainable management of forest resources, prevention and control of major disasters, and efficient use of resources. We should create an all-weather network for more accurate observation, monitoring and prediction in regard to natural resources and ecosystems. We must further implement the Internet Plus initiative in the forestry sector and accelerate the development of smart forestry, so as to modernize forestry with IT application. We should vigorously support soft science research in the forestry sector, give better play to think tanks in this sector in building an ecological civilization, and build our capacity to inform and advice policy-making.

Fifth, we should strengthen the governance system of the forestry sector. A system for natural resources management should be established to ensure full-cycle management, including prevention of violations at the source, strict regulation of resource use, and harsh punishments for any case of violation. Guided by the concept that mountains, rivers, forests, farmlands, lakes, and grasslands are a biotic community, we should create an integrated system for ecological remediation, conservation and regulation. We also need to improve the system for public participation in ecological civilization development in the forestry sector to ensure that everyone performs their duties. Forestry should play an active role in global climate change response and in our efforts towards the goals set in the 2030 Agenda for Sustainable Development, make its contribution to global collaboration and cooperation in ecological governance and to the building of a global ecological civilization.

Advice on the Reform to Advance Waste Classification in China

Jiwen Chang

1 China Introduced Waste Classification Arrangements Suitable for Its Conditions Around the Same Development Stage as Developed Countries

Though seemingly a trivial matter in daily life, waste classification is absolutely a must when we reach a certain stage of economic and social development. It is necessary for environmental protection, natural resources conservation, a better living environment, greater capacity for green and sustainable development, and self-improvement of a civilization. Therefore, waste classification is no trivial matter for a country or nation. Waste classification represents a consensus about how people should live their lives in a modern ecological civilization, and an integral aspect of building an ecological civilization in China. We must pay due attention to it. On an inspection trip to Shanghai in November 2018, General Secretary Xi Jinping remarked that waste classification was a new fashion. He also stressed in June 2019 that waste classification was closely related to people's living environment, it was about using resources in a thrifty way, and it was an important measure of how advanced a society is.

The timing of introducing mandatory waste classification is usually determined by the overall economic and social conditions of a specific country. It won't work if it is introduced too early because the general public will lack the awareness and willingness to follow the rules. It won't do to wait till it is too late either for it will increase the risks of severe environmental pollution and shortage of natural resources. It is thus necessary grasp the best timing for introducing waste classification and set waste classification goals in a phased manner according to a country's specific stage of development. In developed countries like the United States and some European countries, mandatory waste classification was introduced and institutionalized at a

J. Chang (✉)
Research Institute of Resources and Environment Policies, Development Research Center of the State Council, Beijing, China

J. Pan et al. (eds.), *Beautiful China: 70 Years Since 1949 and 70 People's Views on Eco-civilization Construction*, https://doi.org/10.1007/978-981-33-6742-5_34

crucial stage of industrialization when economy and society were both in transition. In Japan, waste classification was first introduced in 1975 in Shizuoka prefecture, dividing waste into combustible and non-combustible waste, and the practice was gradually rolled out nationwide later. After four decades, there are detailed waste classification rules in Japan today. Germany introduced rough waste classification rules in 1904, but was still faced with a shortage of landfill in the 1980s. In response, the German government issued the Packaging Act in 1991, and a system for waste recycling and reprocessing was put in place in its industrial sector to improve classification. Both Japan and Germany now have sound legislations on circular economy and circular society which require strict waste classification. In the transition period of the industrialization process, people are likely to be ready for waste classification as they probably have grown highly aware of environmental problems after widespread ecological education and after seeing for themselves the consequences of environmental pollution. Also, at this time point, technological progress and economic development in the previous stages of industrialization would provide a sound foundation for the government to introduce waste classification rules. Most importantly, at this stage, waste classification will help address the shortage of natural resources and reduce pressure on the environmental by producing less waste. Thus, waste classification is more likely to meet with wide support. This is a pattern we see in history and it is better not to go against it.

China has decided to adopt waste classification policies at a stage similar to when such policies were introduced in developed countries. On December 21, 2016, General Secretary Xi Jinping presided over the 14th meeting of the Central Leading Group for Financial and Economic Affairs, and proposed to roll out waste classification policies nationwide. The General Office of the State Council forwarded the *Implementation Plan for Household Waste Classification* issued by the National Development and Reform Commission (NDRC) and the Ministry of Housing and Urban-Rural Development on March 18, 2017. According to the document, waste classification would be introduced as a pilot program in 46 cities by establishing a system of relevant legislations and standards by the end of 2020. The purpose was to develop models of household waste classification that can be replicated and promoted elsewhere. The General Office of the State Council released the *Work Plan for Zero-Waste City Pilot Program* in December 2018. The 2019 report on the work of the government also required more efforts for urban household waste classification. Based on the experience of these pilot projects, the Ministry of Housing and Urban-Rural Development and eight other authorities jointly issued a circular in June 2019 to promote urban household waste classification in all cities at or above the prefecture level. According to the circular, by 2025, all Chinese cities at or above the prefecture level shall have a complete household waste classification and processing system in place. The circular marks the start of extensive waste classification in China's urban and rural areas. China is currently at a critical stage of building an ecological civilization, and faces an opportunity for the reform of the system for promoting ecological progress. The CPC Central Committee with Xi Jinping at its core decided to introduce waste classification policies, with specific goals for stages and a road map for achieving the goals at the current stage, which is in line with the laws of environmental protection and economic and social development in China. Developed countries have completed industrial transition at a similar stage. In view of the above, we may come to the conclusion that China has been promoting waste classification in a scientific, timely and rational manner, from pilot programs in chosen cities to nationwide efforts.

2 China Should Advance Waste Classification Step by Step

China has made remarkable progress in policymaking and legislation for waste classification. Developed rural areas and city suburbs (e.g. Jindong district of Jinhua city, Zhejiang province, and Ningxiang city of Hunan province) have established complete waste classification facilities, a sound waste collection and management evaluation system, a system for waste classification evaluation and the corresponding reward and punishment measures, a system for waste transportation and disposal based on classification, and a system for relevant fundraising and economic incentives within a few years. The living environment and people's habits in rural areas were greatly improved. In terms of legislation, large and medium-size cities, including Shanghai, Guangzhou, Shenzhen, Beijing and Ningbo all introduced local waste classification regulations, imposing punishments for violations, or made plans to amend existing local legislation on household waste management. For example, *Guangzhou Household Waste Classification and Management Regulations* was adopted on December 27, 2017 at the 11th meeting of the 15th Standing Committee of Guangzhou Municipal People's Congress, and *Shanghai Household Waste Management Regulations* was adopted on January 31, 2019 at the 2nd meeting of the 15th Shanghai Municipal People's Congress. Both regulations contain detailed provisions on the scope of application, responsible entities, classification measures, duties of authorities, measures for waste reduction at the source, hazard-free disposal, recycling, and rewards and punishments. Beijing will amend its household waste management regulations to warn or fine those who fail to sort out waste, and impose severe punishments and heavy fines on those who mix up waste that has been classified in transportation. In the new era of building a socialist ecological civilization, legislation helps consolidate the achievements of pilot programs, and regulations promotes further changes towards a green lifestyle. All these efforts have promoted waste classification in urban and rural areas, which will significantly advance China's efforts to improve the environment.

While we acknowledge our remarkable achievements, we must keep in mind that waste classification must be advanced step by step, not in excessive haste. It might be easy to build waste classification facilities, but it takes time to change people's mentality and lifestyle. While we are determined to press ahead with waste classification, we must also be patient. In advancing waste classification, we need to adopt a step-by-step approach to seek progress while maintaining existing achievements, and must refrain from formality for formality's sake which would waste both human and financial resources. Experience of other countries show that it takes at least one to two generations to move from pilot programs to waste classification nationwide. As a latecomer, China may draw lessons from developed countries in environmental protection.

So far, an institutional frame is in place in the Party and government system for building an ecological civilization, the related supervisory and judicial systems are set up, a pattern for public participation in ecological civilization development is taking shape, and efforts have been made to promote eco-friendly models of economic and social development. As a result, China has made remarkable progress towards an ecological civilization and Chinese people have reached a consensus on the significance of this pursuit. It may, therefore, take far less time for China to change people's attitude towards waste classification and establish sound systems for waste classification at various levels. It is expected that by 2035, when China completes the transition, it will catch up with moderately developed countries in waste classification in terms of legislation, policies and implementation.

3 Design the Waste Classification System with China's National Conditions in Mind While Learning from Other Countries

A large proportion of Chinese research literature on waste classification is about the measures and experience of other countries. Waste classification, however, not only addresses environmental pollution, but also aims to promote progress in society. Therefore, we must bear in mind China's national and social conditions when determining how China will proceed, i.e. what institutions, systems and mechanisms, we will adopt for waste classification. It is not advisable to simply copy what other countries did and any research findings that suggest us to do so must be treated with caution. To address China's own problems, we need to adopt a scientific approach based on the specific conditions of the places involved, while learning from other countries.

Considering that China is a socialist country and the staff of public institutions such as government organs, schools, hospitals show great initiative to respond to the call of the government, we may start by requiring waste classification in public institutions. We should also consider the management conditions in different types of residential areas. For bungalows, villas and *hutongs*, as each building is occupied by an individual household, it is easy to spot failure to sort waste and grant rewards or impose punishments accordingly. It is thus advisable to launch waste classification requirements first in these residential areas to accumulate experience and influence before promoting the practice to other areas. As for high-rise apartment buildings, a high density of residents means that waste bins may have to compete with parking lots for space. As a result, the sorted waste must be moved away sooner than in many Western countries. In some German neighborhoods, waste is collected every few days, but in high-rise buildings in China waste must be collected on a daily basis to keep the neighborhood clean. Owing to cooking habits in China, Chinese households may produce a larger proportion of kitchen garbage which may have a higher content of salt and oil. According to this, proper methods must be adopted for classification, collection and disposal. Meanwhile, as a country in the primary stage of socialism, China lags far behind developed countries in terms of access to waste classification facilities, people's awareness of environmental compliance and government's environmental enforcement. It is imperative to promote waste classification by both guiding and mandatory measures. Sole dependence on judicial means such as fines or detention to impose mandatory waste classification may lead to widespread violations of the law and make it difficult to advance waste classification efforts. In some neighborhoods, the Western model was borrowed, requiring residents to throw away waste at a fixed time and designated sites, with many of the waste bins removed. Due to a lack of awareness, pressure for time and some other reasons, littering became a prominent problem, and cleaning staff are still required. To address this problem, basic-level Party organizations and community organizations must play a role to leverage the strength of socialism. In addition to waste collection at fixed time and places and efforts to raise public awareness of waste classification, we should also consider the special needs of some people to throw away waste in different time periods, so as to promote waste classification effectively.

4 Coordinated Planning for and Scientific Management of Waste Classification

First, we need to launch waste classification strategies, measures and requirements in urban and rural areas in good coordination. In the new era, different strategies, targets, priorities, policies and measures may be adopted in urban and rural areas, but progress must be generally in sync. The housing conditions in urban and rural areas are different. Rural areas, where each residential building is occupied by an individual household, have an edge when it comes to waste classification, because it is easier to determine whether residents have followed relevant rules and to grant rewards or impose punishments accordingly. In advancing waste classification, rural areas, as mostly acquaintance-based societies, also have advantages in its organizational structure in comparison with urban areas that are mostly stranger-based society. Therefore, in a time of integrated and coordinated urban-rural development, we should adopt differential management measures, organizational forms and incentives to promote waste classification in urban and rural areas. In cities, it may be better to start from public institutions, such as government organs, schools and hospitals, or *hutong* and villa areas to accumulate experience and social influence before promoting it to a wider scope. If conditions permit, waste classification can be advanced in villages in the rural-urban fringe as well.

Second, while designing plans for waste classification, we should also introduce or amend laws and regulations at the national and local levels and establish systems, institutions and mechanisms for government guidance, supervision, funding, management, incentives and constraints, public participation, evaluation, rewarding and punishments. The ultimate goal is to put in place law-based institutions, systems and mechanisms for the management of waste classification that are driven by the government and participated in by all, with coordinated efforts in urban and rural areas, and taking into account local conditions.

5 Reform of Systems, Institutions and Mechanisms for Waste Classification

Proper institutions are required to effectively promote waste classification in China. Institutions for promotion and supervision in this regard shall be put in place in urban and rural areas in light of local conditions. For governance of waste classification in rural areas, the government should play a leading role, with support from village Party committees, organization and coordination efforts of villagers' committees and cooperation of leading figures. In particular, village Party organizations and villagers' committees should play an active role. For example, we may put rural households in charge of cleaning and waste classification in the surroundings of their homes and contracted fields, put villager groups in charge of areas occupied by group members' households, and put villagers' committees and village Party committees in charge of their respective designated areas. Party members and villagers' committee officials should take charge of specific villager groups and households. Only with a clear division of responsibility and engagement of the people, can we raise public awareness of waste classification and build faith in a green lifestyle. For governance of waste classification in urban areas, it is advisable to clarify the responsibilities of authorities in charge of commerce, housing, cleaning, and civil affairs in relation to waste classification, and integrate waste recycling, recycling services of property management companies, waste collection services, services of neighborhood committees and relevant policies. All sectors involved shall pursue the common goal with unity in action and mindset.

We need to improve the regulations and standards for waste classification. We should follow a multi-pronged approach in promoting waste classification. In both urban and rural areas, we may introduce regulations on the responsibilities of individual entities for a good environment, waste classification, separate collection and transportation of different types of waste, waste reduction, and classification-based waste disposal. A system of regulations, including guidelines and policies related to the aforementioned processes, is required. For example, while a universal charging system for waste collection shall be put in place in urban and rural areas, disadvantaged groups shall be entitled to reduction of or exemption from such charges, and people with excellent performance in waste classification should also be rewarded with reductions or exemptions. Waste classification standards vary across cities. Waste is classified into recyclable, non-recyclable, hazardous and other waste in some cities, dry waste, wet waste and other waste in some other cities, and dry and wet waste in still other places, especially in rural areas. Complicated classification standards may be difficult to follow even for professionals. In making waste classification standards, we should follow a fact-based approach and ensure that the standards are neither too general nor too specific. The standards, aiming to save resources and protect the environment, should be easy to understand and follow. To ensure the effect of waste classification, we must make sure that our work is well coordinated, the measures are properly designed, and the systems are sufficiently feasible.

To promote waste classification in China, we need to introduce new long-term mechanisms for funding, operation and cleaning of facilities, awareness raising, evaluation, and reward and punishment. For funding in rural areas, in addition to charging households for waste disposal, the government, the village collectives and high-profile local residents should also play their roles by giving subsidies, rewards, or donations. In urban areas, in addition to waste disposal fees collected from households, government funding should be the mainstay. For the operation and cleaning of facilities, in rural areas, villagers, especially disadvantaged people, may do the work while village authorities may provide the funding. In rich rural areas, such tasks may be contracted to dedicated service providers in a market-based manner, and the same may also apply to urban areas according to unified plans of local governments. To raise public awareness and encourage people to participate in waste classification, in districts, sub-districts, townships, neighborhoods and villages where mandatory waste classification is universal, residents, Party members, officials at all levels and the cleaning staff should all play an active role, disseminating relevant knowledge and call for good practices through education, rewards and warnings. This will enable urban and rural residents in demonstration areas for waste classification to develop good habit of waste classification, and influence other with their green lifestyle. To ensure the implementation of relevant plans, we need to organize assessment of the waste classification performance of relevant authorities and organizations, including governments at the township and sub-district level, property management services, neighborhood committees, village Party committees and villagers' committees. Rewards and punishment shall be based on the assessment results, and the performance in waste classification and environmental remediation shall be linked with salary. These measures will stimulate the enthusiasm among the general public to practice waste classification in more areas and move towards to goal of universal waste classification in cities at and above the prefecture level by 2025.

Evolution of the Green Development Concept and the Path to Green Development

Cunkuan Bao

1 Introduction

China's rapid economic development since the launch of the reform and opening up has come with huge consumption and inefficient use of energy and resources, resulting in serious ecological damages and environmental degradation that has hindered economic development and had a negative impact on people's life. The traditional model of development contributed significantly to material progress, but it also caused serious damage to the environment. Green development is a different model of development, representing a revolution in the understanding of development and a fundamental solution to environmental and resource issues. Unlike the traditional way of production and consumption featuring massive manufacturing, consumption and emissions, green development pursues coordination between resources, production and consumption, so as to achieve economic and social development while protecting the environment and achieve harmony between man and nature. In pursuit of green development, we need to adjust the economic and energy structures, improve our land development plan, adjust the industrial distribution across regions, promote the development of the environmental protection industry and circular economy to build a high-quality economy. In addition, green development encourages simple, moderate, green, and low-carbon ways of life, and opposes extravagance and excessive consumption.

The concept of green development in China covers a wider scope than in the West where it involves mainly industrialization. It is a new concept of development advocating environmental-friendly, circular and low-carbon development and seeking sustainable development and prosperity of an ecological civilization. Five-year plans for economic and social development are major documents that embody

Lou Huajing also made some contributions to this article.

C. Bao (✉)
Department of Environmental Science and Engineering, Fudan University, Shanghai, China

J. Pan et al. (eds.), *Beautiful China: 70 Years Since 1949 and 70 People's Views on Eco-civilization Construction*, https://doi.org/10.1007/978-981-33-6742-5_35

China's understanding of development and best illustrate how green development evolved in China from ideas to actions. One of the strengths of China's political system as a socialist country is that long-term development plans at the national level continues well with coherence. For implementation of the green development strategy, relentless efforts are required to change the way of economic development and people's way of life. China is at an advantage thanks to the continuity in its development plans, which ensures that the green development strategy will be continuously followed over a long period of time. However, top-level strategies alone are by no means enough for green development. In addition to national plans, it also requires concerted efforts from authorities at provincial, municipal and county levels, and a cohort of plans including national development plans, special plans, regional plans, and land development plans. With national development plans in an overarching position, the aforementioned plans designed by governments at all levels, each with specific functions, complement one another and constitute a coherent system. These plans facilitate the implementation of the green development strategy at various levels, thus contributing to the fulfillment of China's overarching goals. At a certain administrative level, special plans provide more details for the development plans in specific areas of work and must be consistent with national development plans and land development plans.

Green development is a central concept in the 13th Five-year plan. This article examines China's national and provincial development plans during the 11th, 12th and 13th five-year plan periods, and analyzes the evolution of the concept of green development from the perspectives of the guiding thought, basic principles, goals and targets, and content of the plans. Also analyzed are the implementation of the green development strategy and the path to green development at different levels of the Chinese government based on plans of JS province and CS city during the 13th Five-Year Plan period.

2 Evolution of the Concept of Green Development Through the 11th, 12th and 13th Five-Year Plans

Based on analysis of China's 11th, 12th and 13th five-year plans for economic and social development from the perspectives of guiding thoughts, development goals and major content, this article examines the evolution of the concept of green development and its policy implications with the belief that green development is embodied in development plans at different stages, and that the focus in the pursuit of green development has changed from resource utilization to resource-environment balance, and ultimately to overall improvement of the environment.

To achieve green development, it is a must to significantly enhance resource utilization efficiency and hold down consumption of resources. The 11th Five-Year Plan set binding targets for resources and environment for the first time. Among the 22 targets for national economic and social development, seven are related to resources and environment, accounting for 31.8% of the total, including water resources (for industrial and agricultural use), energy, land (mainly farmland), forests, etc. It also raised requirements for the reduction of the discharges of major pollutants and encouraged reuse of solid wastes. The targets are mainly about enhancing efficiency, such as reducing water consumption per unit of industrial value-added and raising the efficiency coefficient of irrigation. The plan aimed to build a resource-conserving and environmentally friendly society by promoting circular economy, protecting and restoring the environment, intensifying environmental protection, tightening management of resources, making proper use of ocean and climate resources and taking other measures. Though the 11th Five-Year Plan did not use the wording "green development," it still highlighted the importance of resource conservation and environmental protection in promoting growth model transformation, industrial restructuring and upgrading, and higher-quality development. The seven targets were all met in 2010, and six were exceeded. The targets related to resources and the environment were binding, curbing the emissions of major pollutants, reversing the trend of environmental degradation, and contributing to China's efforts to upgrade its industrial structure and transform the growth model. "Green development" was first mentioned and elaborated in Part VI of the 12th Five-Year Plan for National Economic and Social Development, "Pursue Green Development: Develop a Resource-Conserving and Environmentally Friendly Society." The said plan includes "achieving significant results in resource conservation and environmental protection" as a major goal, aiming to improve the incentive and constraint mechanisms for energy conservation and emissions reduction. It set eight targets related to resources and the environment. To achieve green development, the 12th Five-Year Plan proposed to actively respond to global climate change, strengthen resource conservation and management, vigorously develop circular economy, enhance environmental protection, promote ecological protection and remediation, and accelerate the development of systems for water conservancy and disaster prevention and mitigation. All of the eight targets were met in 2015. In particular, the discharge of major pollutants dropped significantly, the installed capacity of clean-energy power generation saw significant growth, and coal consumption and CO_2 emissions started to decrease. China's economic growth was decoupled from pollutant emissions, creating a favorable condition for further improvement of the environment. The 13th Five-Year Plan for National Economic and Social Development followed the country's overall deployment for economic, political, cultural, social, and ecological progress and the new concept of development. While the 11th and 12th five-year plans focused on building a resource-conserving and environmentally friendly society, the 13th Five-Year Plan aimed to achieve overall improvement in the quality of the environment and ecosystems by accelerating functional zoning, promoting economical and intensive use of resources, strengthening comprehensive environmental governance, intensifying ecological conservation and restoration, actively responding to global climate change, improving mechanisms for ensuring ecological security, and developing green and environmentally friendly industries. It contained 10 targets related to resources and the environment, including those for air quality and surface water quality, which were included for the first time, marking a strategic adjustment to the priorities in China's environmental protection efforts.

The 11th, 12th and 13th five-year plans respectively included 7, 12 (in 8 categories) and 16 (in 10 categories) targets related to resources and the environment, accounting for 31.8%, 33.3% (42.9% in terms of the numbers of targets) and 40% (48. % in terms of the number of targets) of the total. The concept of green development and the targets related to resources and the environment gained importance and preponderance. These targets can be divided into three types: targets of resources and energy, targets of pollution control, and targets of ecological security. Resources and energy targets mainly involved the consumption of resources and energy in production, pollution control targets were about emission reduction and environmental quality, and ecological security targets were mainly about farmland and forests.

Examining these targets, we note that from the 11th to the 13th five-year plan, the concept of green development evolved, showing mainly the following changes:

First, the focus of environmental efforts shifted from monitoring intensity or efficiency to controlling both intensity and total consumption. The 11th Five-Year Plan included, for the first time, a target for energy consumption intensity per unit of GDP as a binding target, requiring it to be down by 20% from the level of 2005. The 12th Five-Year Plan set the target of decreasing energy consumption per unit of GDP by 16% from the level of 2010, and aimed to keep the total energy consumption within reasonable limits, showing that China is attaching importance to the control of both energy consumption intensity and total energy consumption. Based on achievements made in the 11th and 12th Five-Year Plan periods, progresses were significant in energy conservation. As China's GDP increased from RMB 18.2 trillion in 2005 to RMB 67.7 trillion in 2015, up 14.04% annually, its total energy consumption increased from 2.22 billion tons of standard coal equivalent in 2005 to 4.3 billion tons of standard coal equivalent in 2015, up only 6.83% annually, much lower than that the GDP growth. According to the 13th Five-Year Plan, China will decrease its energy consumption per unit of GDP by 15% by 2020, and will impose limits on total energy consumption, consumption intensity, water resource consumption, and land use for construction purposes. This means that China continued with the practice of controlling both total resource consumption and the consumption intensity in the management of water resources, land for construction and other resources. Imposing limits on both total consumption and consumption intensity is a natural consequence of the development of China's policies for conservation and efficient use of resources. It is vital to addressing the environmental and resource challenges that China is facing. Such measures embody China's determination to control pollution with continuous efforts and improve the environment, which will create a huge market for the energy conservation and environmental protection industries.

Second, China sought to improve environmental quality while controlling the total consumption of energy and resources. For the first time, the 13th Five-Year Plan included targets for air quality and surface water quality, specifically targets of the proportion of days with good air quality in cities at and above the prefecture level, decrease in PM 2.5 concentration in cities with poor air quality at and above prefecture level, proportion of surface water quality at or above Grade III, and proportion of surface water rated below Grade V in quality. This change reflects a shift of focus in environmental protection from quantity to quality. It is imperative to properly handle the relationship between quantity and quality in environmental protection. Setting limits on quantity is a major way to improve the environment. As pollutant emissions from various sources have direct or indirect impact on the environment, emissions reduction, though not the only way to improve the environment, contributes significantly to environmental quality improvement. The 13th Five-Year Plan included targets for total emissions of sulfur dioxide, nitrogen oxides, ammonia and nitrogen, and chemical oxygen demand which shall be achieved by putting in place relevant facilities and phasing out outdated capacity at pollution sources in major industries. Meanwhile, emissions reduction shall be achieved without compromising development quality. Emissions reduction is a fundamental and compulsory requirement, while environmental quality improvement is a red line that must not be crossed.[1]

Third, the binding effect of green development targets grew. In the 11th Five-Year Plan, five of the seven resource and environment-related targets were binding, in the 12th Five-Year Plan, seven out of eight such targets were binding, and in the 13th Five-Year Plan, all were binding. Binding targets, which are considered government commitment to society, have binding effect on governments at all levels, especially government performance related to public services, environmental protection and land use. It is evident that the plans for national economic and social development have had an increasingly binding effect in the environment and resource sectors. This significantly improved the constraint mechanism which ensured that governments at all levels would fulfill their environmental protection responsibilities, so that China on the whole can meet or even exceed the green development targets as scheduled.

3 Implementation of the Green Development Concept at Local Levels

Development plans for the 13th Five-Year Plan period of 29 provincial-level regions in China were gathered from the Internet and a brief look through them reviewed that 19 had targets related to resources and the environment, accounting for over 30% of the total. Most of them have specific requirements for environmental quality improvement, mostly the same or even higher than the national targets, including targets for the proportion of days with good air quality in cities at and above the prefecture level, decrease in PM 2.5 concentration in cities with poor air quality at and above the prefecture level, proportion of surface water quality at or above Grade III, and proportion of surface water below Grade V.

Nonetheless, we still found the following problems in their pursuit of green development. First, the binding force of green development targets needs to be strengthened. While local authorities have set most environment and resource-related targets as binding ones (Beijing and Jilin each has one anticipatory target related to resources

[1] http://politics.people.com.cn/n1/2016/0122/c1001-28075425.html.

and the environment), the targets are not specific enough. For decrease in energy consumption per unit of GDP, CO_2 emissions reduction per unit of GDP, and major pollutants emissions reduction, only Hunan and Chongqing set specific targets, while all other provincial regions only required fulfilment of the national targets, showing that local authorities are hesitant and conservative about setting targets. Second, there is a lack of coordination between the national and provincial green development targets. Of the 12 regions with targets for the proportion of non-fossil fuel in primary energy consumption in their development plans, only seven aimed higher than the national target. Third, there is a lack of coordination and collaboration between different regions in the pursuit of green development. In designing development strategies and environmental protection measures, most provinces have only considered their own conditions, not paying due attention to coordination and collaboration with other provincial-level regions.

In the following section, we will further analyze the implementation of the green development strategy at the provincial and city levels based on the development plans of JS province and CS city in the province.

JS province strongly advocates green development. The guiding thought of its development plan is to fully implement the new concept of development and the strategy of sustainable development, and stick to green development to deliver a better life to the people, which is in line with the strategic significance of green development. The plan includes the same number of targets related to resources and the environment as in the national development plan, with slight differences in wording. As for the specific targets, five are the same with corresponding national ones, three are higher, and two are lower. In setting specific tasks, the province attaches great importance to improving environmental quality, and gives priority to air pollution, water pollution and soil pollution control. To achieve the targets, the province introduced measures to control the discharge of multiple pollutants and to coordinate between pollution control efforts of different provinces. The plan also proposed to strengthen the reform of the system for building an ecological civilization and improve the mechanisms for performance evaluation and accountability in this regard. JS, an economically developed province in China with high GDP and a high growth rate, contributed about 10% of the country's GDP each year during the 11th and 12th five-year plan periods. Over the past decade, however, JS had been seeing a greater proportion of the secondary industry than the national level in its overall economic structure, while that of the tertiary sector had been lower than the national level. The economic growth model driven by manufacturing is a major reason why the proportion of fossil fuel in JS's energy portfolio was significantly higher than the national level. To pursue green development amid rapid industrial growth, JS province faced arduous tasks to improve its energy portfolio, accelerate the development of non-fossil energy, promote industrial upgrading and improve its industrial structure.

The development plan of CS city for the 13th Five-Year Plan period follows the concept of green development, requiring implementation of the new concept of development in all aspects to seek economic, political, cultural, social, and ecological progress. The fundamental principle is to pursue green growth and implement the green development strategy. However, the significance of green development is not fully reflected in the targets set. There are six targets related to resources and the environment, accounting for 18.2% of the total, lower than the level of JS's plan and the national plan. The plan includes targets for industrial water consumption, total consumption in production, improvement of air quality and surface water quality, pollutant emission reduction, and wetland conservation, the last of which is a special characteristic of CS city. Many environmental factors, such as farmland, forests and energy structure are left, and the targets, especially the targets for air quality, surface water quality, and major pollutant discharge reduction, are not specific enough. To achieve its targets, CS, with a strong economy, emphasized green development and made great efforts to transform and upgrade its economic structure by restructuring energy-intensive and highly polluting traditional industries, moving faster to phase out outdated capacity, pushing major industries towards the high end of the industry chain, improving technology, developing environmentally friendly emerging industries, and fostering new growth drivers. There is a gap between what CS said about green development in its development plan and what it actually did. Though green development is highlighted in the city's development plan for the 13th Five-Year Plan period as a guiding thought, a basic principle, a development strategy and an important goal of development, resource and environment-related targets make up only a small proportion of all the targets set, covering a limited scope and having inadequate binding force.

4 Conclusions and Recommendations

The concept of green development is proposed to address China's problems in resources, the environment and development in the new era. It is an innovative development model that is conducive to resource conservation, environmental protection and ecological improvement while taking into full consideration the carrying capacity of the environment and resource endowment. In China's pursuit of green development, development plans play a leading role. If China is to actually achieve its targets for green development, what is specified in the national plan must truly count at local levels, and this requires coordination between various policies and measures and collaboration between different levels of the government. Therefore, to continue its pursuit of green development in the 14th Five-Year Plan period, China is recommended to:

(1) Strengthen macro control and guidance for local governments in regard to green development. In accordance with the overall requirement, the government needs to strengthen the macro control and differential constraints on provincial-level governments in the preparation of their own development plans to ensure that the national binding targets for resources, the environment and ecology can be met.
(2) Develop a stronger sense of responsibility for green development in local governments, and stimulate their enthusiasm and creativity for this purpose.

Measures should be taken to help local governments develop a stronger sense of responsibility for the environmental quality within their jurisdictions. To implement specific tasks and measures for the achievement of the national targets of green development, local governments shall be encouraged to take the initiative to understand fully the requirements proposed by higher-level authorities in advance and allocate relevant tasks and targets accordingly. Local governments are advised to seek advice from higher-level environmental protection authorities before making their own development plans, so as to ensure the consistency between targets and tasks at various levels. Local governments at all levels should work to transform pressure for the solution of problems related to resources and the environment into motivations for green development and improvement of environmental quality.

(3) Promote collaboration between different administrative areas for green development. In addition to local authorities' efforts for green development based on local conditions, it is also necessary to strengthen cooperation between regions, establish a long-term mechanism for inter-regional collaboration for pollution control, and conduct inter-regional cooperation in ecological remediation and environmental governance.

(4) Let better developed regions spearhead green development efforts. Better developed regions should pioneer in the pursuit of green development to accumulate experience and serve as demonstrations for other regions.

(5) Unleash the enthusiasm of local governments, especially basic-level governments, for green development in addition to making plans. In particular, we must make full use of people's wisdom and creativity and establish a long-term mechanism for market-driven green development.

Develop Zero-Waste Cities for the Building of an Ecological Civilization

Huiqiang Cheng

Zero-waste city represents a model of sustainable urban development, which aims to build an ecological civilization and follows the new development concepts featuring innovation, coordination and green development. By promoting a green development model and green lifestyle, cities strive to reduce solid wastes at the source, enhance the efficiency of resource utilization, and make full use of wastes. With the launch of pilot zero-waste city programs, China entered the advanced stage of a circular economy as part of its efforts to build an ecological civilization.

1 History of Zero-Waste City and International Experience

Zero-waste city is a concept derived from the zero waste strategy first proposed by Dr. Paul Palmer who founded Zero Waste Systems Inc. in 1972 for the recycling and reuse of chemicals. This concept drew extensive attention in the late 1990s. As a goal of urban waste management and reduction, zero waste means turning waste produced in production and life into materials that can be used in other industries, forming a cycle that maximizes use. The first working definition of zero waste was given by Zero Waste International Alliance in 2004 and revised in 2009: Zero waste is "a goal that is ethical, economical, efficient and visionary, to guide people in changing their lifestyles and practices to emulate sustainable natural cycles, where all discarded materials are designed to become resources for others to use. Zero waste means designing and managing products and processes to systematically avoid and eliminate the volume and toxicity of waste and materials, conserve and recover all resources, and not burn or bury them."

H. Cheng (✉)
Research Institute of Resources and Environment Policies, Development Research Center of the State Council, Beijing, China

J. Pan et al. (eds.), *Beautiful China: 70 Years Since 1949 and 70 People's Views on Eco-civilization Construction*, https://doi.org/10.1007/978-981-33-6742-5_36

Canberra, Australia, became the first city in the world to set zero waste as an official goal when it passed *Act No Waste by 2010* in 1995. Thanks to economic and social development and improvement of the waste management system, more and more countries have adopted zero waste as a goal in urban planning. The world has seen the emergence of organizations such as Zero Waste International Alliance, Zero Waste Europe and the like. Eight cities, namely San Francisco of the United States, Vancouver of Canada, Kamikatsu of Japan, Masdar City of the United Arab Emirates, Capannori of Italy, Sydney of Australia, Ljubljana of Slovenia, and Auckland of New Zealand, have been going zero waste. The United States Conference of Mayors adopted a resolution in support of municipal zero waste principles in 2015. The theme of the 2017 UN Environment Assembly was "Towards a Pollution Free Planet." In 2018, 23 cities from around the world issued a joint declaration on building zero waste cities.

China can mainly draw the following lessons from the experience of other countries in building zero-waste cities:

(1) Waste management is the foundation of zero-waste cities. Most cities, before committing to the goal of zero waste, had performed sound waste management for decades and even over a century, and established a complete government-led waste management system in which producers took extended responsibility, household waste was classified, and dedicated waste disposal services were in place to collect, transport and dispose of wastes. With a well-established waste management system, most cities we studied are able to cover all relevant expenditures with the waste disposal fees they collect, forming a fully viable waste management model.

(2) The hierarchy of "rethink, reduce, reuse, recycle, and material recovery" should be followed. Sample cities encourage waste producers (e.g. enterprises and households) to take the responsibility to avoid and reduce waste generation and there are waste recycling or reuse centers. Household waste can be donated or traded at secondhand markets, on e-commerce sites and through other channels. Reuse and recycling rates are set for construction and industrial waste to ensure their reuse and recycling. The cities have also established composting plants, incineration plants, landfills and other facilities to dispose of waste. Some cities collaborate in waste management to cut cost.

(3) Market players and specialized institutions should be encouraged to participate in waste management. In view of the complicated process of waste collection, transportation and disposal, the governments of sample cities made good use of market capital and specialized services for better waste management. San Francisco, for example, engaged Recology, a waste management company with over 100 years of history, to place trash cans, and collect and handle household waste, including recycling, incineration and landfill. The city also worked together with Recology to design waste management schemes. In Vancouver, the government is responsible for collecting and disposing of unrecyclable household waste, while Recycle BC, a waste management company, is engaged for recyclables. Private enterprises and not-for-profit organizations

are also welcomed in the various stages of waste collection, transportation and disposal.

(4) Mandatory and binding tools such as regulations and laws are important ways to build zero-waste cities. Sample cities introduced different regulations for different types of waste. For example, construction waste cannot be sent to a random landfill, and must be transported to dedicated disposal plants, with compulsory targets set for the reuse and recycling of construction waste. Disposable products such as cups, straws, tableware and plastic bags are strictly prohibited, and compost like kitchen garbage must not go to landfill. Only compostable plastic bags are allowed, and household waste must be carefully classified. These measures were generally introduced step by step in the sample cities so that the market and residents have time to adapt to the requirements, but on the whole the scopes of prohibition and mandatory measures expand.

(5) Producer accountability system is widespread. Most sample cities adopt extended producer responsibility system, and keep expanding the areas where manufacturers are required to take the responsibility, including product design, choice of material, and end-of-life recycling and disposal. In accordance with the requirements of the Ministry for the Environment of New Zealand, for example, enterprises in the tire, electronic devices and packaging industries in Auckland are required to take responsibility for recycling and disposal of their products. Such practices significantly contribute to waste recycling and disposal, and reduce waste at the source, as producers are encouraged to use or manufacture products with small environmental impact.

(6) Efforts should be made to explore and apply new technologies related to waste disposal. New technologies have made waste disposal easier and made zero waste more likely. For Masdar, a new city under construction in the United Arab Emirates, a low energy-consuming underground flatbed freight transit system was designed for waste transportation, raising the transportation efficiency and bringing down labor cost. Some other cities have been studying and applying degradable materials and adopting technologies to enhance the efficiency of waste composting and incineration.

(7) Information dissemination and training should be organized to raise public awareness of waste recycling. Dissemination of relevant information and public awareness raising are the most fundamental and yet the most important parts of the efforts for urban waste management. In Sydney, the government provides comprehensive information about waste management and community activities through websites, and offers courses on environmental protection to provide knowledge about waste classification and recycling starting from elementary school. San Francisco has a website and an app to provide information about waste classification and disposal, and a database for people to search for waste drop-off sites and make garbage collection appointment, and provides households and businesses with extensive door-to-door training on waste management in multiple languages. Extensive publicity and training programs have significantly raised public awareness of waste recycling and environmental protection, laying a solid foundation for achieving the goal of zero waste.

2 Building Zero-Waste Cities Is a Sign of the Advanced Stage of a Circular Economy in China

The launch of zero-waste city programs in China is on the one hand a choice made at China's own initiative in the face of the general development trend, and on the other hand a natural result of the circular economy development in the country. Generally, circular economy development has gone through three stages in China:

The preliminary stage (1949–2004): This stage is characterized by comprehensive use of resources. Resource recycling focused mainly on waste recycling and reuse in order to mitigate shortage of resources. China's economy at this stage is not a circular economy in its strict sense, as efforts mainly focused on reuse and turning waste into resources, not reduction at the source.

The intermediate stage (2004–2013): China adopted developing circular economy as a strategy for the transformation of its development model. During this period, the State Council held a national conference on circular economy in 2004 and issued several milestone policies, including the *Opinions of the State Council on Accelerating the Development of Circular Economy* and the *Circular Economy Development Strategy and the Near-Term Action Plan*, and introduced the Circular Economy Promotion Law. Two circular economy pilot programs and several special pilot programs were launched, and 100 cities were established as national demonstration cities of circular economy. This stage saw the development of circular economy in China from theory to practice and from pilot programs to demonstrations, with corresponding guiding strategies, plans, policies, legislations, projects, technological progress and publicity efforts.

The advanced stage (2013–present): At this stage, circular economy offers strong support for building an ecological civilization. The 19th National Congress of the CPC pointed out that "building an ecological civilization is vital to sustain the Chinese nation's development." From recycling to circular economy, and then to ecological civilization and green development, China has reached a new stage in its efforts to build an ecological civilization. In particular, the initiative to build zero-waste cities represents the expansion of the concept of circular economy in both theory and practice, which is a symbol that China is entering the advanced stage of circular economy development.

Zero-waste city is an advanced concept of urban management and development showing distinctive features of common-pool resources. In terms of the flow of resources, zero-waste cities manage their material flows through the cycle of production, products, waste and utilization, aiming to make full use of resources at every step in the production and consumption processes. Governments, enterprises, market entities and individuals are all part of the micro-level governance system in waste cities where micro- and macro-level governance are integrated for fine urban management.

3 Zero-Waste City Building Will Boost Circular Economy and Contribute to Progress Towards an Ecological Civilization

Building zero-waste cities is a major decision made by the CPC Central Committee and the State Council in order to fight pollution and build a moderately prosperous society in all respects. It will contribute to waste reduction at the source, waste recycling and environmentally sound disposal in cities, and improve urban waste management. It will also promote green production, green lifestyle and green consumption, and contribute to the building of an ecological civilization.

(1) The campaign to build zero-waste cities is an innovative integration of various relevant policies so as to form a system for coordinated development. There used to be separate pilot projects for waste management initiated by different authorities, such as the "Double Hundred" project for comprehensive use of resources implemented by the National Development and Reform Commission (NDRC), and the pilot project initiated by the Ministry of Agriculture and Rural Affairs for comprehensive use of crop straw. More departments now cooperate to align policies and seek coordinated development. The NDRC and the Ministry of Industry and Information Technology, for example, cooperated in the development of industrial bases for comprehensive use of bulk solid waste. Pilot zero-waste programs approach urban solid waste management from the perspective of the overall development of cities, and aim to create a new development pattern that coordinates between solid waste management and urban transformation and development, facilitating industrial restructuring and the development of emerging industries. These programs offered a platform for practice so that we can deepen the reform of systems for solid waste management, remove long-standing barriers that resulted in fragmented administration and overlapping responsibilities of various relevant authorities, and explore systems and technologies for comprehensive management of solid waste suitable for China's conditions.

(2) Building zero-waste cities helps resolve the solid waste problem in cities and improve environmental quality. China is the largest producer of solid waste in the world, with an annual output of approximately 10 billion tons and a backlog of 60–70 billion tons. Among the over 600 large and medium-size cities in China, two thirds face the challenge of excessive waste. A survey of 202 such cities shows that comprehensive utilization of industrial solid waste cover 42.5% of the total amount of waste, and the rate for hazardous industrial solid waste is 48.6%. To build zero-waste cities, measures need to be adopted to address solid waste throughout its lifecycle, including production, consumption, management and disposal, to improve urban environmental quality in all aspects.

(3) The building of zero-waste cities will foster new growth drivers in the solid waste treatment industry. The efforts to build zero-waste cities will help shape a large-scale solid waste treatment industry. During the 13th Five-Year Plan

period, the solid waste treatment industry is expected to expand from 10% of the entire environmental protection market to about 25% and investment in China's solid waste treatment industry will exceed RMB 3.5 trillion. The market size of construction waste treatment in China, for example, exceeded RMB 80 billion in 2017, twice the size of 2010, registering an annual growth rate of more than 10% on average. If the market continues to grow at the current rate, by 2020, China's construction waste treatment market will worth more than RMB 100 billion.

4 Advice on Building Zero-Waste Cities in China

Zero-waste cities cannot be built overnight, as they involve various processes. We shall start with pilot programs, make overall plans, and advance the efforts in an integrated and coordinated manner. Recommendations are as follows:

(1) Integrate the building of zero-waste cities into the efforts to build an ecological civilization in urban areas. Zero-waste city represents a full-blown green transformation involving the development of circular economy in the development of an urban ecological civilization. It requires both transition from a linear model to a circular model in production and the transition from excessive consumption to green consumption. Therefore, we need to integrate zero-waste targets into the evaluation system for building an ecological civilization, and consider the building of zero-waste cities an important way to achieve ecological progress in cities. We must coordinate our efforts to improve overall management of solid waste with the supply-side structural reform in cities as well as the overall urban development and management, so as to accelerate the formation of a sound spatial distribution pattern of various functions in cities, a better industrial structure, and eco-friendly models of agricultural and industrial production and consumption, so as to boost green development in cities.
(2) Promote green production in agriculture and industry to reduce solid waste and maximize recycling. To reduce waste at the source, green mining shall be promoted to reduce new solid mining waste and the stock of such waste to be disposed of. To reduce waste generated in production activities, we need to reduce resource inputs and the content of hazardous substances in raw materials by promoting shared standards and eco-friendly product design, and encourage recycling through green design and green supply chains. For waste reduction in the utilization of resources, we should promote recycling of bulk solid waste, and gradually achieve comprehensive utilization of animal waste at or near the places where they are produced. In waste disposal, efforts should be made to maximize recycling, and non-recyclable waste should be disposed of in an environmentally sound manner.
(3) Use market mechanisms to unleash the vitality of market players and foster new models of industrial development. Flexible and diverse market forces are

conducive to effecting changes in the behavior of waste producers. We may consider providing waste treatment enterprises with tax reduction or exemption, or low-interest loans and give them preference in land allocation so as to encourage enterprises to participate in waste disposal. Landfills shall be charged according to the amount of waste buried to raise landfill cost, which will help reduce the amount waste going to landfills. We need to promote coordination of policies and measures such as credibility assessment based on environmental protection performance, green finance and pollution liability insurance, in relation to solid waste management. We should encourage the development of online solid waste treatment businesses and combine online platforms and offline recycling services. We also need to encourage specialized third-party organizations to participate in solid waste recycling, environmental pollution control and relevant consulting services and create new business forms in relation to solid waste treatment and environmental services.

(4) Advocate green development and promote green lifestyle and green development. The building of zero-waste cities represents a profound change in our approach to development. It requires not only a transformation of how things are produced, but also efforts to encourage simple, moderate, green, and low-carbon ways of life. A wide array of measures may be adopted to promote green, circular and low-carbon lifestyles and consumption. Charging for excessive waste can reduce household waste. Deposit for plastic bottles and other similar measures may encourage consumers to dispose of recyclables properly and boost recycling. We should encourage the economy of sharing so as to reduce waste of resources, move faster towards green packaging in the express delivery industry, as well as green restaurants and catering services, and the like.

In summary, the endeavor to build zero-waste cities takes waste management as an integral part of urban development and is likely to initiate profound changes in our ways of life, production and consumption, and help from urban planning, industrial structures, spatial structures and development approaches that are conducive to green development and progress towards an ecological civilization.

Create a Strategic System for Ecological Civilization Development in the Guangdong-Hong Kong-Macao Greater Bay Area

Xiuyu Zhang

General Secretary Xi Jinping emphasized the significance of the Belt and Road Initiative and the Guangdong-Hong Kong-Macao Greater Bay Area (hereinafter referred to as the Greater Bay Area) in creating a new model of all-round opening up in China.

As specified in the *Development Plan for the Guangdong-Hong Kong-Macao Greater Bay Area* (hereinafter referred to as the Plan), a basic principle for the development of the Greater Bay Area is to pursue green development and ecological conservation. The plan requires that efforts be made to "vigorously push forward ecological conservation, establish the approach of green development, uphold state policies on saving resources and protecting the environment, implement the most rigorous ecological protection system, adhere to the most rigorous farmland protection system and the most economical land use system, promote the adoption of green and low-carbon production methods and lifestyles as well as models of urban development and operation, provide a good ecological environment for residents, and facilitate sustainable development of the Greater Bay Area."

As pointed out by General Secretary Xi Jinping, human beings are a community with shared future, and it is the common aspiration of mankind to build a green home. Ecological and environmental crises are common challenges for the world. All members of the international community should work together to build an ecosystem that puts mother nature and green development first and build a sound global environment. A strategic system for ecological civilization development in the Greater Bay Area will facilitate the Belt and Road Initiative and sustainable development of the Greater Bay Area. It will also enable further implementation of Xi Jinping Thought on Ecological Civilization, therefore it is highly relevant to upholding harmony between man and nature, building a socialist ecological civilization, and finding China's approach towards global governance.

X. Zhang (✉)
Ecological Civilization Research Center, South China Institute of Environmental Sciences, Ministry of Ecology and Environment, Guangzhou, China

J. Pan et al. (eds.), *Beautiful China: 70 Years Since 1949 and 70 People's Views on Eco-civilization Construction*, https://doi.org/10.1007/978-981-33-6742-5_37

1 Definition of a Bay Area

In geography, a bay area is an area where river, marine and terrestrial ecosystems meet, with abundant marine, biological and environmental resources endowment, unique geological landscapes and special ecological value. In socioeconomic terms, a bay area refers to a socioeconomic agglomeration of ports and towns facing the same sea, forming strong synergy. In general, cities in bay areas benefit from their unique natural endowments and enjoy a good environment, an open culture, sound industrial development, and a concentration of talent. Bay areas thus play an important part in regional, national and even global economic development, and are highlights in the global economic landscape. Famous bay areas in the world are mostly major growth poles of the world economy, characterized by technological innovation. The Silicon Valley, which is in the San Francisco Bay Area of the United States, is the world's center of technological innovations. Wall Street of the New York Bay Area is the world's financial center. The Tokyo Bay Area is the mainstay of Japan's economy, contributing one third of the country's GDP.

2 Background of the Greater Bay Area

The Guangdong-Hong Kong-Macao Greater Bay Area is a national strategy of China proposed and headed by General Secretary Xi Jinping himself. It is a move in the new era towards a new model of all-round opening up of China, and a new practice of "one country, two systems." This strategy was included in the report to the 19th National Congress of the CPC and the report on the work of the government. It is expected to help promote exchange and cooperation between the Chinese mainland and Hong Kong and Macao SARs. It is of great significance to Hong Kong and Macao in facilitating their participation in national development strategy, enhancing their competitiveness and maintaining prosperity and stability in the long run.

The NDRC issued the *Overall Plan for Pilot Reforms for Comprehensive Innovation in Some Regions and Sectors* in September 2015 and Guangdong was one of the, focusing on promoting cooperation in innovation between Guangdong, Hong Kong and Macao. The report of Guangdong government on its work in 2016 included a section about improving cities in the Pearl River Delta and joining hands with Hong Kong and Macao to develop the Greater Bay Area.

At the fifth session of the 12th National People's Congress held in March 2017, Premier Li Keqiang proposed in the report on the work of the government that China would promote closer cooperation between the mainland and Hong Kong and Macao SARs, draw up a plan for the development of a city cluster in the Greater Bay Area, give full play to the distinctive strengths of Hong Kong and Macao, and elevate their positions and roles in China's economic development and opening up.

President Xi Jinping attended the signing ceremony of the *Framework Agreement on Deepening Guangdong-Hong Kong-Macao Cooperation in the Development of the Greater Bay Area* in Hong Kong in July 2017, where the agreement was signed by Hong Kong SAR's Chief Executive Lam Cheng Yuet-ngor, Macao SAR's Chief Executive Chui Sai On, NDRC representative He Lifeng, and Governor Ma Xingrui, of Guangdong province.

In April 2018, while attending the Boao Forum, Guangdong governor Ma Xingrui said that the plan for development of the Greater Bay Area would soon be released. He expressed strong confidence in making the Greater Bay Area a world-class bay area and making Hong Kong, Macao, Guangzhou and Shenzhen a city cluster that functions as a comprehensive technological innovation center.

The CPC Central Committee and the State Council issued the *Development Plan for the Guangdong-Hong Kong-Macao Greater Bay Area* in February 2019, officially launching the Greater Bay Area development initiative as a new attempt at greater openness in the new era and a new practice of "one country, two systems." The Plan was formulated fully in line with the spirit of the 19th National Congress of the CPC, fully and accurately aligned to the principle of "one country, two systems," and with an aim to fully leverage the compound advantages of Guangdong, Hong Kong and Macao. It requires deepened cooperation between the Chinese mainland and Hong Kong and Macao SARs and enhancement of the Greater Bay Area's supporting and leading role in China's economic development and opening up.

3 Natural and Economic Conditions of the Greater Bay Area

With a long coastline, China has a lot of ocean-bound rivers and bays. Since the launch of the reform and opening up, the country has seen continuous development of modern industries in its coastal areas and steady progress in exploring the oceans. Bay areas in China, in particular, have made remarkable achievements in economic development. The economic growth is especially fast in the Greater Bay Area comprising the nine mainland cities of Guangzhou, Foshan, Zhaoqing, Shenzhen, Dongguan, Huizhou, Zhuhai, Zhongshan and Jiangmen, as well as Hong Kong and Macao SARs. Up to 56,000 km^2 in area, the Greater Bay Area has a population of 70 million, and its GDP exceeded RMB 10 trillion in 2017, making it the best developed region in China in terms of economic strength. The Greater Bay Area is the world's fourth largest bay area following the New York Bay Area, the San Francisco Bay Area, and the Tokyo Bay Area. With a well-developed economy, great biodiversity, an inclusive and open culture, and abundant science, technology and education resources, the Greater Bay Area has been an important powerhouse in China's economic development. It is also an important platform for the country to build a world-class city cluster and participate in global competition.

4 Environmental Issues of the Greater Bay Area

The rapid social and economic development in the Greater Bay Area brought about mounting pressures on resources and the environment, mainly manifested in the following aspects: First, some cities failed to coordinate well between plans for terrestrial and marine development, and developed no unique feature in spatial planning or municipal functions. Marine development activities concentrated in offshore areas, and the available coastline, beaches and shallow-water biological resources are decreasing. Second, the efforts for industrial development focused on resources and producing primary products with low value added, and many places had similar industrial structures. Third, offshore water quality deteriorated, leading to declining ecological functions and frequent marine ecological disasters. If we look back at the development of some famous bay areas in the world, we can see that they also once encountered environmental problems and unfortunately took wrong turns along the way. Practice has proven that a forward-looking approach in designing the framework for ecological endeavors is the way to innovative, coordinated, green, open and inclusive development in the Greater Bay Area.

5 Framework for Building an Ecological Civilization in the Greater Bay Area

1. **Implement the Plan to build a green Greater Bay Area**

The Plan connects the green development of the Greater Bay Area with major national strategies like the Belt and Road Initiative. The area will undergo a transition from factor- and investment-driven growth to an innovation-driven model, and significantly intensify its efforts to develop environment-friendly industries. With an accurate understanding of the functions of cities in the Greater Bay Area, the Plan aims to establish a comprehensive support system for green innovation to promote the cultivation of talent, R&D, products and the development of the market, seeking breakthroughs in key cutting-edge technologies related to resources, the ecosystem and the environment. Mass entrepreneurship and innovation is encouraged in various sectors and stages of development as a constant driving force for green transformation. What is required in the supply-side structural reform is perfectly in line with what our ecological endeavors call for. Focusing on high-quality industries in the Greater Bay Area, the Plan promotes green production and requires all factors to be environmentally friendly. We must vigorously promote green industries such as environment-friendly modern agriculture, marine biological industry and ecotourism, and create a green financial system to make environmental protection a new growth driver in the Greater Bay Area.

According to the Plan, the Greater Bay Area shall actively adapt to climate change and strengthen exchanges and cooperation on low-carbon development, energy conservation, and environmental protection technologies. To this end, the area needs to advance pilot projects for low-carbon development, promote the evaluation of green and low-carbon development, build a green development demonstration zone, and accelerate the development of a green industrial system. It should also take forward the revolution in energy production and consumption, build an energy sector that is clean, low-carbon, safe and efficient, and encourage conservation and recycling across the board. Efforts are required to promote green lifestyles, build more pedestrian paths for the general public such as urban greenways and pedestrian trails in forests and marshes. The Plan also requires promoting the experience of the pilot project for incentives for carbon emissions reduction, and continue with studies and demonstration programs for mutual recognition of carbon labelling across Guangdong, Hong Kong and Macao. The Plan offered an unprecedented development opportunity for the Greater Bay Area, and we should seize this opportunity to build a green bay area.

2. **Improve the structure to ensure ecological progress in the Greater Bay Area**

Following the basic rules for development of bay areas, the Plan requires a spatial layout for the Greater Bay Area that gives priority to ecological civilization development. In line with the requirements proposed in the 13th Five-Year Plan for National Economic and Social Development and the functional zoning system designed for the Greater Bay Area, efforts should be made to coordinate and align various urban plans and special plans for well-measured and orderly development of the area, so as to safeguard its ecological integrity and stability. In pursuit of development, we must pay attention to ecological conservation. We should strictly control the development intensity and scale in offshore areas, and promote moderate development activities in deep and open ocean areas. We shall never sever the ties between land and sea or engage in blind marine development without considering what needs to be sacrificed. The efforts for economic and social development shall be made in good coordination with the endeavor to improve resource allocation and create a living environment. We shall leverage the complementary advantages of terrestrial and marine ecosystems and promote their interaction and coordinated development to foster a development pattern featuring balanced and mutually reinforcing development of ocean and land culture in the Greater Bay Area.

The Greater Bay Area needs to take active measures to create a comfortable living environment with beautiful landscape. It needs to develop an ecological network comprising green networks and green blocks, and create a pattern centred on ecological corridors and blocks of green space. Urban landscaping should take the natural ecosystems into full consideration to improve the regional ecosystem which covers forests, waters, mountains and cities. There should be a multi-level and multi-dimensional ecological landscape structure with multiple functions based on networks, highlighting the distinctive characteristics and natural beauty of the cities. It should take forward the building of environmentally friendly communities to create a beautiful living environment. Eco-friendly ideas should be applied to the planning, construction and management processes of residential areas to make them eco-friendly. The distribution of public utilities shall be coordinated between urban and rural areas based on service range and population size. We need to increase the coverage and density of municipal infrastructure and improve their quality to satisfy residents' daily needs. We must promote energy conservation and recycling and improve the environment in residential areas to build an eco-friendly bay area that is energy-efficient and beautiful.

3. **Pursue innovation-driven development to make the Greater Bay Area a technology hub**

In accordance with the Plan, a technological innovation circle comprising Guangzhou-Dongguan, Huizhou-Shenzhen-Hong Kong-Macao, and Zhuhai-Zhongshan-Jiangmen-Foshan city clusters will be established in the Greater Bay Area. Advance should be faster to form innovation cooperation zones in Nansha of Guangzhou, Qianhai of Shenzhen, and Lok Ma Chau Loop and Hengqin of Zhuhai, and develop major innovation platforms such as in the Songshan Lake area in Dongguan, the Sino-German Industrial Services Zone in Foshan, and Huizhou Tonghu Ecological Smart Zone. Taking the opportunity of building Shenzhen and the Pearl River Delta into national innovation demonstration zones, efforts should be made to foster innovation clusters in major equipment manufacturing, telecommunication and information, smart manufacturing, biomedicine and other industries. The Greater Bay Area must pool and make full use of technology and innovation resources from around the world, improve the development models for cross-regional cooperation and innovation, and develop an open and international innovation system. We need to build a community for innovation and cooperation in the Greater Bay Area, and accelerate the development of the Pearl River Delta National Innovation Demonstration Zone, so as to push the Greater Bay Area towards the position of an international technology and innovation hub.

In the new era, cities in the Greater Bay Area must strive for a better regional innovation system with sound division of labor, leverage the advantages of the four innovation centers of the Hong Kong SAR, Guangzhou, Shenzhen, and Macao SAR-Zhuhai to promote innovation and enhance international cooperation in this regard; take an active part in the global innovation network, and build the Greater Bay Area into a global technology and innovation hub. Support should be provided for the Hong Kong SAR to become a world-class innovation center in the services sector. The Greater Bay Area should serve to gather first-class innovation resources to form networks and invite more first-class organizations and research institutes from the Chinese mainland. Efforts should be made to support the implementation of national projects and the development of innovation platforms in the region, speed up the attraction, integration and development of innovation resources, and improve the network for integrated innovation and cooperation. Relevant entities in the region are encouraged to undertake major science and technology projects to contribute to the development of an international science and innovation platform. The Greater Bay Area needs to build laboratories and engineering centers to improve its primary innovation capacity. Universities, research institutes and enterprises are encouraged to join hands to build cooperation platforms. Efforts are required to accelerate the commercialization of research findings produced in Hong Kong Science and Technology Park, Hong Kong Applied Science and Technology Research Institute, and various universities in the area, so as to make the area a technology hub.

4. **Highlight environmental protection to ensure a good environment**

The Plan requires strong faith in the idea that lucid waters and lush mountains are invaluable assets, and, in that vein, the most rigorous ecological protection rule. According to these, the Greater Bay Area needs to launch major projects for conservation and restoration of ecosystems as soon as possible, develop biodiversity conservation networks, and enhance the quality and stability of its ecosystems. The red lines for ecological conservation must be strictly observed, and control over the use of natural ecological space must be strengthened. We need to step up the protection of mountain, hill, and forest ecosystems in the vicinity of the Pearl River Delta, and develop an ecological barrier of an uninterrupted belt of forest-covered mountains in the north. We must enhance the protection and management of coastlines, strengthen the protection of shoreline resources and the conservation of natural properties, and establish a sound and comprehensive mechanism for dynamic monitoring of coastlines. We should enhance the protection and restoration of near-shore marine ecosystems, take forward the "Blue Bay" action plan for ecological remediation, protect coastal mangroves, and create a coastal ecological belt. Guangdong, Hong Kong SAR and Macao SAR should enhance their cooperation in ecological conservation, jointly protect key wetlands of national and international importance and cross-boundary coastal wetlands in an all-round manner.

The Plan requires that we take forward cooperation in the management of water resources, water environment and water-related projects in the Pearl River estuary area, focusing on pollution control along the east and west banks of the Pearl River. It also requires efforts to strengthen the management of projects concerning land pollution discharge, water-related projects as well as the environment of coastlines and beaches. To this end, we must expedite the development of a system for the control of the total amount of major pollutants discharged into the sea as well as a system for real-time online monitoring of the marine environment in the Greater Bay Area, control the total discharge of pollutants into Dongjiang, Xijiang and other smaller rivers in the Pearl River Delta to ensure that the water quality in the water functional zones is up to standard. We must strengthen conservation of the water environment and aquatic biological resources of major rivers such as Dongjiang, Xijiang and Beijiang, step up efforts to restore heavily-polluted river systems including the Shenzhen River, continue with comprehensive environmental improvement efforts targeted at heavily polluted water bodies in urban areas, link up water networks within the Pearl River Delta, and develop a green aquatic ecological network across the entire bay area. The Greater Bay Area also needs to strengthen joint prevention and control of air pollution, launch more stringent clean shipping policies, make concerted efforts to reduce the emissions of various pollutants, and take coordinated prevention and control measures against pollution caused by ozone and fine suspended particulates (PM 2.5). Efforts should be made to build the capacity for regional cooperation in the handling and disposal of hazardous waste, exercise stronger control over the cross-boundary transfer of waste, and improve the environmentally sound disposal, reduction and recycling of solid waste. Guangdong, Hong Kong SAR and Macao SAR should launch technological exchanges and cooperation on soil improvement, proactively advance the demonstration projects for this purpose, strengthen the safe use of contaminated farmland and land plots, prevent and control non-point source agricultural pollution, and ensure the quality of agricultural produce and a safe living environment. The Plan demands the establishment of a blacklist system for pollution control, and requires efforts to improve mechanisms for credibility assessment based on environmental protection performance, mandatory information disclosure as well as severe punishments and penalties for violations. More should be done to solve environment-related problems that accumulated over time and aroused extensive concern among the general public.

5. **Coordinate the efforts to protect terrestrial and marine environment**

To implement the Plan, the Greater Bay Area needs to make a better scheme for jointly building a quality living circle across Guangdong, Hong Kong and Macao, and develop a green development demonstration zone. Coordinated efforts should be made to advance the development of land-ocean integration of ecological spaces, an eco-friendly economy, a sound environment, an ecological culture and ecosystem-related systems, in a bid to make the Greater Bay Area a world-class city cluster with well-developed economy and high environmental awareness, a green place to live and work in and travel to, and a model of ecological conservation along the twenty-first Century Maritime Silk Road. Cooperation in tourism shall be strengthened to leverage the distinctive characteristics and strengths of cities in the Greater Bay Area. Giving priority to promoting integrated tourism, the Greater Bay Area should develop a system that put together diverse tourism products that meet tourists' needs for cultural and historical attractions, leisure activities, healthcare, and cruise and yacht tours. Joint efforts should be made to attract tourists from home and abroad and make the area a world-class travel and leisure destination.

We should work out development plans for the Greater Bay Area for ecotourism based on its quality ecotourism resources so as to provide more multi-destination premium travel packages. The visa-free policy should be improved. Support should be given to Hong Kong SAR for the development of a diverse tourism platform and to Macao SAR to build it into a world-class tourism and leisure center. The region should promote development of international cruise terminals in Hong Kong SAR, Guangzhou and Shenzhen and further increase the number of international liner routes to facilitate the healthy and orderly development of cruise tourism. It also needs to speed up the implementation of measures for free yacht travel with support given to Shenzhen, Zhuhai, Zhongshan and Huizhou for the establishment of national demonstration zones of integrated tourism. The Qianhai and Shekou area of Shenzhen, the Nansha area of Guangzhou, and the Hengqin New Area of Zhuhai of China (Guangdong) Pilot Free Trade Zone will be developed into demonstration zones for cooperation among Guangdong, Hong Kong and Macao in tourism, and Hengqin will be developed into an international tourism island. Efforts should be made to develop small coastal towns with distinctive features and create a hospitable environment in the bay area.

6. **Build a bay area with abundant cultural resources**

The Plan calls for cooperation in education and sharing of education resources among Guangdong, Hong Kong and Macao and strengthening joint running of education institutions. It supports the development of the Guangdong-Hong-Kong-Macao University Alliance, and encourages joint academic programs, laboratories and research centers in fields with competitive edges. Higher education institutions of the three places are encouraged to explore cooperation and exchange activities such as mutual recognition of academic credits and degrees and the sharing and commercialization of research findings. Efforts are required to explore cross-boundary sharing of libraries, laboratories and large scientific facilities in higher education institutions. Support will be given to Shenzhen to develop into an international demonstration zone for education, bringing in world-renowned universities and specialized colleges, and seek to build world-class universities and academic programs. Cooperation should be strengthened in vocational education and talent training, encouraging exchanges and cooperation among training bases for vocational education. Through joint efforts, a bunch of technical personnel training bases featuring cooperation between enterprises, vocational schools and universities, and vocational education parks with distinctive characteristics will be established. Support will also be given to Macao to establish a training base for talent proficient in Chinese and Portuguese and to Hengqin to create a national education and training base for tourism professionals.

In line with the Plan, Guangdong, Hong Kong SAR and Macao SAR need to strengthen cooperation in the preservation of cultural heritage, and jointly organize different types of cultural heritage-themed exhibitions and shows. Joint efforts are required to develop an ecological reserve characterized by Lingnan culture. Support will be given to Guangdong to develop projects such as an underwater cultural heritage preservation center. Guangdong, Hong Kong and Macao should work together to establish a system for the sharing of cultural resources and information, explore ways to share public cultural information and promote joint efforts in cultural production. They should also make concerted efforts to make the area an international leisure, culture and tourism center. Efforts should be made to promote cultural exchanges among the youth of Guangdong, Hong Kong SAR and Macao SAR with support given to projects such as the Funding Scheme for Youth Exchange between the Mainland and Hong Kong and the Thousand Talent Program of Macao. Efforts are also required to promote cooperation in youth study tours, and jointly develop demonstration bases for study tours. The Plan encourages the organization of Greater Bay Area youth summits and the establishment of platforms for cultural exchanges and cooperation so as to develop a bay area with rich culture.

7. Promote law-based governance in the Greater Bay Area

The Plan requires enhanced cooperation in social governance. Efforts should be made to promote law-based governance, strengthen coordination between anti-corruption mechanisms across the Greater Bay Area, establish quality, efficient and clean governments, and enhance the efficiency of government services and people's sense of gain. In urban and rural communities across the nine Pearl River Delta cities with many Hong Kong and Macao residents, service functions should be improved in a targeted manner to provide timely, efficient and convenient social services to Hong Kong and Macao residents. The Constitution of the Chinese mainland and the Basic Laws of the two SARs must be strictly observed. Guangdong, Hong Kong SAR and Macao SAR should strengthen juridical cooperation on the basis of respect for one another's jurisdiction. A cooperation mechanism for public security and social governance should be established. The three places should also strengthen cooperation in the examination and issuance of early warnings for conflicts and disputes as well as in contingency management. They need to join hands to curb illegal immigration, make greater efforts to crack down on cross-boundary crimes, and coordinate responses to conventional and unconventional security threats. The mechanism for response to emergencies needs to be improved, and a coordination platform for emergency response across the Greater Bay Area should be established. The three places should work together to formulate emergency response plans for major emergencies such as serious accidents, natural disasters, public health emergencies and public security emergencies, and conduct drills for emergency response from time to time to enhance the capability to deal with emergencies through collaboration.

Measures should be taken to prevent and mitigate major risks, especially financial risks. Efforts are required to strengthen the mechanism for local governments' responsibilities for financial risk management, and forestall and defuse risks in key areas properly. We must resolutely crack down on illegal financial activities, improve the regulatory system for weak points in the financial sector, and forestall systemic financial risks. Guangdong province must rigorously implement the relevant provisions of the Budget Law, strengthen the management of ceilings for local government debts, and effectively regulate debt financing by government bodies. Fiscal control should be intensified to effectively keep in check insolvent development projects. Supervision and accountability efforts should be strengthened to resolutely prevent financing guarantees that go against laws and regulations. In summary, the rule of law should be promoted in the Greater Bay Area.

8. Integrate the systems to ensure shared development in the greater bay area

A fundamental difference between the Greater Bay Area and its counterparts in the rest of the world is the existence of different political systems. It is thus imperative to establish a comprehensive and sound mechanism for coordination and implementation, and strengthen the exchanges and cooperation between governments of Guangdong, Hong Kong SAR and Macao SAR, between central and local governments and between the governments, enterprises and the general public. A clear division of labor, coordinated efforts and detailed work plans are required to ensure effective implementation of relevant decisions and policies. We should encourage the development of funding systems and long-term mechanisms with funds from the central and local governments and the non-governmental sector. Efforts should be made to cultivate talent and build think tanks to support the development of the Greater Bay Area and the green Belt and Road, and create new systems and mechanisms for making concerted efforts to foster a sound environment shared by all. We need to improve mechanisms for talent training and introduce new mechanisms in this regard, focusing on the training of versatile talent with a global vision who are familiar with international rules and environmental protection, so as to provide more high-caliber professionals for the development of the Belt and Road Initiative.

Guangdong, Hong Kong SAR and Macao SAR are now embarking on a new journey to develop the Greater Bay Area. Such strategic efforts are a natural result of China's transition from a big economy to a strong one and the need to build its global competitiveness and influence. It is also what Guangdong, Hong Kong and Macao must go through in order to push forward the profound economic and social transformation in pursuit of sustainable development. The development of the Greater Bay Area will help explore new mechanisms for regional cooperation within the framework of "one country, two systems." It will also leverage the unique strengths of Hong Kong and Macao SARs to promote their in-depth cooperation with the Chinese mainland, create new space for their development, and maintain long-term prosperity and stability in these two regions. Building the Greater Bay Area will also boost innovation and development in the Pearl River Delta and contribute to shared development in the entire region covered by this strategy.

Difficulties to Be Addressed with China in the Lead for the Building of an Ecological Civilization in the New Era

Dongjun Meng

General Secretary Xi Jinping emphasized in the report to the 19th National Congress of the CPC that we should cherish the environment as we cherish our lives. He pointed out that taking the driver's seat in international cooperation to respond to climate change, China has become an important participant, contributor, and torchbearer in the global endeavor for the building of an ecological civilization. At the National Conference on Eco-environmental Protection, Xi Jinping proposed that we should build a global ecological civilization. He required that China should create solutions to problems in environmental protection and sustainable development in the world and take an active part in international cooperation on response to climate change. He called for joint efforts to build a global ecological civilization. Addressing the Belt and Road Forum for International Cooperation, Xi Jinping stated that China would set up a big data service platform on ecological and environmental protection, and proposed the establishment of an international coalition for green development along the Belt and Road, saying that China would provide support to related countries in responding to climate change.

For the building of an ecological civilization in the new era, further efforts are required to explore new theories and practices. We need to determine the direction and resolve difficult problems to produce solutions for environmental protection and sustainable development around the world. China should take the driver's seat in international cooperation in responding to climate change, encourage major countries to participate, take a leading role in global environmental governance, and promote joint efforts for green development.

D. Meng (✉)
Ecological Civilization Research Center, China Academy of West Region Development, Zhejiang University, Hangzhou, China

J. Pan et al. (eds.), *Beautiful China: 70 Years Since 1949 and 70 People's Views on Eco-civilization Construction*, https://doi.org/10.1007/978-981-33-6742-5_38

1 Xi Jinping Thought on Ecological Civilization Is a Guiding Theory for Building a Global Ecological Civilization

The building of an ecological civilization is a strategic task involving the fundamental change of the way of production and life, which is vital to sustain the Chinese nation's development.

After proposal of the concept of ecological civilization, definition and exploration of the paths to and practices of buiding an ecological civilization, Xi Jinping Thought on Ecological Civilization became a complete and mature system. It originated from Xi's famous saying that "lucid waters and lush mountains are invaluable assets." Then the concept of ecological civilization was formally proposed at the 17th National Congress of the CPC, scientifically defined and included into the Constitution of the Party at its 18th National Congress. The Third Plenary Session of the 18th CPC Central Committee specified the path towards an ecological civilization. The *Opinions on Accelerating Ecological Progress* specified the tasks, requiring coordinated efforts in pursuit of economic, political, cultural, social and ecological progress. The *Overall Plan for Reform of the System for Developing an Ecological Civilization* served as the top-level design and made specific plans in this regard, which ensured that relevant reform was to be conducted in a more systematic, holistic, and coordinated way. The 19th National Congress of the CPC required that we speed up the reform of the system for developing an ecological civilization, and build a beautiful China. It was specified in the Constitution of the Party that the Party shall "fully understand that lucid waters and lush mountains are invaluable assets."

As China entered the new era, General Secretary Xi Jinping decided that China is in a critical period with pressure from various sources that requires more quality ecological goods to meet people's ever-growing demands for a beautiful environment, He also decided that China is capable of addressing prominent ecological and environmental issues during this period. He believed that China's efforts to promote environmental protection have gone through a historic turnaround with wide-ranging implications. While highlighting the significance of eco-environmental issues, Xi Jinping Thought on Ecological Civilization specifies the new principles, systems, targets, tasks and requirements for ecological civilization development in the new era. Epitomizing the latest thoughts of Xi Jinping on ecological civilization, it is a comprehensive, systematic, profound and scientific elaboration of the concept of ecological civilization, and the fundamental principle for building an ecological civilization with Chinese characteristics in the new era. Xi Jinping Thought on Ecological Civilization was established at the 19th National Congress of the CPC, and improved at the National Conference on Eco-environmental Protection. Xi Jinping Thought on Ecological Civilization has become a guiding theory in the effort of building a global ecological civilization.

2 For a Global Ecological Civilization, It Is Imperative to Reform the Systems and Introduce New Mechanisms and Institutions

China's attempts at systematic reform and innovation of institutions and mechanisms is a symbol that the country has become a leader in the building of an ecological civilization. In this regard, we need to strengthen the systems in many areas, such as pilot programs for the integration of separate development plans into a single master plan, the preparation of a natural resources balance sheet, the confirmation and registration of natural resource ownership, the delineation of ecological conservation red lines, the audit of outgoing officials on management of natural resources, ecological compensation, target-based performance evaluation in relation to ecological civilization development (based on GEP), a national park system, consistency of environmental laws and regulations and environmental quality standards with international counterparts, judicial protection of resources and the environment (through settlement and adjudication mechanisms for environment-related disputes), education programs to raise ecological awareness (including programs for the public and courses in universities). The government should play a guiding role, while enterprises should take the lead in the market and the public should also play a part. We must take a global perspective and adopt a forward-looking approach. The policies must be feasible and replicable, which can guide the implementation of specific tasks, and should be in keeping with international practices so that they can be promoted to the rest of the world. We should be brave and confident to take on responsibilities to develop a comprehensive and sound system of institutions and practices and take a leading role in global endeavor for ecological progress.

3 For a Global Ecological Civilization, It Is Imperative to Promote Green Technologies and Their Commercialization

Technological innovation and application of technologies, which are found in almost every step of our ecological endeavors, have improved the mode of production and significantly facilitated the development of an ecological civilization. It would be impossible to build an ecological civilization without technological progress. We must follow the development trend of the fourth industrial revolution, and adopt technologies such as big data, artificial intelligence, new materials and new energy and the interdisciplinary and platform-based approaches to create market-oriented systems for innovations in green technologies, promote the development of industries featuring energy conservation, environmental protection, clean production and clean energy to encourage the coordination between industrial development and ecological conservation. The Green Technology Bank (GTB) initiative launched by China is a good example in this regard. The GTB is a comprehensive service platform with a pool of advanced green technologies that promote sustainable development by saving energy, maintaining a sound environment, and ensuring security, efficiency and health. It advocates a better integration of technologies and finance, capitalization of technological achievements, and acceleration of technology transfer and industrial application. The platform facilitates both sustainable development of China and South-South cooperation in the field of green technology. Committing to primary innovation in the field of green technology and industrial innovation, application and international transfer of green technologies, the GTB is a major measure for building a technology and innovation hub with global influence, and an important strategy for China to contribute to the global narrative.

4 For a Global Ecological Civilization, It Is Imperative to Promote Innovation in Green Finance

Green finance has injected new momentum to our ecological endeavors. Over the years, we have been looking for standard practices and a sound organizational structure for green finance, and making efforts in regard to related goods and services, supporting systems and risk prevention. China issued guiding opinions on green finance and the *Environmental Risk Management Initiative for China's Overseas Investment*. It launched green finance pilot programs, and established a financial system for building an ecological civilization. We will focus on green assets as securities, develop market channels for green credits and green trading; promote the development of green PE and VR, and use technologies to reduce the cost of green products; and promote innovation in green finance, focusing on using fintech to reduce costs of financial assets accreditation and assessment, etc.

5 For a Global Ecological Civilization, It Is Imperative to Develop Market-Oriented Ecological Industries with Enterprises as the Main Players

To build an economic system aimed at integrating industries and ecosystems, we must allow the market, in which enterprises are leading players, to play a decisive role in resource allocation. Integration of industries and ecosystems is a trend in industrial development and many enterprises have made active efforts in this regard. We have been exploring practices for integration of ecological conservation and industrial development by developing eco-friendly products and services and combining the efforts in this regard with poverty alleviation, rural vitalization and sharing economy, etc. Liabilities accounting relating to natural resource assets remains a challenge across the world. In this regard, Chinese government and enterprises have moved one step ahead and started natural resources balance sheet compiling pilot. In accordance with the Guiding Opinions on Integrated Efforts for Reform of the Property Rights System for Natural Resource Assets, by 2020, a property rights system for natural resource assets featuring confirmed ownership, clarified rights and responsibility, strict protection, smooth transfer and effective supervision should have been basically in place. China has established a big data platform for comprehensive management of resources and the environment, which uses big data and cloud computing for classification of natural resources and data collection, and adopts accounting standards and econometric methods for accurate and systematic value management of the eco-environment and natural resources. The platform enables green development by improving the stock, increasing the flow and controlling the variables so that development and ecological conservation will enforce each other. It also uses big data for predication and assessment of ecological projects. The platform has been piloted in Haikou of Hainan province and Guanshanhu of Guiyang province, with positive comments from local governments, researchers from China and abroad and UN officials. It is advisable to promote the platform worldwide for global ecological progress.

General Secretary Xi Jinping is determined to "take full advantage of the solid material foundations built up over 40 years of reform and opening up, and step up our efforts to build an ecological civilization." He called for resolute efforts to address ecological and environmental problems, fight pollution and push China's development of an ecological civilization to a new stage. In its efforts to build an ecological civilization in the new era, China will make breakthroughs in institutional innovation, systems and mechanism, green technology, green finance and ecological industries, etc. and take a leading role. China will work to implement the 2030 Agenda for Sustainable Development and establish cooperation mechanism for Belt and Road ecological and environment protection. In the endeavor for a global ecological civilization, China will stay true to its original aspirations and play an active and leading role, contributing China's wisdom and solutions to facilitate the pursuit of global ecological progress.

Basic Theories of Ecological Civilization and Ecological Culture System

Developing Socialist Ecological Civilization with Chinese Characteristics

Yue Pan

As the world is going through profound changes never seen in a century, the human civilization is also switching from industrial civilization to ecological civilization. Upholding the basic governance philosophy of ecological civilization and guided by the Xi Jinping Thought on Socialism with Chinese Characteristics for a New Era, especially the Xi Jinping Thought on Ecological Civilization, the CPC Central Committee with Xi Jinping at the core has materialized historic, epochal and all-round changes in China's ecological civilization by promoting the green initiative of building a beautiful China. They have laid a solid ecological foundation for the great rejuvenation of the Chinese nation, made ecological practices for building a community with a shared future for mankind, and provided a Chinese plan for developing countries' leapfrog during transformation.

1 Theoretical Sources for Ecological Civilization with Chinese Characteristics

The first theoretical source is Marxist ecological outlook. General Secretary Xi Jinping called for "studying and practicing the Marxist thought on the relation between man and nature." According to Marxism, man-nature harmony can only be achieved after man-society relations fundamentally change. The human-centered capitalist industrial development pursues maximal profits as the ultimate goal through uninhibited capital expansion, which is bound to cause tension between man and nature, alienate interpersonal relation, drive the global transfer of ecological environmental cost, and eventually lead to regional and global wars, which is how colonists and imperialists plunder resources. According to Marxism, "This communism, as fully developed naturalism, equals humanism, and as fully developed humanism,

Y. Pan (✉)
The Central Institute of Socialism, Beijing, China

J. Pan et al. (eds.), *Beautiful China: 70 Years Since 1949 and 70 People's Views on Eco-civilization Construction*, https://doi.org/10.1007/978-981-33-6742-5_39

equals naturalism; it is the genuine resolution to the conflict between man and nature and between man and man." The CPC believes in Marxism, adheres to the great communist aspiration and strives for a "real community," so it naturally criticizes the "spurious community" that alienates the people and causes human destruction of nature, and strives to restructure the relation between man and man, man and nature, and man and society, including ideas and systems, natural and cultural environments, and environmental equity and social justice.

The second theoretical source is traditional Chinese ecological ethics. As General Secretary Xi Jinping pointed out, "the Chinese nation always respects and loves nature. The 5,000-year-long Chinese civilization has nurtured a rich ecological culture." Aspiring for harmonious co-existence, ancient Chinese thought of heaven, earth and mankind as a unified and harmonious entity, so they fetch from nature at the right time and utilize natural resources properly, which gradually gave birth to the "ethics of unlimited responsibility" between man and nature, man and society. Every generation of Chinese would pay debts owed by unknown ancestors and would bear responsibilities for unknown offspring. From time immemorial, China, with merely 3 million km^2 of livable land or so, has fed millions to over a billion of its people without ever resorting to expansion and looting, transfer of ecological cost, or imperialism and colonialism. What it has been relying on is the Chinese ecological ethics comprising the Confucian philosophy of "man and heaven in one," the Taoist doctrine of "following nature's course," and the Buddhist tenet of "all are created equal," as well as the ecological principle of "balance, moderation, order and restraint" that has been upheld by all dynasties and administrations in Chinese history. As the inheritor and promoter of China's fine traditional culture, the CPC is bound to push forward the evolution and innovation in our traditional ecological culture in support of the modern transformation of Chinese civilization.

The third theoretical source is the western concept of sustainable development. The excessive development of industrialization in the middle of the twentieth century caused unprecedented tension between man and nature, and the constant, large-scale environmental crisis across developed countries forced the western world to reflect on the traditional industrial development mode. In the 1960s, the publication of *Silent Spring* triggered a worldwide debate on development concepts; in the 1970s, the Club of Rome explicitly called for "rational, sustained and balanced development"; in the 1980s, the United Nations officially put forth the idea of sustainable development; in the 1990s, the United Nations Conference on Environment and Development (UNCED) initiated the sustainable development strategy; in 2000, the UN *Millennium Declaration* included "sustainability" as a key global agenda. In the West today, both the left-wing and right-wing are reforming the traditional political and economic institutions and rectifying traditional industrial civilization with ecological industrial civilization. Both developed and developing countries are reaching a sustainability-based "green consensus" in spite of their ideological differences, and setting the goal of all-round human development guided by the three principles of equity, sustainability, and commonness. The CPC, which is good at learning other countries' strengths while holding fast to its conviction and aspiration, will definitely draw theoretical and policy inspirations from the western concept and practices of sustainable development, and take an active part in global ecological environmental governance.

2 Basic Connotations of Ecological Civilization with Chinese Characteristics

1. **Ecological civilization is a new economic and social normal**

Ecological civilization requires the liberation and development of green productive forces. General Secretary Xi Jinping stressed that "protecting and improving the ecological environment is protecting and improving productivity," a statement that has enriched and developed the Marxist theory on productivity by classifying natural environment as a kind of productive force. Agricultural civilization creates the economic model of static circulation, industrial civilization pursues economic growth only, but ecological civilization combines static circulation with economic growth—it boosts circular economy to raise the efficiency of resource use, promotes clean production to reduce pollution during production, advocates green consumption to mitigate ecological impacts, and develops new energies for a thorough shift of production mode. Only when we work on productivity can we straighten out the man-nature relation at source, and make it possible to build a green economic system and environment-friendly society that is based on the environment and resource's carrying capacity, follows natural laws, and achieves maximal ecological benefits at minimal ecological cost.

Ecological civilization requires an ecological approach to industrial development. An ecological layout shall be formed through industrial transformation and upgrade to combine economic and ecological benefits. We should develop ecological agriculture, recover the ecological chain and improve resource use ratio, so as to coordinate the development of planting, forestry, animal husbandry and fishery. We should develop ecological industry, reform traditional productive technologies and create new technologies of energy conservation, circular resource utilization, and development and utilization of new energies and materials. We also should develop ecological service industry, and establish a modern service industry chain focused on high-end finance and commerce, trade and logistics, conference and exhibition, cultural creativity and ecological tourism by adopting clean service paths, ecological service providers and green consumption mode.

Ecological civilization advocates a green lifestyle. Lifestyle in industrial civilization, characterized by materialism and high consumption, upholds that more consumption contributes to economic development. In contrast, lifestyle in ecological civilization, characterized by practicality, economy and green consumption, aims to meet basic living needs while aspiring for mental and cultural advancement. It is estimated that if Chinese people copy the modern American lifestyle, we would need another 1.2 earths to meet the resource and energy demand. The capitalist industrial society isn't an ideal society for the Chinese people, who, enlightened by ecological civilization, has continuously intensified the public awareness of green development through different channels – from forest parks to religious venues, from street sculptures to toilet revolution, from public entertainment to community sanitation, from kindergartens to college education in environmental protection, from green labeling to green technology, and from respecting history to restoring legacy. Thanks to this societal green campaign, the Chinese people have formed a healthy lifestyle that recognizes simplicity, moderation, green and low-carbon mode, and environmental protection, fostered an atmosphere where every citizen contributes to ecological civilization, and turns the ecological consensus into a code of conduct observed by all.

2. **Ecological civilization is a new institutional system**

It means a new legal system. Ecological progress has been put on a law-based and institutional track across the board. To date, China has implemented 49 important laws and released 49 administrative regulations on environmental protection, as well as 90 departmental rules related with pollution control, ecological protection and nuclear radiation, 33 judicial documents on environmental protection, and 2,019 national environmental standards currently. By revising and enacting the most rigorous *Law on Environmental Protection* in history, and formulating such regulations as *Methods of Evaluation and Assessment on Ecological Progress*, *Methods of Holding Party and Government Leaders Accountable for Ecological and Environmental Damages (Trial)* and *Pilot Plan for Conducting Off-office Audit of Leaders on Natural Resource Assets*, the country has established both materially and procedurally the compensation system and life-long accountability system for environmental damages. As a result, ecological progress becomes an ideal testing field for the rule of law. For instance, the environmental protection department tested and launched China's first method of public participation in environmental protection and the first regulation on environmental information disclosure, held the first hearing on environmental impact assessment, and lodged the first public interest litigation concerning environment. These efforts are aimed to not only fundamentally solve the chronic problem that the cost of breaking environmental law is too low while obeying it too high, but also keep integrating China's law-based governance with green development from the perspective of national construction.

It means a new policy system. A package of effective economic policies has been formed for ecological progress. As an important part of macro-economic policies, environmental-economic policies, following market economy rules, are regulating or affecting the behaviors of market entities by means of price, tax, finance, credit, charge and insurance, so as to coordinate economic development with environmental protection. Based on different targets, there are policies on pollution control like charges for pollutant discharge, those on environmental infrastructure like charges for sewage and garbage disposal, and those on environmental protection like ecological compensation and regional equity. Based on the type of policy, there are market approaches such as pollutant discharge trading, policies on environmental taxes and fees such as environmental tax, charges for pollutant discharge, and payment by users, financial and capital moves such as green credit and insurance, fiscal incentives like subsidies for the development and use of environmental protection technologies, and fiscal transfer payment aimed at ecological compensation.

It means a new institution and mechanisms. Scientific and effective governance system and assessment and accountability mechanisms are being established for ecological progress. The ecological and environmental protection department is restructured to change the multi-lead system of environmental governance. Vertical management of the monitoring and inspecting performance of environmental authorities under provincial level is promoted to reduce local government's interference in environmental management. An interactive mechanism of joint prevention and control of water pollution among seven river basins has been established, as well as the coordination mechanism of joint prevention and control of air pollution among Beijing, Tianjin, Hebei and the surrounding areas. The central environmental protection supervision system is in place that covers 31 provinces, autonomous regions and municipalities, which will be assessed and held accountable in this regard rather than being assessed on GDP only like before. In this way, the objective system, assessing methods and reward and punishment mechanisms have been formed that reflect the requirements of ecological civilization, which have a direct bearing on the adjustment of Party and government leadership and on the selection and appointment, training and education, reward and punishment of the officials.

3. **Ecological civilization is new cultural ethics**

Ecological civilization emphasizes the combination of social and ecological ethics. The general belief in industrial civilization is that man is the subject with value while nature and all other lives are valueless objects, whereas ecological civilization holds that both man and nature are valuable subjects. The relation between man and nature restraints the relation between man and man, so properly handling the former is good for coordinating the latter and promoting peace and progress of human society.

Ecological civilization emphasizes the dialectical unity of ecological rights and obligations. Man has the right to enjoy wealth and pursue freedom and happiness, but only within the environmental carrying capacity. With ecological civilization, the gap between regions, between urban and rural areas and between different groups has to be narrowed. For instance, through the regional environmental compensation mechanism, developed regions should allocate more money to support ecological protection in less developed ones; an eco-compensation system in which the urban areas allocate more green subsidies to rural agriculture has been established; rich residents are also guided to pay a higher environmental fee to improve the environment and protect public facilities to benefit the disadvantaged groups suffering most from environmental deterioration; and a natural capital market has been formed to turn good natural environment into an economic advantage. On the global level, developed countries and high-consumption groups, as the main consumers of global resources, are obligated to bear greater ecological responsibilities.

Ecological civilization emphasizes common values. As far as man-environment relation is concerned, people living in agricultural civilization form living community for survival, those in industrial civilization form interest community in pursuit of wealth and resources, but those in ecological civilization form the community of a shared future for the harmonious co-existence of man and nature and the sustainable development of human society. The man-nature co-existence, society-ecology coordination and state-to-state sharing advocated by ecological civilization not only make its inherent principles, but also represent the common values of the community with a shared future for mankind.

3 Chief Characteristics of Ecological Civilization with Chinese Characteristics

1. **Establishing a green political agenda in the new era**

As the main social contradiction has changed in the new era, achieving socialist ecological civilization is an important governance philosophy of great strategic significance upheld by the CPC in its endeavor to meet people's aspiration for a good life. Ecological progress is a major political topic that concerns the Party's mission and purpose, a social issue concerning people's livelihood, and a millennial initiative concerning the sustainable development of the Chinese nation. There are specific strategic deployments for it. China has incorporated ecological progress in its Five-Sphere Integrated Plan for socialism with Chinese characteristics, put "green" into the five development concepts, made "beautiful China" a goal for building a great modern socialist country, and given priority to establishing and improving the ecological civilization system, thus outlining a whole plan covering objectives, principles and road map. There are systematic strategic measures to achieve it. China has waged a series of battles to safeguard our blue sky, lucid water and clean soil, and considered "pollution prevention and control" a tough battle in securing a decisive victory in building a moderately prosperous society in all respects. It has also implemented a string of strategies on all fronts, including poverty alleviation through ecological progress, Healthy China, Rural Revitalization, and building national experimental zone for ecological progress.

2. **Demonstrating the essence of socialism**

Socialism represents a better society superior to capitalism, and ecological civilization represents a more advanced one superior to industrial civilization. Since capitalism is the institutional origin of the global ecological and environmental crisis, a more advanced institution or civilization is inevitable and is the only way to go past and above industrial civilization as a whole. Moreover, socialism holds the moral bottom line of equity and justice. If making a group of people, some regions or a generation wealthy first sacrifices the environment of the majority, of other regions and of future generations, such environmental inequity is sure to exacerbate social inequity. That's why ecological civilization must coordinate the interests of different groups, regions and generations. It is not the simple summation, but an inherent unification of socialism and ecological progress; nor is it a fashionable outcome of humanists and environmentalists, but the inevitable choice given the actual national conditions. Ecological civilization and socialism will promote each other across the board.

3. **Centering on the people**

"Ecological civilization is a cause jointly undertaken and shared by the people. Building a beautiful China should be a voluntary action of all," said General Secretary Xi Jinping. Being fundamentally people-oriented, ecological civilization is set to "create a sound production and living environment for the people" and "continuously meet people's growing needs for a beautiful ecological environment." This is the most prominent, fundamental and universal livelihood campaign with people at its core. Therefore, when it comes to environmental issues, we must endow the people with the right to access to information on ecological or environmental crisis; the right to exercise supervision, particularly over government-funded public projects that impact the environment; and the right to participate, especially in making strategic decisions concerning environmental security. Only by relying on and mobilizing the people can we truly co-build an ecological civilization and share its benefits, making sure that everyone will participate in, promote and benefit from this undertaking.

4. **Reflecting an ideological consciousness**

The topic of ecological and environmental protection has become an ideological tool in the game of diverse interest groups. To win this ideological battle, we must firmly control the dominant power and discourse power in ecological progress and occupy the ideological high ground with advanced ecological concepts. To boost global economic recovery and cope with such challenges as climate change and energy and resource crisis after the global financial crisis in 2008, the international community, especially western developed countries, all put forth and implemented "green new deal," "green economy" and "green growth," triggering a new struggle for international discourse power. In advancing our ecological progress, China should foster a green concept that transcends class interests and political mindset, aim for equity, justice, democracy and the rule of law, and forge consensus, dispel divergence and realize solidarity to the largest extent. This not only relates to our ecological and environmental security, but also our ideological and cultural security.

5. **Promoting exchange and mutual learning between different civilizations**

On the one hand, ecological civilization provides a platform of theoretical fusion and practices for various socialist schools. Traditional socialist schools, such as scientific socialism, democratic socialism and a number of localized socialism, are criticizing the many flaws of capitalist industrial civilization and exploring for theories and practices of ecological civilization in the face of global ecological and environmental crisis. Emerging schools like ecological socialism and ecological Marxism have considerably supplemented traditional socialism by creating a vast room of imagination regarding how to respond to global ecological crisis. On the other hand, ecological civilization provides a platform of dialogue for different civilizations, religious beliefs and ideologies. Nowadays many countries are reflecting on traditional industrial civilization while seeking ecological wisdom from various ancient civilizations. They found that the Christian request for man to respect and protect biological groups should be made part of the natural law; Islamism takes man as part of nature, advocates moderate utilization of nature and opposes extravagance; and Hinduism values the inner spirit and initiated the "Chipko Movement" to practice its ecological ethics. Similar examples can be found from Chinese civilization as well. While there is the western mechanical outlook on nature, humanistic view of ethics and dualistic theory of evolution, the Chinese nation believes in a neutral and organic nature, harmony with nature, and the interaction of man and nature. It is such complementarity that has made ecological civilization a platform of exchanges among different civilizations beyond time and space. As such, it gives a stronger impetus to rational exchanges and dialogues in human society by overcoming barriers in system, race, belief and political ideology.

4 Path of Developing Ecological Civilization with Chinese Characteristics

1. **Comprehensive promotion based on institutional strengths**

As a new trend in the world, ecological civilization would have first appeared in developed countries, where ecological crisis broke out first. It hasn't for three reasons. First, western countries' potent technologies and capital mitigated their ecological crisis; second, the immense inertia of western industrial civilization would continue for a long time; and third, the capitalist west has continuously transferred ecological cost to less developed regions and hasn't become the first to step into ecological civilization as a whole. This created a chance for the Chinese nation to transform its development by leaps and bounds. There is no doubt that socialism with Chinese characteristics has more institutional strengths than capitalism in fundamentally resolving ecological issues. In its Five-Sphere Integrated Plan, ecological civilization is at the same time an independent sphere and one embedded in political, economic, cultural and social development. The five spheres serve as each other's foundation and conditions through mutual promotion. Economic development lays economic foundation, political development provides institutional guarantee, social development engages plural entities, and cultural development fosters a green atmosphere. In return, ecological progress forces the realization of green economic growth, scientific governance mechanism, harmonious and joint social governance, and cultural ethics that cultivate ecology-conscious citizens. Therefore, on the strategic level, socialist ecological progress with Chinese characteristics is not a phased or partial reform of industrial civilization, but an overall transcendence of traditional industrial civilization, especially capitalist industrial civilization. On the institutional level, it neither upholds a strict vertical and lateral division of government duties nor leaves all government functions to the market, but deals with major tasks by pooling together resources and forces in a coordinated way and tackles difficulties through strict Party governance, making sure the green political agenda is followed through. On the cultural and ethical level, ecological civilization doesn't equal environmental technologies, but is an overall campaign aligned with national strategies. It conducts targeted poverty alleviation in an ecological way, ensures environmental equity and justice, and manifests socialist principles and superiority.

2. **Systematic development by intensifying state governance capability**

We should establish and improve the socialist ecological civilization system with Chinese characteristics as required by General Secretary Xi Jinping. First, we should create an ecological cultural system to enhance the whole society's scientific and moral capability of ecological civilization under the guidance of ecological values. Second, we should create an ecological economic system that incorporates environmental impact assessment in the formulation and implementation of industrial development strategies and regional and local development plans, thus integrating industry and ecology into an organic whole. Third, we should put in place the ecological target and responsibility system, which sets the bottom-line environmental requirements for different regions, river basins, sea areas and stages with the goal of improving ecological quality. Fourth, we should establish an ecological institution, whereby the legislative, judicial and executive sectors of ecological environment will be more coordinated and complementary through the modernization of our governance system and capability. Fifth, we should foster an ecological security system that, focused on sound ecosystem circulation and effective prevention and control of environmental risks, ensures the regional or national natural foundation to guarantee human survival and development and sustainable economic and social development, and maintain the integrity, stability and functionality of ecosystems.

3. **Engagement in global ecological governance as a major and responsible country**

General Secretary Xi Jinping said "man and nature are in a community of life" and called for building "a community with a shared future for mankind." That is the latest development of the Marxist community theory in China, which points out the consistency between ecological civilization and the human community of shared future and orients the former directly to the latter. A basic fact is that the natural existence of mountains, water bodies, forests, farmland, lakes and grassland make the precondition and foundation for building the community with a shared future for mankind. Only through the community of life featured by man-nature co-existence can the human community of a shared future sustain, can we achieve the co-existence between man and man, nation and nation, state and state. China has honored its commitment to being a "participant, contributor and leader in global ecological progress" with concrete actions. Aiming for the fundamental and long-term common interests of the whole world, it insists on taking the path of co-existence, inclusiveness and win-win cooperation, blazing out a green road to the community with a shared future for mankind. We have continuously confirmed and performed our ecological responsibilities as a major and responsible country. China took the initiative to issue the *National Plan on Implementation of the 2030 Agenda for Sustainable Development*, implemented the *National Plan for Climate Change Response (2014–2020)*, and stepped up the efforts to boost ecological progress and global green cooperation after the US withdrew from the *Paris Agreement.* We have also implemented the Belt and Road Initiative (BRI) guided by ecological and green concepts. In countries and regions along the Belt and Road, we have actively carried out multilateral or bilateral dialogues, exchanges and collaboration on ecological environmental protection, strengthened information services in that area, and promoted environmental standard, technical and industrial cooperation. We have provided the Chinese plan for global ecological and environmental governance, set up the platform to provide public products for that purpose, and shared China's green wisdom and experience with the international community, in a bid to reinforce the recognition of common interests and emotional identification in creating the community with a shared future for mankind.

Traditional Ecological Awareness in Confucianism

Shucong Zhao

"Studying Confucius and Confucianism is an important way to understand the nature of the Chinese nation and the history of the spiritual world of modern Chinese."

Confucianism is the prevalent and underlying thought of the Chinese nation which still plays an important role in all aspects of social life today. The rich ecological awareness contained in this broad and profound thought can be called traditional Confucian ecological awareness, which has contributed significantly to the continuation of Chinese civilization over thousands of years. Making a systematic and in-depth study of Confucian ecological awareness is of great importance in that it will enlighten us in what kind of socialist ecological civilization to build and how to build it, guide us in practicing the Xi Jinping Thought on Ecological Civilization with Chinese characteristics for a new era, and reinforce our cultural confidence.

According to Confucianism, the human society exists in a natural environment, from which it receives all the materials required to make living necessities, and humanity itself is part of this environment. Based on this philosophy, Confucianism has developed a range of ecological principles including understanding and respecting nature, protecting and making moderate use of it, planting trees for human benefits and exercising restraint, people-centered ecological education, and affinity and harmony between man and nature.

1 Understanding and Respecting Nature

"The Master said, 'Young friends, why not study poetry. Odes may help spur aspirations, increase observation, teach you how to get along with others and voice your complaints. Near at hand, they can guide you to serve your fathers; more distantly, they can guide you to serve a ruler—and you can learn the names of many birds and

S. Zhao (✉)
Chinese Society of Forestry, Beijing, China

J. Pan et al. (eds.), *Beautiful China: 70 Years Since 1949 and 70 People's Views on Eco-civilization Construction*, https://doi.org/10.1007/978-981-33-6742-5_40

beasts, trees and grass." ("Yanghuo" in *the Analects of Confucius*) The Confucian ecological awareness is based on understanding nature. *The Book of Changes* states "there is first heaven and earth, then the various species, then males and females," and Xu Shen's *Origin of Chinese Characters* holds that "the Way gives birth to heaven and earth and all living things." It is the Confucian belief that the Yin and the Yang are symbiotic while the five elements subdue each other, and man derives from nature and constitutes a part of it. Such a perception could hardly arouse the ambition to conquer nature, but would generate awe for nature. "Yanghuo" in *The Analects of Confucius* also says "Does heaven speak? Yet the seasons pass and the creatures of the world are born," which reveals Confucius' awe for the laws of nature, which is the prerequisite and basic rule for human activities. Mencius proposed that the Way of Heaven is the ultimate origin and basis of the Way of Man. "Liqi" in *The Book of Rites* states "Rite shall comply with the course of time, the properties on land, the wishes of spirits and the thoughts of man in order to governance the universe." Li Bai, a renowned poet of the Tang Dynasty, wrote "heaven and earth don't speak, but the four seasons pass and the living things grow without fail," which also expressed his awe for the irresistible laws of nature. The industrial revolution and sci-tech progress both rely on our knowledge of nature, so does ecological civilization. The Confucian perception of nature has scientific elements in it that make man stand in awe of nature rather than attempt to conquer it. Engels wrote in his *Dialectics of Nature*, "Let us not, however, flatter ourselves overmuch on account of human conquest over nature. For each such conquest takes its revenge on us. Each of them, it is true, has in the first place the consequences on which we counted, but in the second and third places it has quite different, unforeseen effects which only too often cancel out the first." This is a manifestation of his awe of nature.

2 Protecting and Making Moderate Use of Nature

Ecological civilization, at the bottom, is a system of man's understanding of the relation between their own behaviors and the natural environment. It is recorded in *The Analects of Confucius* that Confucius only fishes but never nets; only shoots flying birds in the day but never resting ones at night as they may be breeding babies. *Mencius* records more explicit remarks that "growing forests should not be logged so that grass and trees can grow; fishing should be banned when fishes and turtles are growing; fawns and bird eggs should not be hunted or broken so that young birds and beasts can grow." ("Lianghuiwang I" in *Mencius*). *Xunzi* also states that "A wise lord would implement such a system that axes are banned in the forest when grass and trees are blooming or bearing fruits, and nets and poison are banned in rivers and lakes when fishes, crocodiles, loaches and eels are spawning, so they can live and grow. Plow in spring and summer, harvest in autumn and store up in winter, strictly ban fishing in ponds and rivers, and log and plant trees all at the right time, so grain and fish will abound, mountains will not go bare, and people have more than enough to eat and use." ("Wangzhi" in *Xunzi*) Even when our ancestors had to take materials from nature for survival, the Confucian ecological ideology contained many precepts on protecting nature. This view on natural protection has made immense contributions to the continuity and inheritance of Chinese civilization. Engels also wrote in his *Dialectics of Nature* that "The people who, in Mesopotamia, Greece, Asia Minor, and elsewhere, destroyed the forests to obtain cultivable land, never dreamed that they were laying the basis for the present devastated condition of these countries, by removing along with the forests the collecting centers and reservoirs of moisture. When, on the southern slopes of the mountains, the Italians of the Alps used up the pine forests so carefully cherished on the northern slopes, they had no inkling that by doing so they were cutting at the roots of the dairy industry in their region; they had still less inkling that they were thereby depriving their mountain springs of water for the greater part of the year, with the effect that these would be able to pour still more furious flood torrents on the plains during the rainy seasons." Although the Chinese civilization had once in a while made excessive use of natural resources during its long development, the consciousness to protect and moderately utilize nature has always been dominant thanks to the mainstream position of Confucianism. In Confucius' opinion, destroying nature and ecology through excessive felling, fishing and hunting is also a breach of filial piety. His student Gao Chai proposed to not "kill emerging insects from hibernation or break growing plants," and Zeng Shen advocated "logging trees and killing beasts in the right season." These ideas, being prevalent in ancient Chinese society, could be found in large volumes of imperial edicts, village rules and pacts, and family precepts. Even today, environmental damage, unbridled forest logging, and air, water and land pollution remain the most detested behaviors. General Secretary Xi Jinping's remark that "man can utilize and reform nature, but more importantly, it should take care of it and never put itself above it as humanity is essentially part of nature" is consistent with the Confucian ecological philosophy.

3 Planting Trees and Exercising Restraint

There is a juniper at the Temple of Confucius in Qufu, Shandong province that is allegedly planted by Confucius himself and is taken as a symbol of the ancient sage's advocacy for afforestation. *Red Star over China* describes how Chairman Mao felt when he visited this tree in Qufu during a trip in the spring of 1920. "I stopped at Qufu and visited Confucius' grave, I saw the small stream where Confucius' disciplines bathed their feet and the little town where the sage lived as a child. He is supposed to have planted a famous tree near the historic temple dedicated to him, and I saw that. I also stopped by the river where Yan Hui, one of Confucius' famous disciples, had once lived, and I saw the birthplace of Mencius." The Forest of Confucius is the most enduring forest on a family cemetery in the world, which houses more than 100,000 valuable trees and 342 types of herbaceous plants today, becoming a treasure house of ancient and peculiar plants. It is recorded in Li Daoyuan's *Commentary on the Waterways Classic* that "after Confucius passed away, his disciplines planted rare plants from various places, many of which the Shandong people cannot name through generations." A prominent example was the Pistacia chinensis personally planted by Zi Gong, which has been taken as the model for protecting nature and respecting teacher even though only a stump is left today. General Secretary Xi Jinping pointed out, "afforestation is an important way to realize blue sky, green land and lucid water. It is the most inclusive livelihood campaign." "In comparison with our goal of building a moderately prosperous society in all respect and with the people's aspiration for a beautiful ecological environment, we still have a lot to do on ecological progress and environmental deterioration remains serious. The shortage of forests and greening is a pressing issue that has to be addressed." "Afforestation by one generation will benefit many generations to come. Let's roll up our sleeves and carry on this cause year after year, generation after generation." Mr. Liang Xi, who has dedicated his life to China's forestry development, dreams of a day when "every mountain is green, all rivers and lakes are clean, flowers are blooming all year long and birds are singing in every valley" across the country. Having given top priority to afforestation since its founding China has planted and preserved the largest volume of artificial forests worldwide, a perfect illustration of the Chinese genes of "planting trees for human benefits."

Self-restraint is not only an important aspect of Confucianism, but also Confucius' norm on human protection of nature. At present, we are faced with tremendous ecological and environmental pressure with tightening resource limitations, which is all the more reason why we should exercise restraint and save resources. Confucius' remark that "the whole world will call you benevolent when one day you can restrain yourself and do everything according to rite" highlights the necessity for man to restrain themselves, an ability also demonstrated in the fact that Confucius "only fishes but never nets; only shoots flying birds in the day but never resting ones at night as they may be breeding babies." As General Secretary Xi Jinping noted, "successful ecological and environmental protection eventually depends on economic structure and development mode." "We cannot take the old American or European path in building a modern country; that way several more earths wouldn't provide enough resources to meet our consumption." "Resource conservation is the fundamental environmental policy. Most environmental damages are caused by excessive exploitation and extensive use of resources. There won't be any fish left if we drain the pond for fish." That's the underlying principle for ecological progress in contemporary times.

4 People-Centered Ecological Education

Ecological progress is both a goal and a process, especially the process of people receiving ecological and cultural education. Confucius had about 3,000 disciples, but only 72 turned out sages. It is said that Confucius liked reading and teaching under the gingko tree, so his descendants called the place "gingko rostrum." In front of the ritual hall of the Temple of Confucius in Qufu, a male and a female gingko trees planted in the Song Dynasty are still standing tall in memory of the ancient teacher, and two rows of tall and strong gingko trees are also growing in front of the family temple of the southern sect of Confucian family in Quzhou. In Anqiu, Shandong where Confucius' son-in-law Gongye Chang studied, a male and a female gingko saplings are growing tall into the sky, which are said to be brought there by Confucius when he visited his son-in-law who then planted the trees personally. Speaking about the relation between education and environment, Confucius said, "Living with the noble is like living in a room filled with fragrant orchid. Over time, you won't notice the fragrance because you've become part of it. Living with the despicable is like living in a salted fish store. Over time, you won't notice the stink because you've become part of it." (*Family Instructions of Confucius* by Wang Su). He also compared pine and cypress to human tenacity and integrity by saying "the coldest moment in winter withers all other plants but pines and cypresses." In ancient China, prestigious academies were often founded in picturesque forests as our ancestors had long realized an appealing environment was in itself good education.

The Analects of Confucius records, "When the stables burnt, Confucius returned from court asking, 'Was anyone hurt?' He did not ask after the horses." The ultimate purpose of Confucian ecological awareness falls on man—to promote human development and guide human activities. Believing that nature provided for human existence, Confucius called for protecting it, not draining the pond for fish, and allowing nature to rest and recover, with the ultimate goal of serving human development. *Mister Lv's Spring and Autumn Annals* states, "how can you not get plenty of fish if you drain the pond? But there will be no fish next year. How can you not get games if you burn the forest? But there will be no game next year." The purpose of protecting nature is to sustain human development. The report at the 19th CPC National Congress points out, "Man and nature form a community of life… Any harm we inflict on nature will eventually return to haunt us." General Secretary Xi Jinping also noted, "What we work for is harmony between man and nature, and economy and society. In plain words, we strive for lucid waters and lush mountains that can also be invaluable assets." This in essence means putting people first.

5 Affinity and Harmony Between Man and Nature

Daide's Book of Rites records Confucius as saying "not breaking growing plants can be called benevolence." Such a protective attitude toward plants is closely related with principal Confucian moralities, such as filial piety, forgiveness and benevolence, implying that the attitude toward nature is inseparable from the attitude toward man. According to Confucianism, man and nature essentially have many in common. Wang Yangming wrote in his *Big Studies* that "the great man identifies with all creations…. It hurts him to hear birds and beasts wail in fear as his benevolence goes to them; birds and beasts can feel, but it hurts him to see plants being felled as his benevolence goes to them too; plants have life, but it also hurts him to watch tiles and rocks being destroyed as his benevolence goes to them as well." The "big man" refers to those who are noble and benevolent, who are conscious of man and nature in one. The word "benevolence," as the crystallization of Confucius' ecological thoughts, appears in *The Analects of Confucius* more than 100 times. On the one hand, Confucius told a person's character and integrity through his attitude toward nature, which was why he said "the benevolent find pleasure in hills; the wise in water." On the other hand, he took achievements in natural preservation as the indicator of "benevolent governance." Confucius went to inspect the performance of Zilu, his top disciple who had been serving as an official in Pu, State of Wei, for three years. On his way there, Confucius couldn't help exclaiming "bravo" three times because of what he saw – firm houses and lush trees along the way, which implied the people there were hard-working without loafing around. He also hailed the "benevolence" in trees, forests and nature. "Of things that follow laws of earth, it is only pine and cypress that are brightly green in winter as in summer. Of people who follow the rule of Heaven, only Yao and Shun, as the head of all creatures, are the greatest." Zhu Xi also wrote in his *On Benevolence*, "Heaven and earth takes creatures as the center, Human beings with benevolence is the center of all creatures." ecological progress in contemporary times is the maximal manifestation of "benevolence." The report of the 19th CPC National Congress states, "The modernization that we pursue is one characterized by harmonious coexistence between man and nature. In addition to creating more material and cultural wealth to meet people's ever-increasing needs for a better life, we need also to provide more quality ecological goods to meet people's ever-growing demands for a beautiful environment. We should, acting on the principles of prioritizing resource conservation and environmental protection and letting nature restore itself, develop spatial layouts, industrial structures, and ways of work and life that help conserve resources and protect the environment. With this, we can restore the serenity, harmony, and beauty of nature." As long as we adhere to these principles, the grand goal of building a prosperous, strong, democratic, harmonious and beautiful socialist modern country will definitely become reality.

Promoting Ecological Progress with Big Vision, Broad Coordination and Continued Efforts

Guangyao Zhu

Promoting ecological progress concerns the people's well-being and China's future. On the strategic height of the Five-Sphere Integrated Plan for socialism with Chinese characteristics, the 18th CPC National Congress made strategic deployments regarding ecological civilization and the great rejuvenation of the Chinese nation. Since then, the new generation of central leadership with Xi Jinping at the core has earnestly advanced ecological progress with firmer resolve and more pragmatic measures. They have given new connotations to and made new requirements on the cause of pushing the Chinese nation to the new era of ecological civilization and realizing the dream of a beautiful China.

Developing ecological civilization is a great undertaking that requires both theoretical and practical innovations. In the process, we must form the concept of "big vision, broad coordination and continued efforts" and actively practice and promote this undertaking, so as to well play our role as a think tank, service provider and bond in pushing our ecological progress forward.

1 We Must Have a Big Vision in Exploring Ecological Civilization

Ecological civilization is the ecological manifestation of social civilization. It is not only a complex social program, but also a systematic multi-disciplinary science. It not only needs top-level design, but also powerful theoretical support. Therefore, when studying and deepening ecological civilization, we must adopt a broad vision

---

The original article was published on *Ecological Civilization in China* in March 2012. Deletions and revisions are made when it is collected in this book.

G. Zhu (✉)
Former State Environmental Protection Administration, Beijing, China

J. Pan et al. (eds.), *Beautiful China: 70 Years Since 1949 and 70 People's Views on Eco-civilization Construction*, https://doi.org/10.1007/978-981-33-6742-5_41

encompassing the relations between time and space, economy and society, science and morality, and man and nature, and stand at the strategic height of sustainable development. On that basis, we will keep exploring and better understanding the significance, profundity and path of realization of ecological civilization, and lay a solid theoretical foundation for it.

1. **Understanding ecological civilization from the historical perspective of national rejuvenation**

Any form of civilization in human history is just a historical phenomenon and process that will be replaced by a new civilization. Ecological civilization is a continuation/elimination of industrial civilization and a higher state of civilization after the primitive, agricultural and industrial civilization. The 300-year-long industrial civilization of developed countries has created resplendent material wealth, but also brought with it unprecedented plights and challenges to the mankind, such as aggravated conflicts between man and nature, and unseen resource shortage, environmental deterioration and ecological crisis. To lift the mankind out of this crisis, industrial civilization has to be profoundly reformed or, in other words, replaced by ecological civilization. This is an irreversible and unavoidable choice in the process of human development.

Throughout its splendid history of 5000 years, the Chinese nation passed a long period of agricultural civilization but never had an industrial civilization in the real sense, and it suffered tremendously from the plundering and bullying by industrialized countries in modern times. National rejuvenation is the aspiration of generations of Chinese. Our efforts for ecological progress must be based on our national conditions and follow the path of socialism with Chinese characteristics. Environmental protection undertaken by one generation will benefit many generations to come. As General Secretary Xi Jinping pointed out at the 6th group study of the Political Bureau of the CPC Central Committee, driven by a strong sense of responsibility towards our country, our people, our future generations, we must devote ourselves to the promotion of ecological progress with bitter resolve. We will make great efforts to move toward the new era of socialist ecological civilization, create a favorable production and living environment for the people, and write a new chapter of the Chinese Dream in the new era.

2. **Promoting ecological progress following the Five-Sphere Integrated Plan**

The 18th CPC National Congress included ecological progress in the Five-Sphere Integrated Plan, and vowed to energetically advance ecological progress, build a beautiful China, and achieve national sustainability. This demonstrates the CPC's deepened understanding of the laws concerning socialism with Chinese characteristics, as well as its firm resolve to intensify ecological progress. The Five-Sphere Integrated Plan has presented to us a better strategy, more definite goals, richer connotations and a broader path for developing socialism with Chinese characteristics.

In the five-sphere overall layout, economic progress is the center and foundation, political progress the direction and guarantee, cultural progress the soul and vein, social progress the support and objective, and ecological progress the cornerstone and condition. These five aspects complement and promote each other and constitute the comprehensive blueprint for socialism with Chinese characteristics.

China and the world need each other for their own development, especially sustainable development. As the CPC Central Committee has proposed with vision to strongly boost ecological progress, China will work hard to address global dilemmas like climate change, which is of great significance for resolving the serious ecological crisis endangering the security of human society and realizing sustainable development throughout the world, especially achieving harmony between man and nature. As stated in the report of the 18th CPC National Congress, in striving for ecological progress, we will create a sound working and living environment for the people, and contribute our share to global ecological security.

2 We Must Exercise Broad Coordination in Advancing Ecological Progress

Promoting ecological civilization is a new strategic task involving all aspects of the society. Therefore, we need to pool together the Party and government at all levels, all social strata, all walks of life and international forces to work together against obstacles and form a synergy to continuously enhance the drive, cohesion and combat force for promoting this great cause.

1. **Working in a coordinated way to form a "united front" of ecological progress**

Given the current institutional situation, an ecological progress committee or leading group can be formed on the top level to exercise unified leadership and coordination and step up inspection and promotion. An interactive mechanism has been established in some places, whereby the local CPC committee and government take the lead, local People's Congress and People's Political Consultative Congress supervise the work, relevant departments take actions in coordination, and the whole society takes a part. Such a mechanism should be further improved. Ecological progress concerns all aspects of the "man-society-nature" cycle and makes an integral part of the five-sphere layout. If we only focus on one specific field, department or task, we will lose the whole picture and commit omissions. It is such a complicated systematic program that we can only push it forward in an orderly manner by coordinating all parties and rationally allocating social resources under the lead of the CPC committees and governments of all levels. Departments, industries and enterprises are important forces promoting ecological progress. We should fully mobilize and utilize different working conditions and foundations, exercise strengths, and stage a "chorus" of ecological efforts. To that end, supportive policies and incentive mechanisms are needed. Measures such as fiscal input, ecological compensation and tax and fee concessions should be adopted to encourage local governments, businesses and industries to consciously perform their social responsibilities and take an active part in ecological progress. Relevant authorities can set up a dedicated unit to coordinate and guide their ecological efforts toward inter-departmental and whole-of-society interaction.

2. **Establishing coordinated institutions for ecological progress**

A scientific and standard performance assessment system is an important basis and institutional guarantee for boosting ecological progress. Since the 18th CPC National Congress, relevant central authorities and some local departments, as per the requirement of ecological civilization, have begun to incorporate ecological indicators in the assessment system for local CPC Committees and governments. An accountability system has also been put in place to hold accountable the units and individuals whose blind decisions regardless of environmental carrying capacity have caused grave consequences. Considering the differences in economic and social development across the country, we should establish and improve sound assessment methods and rewarding and punishing mechanisms based on resource consumption, environmental damages and ecological benefits. According to the general layout and requirement of national land resources and space development, systems concerning resource and environmental conservation, paid use of resources, environmental accountability and damage claims, as well as environmental laws and regulations should all be perfected. Attention should also be paid to inter-regional cooperation. We should create solid systems for major project consultation, joint law enforcement and information sharing and disclosure among regions and prevent contradiction and mutual obstruction among the policies and rules released by various regions, industries and departments.

3. **Conducting coordinated publicity and mobilization to foster a powerful social atmosphere in favor of ecological progress**

Strengthening ecological publicity and education and promoting ecological civilization is the first step to boost ecological progress. Publicity and education are intended to lay the social, moral and cultural foundation for ecological campaign and to transform the way of production and life. We need to make full use of all publicity channels and create new approaches to popularize relevant knowledge, enhance people's awareness of conservation, environmental protection and ecology, and accelerate the shift to a simple and moderate, green and low-carbon, and healthy and culturally advanced lifestyle, making resource and environmental conservation the mainstream value in the society. Upholding an ecological culture and building an ecological cultural system with Chinese characteristics provides the ideological assurance for elevating the people's environmental ethics and levels of morality and consolidating the social basis for ecological progress. New thoughts and approaches must be explored in publicity and education carried out by social groups and organizations, which shall strive to disseminate and practice ecological ideas, foster ecological morality and promote environmental culture in their work.

4. **Coordinating with international forces to create a new situation of win–win interactions**

As globalization is deepened day by day, ecological protection and environmental governance are no longer the work of one country or one organization. It takes the united efforts of all countries in the world to bring about a healthy ecosystem on earth and transform the human civilization. As the world's largest developing country, China should combine its ecological efforts with global eco-environmental issues that mankind is most concerned with. On the principles of common but differentiated responsibilities, equity and respective capability, it should engage in extensive cooperation with the international community to actively cope with global climate change and jointly advance sustainability.

3 We Must Keep Up Ecological Efforts Across Administrations and Generations

Promoting ecological civilization and building a beautiful China is a long-term and arduous task that cannot be achieved overnight. We must clearly and fully understand how onerous, complicated and prolonged this task is, so that the whole Party and whole nation will make consistent efforts toward it without fail. Every administration, every generation should work hard and pass the baton on to the successors, and forces both inside and outside China should be mobilized toward the goal of lucid waters and lush mountains. That's the only way to achieve a beautiful China characterized by advanced production, affluent life, and a sound ecology of blue sky, green land and clear water.

To make continued efforts for ecological civilization, what's most important is that CPC committees and governments of all levels should maintain continuity in their work. They must undertake ecological civilization as a great cause from term to term, and take no unwise or short-sighted actions for immediate GDP growth. Rather, they should truly take people's welfare and national rise as their own responsibility, and view a beautiful ecological environment as the greatest achievement. Resting on the ancestors' laurels and leaving trouble to the offspring for personal performance and immediate benefits is absolutely out of the question. General Secretary Xi Jinping stressed the importance of holding the ecological red line, adding that any overstepping on environmental protection would be punished. This is the new, higher requirement made by the CPC Central Committee on the whole Party and whole nation regarding environmental conservation. In light of the implementation of the *National Plan for Main Functional Zones* and the *Opinions of the State Council on Key Work of Strengthening Environmental Protection*, we should carry out studies and demonstrations to release the implementing plans as soon as possible. To hold the ecological red line, ecological issues must be addressed, inspected and dealt with in a resolute, open and transparent manner. Only when relevant people and units are held accountable in strict accordance with the law can we build a solid defense line against ecological violations. It is necessary to educate the whole society, especially the young people, on ecological civilization to raise their awareness thereof, so that it will become their conscientious action and the historical responsibility for ecological progress will be passed down from one generation to another.

To make continued efforts for ecological civilization, close attention must be paid to the overall plan or top-level design of each local area and each industry. Developing ecological civilization is an arduous, long-term initiative that spans across generations, to which the key lies in practice, innovation and persistence. China is currently at an important period of economic and social transition, when the factors affecting ecological efforts are getting increasingly complex, deep-rooted conflicts left from earlier times are coming to the fore, and people are having higher demands on ecological quality. To avoid short-sighted actions that only treat superficial symptoms through stopgap measures without taking out the root cause and resolve piled-up problems at the source, we must make overall plans on a strategic level and roll out a top-level design that lists the basic strategy, tasks and goals, policy measures and administrative system conducive to ecological progress and the Five-sphere Integrated Plan. This design should combine the general deployments and strategic objectives decided by the CPC Central Committee with the specific conditions in each region and industry, and combine periodic goals with long-term plans. It should also be executed step by step with targeted guidance. The overall plan should be decided and enacted by the standing committee of the People's Congress of the same level, so that relevant work will be carried on consistently and won't be derailed either because the top leaders have new priorities or because the leadership has been changed.

To make continued efforts for ecological civilization, we must carry forward the fine Chinese traditions, draw on the outstanding achievements of foreign cultures, and promote modern ecological culture. Developing ecological civilization is both the transformation and the inheritance of civilization. The Chinese nation is the only one in the world that has survived and thrived as a state for thousands of years with the same culture, language and origin. Its incisive ecological wisdom is ingrained in our culture, ethics and morality. The Confucian concept of "unification between nature and man," the Taoist idea to follow the laws of nature and the Buddhist belief in "equality of all creatures" all manifest the ecological philosophy of harmony between man and nature, which continues to shed light on our ecological work today. Carrying the characteristics of individual nations as well as the common traits of all mankind, ecological civilization—and its ultimate goals—cannot be realized in or by one country or region, but requires the fundamental changeover from industrial to ecological civilization throughout the world. Therefore, it is necessary to absorb the cultural nourishment from different nations and launch a global relay on the cultural level.

At the moment, sustainable development is becoming a national and international hot spot and has turned into a torrential, irreversible trend in some places, and ecological civilization has become the most sonorous slogan across the Chinese land, giving us new and higher requirements with its new situations and new tasks. We must further sharpen our consciousnesses of maintaining political integrity, thinking in terms of the big picture and fulfilling our mission, and bear firmly in mind the "big vision, broad coordination and continued efforts." We need to live out the spirit of "solidarity, enterprising and dedication," aim higher and make greater contributions to ecological progress.

Mankind Will Awaken to Ecological Civilization

Xinsheng Zhang

1 The Appearance of a New Civilization Form is First and Foremost Marked by Advanced Productive Force

Development without protection is fishing by draining the pond; protection without development is seeking fish from trees. As ecological civilization is a new form of civilization marking the orientation of the twenty-first century, its primary criterion is without doubt the advanced productive force.

1. **The new round of industrial revolution is premised on ecological priority and green development**

As ecological civilization is a new civilization form after agricultural and nomadic civilizations, industrial revolution and industrial civilization, it features a productive force stemming from industrial civilization but more advanced than it. What's critical now is that we should acutely sense the historical opportunities presented by the new round of technological and industrial revolution, and seize the opportunities for industrial upgrade and transformation that come along as we move from automation to a series of major sci-tech advancements such as intelligent applications, the Internet, Internet of Things, big data, cloud computing, 5G and gene technology. As a matter of fact, we are already moving into a new civilization form that will eventually materialize through a new technological and industrial revolution. As we awaken to the concept of ecological civilization, which further liberates our mind, the unleashing of productive forces will inevitably be premised on no other than ecological priority and green development.

The world is going through profound changes never seen in a century, and China is standing at an important period of strategic opportunities when it can make a huge difference. A new round or the third industrial evolution has emerged, bringing

X. Zhang (✉)
ECO Forum Global Guiyang, Guiyang, China

J. Pan et al. (eds.), *Beautiful China: 70 Years Since 1949 and 70 People's Views on Eco-civilization Construction*, https://doi.org/10.1007/978-981-33-6742-5_42

with it tremendous historical opportunities. There are two important norms to define an "industrial revolution," namely revolutionary changes in energy and information communication. The *Paris Agreement* of 2015 commits to holding the increase in the global average temperature to well below 2 °C above pre-industrial levels by the middle of the twenty-first century, for the achievement of which the world will comprehensively stop the use of fossil energies around 2050 and promote the global use of renewable energies. The three basic scientific discoveries of the twentieth century will lead to unimaginable inventions and creations, particularly subversive technologies, in the first half of the twenty-first century. Revolution in information and communication technologies is self-evident in comparison to the previous two industrial revolutions, as the Internet, IoT, big data, cloud computing, AI and 5G are dramatically changing the way we produce and live worldwide. As to energy, the first industrial revolution was powered by steam, the second one by electricity. It seems that the transformation from automation to intelligentization, which some western organizations call a new round of industrial revolution, is too short in time and too narrow in scope for that term. Besides, we believe there is one important criterion for the "third industrial revolution" that the west has neglected or internationally avoided. The first two industrial revolutions were led by developed European and American countries while the developing countries were just followers, some even being reduced to colonies and suffering from serious exploitation. In contrast, the new industrial revolution is led by both oriental and western countries, with developing countries and emerging economies displaying robust momentum and vitality and engaging in and leading the movement even more deeply.

Therefore, the so-called third industrial revolution should be more objectively defined in four aspects—green ecology, energy, information and communication technology, and multi-polarization of global leaders. China has not only revealed first that ecological civilization is the direction of the twenty-first century, but also successively come to the fore in all those four aspects. How to grab the historical opportunities during this civilization transition driven by sci-tech revolution and stay in a leading position—this is vital for China in implementing the strategy and path of ecological civilization.

2. **Circular economy and sharing economy are important roads to change the production mode and consumption pattern in ecological civilization**

The report of the 18th CPC National Congress pointed out, "to move toward ecological civilization and achieve green transformation and development, we must change of production mode and consumption pattern." To put it briefly, production mode and consumption pattern in ecological civilization should make five important shifts—from fossil fuel to renewable energy, from economic capital to natural capital, from resource exploitation to sustainable utilization, from carbon-intensive to carbon-sensitive development, and fundamentally from the traditional development mode featuring massive input, massive use of resources and massive consumption to the production mode and consumption pattern featuring low energy consumption, circularity and renewability. The green shift and green growth of economy will bring tangible and inestimable benefits. We must change our old way of thinking and the old development mode and shift to green economy and blue economy that gives priority to ecology—that's the only way to achieve sustainable development.

In green economic mode, carbon emission is reduced, natural resources are rationally used and environmental pollution is addressed. The use of renewable energy promotes the development of green energies, improves infrastructure and gives a boost to green food and eco-friendly agriculture. The financial sector has turned green too. Intelligent and innovative eco-cities have appeared, urban–rural interaction is promoted, trade and traffic infrastructure is bettered, Internet and big data reform display evident advantages, and green and clean technologies are applied extensively.

All industries should deepen their understanding of green transformation, press ahead toward green economy, green society and eco-city, and eventually realize ecological civilization. Public participation is very important in this process.

It is critical for China, a major manufacturing power, to practice circular economy at this stage, which is, inter alia, the most common and prominent way for us to change the production mode given the current national conditions. Much has been achieved in this field in the country. The basic guiding principle for circular economy is to "minimize the input of natural resources at the front end of economic activities while strengthening the research and development of technologies for back-end treatment, so as to reuse wastes and minimize waste discharge to the point of zero." Following ecological laws, circular economy is guided by the notion of minimization, reuse and recycling and by the principle of low consumption, low emission and high efficiency, which is obviously different from the traditional productive mode.

Mohandas Karamchand Gandhi, the holy figure of India, once said, "The world has enough for everyone's need, enough for anyone's greed." The capitalist system is bound to create false or distorted demand to prompt irrational, endless excessive consumption in order to avoid or delay its periodic crisis while pursuing maximal profits. One needs only one bed to sleep on no matter how big the house is, and only three meals a day no matter how wealthy he is. Natural resources on earth and the carrying capacity of the ecological environment are limited. The expanding economic scale driven by increasing population and technological progress will approach or even exceed the limit of earth's carrying capacity. In fact, the call for the society to shift its consumption pattern is more a matter of values.

In recent years, explorations have been made both in China and across the world for the strategy and path of sharing economy, leading a revolutionary change of consumption pattern as more and more people, especially the young, prefer the right of using and sharing to ownership. By revitalizing idle resources, sharing economy realizes optimal allocation of social resources—that's its core value and chief objective. As a result, it is an important means to addressing the unbalanced and inadequate development, meeting "people's growing demand for a better life," and realizing the coordinated development of the economic, social and natural systems. It won't be a plain sailing for something new to stand out, but the wave of change of production mode and consumption pattern initiated by sharing economy has primarily shown its prospects and vitality.

3. **Natural capital holds the key to turning "lucid waters and lush mountains" into "valuable assets"**

Plenty of meaningful practices of various forms have been carried out to make "valuable assets" out of "lucid waters and lush mountains." In doing so, a fundamental, inevitable question is how to turn natural resources into natural assets and then to natural capital—that's a cutting-edge focus of international studies and explorations too.

The traditional mode of human development has generated enormous ecological debts, so ecological restoration is a crucial task as we work toward green transformation and development now. Yet it would be extremely costly to make up for past ecological damages, much less to recover the entire ecological system on earth, which would require an astronomical amount of capital and resources. If natural resources cannot be translated into natural capital, our ecological efforts would be of little avail no matter how hard we try.

The vision of International Union for Conservation of Nature (IUCN) is "a just world that values and conserves nature and natural resources." This stresses the endowment of values upon nature and natural resources. We used to think that nature doesn't have a price, but it does; what's difficult is how to price it. For a long time in history, the intake of natural resources by agricultural and nomadic civilizations was well within the natural limit, so man and nature could co-exist in peace. But when it came to commercial civilization, the wealth created by industrial revolution was unprecedented, so were the plunder of natural resources and the environmental price paid for it. We are carrying a huge natural resource debt, so much so that the ecological bank has almost reached the edge of bankruptcy due to the deficit of environmental and natural resources, and the regulatory capabilities of earth's four ecological systems, be it the land or the sea, is approaching an inflection point. Putting a value on natural resources is a strategic and basic step to turn the tide, which, however, isn't easy as we must tackle several major difficulties, namely the measurement, accounting and trading of ecological capital. There are several points to be considered regarding the accounting of natural capital, such as designating a specific, common unit for different types of natural capital, collecting and calculating data with the help of Internet, IoT and big data technologies to determine the ownership of natural capital, and delimiting an area where the natural capital can be priced and traded in the market. On that basis, natural capital, like material capital, can participate in all kinds of economic activities in market economy.

Developing calculating rules that conform to economic laws can fundamentally clear the way for implementing the strategy to turn lucid waters and lush mountains into invaluable assets.

4. **An assessment system for green development shall be established based on natural and ecological endowments and national land use plans and strategies**

As the leading force of new productive forces, ecological civilization needs its own assessment system. In industrial civilization, GDP and a string of related standards are used for assessment. The new form of civilization—ecological civilization—also needs an assessment system that matches its theories and strategies, and GEP (Gross Ecosystem Product) is exactly an exploration and innovation toward that goal. The UN-released *2030 Agenda for Sustainable Development*, the "Rio + 20" resolution and the 18th CPC National Congress all highlighted the importance of correcting the mistakes arising from the excessive emphasis on GDP. Long years of practices indicate that a fundamental measure to correct those mistakes is establishing a new assessment system consistent with the laws of ecological civilization. Other factors aside, there are several main reasons why it is so difficult to promote "green GDP" in recent years. First, officials still think in the framework of GDP-based assessment. Despite many attempts, "green GDP" remains hard to promote basically because people haven't jumped out of the old GDP-oriented framework or the theoretical and assessment system of commercial civilization. Second, an important way to calculate "green GDP" is "deduction," meaning output value with negative ecological and environmental effects is deducted from GDP. This is one of the reasons why the local areas and authorities are reluctant to practice "deduction."

An assessment system complying with the laws and values of ecological civilization should be able to reflect natural and ecological endowments and also accord with the land and space plans made according to national strategy. Based on the explorations and practices in several locations in China over the past period of time, it seems GEP can reflect the values and norms for the assessment of ecological civilization while achieving innovation and breakthrough at the same time. Although it's hard to replace GDP completely at present, the inherent natural resource endowments of different provinces and regions and the strategic requirements of national land and space plans can be evaluated with GEP. Take Qinghai, a province located on the Qinghai-Tibet plateau in western China, for example. It's hard for it to catch up with the coastal provinces in the east on GDP no matter how hard it works because the two regions feature completely different geographical conditions and natural endowments, yet it may beat the eastern coastal provinces by a large margin on GEP contribution.

Therefore, GEP not only overcomes the difficulty in practicing "deduction," but also better reflects different natural endowments and conditions of each area as well as the varying requirements of national space plans. In this way, GDP and GEP are both adopted as indicators, so that each region and province can exert its strengths based on local conditions and be "horses for courses" during this period of transformation and transition. In this process, the stimulant role played by the GEP "baton" must not be neglected as it provides not only the critical basis for the assessment of distribution system, but also an important basis for assessing the performance of local governments while fully demonstrating equity. GEP looks at the basics. It assesses the total output value of ecological system (originally called the ecological system's service function), which essentially refers to the total output value of the ecological product it has produced, and that should be entitled to a reasonable return rather than being reduced to the idea of "ecological compensation" in the GDP or green-GDP-based assessment system.

This being said, it's easy to conclude that ecological civilization needs its own assessment system, that is, the "green baton." There is an old adage in Suzhou, "one has rowed the boat for quite a while, only to find the mooring rope still tied." If the "mooring rope" of the "green baton" remains tied, our ecological efforts would be inefficient; but if the "mooring rope" of the "green baton" is untied, our ecological efforts would be much more efficient.

2 Changing Values and Reshaping Ethics

The current inertial growth pattern and trend has to be changed, starting with changing our way of thinking, like how the Enlightenment Movement in Europe emancipated people from the shackles of religious doctrines.

The *2030 Agenda* underscored economic, social and environmental development for the fulfillment of sustainable development goals, but left out cultural and political development in comparison to the five spheres stressed in ecological civilization. In the transformation of civilization, what's more essential and crucial is the establishment of values and ethics. That's the underlying reason why the commercial civilization, which is dominated by the west, has found itself going down an ever narrower path, even going astray in some areas.

Culture deep down is a matter of values and ethics. Adam Smith, the founder of western modern economics and the great British economist, emphasized in his *The Wealth of Nations* the existence of two preconditions for market economy to function and the "invisible hand" to play its part. First, a society has to have a sound government and sound social organization; second, a society has to have solid morals and ethics. Only when these two important preconditions are in place will market economy function in such a manner as not to harm the health of the nation and the society—preconditions that the mainstream western society has almost forgotten or has been trying to avoid since neoliberalism caught on after American President Reagan and British Prime Minister Thatcher took office. The collapse of America's Nasdaq in early twenty-first century was blasted by the exposure of the Enron scandal, and it was no coincidence that the global financial crisis in 2008 was blasted by the bankruptcy of Lehman Brothers caused by its own fraud. In both cases fraud and deception were the root cause, and the US government played "double standards" both times, so it all boiled down to morality and ethics in the end.

Human realization of nature's value and their nature-related values can trace back to Renaissance that took place in Europe between the 14th and thesixteenth century, and the Enlightenment Movement later that gave birth to the humanistic spirit that put man instead of god in the center. Humanitarianism emancipated the mind and laid an important foundation for the new round of industrial revolution and commercial civilization. Ecological civilization, a new form of civilization, also needs a new industrial revolution, which is preluded by the new mental emancipation and new values. This is bound to bring about the switch from "human-based development" to "nature-based + human-centered development" as well as harmony between man and nature.

3 Respecting the Scientific Laws of the Natural Community of Lives and Nature-Based Solutions

In the long history of humanity, man has enjoyed nature's selfless supply of resources and taken it for granted. In the age of agricultural civilization, human demand didn't exceed the threshold of ecosystem's self-regulation and self-recovery, so man and nature could co-exist in peace. However, since the beginning of the Industrial Revolution, the accelerated pace of human conquest and change of nature has exposed mankind to a grave situation featured by unprecedented resource shortage, deteriorating service functionality of ecosystems, worsening environmental pollution, and diminishing bio-diversity—a situation that has spread from local areas to the whole world.

People have increasingly realized that they can neither absolutely dominate nature nor completely succumb to it. The earth will continue to exist without humans, but humans cannot survive a minute without earth. Man and nature should be an organic whole in which they prosper or decline together through interdependence and mutual promotion. Like all other creatures on earth, humans are part of the natural world and they all live on the same planet that supports bio-diversity and plural ecosystems.

Noting the symbiotic relation between man and nature, Marx held that "man itself is the outcome of nature"; they are "active natural existence" that's at the same time "restricted and limited." Engels also stressed that "we, with flesh, blood, and brain, belong to nature, and exist in its midst." Marx and Engels not only viewed man and nature as a community of life, but highlighted the important precondition of respecting nature for the survival and development of human society, as "protecting nature is protecting ourselves." Only when nature is healthy can humans be healthy; only when bio-diversity is protected can human life be sound. The loss of bio-diversity will lead to human extinction as well.

Therefore, the concept of a community of life actually points to the fundamental solution to the relation between man and nature, development and protection—the solution that's based on nature and natural laws, which is termed as "the Way follows nature's course" in Chinese philosophy. There are twofold meanings to this solution. In the narrow sense, the nature-based solution, like other solutions based on engineering, technology, economy or society, is just one of the many parallel options. In the broad sense, however, technological, economic and social solutions should all observe natural laws, or "follow nature's course." In other words, no solution should violate such laws that come above anything else.

The report of the 19th CPC National Congress states that "smooth progress has been made in major ecological conservation and restoration projects" when summarizing China's progress in building an ecological civilization in the past five years, and vows to "implement major ecosystem conservation and recovery projects" when making work arrangements for the future. At present, China is endeavoring to restore ecosystems, recover biodiversity and re-establish the habitat environment they rely on, covering mountains, rivers, forests, farmlands, lakes, and grasslands both in coastal and inland areas. According to past experience and lessons, a solution based on natural conditions and laws must be applied to such a massive and systematic ecological restoration program as it is not only more cost-effective than other solutions, but also accomplishes win–win results for man and nature, development and protection by addressing the root causes.

4 Combining Knowledge with Action

Practice is the only criterion to verify truth. For many years we have explored wide and far for the vision of ecological civilization. Apart from theoretical discussions, we have spent ten years implementing this vision in Guizhou, the province of the most backward economic and social development in China, and worked with local governments, Party organs and other related parties to create the ECO Forum Global Guiyang.

Marx said, "Philosophers only know how to explain the world in different ways, but the key is how to change the world." Bearing this in mind, we, while believing in the huge guiding force of thoughts and visions, also advocate concrete actions by hosting the ECO Forum Global, which has yielded bountiful fruits through ten years' efforts. ECO Forum Global Guiyang was the first to practice the assertion that "civilization prospers when ecological prospers; civilization declines when ecological declines," the first to call attention to climate change in China, the first to propose green financial measures, the first to promote carbon–neutral infrastructure construction, and the first to explore the legal guarantee for ecological advancement in China...It has contributed a lot to China's explorations for the strategy and path of building ecological civilization.

"Combining knowledge with action" is the only way to solve the conflict between development and protection. For a very long time in the past, Guizhou has been one of the poorest regions in China with its per capita GDP ranking last nationwide. In the meantime, it is an important ecological barrier in western China, and maintaining its somewhat fragile ecosystem is an arduous task since the province is filled with mountains and hills typical of a karst landform. Thanks to the unremitting efforts of the CPC Guizhou provincial committee and the Guizhou government, eco-civilization is deeply driven home across the province today. More importantly, ecological conservation and economic development have made headway in parallel guided by concept. In 2018, forest coverage rate in the province increased to 57%. Guizhou piloted the "river chief system" and found a way of environmental protection on that basis in view of local conditions. The Fanjing Mountain was successfully listed as a world natural heritage, driving the total number of such heritages in Guizhou to four. Also in 2018, the provincial GDP climbed up 9.1% from the previous year, the highest growth rate in the country and ranking top three for several years in a row. It was also designated by the central government as one of the first three pilot zones for ecological civilization development. According to Guizhou's overall big data strategy, it has become the first national big data experimental zone in China, firmly seizing the opportunity of transformation during the new industrial revolution. During his inspection tour to Guizhou, General Secretary Xi Jinping said "Guizhou is blazing out a development path different from that in eastern China and that of other western provinces." In combining knowledge with action, the key lies in action, or practice. Guizhou's practices have demonstrated what powerful reform the implementation of ecological civilization could bring about.

5 Building a Green "Belt and Road" by Promoting Ecological Civilization Worldwide

The Belt and Road Initiative is aimed to build interconnected infrastructure hardware, while the building and international promotion of ecological civilization is the software that underpins the hardware. They jointly create the green "Belt and Road."

Ecological civilization has not only awakened the world's ecological awareness, but also given it a better idea of the Chinese version of the UN's *2030 Agenda* and manifested its prospects through the "five transcendences" exclusive to ecological civilization. As ecological civilization is centered on reflecting the natural and ecological laws of earth, its "five transcendences" have innate eloquence and penetration that transcends the differences in nationality and territory, in stage of economic and social development (e.g. developing and developed countries), in political and social system, in thinking and ideology, and in culture and religious belief.

Obviously ecological civilization transcends the different civilizations currently existing in the world, which makes Samuel Huntington's "conflict of civilization" theory unfounded. Peace is the common aspiration of the majority of global citizens, and ecological civilization will enjoy the most extensive support because it can promote equal dialogue, exchange and mutual learning among different civilizations, and take the world to the new era featuring extensive consultation, joint contribution, mutual help and shared benefits.

6 Only Ecological Civilization Can Resolve the Three Basic Contradictions Faced by the Mankind in the Twenty-First Century and Bring New Hope and Vision for the Human Society

Amid the profound changes of the twenty-first century, there are three most fundamental and inevitable contradictions facing the humankind—(1) the increasingly prominent relation between man and nature, development and protection; (2) the new round of technological revolution and the relation between man and technology; and (3) the relation between an established major power and an emerging one.

Obviously the UN *2030 Agenda* and *Paris Agreement*—two milestone resolutions for human development—are not enough to cope with the three major contradictions of the twenty-first century.

This paper has expounded that ecological civilization is the ultimate way to resolve the contradiction between man and nature, development and protection. One more point will be emphasized here—according to the research reports issued by the UN and many other scientific institutions, it will be difficult to keep global temperature rise within 2 °C by the middle of twenty-first century, let alone 1.5 °C with the current global governance approach. If we cannot effectively curb the deterioration of earth's ecosystem and the loss of biodiversity, and if mankind has truly entered the "sixth period of extinction of species" with the sea level continuously rising, as many research reports indicate … should such trend continue, particularly as many western politicians always transfer their own ecological problems to other countries and fool their own people with populism, the human society will inevitably face a catastrophe imposed by nature. This knowledge will force the mankind to become ecologically conscious, restrain themselves, join hands in earnestly implementing the *2030 Agenda* and *Paris Agreement*, and opt for a future of ecological civilization.

As mentioned earlier, the new civilization form will be driven by a new round of technological revolution, while the relation between man and technology should trace back to the ecological morality, ethics and values discussed earlier in this paper. Only the Five-Sphere Integrated Plan stressed by ecological civilization and the imperative ethics and morality can more clearly define and guide the relation between man and technology. In other words, man will eventually control technology rather than be alienated, engulfed or destroyed by it. AI, stem cell, trans-genetics technology or gene duplication—moral and ethic issues have to be addressed in the development of these technologies, and only the values of ecological civilization can lead us to find the solution. That's why we need to stress the Five-Sphere Integrated Plan.

How do we handle the third major relation—the relation between an established major power and an emerging one then? By revealing and analyzing objective natural laws, ecological civilization has discerned one of the most important characteristics of the universe and its species—ecosystem pluralism and biodiversity. The world is a community where all creatures depend on and act on each other, which naturally gives rise to diversity and pluralism. The "Thucydides's trap" initiated by Professor Graham Allison was still confined in the framework of agricultural and commercial civilization and still continued the latter's logic, no wonder he came to the conclusion that a powerful country is bound to pursue hegemony, and an established major power and an emerging one are bound to have a war, after which the world will enter a new cycle of uni-polar domination. In contrast, ecological civilization reveals the law that all creatures grow together without harming each other and all paths develop in parallel without contradicting each other, thus giving rise to a world of multi-polar governance. As a matter of fact, early signs of a multi-polar world of joint governance have already appeared although only 20 years past the twenty-first century. The

US-dominated global governance landscape that took form after the Cold War has undergone notable changes, and developing countries are having a greater say in international politics with the rapid development of emerging economies like China, India, Brazil and South Africa. The expansion from G7 to G20 and the formation of BRICS are clear manifestation of multi-polar governance. In today's world, there are at least three poles—North America, West Europe and Asia. The world is marching steadily toward multi-polar governance.

As a major reform of civilization, ecological civilization is the inevitable trend of world evolution. Fortunately, China's role and contribution in this new transition is drastically different from the last time. Under the feudal autocracy of the Qing Dynasty, China in modern times missed the historic opportunity of transformation and slipped to a position of being bullied and humiliated. In comparison, the new round of technological revolution and the new civilization form have appeared along with the prophecy that "the mankind will eventually walk into the new era of ecological civilization" made by CPC and the Chinese government. How to grasp this strategic opportunity, make the best use of it in China and spread the best practices to the world is a mission that has fallen on the shoulders of China as well as all people in the world that pursue peace and development. The prophecy, like a beacon, will lead China and the whole world toward a beautiful future in the midst of bewildering and profound changes and uncertainties, will be a millennial cause of lasting development jointly pursued by mankind, and will bring the great vision of the harmony between man and nature. To be able to unite and lead the world, we need ideal and vision, which is why ecological civilization is even more significant than the *2030 Agenda* and *Paris Agreement.*

I believe justice, kindness and wisdom will defeat evil, greed and ignorance. Guided by the Xi Jinping Thought on Socialism with Chinese Characteristics for a new era, especially the Xi Jinping Thought on Ecological Civilization, we will strive to build a beautiful China, build a community with a shared future using Five-Sphere Integrated Plan and "Five Transcendences," and make sure man and nature will co-exist and co-develop together in harmony. "Civilization prospers when ecological prospers; civilization declines when ecological declines." The world trend forges ahead with unstoppable momentum. Those who follow it will thrive and those who go against it will perish. Thirty years from now, when China has achieved the "two centenary goals," the human community will understand more deeply that only when the whole world joins hands, awakens to ecological civilization and ultimately marches toward that new era is it possible to bring our beautiful vision into reality. That's not a choice—it is the only way forward for mankind.

Several Issues About Promoting Ecological Civilization that Should Be Studied Further

Yuqing Wang

The CPC and the Chinese government have attached great importance to ecological civilization, and General Secretary Xi Jinping has personally directed and promoted this cause, which has brought out historic and comprehensive changes in the country. The environmental situation in China has notably improved thanks to the concerted efforts of all parties. In Beijing, people are enjoying more days of blue sky and white cloud; around the country, the ecological environment is getting better as bald mountains become green and rivers lucid. However, for a big country with a population of nearly 1.4 billion people, the environmental pressure is huge on its way to developing the economy and pursuing a happy life, and ecological improvement will get increasingly difficult as we move forward. There is a long way to go in our endeavor to build a beautiful China, and it is absolutely necessary to carry out relevant in-depth studies, clarify various relations and keep deepening our understanding.

1 How to Understand Harmony Between Man and Nature

Civilization is the summation of the material and cultural (including institutional) achievements made in the process of human renovation and protection of nature and organization and reform of society. It is the symbol of social progress. Ecological civilization can be viewed as an aspect of human civilization, namely the degree of civilization that is reached by mankind in handling the relation with nature. It is manifested in a state of harmonious co-existence and benign interaction between human society and nature that supports the development of human society. When discussing the concept and contents of ecological civilization, people generally take "harmony between man and nature" as the core, but discussions are neither extensive

Y. Wang (✉)
State Environmental Protection Administration, Beijing, China

J. Pan et al. (eds.), *Beautiful China: 70 Years Since 1949 and 70 People's Views on Eco-civilization Construction*, https://doi.org/10.1007/978-981-33-6742-5_43

nor in-depth enough regarding what "harmony between man and nature" really means and how to achieve it.

We hope science and technology can help us raise the resource and energy efficiency and create more wealth to change this situation, but sci-tech advancements are usually a double-edged sword on the environment with both pros and cons. Micro plastic particles in the ocean, natural disasters caused by climate change due to greenhouse gases...these problems have only drawn close attention in the past few decades. Natural resources include renewable, non-renewable and sustainable resources, and the utilization of them shall be intensive, efficient and rational based on their respective properties. For renewable resources, mainly bio-resources, we must enhance their biological productivity on the basis of protection and nurturing, and the exploitation and utilization thereof shall not exceed their speed of growth, thus ensuring sustainable use. For non-renewable resources, mainly minerals, we must economize on them and carry out cyclic utilization. Humans should learn from the natural ecosystems in their productive and living activities, so as to realize the process of material consumption that generates almost zero waste thanks to the circular chain of producers, consumers and decomposers, with solar energy being the only resource used—a world where all creatures truly live on the sun. The waste-less city that's being promoted now is working in this direction.

In addition to rational utilization of various resources, the diverse functions of life-supporting systems must be maintained to realize harmony between man and nature, including climate adjustment, water and soil conservation, pollution control, wind prevention and sand fixation, disease and insect control, natural material circulation (including the circulation of water, carbon, nitrogen and oxygen), among others. This mainly refers to maintaining the functions of different types of ecosystems, including forests, grasslands, wetlands, rivers, lakes and oceans. To that end, the mankind should cherish land, avoid increasing farmland and building all kinds of facilities with the increase of population, which will encroach upon the land for natural ecosystems. Human activities inevitably affect natural ecosystems, but these effects and interference shouldn't exceed the limit of ecological stability, meaning they cannot cause structural imbalance and serious functional deterioration of ecosystems in a large scope. We must work hard to increase natural capital and keep up the productivity and biodiversity of ecosystems. Roy Morrison wrote in his *Ecological Democracy* that ecological civilization, on the operational level, can be simply defined as follows: "economic growth means ecological improvement and the regeneration of natural capital." At the moment, it seems that human efforts can improve other indicators, but the trend of diminishing biodiversity is hard to curb. According to a research report by World Wildlife Fund in 2016, the per capita ecological footprint of mankind reached 2.6 global hectares in 2012, which had to be sustained by the natural resources and ecological services of 1.6 earths, and the number of vertebrate population dropped by 58% from 1970 to 2012. To realize harmony between man and nature, the human society must make substantial and fundamental changes.

2 Nature-Centrism and Human-Centrism

This is a central issue in ecological ethics or environmental philosophy and an important theoretical subject in ecological civilization, while it also concerns how to view nature's practical and intrinsic values. From the early twentieth century when ecological ethics was founded in the west to the 1980s when China introduced the discipline and began to study it, and to the first decade of the twenty-first century, the debate over nature-centrism and human-centrism has never stopped. I'm no expert on this field, so I'm not in a position to make comments, but when I recently read Xiao Xianjing's *Environment and Society: Human Perspectives on Environmental Issues*, a book published in 2006, I found a simple and clear overview of this topic. I agree to most of its views, but have my reservations on some of the points. Regarding the relation between man and nature, I basically agree with the view of modern human-centrism mentioned in the book. When we talk about harmony between man and nature, "man" must be the main entity or main aspect of the contradiction, meaning the man-nature relation should exclude the social relation among people. Although the man-man relation, if not properly handled, would affect the man-nature relation, the point in question here is what attitude the mankind as a group with independent consciousness should adopt toward nature for its overall interests, not the specific attitude or behavior of some individual, class or group toward nature.

From an axiological point of view, modern human-centrism holds that all human activities stem from and serve human interests, which is not wrong. The key lies in how to understand "interest." There is nothing wrong with this view as long as the interests refer to the long-term, fundamental and overall interests of humankind rather than immediate, narrow and partial interests. Some people are worried that the concept of human-centrism would base everything on human interests, thus limiting the scope of protection and focusing only on protecting the human-related ecological environment. But the question is, are there any species or their living environments on earth that are unrelated with human race today? The UN's Millennium Ecosystem Assessment (2005) generalized the various benefits drawn by man from ecosystem with ecosystem services, which consisted of four categories—supply service, regulatory service, cultural service and supportive service—that were further divided into 27 items. This assessment covers all aspects of nature's intrinsic and instrumental values mentioned in ecological ethics. If we can realize these service functions of natural ecosystems and cherish, protect and make proper use of them, I believe many scholars would no longer have concerns about the idea of modern human-centrism.

Environment and Society: Human Perspectives on Environmental Issues mentions that Paul Taylor, a proponent of modern bio-centrism, put forth in 1986 four principles on respecting all lives as well as five priority principles that must be observed when the protection of species based on the four principles conflicted with the protection of human values, rights and interests. All those principles are worth popularizing and practicing now as they are meaningful rules regarding our respect for and protection of nature, especially living entities. I find Fu Hua's *An Inquiry into Ecological Ethics*, which was published quite early (2002), a good book because it explicitly generalizes the basic western and Soviet views about human-centrism. When talking about the relation between man and nature, although it admits man's position as the main entity, it also stresses that we can only reform nature if and when we follow its course, and activities to reform nature must be based on the knowledge and understanding of natural laws. It also specifies that it is man's mission to scientifically reform nature and also correctly understand and reform themselves. These ideas are worth drawing on.

Like the topic above, the instrumental and intrinsic values of nature have also been the focus of long-standing debate in ecological ethics and environmental philosophy, each party having its own reasons. The debate, in my opinion, largely results from the different definition of "value" or the different understanding of its essence. The *Encyclopedia Britannica (Chinese Edition)* defines value as "anything useful," which falls under the theory of entity, but it also explains that the judgment of value is a human cognitive activity that exists because of a person's wish for something. Some experts argue that the "relation theory" is closer to Marx's definition of value. For general workers, it is of course important to recognize nature's intrinsic values, but what's more important is where these values are manifested, how to protect them in practice so the value can be maintained and increased, and how to make proper use of them for human benefits. A quantitative statement would be better for it can be used to evaluate the environmental costs of various construction and development activities to decide which can continue and which must be stopped. As mentioned earlier, the ecosystem service value adopted by UN's Millennium Ecosystem Assessment is a practical guidance on the evaluation of those activities.

The debate in ecological ethics over human-centrism and nature-centrism and whether nature has intrinsic values is a good thing in that it promotes the development of the discipline and broadens the mind and horizon of researchers. However, some basic understandings of the relation between man and nature must be extracted and generalized as the ideological basis to advance concrete work.

I want to add a few more. First, man is the outcome when natural species evolve to a certain stage, but is different from other species in that it has both natural and social properties. Likewise, the natural world today is the outcome of human activities, and there is almost no part of earth but has been affected by man. Second, the humanized natural world and the nature-supported human society are a pair of entities in a contradiction that are both closely connected with and restricted by each other through continuous changes. The natural world is changing according to its own laws and pace, but human activities exert immense effects on natural changes and the effects are likely to intensify in the foreseeable future. In the natural world, the reproduction of various creatures is subject to natural conditions, but mankind as a thinking creature is much more capable than other creatures. Human beings exert more effects on natural ecosystems than other creatures but are hardly restricted by other creatures. Therefore, the mankind as an intelligent creature has to restrain its own behaviors. The human society would have come to an end if it had to rely on natural laws to curb the excessive human activities. Third, the ideas of humans, the main entity of the contradiction, also change with social development. When there isn't enough to eat, people aspire to live an affluent urban life and forest, wilderness and mountains become obstacles on their way to wealth; but when they have lived a comfortable urban life, they want to relax themselves in nature, although it is already impossible for most people to live in a "pure" natural environment for a long time as they have become socialized. Fourth, with social and technological progress, it is hard to imagine what the humankind and the natural world it lives in will be like in hundreds of years, but we must keep in mind that nature remains nature whatever changes come its way. To quote from a documentary, "nature doesn't need man, man needs nature." I put down these points to remind us all that nature and society are closely intertwined, even integrated as a whole, and human activities have tremendous and far-reaching effects on nature, of which we know little about. Therefore, it would be wise for man to stand in awe of nature and be very cautious when developing and utilizing it.

3 Ecological Civilization and Sustainable Development

We have briefly reviewed how the international community has gradually realized and recognized sustainable development and the policy measures adopted by China. In fact, ecological civilization is closely related with sustainable development, with their basic concepts being connected and consistent. The former is the result of combining sustainable development and China's national conditions.

First of all, both ecological civilization and sustainable development are initiated to cope with the environmental crisis in the process of human development. Sustainable development is defined as "development that meets the needs of the present without compromising the ability of future generations to meet their own needs." Second, the two ideas have similar contents. The three pillars of sustainable development that support and strengthen each other are economic development, social development and environmental protection. Ecological progress is an integral part of socialism with Chinese characteristics along with economic, social, cultural and political progress, and is embedded in the other four aspects. Developing ecological civilization and pursuing sustainable development are both significant decisions made in consideration of the overall interests and long-term development of human society. However, they are different in terms of the perspective of viewing things and the focus of resolving problems, the latter focused on ensuring stable and sustained economic growth. It is stated in the report *Our Common Future* that "the environment is where we all live; and development is what we all do in attempting to improve our lot within that abode. The two are inseparable." In the same year that United Nations Conference on Environment & Development in Rio de Janeiro released the *Agenda 21*, the World Bank's *World Development Report 1992* defined development as the cause of reforming people's lives and the cyclic progress of the economy and society. Ten years later, the UN held the World Summit on Sustainable Development in Johannesburg in 2002, at which a political statement and the execution plan (for *Agenda 21*) were issued, which prioritized eliminating poverty and developing economy and took poverty as the most severe global challenge. According to Xiao

Xianjing, "economic sustainability holds the key to the sustainable development strategy. Natural sustainability is achieved through sustainable economic operation, and a sustainable nature provides the material foundation for sustainable economic development, which is the only way to ensure sustainable social development." Wang Nuo mentioned that when they studied the concept of sustainable development, some expert proposed to define it as "limited development that doesn't exceed the environmental carrying capacity," which was not accepted, indicating that the concept may become utilitarian or misleading if not understood correctly.

Although ecological progress starts with environmental protection, it is elevated to the height of civilization, thus standing on a higher level than sustainable development when it comes to the handling of the man-nature relation. Civilization prospers when ecological prospers. Building ecological civilization concerns the country's future, the people's well-being, and the Chinese nation's lasting development. The importance of environment was never uplifted to this height when we were studying sustainable development. That ecological progress is merged with the other four spheres of development underscores the fact that this is a comprehensive and important campaign that cannot be summarized as environmental protection, but can only be achieved through the participation of the whole people and whole society and the concerted efforts of all sectors. The saying that "lucid waters and lush mountains are invaluable assets" reveals the dialectical relationship between economic development and environmental protection—protecting the environment is protecting productive force, and improving the environment is developing productive force. Regarding the relation between ecological progress and social development, a good ecological environment is the fairest public product, and the most accessible welfare for the people. Environment is an aspect of livelihood, lush mountains create beauty, and blue sky is an element of happiness—they can make people truly feel the superiority of socialist system and the environmental benefits of economic development, giving them a solid sense of gain. Regarding the relation between ecological progress and political development, the most rigorous rules and laws have been adopted to protect the environment. China has established a complete system of ecological civilization featuring clear ownership, multiple participants and equal emphasis on incentive and restraint, with a view to clearing the systematic and institutional obstacles to the development of ecological civilization. As to the relation between ecological progress and cultural development, it is a revolution involving the way of production, way of life, way of thinking and values. It is necessary to intensify the publicity of and education in ecological civilization, enhance the public environmental awareness, push to form a healthy lifestyle and consumption mode that advocates thrift, moderation, and green and low-carbon ways, and foster a favorable atmosphere in the whole society.

The series of insightful remarks and statements made by General Secretary Xi Jinping on ecological civilization demonstrate that this vision advocated by China as a developing country is a guiding thought and strategy that's more comprehensive, more incisive and more suited for China's national conditions when guiding us to deal with the man-nature relation in the process of development. This vision should be promoted in the whole world. More importantly, it offers the Chinese wisdom and plan on global environmental governance for the majority of developing countries, UN bodies and important think tanks. Therefore, we must conduct in-depth study of the sustainable development strategy that's advocated by the mainstream of international community when handling the relation between environment and development. How it relates with the China-initiated ecological civilization? How do we coordinate them? These questions are the common concerns of people around the world engaged in environmental protection and related sectors, and they are worth close attention from our research institutes and publicity departments.

Following the understanding of ecological civilization mentioned earlier, I would like to offer several points about the connotations for reference.

(1) Ecological civilization is a form of civilization featuring comprehensive and sound development. It doesn't reject economic development, nor does it pursue development at the expense of environment. Rather, it aims to improve man's ability of adapting to, utilizing, protecting and restoring nature through new ideas, technological progress, improvement of productive force, and fundamental changes in the way of life and production, so as to promote healthy economic and social development on the basis of harmony between man and nature.

(2) Ecological civilization is a sustainable form of civilization that encompasses human sustainability and natural sustainability, the two of which are different but inseparable with mutual effects on each other. Human sustainability is in itself a massive undertaking that, if not handled properly, would undermine natural sustainability. Natural sustainability, on the other hand, requires that all human activities to utilize the environment and exploit resources must be carried out on the premise that the environment can bear the impacts and is recoverable, resources are renewable, and the development opportunities of later generations are not compromised. It is a sustainable exploitation and utilization of resources that keeps the ecosystems functioning well.

(3) Ecological civilization is a form of civilization where economic development and human development are in unity. It insists on human-centrism, makes sustainable human development the fundamental value, and changes people's values and concept of life instead of prioritizing ecology over people. "We can live a happy life even if we don't have much. It's sad if we have everything but no time to enjoy life." Moderate material consumption, colorful cultural life, healthy lifestyle and harmonious families and friendly neighbors—this should be the life that modern people aspire for.

The term ecological civilization consists of two words. Ecology usually refers to the natural ecology that has nothing to do with man for the most part of earth's billions of years of evolution, and it only became related with man in the past millions of years after mankind came into being. The continuous evolution and progress of man gave birth to "civilization," which is a concept of the human society. Civilization is a state away from ignorance and barbarism; in a way it is also a sign that man is getting farther away from the natural ecology. That we have put these two words together means that we should view our planet in a new ethical light, respect all living creatures, and build our common home into a new civilization.

Basic Path of Building Ecological Civilization

Yi Qian

1 Origin of Ecological Civilization

Civilization is the progress and development of human society and human nature that symbolizes the advanced cultural state and values of mankind. As the focus of ecological study has shifted to institutions and theories, the concept of ecology has also shifted from an ecological system to one that respects and integrates the laws of ecosystems.

In the 1970s, the Soviet academia first broached the idea of ecological civilization, but it didn't define it as a form of civilization after industrial civilization, but as the elevation of ecological culture and attainments. Later scholars in many countries started to expound this concept and began to view ecological civilization from the perspective of ecological science and philosophy. In this way, ecological civilization is considered a new civilization form after industrial civilization and indicates a state of material and cultural advancement.

From primitive civilization, agricultural civilization to industrial civilization, man has evolved from living by instincts to relying on physical strength and skills, from worshiping nature to depending on, changing and struggling with nature, from relying on the natural food chain to self-sufficiency and affluence, from simple fishing and hunting to simple production and then complicated production, from meeting the need for reproduction to working for low-level survival and then high-level overdraft, and from zero environmental pollution to causing serious environmental deterioration and global environmental pressure. While industrial revolution has brought technological development and material abundance, it has also increasingly sharpened the conflict between man and nature and undermined human health and survival, even the fate of earth. It is against such a background that ecological civilization has come into being, which is characterized by a wide range of good points—reliance on intelligence, harmony between man and nature, economic optimization, resource conservation

Y. Qian (✉)
Chinese Academy of Engineering, Beijing, China

J. Pan et al. (eds.), *Beautiful China: 70 Years Since 1949 and 70 People's Views on Eco-civilization Construction*, https://doi.org/10.1007/978-981-33-6742-5_44

and sustainable utilization, environmental improvement, and protection of the health and safety of people and other creatures. Ecological civilization, an advanced state, aims at a balanced development featuring the harmonious co-existence between man and nature, man and man, the current generation and future generations, and man and social economy, and a sound development featuring economic efficacy, social equity and ecological soundness. Ecological civilization involves a spate of aspects, including world outlook, values, ways of production and life, development mode, consumption mode, social system and legal system, among others.

2 Main Ideas of Ecological Civilization and Existing Experience

To build ecological civilization, we, following ecological laws and the principles of harmonious and enduring ecosystem, will build a resource-saving, environment-friendly and ecologically healthy society based on the carrying capacity of resources and environment and aimed at enhancing the capability of sustainable development and maintaining ecological justice, by making comprehensive use of political, economic, cultural, social and natural approaches.

Industrial activities can be considered the earliest and most direct reflections of ecological civilization. From the 1970s, the mankind has been advocating clean production and circular economy, initiated ecological industry and built eco-industrial parks. They have studied industrial symbiosis and metabolism, life cycle management and ecological design of industrial products, and worked hard on de-materialization, improvement of ecological efficiency, re-manufacturing, and green chemical industry and construction. They have also promoted eco-agriculture and eco-cities. All these efforts are aimed at the rational use of resources and environment within nature's carrying capacity, so that man can get on the path of building a resource-conserving, environmentally friendly society and a sound ecology.

China is making all-round efforts to create a new era of ecological civilization.

More than 3000 years ago, *The Book of Changes* elaborated on the unity and inseparability between man and nature, and "unification of nature and man," which advocated respect and love for, emulation of and harmony with nature, became the essence of Chinese philosophy.

The per capita resource occupation in China is far less than world average and the resources are unevenly distributed. China has enjoyed rapid economic growth since the reform and opening-up, but also suffered from serious resource shortage, ecological damages and environmental pollution. It has a pressing need to develop economy, improve people's lives and recover the environment on the premise of resource conservation, pollution prevention and control and ecological protection. Although the country hasn't accomplished industrialization, it has to quickly foster the new concept of ecological civilization in a step-by-step approach, so as to get on the track of new-type industrialization and walk into the new era of ecological civilization.

General Secretary Xi Jinping has repeatedly emphasized that "lucid waters and lush mountains are invaluable assets" and "protecting and improving the environment is protecting and increasing productivity."

Calling the nation to strongly promote ecological progress, the 18th CPC National Congress vowed to "give priority to ecological progress and integrate it in every aspect and the whole process of economic, political, cultural and social development."

3 Six Parallel Areas of Ecological Efforts

1. **Production**

The promulgation of the *Law on the Promotion of Clean Production* reflects China's comprehensive considerations for resources and environment in the field of production, and the launch of pilot programs of circular economy in provinces and cities has initiated a surge of circular economy across the country. The purpose of clean production is to establish an ecological industrial system comprising products (including services) and production process and centered on eco-friendly restructuring and transformation, and to design products—the core of linear material metabolism—in an ecological way. Circular economy goes beyond clean production on a higher level and in a larger scope by pushing for the transformation toward an ecological industrial system and the shift of consumption mode. In the new era of ecological civilization, we must keep promoting clean production and circular economy, developing ecological industry and pushing the ecological renovation of industrial parks. We must also strongly advance ecological agriculture, and promote green utilization of traditional energies and the exploitation and use of renewable energies.

2. **Consumption**

Consumption is a downstream sector, where the waste would lead to the hundreds, thousands or more times of waste in resource exploitation. China has a huge base of consumers with a low-level yet polarized consumption pattern. The outline of the 12th Five-year Plan called for the formation of a civilized, economical, green and low-carbon consumption concept and a green lifestyle and consumption mode suitable for our national conditions. This includes using water-saving products, energy-saving cars and energy/land-saving housing, reducing the use of disposable goods, restricting excessive packaging, controlling irrational consumption, and promoting green government procurement.

We should oppose extravagance and waste and the pursuit for material luxury, establish a moderate consumption scale, scientific consumption structure, fair consumption principles and civilized consumption behaviors, and pursue ecological consumer goods and pollution-free results. Despite China's vast territory and abundant natural resources, we have a smaller per capita territory and resource occupation than many other countries due to the large population, so we cannot look up to developed and wealthy countries when it comes to consumption.

3. **Urbanization**

Urbanization is an important sign of modernization, and this process means the contraction of natural and primitive ecosystems as well as the expansion of artificial ones. A city is a typical artificial ecosystem and a complicated one integrating nature, economy and society that has more needs for artificial management and regulation. City planning and design should be guided by the ecological civilization concept. To be more specific, city scale must be controlled instead of blindly expanded; various types of buildings should be rationally arranged to reduce the demand for intra-city traffic; public transit should be prioritized and walking and cycling encouraged; public buildings, especially government office buildings and plazas, should restrain from excessive size and luxury; and the so-called image and signature projects impractical but costing should be rejected.

We can learn from Germany's de-centralized city development mode and develop a number of smaller-scale cities which are balanced distribution. The administrative resources and service functions in cities should be evenly distributed, and efforts must be made to revitalize small and medium-sized towns and boost the equal development of different regions and between urban and rural areas.

Urban infrastructure, including water supply and drainage system, traffic system and garbage collection and recycling system, should also be designed and built to support ecological progress. Measures should be taken to develop "urban mines" and turn urban wastes into "mineral resources," achieving tri-wins in resource recycling, environmental protection and economic benefits.

4. **Ecosystem protection**

Ecosystem protection, as one of the key tasks of ecological progress, is mainly carried out through intensified protective measures and major eco-recovery projects to produce more ecological products, and by stepping up the comprehensive governance of desertification, stony desertification and water and soil erosion to improve regional ecological environment. We can monitor ecosystem changes to better the layout of land and space development; study the principles of those changes to support ecosystem protection; develop new technologies and better modes of ecological construction to realize efficient utilization of ecosystems; study global climate change and environmental issues to be better at coping with climate change; and disseminate knowledge in natural and ecological protection to support the national environmental protection campaign. Given China's vast territory, the natural ecosystems have different traits from one place to another, so protective measures should be targeted while the damaged systems must be recovered.

5. **Culture and education**

To build ecological civilization and realize coordinated development between man and nature, we have to enhance the public's ecological awareness and drive home ecological knowledge in the whole society, for which culture and education is the primary approach. Ecological education in schools, including primary and middle schools as well as colleges, should be strengthened, so should social education, covering governments, enterprises and the public, with the media, training and self-education being effective approaches. It must be made clear that professionals engaged in science, engineering, law, management, finance and social governance all bear a responsibility for ecological progress and therefore should be educated thereof. In this way, we can foster a social ethos where everyone feels responsible for protecting the environment and promoting sustainable development.

Most importantly, we should intensify quality education across the board, advocate environmental ethics, and urge the public to respect and protect nature, care for themselves and the whole mankind, and consider both the present and the future. We must continuously publicize the sustainable development strategy and ecological progress, drive home the idea of "unification of nature and man" that is a key message of traditional Chinese culture, and also introduce new ideas and theories from developed countries, such as "industrial ecology" and "eco-city."

6. **Legal system and administration**

The legal system for ecological progress must be founded on the basis of a series of established and interrelated legal systems, in accordance with the concept and basic laws of ecological civilization as well as the general evolution laws of environmental factors, and on the principle of ecology first, non-deterioration, ecological democracy and common responsibility. The legal systems can be categorized into preventive, controlling and relief ones. The current laws should be revised to incorporate ecological requirements and be more interrelated with other laws, including criminal ones, law enforcement and punishment of violations should be reinforced, and a lifetime accountability system for resource waste and environmental damage should be established. We must perfect the economic and social development assessment system to include indicators that reflect the status of ecological progress, such as resource consumption, environmental damage and ecological benefits, thus making it an important guide and restriction for building ecological civilization. Environmental protection systems should also be established, such as the system regarding the property rights and the use of natural resource assets, regarding the red line of ecological protection, ecological compensation mechanism, environmental damage claims and accountability, and ecological work assessment and statistics. The assessment indicators should include sindicators to build an ecological civilization across the board.

4 Conclusion

Ecological civilization is proposed based on the denial of the industrial outlook on nature and values. It is a new civilization concept marking the direction of human civilization development. Compared with traditional civilizations, ecological civilization admits the close relation between man and nature, and stresses promoting their harmonious co-existence by changing the ways of production, life and thinking. Building an ecological civilization is the only way to changing development mode and implementing the strategy of sustainable development, and it must be merged with efforts of economic, political, cultural and social development. China has made some progress in this area, but we are still in the initial stage with arduous tasks and plenty of difficulties ahead. Therefore, let's start right now and spare no effort to do something every day. Ecological civilization is a bright prospect, and the dream of a beautiful China with man-nature harmony will definitely come true.

Ecological Civilization is the Chinese Wisdom and Chinese Plan for Sustainable Development

Chunyi Wang

Man has only one earth, which we didn't inherit from our forebears but borrowed from our offspring. The current generation should never live on what our ancestors have created while leaving troubles to later generations. The four great ancient civilizations of Egypt, Babylon, India, and China all began in regions with thick forests, abundant water, and fertile soil. About 3000–4000 years ago, man established resplendent civilizations by large rivers, including China by the Yellow River, ancient Egypt by the Nile, ancient Babylon by the Tigris–Euphrates river system, and India by Indian River, which have been internationally acknowledged as one of the four major civilizations. Natural disasters and human activities caused the dramatic deterioration of ecology and sent the civilizations of ancient Egypt, ancient Babylon, ancient India and Maya from peak to perishing. Man has only one earth and protecting it is our common responsibility. There are many different nations, states, interest groups, religious beliefs and social systems in the world. What common interests and consensus do we have to maintain such a world of great diversity? Ecological civilization and sustainable development are the only way to make humans live together in harmony and rationally choose a common future. Man and earth are a community with a shared future, in which all countries should join hands to build ecological civilization. Major countries in particular should fulfill their responsibilities as such.

1 The Essence of Ecological Civilization is the Harmonious Co-Existence Between Man and Nature

A conclusion has not been reached yet on what ecological civilization is. In my opinion, man is the main subject of ecological civilization, and the harmonious co-existence between man and nature is its essence. As far as the properties of culture and

C. Wang (✉)
China Ecological Civilization Research and Promotion Association, Beijing, China

J. Pan et al. (eds.), *Beautiful China: 70 Years Since 1949 and 70 People's Views on Eco-civilization Construction*, https://doi.org/10.1007/978-981-33-6742-5_45

civilization are concerned, this concept can be interpreted and summarized with one word—benevolence. Man should respect, comply with and protect natural ecology, cherish earth and treat nature well instead of putting itself above nature in conceit and pride. Only when the current world is beautiful and harmonious can benevolence be achieved, said Alfred North Whitehead, founder of process philosophy. In past global economic development, man's excessive exploitation and greedy plundering of natural resources in the sole pursuit for wealth has inflicted countless ecological disasters on the world. Human-centrism severs the unity, integrity, process and organic entirety of man and nature and gives the excuse for man to destroy natural diversity and richness and the connection between openness and circulation, infinity and limitation. The human-centered tradition is the culprit of ecological crisis in modern west. President Xi Jinping has deeply revealed the essence, connotations and values of ecological civilization—the harmonious co-existence between man and nature. Ecological civilization is the crystallization of all human efforts and achievements made in the process of reforming nature to benefit themselves with the aim of man-nature harmony. Man and nature are a community of life, and only when common growth and prosperity is achieved can co-existence be possible. Ecological civilization is an outcome when human civilization has evolved to a certain stage. From primitive civilization to agricultural civilization and then industrial civilization (post-industrial civilization), human development has been seriously restricted by resource and environment, and even human survival is threatened by global ecological and environmental issues. Some people argue that human civilization has evolved from "yellow" to "black" and then "green." Developing ecological civilization doesn't mean abandoning industrial civilization and going back to the primitive way of life and production, but to build an ecological society of advanced production, affluent life, sound ecology and healthy life in consideration of the carrying capacity of resources and environment, following natural laws, and aiming at sustainable development and harmonious co-existence between man and nature.

2 Ecological Civilization is a Fine Outcome of Human Ideology and Culture

1. Fine traditional culture of China

Ecological civilization stems from ecological thinking. The ecological thinking of Chinese nation is contained in its cultural thoughts, ethical systems and historical traditions. Confucianism, Buddhism and Taoism have all answered a fundamental question in Chinese philosophy—what's the relation between man and nature and what's man's position in nature—despite their different attitudes toward nature. All of the three philosophical schools take man as the "main subject that aims to achieve harmony and unity between man and nature" and take "unification of nature and man" as the basic principle on handling the man-nature relation. The Confucian thought on ecology is essentially boiled down to complying with nature and treating it well. Confucius said "The benevolent love others," urging people to extend their benevolence to all things and creatures on earth. Buddhism deems "Buddhahood" as the origin of all things, whose differences are only the different representations of Buddhahood; it also advocates equality of all, believing that "mountains, rivers and plants can all aspire for Buddhahood" and "a pure heart is a pure land." Taoism stresses that respecting natural laws is the paramount principle and following the courses of nature, heaven and earth is the underlying guidance for human behaviors. These thoughts and assertions are valuable wisdom and spiritual assets left by our ancestors during their exploration for man-nature relation, and are of great ideological importance for us to understand the essence of ecological civilization today.

2. **Philosophy and fine ideology and culture of the west**

   (1) The basic idea of Gaia hypothesis is that the interactions between life and environment can condition earth to be suitable for the lasting survival and development of life. The Gaia hypothesis was proposed by English atmospheric scientist James E. Lovelock at the end of the 1960s and was later developed by him and American biologist Lynn Margulis. This hypothesis gradually gain much attention in the western scientific circle, exerting growing influence on people's perception of earth. The key idea of Gaia hypothesis is that earth is an organic entity of life. James Lovelock described earth as "living" and self-regulating. To keep this entity healthy, "Gaia" would respond and counteract if harmful factors appeared within the entity.

   (2) Process philosophy: Famous English philosopher Alfred North Whitehead founded the process philosophy, also called the philosophy of organism, in the US in the mid-twentieth century, which is expounded in his *Process and Reality* and other books. Whitehead described and explained every physical and realistic state in relations, such as relation between whole and part, individual and community, subject and object, system and environment, substance and spirit, end and means, and reality and surreality, and based his ideas on organic entirety, inherent process, generation and circulation, and pluralistic co-existence. Some scholars call his process philosophy the "philosophy of internal relations," including the internalization of man-nature relation, which leads to the conclusion that caring for nature is caring for man themselves, and they both exist in the process of changes. Whitehead's process philosophy has established a harmonious value—achieving a harmonious universe of beauty and order and eventually realizing harmony between man and nature, man and man, and man and itself. This is where process philosophy carries the elements of ecological civilization, as its ideas of entirety, process, generativism and internal relations under the thinking framework of process dialectics have enriched the philosophical essence of China's ecological civilization.

(3) Deep ecology: Deep ecology is an environmental philosophy that emphasizes that all creatures have the right of survival and inherent values regardless of their use value to mankind, and that the natural world is a subtle balance of various complicated relations where all organisms are related with other creatures in the ecosystem. In 1973, Norwegian philosopher Arne Naess first put forth the term "deep ecology" in his article titled *The Shallow and the Deep, Long-Range Ecology Movement: A Summary. The article criticizes human-centrism that instrumentalizes nature and the opposite term "shallow ecology" that only nurtures the wilderness and protects biodiversity for human benefits and therefore limits nature's value to its use value to mankind. In response,* Arne Naess proposed deep ecological outlook that affirms the basic living rights and inherent values of all creatures. In 1984, Arne Naess and George Sessions jointly drafted the eight basic principles of deep ecology, which they stressed were not rigid doctrines, but only outlines for discussion and were open to revision and perfection by people accepting them in the broad sense. The Schumacher College in Britain opened a course on deep ecology in May 1995, and the course takers discussed and summarized the eight basic principles as follows:

a. The well-being and flourishing of human and non-human life on Earth have value in themselves. These values are independent of the usefulness of the non-human world for human purposes.
b. Richness and diversity of life forms contribute to the realization of these values and are also values in themselves.
c. Humans have no right to reduce this richness and diversity except to satisfy vital needs.
d. The flourishing of human life and cultures is compatible with a substantial decrease of the human population. The flourishing of non-human life requires such a decrease.
e. Present human interference with the non-human world is excessive, and the situation is rapidly worsening.
f. The dominant socio-political living situation must therefore end. This will affect basic economic, technological, and ideological structures. The resulting state of affairs will be deeply different from the present.
g. The ideological change is mainly that of appreciating quality (dwelling in situations of inherent worth) rather than adhering to an increasingly higher standard of living. There will be a profound awareness of the difference between big and great.
h. Those who subscribe to the foregoing points have an obligation directly or indirectly to participate in the attempt to implement the necessary changes.

(4) Organic Marxism: From the 1970s, western Marxists, such as Andre Gorz, criticized capitalism with ecological thinking, and others like James O'Connor and John Foster studied and expounded ecological materialism and other thoughts by such classical writers as Marx and Engels,

giving rise to ecological Marxism. Meanwhile, process philosophers and constructive post-modern thinkers in the US, represented by Doctor John B Cobb Jr., deeply reflected on the ecological crisis caused by unbridled global development. During the study, they combined Whitehead's organic philosophy with classical Marxism, especially traditional Chinese culture, and put forth the nascent school of Organic Marxism. According to Doctor Cobb, of all Marxist schools in the west, organic Marxism was the only one so far with distinct cultural elements from China. In this sense, organic Marxism is neither a western nor a Chinese school, but a new international Marxism. According to organic Marxism, the real root of ecological crisis is modernity, especially the obsession with unlimited economic growth it contains, namely the pursuit for high-speed growth, which is exactly what organic Marxism opposes. In his *Organic Marxism*, Philip Clayton rolled out four principles of organic Marxism and sustainable development, including common good, organic and ecological thinking, caring for disadvantaged strata, and long-term and comprehensive viewpoint—all containing profound elements of ecological civilization. He also summarized the "ten ideas of saving the planet" generalized by Dr. Cobb, which could shed light on China's efforts of ecological civilization. The book *For the Common Good* co-authored by Cobb and Herman Daly elaborates on such new concepts as "common good" economics, ecological economics and sustainable economics from the broadest perspective of the entire biosphere on earth. There is no doubt that organic Marxism focuses on global ecological crisis, social crisis, cultural crisis and the reconstruction of environmental philosophy.

In general, the western thoughts such as Gaia hypothesis, process philosophy, deep ecology and organic Marxism mostly revolve around organic entirety, internal correlation, generation and development and pluralistic co-existence, and take an organic view of the earth as a network of lives. These thoughts carry deep-rooted ecological ideas, view nature and man in a whole, ongoing and sustainable light, and are ecological, organic and even grassroots in essence. That's why it is said that China's fine traditional culture provides the soil and root for ecological civilization, while the west's fine thoughts and culture nurture it with plantation and nutrition.

3 Ecological Civilization is the Chinese Expression of Sustainable Development

Sustainable development is "development that meets the needs of the present without compromising the ability of future generations to meet their own needs." As a new outlook on development, it is created in adaptation to the changes of time and the needs of social and economic development, and is the outcome of people's reflection on their path since entering industrial civilization. The 1972 UN Conference on the

Human Environment in Stockholm, *Our Common Future* published in 1987, the 1992 UN Conference on Environment and Development in Rio de Janeiro, the 2002 UN World Summit on Sustainable Development or Rio + 10 in Johannesburg, the 2012 UN Conference on Sustainable Development or Rio + 20, and the *Transforming Our World: The 2030 Agenda for Sustainable Development* adopted by UN General Assembly in New York in 2015—all of them are milestones in this process, witnessing how sustainable development has evolved from a concept to a common development strategy and practice upheld by all countries in the world.

Though the modern idea of sustainable development originated in the west, China has gradually accepted it while reflecting on its own development process and later elevated it to be a national strategy. During its reform and opening-up, China has not only witnessed and experienced global sustainable development, but also, as a major developing country, actively promoted this strategy and made immense contributions to it. Especially in the new era, China has provided Chinese wisdom to help achieve global sustainability, promoted win–win cooperation with ecological thinking, and endeavored to build a community with a shared future for mankind.

In some way, China's ecological civilization is consistent with the sustainable development advocated by the international community in that they both take an ecological view of the biosphere on earth, which gives birth to an ecological consensus on global development. They are both related and different from a rigorous academic point of view. Sustainable development is the result of deep reflections, made with ecological, process, organic and dialectical thinking, on the problems occurring in the highly advanced industrial (post-industrial) society. A basic consensus can be reached among all countries on sustainable development and environmental protection because sustainable development reflects the basic laws of biosphere. As a new concept that looks at the relations between man and nature, environment and economic development, and inter-generational and intra-generational equity in an ecological light, it in a way transcends such barriers as ideology, race and religious belief. The whole world implementing this concept is good for promoting ecological civilization in the human society on a higher level.

In comparison with sustainable development, ecological civilization is more macroscopic. Standing at a new height to grasp the development of human civilization, and holistically coordinates ecological, resource, environmental and development issues, and lets the ruling Party and government to deploy and plan ecological progress on all fronts, covering economy, politics, culture and society. Making ecological progress a component of the Five-Sphere Integrated Plan, China aims at the harmony between man and nature, environment and economy, and man and society on a higher level. The concept is a more scientific guide on enhancing our ability of sustainable development and achieving lasting development of the Chinese nation, and also a huge contribution to world sustainability, both theory and practice. The internationally advocated sustainable development is highly aligned with China's ecological civilization for their consistence in background, problems to be solved, path of implementation and goals to be achieved. They are essentially related as ecological civilization can be deemed as the expansion and sublimation of sustainable development and the Chinese wisdom and plan to bring it into reality. Going forward, the world will continue down the path of sustainable development. As General Secretary Xi Jinping said at the National Conference on Eco-environmental Protection, "We should work together on global ecological civilization construction and get deeply involved in global environmental protection governance to come up with a worldwide solution for environmental protection and sustainable development, while guiding international cooperation to tackle climate change."

4 Ideas and Practices of Ecological Civilization in China

The 40-year reform and opening-up in China has brought rapid industrialization and urbanization and much better lives to the people, but also resulted in a huge ecological deficit, with problems like air and river pollution, excessive heavy metal content in soil and food safety issues. For instance, small-scale pollution and black and fetid water bodies in certain regions haven't been fundamentally resolved, smog in some medium and large cities has incurred strong public complaint, a few enterprises are still discharging effluent against law, and the large-scale prospecting and mining activities in certain areas over the years have caused local vegetation damages, water and soil losses and surface subsidence in protected areas. All these have led to an arduous task of ecological restoration, affected social stability and obstructed sustainable economic and social development. It will take some time to solve the environmental problems piled up in the fast development process, and there is still a long way go to in environmental protection. Reality and international experience have proved that the traditional extensive development approach cannot sustain and the resources and environment have reached or approached the upper limit of their carrying capacity. Therefore, stepping up the efforts of ecological civilization and building a beautiful China is the only right way forward to keep the Chinese nation prosperous forever.

What ecological outlook a political Party holds depends on its philosophical thinking and its view of the biosphere on earth. The CPC is the only party in the world that has proposed, on the level of the Party and the state, to build ecological civilization. The report of the 17th CPC National Congress officially vowed to build ecological civilization, which later evolved into building a resource-saving and environment-friendly society. The 18th CPC National Congress included ecological progress as a link of the Five-Sphere Integrated Plan and put the establishment of ecological civilization into the Party Constitution, and proposed to build a beautiful China and strive for lasting national development. The 19th CPC National Congress made ecological civilization an important part of socialism with Chinese characteristics for a new era. As China has steadfastly advanced ecological progress and deepened institutional reform for that purpose, the whole country, including all local areas and all walks of life, has made active explorations and innovations to address protruding environmental problems concerning water, soil and air. Measures have been taken to protect the mountains, rivers, forests, farmland, lakes and grasslands in a holistic way, push the supply-side structural reform, pursue green development and take an active part in coping with global climate change, so as to contribute to the global ecological progress and sustainable development. In March 2018, ecological civilization was written into the Chinese Constitution to become the will of the nation and the people, guiding, as a basic state policy, the efforts to correctly handle the relation between economic development and environmental protection and to promote sustainable development. At the National Conference on Eco-environmental Protection held in May 2018, General Secretary Xi Jinping made a scientific summary of the Party and the country's strategic deployments and practical innovations for promoting ecological progress, saying that the Party's thoughts on this matter had been continuously enriched and perfected since the 18th CPC National Congress. Ecological progress is one of the five spheres; harmonious co-existence between man and nature is a basic element in the strategy of upholding and developing socialism with Chinese characteristics in the new era; green is a key aspect of the new development concepts; and pollution prevention and control is one of the three tough battles. The Five-Sphere Integrated Plan outlines the general layout of socialism with Chinese characteristics; upholding and developing socialism with Chinese characteristics in the new era is a basic strategy that must be implemented across the board to guide the Party's and the nation's undertakings; the new development concepts reflect our thoughts, direction and priorities of development; and the three tough battles are critical for securing a decisive victory in building a moderately prosperous society in all respects. In each and every one of them, ecological progress is a vital part, which indicates its importance in the undertakings of the Party and the nation. It is generally believed, both in China and abroad, that the past few years have witnessed the strongest measures, fastest progress and best effects in China's ecological progress, giving the people a keener sense of gain.

Since the 18th CPC National Congress, the CPC Central Committee with Xi Jinping at the core has put ecological progress in a prominent position in state governance. It has carried out a series of fundamental, innovative and insightful work, incisively answered a string of major theoretical and practical questions, such as why we build ecological civilization, what kind of ecological civilization to build and how to build it. These have given birth to the Xi Jinping Thought on Ecological Civilization, which is an important part of Xi Jinping Thought on Socialism with Chinese Characteristics for a New Era and a scientific guide on ecological progress in the new era.

The series of new concepts, new requirements, new goals and new deployments put forth by President Xi Jinping are the fundamental guidance and principles for enhancing ecological civilization and building a beautiful China. "Lucid waters and lush mountains are invaluable assets"; "We should protect the environment like we protect our eyes and treat the environment like it is our lives"; Protecting and improving the ecological environment is protecting and increasing productivity; "Mountains, rivers, forests, farmland, lakes and grasslands constitute a community of life"; and "creating a sound living environment and ecological products for the people"—these remarks have been deeply driven home with widespread popularity and support.

The Xi Jinping Thought on Ecological Civilization has rich contents. A civilization may thrive if its natural surroundings thrive, and will suffer if its natural surroundings suffer. Building ecological civilization is a fundamental strategy concerning the lasting development of the Chinese nation. In promoting this campaign in the new era, we must bear the following key points in mind.

One ultimate goal: Supplying more high-quality ecological goods to meet the growing demands of the people for a pristine environment.

Two development stages: By 2035, the environmental quality will be fundamentally improved and the goal of beautiful China basically achieved; by the middle of the twenty-first century, material, political, cultural, social and ecological progress will see all-round enhancement, green development mode and lifestyle will be established, man live in harmony with nature, and the national system and capability of environmental governance will be comprehensively modernized. A beautiful China will have been built.

Three basic judgments: Our efforts to build an ecological civilization are now in a crucial phase in which we must carry forward despite heavy strain and immense pressure, a decisive stage in which we will supply more high-quality ecological goods to meet the growing demands of the people for a pristine environment, and also a period of opportunity in which we have the conditions and abilities necessary to resolve prominent environmental issues.

Four clear-cut requirements: Promoting green development across the board; giving priority to addressing prominent environmental problems in the livelihood sector; effectively preventing environmental risks; and raising environmental governance level.

Five core systems: (1) The ecological culture system with ecological values as the norm; (2) the ecological economy system featuring industry-ecology integration; (3) the target and responsibility system centered on improving the environmental quality; (4) the ecological institution guaranteed by the modernization of governance system and capability; and (5) the ecological security system focused on sound ecosystem circulation and effective prevention and control of environmental risks.

Six major principles: (1) Uphold harmonious co-existence between man and nature, and prioritize resource conservation, environmental protection and letting nature restore itself. We should protect the environment like we protect our eyes and treat the environment like it is our lives, so that the beautiful nature and ecology will last forever in serenity and harmony. (2) Our natural environment is precious. Lucid waters and lush mountains are invaluable assets. Implement the concept of innovative, coordinated, green, open and shared development, and accelerate the formation of spatial patterns, industrial structures, modes of production, and lifestyles conducive to resource conservation and environmental protection, so as to leave time and space for natural and ecological restoration. (3) There is no welfare more universally beneficial than a sound natural environment. We should protect ecology for people's benefits, work hard to address protruding environmental problems detrimental to public health, and keep meeting people's growing needs for a beautiful environment. (4) Mountains, rivers, forests, farmlands, lakes and grasslands constitute a biotic community with a shared future. Ecological civilization must be developed on all fronts, in all regions and in full process in a holistic and coordinated approach with comprehensive and multi-pronged measures. (5) The strictest regulations and laws must be applied in protecting the environment. Institutional innovation should be accelerated and implementation reinforced to put an impregnable defense line on the utilization of natural resources. (6) Joint efforts must be made in building a global ecological civilization. China should deeply participate in global environmental governance, contribute solutions to world environmental protection and sustainable development, and lead international cooperation on climate change.

The six principles are the major guidelines with philosophical, global and realistic significance that we must abide by in developing ecological civilization; the five systems are the pillars that underpin and ensure ecological progress; the four requirements, which are specific and operable, point out our goals and tasks as well as the specific areas of ecological progress; the three judgments present the current situation of environmental protection in China – we must grit our teeth and conquer the challenges that lie ahead; the two stages outline a definite "timetable" and a blueprint that we should follow through; and the one goal is to meet people's aspiration and growing needs for a good life. These draw up a clear and definite route as well as a complete and operable construction drawing as to how to build an ecological civilization.

5 China is an Important Participant, Contributor and Torchbearer in the Global Endeavor for Ecological Civilization

The mankind lives in a community with a shared future and protecting the ecological environment is a common challenge and common responsibility faced by the whole world. Our success in ecological progress will do credit to socialism with Chinese characteristics; our failure in it will give those with ulterior motives an excuse to criticize and attack us. Ever since the human world stepped into industrial civilization, the rapid development of traditional industrialization has generated immense material wealth, but also expedited the exploitation of natural resources, disrupted the cycle and balance of ecosystems on earth and caused tension between man and nature. Since the 1930s, a number of environmental incidents have occurred in Western countries, causing huge losses, which have shocked the world and triggered a profound reflection on the development model of capitalism. In the process of more than 200 years of modernization, industrialization was realized in no more than 30 countries with a total population of no more than 1 billion. If we promote environmental progress and build China, the largest developing country with a population of more than 1.3 billion, into a modern socialist country that is prosperous, strong, democratic, culturally advanced, harmonious and beautiful, the influence of our achievement will be worldwide.

China was the first to issue the *National Plan on Implementation of the 2030 Agenda for Sustainable Development*, implement the *National Climate Change Programme (2014–2020)* and deposit the instrument of ratification of *The Paris Agreement* with the UN. China's elimination of ozone-consuming substances accounts for over 50% of the total elimination in developing countries, making the greatest contributions to global ozone preservation.

China has deeply engaged in and led the international cooperation in response to climate change, and is one of the first countries to approve the *Paris Agreement* through legislative procedure and deposit its national climate change plan with the UN. As a responsible and major country, China has actively participated in global environment governance and honored its commitment to emission reduction. It pledged as early as in 2015 to peak its CO_2 emission around 2030 and strive to achieve that at an early date. To that end, the Chinese government has adopted a raft of powerful environmental protection measures, such as tightening bulk coal management and energy conservation and emission reduction in key industries, and optimizing the energy mix and decreasing the ratio of coal consumption. From 2013 to 2018, China's energy and water consumption per unit GDP dropped more than 20%, the emission of main pollutants kept falling, the number of heavily polluted days was halved in key cities, 163 million *mu* new forests were planted, and the area of sandy land shrank nearly 2,000 square kilometers per year.

At the 2015 UN Climate Change Conference held in Paris, President Xi Jinping called on all countries to abandon the narrow-minded "zero-sum game" and urged them, especially developed ones, to work for mutual benefits by sharing more and bearing more responsibilities. While the international community is still arguing about the division of duties, the Chinese government has begun to actively cope with climate change with amazing resolve and intensity. While the US announced to withdraw from the *Paris Agreement*, China reaffirmed its position of upholding its goals and commitments regarding climate protection, invigorating the world with its sincerity and determination. In the twenty-first century, China has been continuously making contributions to the global low-carbon economy, not only making its policies more transparent to enhance mutual trust, but also providing other developing countries with strong financial support to transform and upgrade their development mode. As early as in December 2015, the People's Bank of China issued green financial bonds. At present, China boasts the largest green bond market in the world and has created the largest carbon emission trading system as well. In September 2017, China kicked off a research program on the timetable for the halt of production and sales of cars burning traditional energies, and it is now the world's largest new energy vehicle (NEV) producer and consumer. Through ceaseless innovation and development, China has de facto become the global benchmark in such areas as environmental protection, renewable energy utilization and green finance.

After initiating the Belt and Road Initiative over six years ago, China has signed 173 cooperation documents or agreements (note: by the end of March 2019) with 125 countries and 29 international organizations. As President Xi Jinping said, the BRI originated in China, but its opportunities and fruits belong to the whole world. China has no geopolitical calculations, seeks no exclusionary blocs and imposes no business deals on others. The claim by certain foreign individuals and media that the BRI, while bringing development opportunities to regions and countries along the route, will threaten local ecology and environment is totally groundless. The BRI advocated by China is a path of green development. When working with BRI countries and regions, Chinese enterprises strictly implement the local environmental requirements and provide them with advanced technologies. It is by means an output of backward production capacity, but an initiative to build a beautiful and clean world with sound ecology and common ecological and environmental security. Now that China has become a world leader in energy conservation, environmental protection and renewable energy utilization, it is trying to spread its ecological concepts to the world, offer Chinese wisdom and plan for building a community with a shared future for mankind and resolving the world's environmental problems. China strives to be a major participant, contributor and torchbearer in the global endeavor for ecological civilization.

In his speech titled "Working Together for a Green and Better Future for All" at the opening ceremony of the International Horticultural Exhibition 2019 Beijing, President Xi Jinping said that we need to "advocate harmony between man and nature," "pursue prosperity through green development," "follow a philosophy that cares for nature," "adopt a scientific approach to ecological conservation," and "join hands to meet common challenges." He vowed to provide supportive concepts and Chinese wisdom for developing global ecological civilization. Building a beautiful home is the common dream of mankind, and no country can stay intact by itself in this community of intertwined interests and shared future in face of ecological and environmental challenges. Only through concerted efforts can we effectively deal with global environmental issues such as climate change, marine pollution and biological protection, achieve the 2030 sustainable development goals set by the UN, drive home the concept of green development, and ensure steady and sustained progress in ecological civilization development worldwide.

Ecological civilization is the wisdom, experience and innovation contributed by China to global ecological governance and protection. It belongs to China as well as the world and all mankind.

The Rise of a Green Society

Dayong Hong

Ever since its emergence, human being has been in a relation of unity of opposites with the environment. On the one hand, obtaining resources from the environment is the basic condition of human survival and development, and man's productive and living activities always exert impacts on the environment, as reflected in the consumption, deterioration and damage of it. On the other hand, excessive exploitation of resources and environmental damage will eventually affect human survival and development. Therefore, in different development stages of productivity and different social systems, man, based on its understanding of its relation with environment that is acquired from productive and living practices, has always incorporated environmental factors in its activities of social construction in various ways, aiming at the coordination between human society and environment. In this sense, there is always an environmental dimension to social construction though it varies in different historical periods and backgrounds. For one thing, the human perception of environmental status and impacts varies in different stages of social development; for another, the inclusion of environmental factors in social construction varies in breadth, depth and intensity. This paper tries to analyze how China has incorporated environmental factors in its social construction and gradually evolved into a green society since reform and opening-up. A green society refers to a process and state where man, on the basis of understanding the interaction between society and environment, consciously pushes for social reform for their coordination. This is an important aspect, even a key connotation with dispersive and penetrative effect, of social construction in contemporary China.

D. Hong (✉)
Department of Degree Management & Postgraduate Education, Ministry of Education, Beijing, China

J. Pan et al. (eds.), *Beautiful China: 70 Years Since 1949 and 70 People's Views on Eco-civilization Construction*, https://doi.org/10.1007/978-981-33-6742-5_46

1 Social Construction: From Developing Environment to Protecting Environment

If the social security and the improvement of livelihood is the focus of social construction, the chief task at the beginning of reform and opening-up is eliminating poverty and improving people's lives. In 1978, China's per capita GDP was only RMB 385, ranking way back in the world, and 97.5% of its rural population lived in poverty with insufficient clothing and food according to the current standard. Under such circumstances, translating environmental resources into goods and services on a larger scale, at a faster pace and with a higher efficiency to lift people out of poverty was naturally the top priority, so the development and utilization of environment was the main social activity.

The book *Poverty of Abundance* published in 1986 was very influential at the time. It analyzed why the western region fell behind the eastern region and pointed out the path for it to step out of poverty—by improving human quality, changing the basic social structure, taking a new look at environmental resources, and changing the way of resource exploitation and utilization. Marveling at the abundant resources in the west, the author explained why the people there "failed to do things in spite of the resources that could have made them succeed in everything"—it was because of the "traditional socioeconomic structure and the people with low quality of commodity production couldn't effectively develop and utilize the resources and create more wealth. In return, the low level of resource development and utilization and human quality dragged the change of basic social structure, resulting in a vicious cycle of childish social system and its poverty." The author deemed agriculture and animal husbandry as the root cause of western region's poverty and backwardness. "In the relation between human production and natural resources, the modern production mode makes multi-layered development and multi-time utilization of natural resources." Therefore, it is necessary to change people's ideas and their ways of production and life, and develop and utilize environmental resources according to commodity economy laws to shake off poverty. Though focused on the western region, this book is to a large extent applicable to China's development in general, and its views on the relation between man and environmental resources are typical in the early stage of reform and opening-up.

Indeed marketization, industrialization and urbanization have dramatically changed people's development and utilization of resources, boosted economic growth and improved people's lives. China's achievements in poverty alleviation and economic growth since its reform and opening-up have benefited all people in the country and astonished the world. Yet we must be mindful of the louder voices for environmental protection and improvement instead of simple development and utilization as reform and opening-up goes deeper. In addition to the basic state policy of environmental protection (1983) and the sustainable development strategy (1995), the CPC Central Committee proposed to build an environment-friendly society at a symposium on population, resource and environmental work held in March 2005, and the Fifth plenary session of the 16th CPC Central Committee held in October that year further specified the goal of building a "resource-saving, environment-friendly society." In 2007, the 17th CPC National Congress vowed to "develop ecological civilization." In 2012, the 18th CPC National Congress stated that man and nature constitute a community of life and man must respect, accommodate to and protect nature. Only by observing natural laws can man avoid detours in the development and utilization of natural resources. We should, acting on the principles of prioritizing resource conservation and environmental protection and letting nature restore itself, develop spatial layouts, industrial structures, and ways of work and life that help conserve resources and protect the environment. With this, we can restore the serenity, harmony, and beauty of nature. We must put ecological progress in a key position, put it into practice in every aspect and the whole process of economic, political, cultural and social development, in the endeavor to build a beautiful China and realize lasting development of the Chinese nation. The 19th CPC National Congress in 2017 singled out pollution prevention and control as one of the three tough battles that must be won in building a moderately prosperous society in all respects.

In particular, General Secretary Xi Jinping's important remarks on environmental protection have vividly and profoundly expounded the significance of this work during social construction. For instance, "the environment has no substitute, and we may not realize how truly precious it is until it is damaged to the point of being unsalvageable"; "a civilization may thrive if its natural surroundings thrive, and will suffer if its natural surroundings suffer"; "we should protect the environment like we protect our eyes and treat the environment like it is our lives"; "protecting and improving the environment is protecting and increasing productivity"; "lucid waters and lush mountains are invaluable assets"; and "a good ecological environment is the fairest public product, and the most accessible welfare for the people," and so on and so forth. Under the firm leadership of CPC Central Committee with Xi Jinping at the core, China has accelerated the top-level design and institutional establishment for ecological civilization since the 18th CPC National Congress. It has set to protect the environment with the most rigorous systems and regulations, made faster institutional innovations, reinforced execution and taken a series of fundamental, innovative and forward-looking steps. As a result, we have got onto a fast lane of environmental governance addressing symptoms and root causes simultaneously, along which historic, dramatic and comprehensive changes are taking place. As such, environmental factors have been integrated in social construction in a new light and have driven the profound changes of the society.

2 Main Endogenous Drive for Social Transformation

If our initial steps of environmental protection were somehow influenced by the international wave of environmental protection, our sustained and deepened efforts today for environmental protection and ecological progress are a conscientious initiative in response to domestic development demand. Particularly, in comparison to the most developed countries' retrogression in environmental protection and the consequent countercurrent, China's green social construction has further asserted its independence rather than being forced by external influence. What exactly is the endogenous drive of such a green turn in Chinese society? This involves multiple forces of all fronts and levels, with the following five points being especially crucial.

First, the conflict between development and environment becomes increasingly pointed and environmental quality faces serious threat. After decades of traditional high-speed growth, the huge debts we owe to ecological environment have become a marked weak link. In 2012, China's economic aggregate accounted for about 11.5% of world total, for which, however, we consumed 21.3% of global energy, 45% of steel, 43% of copper, 54% of cement and emitted the most SO_2 and NO_x. In 1985, China discharged 34,154.2 million ton wastewater, which kept rising to 71,109.54 million tons in 2016; it emitted 13.25 million ton SO_2 in 1985, which peaked at 25.888 million tons in 2006 and fell to 11,028,643 tons in 2016.

According to the *Bulletin of Environmental Situation in China* issued by the Ministry of Ecology and Environment in 2017, only 99 of the 338 cities (29.3%) of and above prefecture level passed muster in urban air quality, while the rest 239 cities (70.7%) all fell short of the standard. Of the 112 important lakes (reservoirs) nationwide, 6 had Grade I water quality, 27 Grade II quality, 37 Grade III, 22 Grade IV, 8 Grade V and 12 Grade V-minus, accounting for 5.4%, 24.1%, 33.0%, 19.6%, 7.1% and 10.7% respectively. Regarding the quality of groundwater, monitoring points that recorded excellent, good, fairly good, fairly bad and bad groundwater quality accounted for 8.8%, 23.1%, 1.5%, 51.8% and 14.8% respectively. In fact, in addition to grave air and water pollution, other forms of pollution such as solid waste and soil pollution are serious too; and there wasn't just environmental pollution, but also ecological destruction. Of all 2591 counties across the country, those with "excellent" and "good" environmental quality take up only 42.0% of national territory, the "mediocre" ones take up 24.5% and "fairly bad" and "bad" ones 33.5%. Moreover, the worsening environmental quality has led to mounting threats to food and drug safety, life and health. These make the background and key motivation for us to restructure the society.

Second, people's needs have changed dramatically. The biggest achievement in our social construction since reform and opening-up is that we have substantially raised people's living standards, basically satisfied their material needs, and kept them generally well-off. According to the current standard, the ratio of poor rural population fell from 97.5% in 1978 to 3.1% in 2017, and a national subsistence allowance system for both urban and rural residents has been put in place to institutionally guarantee the basic life of every citizen. In 2017, the per capita disposable income in China reached RMB 25,974, and the Engel coefficient fell to 29.3% in general from 57.5% (urban) and 67.7% (rural) in 1978, hitting the affluence line set by the UN. More importantly, the more assets people have, the more capable they are to fend off risks. For instance, the total household deposits increased from RMB 21.1 billion in 1978 to RMB 62.6 trillion in 2017. On that basis, people's needs are more diverse and high-level, with greater emphasis on security, comfort and sustainability. Eating safe good, drinking clean water, breathing clean air, and living a sustainable life in a pleasant environment have been increasingly acknowledged as the basic need.

Therefore, the public is paying more attention to environmental quality, which has become a familiar and important topic among the public and media. I participated in organizing a national environmental awareness survey in 1995, and the results showed that 23.6% of the respondents had no idea of environmental protection, 16.5% of them had "very little" knowledge about it, 66.9% had "little knowledge," 16.1% had "a fair amount of knowledge," and only 0.5% considered themselves as having "a lot of knowledge" about environmental protection. Meanwhile, most urban and rural residents were unfamiliar with relevant policies and regulations. Only 31.8% considered themselves as "know something or know very well" (only 0.5% thought they know very well), 42% "heard of them," and 26.2% "never heard of them." However, according to the results of 2010 Chinese General Social Survey (CGSS) that I participated in designing, 70% of the respondents realized China was facing "very serious" or "quite serious" environmental problems, 65.7% of them were "very concerned" or "quite concerned" with environmental issues, and only 3.1% "didn't care at all."

Third, social tension and conflicts arising from environmental damages (risks) become increasingly marked. Easing social relation and dissolving social conflicts is an important part of social construction. In a period of social transition, labor disputes, land acquisition and demolition and social security used to be main reasons for social conflicts. As people live a better life and become more aware of environmental rights and interests, actual environmental damages and possible environmental risks also become important reasons that aggravate social conflicts, making the green transformation of society an inherent need for promoting social harmony. According to data published by the state, both the number of people petitioning against environmental pollution and the frequency of such petitions have increased in a certain period of time. In 1987, 77,673 people petitioned against environmental pollution, which increased to 139,424 in 2000. No statistics were available in 2001–2010, but the number was 104,323 in 2015. Some petitions were made by large groups, the number of which was 47,714 in 1996, 88,237 in 2005 and 48,010 in 2015. The actual number of petitioning groups may be larger as the statistical specifications have changed.

In the twenty-first century, environmental incidents usually involving tens of thousands of people captured extensive attention. Such examples as the PX event in Fujian's Xiamen in 2007, cadmium pollution in Hunan's Liuyang, "lead in blood" event in Shaanxi's Fengxiang and the opposition to the construction of waste incineration power plant in Guangdong's Panyu in 2009, PX event in Liaoning's Dalian in 2011, opposition to PC project in Tianjin, resistance against Oji Holdings Corporation's project in Jiangsu's Qidong and resistance to Hongda Group's molybdenum copper alloy project in Sichuan's Shifang in 2012. The list goes on. Some of these projects triggered protests and conflicts in the stage of environmental impact assessment, such as the event in Jiangsu's Qidong, some caused conflicts during construction, and some after completion and operation, such as the PX event in Dalian. Studies show that during the decade from 2003 to 2012, 230 environmental incidents of a rather large scale were reported by the media, with 58 in 2011, displaying an obvious upcurve. Such a trend has been corroborated by leaders of environmental protection authorities, posing a contrast to the decline of conflicts in other forms. It is reported that Zhou Shengxian, former Minister of Environmental Protection, once said, "While the total number of petitions, and the number of group petitions, irregular petitions and mass events have been falling, the number of environment-related petitions and mass events has increased by more than 30% every year."

Fourth, the CPC and the government have adjusted their work priorities and acted proactively. The socialism with Chinese characteristics, as the basic system of China, is essentially distinguished by the adherence to the CPC's leadership. The CPC in return upholds the people's principal position, implements its fundamental purpose of serving the people whole-heartedly, works to meet their aspiration for a good life, and adjusts its work directions and priorities continuously according to the changing social contradictions. At the beginning of the reform and opening-up when productivity was low and people generally lived in poverty, accelerating the emancipation and development of productivity, stepping up the development and utilization of resources and environment and taking economic growth as the center were naturally and imperatively made priorities. But even in such a case, the CPC and the government still attached importance to environmental protection and kept improving the team, organization and institution for that purpose while boosting economic growth.

Since we entered the twenty-first century, especially after the 18th CPC National Congress, the CPC Central Committee, having deeply analyzed the changing trend in main social contradictions, pointed out in the report of the 19th CPC National Congress, "As socialism with Chinese characteristics has entered a new era, the principal contradiction facing Chinese society has evolved. What we now face is the contradiction between unbalanced and inadequate development and the people's ever-growing needs for a better life. China has seen the basic needs of over a billion people met, has basically made it possible for people to live decent lives, and will soon bring the building of a moderately prosperous society to a successful completion. The needs to be met for the people to live better lives are increasingly broad. Not only have their material and cultural needs grown; their demands for democracy, rule of law, fairness and justice, security, and a better environment are also increasing. At the same time, China's overall productive forces have significantly improved and in many areas our production capacity leads the world. The more prominent problem is that our development is unbalanced and inadequate, which has become the main constraining factor in meeting the people's increasing needs for a better life."

Based on this judgment of the overall situation of socioeconomic development and the understanding of the changed principal social contradiction, the CPC and the government, bearing in mind people's general and long-term interests, have placed more stress on high-quality development, promoted the Five-Sphere Integrated Plan, and made careful deployments for the tough battle of environmental protection. The Third plenary session of the 18th CPC Central Committee decided that we must establish a systematic and complete institution for ecological civilization, carry out the most rigorous source protection system, damage compensation system and accountability system, and perfect the environmental governance and ecological restoration system to protect the ecological environment. China's institutional work for ecological civilization has made rapid and substantial progress since 2015.

After the new *Law on Environmental Protection*, considered "most stringent in history," officially came into force in January 2015, a long list of important documents have been rolled out, including the *Opinions on Accelerating the Development of Ecological Civilization, Methods of Public Participation in Environmental Protection, Plan for Inspection and Supervision of Environmental Protection (Trial), Methods of Holding Party and Government Leaders Accountable for Environmental Damages (Trial), Master Plan for Institutional Reform for Ecological Civilization, Guiding Opinions on Pilot Implementation of Vertical Management Reform of the Monitoring, Inspection and Law Enforcement by Environmental Protection Agencies under Provincial Level, Air Pollution Prevention and Control Action Plan* (commonly known as "10-Chapter Air Pollution Action Plan", *Water Pollution Prevention and Control Action Plan* (commonly known as "10-Chapter Water Pollution Action Plan"), *Soil Pollution Prevention and Control Action Plan (commonly known as "10-Chapter Soil Pollution Action Plan"), Action Plan for Soil Pollution Prevention and Control, Reform Plan for Environmental Damage Compensation System*, and *Several Opinions on Delimiting and Observing the Red Line of Ecological Protection*. Ecological civilization is also included in the 13th Five-year Plan. In particular, after the inspection and supervision mechanism was established based on the *Plan for Inspection and Supervision of Environmental Protection (Trial)*, the central environmental protection authority has completed the first round of inspection in full coverage. According to the *Bulletin of Environmental Situation in China* issued by the Ministry of Ecology and Environment in 2017, this round of inspection saw more than 18,000 Party and government leaders held accountable, handled 135,000 environmental complaints made by the public, and directly pushed for the solution of over 80,000 environmental problems immediately concerning the people. In 2017 alone, environmental inspectors held talks with 30 cities (counties and districts) departments and units, imposed administrative punishment on 233,000 cases nationwide, and imposed fines totaling RMB 11.58 billion, up 265% from 2014 before the new *Law on Environmental Protection* was promulgated. As a matter of fact, the increasingly rigorous and meticulous institutional design and execution has transmitted the pressure of environmental protection from the central level to local level, from government to enterprises, and from the state to individuals. The green storm waged by the Party and the government to supervise government and business behaviors in environmental protection is blazing a Chinese path of composite environmental governance. Such proactive adjustment and action is intrinsically inevitable for the Party and the government that always center their work on people's interests.

Fifth, in face of environmental deterioration, people's changed consumption preference and government regulation and governance, businesses have seen new opportunities to make profits and displayed a more obvious green tendency. The accelerated development of new industries, business forms, modes and products is boosting social changes while meeting new needs. China has kept increasing its environmental input since the reform and opening-up, which for the first time accounted for more than 1% of GDP in 1999 and rose to 3.5% during the 12th Five-year Plan period, giving a strong boost to the rapid development of environmental protection industry. In 2000, the industry registered a yearly output value of RMB 108 billion, which rose to RMB 1100 billion in 2010.

Apart from environmental protection industry, resource-oriented enterprises that specialize in resource mining and produce mineral and primary products command an important position, but their production is accompanied with serious environmental impacts. In 2013, the industrial solid waste, waste water and waste gas emitted by resource enterprises accounted for 97.1%, 77.7% and 92.4% of all industrial discharge. But recent studies show a growing trend of green behaviors among them, although still with room for improvement. For instance, 84.5% of the resource enterprises surveyed included environmental protection in their target system, 80.4% attached importance to their environment-friendly image, 70.3% regularly trained their employees in environmental awareness and environmental management skills, 69.4% seen their employees actively participate in environmental practices, 83.5% considered energy conservation and circular energy use in production design, 77.6% prioritized renewable and recyclable production materials, 79.4% had material and scrapped recycling system during production, and 80.8% adopted environment-friendly production process. All these indicate that businesses' green actions in rational pursuit for profits may also become an endogenous drive for social transformation.

3 Effects and Prospects of Green Social Construction

Green social construction generates multi-faceted effects, with environmental effect being a main facet. From the perspective of environmental protection, an official at the Ministry of Ecology and Environment described with five "unprecedented" the profound social changes aimed at better environmental quality since the 18th CPC National Congress—the depth of understanding is unprecedented, the intensity of pollution prevention and control is unprecedented, the high frequency of the issuance of new regulations is unprecedented, the vigor of law enforcement is unprecedented, and the high speed of environmental improvement is unprecedented.

Relevant materials indeed denote remarkable progress in environmental protection and ecological progress in the five years after the 18th CPC National Congress. For example, the forest coverage rate in China has increased continuously from 21.38% in 2012 to 22.3% in 2016; the average concentration of inhalable particle (PM 10) in 338 prefecture and higher level cities fell by 22.7% from 2013; the average concentration of fine particulate matter (PM 2.5) in the Beijing-Tianjin-Hebei region, the Yangtze River Delta and Pearl River Delta was down by 39.6%, 34.3% and 27.7% from 2013; and after the promulgation of "10-Chapter Water Pollution Action Plan," Grade I, II and III surface water rose from 6% to 67.8% and that of Grade V-minus surface water fell from 9.7% to 8.6%. The *2017 Bulletin of Environmental Situation in China* made the following conclusion: "the air and water quality in China is further improved, soil environment risks are curbed, ecosystem is generally stable and nuclear and radioactive security is effectively guaranteed. People have truly felt the positive changes in environmental quality."

In terms of energy conservation and emission reduction, China's GDP has increased faster than its total energy consumption after 2008. If we trace farther back, our energy consumption per RMB 10,000 GDP has kept falling since the reform and opening-up and hit 0.588 ton standard coal in 2016 according to data from the National Bureau of Statistics. During the implementation of the 12th Five-year Plan, the carbon intensity in China dropped 20%, better than the 17% target set in the plan. Our SO_2 emission has been on a continuous, accelerated downcurve after peaking in 2006.

Yet we should have a right attitude toward the notable environmental improvements resulting from green social construction. On the one hand, we must fully realize that this is the right direction, steadfastly promote green transformation in society, continue to work for harmonious co-existence between man and nature, and provide more quality ecological products to meet people's ever-growing needs for a beautiful environment. On the other hand, we should remain science-based and calm, make continuous self-reflections, and fully realize that even we have achieved interim positive results, local and green social construction is a complicated process. Without comprehensive, in-depth and sustainable green social construction, the improvements so far won't sustain, and man-nature harmony will be hard to achieve or effectively maintain.

Why? While building a green society is without any doubt the only right choice in light of the country's situation, the natural laws and the people's needs, it is nonetheless a tough and long-term campaign. I can think of at least the following reasons. First, human interference and impact is still limited in comparison to the environmental system's own evolution. Some environmental problems have complicated causes, either human or natural or both. For instance, our efforts to control air pollution can hardly affect climate change and the long-term, complicated evolution of earth's environmental system, both of which are usually important factors that either aggravate air pollution or offset the effect of air pollution control. It's safe to say that our knowledge of the laws governing the operation and evolution of the earth and the universe is very limited, so instead of indulging in our "victory" over nature, we must more consciously respect, accommodate to and protect nature and abide by its laws. Second, compared with fluid air and water pollution that is highly visible with definite pollutants, pollution that's invisible and stagnant with complex causes tends to be neglected and therefore harder to address given its continuous accumulation, extreme complexity and delayed social impacts, such as soil pollution, gene pollution and biodiversity losses. As a matter of fact, this type of pollution may threaten human health and social sustainability on a deeper level and in a wider scope, but we are barely able to cope with it yet. Third, with regard to social transformation, the current focus is on changing the administrative system and behaviors and adjusting the spatial layout, industrial structure and production mode, which is onerous and difficult especially as sustained change faces objective challenges. In contrast, however, adjusting lifestyle, guiding public behaviors and converging social consensus is something more difficult where quick progress is almost impossible. In this field, we haven't done enough and effective measures are limited. Fourth, compared to the establishment of rules and regulations, their execution is more complicated and difficult. In particular, internalizing them into social members' conscientious acts is a long and complex process, even with distortion, conflict and relapse. This is an important area we should pay special attention to and from which we can draw wisdom of such social sciences as sociology and psychology. We must pay more attention to people's green practices in everyday life and, aim at a green lifestyle, rebuild living infrastructure, restructure opportunities and space of daily life, and arrange convenient and effective living guidance, in a bid to promote green social construction on a deeper level. Otherwise the superficial reforms would be delayed, dysfunctional or even subverted due to deep-going reasons. Fifth, although the main contradiction in Chinese society has changed, our fundamental national context that China is still in the primary stage of socialism and will stay so for a long time remains unchanged, and our international standing as the world's largest developing country remains unchanged. Therefore, the contradiction between development and environment will continue to exist for a long time, and we still need to balance them and achieve higher-quality development while boosting environmental protection. Given the unbalanced development between urban and rural areas, different regions and groups and the diversification of social values, environmental equity will be a prominent challenge in the process of green social construction. Facing up to this challenge and properly handling it will increase the endogenous drive for green social construction, whereas neglecting and mishandling the challenge will undermine the green consensus and increase the endogenous resistance to green social construction.

The current green social construction in China is just the first ray in the morning, while the journey ahead remains long and hard. A real green society not only needs an extensive and dominant green consensus and scientific, comprehensive, systematic and meticulous institutional arrangements, but also rigorous, effective and regular execution. It has to develop suitable technologies and transmit knowledge and information, form many active green social organizations and carry out green activities, and systematically restructure and reshape the public living practices. In the age of globalization, green social construction in one region is bound to be affected by the external environment and has to have effective interactions and coordination with it. Considering the current development status, the ultimate green society will definitely be a global one, which requires coordinated and profound reform across the globe and calls upon all people in the world to truly stand in awe of nature and cherish the community with a shared future that we all live in.

Environment is a Major Political Issue Which Bears upon the Mission and Purpose of the CPC

Guang Xia

At the National Conference on Eco-environmental Protection, General Secretary Xi Jinping said "environment is a major political issue which bears upon the mission and purpose of the CPC." This was a solemn and thought-provoking remark in that it not only directly connected environmental protection and ecological progress with the Party's properties for the first time, but also elevated them to the highest political level.

Since the 18th CPC National Congress, General Secretary Xi Jinping has repeatedly elaborated on environmental protection and ecological progress from a political perspective, which has become an important part of the Xi Jinping Thought on Ecological Civilization. At the meeting of the Standing Committee of the Political Bureau of the 18th CPC Central Committee held in 2013, Xi made it clear that "we cannot view strengthening ecological progress, advancing environmental protection and promoting a green and low-carbon lifestyle merely as an economic issue. It is a major political issue." At the National Conference on Eco-environmental Protection, he further stressed that "we should give full play to the Party's leadership and the political advantage of our socialist system that it is able to pool forces to accomplish grand undertakings," and "all regions and departments should reinforce the 'Four Consciousnesses,' resolutely safeguard the CPC Central Committee's authority and centralized, unified leadership, and steadfastly shoulder the political responsibility for ecological civilization development." These simple words carry far-reaching and deep-going significance.

"Ecological environment is a major political issue concerning the CPC's mission and purpose"—deeply understanding this proposition, including its contents and reasons, is of great importance for sharpening our political consciousness, increasing our initiative in implementing the Xi Jinping Thought on Ecological Civilization and the requirement of the National Conference on Eco-environmental Protection, and issuing political measures for environmental protection.

G. Xia (✉)
Ministry of Ecology and Environment of the People's Republic of China, Beijing, China

J. Pan et al. (eds.), *Beautiful China: 70 Years Since 1949 and 70 People's Views on Eco-civilization Construction*, https://doi.org/10.1007/978-981-33-6742-5_47

1 How to Understand "Environment is a Major Political Issue Which Bears upon the Mission and Purpose of the CPC"?

Politics refers to the process in which people carry out activities revolving around public power and the government uses public power to distribute resources with authority.[1] In short, politics means using political power to run the country. Only those matters that concern the overall national development and call for the exercise of state power are major political issues. As such, we can understand why "environment is a major political issue which bears upon the mission and purpose of the CPC" on three dimensions—mission, purpose and state.

1. **Correctly handling the relation between man and nature is an important subject in the Party's effort to govern and revitalize the country**

Ever since the CPC came in power, it has been committed to seeking happiness for Chinese people and great rejuvenation for the Chinese nation. This is the CPC's grand mission but also a massive and onerous campaign, in which countless complicated conflicts and relations have to be addressed, posing high requirements on every aspect of state governance and allowing no obvious short link in this system. Correctly, the CPC attaches great importance to handling the relation between man and nature. As General Secretary Xi Jinping emphasized, all-round moderate prosperity is a political commitment; we cannot vow to build the moderately prosperous society on one hand and leave a poor ecological environment on the other—which won't be accepted by the people nor stand up to the test of history. That the Xi Jinping Thought on Socialism with Chinese Characteristics for a new era explicitly puts forth the Five-Sphere Integrated Plan encompassing economic, political, cultural, social and ecological progress underscores the systematic and complete system of CPC's state governance. It is an advanced philosophy and approach of state governance in the modern world.

(1) Historical price

Governing a gigantic country is a difficult journey teeming with hardships and setbacks. The Five-Sphere Integrated Plan has only been formed after long years of explorations and summary at a high historical price. To be objective, China's state governance system, for a very long time in the past, lacked the part about correctly handling the relation between man and nature. It was generally believed that natural elements like resources, environment and ecology were all abundantly at our disposal, and our main task was to make full use of them for national development. As a result, environmental protection has been lagging behind the rapid economic and social development, even leading to the "price scissors"—the faster the economic society develops, the greater environmental pressure and the more environmental problems. Even today, we still have to face the huge imbalance between

[1] Yang (2012).

socioeconomic development and environmental protection—while people's material life is significantly improved and urban and rural construction progresses by leaps and bounds, the ecological environment is getting increasingly overburdened. Such a governance system is unsound, which results in unbalanced and unsustainable development that obviously deviates from the Chinese dream of great national rejuvenation.

Such a development state brings many adverse effects on our economic, livelihood and political development (eg. undermine the "Four Confidences"). For a long time in the past, there has been a sore spot in the heart of the Chinese—although China's overall national strength is among the top in the world and our remarkable achievements in economic growth and urban and rural construction have been widely acclaimed, we are still labeled as a country with poor environmental performance. There was a time when ecological environment in China was associated with foul air and sprawling sewage. Foreign media called Beijing one of the capital cities with the worst air quality, and even asked "will China be suffocated in its self-imposed pollution?" Chinese people loathed remarks like that but couldn't utter a word of objection. The severe environmental pollution and ecological damage have dampened our sense of pride on our development achievements. When speaking of our development path and pattern on the international stage, we were not completely confident and assured because "our economic growth came at a great price of resources and environment." Confidence has to be rooted somewhere. If our own ecological environment are not protected, if clean water and air are not secured everywhere, how can our development pattern be convincing and compelling in the world? It's safe to say that the environmental issue is a major short link that undercuts people's confidence in the Chinese development path.

(2) Why do we lag behind?

Why was environmental protection lacking in the governance system of the past? It had something to do with the principal social contradiction at the time, which was the contradiction between people's ever-growing material and cultural needs and the backward social productivity, so the focus of state governance was naturally on economic and social development. At one time there were extensive discussions about depriving China of its "global membership" for its backward economy. Given such a strong sense of crisis, economic development became the overriding priority, and "only development counts" actually meant "only economic development counts." Under such circumstances, the institutional setup of state governance and resource allocation were disproportionately tilted toward economic development, while environmental protection was marginalized. The result is that economic development has seen all-round expansion in scale, structure, distribution and speed, putting tremendous pressure on the ecological environment, while environmental protection always stands at disadvantage to economic development even though more attention has been paid to it, greater efforts made and impressive progress achieved. The system and capability of environmental governance falls seriously behind the actual needs for environmental protection and improvement; the laws, regulations, standards and policies in that regard are not established and perfected until long afterwards, and the same happens to the regulatory mechanism and administrative forces needed to ensure rigorous execution. The development of environmental protection agencies is far behind economic development authorities, and the Ministry of Environmental Protection wasn't formed until 2008.

(3) Important turning point

The consistent backwardness of environmental protection began to change after the 18th CPC National Congress, when our principal social contradiction was between people's ever-growing needs for a better life and the unbalanced and inadequate development—this is the background for the rise of environmental protection. After the Five-Sphere Integrated Plan was determined, efforts of environmental protection have obviously been stepped up with the issuance of tens of institutional reform plans for ecological civilization and the implementation of the stringent new *Law on Environmental Protection*. Thanks to the central environmental protection inspections and the accountability system, a large number of enterprises that are illegally built, poorly managed and highly polluting have been shut down, clean energies have replaced bulk coal burning on a large scale, and masses of heavily polluting production capacity have been eliminated. The economic structure and development quality is improved, so is environmental quality. Ecological progress has gradually become a widely understood and accepted political consensus among Party and government leaders of all levels as well as among the public. In sum, the series of fundamental, innovative and forward-looking work conducted since the 18th CPC National Congress has brought about historic, dramatic and comprehensive changes in environmental protection, indicating that our governance system is getting more complete and sound. General Secretary Xi Jinping stressed at the National Conference on Eco-environmental Protection that we must move faster to build systems for an ecological civilization with ecological culture, ecological economy, targets and responsibilities, institutional guarantee and ecological security as the main contents. In other words, we must make comprehensive arrangements for ecological progress as a key domain of the Party's state governance, and work to modernize the governance system and capability.

As a whole, continuously deepening our understanding of China's national conditions, correctly handling the relation between man and nature and speeding up ecological progress are a new proposition on our way to achieving the historical mission of Chinese nation's great rejuvenation, and is bound to become a key domain of the CPC's state governance. We have been negligent and delayed on this front in the past, but now we are catching up and shoring up this weak link—this is a major political issue and task that calls for the united efforts of all Chinese.

2. **Ecological civilization is an inherent part of socialism**

Article 1 of the Constitution of the People's Republic of China provides, the People's Republic of China is a socialist state under the people's democratic dictatorship led by the working class and based on the alliance of workers and peasants. The nature of China decides that the primary function and mission of state governance is safeguarding its socialist nature—that's the overriding political mission. General Secretary Xi Jinping pointed out that socialism with Chinese characteristics is socialism and nothing else. Therefore, ecological progress, a key domain of the CPC's state governance, must have special significance and purpose for safeguarding our state nature.

When we view ecological progress from the perspective of building socialism with Chinese characteristics, we must figure out two things. One, what's the relation between ecological civilization and socialism? Two, how to safeguard our nature as a socialist country by developing ecological civilization?

(1) Ecological civilization and socialism

At first sight, ecological civilization and socialism don't seem to have any direct connection as one is practical work and the other is an ideological issue, but a closer look would reveal their inherent connections.

A country's development path isn't chosen randomly, but is closely related with the country's history and national conditions. When the People's Republic of China was founded in 1949, against the background that the old China was a semi-feudal, semi-colonial country that had suffered too long and too much from the suppression and bullying of feudalism and imperialism, China chose to embark on the socialist path to meet people's ardent desire for an independent, equal and affluent life. Only socialism represented by the Communist Party of China was able to meet this desire. In other words, choosing the socialist path at that time was mainly out of "social justice."

In addition to this historical reason, today due to the special relation between man and nature in China, we insist on the socialist path with Chinese characteristics.

At the moment, China is one of the countries witnessing the greatest tension between man and nature in the world. This is reflected in three ways. First, China, with a territory of 9.6 million square kilometers, houses 1/5 of the world's population, but the natural conditions on its land are far from favorable as only slightly more than half of the territory is suitable for human production and life, while the rest is covered by desert, plateau, Gobi desert and high mountains. With so many people crowded on so limited land, the tension between man and nature can well be imagined. Second, China is short of resources. Necessary resources to meet the needs of modern life, such as land, water, energy, grassland, forestry and metal, are all scarce, with per capita resource ownership much less than world average. Meanwhile, the rising living standards have been driving up the per capita resource consumption, which forces China to import resources in large quantities and expands the effects caused by man-nature tension at home to a wider scope. Third, China has a strong economy with the fullest economic categories and the largest manufacturing capacity in the world, which, however, has taken up enormous ecological space and environmental capacity. The environmental consumption and pollution caused by many productive activities has far exceeded the environment's carrying capacity, not only putting extreme pressure on human survival and life, but also imposing serious resource and environmental restrictions on future economic development. This has aggravated the man-nature tension to an unprecedented point.

China's almost unique national condition featuring "large population, resource shortage and strong economy" in parallel has led to the tensest relation between man and nature and consequently given rise to the unique proposition or conclusion that we must adhere to the socialist path with Chinese characteristics. Socialism in essence means collectivism and seeks the common interests of the greatest majority of people. Exactly because of the extreme strain between man and nature in China, the whole society should collectively own the land and natural resources to avoid the situation where a handful of people possess and monopolize scarce natural resources while the majority falls into a development dilemma. On that basis, a powerful leadership is needed to lead and organize the whole people to collectively use the land and natural resources in a fair, rational, sustainable and planned approach, on the principle of meeting the general interests of the whole society rather than those of a small group of people. In other words, we must adhere to the public ownership of territorial resources and the CPC's leadership, which is equivalent to adhering to socialism with Chinese characteristics. That's the only way to provide the whole society with universal development opportunities, and ensure national stability and sustained development. Therefore, if we chose socialism out of "social justice" at the beginning, then our adherence to socialism with Chinese characteristics today is also for "environmental (ecological) justice."

(2) Ecological civilization and state nature

How do we maintain our nature as a socialist country by developing ecological civilization then?

Usually ecological progress is regarded as a special undertaking that's not directly connected with a political topic like maintaining the state nature, but such a connection does exist whether we think of it or not. As a matter of fact, with the central environmental protection inspections and the implementation of other strong measures, ecological progress, particularly environmental protection, is displaying ever more obvious political properties.

The central environmental protection inspection has revealed a large number of appalling cases of environmental pollution and ecological damage, which, on the surface, reflect natural damages caused by man, but expose on a deeper level a few people's infringement on the public interests of others and the society as a whole. The "few people" refer to special interest groups. Enterprises and places that discharge excessive pollutants are transferring the cost of pollution prevention and control to the society while making more profits themselves. The society would be able to bear it if there are very fewer polluting enterprises, their discharge can be digested by nature itself, and the environmental damages are relatively small. But when a lot of enterprises are doing the same, social damages would surface and conflicts of interests would break out. Why is it that the serious pollution in certain places has not be fundamentally addressed for so long? Why are many places and enterprises still pretending to take corrective measures under the vigorous central supervision? It is because of the opposition between individual invested interests and the overall social interests. Essentially, promoting economic development regardless of environmental pollution and ecological damage means a few interest groups illegally occupying the natural capital that belongs to the whole society, or in other words, they gain while the society loses. That is environmental (ecological) injustice and is against socialism.

The opposition between capital interests and social interests appears in every country, so every country proposes to "contain capital." With the development of market economy in China, it becomes increasingly imperative to keep capital from expanding so excessively as to threaten public interests—that's why our market economy is prefixed with socialist. The environmental issue is a manifestation of the conflict between capital interests and public social interests, and strengthening environmental protection and ecological progress is actually an endeavor to regulate and control this conflict of interests, the intensity and extent of which is related to how severe the conflict is and how much the society can bear. The central authority has adopted stringent supervision and law-based governance because the current conflict is much graver than the society can bear. Therefore, the nation's stringent measures, including environmental supervision, has an intrinsic connection with safeguarding and maintaining its socialist nature. This is the political position we should uphold when understanding the significance of environmental protection and ecological progress.

(3) New struggle

Protecting the public environmental interests from being eroded by capital interests is a great struggle with many new historical features, during which we must dialectically handle two special aspects.

First, we must stand from the people's perspective in environmental protection and ecological progress. General Secretary Xi Jinping has stressed repeatedly that "public sentiment is the greatest test of political efficacy." We must actively respond to their keen expectation for faster improvement of environmental quality, strongly promote ecological progress and provide more quality ecological products to meet their ever-growing demand for a beautiful eco-logical environment. Standing with the people complies with the ruling Party's purpose and is the foothold for viewing the environmental issue as a major political issue. Standing at such a position, any behavior, as far as environmental protection is concerned, that violates the people's overall interests must be opposed and stopped no matter how many local interests it can generate—that's the political position we must uphold. The handling of a series of cases in point, such as the ecological damage on Qilian Mountain and the desert pollution in western China, has put an end to many enterprises that make profits by sacrificing the environment. This represents the prohibition and punishment of local interests for the society's overall interests, and the crackdown on irrational capital interests.

Standing with the people on environmental protection is of great realistic importance now. A local government official once said, "If an enterprise causes pollution during production and the nearby residents complain about that, I would stand on the enterprise's side because it generates taxes." This fully proves that political stance is not a remote concept, but a realistic issue. If we don't elevate the environmental issue to the political level, many officials won't stand with the people when local interests are involved, and if we deviate from the people on environmental protection, an undertaking with such extensive and sustained impacts, how can we call ourselves a country of the people? In the history of many developed countries, serious environmental events caused strong public protests and anti-pollution movements that forced the congresses to pass strict environmental laws and standards. China, as a socialist country, commits to people-centered development from the very beginning. This is decided by our political system and where our political advantage lies; this is also why the environmental undertaking is closely related with our purpose of serving the people.

Second, we must expand the common interests of various parties during ecological progress. The regulation of capital interests in the environmental field is special in that capital brings multi-faceted interests. To be more specific, while capital causes negative effects like environmental damages in the process of pursuing profits and value increase, it also generates positive effects such as employment, income increase and tax contribution. On the one hand, we should prevent interest groups from working together to harm public environmental interests; on the other hand, we should recognize and protect their legal pursuit for profits. Therefore, we should take a two-sided approach toward capital—promoting the positive effects while eliminating the negative ones, instead of throwing the baby out with the bathwater. The interest layout during economic development must be restructured. Protecting the ecological environment is social interests, so is developing economy and improving livelihood. Therefore, on the premise of adhering to the people's stance, efforts must be made to increase the synergy between developing economy and protecting the ecological environment, which is another implication of "ecological environment is a major political subject." During the storm of strict central environmental protection inspection in recent years, some areas have adopted the universal approach of shutting down all enterprises, which is anything but coordinating the two interests. To address this, the Ministry of Ecology and Environment timely formulated the *Working Opinions on Prohibiting a Universal Approach to Environmental Protection* to strictly ban the universal shutdown of enterprises and avoid concentrated stoppage of work and production. It also assigned a large group of officials to carry out deep investigations on the grassroots level and ensure heating for the people last winter in response to serious air pollution. All these actions are based on and driven by a correct political stance.

For Party and government leaders at all levels, properly handling the various interest demands and enlarging their synergy in ecological progress is a new task never encountered before. So it's no surprise that they sometimes seem not competent enough when performing this new duty. What's important is to accumulate new governing skills through study and practice on the premise of upholding the people's stance and serving their interests. As the environmental governance system is modernized and officials become more politically mature, more Party and government leaders with new governing skills and political wisdom will come to the fore, who will blaze out a successful path of handling the complicated contradictions in ecological civilization development.

3. Strong state power has to be exercised for environmental protection

As General Secretary Xi Jinping said, "To resolve environmental issues, we must fully exert the political strength of CPC leadership and China's socialist system which can bring together the resources necessary to take on great tasks, take full advantage of the solid material foundations built up over 40 years of reform and opening up, step up our efforts to build an ecological civilization, wage a determined battle to prevent and control pollution, and push China's development of an ecological civilization to a new stage." This makes it clear that national forces under political leadership should be mobilized and employed in advancing environmental protection.

(1) National strength in environmental protection

Environmental protection is a grand cause with comprehensive, long-term and public significance that calls for immense input of manpower and resources. All the resources mobilized by the state through various means for the purpose of environmental protection can be generalized as "national strength in environmental protection," or "national eco-strength" for short.

The national eco-strength is part of the overall national strength and may, on certain conditions, become the main factor that affects overall national strength due to the "cask effect." With China's rapid development in recent years, people have become more confident in its overall national strength and paid more attention to how much environmental protection has contributed to it. The green GDP estimated by environmental professionals gauges the national eco-strength. In general, although China has become a major strong power in economic, political and military aspect, it remains weak in environmental protection and has a long way to go before realizing the modernization of its environmental governance system and capability. It is stated in the *Opinions of the State Council of CPC Central Committee on Comprehensively Strengthening Environmental Protection and Resolutely Fighting the Tough Battle of Pollution Prevention and Control* that "some areas and departments haven't fully recognized the importance of or performed their duties in environmental protection, and work in this area isn't well coordinated between urban and rural areas." A weak eco-strength has in a way offset the country's overall strength. To put it briefly, our current eco-strength may only be enough to cope with the environmental pressure caused by our economic development scale in the beginning of this century, but it has to be multiplied in order to bear the current or future environmental pressure.

(2) Political mobilization for environmental protection

Politics is the activity of governing a state through the exercise of political power, mainly by means of resource allocation. Political mobilization is a must to enhance China's national eco-strength because environmental conservation is a nonprofit cause for public welfare, so it has to be carried out by "talking politics" or through political mobilization as most resource input in it doesn't generate direct economic returns. Furthermore, national eco-strength is a structural force that, with political leadership in the center, pools the whole society together through political mobilization to jointly advance the cause of environmental conservation. This is exactly what we mean by saying "fully exerting the political strength of bringing together the resources necessary to take on great tasks."

General Secretary Xi Jinping urged us to "consciously coordinate economic and social development with ecological progress." This requires us to coordinate environmental interests with economic and livelihood interests, and local economic interests with overall social interests—a new leadership ability required of Party and government leaders at all levels in the new era, which, to be more specific, is the ability of promoting green development. General Secretary Xi Jinping made the following remarks at the National Conference on Eco-environmental Protection: "Green development is a critical element of modernizing an economy; fostering green modes of development is the key to resolving our pollution problems. We must adjust the economic and energy structures, optimize the distribution of territorial space available for development, adjust the distribution of industries within regions and river basins, and foster growing industries that promote energy conservation and environmental protection, clean production, and clean energy. We must promote all-round conservation and recycling of resources, and ensure that the systems of production and everyday life overlap in their circular use of resources. We must advocate a green and low-carbon lifestyle of moderation and frugality and oppose extravagance and unjustifiable consumption." Premier Li Keqiang called for "promoting green development and preventing environmental pollution at the source." Green development isn't purely a development mode, but also a political attitude and task. In the past we didn't have the awareness or ability to have invaluable assets in lucid waters and lush mountains, now we want to foster and temper such awareness and ability, which is doubtlessly a new challenge to the CPC's governing approach and capability and a new test to the Party and government leaders at all levels. Cadres growing up through such a major test conscious of ecological civilization and green development will be CPC's governing backbone in the new era and will expand into a political force competent enough to implement the Five-Sphere Integrated Plan. This constitutes a dramatic change of epoch-making significance for the Chinese nation and the world in general.

2 Political Strategy of Winning the Tough Battle of Pollution Prevention and Control

General Secretary Xi Jinping also said "The battle to prevent and control pollution is one in which we must act with urgency to overcome burdensome tasks and difficult challenges. In this bitter and unrelenting fight, it is essential that we strengthen the leadership of the CPC." CPC's leadership and socialism system are our political advantage in boosting environmental protection, an advantage that must especially be brought into full play in this special period of fighting the critical battle of pollution prevention and control.

1. **What obstacles to overcome in pollution prevention and control?**

Fighting the critical battle of pollution prevention and control means concentrating powerful forces within a specified period of time to overcome a batch of environmental difficulties that should have been resolved in the medium and long term but haven't, with the focus on protruding environmental problems, so as to adapt the environmental state to the goal of building a moderately prosperous society in all respects. It's safe to say that the battle of pollution prevention and control is a special one waged to achieve specific objectives, for which a swim-or-sink battle of strong firepower and unusual measures is inevitable.

In fighting this battle, we have to crack three "hard nuts." First, we must resolve protruding environmental problems and notably improve environmental quality; second, we must resolve the conflict between development and environmental protection and foster a green development mode and lifestyle; third, we must deepen the institutional reform for ecological civilization and advance the modernization of ecological governance system and capability.

These difficulties that have been piled up over long years are hard to overcome with regular governance measures, but call for necessary political approaches and systematic political strategy. We used to employ many "administrative means," some of which are essentially political approaches.

2. **Political strategy for fighting and winning the tough battle of pollution prevention and control**

The political strategy for fighting and winning the tough battle of pollution prevention and control consists of the following points.

(1) Resolutely safeguarding the CPC Central Committee's authority and centralized and unified leadership. Guided by the Xi Jinping Thought on Ecological Civilization, we must uphold the CPC Central Committee's decisions and deployments on ecological progress and environmental protection, especially the *Opinions of the State Council of CPC Central Committee on Comprehensively Strengthening Environmental Protection and Resolutely Fighting the Tough Battle of Pollution Prevention and Control*, as the paramount political requirement and implement it in the whole process of strict Party discipline and state governance. No place or organization shall selectively implement this requirement with local interests as the yardstick, no leniency or second chance shall be given to violators, and typical cases shall be dealt with and severe punishment imposed to set examples.

(2) Resolutely fulfilling the political responsibility for ecological progress. Chief leaders of CPC Committees and governments at all levels are the principals bearing overall responsibility for local environmental protection work and

quality, and they should spend more time and energy studying and deploying this work. Relevant departments shall earnestly perform their related duties and work in a concerted and coordinated way.

(3) Perfecting the system of environmental goals and responsibilities. All local areas shall work out a list of responsibilities and break down the tasks to each specific department, and the list of duties for relevant central departments and state organs shall also be rolled out as soon as possible. Relevant departments shall earnestly perform their environmental duties and work out annual work plans and measures. Environmental work shall be given more weighting in the assessment and appointment of officials, and those responsible for environmental damage shall be strictly and severely held accountable, for life.

(4) Implementing enduring and forceful central inspection and supervision over environmental protection. The political system shall play a bigger role in safeguarding environmental protection. While improving the central inspection over local areas, the same shall also be done on the ministries and commissions. Efforts should be made to polish the supervision, assignment, inspection, talk and special inspection mechanism, and carry out targeted supervision in key regions, sectors and industries. Environmental protection shall be given a higher position and authority in the nation's macro-regulation with a more rigorous reward and penalty system.

(5) Adopting a green-oriented cadre selection and appointment system. A scientific and reasonable ecological civilization assessment system and the assessment results shall be taken as an important basis for rewarding and punishing the leading groups and officials at all levels. Those performing superbly in promoting green development and environmental protection shall be prioritized in promotion. Contents about ecological progress shall be increased in the training of young officials, and those who are neither active nor effective in boosting ecological progress shall not be promoted to key positions despite their exceptional performance elsewhere.

(6) Intensifying law enforcement and judiciary work on environmental protection. We should further enhance the position and weighting of environmental protection in legal construction, and launch an ecological and green overhaul of various economic and social laws and regulations. The People's Congress' legislative process on ecological progress and environmental protection shall have quantified targets. Law enforcement directly carried out by the government plays the dominant role in environmental regulation, assisted by social supervision, thus combining "government regulation" with "social check." The courts and procuratorates should bear more environmental responsibilities by boosting judiciary practices in that area, ensuring environmental justice during case filing and trial, and supporting the development of environmental laws by creating precedents.

(7) Enhancing the role of the people's congress and people's political consultative conference in promoting ecological progress. Advice on environmental legislation proposed by deputies of the People's Congress and members

of the People's Political Consultative Conference shall be prioritized. The People's Congress deputies shall play a bigger role in environmental inspection and supervision. Those with problematic performance in ecological environment shall not be selected as People's Congress deputies or People's Political Consultative Conference members, and the two bodies shall have more deputies and members from the environmental sector.

(8) Building and fortifying the united front of environmental protection. Following the political wisdom of "increasing friends and decreasing foes," we, under the CPC's unified leadership, should build and fortify a joint force of environmental protection including political forces, economic groups, social organizations and individuals. We must extensively listen to the calls for and suggestions on ecological progress and environmental protection, foster a commonly recognized value of ecological civilization, balance and accommodate the interests of all parties, seek the greatest common divisor of various interests, and maximize common action and minimize resistance to environmental protection. The technological circle should fully exert its technological advantages and take the lead in putting up a "united front of environmental protection." The ecological and environmental authority, as the core and backbone of the united front, shall be responsible for organizing and guiding it and pushing its development while seeking successful cooperation.

(9) Adhering to the mass line of environmental protection. "A giant hawk cannot fly with only one feather; a fine horse cannot run fast with only one leg." In protecting the ecological environment, we must carry on the CPC's traditional approach of trusting and relying on the people. The *Action Plan for Green Environment for All* shall be implemented to foster a green lifestyle by pooling the wisdom and efforts of everyone. Efforts should be made to expand and protect the public's environmental rights and interests and develop environmental NGOs in an orderly manner. The press is encouraged to play their supervisory role and contribute to the people's war of pollution prevention and control.

(10) Timely utilizing the rare political opportunities. Given the special requirements of the critical battle of pollution prevention and control, it's necessary to roll out a plan for the modernization of the environmental governance system and capability, list that in the national plan for organizational reform and budget, so as to significantly enhance our capability of fighting the battle. Governments of all levels shall have more authority as well as bear more duties in environmental protection, environmental organizations shall be enlarged, and the capability and level of technology, monitoring, law enforcement, emergency response and management revolving around environmental protection shall all be improved. The size and quality of environmental personnel should be improved, with a special focus on covering rural areas in grassroots environmental efforts.

Reference

Yang, G. (2012). *An introduction to politics* (4th ed.). Beijing: China Renmin University Press.

Profoundly Understanding the Multiple Relations in Ecological Civilization

Hongchun Zhou

Ecological progress is coordinated and integrated with economic, political, cultural and social progress. Economic development is the foundation, social development the goal, political development the guarantee and cultural inheritance and innovation the soul. In promoting ecological progress, we should bear in mind the current conditions while looking to the future; push it forward comprehensively with clear priorities; practice it in China but also take the lead in the world; and focus on outstanding environmental problems while emphasizing systematic and overall effects. We cannot drain the pond for fish or kill the hen for eggs—that's sacrificing our offspring's well-being for immediate interests and short-term growth; nor can we give up development and willingly stay in poverty for environmental protection. It's hard to realize a sound ecosystem cycle without livelihood improvement, and livelihood won't sustain either if we neglect the improvement of environmental quality. Therefore, we must adopt green development, take the path of sustainable development and walk into a new era of ecological civilization.

1 The Relation Between Five-Sphere Integrated Plan and Spheres Involved

Ecological progress is coordinated and integrated with economic, political, cultural and social development. Ecological civilization, which encompasses vitality, carrying capacity and integrating ability, is the foundation for economic, political, social and cultural progress.

Economic development is the foundation for material civilization. Putting economic development in the center holds the key to solving all problems in China and prospering the nation, and is a fundamental step to lift the nation from poverty

H. Zhou (✉)
Development Research Center of the State Council, Beijing, China

J. Pan et al. (eds.), *Beautiful China: 70 Years Since 1949 and 70 People's Views on Eco-civilization Construction*, https://doi.org/10.1007/978-981-33-6742-5_48

to affluence and bypass the "middle income trap." The ecology functions as a bond between nature and economy, society, culture and politics, with the two ends of the bond promoting and complementing each other. Ecological progress is an initiative that considers the big picture and the long term, for which holistic planning of various terms and aligned policies and measures are needed to advance the overall green transformation on the economic, political, cultural and social fronts. In this way, it will put in place a powerful balance mechanism and point out the direction for achieving the harmonious co-existence between man and man, man and society, man and nature.

Political civilization points out the direction forward for ecological progress, which in turn creates the conditions for our current political development. Political civilization should protect different interest groups and their diverse needs, and ecological progress should be an integral part of political development and serve as an opening for institutional reform. As mass events caused by environmental problems have diminished the credibility of some local governments or fueled public discontent, starting the institutional reform in environmental sector will meet less resistance as it complies with the people's demand for a good life. Steps should be taken to promote public participation in environmental protection, carry out democratic decision-making, management and supervision, so as to gain experience and lay a solid public foundation for political system reform and break new ground for ecological civilization.

Ecological civilization has to be bolstered by cultural civilization. As the fine outcomes of man's reform of the objective world as well as the subjective world, cultural civilization reflects the progress of human wisdom and morality mainly in two domains. First, it is reflected in science and culture, including social culture, knowledge and wisdom, and the development scale and level of such social undertakings as education, science, culture, art, health and sports. Second, it is reflected in ideology and morality, including political ideology, morality, social ethos, and human values, ideal, integrity, awareness, conviction and sense of organization and discipline. Ecological civilization is an important component of cultural civilization, while cultural development provides mental guidance, spiritual motivation and intelligent support for ecological progress.

In developing ecological civilization, we should focus on prominent environmental problems while emphasizing the systematic and overall effects. Bearing in mind the Five-Sphere Integrated Plan and Four-Pronged Comprehensive Strategy of socialism with Chinese characteristics, we should merge ecological progress with economic, political, cultural and social development in the whole process and on all fronts, and give priority to resolving key environmental pollution issues affecting public health. Measures will be taken to win the critical battles for blue sky, lucid water and clean soil, coordinate source governance, strict process regulation and punishment of non-standard pollutant discharge, and advance resource conservation, low-carbon development, spread of green development concept, institutional establishment, technological innovation and investment in a holistic manner. As such, all the links and factors will form a complete system to make sure lucid waters and lush mountains are invaluable assets.

But a good ecological environment doesn't equal ecological civilization. Ecology is a natural state, civilization is social progress, and the combination of them—ecological civilization—is a harmonious state between man and nature. Civilization prospers when ecology prospers. Ecological civilization not only features a sound ecological environment, but also a sound cultural civilization. Some impoverished regions in the world have great ecological environment, but the people there have to slash and burn wood because of extreme shortage of materials, which leads to water and soil loss and ecological deterioration, which in return worsens poverty, hence the vicious cycle of "poverty—ecological deterioration—poverty." Only when ecological environment and cultural civilization proceed in parallel can ecological civilization last long.

2 Relation Between Ecological Civilization and Traditional Civilizations

The human civilization is evolving all the time. In terms of time, human civilization has passed the primitive civilization, agricultural civilization and industrial civilization. Ecological civilization is an inheritance of traditional civilizations while discarding the backwardness and making innovations of its own.

Man has been dependent on nature since the day it was born. In primitive society, man lived by simple productive activities such as collecting, fishing and hunting and could only survive by living in groups, so it evolved together with other creatures and the environment in general. It was only in modern times that man began to shake off the regional restrictions from nature and its ecological conditions. The emergence and use of iron ware marked a substantial leap forward in human ability to utilize and reform nature and pushed human society to the age of agricultural civilization. The extinction of the ancient Egypt, ancient Babylon and America's Maya could perhaps be ascribed to excessive grazing, land reclamation and blind irrigation.

Agricultural civilization represents the time-honored history and splendid civilization of China, and a natural environmental friendliness is the precondition for the uninterrupted agricultural civilization or Chinese civilization for thousands of years. After the People's Republic of China was founded, a major challenge facing us was how to carry on our fine traditions while proceeding with socioeconomic modernization, and blaze out a path of industrialization and urbanization featuring the harmonious co-existence between man and nature. From agricultural modernization in the beginning of new China to urbanization of rural areas after reform and opening-up and to the new model of industrialization and urbanization in recent years, it seems we are still in the mindset of "catching up with" modernization.

The Industrial Revolution in Britain lifted the curtain on the age of industrial civilization, the rise of which prompted the human society to gradually "forget" natural laws and restrictions and pursue a universal (regardless of region), all-time (regardless of season) and unbridled (without restraint) development overriding the restrictions imposed by natural resources and ecological laws. The development of productivity empowered man to conquer nature on an unprecedented scale, which resulted in tremendous wealth as well as serious environmental crisis. The "eight environmental hazards" in the twentieth century, including the smog disaster in Belgium's Meuse Valley, Donora's "killer smog" in the US, London's smog, photochemical smog in Los Angeles, minamata disease in Japan, and rice bran oil event in Toyama and Yokkaichi, threatened public health and prompted the people to deeply reflect on the traditional economic growth pattern.

In 1962, American writer Rachel Carson published a book titled *Silent Spring*, which sounded the alarm of environmental crisis in industrial society by elaborating on how the massive use of pesticide harmed human race with appalling real cases. In the 1970s, two global energy crises further set off the conflict between economic growth and environment. In 1972, the report *The Limits to Growth* published by the Club of Rome warned the world for the first time that "If the present growth trends in world population, industrialization, pollution, food production, and resource depletion continue unchanged, the limits to growth on this planet will be reached sometime within the next one hundred years," which drew close attention from all countries. Also that year, the United Nations Conference on the Human Environment was convened, at which a Declaration was released that put out the slogan of "only one earth," calling on mankind to bear the responsibility and obligation of natural preservation while developing and utilizing it.

In 1987, the then Norwegian prime minister Gro Harlem Brundtland systematically expounded the connotations of sustainable development in the *Our Common Future* report. In 1992, the United Nations Conference on Environment and Development held in Rio de Janeiro, Brazil adopted the *Rio Declaration on Environment and Development* and *Agenda 21*, urging countries to, while boosting economic growth, not just focus on the quantity and speed of development, but also on quality and sustainability.

We cannot simply take ecological civilization as a brand new form of civilization. The concepts it involves carry the ideological quintessence since agricultural civilization. "Harmonious co-existence between man and nature" is a continuation of "unification of nature and man" proposed by ancient Chinese sages, and respecting nature is a theory proposed by Engels in his *Dialectics of Nature*. China's ecological progress will employ modern technologies and the latest achievements of big data and mobile Internet, take the path of sustainable development, and take the country to the new era of ecological civilization.

3 Relation Between Ecological Progress and Social Development

Ecological progress is an important means to improving people's lives. As General Secretary Xi Jinping pointed out, the environment is livelihood. Ecological progress will enable people to fairly enjoy the fruits of development and make the society more harmonious. It seeks benefits and solves difficulties for the people, addresses their most immediate concerns, and makes continuous progress in ensuring people's access to childcare, education, employment, medical services, elderly care and housing. It should be viewed as a key task to materialize, protect and enhance people's fundamental interests, and safeguarding people's environmental rights and interests should be made the highest standard for work performance. Through the united efforts of the whole Party and the whole nation, we will strive to create a dynamic situation where everyone contributes to and shares the benefits in social harmony.

Ecological progress and livelihood improvement promote each other. As people are living an increasingly better life, they are no longer content with being "well fed and well clad," but have higher-level needs such as for a sound ecological environment. They are paying more attention to ecological protection in river basins, sources of drinking water, and life-related topics like smog. Environmental quality directly bears on public life safety as the emergence and spread of some diseases originates in environmental deterioration. Only by intensifying ecological progress and realizing the sustainable use of natural resources and sound cycle of ecological environment can we keep improving the livelihood. If we don't pay close attention to livelihood, a sound environmental cycle is hard to come through; but if we only focus on livelihood and neglect environmental improvement, livelihood won't sustain either. Therefore, we must place equal emphasis on environmental protection and livelihood improvement, and embark on a development path characterized by advanced production, affluent life and sound ecology.

Ecological progress should serve social construction. When approving projects and handling environmental violations and pollution disputes, we must stand fast by the following bottom line—are we safeguarding the people's environmental rights and interests? Are we putting public health and safety first? Democratic rights of the people should include environmental rights, which therefore have to be protected. On the other hand, the people have the right to exercise democratic supervision according to law, and managing public affairs and public welfare undertakings, including environmental protection, is the most effective and widespread approach to letting the people be their own masters. When it comes to specific work, we must make sure the people have access to clean drinking water by strengthening water source protection and the protection and improvement of rural ecology and environment; make sure they can breathe fresh air by reducing the occurrence and adverse effects of smog weather. We must also make sure they have access to safe food by encouraging and supporting ecological agriculture and ecological product development, reinforcing agricultural non-point pollution governance and food safety supervision, and intensifying full-process supervision over the production, processing and circulation of grain, vegetables, meat and aquatic products. Efforts should be made to energetically develop characteristic economic forests and fruits, vegetables, medicine and animal husbandry for "win–win" interactions of "Sannong (agriculture, farms and countryside)."

Ecological progress doesn't equal environmental protection. Ecological progress in the broad sense consists of ecological construction, ecological economy, ecological security and ecological culture. Ecological progress in the narrow sense consists of land and space layout and its optimization, conservation, protection and sustainable use of resources, environmental protection, pollution control, and institutional establishment. Environmental protection is part of ecological progress, even a priority of priorities given the severe environmental situation, but ecological progress cannot be viewed as a simple upgrade of environmental protection, otherwise it might be delinked from other sectors. Correspondingly, ecological progress is by no means the sole responsibility of the environmental department, but an initiative for which all government departments shall bear their respective duties. Spatial optimization, resource conservation, environmental protection, industrial upgrade, green construction and transportation, social transformation, sci-tech innovation, ecological culture, green consumption, green tax and finance…these are all important contents of ecological progress. The organizational reform of the State Council in 2018 created favorable conditions for establishing and improving the leading system and inter-departmental interaction for ecological civilization development.

4 Relation Between Economic Development and Environmental Protection

Environmental protection and economic development are a unity of opposites. Human production and life cause environmental pollution and ecological damage, even ecological crisis. On the other hand, environmental protection needs capital and therefore affects the quantity and speed of economic growth to some extent. Only by developing the economy can we continuously meet people's ever-growing material and cultural needs. Without quantitative economic growth, the leap from quantitative change to qualitative change would never appear and the accumulation of material wealth would be impossible. Without an economic foundation, there is no ecological progress to speak of. However, if we blindly pursue the scale and speed of economic growth without heeding its quality and benefits, if we wantonly claim natural resources and unscrupulously destroy the ecological environment, and if we trade resources and the environment for economic growth, what awaits us will be resource depletion, environmental destruction and non-sustainable development. We cannot sacrifice our offspring's welfare for immediate interests and temporary growth, which is draining the pond for fish and killing the hen for eggs. Nor can we give up development and willingly stay in poverty or wait for assistance for environmental protection. The key lies in the degree of and balance between development and protection.

Coordinating the relation between man and nature concerns the great victory of the socialist cause with Chinese characteristics, the achievement of the two centenary goals, the great rejuvenation of the Chinese nation, and the building of a community with a shared future for mankind. China is looking at the basic national condition of a large population, inadequate per capita resources, and weak carrying capacity of the ecological environment. With our economic growth and continuous population increase, especially the rapid development of industrialization and urbanization, the conflict between inadequate resources such as water, land, energy and minerals and fast economic growth is becoming more conspicuous. A direct consequence is the serious environmental situation. If the traditional development mode continues, it will deplete the resources, overwhelm the environment and make development hard to sustain. To meet people's new expectations for a beautiful ecological environment, the key is to improving their lives, to bolstering the sustained and healthy socioeconomic development, and to demonstrating China's positive and responsible image.

We cannot set ecological civilization against economic development. There are some people who put the sole emphasis on "lucid waters and lush mountains" in total disregard of economic foundation. Ecological progress is not giving up development, but to pursue high-quality development at low resource consumption and high efficiency. Development doesn't mean economic development alone or simply GDP growth. Resource conservation, environmental protection, sci-tech innovation, cultural prosperity, social progress and many other factors all make part of the connotations of development. The sentence "lucid waters and lush mountains are invaluable assets" is a full and complete statement that integrates protection and development in an organic whole. Affluent life at the price of ecological deterioration is not ecological civilization; beautiful environment surrounded by poverty and backwardness is not ecological civilization either. We must try and strike a balance between economic development and environmental protection.

5 Paths of Developing Ecological Civilization and Their Relations

The Third plenary session of the 18th CPC Central Committee said ecological civilization should be developed through green, circular, and low-carbon development. In the new era, China has acquired a deeper understanding of what ecological progress implies. Green development is elevated to be a development concept, with building a resource-saving and environment-friendly society being an important aspect. It also coordinates work in such areas as energy conservation and emission reduction, circular economy, improvement of environmental quality, and low-carbon economy.

Circular economy is a term that generalizes the down-sizing and recycling activities in the process of production, circulation and consumption. It gauges development in terms of resource efficiency. In traditional economic growth mode, the earth was regarded as an infinite resource repository and junk yard. While one end of the economic system exploits multitudes of resources from the earth to produce products, the other end discharges massive amounts of waste water, waste gas and garbage to the environment. This is a linear growth pattern following the route of "resources—products—scraps." When the industrial chain is extended in the principle of "material metabolism," "process coupling" and "resource sharing," economic growth has shifted from relying on the exploitation and utilization of natural resources to the utilization of both natural and recycled resources. This is an intensive growth pattern following the cycle of "resources—products—scraps—recycled resources."

Low-carbon economy is an economic growth pattern characterized by low energy consumption, low pollution and low emission. Its core is the innovation of energy technologies, innovation of systems, and the fundamental change of human view on survival and development, involving the reform of production mode, lifestyle and values. The British energy white paper—*Our Energy Future: Creating a Low Carbon Economy*—published in 2003 defined low-carbon economy as follows: higher resource productivity—producing more with fewer natural resources and less pollution—will contribute to higher living standards and a better quality of life, the opportunity to develop, apply and export leading-edge technologies, creating new businesses and jobs. Through favorable tax and financing policies, we should guide private investors to increase the input in the R&D of low-carbon technologies, and energetically develop industries focused on biomass energy, solar power, wind power, energy-saving equipment, hydro power, and environmental protection.

Green development is one of the new development concepts. Green development is a kind of development respecting the restraint imposed by resource and environment's carrying capacity, and a sustainable mode of production and consumption. There is no substitute for the current ecological environment. When we use it, we take it for granted. But once damaged, it will struggle to recover. As the Chinese economy is shifting from high-speed growth to high-quality development, green development has become a common consensus. In building a moderately prosperous society in all respect, we cannot give up development in order to protect the environment and guard the "lucid waters and lush mountains." Nor can we repeat the old path of "governance after pollution" or "pollution and governance in parallel," still less live on what our ancestors have created while burning the bridges for later generations. Instead, we should adjust the industrial structure, foster new drivers and change the traditional production and consumption pattern featuring "massive production, massive consumption and massive emission." The path of green, low-carbon and sustainable development has to be upheld to realize socialist modernization with Chinese characteristics where economic and social development complies with environmental protection and man and nature co-exist in harmony.

The relation between current construction and long-term development should be properly handled. We should bear in mind the current conditions as well as future development, short-term planning as well as long-term arrangements. We should also take forceful and effective actions to resolve pressing environmental issues, as well as keeping environmental pollution from affecting public health and ecosystem on the principle of "prevention first." On the one hand, we, eyeing the goal of building a moderately prosperous society in all respects, shall perform the ecological civilization tasks for the next five years set out in the report of the 19th CPC National Congress. That is to develop spatial layouts, industrial structures, and ways of work and life that help conserve resources and protect the environment. On the other hand, eyeing the Chinese dream of building a socialist modern power and reviving the Chinese nation, we shall work out the strategic goals, principles and paths for ecological advancement for the medium and long term, and strive for a modernization featuring man-nature harmony. Taking into account the actual conditions of different areas and sectors, we should deepen and perfect the macro strategy and break it into specific tasks, develop feasible construction drawings and road maps for ecological progress, and make sure all the targets are accomplished on schedule.

6 Relation Between Technological Innovation and Social Governance for Ecological Progress

Science and technology constitute the primary productive force. The drive of development decides the speed, efficiency and sustainability of development. Given China's economic development and size, it would be very hard to achieve sustained and healthy economic development and "double our GDP and per capita income of urban and rural residents from 2010 to 2020" if we cannot ensure a sound development drive.

The technological revolution in the twentieth century ushered in the age of traditional industrial civilization, while the R&D and application of a great deal of green technologies is kicking off the new era of ecological industrial civilization. If we still cling to the view of omnipotent science and technology, we may have the illusion that the current lifestyle is ok, science and technology can solve everything, and resource shortage, diminishing species, climate change and environmental pollution are no big deal and can be effectively resolved with sci-tech progress. If we still cling to that view, we will never go beyond materialism or fundamentally change the direction of civilization development.

In building ecological civilization, we must fully rely on and actively exert technological innovations. The promotion and application of green technologies shall be combined with intensified independent R&D. Also shall be combined is integrating existing technological resources with deploying future technological innovations, primitive innovations and system integration innovations with introduction, digestion, absorption and re-innovation, and strengthening environmental governance and recovery with adjusting economic and industrial structure. Special efforts shall be made to make breakthroughs in major scientific issues and key technologies that obstruct the shift of economic development pattern and sustainable development, thus providing ceaseless knowledge, scientific basis and technological support for ecological civilization development. The role of science and technology is reflected in three ways. First, it enables people to systematically and profoundly understand the natural laws and the laws governing the interactions between man and nature. It makes us realize China's status quo in natural resources and ecological environment and their changing trend. It also helps us recognize the laws concerning the evolution and regulation of the complicated social system. As such, we are in a position to timely adjust the relation between man and nature. Second, the promotion and application of science and technology in resources, environment and sustainable development can push the shift toward a resource-saving, environment-friendly society. Third, it serves as supportive conditions for changing the economic development mode, boosting economic restructuring and industrial technology upgrade, and realizing sound and fast development.

Modern technologies are powerful instruments for ecological progress and the improvement of environmental quality, but they are not the whole approaches to ecological progress. In fact, science and technology are a "double-edged sword." On the one hand, with the traditional industrialization in the twentieth century, natural resources have been subject to high-intensity and predatory exploitation and utilization, which has caused serious ecological damages and environmental pollution. This is to a large extent driven by modern technological revolution. On the other hand, science and technology are playing an increasing role in resource conservation, ecological protection and environmental improvement, but the view of omnipotent science and technology must be discarded in building ecological civilization. While stressing the positive effect of sci-tech progress in ecological progress, we cannot become superstitious about modern science and technology. Nor can we equate ecological progress, a complicated socioeconomic progress, with a simple process of technology-based governance while neglecting the role and influence of culture.

Science and technology are in no way omnipotent. Bearing this in mind, we won't think there is an alternative to everything, but would admit "we have only one earth" and consequently protect this only planet we call home. The theories and methods of ecosystem management point out the way to straighten out the macro system encompassing population, resources, environment, economy and society. Employing the approaches of systematic engineering and the principles of human ecology, it studies the dynamics and cybernetics of the composite ecosystem. Ecosystem management coordinates the complicated relation of systematic coupling between man and nature, economy and environment, part and whole on such dimensions as time, space, quantity, structure and order. It promotes efficient use of matters, energies and information, fully merges technology and nature, and maximizes human creativity and productivity, so as to protect the ecosystem functions and people's physical and mental health to the largest extent, and ensure the sustained and healthy development of economy, nature and culture. Promoting ecosystem management on the basis of China's advantageous management system will help generate satisfactory results across the board through our ecological civilization efforts.

7 Relation Between Ecological Security and Ecological Culture

The Chinese plan for building ecological functional zones proposes to develop an ecologically secure spatial layout. This consists of "two barriers"—the Qinghai-Tibet Plateau and the Loess Plateau-Sichuan-Yunnan ecological barriers, and "three belts"—the northeast forest belt, north China sand belt, and south China hilly belt. They, along with important water systems including rivers and seas, constitute the general skeleton of the spatial layout, interspersed with important areas where development is prohibited by the state.

Ecosystem is the material foundation for human survival and development. Ecological civilization is a conscientious human action following the rules of ecology and sociology—it is the practice and outcome of pursuing harmonious co-existence between man and nature when productivity is highly advanced and material very abundant. Human survival and development is an integral part of natural evolution. Engels said, "we, with flesh, blood, and brain, belong to nature, and exist in its midst." Man and nature have a dialectical relation. On the one hand, man takes resources and space from nature, enjoys the services and functions provided by ecosystems, discharges wastes into nature, and its activities affect nature's structure, function and evolution. On the other hand, nature provides the resources and environment needed for man's survival and development, accommodates and digests the wastes generated by human activities; but natural changes, such as natural disasters, environmental pollution and ecological damages, also act upon man and restrict its survival and development.

We must combine the concept with conscientious practices. To build ecological civilization, we must uphold green development and keep enhancing the public awareness of ecological civilization, a concept that takes a long time to form. We must strengthen the publicity and education of this concept in breadth, intensity and depth to truly make the public more aware of conservation, environmental protection and ecology, create an ecology-caring social climate, and turn the concept into a common consensus and code of conduct for every social member. In the meantime, we have to "change both the way of thinking and the way of action; the way of production and the way of life; the way of economic development and the way of social development." From the government and society to enterprises and individuals—they should all start with details and strive for long-term development. All entities shall implement the idea that everyone is responsible for protecting the environment, and act consciously to implement and demonstrate the ecological outlook and promote the harmonious development of man and nature.

Steps should be taken to include ecological civilization in the educational system and social construction, spread knowledge about ecology and environment among the public, and make such educational or informative activities are extensive, enduring, targeted and interesting. Topics such as respecting nature, protecting the ecological environment, conserving resources and coordinated co-existence should be advocated in a universal, regular and permanent manner. The moral standard and institutional incentive system for ecological equity and justice shall be established across the society and observed and supervised by all. We must take advantage of the market mechanism and perfect a long-term mechanism aimed at minimal resource consumption and maximal benefit and service. No one has the right to occupy more resources than others, and everyone can find its right place and exert its full potential, thus integrating equity, efficiency and justice. We will carry out a series of ecological civilization activities to urge the people to actively participate in ecological progress and environmental protection. Various social groups and organizations will be brought into full play to foster the production mode, lifestyle and consumption mode conducive to resource conservation and environmental protection.

Ecological progress is a systematic undertaking directly concerning national welfare, people's livelihood and socioeconomic development. It needs not only the close attention, firm leadership and effective guidance from the CPC and governments at all levels, but also the close coordination of all sectors and the concerted efforts of the whole society. As Chinese economy keeps developing at a fast pace and people are living a better life, they have high demands on production safety, food safety, water security and environmental quality. To resolve or mitigate resource and environmental issues, not only should the management be more professional in such sectors as water conservancy, forestry, agriculture, industry, environmental protection and land, but social management should be enhanced too. Sector management should focus on the normal state and be law-based, unified and scientific. Social management, on the other hand, should focus on the crisis state of special, emergency and serious conditions, and the ability and level of it should be reinforced in various ways.

8 Relation Between Government and Market in Ecological Progress

On March 5, 2019, General Secretary Xi Jinxing generalized for the first time the "four ones" about ecological progress when participating in the discussions of the Inner Mongolia delegation. To be specific, ecological progress is one aspect of the Five-Sphere Integrated Plan; adhering to harmonious co-existence between man and nature is one basic policy of upholding and developing socialism with Chinese characteristics in the new era; green development is one of the new development concepts; and pollution prevention and control is one of the three tough battles. This generalization has enriched the Xi Jinping Thought on Ecological Civilization and is a basic principle and action guide for ecological civilization development in China.

When building ecological civilization, we should give full play to the market's basic role in resource allocation to make economic growth more vibrant and efficient. In the meantime, we should give full play to the government's role in macro-management and regulation to overcome or make up for the market's defects and shortcomings and solve problems that the market cannot. As industrialization, urbanization, marketization, information application and global economic integration are all picking up speed and our reform has come to a critical juncture, the government should well play its role in boosting employment, adjusting allocation, perfecting social security, ensuring social equity, coordinating urban–rural and regional development, protecting the ecological environment, and maintaining economic stability.

Strengthened government control can effectively conserve resources and reduce the immense hazards imposed by backward market economy on the whole society. It also avoids the waste of resources resulting from repeated construction and blind investment, and restricts construction projects that cause high consumption and high pollution and seriously affect the ecological environment. Great efforts must be made to accelerate the shift in government functions, clear the institutional barriers to fundamentally change the economic growth mode, and effectively address the over-reaching, omission and mismatch of government functions. Upholding the idea of open, transparent and honest management and power-duty consistency, the government should comprehensively and correctly perform its duties in "economic adjustment, market supervision, social administration and public service." It must represent equality and efficiency in resource and environmental planning, price mechanism, investment policy and administrative system, and make sure government intervention is just, open and equitable. Government decisions should be more scientific, so they can rectify market malfunctions while preventing government malfunction, and avoid disorderly or excessive competition and possible inefficient, unfair or unequal phenomenon such as monopoly. Special efforts should be made regarding the allocation of significant rare resources and state-owned public resources. The government should work hard to achieve reasonable and just social objectives and sound economic order, and modernize its governing capabilities.

We can use the market mechanism to build ecological civilization at a lower cost and higher efficiency. Ecological progress doesn't equal government leadership, fiscal input and government purchase of services. Instead, the market's basic role in resource allocation should be brought into full play. We must take advantage of the price leverage, so that the price of resources fully reflects the real cost of ecology, resources and environment, and polluters, resource developers and users will bear the losses of environmental and ecological damages as well as the cost of resource depletion. An environmental resource market shall be created, such as the water market and discharge trading market, to effectively keep down the cost of pollution control. Public fiscal fund shall be spent in such a way as to steer private capital to ecological progress, and all kinds of entities shall be mobilized to actively participate in this cause. Enterprises should consciously act as entities responsible for pollution control, have a keener sense of social responsibility for environmental protection, and achieve harmonious development between man and nature through economic means.

Regarding the relation between domestic governance and international cooperation, we have to bear in mind both the domestic and international situations, and foster a domestic environmental governance system with the government taking the lead, enterprises as main entities, and social organizations and general public taking part. We must work to resolve the environmental problems in China based on national conditions, and stick to the path of ecological progress characterizing advanced production, affluent life and sound ecology, with the goal of building a beautiful China. We will actively engage in global environmental governance, and adopt a global perspective toward ecological progress at home. We will urge ecological exchanges and cooperation of all countries, try to secure more international discourse power, and contribute to tackling global ecological challenges and boosting world sustainable development. China as a responsible developing major country will take an active part in global ecological governance and undertake international responsibilities commensurate with its national conditions, development stage and actual capabilities. Making full use of multilateral cooperation mechanisms such as the Belt and Road Initiative, it will communicate and cooperate with the international community in management mode, advanced technologies, experience and achievements, in the joint endeavor to explore a path of ecological civilization for the whole world.

On General Sustainable Development

Kaizhong Yang

1 Concept of Sustainable Development

The proposition of the sustainable development concept is a big step forward in mankind's "environmental philosophy." Before the 1950s, there was the famous Malthusian Theory of Population, but in general, man tends to believe, consciously or unconsciously, that the environment is able to provide human society with natural resources and environmental labor infinitely, and that its functions won't be harmed no matter how we dispose of and utilize it. Such an idea is detrimental to human welfare and happiness and makes man, the entrusted manager, feel that they do not have to shoulder responsibilities for nature, society and later generations and no use to manage the man-environment system. However, the increasing breadth and depth of human intervention on nature has brought about more serious environmental damages. Since the 1960s, man has begun to realize that the natural resources and environmental labor provided by the earth are not infinite, but are as rare as capital and labor. How to dispose of and use what the earth provides relates with human welfare and happiness, and the current generation has the responsibility to rationally manage the earth environment for human interests. However, for some time we didn't fully grasp the essence of environmental problems, and adopted the simple approach of "post-pollution governance" to environmental management. After the 1970s, man has gradually realized that environmental problems are rooted in the way how society and economy operate, and resolving them largely depends on adjusting the way of operation. Therefore, environmental management shall be incorporated in economic and social development policies, which requires that the relation among population, economy and environment should be shifted from "confrontational" (development

The article was originally published in the first and second issue of *China Population, Resources and Environment*, 1994. It was slightly edited when included in this book.

K. Yang (✉)
Research Insitute for Eco-civilization, Chinese Academy of Social Sciences, Beijing, China

J. Pan et al. (eds.), *Beautiful China: 70 Years Since 1949 and 70 People's Views on Eco-civilization Construction*, https://doi.org/10.1007/978-981-33-6742-5_49

comes with destruction of the environment) to "complementary" (development and environmental protection proceed in parallel). Based on explorations in the 1970s and 1980s, the World Commission on Environment and Development (WCED, aka Brundtland Commission) published the *Our Common Future* in 1987, debuting the concept of sustainable development. From the *Montreal Protocol on Substances that Deplete the Ozone Layer* to the International Geosphere-Biosphere Programme (IGBP) of International Council for Science (ICSU), from the Human Dimensions of Global Change Programme (HDGCP) co-chaired by the International Federation of Institutes for Advanced Study (IFIAS), International Social Science Council (ISSC) and United Nations University to the United Nations Conference on Environment and Development in Rio de Janeiro in 1993—all these programs and events have shown that sustainable development has become a cross-disciplinary subject encompassing natural, life and social sciences and a guiding action program for the international community.

Sustainable development is a norm governing the changing relation between man and environment, but no agreement has been reached on its connotations and extensions yet. A generally accepted definition comes from the Brundtland Report—development that meets the needs of the present without compromising the ability of future generations to meet their own needs. This definition, however, simply emphasizes the temporal dimension but neglects the spatial dimension of sustainable development. The spatial dimension ranges from the global level to regional level horizontally, and from natural circles to different sectors of human activities vertically. These spaces are relatively independent while mutually affecting each other. The interaction between vertical spaces is self-evident, while that between horizontal spaces takes the form of regional effects of global changes as well as global effects of regional changes.

Therefore, sustainable development should reflect both the relative independence of the globe, regions and sectors and their mutual effects. Spatial dimension is its qualitative feature, so it must be reflected in the definition. The author believes sustainable development can be better defined as follows: it meets the needs of the present generation without compromising the ability of future generations to meet their own needs; it meets the interests of both local population and global population. This definition consists of four important interrelated aspects.

(1) General sustainable development: general methods and actions of theoretical studies of sustainable development;
(2) Sectorial sustainable development: it meets the needs of a specific sector without compromising the ability of the whole world to meet its needs. Examples include ecological or sustainable agriculture, ecologically sustainable industry, and sustainable housing and transportation;
(3) Regional sustainable development: it meets the needs of local population without compromising the ability of global population to meet their own needs. Here region may refer to a bloc of states, a state, or areas, including cities, within a state.
(4) Global sustainable development: it meets the needs of the whole world.

2 Sustainable Development Goals

Economic growth and income redistribution are the basic goals of traditional development mode. This kind of bi-dimensional standard neglects natural capital and doesn't consist with sustainable development, which has three fundamental principles—efficiency, equality and ecological sustainability in the broad sense.

1. **Efficiency**

Sustainable development isn't asceticism. Its basic goal is meeting human needs. Efficiency, which means using minimal cost of rare resources for maximal benefits, is still one of its basic goals. Unlike in traditional development, benefits here don't just include the consumption of man-made products and labor (human system), but also environmental capital, and cost doesn't just include the input of man-made material capital, but also the consumption of environmental capital.

Efficiency can be improved in two ways. First, between the consumption of man-made products and environmental capital, we can increase one and decrease the other, resulting in net benefits increase. Second, we can increase both the consumption of man-made products and environmental capital. Some scholars call the former Weak Sustainability and the latter Strong Sustainability (Giaoutzi and Nijkamp 1993). Weak sustainability can be further divided into two scenarios—increase of the consumption of man-made products but decrease of environmental capital, or decrease of the consumption of man-made products but increase of environmental capital increases, the latter very rare in reality. Generally speaking, due to different development stages of productivity and human needs, developing countries usually opt for "weak sustainability" while developed ones prefer "strong sustainability."

2. **Equality**

The rich side (rich classes, regions and countries) can keep investing in new areas in pursuit of maximal profits regardless of environment's carrying capacity, whereas the poor side (poor classes, regions and countries) abuses natural resources and labor in order to stay alive. Empirical studies (eg. Gallopin and Barrera) show that the richest and the poorest strata do more harm to environment than the middle class, indicating that a high level of bipolarization is sure to lead to environmental deterioration. Furthermore, it will lead to sharp conflicts between the rich and the poor and eventually result in social revolution, thus destroying productive forces and sabotaging sustainable development. However, these risks can be avoided if the land system is reformed and other measures are adopted to ensure fairer distribution of natural capital and social benefits. Therefore, promoting fair distribution should be made an important goal and norm of sustainable development.

Fair distribution doesn't only involve space, but also time. The common approach of discount rate tends to ignore long-term impacts and may eventually cause marked or irreversible hazards to the environmental system. Sustainable development stresses inter-generational equality, so a low discount rate is advised to give a large weighting to long-term environmental damages. However, as Fisher and Krutilla (1975) pointed out, a low discount rate may lead to a high investment rate, which gives rise to even greater environmental damages. Another method (Fisher and Krutilla 1975) is appropriately underestimating irreplaceable environmental capital. If economic development endangers such resources, the long-term damage cost will quickly increase. Syrah Waltb et al. proposed to analyze the "minimal security standard" decided by experts to prevent irreversible damage to important resources.

3. **Ecological sustainability**

The maintenance of the environmental system depends on the magnitude and pace of the flow of its material and energy input and output. Within a certain limit, human activities can change this flow to meet the society's demand for natural resources, environmental comfort and waste disposal capability. But there is an upper and lower limit to how much an environmental system can bear, and both the environment and mankind would be endangered if the limit is exceeded. Therefore, protecting the environment and maintaining a sound human ecology—ecological sustainability—is a basic goal and principle of sustainable development.

The most important standards of ecological sustainability are as follows:

(1) Environmental parameters

In theory, the evaluation indicators of ecological sustainability can be the total natural capital inventory of environmental system, or different segments of the system, such as non-renewable resources, renewable resources and biodiversity. In practice, however, the parameters selected for evaluation usually have to be revised based on the time and place. The higher the overall level is, the harder it is, even impossible, to determine the weighting of lower-level indicators. Therefore, as experience in environmental impact assessment has proven, what's critical is putting a few "environmental parameters" in the scope of evaluation. To do that, it is necessary to design a set of procedures to help different social groups identify what they think are the most important parameters, and create corresponding forums and consultation channels for them to discuss and agree on commonly concerned issues and important environmental parameters.

(2) Critical level

As we all have a judgment of our responsibility for future generations and different evaluation of risks and technology alternatives, our understanding of the sustainability of the selected environmental parameters is bound to differ. There is the minimal security level, natural level or current level, the last one being the most common reference. The strong sustainability strategy stresses the complementarity between man-made capital and natural capital, while the weak sustainability strategy is more optimistic about replacing environmental facilities with man-made products. The former imposes stricter restrictions on to what extent can natural capital be used.

(3) Acceptable risk level

The human understanding of environmental changes is unavoidably incomprehensive and inadequate, which means the evaluation also includes uncertainties. That's why we should systematically specify the acceptable risk levels and uncertainties. For example, do we allow the absence of long-term risks? Will it be safe then? Are certain risks acceptable with the foreseeable technological progress?

(4) Spatial level

The micro-space and macro-space of human activities are distinguished as well as connected. How the sustainable utilization of natural capital is defined on the dimension of human activity space is of great importance. If ecological sustainability goal is only related with overall project level, a project is acceptable as long as its non-sustainability is compensated by other projects' sustainability. Yet if the sustainability goal is defined as specific project level, any non-sustainable project would be unacceptable. Some people support defining the goal as specific project level and some overall project level. Either way, sustainability policies based on specific project level mean more projects will be rejected as opposed to the overall project level. Therefore, which definition to adopt in planning should depend on the time and place. For example, developing countries may prefer the overall project level, but developed countries tend to and are obligated to employ sustainability standards based on specific project level.

Efficiency, equality and ecological sustainability—these goals and principles that supplement each other shall be observed by all countries and regions around the globe. However, given different development stages and tasks, all countries and regions should be allowed and encouraged to bestow various weighting on those goals and other internal factors in view of their own national and regional conditions. In this way, they can combine the principles and flexibility of sustainable development while simultaneously combining the general principles of sustainable development with the actual conditions in each specific country and region.

3 Strategic Factors of Sustainable Development

Factors that affect sustainable development include the interaction between the human system and environmental system and their respective internal drives. Some of these factors are controllable, some are not, and those controllable and critical ones are the strategic factors of sustainable development. There are five basic strategic factors—ecological modernization of economic structure, prevention-based environmental management, economic mechanism, controlled demographic change, and fight against poverty.

1. **Ecological modernization of economic structure**

The basic framework of traditional economy is that resources obtained from the environmental system are transformed into goods to meet the needs for man-made products while "three kinds of wastes" are discharged into the environment. This kind of cowboy economy is the fundamental reason for non-sustainability. Therefore, reforming the economic structure to gradually delink economic growth from environment-affecting input (resources) is the paramount strategic factor of sustainable development. It is called the ecological modernization of economic structure.

Based on the comparison between economic growth and the increased use of environment-affecting input, ecological modernization of economic structure can be divided into two patterns:

(1) Absolute ecological modernization: the use of environment-affecting input keeps falling when economy increases;
(2) Relative ecological modernization: the use of environment-affecting input remains unchanged or increases slower than the economy.

The two patterns correspond to strong sustainability and weak sustainability respectively.

M. Janicke et al., focused on factors like energy, iron and steel, cement and shipment that have strong negative effects on the environment, and studied the relation between economic structural changes and environmental impacts in 31 member states of the Council for Mutual Economic Assistance and Organization for Economic Co-operation and Development (OECD). They found the following three patterns:

(1) Absolute structural improvement: GDP has positive growth, the use of environment-affecting input has negative growth, and absolute environmental benefits are expected;
(2) Relative structural improvement: the use of environment-affecting input remains unchanged or increases slower than GDP, and relative environmental benefits are expected.
(3) Structural deterioration: the use of environment-affecting input increases faster than GDP.

The first and second pattern corresponds to the absolute and relative ecological modernization of economic structure respectively, but the third pattern corresponds to non-sustainability. The ecological modernization pattern of economic structure depends on consumer demand and R&D capability. As developed countries and regions are in a stage where they pursue "better quality of life," information application and services, the structural change is dominated by the tertiary industry that provides labor and improves the "quality of life." Possessed with abundant materials, capital and talents as well as strong R&D capability, they boast advanced technologies as the global center of technological innovation. As such, developed countries and regions naturally choose and implement the absolute ecological modernization of economic structure. In contrast, developing countries and regions are still in the stage where their chief pursuit is living necessities and industrialization, so the structural change is dominated by industries that produce tangible goods. Troubled by insufficient materials, capital and talents and weak R&D capability, they feature backward technologies and rely on import for technological progress. As such, developing countries and regions can only choose and implement the relative ecological modernization of economic structure.

2. **Prevention-based environmental policies**

Up till now, environmental policies in China are still stop-gap measures that only address symptoms. Rolled out to target specific pollutants, these policies don't tackle the root causes and the detailed objectives, indicators and systems of various programs are hardly coordinated, so pollution may transfer among various environmental media or regions. Therefore, although existing policies are necessary to address pollution that has already taken place, prevention-based environmental policies, as a new approach, have to be launched to root out the problem.

The accumulation of environmental damages, acquisition of technology and knowledge and enhancement of public awareness—the sequence of these three interrelated and simultaneous policy processes, especially the sequence of them reaching criticality, is decisive for the prevention-based environmental policy system.

We can accelerate the accumulation of technology and knowledge and enhance public awareness by investing more in environmental technology R&D and environmental education, but that still largely depends on the specific environmental problems that need to be addressed. Environmental impact assessment (EIA) enables people to learn more about the possible environmental impacts of construction projects and urges proper measures before any damage is done. It usually allows imposing the economic burden on the project initiator, which complies with the precondition or principle for implementing prevention-based environmental policies—whoever causes the pollution shall pay for prevention and control. This can be regarded as an important part of prevention-based environmental policies.

The threshold of technology, knowledge and public awareness changes as people come to know and understand more about environmental damages that have happened or will happen. Therefore, environmental standards should be dynamic and adjusted in different countries, sectors or even different regions of the same country.

3. **Ecological modernization of economic mechanism**

The planned economy featuring unmatched responsibilities and rights falls to encourage enterprises enough to pursue economic benefits. Establishing market economy is one of the preconditions for sustainable development, yet even the most advanced market economy today isn't perfect, otherwise market economies wouldn't face such serious environmental problems. Therefore, following the requirement of ecological sustainability, reforming the current market economy operation mechanism and conducting ecological modernization of it is another crucial strategic factor of sustainable development.

The ecological modernization of economic mechanism includes enabling the economy to make ecological self-adjustment and redirecting economic policies according to ecological requirements.

Producers and consumers can benefit from the use of natural resources and environmental labor but don't have to pay everything they have for it; rather they transfer the subsequent results such as resource shortage and environmental pollution to the society, to future generations and to nature. The existence of such externality is the underlying cause for the loss of balance in man-environment system. For sustainable development, we must internalize the externality of production and consumption and let the producers and consumers that create the environmental problems pay the price. We should also incorporate ecological views into all production and consumption-related decisions, and make environmental protection a conscientious action of all producers and consumers. That's what we call economy's ecological self-adjustment.

How to achieve such self-adjustment is both a theoretical and a practical challenge. Current discussions are focused on the following directions: (1) setting fully effective price and cost signals for natural resources and environmental labor based on the marginal opportunity cost of resource depletion; (2) carrying out ecological accounting on corporate level and national accounting level; and (3) finalizing plans for carrying environmental responsibilities and compensating for environmental damages through the court and legislation.

Traditional economic policies are based on the increase and distribution of man-made products and labor. They are biased and their goals, indicators, approaches and organizations have to be redefined and supplemented according to the requirement of sustainable development. To be more specific, first, the protection and increase of natural capital should be included in the goal and indicator system of economic policies, with sustainable development as the guiding principle for economic development. Second, unreasonable interference in agricultural product price must be abolished and resource exploitation tax and pollution tax should be levied to complement economic policies. Third, macro-regulators should be made more conscious of sustainable development, the capability of environmental protection departments should be reinforced and EIA should be included in all major economic policies.

4. **Controlled demographic change**

The demographic change follows the Logistic curve and consists of five stages. In the first stage, a Malthusian Equilibrium is reached in population as the high death rate and high birth rate offset each other; or population doesn't flow and is evenly distributed under natural economy. In the second stage, population increases quickly because of high birth rate but falling death rate; or population is rather concentrated as natural economy disintegrates. In the third stage, birth rate falls quickly; or the speed of population concentration quickly slows down due to dispersion. In the fourth stage, death rate falls to a very low level and birth rate is basically on a par; or population concentration and dispersion tend to reach an equilibrium. In the fifth stage, population size is quite stable as birth rate and death rate, or population concentration and dispersion are in equilibrium. Experience shows that in the second and third stage, demographic change may impede economic growth and cause poverty and environmental deterioration. Therefore, abandoning the mode where a family has absolute freedom in child birth and the governments implement reasonable interference in demographic change is an important factor of sustainable development.

The goal of controlling the demographic change process is to, according to changes in capital accumulation, technological progress and natural capital, arrive at an appropriate size and growth rate of population that adapts to the environment's and economy's demographic carrying capacity and maximizes per capita social benefits. Two ways to achieve this goal are family planning and population re-distribution. The former aims to coordinate population with natural, economic and social development in different periods by influencing the family child-bearing behavior. The latter aims to coordinate population with natural, economic and social development in different places by influencing people's geographic movement, of which a sustainable urban and regional population size and urbanization and immigration policies are typical examples.

5. **Fight against poverty**

Poverty could be the result of uneven distribution of production factors and poor geographic environment under a certain socioeconomic mode, or it could be the outcome of environmental deterioration resulting from population increase and predatory exploitation of natural resources. Yet it is also a main driver of the degradation of man-environment system. In order to survive, the poor have no choice but excessively exploit the environment and resources day in, day out, which worsens the environment; that in turn makes the poor even poorer and their lives ever harder, which then further degenerates the environment. Therefore, fighting against poverty is an indispensable strategic factor of sustainable development.

References

Fisher, A. C., & Krutilla, J. V. K. (1975). Resource conservation, environmental preservation and the rate of discount. *Quarterly Journal of Economics* (89).

Giaoutzi, M., & Nilkamp, P. (1993). *Decision support models for regional sustainable development*. Aldershou, U.K.: Gower.

Ecological Civilization is Not Equal to Green Industrial Civilization

Yongsheng Zhang

The industrial civilization doubtlessly represents huge progress of mankind, but as traditional industrial mode is unsustainable, ecological civilization has become the only choice to achieve sustainable development. When we discuss ecological civilization, one typical thinking is to equate it with green industrial civilization, which hopes to resolve environmental problems through technological progress under the traditional industrial mode (Hallegatte et al. 2012; UNEP 2011; Acemoglu et al. 2012). As a matter of fact, non-sustainability is the inherent limitation of traditional industrial mode. Despite the significance of green technological progress and higher efficiency, they cannot fundamentally solve the sustainability issue. Therefore, we have to think out of the framework of green industrial civilization, and push for the fundamental change of development pattern according to the internal requirements of ecological civilization. This paper has no intention to make a comprehensive discussion of the mega concept of "civilization," but will only, from the economics point of view, reveal the internal logic and implications of the respective economic mode that underpins ecological civilization and industrial civilization.

1 The Reason Why Cannot Traditional Industrial Mode Sustain

1. Mass industrial production entails mass consumption

When it comes to what to produce and how to produce, the traditional industrial logic has the following intrinsic characteristics. First, production focuses on material wealth and the values created are to meet people's material needs, which, however, only represent part of their diverse needs. Second, mass production is carried out

Y. Zhang (✉)
Development Research Center of the State Council, Beijing, China

J. Pan et al. (eds.), *Beautiful China: 70 Years Since 1949 and 70 People's Views on Eco-civilization Construction*, https://doi.org/10.1007/978-981-33-6742-5_50

to produce identical products to largely improve productivity. Third, an industrial vision is limited to the relation between man and things without much heed to the more complicated interactions between man and nature. These characteristics decide the logic of industrial expansion and its consequences.

Industrial mass production results in high material productivity, which requires mass consumption. However, in the age of farming and handicraft of low productivity before industrial revolution, people had formed the habit and cultural and religious tradition of frugality. Although industrialization is centered on material wealth production, there is a limit to how many products people need, so unlimited expansion is impossible. If there isn't a market for mass production, the modern economy built on industrialization cannot sustain. As a result, the traditional consumer psychology and consumption mode poses the biggest obstacle to mass industrial production.

2. **Mass consumption entails systematic reshaping of consumer psychology**

What should be done then? The way out is to reshape the frugal habits formed in the long period of farming and handicraft into mass consumption that fits mass production, so as to enlarge market demand. This requires a new wealth narrative matching the mass industrial production—It seems that people's material demand is limited, but if material wealth becomes a sign of career success and social status, then, there would be no material limit since people's pursuit for career success and social status is unlimited.

The American Dream is a representative of this new wealth narrative, where wealth is the sign of professional success and social status. In the 1930s, the capitalist Europe and America had a major crisis never seen before, which marked a milestone in reshaping the social psychology and people's consumption mode. To resolve the surplus crisis, Keynesianism and easing monetary policies were adopted on the macro level to stimulate effective demand, while ad marketing and financial measures like consumer credit and credit card were adopted on the micro level. This gave rise to a whole set of institutional arrangements in support of mass consumption for the purpose of reshaping the relation between man and commodities and reshaping the social psychology of consumption. In the end, consumerism became the cornerstone of modern economic operation and consumer society became the sign of industrial society (e.g. Ng 2003; Scitovsky 1992; Skidelsky and Skidelsky 2012; Goodwin et al. 2008; Mukerji 1984; Atkinson 2012; Veblen 1899). As late-comer countries join the camp of industrialization and globalization one after another, material desire and consumerism spread across the globe, as represented by the "November 11" shopping carnival in China.

3. **The formation of a consumer society is a long-term and conscientious social reform**

Industrial society of high productivity has to be built on consumerism, and the rise of consumerism is the systematic action of the entire economic system rather than the behavior of an individual entity. Only when everyone "buys more and sells more" will there be a market for all producers—and everyone is happy. Such a massive reform of social psychology doesn't and cannot depend on a planned "social reform program" or "conspiracy" by the government or individual enterprise. It is the inherent requirement of industrialization and capital. It is self-enforcing and is the outcome of the common action of a lot of dispersive market entities. Unlike previously believed, modern consumption is actually the direct result of a long-term, conscientious but barely known social reform (Atkisson 2012). As Nobel prize winners Akerlof and Shiller (2015) pointed out, what consumers really like is actually different from what they are manipulated to like in the market.

Modern economics have given a push to this unsustainable development mode. First, consumerism is embedded in the analytical architecture of standard economics. In this architecture, consumer's efficacy is a function to the volume of commodity consumption. The more consumption, the higher efficacy, so more consumption is always welcome, although in reality people's material consumption does have a limit. Besides, producers have to sell more to earn more. Second, in the industrial logic that advocates buying and selling more products, human well-being and ecological environment are not in the narrow analytical architecture of standard economics. Once we jump out of this architecture and scrutinize economic conclusions in the broader "man-nature" perspective, with human well-being included, many conclusions of standard economics are no longer tenable or somewhat misleading. Therefore, only by reflecting on and restructuring standard economics can we truly understand sustainable development.

4. Benefits and environmental consequences under industrial logic

Industrialization has dramatically increased the productivity of material wealth and advanced human civilization, but the economic prosperity built on the consumption of material wealth has brought about impacts on human well-being and environment. First, the expansion of consumption usually goes counter to human well-being. The ultimate goal of development is to increase people's well-being or sense of happiness, while economic growth and consumption are just means to that end. A lot of studies show that in many countries including China, economic development under traditional industrial mode hasn't consistently enhanced people's sense of happiness as expected (e.g. Easterlin et al. 2012; Ng 2003; Scitovsky 1992; Jackson 2016; Skidelsky and Skidelsky 2012). Smith (1759) pointed out that the high productivity of market economy was driven by the misleading belief that material wealth could bring happiness. "This deception is what starts men working and keeps them at it… having entirely changed the whole face of the globe."

Second, environmental crisis makes the world unsustainable. The impacts imposed by traditional industrial mode on the ecological environment are reflected in aggregate and logic. As this mode depends on the continuous expansion of consumerism, man will inevitably surpass the environmental carrying capacity one day. Compared with this consequence, the damages done by the infringement of traditional industrial logic upon the environmental system are less noticed. Ecosystem is a symbiotic system (including the natural system and the relation between man and nature) where a number of agents depend on each other, but according to traditional industrial logic, powerful industrial technologies and organizations shall be mobilized to extract certain links that are considered "useful" from this symbiotic system for mass industrial production. When this interdependent system is broken, the entire natural ecosystem may collapse as a whole.

2 Can Green Industrial Civilization Resolve Sustainability Crisis?

At the moment, the most common thought about achieving sustainable development is through green industrial civilization, which means resolving problems with new green technologies in the current industrial mode (Acemoglu et al. 2012). There is no doubt that green technology innovations are extremely important for boosting productivity and national competitiveness, but this paper focuses on the other side of technological progress. With technological progress and improved efficiency, will people choose to use fewer resources to produce the same products to improve the environment, or will they use the same or even more resources to produce more products, thus aggravating environmental degradation?

Unfortunately, the intrinsic logic of industrial capital decides the option for the latter. Therefore, unlike what people intuitively think, technological progress doesn't always bring sustainable results; it even aggravates environmental crisis under certain circumstances (Polimeni et al. 2009; Huesemann 2001; Huesemann and Huesemann 2011). The counter-intuitive phenomenon revealed by Jevons (1865) that a higher efficiency in Britain's coal industry led to more coal consumption was not an accidental case, but stemmed from the deep-rooted logic behind development—the so-called development paradox.

Why is it that green technological progress may add to and exacerbate the aggregate of environmental pollution? It's simple. Although green technology and higher efficiency will reduce the environmental intensity of unit product, the endogenous drive—capitalists' pursuit for maximal profits and consumers' pursuit for maximal utility—is sure to pump up the aggregate, and the resulting aggravation of environmental pollution will eventually override the reduction in environmental intensity.

Specifically, for consumers who aim at maximal utility, the more they consume, the higher the utility; for producers who aim at maximal profits using green technologies, either self-developed or purchased, the more they produce, the higher the profits. In this process, the producers and consumers follow a highly consistent logic—the more, the better. Therefore, production and consumption keep expanding when resource restrictions are fixed. In the long term, the day will never come when technological progress will reduce resource consumption. In standard economics, the impacts exerted by economic activities on ecological environment (the man-nature relation) are not included in its analytical architecture, so as long as there are resources and technologies available to expand production and consumption, there isn't a compromising force to restrain such expansion.

Therefore, in the framework of traditional industrial civilization, technological progress cannot fundamentally achieve sustainable development, nor can the so-called green industrial civilization. Only by changing the contents of production and consumption and decouple their growth from environmental damage is it possible to avoid environmental non-sustainability caused by economic expansion. The only way for traditional industrial civilization to transform into the so-called green industrial civilization is through the broader framework of ecological civilization. The post-industrial society generally refers to the post-industrialization of production side as its consumption side is no different from that of the industrial society, so it cannot be called a transformation of industrial civilization.

3 Ecological Civilization Marks a Fundamental Change of Development Pattern

The shift from industrial civilization to ecological civilization is an inevitable choice in response to the non-sustainability crisis under traditional industrial mode. It indicates a fundamental change in the underlying logic that defines human behaviors, and the change, which is reflected in vision and value, initiates a profound turnabout in the development pattern.

The change in underlying logic leads to a change in vision. The relation between industrial civilization and ecological civilization resembles that between geocentric theory and heliocentric theory. Economy is a subset in natural ecosystem. Man's past knowledge and renovation of the world was mostly undertaken under the narrow industrial vision and logic, while the impacts imposed by economic activities on nature are not included in the current analytical architecture of standard economics. Once put in a broader perspective, many theories, decisions and behaviors under the narrow economics framework have to be reevaluated, and the cost-profit concept has to be redefined. Some behaviors that used to be deemed as efficient and cost-effective don't remain that way, while new opportunities will be recognized. For instance, people used to think that the existing industrial mode is most efficient while green economy is a "luxury" that can only be afforded when economic development reaches a certain level. This conclusion would nevertheless be rewritten when reviewed in a new perspective—green economy is most efficient and the existing development mode is too costly.

The change in underlying logic leads to a change in value. This means the standards for a "good life" and modernization have to be redefined. The development of human society is driven by the pursuit for a happy and good life, but the definition of a happy and good life varies in different concepts or value systems, thus resulting in different contents of development. As mentioned earlier, the shift from agricultural civilization to industrial civilization is a process of paving the way for mass industrial production by reshaping the social psychology (especially consumer psychology) on a large scale. Likewise, the shift from industrial civilization to ecological civilization also has to be led by a massive reshaping of values and social psychology, so as to push the profound change in consumption contents and lifestyle. Throughout history, "a good life" has always been a dynamic concept. For instance, for several centuries, what Chinese people aspired for was not the acquisition of material wealth, but the cultivation of knowledge and fine art on the basis of moderate prosperity (Etzioni 2012). Therefore, when the modern development concept is combined with China's 5000-year-long culture and philosophy, they will hopefully give birth to a good lifestyle characterized by a sustainable man-nature relation as well as rich meanings of life.

The change in underlying logic leads to the shift of development pattern. The aforementioned new visions and values constitute the new coordinates of human behaviors in ecological civilization. They decide how man will learn about and reform the world with new ideas and visions different from those in the traditional industrial age to create a happy and good life. With this shift, the traditional development pattern in industrial age needs to be comprehensively and profoundly transformed. A lot of ideas have to be redefined, including what to develop and how, and a systematic transformation has to be brought about on that basis, covering a full range of aspects such as development concept and content, resource concept, organizational mode, business mode, institutions and mechanisms, and policy system.

4 Conclusion

China has scored an unprecedented achievements after its reform and opening-up. China is also one of the biggest beneficiaries of traditional industrial mode, but this mode is unsustainable and its development theories are unable to produce effective explanations and solutions. Therefore, based on its development experience and cultural traditions, China has fostered the vision of ecological civilization. The emergence of new development theories is inevitable, just like how heliocentric theory came into being when geocentric theory was found unable to explain many phenomena. However, the current disciplinary system, academic system and discourse system, which mostly derive from the traditional industrial mode and serve its purposes, cannot meet the requirements of global sustainable development and must be rethought and restructured under the ecological civilization framework.

In the new development concepts, as the purpose of development shifts to meeting people's "ever-growing needs for a better life," a lot of new opportunities of economic growth will appear and promote rather than obstruct economic growth, only the contents of growth will change considerably. The new needs have to be met through the new supply-side structural reform and will become a new important part of economic drive. Yet the shift from traditional industrial mode to ecological civilization-based green development is also a huge challenge as the development concepts and their supportive system that have been established over a long time all have to be changed. The reshaping of social psychology and market system in the industrial age is a self-enforcing outcome of different market entities for common interests. In comparison, ecological civilization is a stage when traditional industrial mode, after over 100 years' development, has led to environmental and social crisis, which prompts transformation. The question isn't whether we should move from industrial civilization to ecological civilization, but how to do it. If China can find the path to realize ecological civilization, that would be a significant contribution to mankind.

References

Acemoglu, D., Aghion, P., Bursztyn, L., et al. (2012). The environment and directed technical change. *American Economic Review, 102,* 131–166.

Akerlof, G., & Shiller, R. (2015). *Phishing for phools: The economics of manipulation and deception.* Princeton University Press.

Atkisson, A. (2012). *Life beyond growth: Alternatives and complements to GDP-measured growth as a framing concept for social progress (2012 Annual Survey Report).* Tokyo: Institute for Studies in Happiness, Economy and Society.

Easterlin, R. A., Morgan, R., Switek, M., et al. (2012). China's life satisfaction, 1990–2010. *Proceedings of the National Academy of Sciences of the United States of America, 9*(25), 9775–9780.

Etzioni, A. (2012). *The good life: An international perspective.* https://www.researchgate.net/publication/256065268_The_Good_Life_An_Intemational_Perspective.

Goodwin, N., Nelson, J., Ackerman, F., et al. (2008). Consumption and the consumer society. In M. E. Sharp (Ed.), *Microeconomics in Context.*

Hallegatte, S., Heal, G., Fay, M., et al. (2012). *From growth to green growth: A framework* (World Bank Policy Research Working Paper 5872). Washington, DC: World Bank.

Huesemann, M. H. (2001). Can pollution problems be effectively solved by environmental science and technology? An analysis of critical limitations. *Ecological Economics, 37* (2) 2, 271–287.

Huesemann, M., & Huesemann, J. (2011). *Techno-fix: Why technology won't save us or the environment.* New Society Publishers.

Jackson, T. (2016). *Prosperity without growth: Foundations for the economy of tomorrow.* Taylor & Francis.

Jevons, W. S. *The coal question.* 1865/1965.

Mukerji, C. S. (1984). *The eighteenth century* (Vol. 25, No. 1, pp. 88–94). University of Pennsylvania Press.

Ng, Y. K. (2003). From preference to happiness: Towards a more complete welfare economics. *Social Choice and Welfare, 20,* 307–350.

Polimeni, J. M., Mayumi, K., Giampietro, M., et al. (2009). *The myth of resource efficiency: The Jevons Paradox.*

Scitovsky, T. (1992). *The joyless economy: The psychology of human satisfaction.* Oxford University Press.

Skidelsky, E., & Skidelsky, R. (2012). *How much is enough? Money and the good life.* Penguin UK.

Smith, A. (1759). *The theory of moral sentiments.* Cited in Ashraf, N., Camerer, C., & Loewenstein, G. (2005). Adam Smith, behavioral economist. *Journal of Economic Perspectives,* [1981], 19 (5), 131–145.

UNEP. (2011). *Towards a green economy: Pathways to sustainable development and poverty eradication.* Geneva: United Nations Environment Programme. https://www.unep.org/greeneconomy/Portals/88/documents/ger/ger_fmal_dec_2011/Green%20EconomyReport_Final_Dec2011.pdf.
Veblen, T. (1899). *The theory of the leisure class.* Prometheus Books.

Ecological Civilization—The Chinese Path

Mouchang Yu

President Xi Jinping said, "The path we take determines our destiny," which has proved a truth in history. From the perspective of human civilization development, the Chinese path—the 5000-year-long history of Chinese civilization—has taken three major steps that decide China's future. The first step is in the agricultural age when the Chinese agricultural civilization, guided by core social values then, climbed to the world's top in its accomplishments and made great contributions to mankind. The second step is in the industrial age when China, due to the inertia of agricultural civilization, failed to shift to a new track, and therefore was reduced to a dumping field of products from advanced industrialized countries and was plundered by them for resources, leading to a century of humiliation and misery. The third step is in the age of ecological civilization when China, under the firm leadership of the CPC, has pioneered on the path of ecological civilization for the great rejuvenation of the Chinese nation. China will be great again!

1 Ecological Culture—A Prelude to Ecological Civilization

Culture is the precursor and soul of social development and the leading force of economic development. The rise of ecological culture symbolizes the arrival of a new age for human race—the age of ecological civilization. In the mid-twentieth century, the eight environmental hazards astonished the world and triggered a great environmental protection movement. Scholars in developed countries began to reflect

M. Yu (✉)
The Institute of Philosophy, Chinese Academy of Social Sciences, Beijing, China

J. Pan et al. (eds.), *Beautiful China: 70 Years Since 1949 and 70 People's Views on Eco-civilization Construction*, https://doi.org/10.1007/978-981-33-6742-5_51

on the events, and studies of ecological culture thrived, such as environmental philosophy, environmental politics, environmental ethics, environmental law, environmental economics, ecological theology, ecological feminism, ecological Marxism, and so on. A new culture—the ecological culture—was on the horizon.

In the 1980s when the study of ecological culture just started in China, the author delivered speeches and published articles and books on this topic and proposed viewpoints on behalf of Chinese academia. I wrote in my articles that culture is a way of survival that distinguishes man from animal. Modern culture is one where "man reigns over nature" whereas new culture is one where "man and nature develop in harmony." In the narrow sense, ecological culture refers to the social ideology, human spirit, and socioeconomic systems and institutions under the guidance of ecological values, covering ecological politics, ecological philosophy, ecological ethics, ecological economics, ecological law, ecological art, and ecological aesthetics, among others. In the broad sense, it refers to a new way of human existence encompassing the ecological production mode and lifestyle – a way of harmonious development between man and nature.

The development of human culture follows a historical trajectory. The evolution of cultural forms—from natural culture, humanistic culture, and scientific culture to ecological culture—corresponds to the evolution of human civilization forms—from pre-historic civilization, agricultural civilization, industrial civilization to ecological civilization. In the mid-twentieth century, the global ecological crisis represented by the eight environmental hazards triggered an environmental protection movement across the globe. The rise of ecological culture marked the decline of industrial civilization and the arrival of a new era—the era of ecological civilization.

In 1996, Yu Mouchang published a book titled *New Century of Culture—Theoretical Explanation of Ecological Culture*, in which he used two tables to present the form of human culture, civilization and society, the central industries and axis of society, and the historical development of production mode, energy, social entities and philosophical expressions. For instance, the central axis of the four forms of human society so far, or the factors determining social development, are as follows: morality of the primitive society, power and privileges of the slave and feudal societies, capital of the capitalist society, and wisdom of the ecological civilization society. The society of ecological civilization is essentially ecological socialism, where ecological industry is the central industry, information and intelligent applications define the production mode, intelligent machinery is the tool, information and intelligence is the main direction of resource development, solar energy is the main energy form, knowledge is the main social property, information is the form of science, intellectuals are the backbone of the society, and the expression of philosophy is respecting nature. This is a summary drawn from the analysis of the historical development of human culture.

"Culture is the lifeblood of a nation, and it gives the people a sense of belonging. To complete the building of a moderately prosperous society in all respects and achieve the great renewal of the Chinese nation, we must create a new surge in promoting socialist culture and bring about its great development and enrichment, increase China's cultural soft power, and enable culture to guide social trends, educate the people, serve society, and boost development." This is written in the report of the 18th CPC National Congress.

Today the study of ecological culture is elevated to the national level in terms of leadership, scale and capability. Research institutes of ecological culture have been established on national and sub-national levels, departments and courses regarding ecological culture have been established in colleges, and bachelors, masters and doctors are being cultivated in related disciplines. This is an imperative step to implement the ecological civilization strategy outlined by the 17th, 18th and 19th CPC National Congresses. Relevant studies have made impressive progress and results. Monographs, theses, and literary, artistic and film and TV works on ecological philosophy, ecological politics, ecological economics, ecological law, ecological Marxism and many other subjects have emerged like mushrooms after a spring rain, serving the nation's ecological progress.

2 Ecological Civilization—A Path of Socialism with Chinese Characteristics

In the early twenty-first century, along with China's remarkable achievements in industrialization were serious environmental pollution, especially the pollution of air, water and heavy metals. A number of grave pollution events happened at that time. There were severe heavy metal pollution events such as the "cadmium pollution event" in Hunan's Liuyang (2008–2009) and the "lead in blood event" in Shaanxi's Fengxiang (2009). Water pollution gave rise to several "cancer villages" (2000–2006) in different regions, including Henan's Shenqiu, Anhui's Mengcheng and Lingbi, Shandong's Wenshang, Shangba Village in Guangdong's Shaoguan, and Jiangxi's Dexing and Leping, where the incidence rate and mortality rate of cancer increased shockingly. Food safety events took place constantly, such as the "Sanlu formula event" (2008), poisonous dumpling event (2008), clenbuterol event (Guangdong's Heyuan, 2006), "toxic cowpea event" (Wuhan, 2010), and "gutter oil event" (2010). Air pollution was also serious with thick smog spreading from Beijing and other big cities in the north to the central, eastern and southern part of China (2013). These pollution events were more serious than the eight environmental hazards regarding the scope, gravity, number of deaths, hazard to human health and economic damages. The Chinese government and people were so shocked that they embarked on the path of building an ecological civilization.

In 2007, General Secretary Hu Jintao said, "the 17th CPC National Congress emphasized the need to develop ecological civilization. It was the first time the CPC put it forth as a strategic task." "We will develop ecological civilization and basically establish the industrial structure, growth pattern and consumption mode characterized by energy and resource conservation and environmental protection. The circular economy will reach a large scale, the share of renewable energies will increase considerably, the discharge of main pollutants will be effectively controlled and quality of eco-environment notably improved. The idea of ecological civilization will be firmly driven home across the society." To implement the spirit of the 17th CPC National Congress, a surge of ecological culture studies swept across the nation. Major progress was made in the study of ecological philosophy, ecological ethics, ecological economics, ecological politics, ecological law, ecological literature and art, and ecological culture, making psychological and theoretical preparations for kicking off ecological civilization development. Meanwhile, 30 years of reform and development witnessed China making tremendous achievements in its industrialization and becoming the world's largest machinery manufacturer and industrialized country, making material preparations for ecological civilization development. The 30 years of reform and development also witnessed substantial progress in China's scientific, technological and educational sectors, reaching world leading level in certain fields, which made the sci-tech and talent preparations for ecological civilization development.

In 2012, the 18th CPC National Congress put forth the Five-Sphere Integrated Plan as the national development strategy, whereby ecological progress is deeply integrated with economic, political, cultural and social advances. The *Master Plan for Institutional Reform for Ecological Civilization* was formulated, and the *Opinions on Accelerating Ecological Civilization Development* (2015) was issued by the CPC Central Committee and the State Council. In implementing the strategy, the Master Plan and the Opinions, construction of ecological counties, cities, provinces and demonstration zones was carried out in full swing. Building an ecological civilization is a great endeavor by the whole Chinese people and a great initiative by China.

In 2017, General Secretary Xi Jinping summarized our experience in ecological progress in the report of the 19th CPC National Congress. He said, "In the past five years, we have implemented the Five-Sphere Integrated Plan, made new advances on all fronts, and made notable progress in building an ecological civilization." On May 18, 2018, General Secretary Xi delivered an important speech at the National Conference on Eco-environmental Protection. He pointed out that great efforts have been made to advance ecological progress, the whole Party and the whole nation have become much more conscientious and proactive in practicing the green development concept, and a lot more attention has been paid to environmental protection. In promoting green development, we have focused on protruding environmental issues, intensified ecosystem protection, reformed the environmental regulation system and expedited the institutional reform for ecological civilization. The new institutional system is taking shape at a faster pace, the system of main functional zones is being improved step by step, and national parks are promoted on a trial basis. Solid progress has been made in all-round resource conservation, resulting in a largely reduced energy and resource intensity. Major environmental protection and recovery projects have progressed smoothly with continuously rising forest coverage rate, environmental governance is obviously strengthened, and the environment is of a better quality. Taking a leading position in international cooperation on climate change, China has become an important participant, contributor and torchbearer in the global endeavor for ecological civilization. Building ecological civilization is a millennial cause aimed at the lasting development of Chinese nation.

The nation is now standing tall in the world's east in a brand new outlook, making greater contributions to humankind as it approaches the center of global stage.

Under the CPC's leadership, the ecological civilization drive has become a state action and a grand endeavor of the whole nation. Chines people are walking toward the glorious road of building an ecological civilization.

3 Ecological Civilization—An All-Round Transformation of Human Society

The academic circle has voiced different kinds of views and statements on what ecological civilization is. We believe ecological civilization is a new form of human society. The global ecological crisis breaking out in the middle of the twentieth century severely challenged human existence and became the focal social problem, heralding the arrival of a fundamental reform in world history. The crisis, featuring environmental pollution, ecological damage and resource shortage, indicated that the industrial civilization was declining while a new civilization—ecological civilization—was rising. By building an ecological civilization, we will realize a comprehensive social transformation and integrate ecological progress in social, institutional, material and cultural efforts. A new, sustainable human society will be created—a society of ecological civilization featuring social harmony between man and man and harmonious co-existence between man and nature.

According to the book *Ecological Civilization: A Comprehensive Transformation of Human Society*, ecological civilization is a new and sustainable form of human society that aims for man's reconciliation with ecology and society through all-round transformation.

(1) Transformation of values: we should abandon human-centrism and individualism and foster the core social values of man-nature harmony.
(2) Transformation of philosophy: we should move from the philosophy of man dominating nature to man existing in harmony with nature. A transformation of philosophical pattern should be effected from the modern "subject-object dichotomy" to the environmental philosophy of man-nature unity.
(3) Transformation of social politics: we should change from the "capitalist despotism" of industrial civilization to "people's democracy" of ecological civilization.
(4) Transformation of production mode: we should create advanced ecological technologies and processes and shift from a linear, non-circular economy to an ecological, circular and low-carbon economy.
(5) Transformation of way of life: we should depart from high consumption to green consumption and live a simple, low-carbon and equitable life. That's a sustainable lifestyle of a higher level.

(6) Transformation of culture—an ecology-oriented shift of ethics and morality: the development from social ethics to environmental ethics is a progress in human morality.
(7) Transformation of culture—an ecology-oriented shift of education and scientific technology: the purpose of science and education will expand from humanity, economy and society only to also include the protection of nature, ecology and environment.
(8) Transformation of culture—an ecology-oriented shift of literature and art: fine arts have shifted its focus from man alone to man-nature harmony.
(9) Transformation of medicine—promotion of traditional Chinese medicine: we should go beyond the modern "bio-medical mode" and carry on the 5000-year-old traditional Chinese medicine.

China is making new contributions to mankind by building an ecological civilization. Advanced industrialized countries have failed to shift to the path of ecological civilization due to "path inertia." In contrast, China is prompted to adopt the path of ecological civilization first because of its serious ecological problems, and will lead the world into a bright ecological future.

4 Following Xi Jinping Thought on Ecological Civilization and Striving for a New Victory in Ecological Civilization Development

On May 18, 2018, General Secretary Xi Jinping delivered an important speech at the National Conference on Eco-environmental Protection. Xinhua News Agency reported the speech and called for "earnestly studying and apprehending Xi Jinping Thought on Ecological Civilization." The report said, the Xi Jinping Thought on Ecological Civilization with rich connotations answers several cardinal theoretical and practical questions—why we build ecological civilization, what kind of ecological civilization to build and how to build it. It is a significant theoretical and practical innovation of the Party and the fundamental guidance for promoting ecological progress in the new era. We will abide by the thought, insist on man-nature harmony, and uphold the concepts that lucid waters and lush mountains are invaluable assets and protecting and improving the eco-environment is protecting and developing productivity. We will guide the socialist drive with Chinese characteristics with socialist outlook on ecological civilization, and press ahead till we secure full success in this undertaking.

1. **Adhering to harmony between man and nature**

Harmonious co-existence between man and nature is the basic policy of ecological progress. In 2017, General Secretary Xi Jinping said in the report of the 19th CPC National Congress that ensuring harmony between human and nature and building ecological civilization is vital to sustain the Chinese nation's development. It is the basic policy that underpins our endeavors to develop socialism with Chinese characteristics.

In 2015, General Secretary Xi initiated the vision of "a community with a shared future for mankind" in his speech at the UN General Assembly. "In today's world, all countries are interdependent and share a common future. We should renew our commitment to the purposes and principles of the Charter of the United Nations, build a new type of international relations featuring win-win cooperation, and create a community with a shared future for mankind." This is the "Chinese approach" to handling international relations.

History has proved that man and nature share a common future. The industrial civilization, in which man dominates nature, leads to global ecological crisis. We should establish and implement the philosophy of harmony between man and nature because they essentially exist in a community of cooperative and harmonious co-existence. Man lives on nature and develops by exploiting and utilizing it; man also changes nature and creates a new natural world. The normal state, essence and law of man-nature relationship is harmony, and the direction or goal of this relationship is common survival, common prosperity and common evolution. This is the foundation for the vision of "community with a shared future for mankind."

Harmony between man and nature is our vision and the direction and goal of our endeavors. On the level of state relations, conflict and confrontation, disputes and even wars have seriously damaged the interests of all parties. The development of modern warfare, especially nuclear war, may even lead to the extermination of the human race.

"Harmony" is the quintessence of Chinese culture. China believes in "harmony without uniformity." In Chinese philosophy, as long as we uphold the vision of "community with a shared future," the China-US trade war can be appropriately resolved, so can all disputes and clashes in the world, big or small. "I sincerely hope that the international community will make concerted effort to make peace, relate to each other in cooperation rather than confrontation; beat swords into plowshares and build a community of common destiny of all mankind," said General Secretary Xi. In face of various challenges, only by upholding man-nature harmony and a common human community will we be able to usher in a new century for China and a new era for humankind. This is the best "Chinese plan" for the world.

2. Adhering to "lucid waters and lush mountains are invaluable assets"

On August 15, 2005, Xi Jinping, then Secretary of the CPC Zhejiang Provincial Committee, inspected the Yucun Village in Anji County, Zhejiang, where he affirmed the shutting down of quarries and cement plants in the village for environmental protection. He told the officials, "Don't repeat the old path. Don't cling to the old development mode. You said you were determined to shut down some mines. This is a wise move. We used to say that we want lucid waters and lush mountains as well as economic benefits, but the waters and mountains can bring economic benefits themselves. They are invaluable assets in their own right." The General Secretary proposed a new development concept and path for China. Guided by it, we have made concerted efforts to create a better ecological environment with clearer sky, greener mountains and cleaner waters, and jointly walked into the new era of socialist ecological civilization.

The new development concept put an end to some long-standing disputes—what's the relation between economic development and environmental protection? Is economic development bound to destroy the environment? It drew the definite conclusion that lucid waters and lush mountains are invaluable assets as they can bring economic benefits. As long as we consistently carry out this new concept, we can achieve economic and social development and environmental protection in parallel, and economic, social and ecological benefits in parallel.

3. **Adhering to "protecting and improving the eco-environment is protecting and developing productivity"**

On April 10, 2013, General Secretary Xi Jinping said during an inspection tour in Hainan Province, "throughout the world development history, protecting and improving the eco-environment is protecting and developing productivity." On May 23, 2016, he said "ecology contains resources and productivity" when inspecting Yichun, Heilongjiang Province.

The idea that ecology contains productivity overset the definition of productivity in modern economics, which is that there is only one kind of productivity—the social productivity of human labor. But that definition is incomplete. Ecology contains productivity—this idea proposed by General Secretary Xi is a new economic theory, a theoretical innovation and substantial progress in economics.

Following the idea that "ecology contains resources and productivity," we need to correctly handle the relation between economic development and environmental protection, put economic statistics on natural resources and environmental quality that go into material production, and implement the policy of paid use of natural resources. Ecological assessment should be performed on the exploitation and utilization of resources and environment to compensate for their damages and strive for ecologically balanced development. According to the new economic theory and on the "circular economy" principle, we should invent ecological technologies and processes to recycle or reuse resources invested in economic production for maximal output while reducing the discharge of wastes, so as to realize maximal economic, social and ecological benefits all at once. We need to protect and develop ecological productivity, so that we can create adequate economic products and values to support modern life while conserving natural and ecological values, sustaining the development of man, society and nature, and advancing their common prosperity in a community of shared future. This is both imperative and feasible. It is the basis of economic development in ecological civilization.

4. **Adhering to socialist outlook on ecological civilization and socialist path with Chinese characteristics**

General Secretary Xi Jinping pointed out in the report of the 19th CPC National Congress that what we are doing today to build an ecological civilization will benefit generations to come. We should have a strong commitment to socialist ecological civilization and work to develop a new model of modernization with humans developing in harmony with nature." The CPC's thought on ecological progress consists of four essential parts – the Five-Sphere Integrated Plan, the basic policy of man-nature harmony, the new green development concept, and the tough battle of pollution prevention and control. He stressed, "the 'four parts' reflect the CPC's grip on ecological progress, the position of ecological progress in the Party's and nation's cause in the new era, and the Party's deployments and requirements to achieve that goal."

In May 2018, General Secretary Xi Jinping expounded important views of socialist ecological civilization at the National Conference on Eco-environmental Protection. "We must consciously coordinate economic and social development with ecological progress. To succeed in this endeavor in the new era, we must adhere to the following principles. First, we must pursue harmonious co-existence between man and nature, and prioritize conservation, protection and the restoration of nature. We should protect the eco-environment as we protect our eyes, and cherish it as we cherish our own lives, so that the beautiful nature and ecology will last forever in serenity and harmony. Second, lucid waters and lush mountains are invaluable assets. We must implement the vision of innovative, coordinated, green, open and shared development, and accelerate the formation of spatial configurations, industrial structures, production models and lifestyles that favor resource conservation and environmental protection, so as to leave time and space for nature to recuperate. Third, a good ecological environment is the fairest public product, and the most accessible welfare for the people. We must pursue environmental benefits for the people with the emphasis on solving prominent problems that threaten their health, and meet their growing expectation for a beautiful environment. Fourth, mountains, rivers, forests, farmlands, lakes and grasslands are a community of life. We must make an overall plan that takes all relevant factors into consideration, and adopt multiple measures simultaneously to advance ecological progress in all respects. Fifth, we should protect the eco-environment with the strictest regulations and laws. We must speed up institutional innovation, strengthen implementation, and make regulations mandatory, so that they become a powerful deterrent. Sixth, we should work together to promote a global ecological civilization, be heavily involved in global environmental governance, help form global solutions to environmental protection and sustainable development, and play a constructive role in international cooperation on climate change." These are basic view of socialist ecological civilization.

The socialist ecological civilization is a scientific outlook on history. In the history of human thoughts, socialist thoughts have gone through several periods—from Utopian socialism in the 16th and seventeenth century to democratic socialism and Marx's scientific socialism in the nineteenth century, and to the current socialist outlook on ecological civilization. The latest one is a new socialist thought, the Xi Jinping Thought on Ecological Civilization.

We will follow the Xi Jinping Thought on Ecological Civilization, steadfastly push forward innovative, coordinated, green, open and shared development, and build an ecological socialist country. This is the socialist path with Chinese characteristics. In Xi's new development concepts, innovation is the primary drive of new development, coordination is the inherent requirement, green is the essential approach, opening-up is the only path, and sharing is the ultimate goal.

General Secretary Xi said, environment concerns people's livelihood and green mountains and blue sky are an indispensable part of a good life. Building ecological civilization is a grand cause that relates to the well-being of Chinese people and the future of Chinese nation. Guided by the new concept of socialist ecological civilization and ecological socialism, we should always put people in the center and take their benefits as the ultimate goal. We should cherish nature more consciously, protect ecology more actively, and work harder to build a harmonious socialist society with Chinese characteristics. Faster steps should be taken to establish an ecological culture system guided by ecological values and a circular, green and low-carbon economic system underpinned by ecological high technologies and centered on ecological industries. Efforts should also be made to establish a system of goals and responsibilities with improving environmental quality at the core, an institutional system guaranteed by governance system and capability modernization, and an eco-security system focused on sound ecosystem cycle and effective prevention and control of environmental risks. We have to expedite the process of building ecological civilization and press forward along the Chinese path toward the new era of socialist ecological civilization. By making united efforts, the sons and daughters of the Chinese nation will definitely bring the Chinese Dream of national rejuvenation into reality.

Theoretical Perception and Logic Expression of Ecological Civilization Strategy in the New Era

Sihua Liu

Both the 18th and 19th National Congress of the CPC exalted ecological civilization to be part of the Five-Sphere Integrated Plan for the socialist cause and a national development strategy, and put it in a significant position in endeavors to build a beautiful China and realize the Chinese Dream of national rejuvenation. When joining the deliberations of the Inner Mongolia delegation during the second session of the 13th National People's Congress, General Secretary Xi Jinping particularly stressed "maintaining the strategic resolve of building an ecological civilization." When addressing the opening ceremony of the 2019 Beijing International Horticultural Exhibition, he further announced the inclusion of ecological civilization in China's overall development plan. Since the development of ecological civilization is included both in the Five-Sphere Integrated Plan for modernization in the new era and in the overall national development plan, I believe it is now regarded as a major development strategy, and the series of assertions made by General Secretary Xi are the guiding thoughts on advancing this strategy across the board. Below is a brief overview and systematic explanations of eight important theoretical and practical innovations in the ecological civilization strategy in the new era.

S. Liu (✉)
The Research Institute of Ecological Civilization and Sustainable Development of Economy, Zhongnan University of Economics and Law, Wuhan, China

J. Pan et al. (eds.), *Beautiful China: 70 Years Since 1949 and 70 People's Views on Eco-civilization Construction*, https://doi.org/10.1007/978-981-33-6742-5_52

1 Strategic Positioning and Direction of the Ecological Civilization Strategy in the New Era

From the report of the 18th and 19th National Congress of the CPC, and the important statements in *Xi Jinping: The Governance of China* to General Secretary Xi's speech at the National Conference on Eco-environmental Protection—they have fully showcased that "ushering in the new age of ecological civilization and building a beautiful China is an important part of realizing the Chinese Dream of national rejuvenation." After the 18th National Congress of the CPC, China elevated "building a beautiful country" to be a major task of overall importance in its socialist modernization drive, and highlighted ecological civilization's strategic position as the "fundamental cause to sustain the development of the country and the nation." This pointed out the strategic direction of building an ecological civilization in the new era. Therefore, Marxist scholars, either focused on the political or academic aspect, are all repeatedly illustrating how ecological civilization holds the key to whether the Chinese nation can re-emerge to glory through socialist modernization in the new era, open new horizons in building a beautiful China, and achieve great rejuvenation. Without a sound ecological environment, a beautiful China would be out of the question and the Chinese Dream would not come true.

2 Strategic Goals and Main Signs of Building an Ecological Civilization in the New Era

As a major strategy concerning national development as a whole, the development of ecological civilization should be guided by definite strategic goals. In the context of building a beautiful China and sustaining Chinese nation's development, we should keep in mind that China is still in the early stage of socialism, so the strategic goals of building an ecological civilization should be consistent with current national conditions. In the new era, China remains "a country of a big population and economy, limited resources and poor ecology and environment." It is faced with a serious situation of tightening resource shortage and restrictions, grave environmental pollution and fragile and degrading ecosystem. Therefore, resource and environment issues, particularly ecological ones, pose the biggest challenge and obstruction to China's economic, social and ecological progress. As such, the strategic goal should be building a resource-saving, environment-friendly and ecologically sound society. This is an overriding issue concerning the direction and path as we strive for economic and social sustainability, lasting development of the Chinese nation, and an early entry into ecological civilization. It is a theoretical innovation of the ecological civilization strategy, marking an improvement of the theory in the new era.

3 New Directions and Roads for Implementing the Ecological Civilization Strategy in the New Era

This is a new topic emerging in the course of building an ecological civilization after the 18th National Congress of the CPC. Marxist scholars, including me, have conducted in-depth theoretical explorations around it. For instance, Fang Shijiao and others deeply expounded this topic in a report of the National Social Science Fund of China in 2015. "Today a profound ecological revolution is needed in the economic and social sectors for human survival and development. We should adopt the new path of green development and advance the ecological or green transformation of the way of production and way of life—namely the way we live, covering economy, science and technology, culture and education, politics and social activities. We should energetically push the socialist economic and social systems with Chinese characteristics to be more ecology-friendly and green, so it will become the internal mechanism, main content, basic path and green outcome of the ecological, economic and social systems." Therefore, "China's economic and social development today must follow the path of ecological, green and innovative transformation. This is an irresistible trend." As such, ecological and green socioeconomic development becomes the basic content and path of building an ecological civilization, and the only path toward the social and economic development coming with ecological civilization. Four points are worth special attention in this process. First, we should make ecological and green efforts the core element and internal drive of building an ecological civilization as they represent the ecological features, connotations and guidance of China's development path. Second, ecological and green endeavors are the chief driver that accelerates the shift from the economic, social and cultural pattern dominated by industrial civilization to one dominated by ecological civilization. Third, to comprehensively push forward ecological civilization in the new era, the key lies in realizing the ecological and green development in all aspects of the modernization drive, including economy, politics, culture, spirit, society, and ecology. Fourth, following the ecological and green direction, the development of ecological civilization has become the drive and engine of overall national development, pushing forward the Chinese economy, society and civilization.

4 New Connotations and Thoughts on Implementing the Ecological Civilization Strategy in the New Era

To highlight the strategic position of building an ecological civilization, the report of the 18th National Congress of the CPC used the word "incorporate," which was replaced by "integrate" in the report of the 19th National Congress of the CPC, giving rise to Xi Jinping's new thought and new strategy of integrated development. In his *Ludwig, Feuerbach and the End of Classical German Philosophy*, Engels hailed development as a "great basic thought" and that "the whole process of great development happens through mutual interactions." In my opinion, these mutual interactions include "integrated development," a great thought initiated by General Secretary Xi. According to Marxist views, integrated development is a basic or even universal law of development, and the all-round advancement of ecological civilization strategy in the new era has blazed a broad path for this law. On the premise of adhering to and developing socialism with Chinese characteristics, the various components and sectors of China's socialist economy and society constitute an indivisible entirety of interdependence, interaction and integration, driving orderly development in general and the evolution of Chinese civilization. This is distinctly reflected in the Five-Sphere Integrated Plan and the integrated development in all five spheres, not only injecting new connotations into the ecological civilization strategy, but also giving new guidance on its practices. To be specific, the development of ecological civilization in the Five-Sphere Integrated Plan has to be integrated with economic and social development. Moreover, it has also to be integrated with development of the other four spheres. Ecological and green development should not only liberate and develop productivity in a green way, but also adapt and coordinate productivity with relations of production, and economic base with superstructure, so the overall ecological system will develop in a green and integrated way.

5 New Vision and Measures of the Ecological Civilization Strategy in the New Era

From the perspective of national development strategy, the development of ecological civilization in the new era involves a series of strategies in multiple domains, such as economy, politics, culture and society. It is a massive strategic landscape and measure. China is in a transition period from "becoming affluent" to "becoming powerful." It has the conditions and abilities to fundamentally resolve the ecological, environmental and resource dilemma, and the way it does that inevitably carries features of the time involving integrated ecological and green development, namely the green strategy and measures. In this sense, the strategic deployments to resolve the dilemma in the new era can be called a new landscape or a new measure. The new strategy encompasses rich contents with four focuses. First, resource conservation and protection is the basic policy. As said before, China has limited resources with a low utilization rate and regular shortage. Economizing on resources, raising their efficiency and promoting resource conservation and protection across the board is a top priority in promoting ecological civilization. Second, the general environmental deterioration in China hasn't been well controlled, and the discharge of main pollutants is far more than the environment can bear. We must firmly grasp this basic reality as a key to fully promote ecological civilization and go all out to win the critical battle of environmental governance and protection. Third, ecological recovery, improvement and protection will remain a strategic focus of building an ecological civilization for a long time. We must steadfastly fight this critical battle and consolidate the ecological foundation for economic and social development in order to ensure a good life for all people and create a new situation of modernization drive that is harmonious and integrated with nature. Fourth, establishing the strategic system of ecological civilization is essentially consistent with the "five systems" proposed by General Secretary Xi: "we must act quickly to establish a cultural system in which environmental concepts and values are norms, an economic system aimed at integrating industries and ecosystems, a system of responsibilities and objectives centered on improving environmental quality, a system of institutions backed by modernized governance systems and capacity, and a safety system that emphasizes the maintenance of well-functioning ecosystems and the effective prevention and control of environmental risks."

6 New Choices and Tasks of the Ecological Civilization Strategy in the New Era

Following Xi Jinping's strategic thought on ecology first and green development, the Yangtze River Economic Belt should give top priority to ecological recovery for a long time to come. It should work hard on eco-protection and avoid massive development in the river basin, so as to blaze out a new path of ecological civilization. At present, the whole country, including economically less developed regions and poor mountain areas, is exploring the new path and has made the strategic choice of ecology first, green development and innovation, and integrated development of economy and ecology. This in essence is the strategic choice for ecology-oriented modernization, "a big green strategy and a big green thought." This new strategy has rich contents and extensive strategic tasks. This paper briefly discusses five aspects of the strategy implementation.

(1) The ecology-first strategy is a contemporary representation of Marx's thought on nature preceding man. Man today should observe the laws and implement the strategy of ecology-first development. The implementation of ecological civilization strategy in the new era should naturally revolve around the strategic tasks of ecology first and advance this strategy on all fronts.
(2) The new era is also characterized by intensive green innovations. We should implement the strategy of development driven by green innovation, make ecological reform and green innovation and transformation the drive and engine of the ecological civilization strategy, and combine the two strategies as an organic whole.
(3) Green industry is the dominant industry in the twenty-first century featuring ecological reform and green innovation and transformation, and green industrial development is the inherent requirement, inevitable choice and strategic task of green and innovative development in the new era. The comprehensive implementation of green industry strategy is a crucial aspect in fully advancing

the ecological civilization strategy. A new situation should be created to effectively foster green industries and promote them to quickly grow into dominant industries.

(4) Developing green energy is the inherent demand of energy development laws. From high carbon to low carbon, low efficiency to high efficiency, dirty to clean, black to green—this is the general trend of energy evolution and the basic law of energy development. Therefore, implementing the green energy strategy that prioritizes clean and green energies is a consensus of all countries, both developed and developing, and the strategic task of China's energy development. Another focal point of advancing the ecological civilization strategy comprehensively is pushing the fundamental shift from a high-carbon energy and economic structure to a low-carbon or zero-carbon one. While seeking low-carbon utilization of high-carbon energies, we should actively develop renewable energies and various gaseous fuels in the attempt to walk into the new age of green energy and economy.

(5) The green science and technology strategy in the new era is an innovative continuation of the "strengthening the country through science and technology" strategy implemented after the reform and opening up. Seventeen years ago, I said the ecological trend of modern science and technology was the essential direction of sci-tech and economic progress, and modern science and technology must develop for the purpose of protecting environment, improving ecology and enhancing nature. At the moment, the ecological and green development of science and technology has grown into an unstoppable trend in all countries around the world, including China. It calls for the "comprehensive advancement of ecological civilization" (said General Secretary Xi Jinping) in the new era, which, like the comprehensive implementation of green science and technology strategy, should also revolve around resource conservation, environmental protection, ecological improvement and natural development, which are the strategic tasks of green sci-tech innovation. Such innovation will be the technological pillar for implementing the ecological civilization strategy by integrating green technologies, industries and energies in driving national development.

7 Fundamental Guide and Basic Principles for Implementing the Ecological Civilization Strategy in the New Era

Marxist scholars in China share the view that the Xi Jinping Thought on Socialist Ecological Civilization is a rigorously logical, complete and scientific theoretical system. It systematically reviews the CPC's theoretical explorations for ecological civilization and is a theoretical representation of the ecological civilization as socialism with Chinese characteristics has entered a new era. We must closely follow the Xi Jinping Thought on Socialism with Chinese Characteristics for a New Era and the Xi Jinping Thought on Socialist Ecological Civilization, which make the fundamental guide for the implementation of the ecological civilization strategy.

The principle of promoting ecological civilization proposed by General Secretary Xi should be adopted to guide our ecological efforts. First, we must uphold harmonious co-existence between man and nature, and prioritize conserving resources, protecting the environment and promoting natural restoration. We should protect the environment in a similar way that we protect our own eyes, and treat it like we treat our own lives, so that the beautiful nature and ecology will last forever in serenity and harmony. Second, lucid waters and lush mountains are invaluable assets. To resolve eco-environmental problems once and for all, we should push forward innovative, coordinated, green, open and shared development, and work faster to establish the geographical space, industrial structure, way of production and way of life that help conserve resources and protect the environment, so as to leave time and space for ecological restoration. Third, we should protect ecology for people's benefits and keep meeting people's growing needs for a beautiful environment. Fourth, mountains, rivers, forests, farmland, lakes and grasslands constitute a community of life. Ecological civilization must be developed on all fronts, in all regions and full process, and in a holistic and coordinated approach with comprehensive and multi-pronged measures. Fifth, ecological environment should be protected with the most rigorous systems and regulations, which should function as a powerful deterrent. Sixth, we should work together to promote a global ecological civilization, be heavily involved in global environmental governance, help form global solutions to environmental protection and sustainable development, and play a constructive role in international cooperation on climate change.

8 Strategic Assurance and Institutional and Mechanism Innovations for Implementing the Ecological Civilization Strategy in the New Era

The key problem now is the lack of adaptation, coordination and congruity between the existing systems and mechanisms and the comprehensive advancement of ecological civilization strategy. Institutional and mechanism innovations are needed to guarantee the strategy implementation. In this process, it is particularly important to put in place well-functioning systems and mechanism driven by green innovations. Institutional innovations should cover the following aspects: greening of industrial structure, development of environmental protection industry and ecological civilization industry, green technologies including green energy technologies, paid use of resources and ecological compensation, green credit and investment system, green price system, green fiscal and tax system, and green wealth and evaluation system. We also need a new economic accounting system suitable for green economic development, including a green GDP accounting system, green accounting system and green auditing system, a new administrative and legal guarantee system for green economic development and environmental protection, and a new green performance assessment system for entrepreneurs, government officials and governments. Institutional and mechanism innovations are also needed to create a green financial system and green service industry system. I believe in the great course of reform and opening up, with continuous ecological civilization, we will eventually realize a beautiful China and a green future.

Toward a New Civilization: Ecological Civilization or Information Civilization

Feng Lu

The accelerated progress of modern science and technology and the appearance of global ecological crisis have both made us realize the end of an old age and the arrival of a new one as we are moving from industrial civilization to a new civilization. There are various views as to what defines the new age or civilization and how we should name it. Two views are worth special attention—the new age or civilization is about ecology or about information. This paper will compare the two views and briefly delineate the vision of the new civilization.

1 Ecological Civilization Theory

The first foreign scholar that put forth the idea of ecological civilization was perhaps professor Iring Fetscher from the Department of Politics of the University of Frankfurt. In the article titled "On the Environment of Human Survival and Dialectics of Progress" published in 1978, Fetscher said, the expected and allegedly urgently needed ecological civilization—unlike Schelsky's technical state—presupposes a social entity that consciously controls the system. That will materialize in a humanitarian and free manner rather than through an expert team serving a global ecological autocracy. Today the age of avid aspiration for infinite progress is coming to an end, and the age when man believes in infinite conquest of nature has come under question. Since a merciful lifestyle featuring peaceful co-existence between man and nature is completely possible, uninhibited technological progress must be controlled and restricted. This article by Fetscher marked a milestone in the history of human thoughts. Targeting a fundamental mistake in modern western thoughts—man can conquer nature unlimitedly with technological progress, he primarily envisaged the ecological civilization. Predicting the forthcoming end of the age of avid aspiration

F. Lu (✉)
Center for Ecological Civilization, Tsinghua University, Beijing, China

J. Pan et al. (eds.), *Beautiful China: 70 Years Since 1949 and 70 People's Views on Eco-civilization Construction*, https://doi.org/10.1007/978-981-33-6742-5_53

for infinite progress and the questioning of infinite conquest of nature, he believed a new age was about to emerge.

Although it was a western scholar who initiated the idea of "ecological civilization," very few western scholars have so far used this idea or considered studies of it important. One of these few is John B. Cobb, Jr., a familiar name for many Chinese today. He has said many times that man can only avoid the fate of extinction by embracing ecological civilization, and the hope of ecological civilization is in China.

Arran Gare from the Swinburne University of Technology, Australia is another of the few western scholars who believed ecological civilization was the future. According to him, the crisis we are facing featuring massive environmental problems is a crisis of modern western civilization, and to get out of this crisis, a radical enlightenment is needed. The enlightenment will resort to the process-relational metaphysics supported by post-reductionist natural philosophy and science, and place mankind in a situation where they, through history, realize self-creation of human nature within nature and move toward a new civilization. In this process, the destructive conflicts between clans, civilizations and nations will be overcome, and the whole mankind will be asked to promote a healthy global ecosystem and admit that human community should succumb to this ecosystem. The purpose of this radical enlightenment should be understood as the development of a global ecological civilization.

American scholar Roy Morrison also recognizes ecological civilization and views it as the civilization of the future. He called his *Ecological Democracy* published in 1995 a book about the "fundamental reform" "from industrial civilization to ecological civilization," in which he poignantly criticized industrialism and industrial civilization. In his opinion, industrial civilization is non-sustainable not only because of its commitment to unlimited growth, but also of its inherent proclivity for war. Ecological civilization, however, is based on a variety of lifestyles that sustain the interrelated natural ecology and social ecology. Such a civilization has two essential traits. First, it takes human life as the same as the thriving biosphere, which is dynamic, sustainable and balanced. Man and nature are not in opposition. Man lives in nature. Second, ecological civilization reflects a fundamental reform of human lifestyle, which depends on our ability to opt for a new society. Roy Morrison said naming a civilization was raising a flag. Developing an ecological civilization is a great reform that is as significant as the transformation from agricultural civilization to industrial civilization.

There are other western scholars who also recognize ecological civilization, but they are only a small group as a whole.

The first scholar initiating this concept in China might be Mr. Ye Qianji. He delivered a report titled *On Ecological Civilization* at the conference on water and soil conservation in the Three Gorges reservoir area held in the second half of 1986. His article "Ecological Needs and Ecological Civilization" was collected in *Ecological Agriculture in China* edited by Guo Shutian and published by China Prospect Publishing House in 1988. In that article, he wrote that "ecological civilization means man benefits from nature and benefits nature in return; man changes and protects nature simultaneously. They are in a relation of harmonious unity." "The proposition of ecological civilization turns activities of material development into activities of reforming and protecting nature. Cultural advancement should foster comradeship between man and man as well as partnership between man and nature."

Ye Qianji's definition of "ecological civilization" obviously differs from that given by Fetscher. In the 1980s, China proposed to build socialist material civilization and cultural civilization, and Ye's proposition of ecological civilization was intended to make up for deficiencies of the above two ideas. In the context then, the society consisted of two parts—material civilization and cultural civilization, but Mr. Ye thought that was not enough and ecological civilization had to be added to make a complete society. According to him, material civilization, cultural civilization and ecological civilization are synchronic dimensions (some call them elements) of a civilized society. Today, political civilization and social civilization are added to the system too, driving the number of synchronic dimensions to five. Fetscher didn't mean "ecological civilization" in this sense, but in a diachronic sense, namely ecological civilization is a new social state, whose emergence and development constitute a new age. The difference in Fetscher's and Mr. Ye's definition cannot be neglected. Only Fetscher's diachronic "ecological civilization" marks a new age in the historical sense.

Before the strategy of building an ecological civilization was officially put forth by the 17th National Congress of the CPC in 2007, very few scholars in China studied this topic.

In 1990, Li Shaodong published an article titled "On Ecological Awareness and Ecological Civilization" in the *Journal of Southwest Minzu University (Philosophy and Social Sciences)*. The article defines civilization as a state of progress in material and cultural development that stands opposite to barbarism, ugliness and backwardness. Ecological civilization tries to introduce the rational perception of ecological environment and its positive practices into cultural development as an important component. Ecological civilization comprises a pure ecological morality, lofty ecological ideal, scientific ecological culture and civilized ecological behaviors. To develop ecological civilization, it is necessary to have a clear guiding thought, intensify the communication of ecological knowledge, foster a favorable social environment and establish an institution for it. Li Shaodong's definition is consistent with that given by Mr. Ye—ecological civilization is a "state of progress" opposed to "barbarism, ugliness and backwardness" that man demonstrates in how they treated the ecological environment. It is parallel with material civilization and cultural civilization and penetrates those two.

In 1992, Xie Guangqian published "An Initial Study of Socialist Ecological Civilization" in *Socialism Studies*; in 1993, Shen Xiaohui published "Toward Ecological Civilization" in *Solar Energy* and Liu Zongchao and Liu Yuesheng published "Global Ecological Civilization: Earth Surface Information Proliferation Pattern" in *Chinese Journal of Nature*. Shen Xiaohui held that major ancient agricultural civilizations all declined because of the degradation of natural system and the destruction of eco-balance between man and environment. But ecological destruction in history was after all local, so was the fading away of civilization. When civilization in one place declined or vanished, new civilization would emerge and evolve elsewhere.

The situation is different in modern times, when ecological destruction and environmental pollution are of a global scale. The deterioration of global environment will bring unimaginable consequences to world civilization. The key to resolving this problem lies in whether man can face up to the environmental consequences that their behaviors have caused or may cause, and shoulder unshakable responsibilities for natural degradation. To save the world and man themselves, the traditional way of life, production and thinking all call for a profound environment-friendly revolution. That is the only way to find a new development path and build a new civilization highlighting environmental protection where man and nature develop in harmony for common prosperity without hurting each other. This is ecological civilization. Shen Xiaohui obviously used the term in a diachronic sense.

The earliest monograph on ecological civilization we could find in the National Library's catalogue is Zhang Haiyuan's *Productive Practices and Ecological Civilization* published by China Agriculture Press in 1992. It must be the first book that has "ecological civilization" in its title, which elevates environmental protection to the level of ecological civilization. The author wrote in the preface, "given the reality of environmental pollution, protecting the environment has become a pressing task for the government and citizens of every country, and the precondition and result of accomplishing this task is developing an ecological civilization in the new era." The book claims to have "answered why we should develop ecological civilization, how to develop it and why we can succeed in developing it". However, it doesn't define the term, but rather focuses on how to protect the environment in productive practices. Therefore, it isn't a monograph specific to ecological civilization.

The earliest monograph on ecological civilization is *On Ecological Civilization* edited by Liu Xiangrong published in 1999, which defines ecological civilization as an advanced form of civilization that pursues not only economic and social progress, but also ecological development; not only the coordinated evolution between man and nature, but also between economy-society and the biosphere. Developing an ecological civilization is the overriding way out of the ecological crisis and a thorough change of civilization. It is an unavoidable state in history and also a conscious choice of man; it is the ideal we aspire for and also the reality that is already happening around us. The ecological civilization in this book is similar to Fetscher's diachronic ecological civilization.

The convening of the 17th National Congress of the CPC in 2007 marked a turning point both for the study and the development of ecological civilization. Before that, only a handful of people studied this topic and their work and the results they got received little attention. The 17th National Congress of the CPC proposed the ecological civilization strategy, the 18th National Congress of the CPC uplifted ecological civilization as part of the Five-Sphere Integrated Plan, and the 19th National Congress of the CPC made the special point that "building an ecological civilization is vital to sustain the Chinese nation's development." Today great importance is attached to the study and development of ecological civilization. In the report of the 18th National Congress of the CPC, former President Hu Jintao called for "striving to usher in a new era of socialist ecological progress." General Secretary Xi Jinping also said, "The mankind has gone through primitive civilization, agricultural civilization and industrial civilization. Ecological civilization is the outcome of industrialization reaching a certain stage, and the new requirement for man-nature harmony." Here the "ecological civilization" is also diachronic. It is clear that China has admitted that ecological civilization will replace industrial civilization, and the new era we are ushering in is a new age of socialist ecological civilization.

2 Information Civilization Theory

In the 1980s and 1990s, Alvin Toffler's futuristic studies greatly influenced the Chinese academia. He put forth the idea of "third wave" in the 1980s. The Tofflers wrote in their book *Creating a New Civilization: The Politics of the Third Wave*, published in 1995, that "A new civilization is emerging in our lives, and blind men everywhere are trying to suppress it. This new civilization brings with it new family styles; changed ways of working, loving, and living; a new economy; new political conflicts; and beyond all this an altered consciousness as well." They called the rise of this new civilization "the third wave." According to them, the human race has undergone two great waves of change. The First Wave of change—the agricultural revolution—took thousands of years to play itself out. The Second Wave—the rise of industrial civilization—took a mere three hundred years. Today "We, who happen to share the planet at this explosive moment, will therefore feel the full impact of the Third Wave in our own lifetimes." "We are the final generation of an old civilization and the first generation of a new one." They believed "industrial civilization is coming to an end" and "the general crisis of industrialism" is all too obvious. The First Wave brought agricultural civilization, the second one industrial civilization. "The first civilization is still symbolized by the hoe, the second by the assembly line, and the third by the computer." "While land, labor, raw materials and capital were the main factors in production in the Second Wave economy of the past, knowledge—broadly defined here to include data, information, images, symbols, culture, ideology and values—is now the central resource of the Third Wave economy." It is clear that the economy of the new civilization is the "knowledge economy" or "information economy" that was heatedly discussed at the turn of the century. The Tofflers called it a "super-symbolic Third Wave economy." If railways and highways represent the infrastructure of industrial civilization, "information highways" or "electronic passages" make the infrastructure of the new civilization. The Tofflers tended to call the new civilization (the third civilization) an information civilization.

Tofflers' view on the new age and new economy was echoed by Peter F. Drucker, father of management, although he called it the age of "Post-Capitalism Society." Drucker said we obviously lived in a process of historical transformation, which has changed the world's political, economic, social and moral domains, and after which our values, convictions, social and economic structure, political concepts and systems, and world outlook would be changed far beyond our imagination. In a post-capitalism society, the truly dominant resource and decisive "factor of production" is not capital, land or labor, but knowledge. That is why it is actually the information society, and its economy is information economy. Since post-capitalism society is still capitalist, it is a society dominated by information capitalism.

Paul Hawken, who has studied and promoted green capitalism and business ecology for many years, asserted in his *The Next Economy*, published in the 1980s, that the industrialist economy is material economy, characterized by massive production, massive consumption and massive waste, which is non-sustainable. Material economic growth depends on cheap mineral resources. With the oil price hike and the emergence of computer technology in the 1980s, material economy is declining and a new economy will appear, namely the information economy.

With the development of information technology (including artificial intelligence), many take information civilization as the new civilization of the future. Some Chinese scholar said, "It is generally believed that primitive civilization, agricultural civilization and industrial civilization are the civilization forms that man has experienced or is still experiencing. Today we have stepped into information civilization, a new civilization form after industrial civilization." Others say the same about ecological civilization—we have stepped into or we must embrace ecological civilization, a new civilization form after industrial civilization.

3 Common Views of the Two Theories

Both the theories of ecological civilization and information civilization have their theoretical and realistic basis. In reality, environmental pollution, ecological destruction and climate change are happening around the world, and IT and AI technologies are exerting growing effects on the economy, politics, military, culture and education of all countries.

Both theories need to search for scientific basis from new sciences. The ecological civilization theory resorts to quantum physics and ecology-integrated complex sciences, while information civilization theory resorts to cognitive science and information science that came long after Alan Mathison Turing.

Both theories must be defended with a new philosophy, which refers to ecological philosophy for ecological civilization theory and information one for information civilization theory. Both philosophies claim to be the new philosophy of the new age and both criticize modern philosophies that support industrial civilization. For instance, they both reject the subject-object dichotomy of human centrism advocated since Descartes and Kant. Subject-object dichotomy, or the division of humans and unhuman objects, is the cornerstone of modern philosophy, and rejecting such dichotomy is a common feature of ecological civilization and information civilization theories.

Aldo Leopold is a precursor of ecological philosophy. His "land ethic" takes land as a community that consists of such members as soil, water, plants and animals, and regards man not as a conqueror, but as an equal member and citizen in the community. Arne Naess, founder of deep ecology, held that there is no insurmountable boundary between man and natural objects or even nature as a whole, and everyone is related with nature. All life forms are entitled to a non-quantifiable universal right. No biological specie has more right to live and blossom than other species. Ecological philosophers don't think there is as much a difference between humans and unhuman species as claimed by Descartes and Kant, but hold that unhuman species are qualified to share the same moral community as man.

Information philosophers reject subject-object dichotomy based on the achievements of AI technologies. Luciano Floridi, one of the most active information philosophers in modern world, thinks the earth is changing into an info-sphere or has been such all the time. In this info-sphere, intelligence is not exclusive to man, or man is not "unique" as far as intelligence is concerned. Nowadays people have come to accept the post-Turing idea of "self"—we are not lonely "agents" on an island like Robinson, but are interconnected and inter-embedded inforgs (a new coinage) in the info-sphere that we share with other information agents, both natural and artificial, who process information logically and independently. Natural agents refer to unhuman species while artificial agents refer to various intelligent machinery or robots. We can see that after the Turing revolution, concepts like subjects and subjectivity emphasized by philosophers like Descartes, Kant and Sartre became unimportant and made place for notions like agents and agency. According to Kant, Sartre and their likes, only man is the subject and only man has subjectivity; but in the opinion of information philosophers, all animals (especially advanced animals including man) and intelligent machinery are agents, and they have agency. Agents are those that can interact with other beings. They admit other beings' equal status as themselves and experience their identity and freedom by placing themselves among other beings.

In this sense, information philosophers and ecological philosophers share the view that both humans and unhuman objects have moral qualifications. While ecological philosophers believe ecosystem has moral qualifications and man should protect the health of ecosystem out of moral consciousness, information philosophers think both unhuman animals and intelligent machinery are morally qualified agents that should be kindly treated by man as well. Some information philosophers also endorse energy conservation, emission reduction and environmental protection. Hawken asserted the transition from material economy to information economy out of considerations for environmental protection. Floridi said one of information philosophy's tasks is to build a moral framework in which the human inforgs, as part of the info-sphere, regard the sphere as a new environment worthy of moral attention and care. This moral framework has to explicitly reveal and cope with the unseen challenges in the new environment. It must be an Environmental ethics about the general info-sphere. This overall environmentalism (general, inclusive as well as artificial) sets to change our self-consciousness and realistic roles, considers what is worthy of our respect and care, and thinks of how to bring natural and artificial objects into an alliance.

4 Conclusion: Vision of the New Civilization

Civilization encompasses all that is created by man as a creature other than unhuman animals and is therefore enormously rich and complicated. Some divide civilization into three dimensions—object, institution and idea. China divides it into five synchronic dimensions—material, cultural, political, social and ecological. According to Samuel Phillips Huntington, civilization comprises such dimensions as language, history, religion, custom and system. An Iranian scholar who has devoted long years to the study of civilization believed civilization includes two inseparable parts. One is a clear world outlook that could be presented as a cultural system, an ideology or a religion; and the other part is reflected via a consistent political, military and economic system, which usually takes the form of an empire or a historical institution. Each way of division only represents one angle for understanding civilization, but none of them has a comprehensive picture of it. Civilization is so complicated, multi-dimensional and multi-faceted that there isn't a single noun or adjective convincing enough for all to recognize it as the right term for the emerging new civilization. People have different judgments of the reality, different expectations for the future, and naturally different definition for a new age or civilization.

According to the ecological civilization theory, in the future civilization, we must change the energy structure so more clean energies like solar and wind power will be employed, change the industrial structure so all heavily polluting industries will be kicked out, change the economic growth mode from linear economy to circular economy, and promote the de-materialization of economy. We must apply ecological rules to institutional development and innovation, go beyond the "capital logic" and abandon the GDP-oriented policy. We must create a thriving and diversified ecological culture, strongly advocate green consumption against materialistic consumerism, and fundamentally change our values and outlook on the world and development like a new enlightenment. In sum, we must completely give up the way of production and way of life featuring massive development, massive production, massive consumption and massive emission.

According to information civilization theory, the future civilization will be one based on information, when knowledge and information are the most important factors of production and living necessities, and de-materialization of economy is one aspect of economic informatization. If we want to buy a new book published in the United Sates, it is unnecessary to ship the book to China all across the vast ocean because we can download the electronic version from Kindle. If, as Floridi and others predicted, the future world will be online and offline at the same time with the virtual world and physical world integrated in one, we wouldn't need to take the plane or train to visit Jiuzhaigou, Mount Huangshan or other scenic areas—it will only take a click of the mouse. Information technology will indeed change our way of production, of living and of thinking profoundly.

The two theories of new civilization delineate the vision of the future civilization or new age, and both will be tested through human practices. Since both theories attempt to replace the modern civilization theory that currently remains in a dominant position, they are bound to be strongly resisted by it. Just as Toffler said, "A new civilization is emerging in our lives, and blind men everywhere are trying to suppress it."

Ecological Civilization in a Historical Perspective

Lihua Wang

Reality is the continuation and development of history, and civilization is the accumulation and update of history. The ecological civilization campaign is the grandest undertaking and the most complicated and massive program in the modern world. China being the first country to propose the development of ecological civilization is both imperative and inevitable. The cause, though started in our generation, will benefit many generations to come. We should try to have a comprehensive and deep understanding of the immense and profound significance of this campaign from a historical perspective, so as to come to a thorough and correct judgment.

1 Understanding Ecological Civilization from the Historical Perspective of Human Survival and Development

The relation between man and nature and that between people make the two basic relations of human society that are constantly in development and changes. Respecting, adapting to and caring for nature is the only right path forward for mankind. Correcting our past mistakes, especially serious mistakes made after the Industrial Revolution, and promoting man-nature harmony is the only right direction to sustain civilization. This is because mankind is the outcome of the evolution of life system on earth—it is part of nature and has to rely on the biosphere for survival.

More than 100 years ago, Marx and Engels stated clearly "the first premise of all human history is, of course, the existence of living human individuals. Thus the first fact to be established is the physical organization of these individuals and their consequent relation to the rest of nature... The writing of history must always set out from these natural bases and their modification in the course of history through the action of men." "... Nature is man's inorganic body—nature, that is, insofar as it is not

L. Wang (✉)
Center for China Environmental History Studies, Nankai University, Tianjin, China

J. Pan et al. (eds.), *Beautiful China: 70 Years Since 1949 and 70 People's Views on Eco-civilization Construction*, https://doi.org/10.1007/978-981-33-6742-5_54

itself human body. Man lives on nature—means that nature is his body, with which he must remain in continuous interchange if he is not to die. That man's physical and spiritual life is linked to nature means simply that nature is linked to itself, for man is a part of nature." Marx and Engels took human life and physical organization and its relation with nature as the first premise for establishing historical materialism. From that they analyzed in depth the emergence, development and evolution of the way of social production, division of labor, ownership of property, class conflicts, urban-rural relation, state politics, military struggles, ideas and ideology...

Millions or tens of millions ago, our ancestors came into being on earth and formed a relation with nature subsequently. Human history has been unfolding along two intertwined centerlines. Within the society, with man-man relation as the centerline and the way of production as the axis, humans have completed the transformation from collecting and hunting to farming and animal husbandry, and from agricultural civilization to industrial civilization. They are now striding toward the information and intelligent age at a fast pace. Within the biosphere, with the evolution of man-nature relation as the centerline and the exploitation of natural resources and progress of material civilization as the symbol, humans have gradually improved their skills and enhanced their labor in scope, scale and intensity. In this process, they have become more capable of disturbing and changing the earth's ecosystem, achieved total victory in competition against other species, and eventually acquired the super-power enough to destroy mother earth. This long history can be roughly divided into four stages.

In the first stage, man, as a chance result of life evolution on earth, wandered around for food like other animals most of the time, with no knowledge, skill or memory. Then natural evolution provided it with a "window period" suitable for survival, in which cosmic rays, sunshine, atmosphere, water, temperature, land, fauna, flora, microorganisms and minerals, as well as other unknown natural conditions happened to facilitate the survival of this common yet special primate. Since then man has kept evolving and reproducing.

In the second stage, man stood out from multitudes of natural species, acquired the ability to make and use tools, learned how to start fire manually, and began to distinguish themselves from objects, have self-consciousness and personal will. They therefore became "cultured animals," which continuously reinforced their competitive edges over other species through cultural rather than physical evolution, and got so outstanding that they became the ruling primate.

In the third stage, starting from around 10,000 years ago, man, with their knowledge about plants, animals, water and soil, and climate that they had accumulated through long experience of collecting and hunting as well as corresponding tools and skills, embarked on the "Agricultural Revolution" in multiple regions successively. They planted crops, bred animals, and began to obtain food, energy and other living materials from nature through proactive intervention in it instead of completely relying on its bestowment. From then on, man has kept changing the world according to their own will and ideal while changing themselves as well, leading to a fundamental change in the relation between man and nature.

In the fourth stage, starting from around the mid-eighteenth century, the human society experienced four technological-industrial revolutions through superposition, acceleration or even leap forward, almost changing everything on earth in a short span of over 200 years. Today human activities have replaced natural forces to be the main driver of environmental changes, prompting geologists to announce the arrival of an "anthropocene."

History tells us that environmental problems have never stopped emerging since the day man came into being, and have keep increasing after ape evolved into man. However, environmental problems in different periods varied vastly in representation and influence, with profound differences between then and now. In the age of collecting and hunting, earth was barely populated, tools were extremely simple, and man with obscure consciousness wandered without a settlement. They were hardly able to adapt to, much less disturb nature, had no advantage when struggling with poisonous insects or beasts for food or shelter, and their lives were in constant danger. Therefore, natural worship and animism were prevalent then. In the agricultural age, man, with a stronger self-consciousness, began to think of their relation with nature and believed in man's conquest of and rapport with nature at the same time. At that time, planting and animal breeding were the main form of economic production, social life followed the natural course, and eco-environment was little polluted. However, blind and disorderly resource exploitation and environmental modification had already resulted in resource depletion, water system disorder, water and soil loss and land fertility exhaustion in local areas, and ancient civilizations perished one after another. In the recent 200 years and more, Industrial Revolution and Technological Revolution have dramatically empowered mankind. Motivated by material consumption and a desire for profits, man, armed with powerful technologies, has treated nature and its inhabitants roughly and recklessly, completely upending the life system on earth. Over time, man seemed to have forgotten who they were and where they came from. As man was mentally estranged from nature and their behaviors devastated nature, the conflict between man and nature quickly exacerbated and civilization development slipped into a serious dilemma. Therefore, it was no surprise that the human ability that was in no way restricted by ecological morality backfired at man itself through nature's harsh revenge, posing ever more serious challenges to social and economic development and life health and safety.

More than 100 years ago, Engels issued a clear warning, "...Let us not, however, flatter ourselves overmuch on account of our human conquest over nature. For each such conquest takes its revenge on us. Each of them, it is true, has in the first place the consequences on which we counted, but in the second and third places it has quite different, unforeseen effects which only too often cancel out the first." "...Thus at every step we are reminded that we by no means rule over nature like a conqueror over a foreign people, like someone standing outside nature – but that we, with flesh, blood, and brain, belong to nature, and exist in its midst, and that all our mastery of it consists in the fact that we have the advantage over all other beings of being able to know and correctly apply its laws."

Nevertheless, the huge success in technologies, industry, capital and market has incurred the prevalence of instrumentalism and consumerism, made man increasingly arrogant toward nature and unscrupulous in environmental behaviors, till at last nature takes revenge in great intensity. In the twentieth century, environmental disasters happened ceaselessly, as represented by the notorious "eight environmental pollution (hazard) events" in the West. Global climate change is particularly worrying concerning the prospects of civilization development. Over 50 years ago, the British historian Toynbee commented, "In more than 200 years from 1763 to 1973, man obtained the power to conquer the biosphere, which was unprecedented. In these bewildering circumstances, only one prediction can be made with certainty. Man, the child of Mother Earth, would not be able to survive the crime of matricide if he were to commit it. The penalty for this would be self-annihilation." When Toynbee made these remarks, people were lamenting the "Silent Spring" and worried about "The Limits to Growth," and the international community began to realize the severity of environmental problems and take countermeasures. Yet the desire for wealth and the inertia of development were so overwhelming that man has continued to harm earth's ecology at a faster pace and on a larger scale since then, and there is no telling when we will effectively halt the trend of environmental deterioration.

If we observe all civilization forms in the long history of human life, we can distinguish their different outlook in different time and space. What is penetrating the whole process is the waning and waxing of human power and natural power. Since the Industrial Revolution, human power has grown quickly and material civilization developed by leaps and bounds, along with the quickly worsening natural ecosystem. For tens of thousands of years, social development and civilization progress have been characterized by man trying to break away from natural restrictions to obtain as many natural resources as possible in pursuit of an ever more affluent, even luxurious, life. But things have changed today. The biggest dilemma in the current world is not the contradiction between ever-growing material needs and backward social productivity, but the contradiction between ever-bloating desire for material consumption and the ecosystem's limited bearing capacity. Man must control their material desire with strong mental will, follow nature's course, respect and care for it, take the initiative to dissolve their contradiction, and strive for lasting development of man-nature harmony. This is the great goal that China's ecological civilization is aiming at, which history will prove is the only right direction for human development in the future.

2 Understanding Ecological Civilization from the Perspective of the 5000-Year History of Chinese Civilization

The history of Chinese civilization is the history of the coordinated evolution between the Chinese nation and the natural environment it lives in.

The heave of the Qinghai-Tibet plateau, the formation of the Himalayas, the cycle of hot and cold weather, the up and down of eastern sea level, the aeolian dust deposition on the loess region and plateau, and the evolution of the Yellow River, Huaihe River, Yangtze River and other major water bodies… these natural movements and changes have created an environment suitable for human survival. From the tertiary to the quaternary geological period (about three million years ago), human activities appeared in China, as evidenced by many ancient human relics discovered by archaeologists. However, the latest molecular anthropology studies, especially Y-chromosome DNA studies, indicate that those ancient humans were only remote relatives rather than immediate ancestors of Chinese people. The immediate ancestors emigrated from the African continent after the end of the ice age in the quaternary period 50,000–100,000 years from now. They were the second batch, or much later, of "African immigrants to China." They were wandering, collecting and hunting in dense forests, until 10,000 years from now, crop planting and fowl and poultry breeding began to appear in different places. That marked the beginning of Chinese ancestors' change of living as they began to control and renovate natural species and the environment.

China crossed the threshold to civilization 5000 years ago. Since then the Chinese nation has continuously exploited natural resources, changed the ecological environment, developed agriculture and animal husbandry, and created the Chinese civilization. From the remote ages of Xia, Shang, Zhou through the Qin, Han, Sui and Tang dynasties to the Song, Yuan, Ming and Qing dynasties, the Chinese nation and the state of China kept evolving and integrating into a massive civilization system. In the past 5000 years, Chinese people have created proud and resplendent cultural achievements, such as bronze, ritual and music, the Great Wall, the Great Canal, poetry and painting, agriculture, traditional Chinese medicine and the four great inventions. There have also been magnificent social movements, bloody wars and turmoil, ever-changing political situations and precarious ups and downs… which often make historians retrospective and thoughtful. All these achievements of civilization and stories of man have taken place in the colorful natural environment, closely related with the multitudes of other species also living on the vast Chinese territory.

The underlying history is perhaps not the joys and sorrows of man, but the profound changes in the relation between man and nature. Ever since Chinese forebears came to the agricultural age and lived in settlements, the rural villages that used to be scattered like stars over the sky have been replaced by crowded cities of various sizes, and people seem to get farther and farther away from nature. The advent and development of agriculture not only changed the way of resource utilization thoroughly, but also augmented the ecological bearing capacity considerably. Over the past tens of thousands of years, hundreds of animals and plants have been domesticated or cultivated, and those represented by the five grains and six animals have lived in mutual benefits and common evolution with man. They have sustained the continuation of Chinese nation and enabled the once rare population to keep propagating and reproducing into a huge population of nearly 1.4 billion, most of whom have lived in the east of the Hu Line where conditions are good for farming.

To obtain food and clothing, man has reclaimed forests, grasslands, lakes, swamps, hills and mountains into dry land, water land and terrace, conducted countless irrigation, shipping and flood control projects, and endlessly exploited and utilized all kinds of underground minerals like salt, coal, petroleum, clay and metals, laying the material foundation for the development of Chinese civilization. Through the long living experience, the Chinese nation, always thinking of the relation between man and nature, has formed a rich ecological culture and unique wisdom. Ancient sages proposed and explained the systematic view of heaven, earth and human being in an organic unity, the ecological ethics of "letting all creatures live" and "treating all beings well," and the view of resource conservation and sustainable development featuring "moderate utilization." The hundreds of millions of Chinese farmers have also upheld the recycling tradition of "trash to treasure" for 4000 years. All these make invaluable ecological cultural heritages. They reflect Chinese ancestors' deep thinking and rational practice of man-nature relation, and imply the secret to the continuity of the Chinese civilization.

It goes without saying that all civilizations hitherto have paid a corresponding environmental price for their development. The Chinese civilization has continued for 5000 years thanks to advantageous natural conditions, but the environment it has lived in has indeed undergone earthshaking changes—some are positive to meet human needs, while some are simply reckless destruction with deplorable consequences. For thousands of years, over 80% of the mountains, forests, grasslands and swamps in China have vanished due to constant logging, reclamation and landfill, millions of kilometers of land have been eroded into deserts, and the once vast and flat loess plateau has disintegrated into crisscrossing valleys and ravines. The destruction of forests and vegetation has degraded the water-containing capability, and water and soil loss has disrupted or even dried up many river systems. The Yellow River has seen overflow or inundation more than 1500 times in 2500 years and changed course several times, and the Yangtze River has caused increasingly serious flooding in recent centuries. Numerous lakes and swamps have been destroyed and wasted till complete disappearance. Persistent poaching and habitat damage have led to the extinction of a lot of wildlife, turning the once widespread reindeer, elephant, giant panda and crocodiles into rarities.

Historical literature shows that the man-nature conflict doesn't just happen in modern times. The Chinese nation has encountered at least two extremely severe environmental and resource crises in history. The first time was at the turn of the Zhou and Qin dynasties. At that time, the Chinese civilization was centered in the middle and lower reaches of the Yellow River, where the growing population and land reclamation gradually exhausted mountains, forests and waters. Important materials such as leather, fur, horn, teeth, feather, arrow, grease, glue, cinnabar and lacquer were in short supply, people were generally worried that logging, hunting and handicraft could not sustain, and kings feared the loss of mountains and rivers would weaken their states. In response, the states released various laws and regulations to control logging and hunting and protect the reproductive ability of natural resources. With the establishment of a unified nation and the social and economic transformation, the crisis was dissolved over time.

The second crisis happened after the middle of the Qing Dynasty. During the Ming and Qing dynasties, Chinese population increased steadily and reached 430 million before the Opium War. The plains were so populous that many vagrants went to live in deep mountains, where they felled forests and reclaimed wasteland blindly. As a result, forests decreased, water and soil was eroded, rocks laid bare, water systems in the plains were disrupted, general ecological environment deteriorated and water and drought disasters took place more frequently. The relation among environment, economy and society lost balance, and the development pattern from population (labor force) increase to farming area expansion to economic growth came to an end. Against such a backdrop, men of vision pointed out the origin of natural disasters, emphasized the importance of protecting mountains and lakes, and even proposed to control population. The central and local governments rolled out many policies and measures to prohibit reclamation and hunting and encourage afforestation, and community rules and regulations were released to protect forests, ban hunting and limit reclamation, yet all failing to curb environmental degradation. In the nineteenth century, China was faced with sluggish economy, constant famines, chronic poverty and weakness and ordinary people living on the line of starvation. Seizing this moment, western colonialists bombed open China's gate with gunboats, plunging the Chinese nation into a miserable abyss of internal and external troubles and the once world-leading Chinese civilization to a historical rock bottom.

The national crisis of contemporary China was of course the result of a full array of factors, but the ecological dilemma resulting from the traditional production mode, including environmental degradation and resource scarcity, was an important reason. Therefore, an important precondition for the rejuvenation of Chinese civilization is to comprehensively adjust the relation between man and nature and step out of the economic crisis and ecological dilemma that have taken form over long years. Ecological civilization is not just a decision for realistic development, but also a call of history. It is the only path to the all-round rejuvenation of Chinese civilization, carrying with it immense historical significance and powerful innate momentum.

3 Understanding Ecological Civilization from the Perspective of the 100-Year-Long Struggles of the Communist Party of China

After the Opium War broke out, China was forced to face strong enemies and tremendous changes never seen in its thousands-of-years-long history. When the country was afflicted by internal strife and external humiliation, worsening environment and constant disasters, countless patriots sought up and down despite bewilderments and setbacks for a way to save the nation from misery and subjugation and lead it to revival. Under the wise leadership of the CPC, Chinese people fought fearlessly against foreign aggression regardless of injuries and deaths, and eventually founded an independent, free and democratic new China on the ruins of semi-feudalism and semi-colonialism. This is the choice of history.

The CPC is a great Party that has the strongest sense of historical mission in all the world of all times. From the day of its founding, the Party has taken it as its duty to fight for the nation's independence and liberation and the country's prosperity and strength. Bearing firmly in mind its aspiration to bring a happy life to the vast majority of people, it has fought tirelessly in the past century to accomplish major historical missions in different periods, leading the poor and weak China to become a well-off and strong country and creating numerous miracles in human history. Building an ecological civilization is the CPC's new mission in the new era. Concerning the long-term wellbeing of the Chinese nation and the realization of the two centenary goals, the concept should be earnestly studied and apprehended in the context of the Party's century-long struggles.

The construction of new China started in dire conditions—chronic poverty and weakness, and desolation and ruins everywhere. As a later comer to the industrialization process, China completed within several decades the civilization transformation that took western countries several centuries to complete. The scale and speed of its economic development and social progress in the past 70 years is unprecedented and has amazed the world.

History is filled with conflicts and twists, and pros and cons, weal and woe always come hand in hand. The hard lesson that "lagging behind makes one vulnerable to aggression" has kept inspiring the Chinese nation to make progress in industrialization and modernization, which has unavoidably created a kind of anxiety. Having suffered too much from foreign aggression and internal conflicts, the new China and its ruling Party wasted no time in its efforts to step out of poverty and improve the economy, but they couldn't accomplish everything overnight. For a certain period, the Chinese society was too keen on GDP growth. It didn't fully understand the bearing capacity of ecological environment and natural resources, didn't well coordinate the conflicts among population, economy and environment, and didn't have a sound system for resource administration and environmental protection. Therefore, while tremendous achievements were made in social and economic sectors, the environment was under unprecedented pressure. Environmental problems that were inevitable in the course of rapid economic development, such as consumption of natural resources and pollution of air, water and land, displayed structural, superimposed, compressed and complex traits. If these problems were not quickly and resolutely addressed, ecological and social risks would overlap even to the point of undermining the natural foundation for the survival and development of the Chinese nation!

The aggravating environmental challenges have caused concerns—are social progress and economic development bound to destroy the ecological environment? What exactly is the purpose of development? Will the 5000-year Chinese civilization sustain? Such awareness of civilization crisis has awakened the society and guided environmental protection practices. As far as environmental history is concerned, the current environmental crisis is the outcome of long-term accumulation, particularly the special historical conditions in modern times; it is also the price that Chinese people have to pay for the benefits of modernization and industrialization. Therefore, we should show both understanding and compassion from a historical and rational perspective. In the meantime, we must be fully aware that environmental issues relate to the very existence of Chinese nation and the wellbeing of generations to come. Resolving the crisis and promoting harmony between man and nature is the only right way forward for Chinese civilization and the fundamental approach to sustain the Chinese nation. Once again the CPC has firmly and resolutely taken up this great mission of the new era.

It is generally agreed that environmental protection in China, in the sense of modern science, started in 1972, when China sent a delegation to participate in the first United Nations Conference on the Human Environment in Stockholm, capital of Sweden. But before that, when the new China was just founded, the country already began environmental work on a large scale, including river governance, afforestation, water and soil preservation, wildlife protection, and the investigation and control of solid, liquid and gas pollution. These efforts were the prelude to the development of environmental protection in China. In the 1970s, environmental protection was put on the state agenda and later designated as a basic state policy, and relevant work was carried out from scratch on all fronts. In 1974, the State Council set up the leading group of environmental protection, followed by the establishment of similar organizations of various types at all levels, which later were continuously adjusted and developed to form a complete system. After China promulgated and implemented the *Environmental Protection Law of the People's Republic of China (Trial)* in 1979, relevant laws and regulations were continuously perfected. It was also from the 1970s that specialized work in various domains was carried out, including environmental scientific research, cultivation of specialized talents, publicity, education and foreign exchanges on environmental protection, and environmental assessment and quality monitoring, so that environmental work became more and more professional, systematic and regular. Environmental protection enterprises, green industries, ecological parks, environmental protection demonstration projects and natural and ecological reserves have also been established in large quantities. Since the first National Conference on Eco-environmental Protection in 1973, the conferences in the following years have discussed major problems and key strategies and witnessed substantial achievements in different periods. They have witnessed how this undertaking has evolved from a departmental work to a national strategy, from the work of environmental workers alone to the common action of the whole nation.

The special conditions in China have given its environmental cause some distinct features. Environmental protection as a mankind's common cause started in the 1970s, with China playing an active part in its commencement. At that time, western countries had completed industrialization and become wealthy and advanced modern societies, whereas Chinese people were still struggling for enough to eat. Therefore, China's economic takeoff after the 1980s was accompanied by environmental protection, during which development and protection was in sharp conflict and decisions and actions were always faced with a hard choice. Western countries solved their environmental problems by transferring their backward and low-end industries with high consumption, high emission and heavy pollution (even polluting wastes directly) to developing countries. China with a weak foundation and backward technologies therefore was forced to take over immense waste pollution and pay an extremely high environmental price when trying to attract business and investment and become the "world factory." If environmental protection in the West is mainly driven by the private sector, that in China is organized and mobilized by state power with extensive public participation, which is the reason why it has encountered much greater difficulties and more complicated conflicts than in the West. In addition, the Chinese approach to environmental protection has also demonstrated a much stronger state and national will and the leaders' broader vision on civilization and humanity. It is the CPC that has initiated to develop the "ecological civilization," elevated it to be the top national development strategy, and made it a grand goal of jointly building the "community with a shared future for mankind." These historical facts must be remembered forever!

The 17th National Congress of the CPC first put forth the strategic task of "building an ecological civilization." The 18th National Congress of the CPC incorporated it as part of the Five-Sphere Integrated Plan for socialist development with Chinese characteristics and demanded "we must give high priority to making ecological civilization and incorporate it into all aspects and the whole process of advancing economic, political, cultural, and social progress, work hard to build a beautiful country, and achieve lasting and sustainable development of the Chinese nation." That primarily outlined a new blueprint for civilization development.

The CPC Central Committee with Comrade Xi Jinping at the core uplifted "ecological civilization" to a higher strategic position by writing it into the *Charter of the Communist Party of China* and the *Constitution of the People's Republic of China.* This new choice of civilization reflects the wish of over a billion Chinese people in the new era and is therefore canonized as the unshakable state will. Over the years, General Secretary Xi Jinping has made many succinct remarks about ecological civilization on multiple occasions, which form a complete system of the Xi Jinping Thought on Ecological Civilization. This system reflects the profound legacy of the 5000-year-old Chinese civilization, manifests the foresight and foresight for Chinese nation's sustainable development and people's happy life, and embodies a deep concern for the common future of mankind. This system conveys philosophical dialectics, rational analysis and steadfast practicality, with which a new horizon is opened in combining the Marxist theory on man-nature relation with China's actual conditions and man's lofty ideals. As such, it commands a high ground in both human and natural morality, following the natural course while meeting people's aspirations and expectations. It is under the guidance of this advanced thought that China's ecological civilization has been progressing by leaps and bounds in recent years, exerting widespread effects on the whole world.

Many centuries later, when future historians look back on the "past," they will find a glorious period of hard work and arduous efforts. In the twenty-first century, an ancient oriental power, having gained rebirth from unimaginable tribulations and ordeals, built a beautiful and harmonious country that is affluent, strong, democratic and culturally advanced, and achieved all-round national rejuvenation. What's more, it led mankind through an epochal transformation into the new era of ecological civilization where man and nature lived in harmony with each other!

Several Thoughts on Promoting University Education in Ecological Civilization

Zongguo Wen

The report of the 19th National Congress of the CPC stated that building an ecological civilization is vital to sustain the Chinese nation's development and is the only way to address the contradiction between people's ever-growing needs for a good ecological environment and the inadequate effective supply of ecological products. At the National Conference on Eco-environmental Protection in 2018, the Xi Jinping Thought on Ecological Civilization was established as a beacon for developing ecological civilization. Revolving around the major requirements of this campaign, universities, as institutions undertaking scientific research and talent cultivation, should on the one hand lead as pioneers and guides in promoting green, circular and low-carbon development in the endeavor to contribute to China's and the world's sustainable development. On the other hand, they should also fully exert their advantages and roles to actively advance education in ecological civilization and explore the path of building green universities, and advocate energy conservation and emission reduction on campus, as part of the efforts to promote regional environmental improvement.

1 Important Mission of Higher Education in Building an Ecological Civilization

Ecological civilization calls for the joint efforts of multiple forces, including governments, universities, the press and social groups. Giving the guiding role of education, higher learning institutions should act as pioneers in ecological civilization education. The serious environmental issues, even ecological disasters today, are largely attributed to the lack of sufficient and systematic ecological education in the past.

Z. Wen (✉)
Center for Ecological Civilization, Tsinghua University, Beijing, China

J. Pan et al. (eds.), *Beautiful China: 70 Years Since 1949 and 70 People's Views on Eco-civilization Construction*, https://doi.org/10.1007/978-981-33-6742-5_55

Therefore, it is especially important to establish a multi-layered system for ecological civilization education with more extensive public participation in this era. Tasked with cultivating high-level specialized talents for the society, higher learning institutions play a non-negligible role in economic and social development. As college graduates shoulder an important task to realize economic and social sustainability, their awareness and ideas of ecological civilization have a direct bearing on that process.

1. **Playing a pioneering role in ecological civilization education by accelerating reform of teaching and research system**

Teaching, publicity and research of the theories, methods and key areas of the development of ecological civilization is the requirement of this national campaign and an important task of universities and research institutes. First, ecological civilization should be reflected in disciplinary education, so that all economic and social sectors will establish the awareness and grasp the basic skills thereof, thus boosting industrial upgrade. Second, great efforts must be made to develop resource and environment-related disciplines and cultivate specialized talents, so as to advance the development of ecological civilization in all sectors and promote emerging industries in the field of energy conservation and environmental protection. Third, universities' disciplinary advantages should be used to popularize ecological knowledge to help foster new values among the public, urge them to be more caring and appreciative of nature, and establish standard ecological ethics and morality. In particular, an ecological civilization education system with Chinese characteristics should be built as soon as possible, covering courses, teaching materials and methods, and teachers, and be continuously improved through teaching practices across the society. Tsinghua University, for instance, has actively promoted education in ecological civilization and sustainable development. It has developed the courses such as Environmental Protection and Sustainable Development (opened in 1997, covering an array of subjects including ecology, resources, environment, ethics, economy, management and engineering technology. It was rated as an excellent course at Beijing municipal level and state level in 2006) and Fifteen Lectures on Ecological Civilization (opened in 2015).

In recent years, more high-quality academic papers on "ecological civilization education" and "green university education" have been published and relevant explorations made. For instance, Huang Baiwei and others at Peking University discovered the sizable carbon footprint of canteens in the university using quantification standards, and proposed to mark the users' dining carbon footprint with ecological labels. In this way, the university reduced food waste and unnecessary food demand, improved the food carbon emission mix, cut carbon emission across the board, promoted green university education, and achieved the goal of building a green eco-campus. Liang Lijun, Liu Chao and others at Tsinghua University studied the university's theories and practices in building a green university. They summarized the Tsinghua model of green campus construction and green university education in the past decade and more, and generalized three concepts and practices—environment-friendly campus, economical campus, and ecological and beautiful campus.

2. **Promoting green transformation of university development through green initiative**

Green concepts and ecological civilization awareness should be applied in teaching, scientific research, social service and cultural advancement. Building a "green university" means implementing the principles and guiding thoughts of sustainable development and environmental protection in all university activities and the whole process of university education around the center of "education of man." As a concrete action to promote ecological civilization in higher education, it is an important part of the ecological campaign. In a sense, the significance of building green universities goes far beyond the initiative itself—it is a more important effect to demonstrate and lead the good practices of building an ecological civilization nationwide.

Prestigious universities around the world all take cultivating new talents conscious of environmental protection as their mission. After Yale University rolled out the "Blueprint for a Green Campus" at its "Campus Earth Summit," Brown University in the United States also initiated the "Green Brown" initiative, University of Edinburgh in the United Kingdom launched the "environmental agenda," and University of Waterloo in Canada the "Green Campus Action." Single-item action plans aimed to promote environmental protection and sustainable development both on and off campus are also carried out. Most of the subjects in University of Michigan's cross-disciplinary courses are about world or regional environment, energy, resources and other sustainability-related issues, as well as ecological civilization and industries. A growing number of Asian universities are conducting green practices. There are the Meijo University, Tokai University and its Faculty of Education in Japan, South Korea's Hanyang University, and nine universities in India including the University of New Delhi and Indian Statistical Institute. Their fellows also include Malaysia's University of Malaya, the Philippines' University of Manila, Thailand's Chiang Mai University and Vietnam's Institute for International Relations in Hanoi.

In China, Tsinghua University was the first to suggest building green universities in 1998, and submitted a plan for "building green demonstration projects" to the National Environmental Protection Administration. In view of the actual development needs of domestic universities, Tsinghua University, with "education of man" as the center, proposed to implement the principles of sustainable development and environmental protection in all teaching activities. It also formulated a three-layered action plan for building green universities, namely cultivating talents with "green education," advancing scientific research and environmental protection with "green technology," and enlightening the people through "green campus" demonstration projects. Of the three layers, green education is the core of building a green university. It is aimed to blend sustainability ideas such as resource recycling and environmental protection with all non-environmental subjects and university education in general, and internalize it as a vital part of the basic knowledge and overall quality of all students on campus.

China attaches great importance to the development of green universities. The *Action Outline for National Environmental Publicity and Education (2011–2015)* issued in 2011 emphasized the need to "promote environmental education in higher learning institutions, incorporate it in the teaching plan as a crucial part of quality education of college students, and build green universities." The report of the 19th National Congress of the CPC urged "initiatives to make Party and government offices do better when it comes to conservation, and develop eco-friendly families, schools, communities, and transport services." At present, the concept of green university is widely and deeply driven home and universities have all taken actions to put the initiative into practice with notable results. Going forward, higher education institutions should keep up the efforts to improve the elements of talent cultivation and produce various talents well versed in ecological civilization ideas and expertise.

3. **Building green and energy-saving campus to create an edifying environment**

The per capita energy consumption in higher education institutions is higher than that of common residents with a massive total energy consumption as well. Therefore, they must pay special attention to water and energy conservation, build ecological and beautiful campus, and function well as the place of education. To achieve the goal of ecological campus, universities, following the basic principles and methods of sustainable development and ecology, should apply environmental education in the whole process of building a new model of campus featuring good ecology, advanced ecological technologies, healthy ecological culture, sound ecological management and high ecological comfort. They should turn campus into an exemplary community integrating such roles as talent cultivation, resource conservation, environmental friendliness, sound ecological cycle, and green culture.

University campus also constitutes a sub-system and a special environmental unit of the urban ecosystem with a high level of openness. Resources and energies go into the campus system and talents come out of it—its volume and intensity of campus waste discharge would affect the overall environmental quality of the city. Upholding the value of ecological civilization, the green university initiative, with improved approach and efficiency of resource and energy use, is sure to generate great radiating and demonstrative effects on building an ecological civilization in the society. Therefore, universities, as institutions boasting a high concentration of high-quality talents, should act as role models by practicing the requirements of ecological civilization and green concepts.

A green university campus doesn't just mean energy-saving buildings, but refers to a new model of campus that integrates a wide range of functions, including the conservation of energy, water, land and materials, reducing or eliminating greenhouse gas emission and negative environmental impacts, promoting biodiversity, and enhancing environmental comfort. For example, water, electricity, coal, gas and petroleum are metered online real-time, digital platforms are set up, and energy and resource rationing is implemented for each department. Moreover, improvements are made in the water supply and drainage systems, central plaza layout, road transport system and application of eco-technologies. Efforts are also made to foster the teachers' and students' habits and competence of environmental protection.

4. **Promoting ecological culture by developing advanced eco-technologies**

Universities are an important inheritor of advanced human culture and spirit. When advancing the historical process of ecological civilization, higher education institutions should not only perform the task of building an ecological civilization, but also promote ecological culture. Teachers and students alike should take concrete actions to promote sustainable development both on campus and in the society, create the atmosphere of caring for ecological environment, and urge an environment-friendly production and consumption mode. In this way, the green campus initiative will be put into practice beyond the campus and a benign cycle and common development would be realized between green universities and local communities.

Higher learning institutions should always reflect the concept of ecological civilization in the creation of eco-technologies. They should organize relevant researchers and teachers to step up the study of issues related with ecological civilization, contemplate technological innovation from an ecological perspective, and apply the research results to the society to advance ecological civilization directly or indirectly.

When promoting ecological civilization, higher learning institutions should combine education in science with that in humanities, cultivate the students' awareness of ecological civilization and sustainable development, in a bid to foster and advocate such core values as equity, responsibility and sustainability contained in green culture. They should also keep bettering the teachers' and students' overall quality to actively contribute to the development of green culture.

2 Suggestions on Improving Education and Research in Ecological Civilization

As an important juncture where talents (primary resources), science and technology (primary productive force), innovation (primary driving force) and culture (primary soft power) are mingled, universities have a special and vital role to play in serving the development of ecological civilization. They are the main base of talent cultivation, where all educational activities are designed and intended for the students' growth. Producing high-quality green-conscious talents is the fundamental objective of building green universities and also the key to achieving the goals of building an ecological civilization. In recent years, many universities have attached great importance to the education and research in ecological civilization. They have established a number of relevant research institutes, carried out plenty of research projects and published a series of papers and monographs. Moreover, they have enlightened the society in a positive way thanks to new media such as WeChat public accounts and MOOC, through which information about ecological education and research has been communicated on a primarily large scale. Under the Five-Sphere Integrated Plan, higher education institutions should understand the significance of ecological civilization in the new era more rationally, and make greater contributions to it with regards to talent cultivation, scientific research, social service and disciplinary development.

1. **Accelerating the development of green courses and talent cultivation**

Building an ecological civilization means not only planting trees, but also educating people. To succeed in this strategy and let trees grow, we must enhance ecological education of the people, which is a basic, strategic and long-term task. Universities, as the highland that cultivates talents, makes scientific research and technological innovations, and inherits and promotes culture and civilization, shoulder the duty and mission of serving the socialist modernization drive by providing those services, and should make a difference in the undertaking of ecological civilization.

Establishing the concept of respecting, accommodating and protecting nature is an important ideological foundation for advancing the development of ecological civilization as it reflects the new values and ecological ethics. Universities should cultivate green talents that not only care about the ecological environment, but also know how to do that, especially those specialized in green scientific research. They should make the students understand the strategic significance of giving top priority to the development of ecological civilization. Regarding the curricular system, the educational authority should enact and implement the environmental education system to strictly ask all universities to open green courses as compulsory public courses for non-environmental majors. Normal universities in particular should open relevant courses to make sure the teachers they produce are equipped with basic knowledge and skills of environmental education.

The ideas and thoughts of ecological education should be contained in the courses, so that the "sustainable development" concept will be implemented throughout the teaching process. Theories and skills of sustainable development and environmental protection should be taught in classes to drive home to the students that we should shift from industrialization to ecological civilization and work for green economy and inclusive growth. In 2015, Tsinghua University opened the universal source titled "Fifteen Lectures on Ecological Civilization" for undergraduates, which was selected by about 9% of them. In 2018, the university published *Theories and Practices of Ecological Civilization* as a new textbook.

2. **Enhancing the awareness and competence of environment-friendly technological innovations**

Scientific research carried out by universities is an important component of the national scientific work and innovation system, playing a critical role in advancing scientific and technological progress of the country. Major sci-tech breakthroughs and innovations have constantly created new engines for economic growth, prompted substantial economic restructuring, and stabilized the economy on a higher level based on a better structure. With scientific development in modern times, the human competition for subsistence resources has become fiercer, and science is regarded as an effective tool by countries to vie for resources and drive economic development. In the twenty-first century that is characterized by the shortage of natural fossils, mineral resources and environmental resources, boosting environment-friendly technological innovation is a major social responsibility incumbent on scientific researchers, especially in China.

In recent years, universities in different places across China have successively formed research institutions of "ecological education" to carry out relevant researches and have yielded satisfactory results. When universities promote green tech innovations in the new era, the first thing to do is enhancing the teachers' awareness of scientific research and supporting their green technology R&D. In this way, they will be able to provide key technologies and constructive references for making decisions on the development of ecological civilization and consequently make more contributions to this cause. General Secretary Xi Jinping said at the National Conference on Eco-environmental Protection that to push China's development of ecological civilization to a new stage, we must step up efforts to make technological breakthroughs in major projects, and study environmental issues vital to economic and social development to come up with targeted measures. Universities should fully exert their advantages in disciplines, subjects and talents, delve deep into the theoretical system and practical technologies for building an ecological civilization, and study the dialectical relation between ecological civilization and environmental protection. They should integrate forces in relevant tech fields and keep pushing forward process and technology progress, with a view to realizing harmony between man and nature, win-win results in development and environmental protection, shared fruits of economic and social development, and growing public happiness.

Meanwhile, when promoting green technological innovation, educational and tech workers should not only meet the nation's actual economic and social needs, but also consciously and prudently evaluate the ecological effects their research results may generate on the resources, environment and energies. Responsible scientific researchers, while pursuing tech innovation based on personal curiosity, should also be highly vigilant against the risks and hazards that may derive from the application of their research results. In particular, they should apply the principles of resource conservation, circular utilization and environment friendliness when developing processes and products, and take the initiative to evaluate their research subjects on ecological effects, moral ethics, and social values. Green tech innovations should be rooted in China's basic national conditions and resource endowments, environmental protection needs, and industrial transformation needs. Many developed countries have adopted environmental measures in the form of standards and norms, which put new restraints on the design and development of modern technologies, and relevant environmental policies have consequently recalibrated technological development to reflect ecological goals and priorities. Scientific researchers in China should be urged to consider innovations on ecological efficiency and efficacy when developing processes and products. They should consider how to better the products and production processes to make the use of resources more ecologically and economically reasonable, and how to design products with more ecological and economic rationality and shift from product-oriented to service-oriented innovations, so as to meet market needs to the largest extent.

3. **Supporting decision-making and providing social services for green, low-carbon and circular development**

Higher learning institutions also do a lot of work to support government decision-making and serve the public and enterprises. In this process, university educators should fulfill their social responsibilities by providing science-based suggestions for decision-making in social management, promoting ecological tech innovations to meet economic and social needs, and foster the public environmental awareness in a scientific and objective spirit.

University educators should proactively take up the social responsibility for making substantial decisions on green, low-carbon and circular development. Due to domestic and international situations at the moment, the decision-making process has to be backed by more solid scientific information, which means university educators should carry a more significant social responsibility on their shoulders. At home, following the requirement of building an ecological civilization, the basic state policy of "resource conservation and environmental protection" has to be strictly observed in formulating various economic and social policies, promoting all the work and developing all kinds of plans. Meanwhile, the guideline of "prioritizing resource conservation and environmental protection and letting nature restore itself" should be upheld as the fundamental direction of our work in resource exploitation and utilization as well as environmental and ecosystem protection. Internationally, circular economy, green development and environmental protection are all important strategies many countries have adopted in their economic and social development, and also key measures for them to occupy the economic high ground in the age after the financial and debt crisis. Compared with other countries, China's development involves more factors and more complicated relations, with protruding conflicts between economic growth and environment, resources and energies and greater difficulties in making scientific decisions. During social operation and management, many scientific researchers don't want to get involved in "politics" or participate in "decision-making," but the government often needs scientific and technological basis when it has to make decisions on complicated issues. Scientific researchers in modern times should not only be specialized in their own fields, but as social elites should also participate in the government's and industry's decision-making and management process.

Universities should also proactively conduct publicity and education and guide public opinions. Many pollutants that are discharged into the environment as byproducts of modern technologies lead to diseases and ecological damages, and measures have to be taken to avoid environmental hazards and risks. In possession of specialized knowledge, scientific researchers can predict the possible application prospects of such knowledge more accurately and comprehensively, and therefore have the responsibility to estimate the positive and negative effects of related technologies. The current industrial production in China has caused more and more serious resource waste and environmental pollution, and grave and frequent eco-security hazards have evoked intense mass events. In face of these realities, researchers in universities should not bury themselves in the ivory tower doing only what they are interested in, but should serve the public and educate them to understand relevant issues. They should guide public opinions on controversial hot topics with an objective and scientific attitude rather than play to the gallery.

4. **Setting up cross-disciplinary platform for ecological civilization education and research**

The development of ecological civilization is a systematic program that calls for the common efforts of multiple parties, including governments, universities, media and the society. We are faced with extremely complicated environmental problems today. Environmental risks and events such as air, water and soil pollution and ecosystem degradation are caused by a complex mixture of factors. Developing the economy, exploiting resources and continuously improving the living standards of urban and rural residents – these tasks have to be performed in parallel. Therefore, there are conflicts among government decisions, company production and public life, which have to be balanced in general, but there is also development consensus among them that has to be considered in a holistic manner. What is of fundamental and paramount importance to the development of ecological civilization is to arouse and sharpen the social awareness of ecological civilization, including conservation, environment and ecology. We must advocate a healthy, rational, low-carbon and environment-friendly lifestyle and consumption mode, and put in place systems and mechanisms conducive to developing green technologies and fostering green spirit, so that every citizen, every family, every organization and every enterprise would consciously pioneer, practice and promote ecological civilization. This is why ecological civilization is a complex and cross-regional issue never seen before, and the education of it requires the common efforts of multiple disciplines and the formation of a multi-disciplinary education research team.

But the reality is that while scholars in different subjects discuss environmental problems and teach the subject, their work varies greatly in the object of study, definition of issues, theoretical tools and academic context. Their teaching and research of ecological civilization is generally self-sustained, overlapped or repeated. Therefore, it is urgent for universities to break academic boundaries and set up cross-disciplinary exchange platforms to forge consensus, communicate ideas, integrate teams and concentrate resources, in order to refine and systematize theories and knowledge related to ecological civilization. In November 2015, Nankai University formed the Institute of Ecological Civilization, which brought together over ten teams specialized in environmental science, chemistry, history, economics, etc. The new institute aims to assist in government decision making, theoretical innovation, history recording, and communication and education regarding environmental protection and ecological civilization in China. In April 2016, more than ten schools of Tsinghua University, including the School of Environment, School of Humanities and Laboratory of Law Carbon Energy, co-initiated the Center for Ecological Civilization. Exerting its cross-disciplinary advantages, the new research center gathers scholars specialized in engineering, science and humanities to work closely together, and they have conducted many explorations in the theories, methods and practices of building an ecological civilization. Many research institutes on ecological civilization have also been formed across the country. They include the Thinktank for Eco-civilization Studies of Chinese Academy of Social Sciences, Institute of Ecological Civilization of Xiamen University, Ecological Civilization Research Center of the South China Institute of Environmental Sciences, Ministry of Ecology and Environment, Institute of Ecological Civilization of Hunan Normal University, Xuzhou Institute of Ecological Civilization Construction, Center for Ecological Civilization Research, Chinese Research Academy of Environmental Sciences, Beijing Academy of Eco-civilization, and Guizhou Eco-civilization Institute.

Building on the existing educational or research institutes of ecological civilization, the China Association of Higher Education formed the committee on ecological education in 2018. The move was to further leverage universities' capability of cultivating talents for the development of ecological civilization, consolidate their strengths and features in basic research, cutting-edge explorations and common technology development, and deepen the coordinated innovation between enterprises and government departments in terms of policy formulation and core and key technologies. Supported by existing ecological civilization research institutes, the newly formed committee will serve as a platform of communication and cooperation in ecological education and research, so as to bring into full play the disciplinary, talent and research advantages of all universities. First, the new committee, focused on hot topics and difficult areas, actively reforms the theories, methods and practices of ecological civilization education, covering science and technology, institution, way of life and way of thinking. As this involves natural sciences, engineering technology, social sciences, humanities and law, the committee has gathered a group of engineering, science and humanities scholars in in-depth cooperation. Leveraging their cross-disciplinary advantages, they aim to establish a complete and scientific educational system, theoretical system and practical path for ecological civilization. Second, the committee, following the steps of building an ecological civilization, aspires to unite the members and universities to set up various platforms to promote communication and cooperation among universities. In the meantime, efforts are made to deepen the university-enterprise cooperation, popularize ecological results, intensify the integration of industry, college, research institute and application, step up the cultivation of composite talents, and eventually improve ecological education. Third, the committee, focusing on ecological education, promotes intercollegiate cooperation, establishes teaching alliances, integrates quality resources and builds sharing platforms. Quality courses are opened and excellent textbooks developed to internalize the ecological idea of respecting, accommodating and protecting nature in every aspect and every link of student education.

Systematic Difference and Correlation in Urban and Rural Ecological Civilization

Zhiyuan Ouyang

Ever since the 18th National Congress of the CPC, ecological civilization has been comprehensively implemented as a major national strategy with great progress made in environmental protection. Under the rigorous supervision and guidance of the central government, pollution sources have been shut down and sludge and garbage piled up for a long time quickly cleared away, but environmental problems have kept re-emerging as they also reflect social issues. This situation happens in urban and rural areas alike, but the latter has obviously done a better job than the former in building an ecological civilization. Therefore, there have been many articles and discussions about taking the approach of "encircling cities from the countryside" to this ecological campaign, a view I also endorse. But it won't do if we stay at words and slogans. Studies of the underlying laws and patterns have to be carried out, otherwise the approach has no theoretical or realistic significance. On the other hand, there are studies of self-organizations of rural environmental protection in recent years, but more should be done to elevate such studies on the theoretical level, such as the systematic and scientific review of the self-organizations, and their significance for overall rural development and ecological civilization. This paper is an attempt toward that goal.

1 Origin of the Issue

The most conspicuous environmental issue now is pollution, which in essence is caused by low-efficiency or inefficient utilization of resources, and pollutants are essentially mutated natural resources. At present, pollution control is usually through the harmless treatment of pollutants at the back end, but this is a very passive approach

Z. Ouyang (✉)
School of Philosophy, Renmin University of China, Beijing, China

J. Pan et al. (eds.), *Beautiful China: 70 Years Since 1949 and 70 People's Views on Eco-civilization Construction*, https://doi.org/10.1007/978-981-33-6742-5_56

as it demands input from the polluters but benefits the whole society. The thorough approach to pollution control is clean production, whereby resources are fully transformed during production under full-process control, thus benefiting both the producers and the society. Clean production has been promoted for many years but with little effect. There are two reasons—front-end investment and willingness of implementation, the second reason being more important as technological improvement is an ongoing process. On a deeper level, it is because enterprises are not strictly required to fulfill social responsibilities. In developed countries, enterprises adopt clean production primarily because of the law, then for economic considerations, and at last because of public pressure. When all these three factors, especially the last one, come into play, enterprises would have no choice but to go clean. Consumers express their expectation for a clean life through their buying behaviors. This expectation is not only for the product to be harmless to the buyer, but also to the environment, even for production to be harmless to the environment too. This gives rise to the environmental labeling system, a certification system that is under rigorous supervision and designed to meet both physical and mental needs. However, it must be noted that not everybody would feel the mental impact of environmental damage as that doesn't just depend on income, but also on culturing and upbringing, which is reflected in the ratio between physical and mental needs. When the general public is on a low cultural level, a green social atmosphere cannot be fostered and the producers feel no pressure. We have introduced the environmental labeling system, but consumers only care about the products per se and hardly ever ask about the emission and pollution control during production when paying the bill. The public wouldn't complain either unless something goes wrong at back-end pollution control.

Pollutant discharge doesn't just take place during production, but also in life. With the rising consumption level, the amount and quality of all sorts of solid wastes has been increasing sharply, hazardous wastes in particular. Wastes are either landfilled, burned or recycled. Landfilling can still be seen in many places as it is the simplest way, but it is leaving ever more hidden hazards with the shrinkage of usable land. Burning can generate energy, but siting for this approach is very difficult due to public concerns over incineration technology, the safety of which is under question as hazardous wastes are posing ever bigger threats. Even if the technology is safe, the burning of a vast variety of materials would lead to a very low utility value of the embers. Recycling is a relatively thorough approach. I don't agree with the notion of "circular economy" because circular utilization of energy is impossible and that of materials has a lot of limitations. Therefore, I suggest developing a resource-saving economy that includes maximal recycling. The key to recycling and reuse is sorting. In developed countries, there is a stringent system for waste sorting and recycling. Wastes are not only sorted into tens of kinds, but down to different sizes in the same kind. Although law plays a part in successful recycling, the basic factor is public cooperation. Some scholars, including me, began to call for waste sorting and recycling many years ago, and it was gradually implemented in the twenty-first century, but even rough sorting has met numerous difficulties. There are two reasons for this. The first is the responsibility for management. Some pilot communities have actively sorted wastes, only to find them being taken away all mixed together, which is a wet blanket on their enthusiasm. The second is public acceptance. Even when management catches up, most residents don't take waste sorting seriously and the sorting facilities in cities are left idle. According to foreign experience, legal measures have to be taken to enforce waste sorting, but the division of duty and degree of punishment are both hard to come through. AI has been introduced to help in certain areas, but public cooperation remains indispensable because, given the special nature of wastes, if AI facilities are installed all over the country, the upkeep cost would be unimaginable.

Ecological problems are also reflected in the excessive expansion of artificial nature, represented by the vanishing boundary between cities and the countryside. In an article published in 2019, I said nature had a function that I called "accrete self-organization," which was the result of natural evolution over a long period. The "accrete self-organization" is one that includes vital activities, a function that, once lost, would throw the world back to primitivity. The nature we see can be categorized into primitive nature, humanized nature and artificial nature depending on how much its "accrete self-organization" function is maintained. Primitive nature and artificial nature represent the highest and lowest end of maintenance, with humanized nature in the middle. Cities are a typical example of artificial nature while the countryside is an example of humanized nature. The essence of ecological civilization is utilizing natural resources on the premise of protecting nature's accrete self-organization function to the largest extent. Industrialization and urbanization are twin brothers. Adam Smith in his *Wealth of Nations* ascribed the greatest improvements in the productivity to the division of labor, which aimed at a high level of concentration of manpower and materials. That is how urbanization brings economic efficiency, but at the price of losing the accrete self-organization function. When concentration is of a small scale, the shortfall can be compensated by the limited humanized nature both within and around the city; but once concentration reaches a certain scale, especially after a city belt is formed, "urban maladies" would rise considerably. Urban malady, essentially caused by the lack of accrete self-organization function, cannot be cured simply through sanitary means. Replacing the term "Cheng Shi Hua" with "Cheng Zhen Hua" (both meaning urbanization, the former highlighting "city" while the latter "city and town") reflects the consideration to keep humanized nature as much as possible. The problem now is that man has thought too much of itself and tried to make artificial nature bigger, stronger, faster, higher, deeper and more convenient, which unfortunately weakens the accrete self-organization function unlimitedly. The "landscape" concept proposed by German geographer Humboldt A. encompasses the existential status and ratio of the three kinds of nature. The landscape in a lot of areas where fertile land used to stretch far and wide has been devastated. "Nostalgia" will become a word of the past if urban expansion is not curbed effectively.

2 Analysis of the Issue

For a very long time, the countryside has become the dump site of the "three wastes" generated by cities. Point pollution in cities evolves to be non-point pollution in rural areas, which, combined with the decline of rural farmland and buildings, makes people deeply disappointed in the rural environment. After the Wenchuan earthquake in 2008, the central government exerted China's institutional strengths and quickly and effectively reconstructed the quake-hit areas through paired assistance. Residents in rebuilt villages displayed a general and surprisingly high recognition of ecological civilization and both the environment and their spiritual outlook underwent dramatic changes. Especially after the 18th National Congress of the CPC, building a beautiful countryside has become a state action and ecological efforts in rural areas have yielded fast and large-scale progress. Villagers consciously protect mountains, waters, forests, farmlands, lakes and grasslands, and pollution control has entered a regular and institutionalized track. Waste sorting and cycling, for instance, which is hard to promote in cities, has proceeded quite smoothly in the countryside. This profound change is directly related with central decisions. What's worth further study is that apart from the Beautiful Countryside initiative, there is another state-level task of building a "Beautiful China" with the same connotations. Since cities have much more input and much better conditions than rural areas, why is it so difficult to promote ecological civilization in cities? One viewpoint is that this is because the development of ecological civilization is directly linked with villagers' income. This makes some sense, but why is voluntary environmental protection that is not directly linked with villagers' income implemented with a much stronger momentum in rural areas, particularly those inhabited by ethnic minorities, than in cities? Another phenomenon is that migrants may not follow environmental rules in cities, but they do when coming back to rural areas where there are standard environmental rules. They even help promote them. Villagers' environment-friendly behaviors are more common and conscious on the frontline of sand control. Some argue that this is because they are face to face with nature, but that cannot explain why people who were also face to face with nature under the planned economy didn't display such traits. To seek an answer, it is necessary to employ the synergetics founded by Hermann Haken in the 1970s. Synergetics derives from natural sciences, but it gives rational explanations of social phenomena.

Synergetics studies how an open system "far from Equilibrium State (a stable, even and regular state)" voluntarily develops an orderly structure in time, space and function through internal synergy when having material or energy exchange with the external environment. Haken divided organization into two kinds—"heter-organization" and "self-organization"—based on its evolution. A system is a heter-organization if it is organized by external orders; a system is a self-organization if its various units, following some tacit rules, automatically organize into an orderly structure by each performing their duties while coordinating with each other in the absence of external orders. The aforementioned accrete self-organization is a typical self-organization. This paper will only give qualitative descriptions of the social phenomenon in question according to basic principles. After the reform and opening up, the old ossified organizations in rural areas disintegrated and villagers assumed a large extent of freedom in economic activities, who, having gained experience in different places, on different occasions over several decades, have established different lives. When they gather together again, each member's attitude toward these lives can be regarded as the "macro parameter" of the system. Given the traditional Chinese culture, working together to improve the environment and getting rich through ecological civilization is very likely to become the prevalent attitude, which is the "order parameter" in synergetics. The order parameter strongly influences every member's thoughts and actions, which gradually evolve into a general will. Against such a background, the members are easily inspired by social changes and supportive policies to work in coordination and push the system toward a common goal, thus triggering the "phase change" from disorder to order. The inspiring external factors are called "control parameter." The word "control," according to Haken, means guiding and adjusting, while the decisive factor on the system's phase change is cooperation among the members, which usually takes the form of building a "community," such as the "community of ecological interests." A community isn't necessarily an economic entity. It could be a group bonded by interests, in which every member is both an actor and a manager. There is generally a leader or leaders, but they don't interfere in the members' actions.

A city is a complex mega-system, and a big city is a super one, in which the categories and scope of social activities are far beyond the reach of rural areas. City is the center of industrialization which has three major features. The first is intensiveness, which refers to the high level of compression of space and the intensive utilization of personnel and equipment. The second is synchronization, which refers to the high level of compression of time and the uniform arrangement of work and life. The third is standardization, which refers to the high level of compression of structure and the front-to-end standardization of everything, from details to the general assembly. Human expertise is extended in one direction amid these three features, which keeps raising the quantity and quality of products. The three features essentially constitute a heter-organization, and consumption that is attached to the production system is a part of this system. They seem unconnected but there is a connection—wild consumption and wild production are in a mutual cause-and-effect relation. This is the "human dependence on things," and the result is metamorphosis of both man and nature. Marx divided individual development into three stages—mutual dependence between men, human dependence on things, and all-rounded development of man. In the first stage, man's essential power is weak, but it is tempered in the second stage and then transitioned to the third stage of all-rounded health, including natural, social and mental power. In face of fierce international competition, industrialization and marketization have to be accelerated either for the sake of national security or public needs. Ecological civilization is to offset the negative effects of industrialization and marketization and create conditions for all-round human development. China has to carry out the dual tasks of industrialization and building an ecological civilization, which are both in unity and opposition, as intensively reflected in cities. It is indeed very hard to forge cooperation in environmental protection while we are already busy dealing with unending problems, and cooperation is a must for environmental protection. Public coordination rarely happens unless major pollution events involving the "three wastes" happen, and it vanishes as long as the moment is gone. Developed countries also made strenuous efforts over a long period before finally achieving a sound environment, but to maintain it, continuous large-sum investment and waste transfer is required.

3 Solution of the Issue

After the reform and opening up was launched in China, the western economics theory about its "urban-rural dual structure" gained much traction at one time. This theory is of some reference value as it reveals certain laws of industrialization, but it has the lethal flaw of ignoring the fundamental role of ecological environment and takes the western model as the only choice. In the past few decades, China has always paid close attention to the "Sannong" issue (agriculture, the countryside and farmers), which indicates that the country has kept a distance from the "dual structure" theory. There are many reasons why rural development has been lagging behind, but the apotheosis of this theory is definitely the fundamental one. Many areas take "wiping out the countryside" as a potential goal, and some even make "building towns without farmlands" as the action guide. It is generally believed that we can depend on the international market for food supply, but the fact is all industrialized powers are at the same time major agricultural countries, particularly major food producers. They haven't adjusted their structure, but why are they so picky on ours? Some would argue that this is because they have adopted the modern way of production, they are advanced and we are backward. Yet the reality is the non-sustainability of industrialized agriculture has long been exposed, while traditional agriculture, with appropriate renovation, would become vibrant and organic. The market has shown who is advanced and who is backward. The real purpose of confusing and deluding us is to destroy our confidence and pave the way for their agricultural export. Luckily, we were not fooled. Instead we have seen bumper harvest of grain for years in a row. However, the current situation is far from optimistic. Large areas of farmland have been left wasted or turned into construction site, which then is also abandoned, and the little remaining farmland is mostly used to grow economic crops. Growing economic crops is not a bad thing as it opens a way for farmers to live a better life. The problem is China is too short of arable land, and it would be dangerous for a country with over one billion people not to be self-sufficient in grain. To turn the tables, we have to first abandon the blind belief in western economics theory and put "Sannong" issue in a prominent position.

As a matter of fact, the real breach to truly resolve the "Sannong" issue is the Beautiful Countryside initiative. There are three reasons for this. First, the countryside is not only the production area of food and biological raw materials, but also the place where natural resources regenerate and renew and man and nature interact directly. Beautiful countryside – the concept in itself is pleasant. The worsening ecological environment, the fast pace of work and the boredom with artificial nature are strongly spurring the urbanites to embrace nature again. There are many ways to do that, but the most common way is going to the countryside. Implementing this initiative from the angle of ecological civilization can effect a change in the public view on consumption, thus breaking the theoretical and ideological fetters imposed by western economics. Second, the traditional approach to solving the "Sannong" issue, namely high input, is the result of thinking in the confines of industrialized agriculture. If we make a Gestalt switch in aesthetics, highlight the cultural values of traditional countryside, we can realize a natural and clean countryside. The rural system is relatively simple, where environmental governance is easier than in cities as it depends on the input of manpower rather than capital. As long as ecological progress can increase the income of rural residents, migrant workers will go back to their hometown, bringing the new ideas and technologies they have obtained in cities. The special land ownership in China decides that our "back-to-countryside" wave will have a stronger momentum than the "anti-urbanization" movement in developed countries. Third, currently rural areas mainly produce non-staple food and sundries to achieve prosperity, but the large-scale expansion of breeding industry is detrimental to landscape and health. Historically speaking, anthropoids were vegetarians. Eating meat has improved human nutrition, urged them to increase the living space and made them much stronger fighters, but vegetarian diet is bound to return in the long term. Grain production has both material and spiritual significance, and both can be achieved through ecological tourism. If one day the scene of "green field, growing rice and chirping cuckoos in misty rain" is back, the charm of the countryside will be irresistible. We must advance the construction of a beautiful countryside through forming a rural community of ecological interests, otherwise we won't be able to stimulate the villagers' enthusiasm in a sustained and effective way.

Social reproduction consists of three closely linked aspects—natural reproduction, demographic reproduction and economic reproduction, which are effectively coordinated by the Beautiful Countryside initiative pillared by the rural community of ecological interests. First of all, it creates a stable and reliable strategic rear base for national economy, the significance of which is explained earlier. Second, from a demographic point of view, the rapid development of intelligent technologies will lead to a huge surplus of labor force and consequently greater employment pressure. I have discussed this in the article *Demographic Dividend or Brain Dividend?* At that time, the beautiful countryside will offer a space for the surplus labor force to exert their strengths. At last, from an ecological point of view, building a beautiful countryside will serve three purposes. The first is to demonstrate and spur environmental governance in cities. If we can succeed in building a beautiful countryside, the city polluters would be too ashamed to play the "cat-catch-mouse game" with environmental inspectors and authorities any more, and the society in general would gradually be inclined toward a clean environment. The second purpose is to build barriers against urban expansion and pollution transfer. Environmental violations usually happen and get away with punishment because of the priority to cities, but the construction of beautiful countryside will take the upper hand both morally and operationally. Third, once a beautiful countryside is built, it will restrain the environmental behaviors of urban tourists, and participatory tourism will change their ideas, which will lead to profound impacts on urban environmental administration. These three purposes, if achieved, will form the situation of "encircling cities from the countryside" in building an ecological civilization. Government authorities may change their strategic thinking, focus on the weak links and key contradictions, make breakthroughs and push the overall situation forward. Great efforts should be made to create favorable conditions for the healthy growth of the rural ecological community. This community should stick to three key points. First, it should establish right ecological values and prevent the erosion by superstitious activities and control by clan forces. Second, it should always prioritize ecological and comprehensive benefits over economic benefits and guard against actions or decisions for quick success and profits. Last but not least, it should reform the artificial nature gradually and appropriately and prevent copying the urban landscape and life in the countryside. Party organizations and members should actively yet subtly play the guiding and regulating role as a "control parameter."

4 Conclusion

The difference in the development of ecological civilization between rural and urban areas lies in the different effects of self-organization and heter-organization. Due to the special conditions of the rural system, it is relatively easy to form a self-organized rural community of ecological interests through coordination under new circumstances. Building a beautiful countryside with the community as a pillar will help solve the traditional "Sannong" issue, address the environmental problems in rural areas and give a strong impetus to ecological civilization in cities. In comparison with general discussions about "encircling cities from the countryside" regarding the development of ecological civilization, this paper is a systematic analysis on the theoretical level; in comparison with specialized studies of rural self-organization of environmental protection, this paper eyes the big picture of urban and rural ecological civilization. The conclusions herein can serve as references for macro decision making in relevant sectors. How to form self-organization for urban ecological civilization in this big picture is another topic for further study.

Ecological Civilization Is a New Path of Civilization Imbued with Eastern Wisdom

Xiaode Zhang

The 18th National Congress of the CPC proposed the path of ecological civilization for China, the first time the banner of ecological civilization was held up in the world and marking a milestone in the twenty-first century. It demonstrates the confidence and commitment of the CPC Central Committee with Comrade Xi Jinping at the core for seeking a path different from the western path of industrial civilization with eastern wisdom to adapt to the requirements of the times.

A new era, like spring, often arrives before anyone notices it. Columbia's discovery of the new continent in the fifteenth century unveiled a new era when the world was dominated by the western industrial civilization. But when Columbia landed in America, he didn't know it was a new continent, but believed it to be Asia till the last minute of his life.

Things like this keep happening. At present, many people still equal ecological civilization with environmental protection without realizing that it marks the beginning of a completely new era. When the 18th National Congress of the CPC put forth the development of ecological civilization in the Five-Sphere Integrated Plan, it was indeed to address the environmental crisis, but building an ecological civilization should by no means be equaled with environmental governance. Guided by China's ecological wisdom of "following the course of nature" and systematic and dialectical Marxist philosophy, the overall strategy proposed at the 18th National Congress of the CPC is a new exploration for the new era and new civilization path that decides the future of China and the world.

X. Zhang (✉)
Department of Economics, Party School of the Central Committee of CPC (National Academy of Governance), Beijing, China

J. Pan et al. (eds.), *Beautiful China: 70 Years Since 1949 and 70 People's Views on Eco-civilization Construction*, https://doi.org/10.1007/978-981-33-6742-5_57

1 Background of the Proposal of Ecological Civilization: The Cost-Externalized Western Industrial Civilization Is Unsustainable

If we look at the modern environmental crisis with the systematic and dialectical viewpoint of traditional Chinese medicine, we would find that the environmental crisis is not the source of the modern industrial civilization crisis, but one of its outcomes. In TCM, you cannot just treat the eyes if you have an eye disease, but should treat the liver where the root cause lies. The Five-Sphere Integrated Plan for building an ecological civilization proposed at the 18th National Congress of the CPC doesn't aim at the local improvement of industrial civilization, but attempts to systematically and comprehensively explore a new civilization unlike the industrial one. The 18th National Congress of the CPC put forth this great ecological initiative against the background that the western industrial civilization is unsustainable. It is true that it still has a lot that we can learn from and draw on, but in general the western civilization mode has completed its mission and fallen into a dilemma of more cost than benefits.

1. **Western industrial civilization is a "cost-externalized industrial mode"**

As developed western countries only encountered resource and environmental restraints after industrialization was accomplished, they at first didn't view resources and environment as a cost of industrialization that they had to bear. From an economic perspective, the western industrialization mode was established on the assumption of unlimited resource supply and sufficient self-purifying capability of the environment. Such an assumption may hold water for a certain period of time, but problems emerge when industrialization consumes more energies and causes more environmental pollution than nature can bear. With the western mode, industrialization doesn't pay for energy consumption and environmental pollution, but transfers the price to the nature and society. In this sense, the western industrialization mode is a "cost-externalized" one that benefits western countries at the expense of others. If such a mode is not overhauled and if environmental pollution is only addressed from the outside, it is impossible to fundamentally solve the conflict between man and nature.

2. **Modern industrialized economy is capital economy that lacks restriction on consumption and pursues unlimited expansion of production**

Driven by market competition and technological innovation, how to generate maximal output with minimal input is the intrinsic mechanism and goal of the whole industrial economy. Guided and controlled by this goal, stimulating the human desire for consumption to the largest extent to continuously expand production becomes the internal drive of western industrialization and capital economy. The basic and ultimate goal of production in any age is meeting people's living needs, but in the modern industrialized system dominated by capital, consumption only exists for capital appreciation and consuming behaviors are dominated by the market. Industrialized production without consumption restraints falls into the abyss of unlimited growth and expansion, which results in a unique phenomenon of industrial economy—driven by competition, the unlimited pursuit for GDP growth and for far more material consumption than physically necessary becomes a global malady that is beyond control.

In the modern system of industrial production, on the one hand, we are worried about resource and environmental crisis; on the other hand, all our policies and systems are designed to meet unlimited material needs and achieve high GDP growth. In the industrial economic system with no consumption restraints, such actions are understandable in developing countries where food and clothing sufficiency is not guaranteed yet. But in fact, even in the United States whose per capita GDP exceeds USD 40,000, the whole society would panic once economic growth slows down. The obsession with economic growth has become a common social ailment, and pursuing unlimited growth a culture in modern industrialized civilization. If this morbid consumption mode aimed at capital increase is not reformed, clean and energy-intensive production alone cannot fundamentally resolve the predicament of modern industrialization.

3. **Western industrialization is exclusively beneficial for a small group of people or countries**

The industrialized growth mode driven by market competition is a polarized one that makes the strong stronger and the weak weaker. The competition-induced polarization and cyclic economic crisis posed tremendous obstacles to economic and social development in developed western countries in the early twentieth century. Having experienced the 1929 economic depression that started in the United States and swept across all developed western countries, a new administrative system came along featuring government-led macro regulation and public good supply, which effectively rectified and solved the polarization issue. However, this system only functions within national borders.

After WWII, as more and more developing countries gained independence and the western market economy was spread to the whole world, the competition-driven polarization effect also appeared worldwide along with economic and market globalization—the polarization, which was increasingly intensified, between strong industrialized powers and developing countries. It is currently impossible to establish a global regulatory system to redress the competition-induced polarization, so industrialization actually becomes a civilization controlled and exclusively enjoyed by a small group of developed countries and multinationals. But while the few industrialized countries have the benefits of industrialization all to themselves, they, leveraging their competitive edges, have not only excessively occupied energies around the world, but also transferred energy consumption and pollution to developing countries through industrial transfer. If earth's resources and environment had the capacity to support human development without limit, emerging industrialized countries could join the industrialized club one by one following the western mode. But the fact is that they are already on the brink of a crisis, so earth no longer has the conditions for emerging market economy countries to follow the old path of polarizing competition once adopted by developed western countries.

Western industrialization only satisfies and accommodates few countries. Industrialization has been going on for more than 200 years, but the western developed countries that have truly benefited from it account for only 16% of the world's population while consuming more than 60% of earth's resources, with their ecological footprint exceeding half of earth's bearing capacity. While the earth is in the middle of a dire environmental crisis, the western path of industrialization, if adopted by the rising China that has 20% of global population, will expose both China and the rest of the world to a common crisis.

As the flaws of the west-exclusive industrialization are exposed, the biggest challenge China faces in its industrialization is whether it is able to create a new civilization path that can benefit the vast majority of people. This is a hard question for China's industrialization and for the evolution of human civilization. The solution to it will be a blessing for China's modernization and human civilization.

4. **Western industrialization kills and smothers the diversification of human culture and civilization and seriously destroys human civilization and cultural ecology**

The development of cost-externalized industrialization not only seriously destroys natural ecology and environment, but also damages human civilization ecology and cultural ecology. The ancient society before industrialization was an ecological world where various cultures and civilizations co-existed. Starting with industrialization in the West, the immense productivity of material wealth motivated by technological innovation and market competition quickly led to the collapse of the ancient productive approach and relations. As far as human civilization is concerned, industrialization replacing and overthrowing the ancient production approach and relations is a price that has to be paid for human progress. The ancient agricultural civilization took place in a relatively closed environment as different ethnic groups scattered around the world completed this process almost in parallel in roughly the same period. In comparison, industrial civilization started in the West and then spread to the whole world. It promoted the fast rise of the West with economic, cultural and military predominance. As a result, the process of western industrialization spreading across the world was not only accompanied by the disintegration of ancient production approach, but also by the destruction of the diversified cultural ecology left from the ancient society. From the 15th to the nineteenth century when the West went ahead on colonialism, the Latin American and African aboriginals and cultures that could barely protect themselves were devastated. After WWII, the western industrialization began to make its way to developing countries in a peaceful way, and the latter's national cultures were gradually eroded or replaced unawares by commodity economy and the western lifestyle.

Especially after the end of the Cold War and the disintegration of the Soviet-led socialist bloc in the 1970s, economic globalization evolved into global economic marketization and cultural westernization. Pursuing industrialization at the expense of diversified ecology of ethnic cultures, like sacrificing the ecological environment for economic growth, is also a kind of ecological damage that is worth close attention. We have realized how important biodiversity is for the self-balance and self-protection of ecology and nature, and then we should also realize how important ethnic cultures are for the balance of human civilization ecology. It would be unimaginable if there were only one culture and one identical lifestyle in modern industrial civilization—what a disaster it would be for human civilization! The existence of only one culture in human civilization means the end of its evolution because evolution, either of natural ecology or material world, is only possible in the interactions of diversity. Singularity means stagnation.

2 The 18th National Congress of the CPC Proposes Ecological Civilization as a New Path Imbued with Eastern Wisdom

1. The shift toward ecological civilization should start with a reform of philosophy and thinking

If we interpret the ecological civilization strategy proposed at the 18th National Congress of the CPC in the context of the time, we would easily understand why China hasn't repeated the western path of environmental governance, but instead blazed its own path with Chinese characteristics by creating a new civilization mode. The underlying reason is that the crisis we are facing today is not simply an environmental crisis, but a civilization crisis. The flaws of western industrial civilization are not simply the flaws of industrialization itself, but expose the flaws of the entire civilization system. When we reflect on the western industrialization mode, we should stand at the height of civilization as we not only need to seek a way out of the environmental crisis, but also reflect on and innovate our values, way of thinking, culture and way of life.

Theories of ecological civilization were put forth in the 1980s, but China is the only country that has elevated the concept to be a state strategy and has held up high this banner. This is full evidence that China is standing on a high ground as we march toward the future. Standing at such a height, what we bring to the modern world is, first and foremost, not technology, not GDP, or a sectoral system. What we present is the possibility of a new civilization based on eastern wisdom, the new natural outlook of "man and nature in one," the new values of mutualism, the new thinking of systematic integration reflecting the generation-inhibition of five elements, and the new philosophy of a community with a shared future for mankind.

Perhaps right now with the aggressive impacts of western technologies resulting from the China-US trade friction, we are not taking seriously these intangible factors that are already influencing the world. But we must not forget that the West, as it gradually walked to the world center since the fifteenth century, didn't begin with technology and economy when changing the world, but with culture and philosophy.

The real inception of modern western civilization was not Industrial Revolution in Britain, but the Reconnaissance, the Reformation and the Enlightenment from the 14th to the eighteenth century. This was no accident, but the law of world historical evolution. In the whole history of human civilization, it was always the ideological, conceptual and cultural revolution that ushered in a new era.

2. **China cannot repeat the western path of governance that benefits oneself at the expense of others**

The proposition by the 18th National Congress of the CPC to resolve environmental problems through building an ecological civilization is not just a concept. As specified by the 19th National Congress of the CPC, it is a new governance path and a new Chinese plan created with Chinese wisdom. The commitment to contributing to world environmental protection also made at the 18th National Congress is not empty talk either, but an endeavor to explore a path of environmental governance different from the western one.

In the backdrop of technology worship, many people doubt that the Chinese plan can contribute to world environmental protection. Up to now, both advanced technologies and mature systems for environmental governance belong to the West. What on earth does China have to contribute? In fact, in modern times, the biggest challenge to resolving the environmental crisis is not the lack of technologies, but the problematic philosophy and guideline of governance. Western developed countries began addressing environmental problems in the 1970s. There is no doubt that they have achieved a lot in that aspect, but their approach is not sustainable.

If we keep walking down the western road, the environmental crisis facing the world will be a dead end. Einstein said, "We cannot solve our problems with the same thinking that created them." This is exactly the case with the environmental path adopted by western developed countries—a path featuring the local governance of environment and external governance of pollution export. In the past half century, although these countries have worked hard on environmental protection in terms of technology, legislation, investment and institutional innovation, these efforts are limited and focused on the environmental front alone, without ever touching the deep-rooted reason, namely the way of life and production characterized by high energy intensity and high consumption. Since they haven't found an effective way to systematically and fundamentally resolve energy and environmental problems, western developed countries have taken the path of pollution export leveraged on their advantages and channels of international trade in order to defuse the energy and environmental crisis. China has been a primary victim to such pollution export in the past few decades. Let's take a look at the China-US trade friction. China is in trade surplus with the United States in terms of export-import volume, but it is in trade deficit with the United States if we count in energy consumption and pollution. During the China-US trade over so many years, more than half of the products whose production is transferred from the United States to China are of high energy consumption and high pollution, which developed countries don't want to produce on their own territory. As a result, we produce fair-priced products with our energy and low-cost labor force while causing pollution at home, only to find ourselves being accused of commodity dumping.

The United States has the highest per capita energy consumption in the world—11-ton standard coal per year. Such high energy consumption and high pollution is neither absorbed internally nor addressed by reforming its political, economic or cultural system, but is solved through foreign trade—exporting pollution and importing energy. The per capita energy consumption in China now is 2-ton standard coal, part of which comes from pollution exported by western developed countries. If we look at industrialized production in those countries, their GDP is green and their air is clean, but those are not the result of them making ecological efforts. It is the result of a superficial, unsustainable governance mode that benefits themselves at the expense of others.

3. **The Five-Sphere Integrated Plan for ecological civilization proposed at the 18th National Congress of the CPC points out a way of fundamental governance entrenched in the eastern dialectics of generation and inhibition**

China cannot copy the unsustainable western governance mode. The earth's bearing capacity does not allow China to repeat that path of external governance. Against such a background, the 18th National Congress of the CPC proposed the Five-Sphere Integrated Plan teeming with eastern wisdom that is of great value and significance for China and the world. Combining the eastern ecological wisdom of following nature's course and Marx's systematic dialectical philosophy, it is an attempt to seek a new governance mode and a new civilization path—the Chinese approach presented at the 19th National Congress of the CPC.

The 18th National Congress of the CPC proposed the "Five-Sphere Integrated Plan" concept of governance that integrates ecological advances with political, economic, cultural and social development, which is highly consistent with the ancient Chinese philosophy of the "five elements." Ancient Chinese believed in the generation-inhibition relation among all substances on earth, which is fully reflected in the "five elements" theory involving metal, wood, water, fire and earth. It is based on this theory that traditional Chinese medicine practitioners have understood the generation-inhibition relation among heart, liver, spleen, lung and kidney and established the systematic and dialectical TCM treatment system with Chinese characteristics. The ecological strategy of Five-Sphere Integrated Plan proposed at the 18th National Congress of the CPC doesn't focus on the environment alone, but takes the whole society as a connected life system. The politics, economy, culture, society and ecology in a country are just like the heart, liver, spleen, lung and kidney in a human body. To address environmental problems, we have to also address other factors that generate/inhibit the environment. It is true that environmental governance in China lags seriously behind and it is an arduous task, but China's efforts to systematically address the maladies of industrial civilization by creating a new civilization have brought new hope for the country and the world.

Western developed countries have indeed done a great job in environmental governance. They have even realized zero emission per capita successively in the twenty-first century. However, we have to be aware that this doesn't mean they have completed the building of an ecological civilization. Ever since the 1970s, the western approach to environmental governance has been focused on symptoms without ever touching the root causes, which is highly costly and unsustainable. Two main environmental approaches have been adopted by them. First, they improved the environment per se without changing the morbid way of production and life. It is like when a patient suffers from a lung disease, the doctor, instead of finding a way to cure the lung fundamentally, leverages on technological progress to make a machine to help the lung operate. Second, dominating international high-end trade, they exported pollution and imported energy, thus transferring its own pollution to others. As a result, while developed countries are moving toward zero emission themselves, they are increasing the total emission worldwide. This kind of governance mode that benefits developed countries at the expense of others is unsustainable.

3 Ecological Civilization Is a Civilization of the New Era When China and the Rest of the World Share Win-Win Development

1. **The "lasting development" initiated at the 18th National Congress of the CPC is a mutualistic outlook on world civilization**

The report of the 18th National Congress of the CPC makes it clear that the mission of promoting ecological civilization is to "achieve lasting and sustainable development of the Chinese nation." This is an innovation and extension of the "sustainable development" concept initiated by the United Nations and a new development of China's long-standing mutualistic world outlook in the age of ecological civilization.

The UN-initiated "sustainable development" refers to sustained development of the current and future generations. In other words, it reflects the sustainability from today to tomorrow, from the current generation to generations to come. The "lasting development" initiated by the 18th National Congress of the CPC focuses on the "Chinese nation," which refers to not only the nation with a 5000-year history, but also the contemporary China founded in 1949 and the future Chinese generations. If the sustainable development of the United Nations aims at 2D sustainability in time and space, China's lasting development aims at 3D eternity from the past, present to tomorrow. It inherits the past, focuses on the present and looks to the future. The Chinese nation has a civilization that has lasted continuously for thousands of years without disruption, one that is most longevous and sustainable. By proposing lasting national development at the 18th National Congress, China is set to inherit and absorb wisdom from its 5000-year civilization to serve the sustainable development and explore ecological civilization in contemporary times. This is also the highlight and innovation of building an ecological civilization proposed by the 18th National Congress.

The mainstream world civilization outlook in modern times is rooted in the western concepts of self-centered competition, survival of the fittest and external colonization.

It was based on this outlook that the Western world rose up and dominated the globe after the fifteenth century; it was also based on this outlook that it embarked on the environmental governance path of transferring its own pollution to other countries. If we in contemporary times continue to follow this outlook, the energy and environmental crisis will never be defused. The crisis we are facing today is essentially one of civilization outlook. To address the contemporary energy and environmental crisis at source, what we need is a new "mutualistic outlook on world civilization."

This new civilization outlook needed in the new era is found in the east. Throughout its 5000-year civilization, China upholds the Confucian philosophy of benevolence and ethics characterized by "Do not do onto others what you do not want others do onto you," and the civilization outlook aimed at world peace. It hasn't abandoned this traditional outlook even after it shifted from the track of ancient agricultural civilization to the track of western industrial civilization. Both the capitalist revolution led by Sun Yat-sen that upheld "the whole world is one community" and the new democratic revolution led by Mao Zedong that aimed to liberate the whole mankind represent the continuation and development of the Chinese civilization outlook for world peace. In the report of the 18th National Congress of the CPC, China pledged to "contribute our share to global ecological security," which is an embodiment of its mutualistic civilization outlook on environmental governance.

With China's rise in the twenty-first century, the "China threat theory" has gained ground among western developed countries, which obviously view China's rise from their colonial civilization outlook. As far as the development of human civilization is concerned, the world needs China's rise to defuse the contemporary civilization crisis because only China's rise can ensure that the eastern country's mutualistic civilization outlook for world peace will become the global mainstream, and the world will move from the US-dominated unipolar civilization to a new civilization of diversity and co-existence. In this sense, the rise of the Chinese nation is by no means a threat to world civilization, but a blessing for it to walk out of the crisis.

2. **Chinese wisdom and Chinese approach that follow the trend of the time are recognized and commended in the world**

The ecological civilization strategy rooted in Chinese wisdom proposed at the 18th National Congress of the CPC was soon recognized by the United Nations. In February 2013, the 27th session of the Governing Council of the United Nations Environment Programme (UNEP) decided to promote China's philosophy of ecological civilization, indicating the international community's recognition and endorsement of this Chinese concept.

When delivering a speech at the opening ceremony of the Paris Conference on Climate Change on November 30, 2015, General Secretary Xi Jinping first announced to the world a new plan based on eastern wisdom for addressing climate change. He proposed three guidelines for resolving the climate issue. First, the Paris Conference should reject the narrow-minded mentality of "zero sum game" and adopt the new mentality that "Do not do onto others what you do not want others do onto you," so as to create a future of win-win cooperation, with each country making contribution to the best of its ability. Second, we should reject the mindset of confrontation and meet global climate challenges in the new mindset of "inclusiveness, mutual learning and common development." Third, elaborating the Chinese outlook on the world, justice and interests, he announced China's nationally determined contributions and responsibilities to climate governance. Xi's speech drew vehement responses and applause from the audience and was highly acknowledged by the Paris Conference.

In May 2016, the UNEP published the report titled "Green is Gold—The Strategy and Actions of China's Ecological Civilization" according to the theory of "lucid waters and lush mountains are invaluable assets" initiated by General Secretary Xi Jinping. The report spoke highly of Xi's green development thought and China's ecological civilization vision. In this sense, General Secretary Xi's theory that "lucid waters and lush mountains are invaluable assets" has not only contributed to China's green development, but also to world ecological civilization.

On January 18, 2017, General Secretary Xi Jinping made an important speech titled "Work Together to Build a Community of Shared Future for Mankind" at the United Nations Office at Geneva. It was the first time he proposed China's human civilization outlook and philosophy to the world, which was "building a community of shared future for mankind and achieving shared and win-win development," a vision that once again was commended and affirmed by the international community and the United Nations.

All in all, the ecological civilization infused with Chinese wisdom is not one where China replaces the United States to be the new unipolar power, but one where China and the world work together for shared win-win development through coordination and innovation.

Building Ecological Culture, Showcasing Ecological Wisdom

Jian Lin

Dire environmental issues have exposed the plight of human culture. The only way to overcome the ecological crisis, protect ecological security and achieve harmony between man and nature is establishing the vision of ecological civilization, changing people's values, creating the eco-culture system and taking concrete measures to intensify ecological governance.

1 Ecological Culture and Ecological Civilization

At the National Conference on Eco-environmental Protection held on May 18, 2018, General Secretary Xi Jinping urged the country to move faster to build systems for an ecological civilization. He further elaborated on the five systems—a cultural system in which environmental concepts and values are norms, an economic system aimed at integrating industries and ecosystems, a system of responsibilities and objectives centered on improving environmental quality, a system of institutions backed by modernized governance systems and capacity, and a safety system that emphasizes the maintenance of well-functioning ecosystems and the effective prevention and control of environmental risks.

The ecological culture that features the harmonious co-existence and coordinated development of man and nature is the new way of life and new value chosen by mankind in the twenty-first century in the face of multiple crises. With man in the center and closely associated with nature, the ecological culture encompasses all activities and achievements by man during their interactions with nature in order to adapt to the natural environment, maintain ecological balance, improve ecological environment, realize the cultural values of natural ecology, and meet man's material and spiritual needs. It represents an ecological awakening and social adaptation, a

J. Lin (✉)
Renmin University of China, Beijing, China

J. Pan et al. (eds.), *Beautiful China: 70 Years Since 1949 and 70 People's Views on Eco-civilization Construction*, https://doi.org/10.1007/978-981-33-6742-5_58

process in which man have come to respect nature and follow its laws, and reflect on and adjust themselves in development.

"Ecological culture is about addressing the complicated relation between man and nature. It is based on the ecosystem, respects ecological laws, and aims to meet people's diverse needs by realizing the various values of ecosystem. As an integral part of material, institutional and spiritual culture, it reflects the coordination between man and nature, man and man, man and society, and economy and environment, and advocates a healthy and advanced way of production, life and consumption."

Essentially, ecological culture reflects ecological productivity, crystallizes the progress of human civilization, provides the intelligent and spiritual drive to push the society forward, and permeates all sectors of the social ecology. In the meantime, its essential trait is determined by man's natural and social traits.

Ecological civilization is the summation of all positive results on material, spiritual and institutional fronts aimed to improve the man-nature relation and establish a scientific ecological functioning mechanism and sound ecological environment. According to the vision of ecological civilization, both man and nature are main entities and have values; not only man, but all living objects rely on nature. As a new vision, ecological civilization is a civilization paradigm proposed by man, after reflecting on the defects of industrial civilization, to achieve coordinated development among population, resources and environment. The center of ecological civilization is the implementation of the value of man-nature harmony in economic and social development and the manifestation of its results. It advocates respecting and protecting nature, following its laws and rationally utilizing it, and taking the initiative to promote ecological civilization, so as to sustain the development of human society in sound ecology and harmony with nature.

Ecological culture includes such topics as ecological values, ecological ethics, ecological spirit and ecological aesthetics.

On the structural level, ecological culture consists of material culture, spiritual culture, behavioral culture and institutional culture.

Ecological material culture, or the material reflection of ecological culture, refers to the material results of human activities affecting the ecosystem. It includes natural ecosystem and artificial ecosystem, such as forests, wetlands, grasslands, natural reserves, urban ecosystems, green industrial systems, ecological material products, and various ecological cultural carriers.

Ecological spiritual culture, or the spiritual core of ecological culture, sums up the human knowledge, sentiment and attitude toward ecology. It includes ecological philosophy, ecological natural sciences and social sciences.

Ecological behavioral culture, which refers to the process of man imposing impacts on the ecosystem with various tools, is the generalization of human behaviors, practices and activities that have an impact on ecology. It includes afforestation, ecological business, industrial development, ecological trade and tourism, and consumption of ecological products.

Ecological institutional culture refers to laws, regulations, policies and systems related with ecology. It reflects the human values and institutional needs in different periods.

Cultural ecology views man's relation with nature and social environment from an ecological perspective. In the ecological system, the change of environmental factors in the biosphere and non-biosphere depends on cultural factors of mankind, especially the scientific level, technology and equipment involved in the exploitation of natural resources. Meanwhile, environmental evolution also exerts inestimable effects on man's cultural development. The anthropologist Steward first put forth the concept of "cultural ecology" in 1954, which studies the process of how man adapts to the environment, or in other words, it studies the generation, development and change of culture by analyzing the interactions of various factors in the natural and social environments people live in. The environmental system is deeply affected by man's cultural and economic activities—they are closely connected in mutual restriction and common changes. Human activities have changed the original state of ecological environment by, for instance, building dams on rivers, reclaiming farmland around lakes and creating land by filling seas. Ecological evolution is deeply imprinted with the footprint of human activities.

From a cultural perspective, environmental problems are deeply linked with people's cultural values, involving their ideas, behaviors as well as cultural ecology and ecological culture. The ecological environment, a natural system that man relies on, is the source of all materials we need for survival. It isn't an external system outside human life, but an indispensable source of resources to support our existence, life, production and development. Man is restrained by environmental rules and all human activities must be within the bounds of environmental bearing capacity. Civilization prospers when ecology prospers; civilization declines when ecology declines. When man lives in a cultured manner, the culture should not be anti-nature. The dialectical unity between culture and nature is the essence of human existence. Man is restricted by natural rules, and man's right to enjoy a good material life and pursue freedom and happiness has to be within the bearing capacity of the ecological environment.

The ecological crisis has exposed the dilemma of human existence and human culture. Only by taking concrete steps to reshape the cultural values and change people's ideas can we overcome the ecological crisis, safeguard ecological security and achieve man-nature harmony. Man is at the same time the user, builder and manager of nature. To manage nature well, they should first manage themselves well. Man should effectively prevent ecological risks, so that people can eat and live in safety, maintain the pastoral scenery, and live in a comfortable environment.

Lucid waters and lush mountains not only provide man with material benefits, but also the resources of living, healthy conditions and spiritual pleasure. In Kubuqi, Inner Mongolia, more than 6000 square kilometers of sandy areas have been well harnessed, which now conserve more than 24 billion cubic meters of water resources and is valued at RMB 349 million for its role in biodiversity protection. The Saihanba National Forest Park, a forest converted from a desert, is the largest of its kind in northern China. Presenting the unique views of both a forest and a grassland, it is reputed as the "source of rivers, hometown of clouds, heaven of flowers, ocean of forests, and paradise for precious birds and beasts alike." In December 2017, Saihanba and Kubuqi were jointly conferred the Champions of the Earth award by the UN Environment Assembly. Some poor areas with favorable environmental resources can explore a path of poverty alleviation through ecological reform and innovation to revitalize the local factors such as land, labor force, assets and natural landscape, and consequently turn resources into assets, capital into equity, farmers to shareholders, and lucid waters and lush mountains into mountains of gold and silver.

Ecological culture is the basis to foster ecological civilization. Establishing, promoting and popularizing ecological culture will advance ecological civilization; in return, practices of building an ecological civilization will enrich the connotations of ecological culture. The idea of ecological culture and ecological civilization is deeply integrated with and comprehensively implemented in every aspect and the full process of economic, political, cultural and social advances. Ecological culture, as the mainstream culture in the age of ecological civilization, provides powerful spiritual drive, intelligent support, behavioral guide and institutional guarantee for ecological civilization.

2 Values of Ecological Civilization

The ecological cultural system follows ecological values as the norm.

The ecosystem has important values, both natural values and life values. Its value to mankind lies in supporting life, providing resources and interacting for symbiosis. Ecological culture, on the other hand, has spiritual and social values.

1. **Natural values**

Nature has value, which is created in the process of material production, defined as "stored achievements," and determined by the nature, structure and function of natural objects. Life and nature are both self-organized systems. The natural process is one of material production and incessant value creation. Man should care for nature and fully realize that it is far from being just an instrument employed by man, but has intrinsic values that don't hinge on human will. We should respect or hold in awe the intrinsic value and independence of nature.

Leopold, founder of ecological ethics, urged man to understand nature with a holistic and organic view and realize its integrity and internal connections. He proposed the standard of "land ethic" and asked man to look at nature and treat it kindly from a point beyond human interests.

Nature's value consists of two parts. The first is nature's usage to man, which is called its use value or external value. The other is the self-sufficiency of the natural world or ecosystem, namely the dynamic balance resulting from the mutual connections and utilization among natural objects, which is called nature's internal value. The second part emphasizes that nature's value is objective and the human dimension is the absolute or only dimension.

This shows that nature has two kinds of values. Its external value refers to the value of natural objects as tools, namely their meaning in meeting human needs. For instance, nature provides the resources to sustain the existence and development of human beings and other living things. Its internal value, on the other hand, refers to the value and meaning of the ecosystem not as a tool, but in its own right. The internal value of nature can be further divided into three levels. On the first level, with all kinds of beings in the ecosystem as objects, nature is valuable and meaningful for keeping the whole ecosystem stable, complete and in order. On the second level, different species in the ecosystem have a relation of values. On the third level, beings of the same species have a relation of values among themselves. Man has the obligation to protect and promote the beings with internal values, and to maintain and promote the integrity and stability of the ecosystem with internal values. On the practical level, nature's values are objective and subjective at the same time.

Famous American ecological ethicist Holmes Rolston said, "The creativity of natural system is the mother of value. All creations by nature only have value in the sense that they are the materialization of natural creativity. Value is wherever spontaneous creation is." Value is "such that it can create differences conducive to organisms, enriches the ecosystem and makes it more beautiful, diverse, harmonious and complicated." According to Rolston, nature evolves in the direction of value generation. We don't endow value on nature; nature gifts us with values. He categorized nature's values into a long list of types, including economic value, life-supporting value, recreational value, scientific value, aesthetic value, life value, value in diversity and uniformity, value in stability and spontaneity, dialectical value (of conflicts), and value of religious symbolism. According to him, the first five types can be grouped under the first and second category, which are subject to man's subjective judgment, while the last five types can be classified in the third category, which are determined by nature itself. "When nature is taking form, the outcome—in the form of the whole nature—is the source of value."

We must recognize the value of natural resources by putting a cost on them, implement the paid use of them and establish a sound system for that—this is an important measure to protect resources. "Oceans, lakes, mountains, forests and wet soils contain more values than individual animals," said Callicot.

In a traditional society, man needs nature more in the existential sense; but in the modern society, nature not only meets human needs for existence, but also for enjoyment and development.

Natural right refers to the right of natural objects or the natural right of beings. It is an intrinsic right of natural beings that exists according to ecological laws and is respected by man. Endowing or recognizing natural right isn't to increase nature's obligations to man, but the other way round – increasing man's obligations to nature, for the purpose of protecting the environment.

"Heaven and earth have no heart. Man is their heart." Man is the conscience of nature. To protect our home on earth and protect biodiversity, man should respect natural rights and affirm that the rights of all living beings are inviolable.

2. Supporting life

Man relies on the ecosystem for survival, including air, water and food, among others. Had there been no plants, insects or organisms on earth, mankind would have lived for only a few months at most.

Albert Schweitzer, the French philosopher and Nobel Peace Prize winner, put forth the "ethic of reverence for life" in his *The Philosophy of Civilization*, and further elaborated on the idea in his *Reverence for Life*. He held life as a great creation of nature, saying "life is sacred," calling for utmost respect for all lives and advocating "worship for life" as the core and basic principle of ethics. "Protecting, improving and developing life" is an important content of the "norms of man and nature" and of "good"; "good is protecting and developing life while evil is destroying and injuring it." "Reverence for life doesn't allow individuals to give up caring for the world. It urges individuals to interact with all lives around them and feel a responsibility for them. The development of lives that we can influence and our interactions with and responsibility for these lives should not be limited to maintaining and promoting their existence per se, but should extend to striving to realize their utmost values in any aspects."

3. Providing resources

The ecosystem provides unlimited resources for human production and life. Ecological factors as resources become the objects of labor and constitute the factors of production.

Man can exploit creatures as renewable resources, but within a certain limit.

Ecological tourism is very appealing as it meets people's ecological needs for clean water, fresh air and beautiful environment. Ecology therefore has the health-reinforcing and recuperative function for man. The natural ecosystem provides tourism, recreation and health services for man, having important values for improving their living standards and health and enriching their spiritual life.

Ecological activities have to comply with scientific ecological outlook and ecological laws and aim at the best "ecological benefits."

4. Symbiotic interactions

The famous American ecologist Eugene Odum pointed out, "the principle of ecosystem development bears significantly on the relation between man and nature. The policy of ecosystem development is obtaining 'maximal protection,' or maximal support for the complicated biological structure, while the purpose of man is 'maximal production' or the highest output possible. Conflict often arises between the two." Sometimes man has to make concessions or comprises, namely adjusting their behaviors. Human activities cannot undermine the basic ecological processes on earth. Many values in nature are irreplaceable, without which man cannot survive. For instance, photosynthesis of green plants and the material circulation, transformation and regeneration in ecosystem starting from that turn solar energy into an effective energy for earth, which is the condition to maintain lives on the planet and the basis and prerequisite for human creation of cultural values.

5. **Spiritual and cultural values**

Ecological culture offers a spiritual home for man, continuously elevates their realm and urges them toward a new civilization.

The development of human society is a process in which we use natural values to create and realize cultural values. Natural ecosystem takes effects spontaneously; human labor turns natural values into cultural values; and culture is the representation of human power. Human culture is intelligent in that it turns natural nature into artificial nature and realizes cultural values from natural values.

3 Manifestation of Ecological Wisdom

Material resources in the ecosystem can be continuously used through circulation, transformation and regeneration. The feedback resulting from the interrelations and interactions of various factors in the ecosystem is coordinated and stable on the one hand, and efficient and consistently evolutionary on the other. This is "ecosophy."

Arne Naess said, "What we need today is an extremely expansive ecological mindset that I call ecosophy. 'Sophy' comes from the Greek term 'sophia,' or 'wisdom,' which relates to ethics, norms, rules, and practices. Ecosophy, or deep ecology, involves a shift from science to wisdom. It is a philosophy of wisdom."

Ecological culture reflects a kind of wisdom that has the following core ideas: universal connection and harmonious co-existence; diversity, peace, balance and inclusiveness; mutual respect, equality and interaction; perpetual cycle of life, truth, good, beauty and sanctity.

1. **Universal connection and harmonious co-existence**

Mankind and all things in the ecosystem are universally connected. They live in harmony together and evolve together in the perpetual cycle of life.

American ecologist Commoner proposed the primary principle of ecology: Everything is connected to everything else. This is the principle of ecological connection. An individual existence would lose its vitality if it is not connected with the system.

According to American scholar Carolyn Merchant, "The premise of ecology is that all things in nature are interconnected. It stresses the interactions in nature as paramount. Every part communicates and interacts with other parts and the whole. Every part and every small environment in the ecological community is in dynamic connection with the surrounding ecosystems. Organisms in any specific small environment would affect and at the same time be affected by the whole network comprising both living and non-living environments. As a philosophy of nature, ecology is rooted in the organic theory, holding that the universe is an organic whole, its growth and development rely on its internal force, and it is the unity of structure and function."

It is necessary to adopt the holistic and organic mindset and stress holistic development. "All organisms in an ecosystem are 'interdependent'... Such integral and organic feature of ecosystem maintains the commonality and complementarity of human aesthetics and guard the peculiarity and differentiation of various species."

Nature is man's "inorganic body." Man relies on nature for existence. On the one hand, man acts on nature, changes it and humanizes it. On the other hand, nature acts on man, makes them learn its "wisdom," improves their quality and essential strength, and naturalizes them. This is a unitary process.

2. **Diversity, peace, balance and inclusiveness**

The ecosystem features diversity, so does human culture. They can accommodate each other.

Biodiversity includes the diversity of ecosystems, species and heredity. It plays an irreplaceable role in maintaining eco-balance and protecting the ecological environment. The diversity of ecosystem refers to the totality of different ecological environments, biocenoses and bio-ecological processes. There are 27 categories of ecosystems with a total of 460 types on land alone.

Although biological beings and ecosystems are of different organizational levels, they cannot be separated, nor can they exist independent of each other.

The second principle proposed by Commoner is: Nature knows best—this is the principle of ecosophy. In ecological evolution, "those possible arrangements that cannot co-exist in totality would be phased out during the long process of evolution. As such, an existing biological structure, or a known structure of natural ecosystem, is, as per common sense, the 'best.'"

Arne Naess said, "What we need today is an extremely expansive ecological mindset that I call ecosophy. Ecosophy, or deep ecology, involves a shift from science to wisdom." Naess is an advocate of cultural pluralism and ideological pluralism, which he believes can enhance the movement of deep ecology as pluralism means inclusiveness.

Without diversity, individual parts cannot form an entity capable of growth, development, self-repair and self-creation; without integration, different components cannot combine into a dynamic functional structure.

Ecosystem is in an orderly structure, in which the interactions among different organizational levels demonstrate its functional order.

3. **Mutual respect, equality and interaction**

Nature and man are equal in the ecosystem. They act upon each other. The rights of nature are equal.

Man should respect nature and show it necessary reverence instead of plundering and destroying it. Man should not only develop and utilize nature, but also protect and build the natural ecology. We should abandon the idea of man controlling and dominating nature, and realize that man and everything in nature are equal and must co-exist in harmony.

The third ecological principle proposed by Commoner is: there is no such thing as a free lunch, which is the principle of "ecological cost." "This warns people that every acquisition comes at a cost because the ecosystem on earth is an interconnected system, in which nothing can just be obtained or lost. It is subject to all measures of improvement. Anything extracted from this system through human power must be put back. A cost for that is inevitable, but may be delayed. The environmental crisis today is warning us that we have owed our debts for too long."

"Life means relying on the surrounding environment at all times," said Edgar Morin. Adjustment and restriction constitute the "living wisdom" of all species in nature. Adaption is the selective reaction by creatures to changes in the environment, whereby the system is adjusted to the expected state and ecosystem becomes an adaptive system that performs both adjusting and restrictive functions.

The future of ecology depends on the weakening of control. We must abandon the idea that ecology is independent and isolated. Ecological activities are part of human activities and they reflect the relation between human society and nature and the mutual effects between man and ecology.

4. Perpetual cycle of life, truth, good, beauty and sanctity

The development of ecosystem is a natural and historical process. Ecology evolves in time and is distributed in a certain space, which embodies its temporal and spatial orderliness as well as time-space unity. Everything must go somewhere—which is the fourth ecological principle proposed by Commoner, namely the law of the conservation of matter.

Ecosystem itself is characterized by truth, good, beauty and sanctity. When we look at the ecosystem from a cultural perspective, those are the goals of an ecological culture.

(1) Search for truth

The co-existence between man and ecological environment is a real state.

Rolston said, "A real person has to live in an environment. A person needs a place to live, a base to create values—a need that has both non-moral and moral elements. Either from the biological point of view or the perspective of material need, there would not be human life had there not been a world full of resources and ecosystems... People cannot become free away from their environment. They can only obtain freedom in their environment and would lose a lot of marvelous values of nature unless they observe nature's rules all the time."

(2) Inclination for good

The idea of man's righteous actions should expand to include care for nature and ecology, the moral concept of rights should expand to cover matters and processes in nature, and the domain of moral actions should expand from between man and man to between man and nature.

Good is preserving, promoting and developing life. As Albert Schweitzer said, "It is good to maintain and further life; it is bad to damage and destroy life. We would be considered ethical if we abandon our bias and our estrangement from other beings, and share with living creatures around us. Only that makes us real humans; only that gives us a special and clearly oriented character that will never be lost and will keep evolving."

(3) Pursuit of beauty

Ecological beauty refers to the adaptation and harmony between creatures and environment in the ecosystem and the unlimited diversity of their relations. The beauty of ecosystem includes that in form, color, sound, charm, and atmosphere. The aesthetic value of ecology is reflected in paintings, decorations, poetry, novels and other artistic works of different periods with nature and ecology as the motif, which inject fresh vitality into human art and literature.

(4) Attainment of sanctity

American scholar John B. Cobb, Jr. pointed out in his article *Ecology, Science, and Religion: Toward a Postmodern Worldview* that in postmodern worldview, adventures in the universe have inherent connections with the world. In other words, the adventures are comprised of all kinds of relations with the activities and values in the world. The greater the value, the richer the sacred life.

According to American theologist Thomas Berry, the modern society needs a fundamental transformation in its religious and philosophical pattern, from the reality and value of human-centrism to that of bio-centrism or eco-centrism.

The unity of truth, good, beauty and sanctity is the harmony between man and nature. That represents the true realm of ecological civilization.

The relation between man and nature restricts the relation between man and man, man and society. Facing the grim reality of tightening resource constraints, grave environmental pollution and ecosystem degradation, we must adhere to green development and the vision of ecological civilization that calls for respecting, accommodating and protecting nature. Efforts for ecological civilization should be given priority and incorporated in all aspects and the full process of economic, political, cultural and social development, with the ultimate goal of building a beautiful China and achieving lasting development of the Chinese nation.

Thoughts on the Philosophical Foundation of Ecological Civilization

Chun Xu

1 Conception of Nature Is the Logical Precondition for the Study of Ecological Civilization Theories

The conception of nature is the foundation of philosophy in every age. We cannot study ecological civilization without discussing the conception of nature, which is the theoretical and logical precondition. People's conception of nature is changing with the change of their ability to understand and change nature. This change, which is driven by scientific and technological progress and by the inheritance and development of academic thoughts, should be studied on the theoretical level.

Since the birth of human civilization, humans have first and foremost engaged in agricultural production in utilizing and reforming nature. Ancient civilizations in both the East and West were characterized by agricultural economy as the fundamental sector. In comparison, the agricultural civilization in ancient China, for a very long period of time, was in a leading position of human evolution. It gave birth to the idea of "man and nature in one" that advocated strictly following natural laws and opposed blindly interfering in nature. At that time, the mainstream philosophy on the relation between man and nature was that man should adapt to nature. The generation of such a simple ecological philosophy was attributed to the fact that agricultural production relied completely and fully on nature.

In western culture, before the time of ancient Greece, people generally held a mythological view of nature and animism. In the sixth century BC, ancient Greek philosophy came into being, which was penetrated from the beginning to the end by the study of nature. The Greeks had the valuable instinct that world was a living, mythical and continuously growing organic entity from which all matters and creatures grew. The Greek ideology and the *Bible* of Christianity were the main thoughts that influenced natural philosophy in the Middle Ages. In that period, Christianity dominated the ideological world in Western Europe, and philosophy, as it was employed to

C. Xu (✉)
Peking University, Beijing, China

J. Pan et al. (eds.), *Beautiful China: 70 Years Since 1949 and 70 People's Views on Eco-civilization Construction*, https://doi.org/10.1007/978-981-33-6742-5_59

verify the doctrines and convictions of Christianity, was at first equated to Christian mythology and later separated from it. In this process, reasoning gradually shifted from religious belief to nature and sentimental existence in the empirical reality, and philosophy that was once mingled with Christian mythology, by obtaining new contents and methods, evolved into natural philosophy and provided the ideological nest for the emergence of modern science. To fully understand the emergence and development of natural sciences in modern times and the formation of modern conception of nature, the ancient age and the Middle Ages should and could not be overlooked. Ancient Greece and the Middle Ages represented two completely different directions, but it was in the latter period when ideas of both periods were at work that the seed of science was planted.

Around the sixteenth to seventeenth century, a new conception of nature as opposed to the Greek one began to thrive and quickly replaced the latter to become dominant. The mechanical conception of nature viewed nature not as an organic entity, but as a machine that, like a watch, a clock or a watermill, had no life or vitality, no sense or reason, and no independent movement or change. The difference was that watch, clock and watermill were invented by man, but nature by God. God invents a set of principles that were applied to manipulate nature's movement, thus turning nature into a passive and controlled machine. As such, the idea of "control of nature" that could trace back to the Jewish-Christian tradition and began to emerge in the late period of the Reconnaissance finally acquired a complete philosophical form through the wonderful combination of mechanism and rationalism.

The concept of natural teleology foreboded the disintegration of the mechanical conception of nature. Teleology is a core issue in western philosophy that involves the explanation and criticism of "purpose" and "purposiveness." Teleology had its inception in the pre-Socrates period. Although it was seriously attacked in the 15th to eighteenth century, teleology didn't go extinct but had been lurking in the philosophies of Descartes and others till it eventually came to light again through Kant's philosophy. The teleology in ancient Greece was natural teleology, which held concepts as the highest guidance on all beings. Other than that, there was also the theological teleology in the Middle Ages. In the eighteenth century, when natural sciences theories were unable to explain the self-discipline of life and uniformity of nature, Kant proposed the natural teleology. To make up for the deficiency of mechanical causality, Kant, when explaining the natural system where living organisms were in harmony and uniformity, introduced the concept of natural teleology different from theological teleology and mechanical causality. According to him, organic beings could be considered as having "intrinsic purpose," which was different from the relative purpose or extrinsic purpose measured by empirical judgment. A natural object can be considered to have an intrinsic purpose or a natural purpose because unlike a sheer machine or a man-made apparatus, it is an "organic being with its own organization." Hegel continued Kant's view on organic life and saw a complementary relation between life and purpose. Hegel's natural philosophy was an overview of the natural sciences achievements made in early nineteenth century, and he regarded nature's development from the lowest to the highest stage as an inevitable process. Hegel deemed this kind of inevitability as absolute purposiveness. If we don't sort through the tradition of teleology in western philosophy, it would be hard to understand Rolston's arguments about nature's intrinsic values.

The twentieth century has witnessed dramatic changes in scientific ideas. In mid-twentieth century, the systematic scientific movement motivated by a series of new achievements in modern natural sciences and dominated by three laterally crossover new disciplines, namely information theory, control theory and system theory, presented a natural vista of how the world self-organization has evolved. At that time, evolution captured the attention of more philosophers, leading to Henri Bergson's *Creative Evolution* and Whitehead's process philosophy. The rise and development of modern sciences has eroded the foundation of the mechanical world-view and formed a new scientific one and a new explanatory framework. In a sense the new worldview can be called an "organic" one. While the organic theory in ancient Greece was no more than an intuitive guess and philosophical speculation and that in the nineteenth century was just a feature of a philosophical section, the modern organic worldview is established on the basis of modern sciences including physics, biology, system theory, control theory and information theory. It is the generalization of modern scientific worldview and it lays the natural conception foundation for ecological civilization.

2 Values Are Important for Selecting a Way of Life Conforming to Ecological Civilization

People's evaluation system and standards for nature are changing as their understanding of natural values keeps deepening and expanding. In the agricultural society, nature was a direct existence and direct source of living resources, so it was easy to form intuitive and primitive natural responsibility ethics. In modern times, however, with the continuous process of industrialization, nature has gradually been reduced into an indirect existence, tool and means through the increasingly advanced scientific and technological approaches. Human-centrism has remained a mainstream ideology in the West, which, by viewing nature as a tool and taking mankind as the only center in value judgment, has led to serious damages of natural environment. The damages, affecting every aspect of natural environment from organic biosphere to inorganic sphere, have generated consequences that man has found hard to bear as time goes by. Against such a background, as people with vision began to reflect on human-centrism, environmental ethics have kept evolving, improving and becoming more systematic as people are increasingly aware of environmental protection, thus laying the value foundation for ecological civilization.

In traditional Chinese culture, "man and nature in one" is the basic spirit of Confucianism and it contains some very important environmental ethics. By putting forth "man and nature in one," the Confucian school, on the one hand, wanted to argue for an ideal interpersonal relation that established a universal human ethics following the virtue of heaven. On the other hand, it intended to create the ethics of man relying on and adapting to nature, so as to achieve sustained and stable economic development in the agricultural society. From Confucius advocating "awe for nature" to Zhang Zai recognizing nature's intrinsic values and calling for the fulfillment of responsibility for nature, a natural and logical result is what we call the environmental or ecological ethics today. Therefore, we should fully absorb the essence of traditional Chinese ecological culture and adapt it to modern times.

In traditional western philosophy and ethics, "value" is an exclusive human trait. Only man has value (intrinsic value) and natural beings only have meaning when they are relevant to man's subjective purpose and only have value (instrumental value) when they have the usages and functions to meet human needs. Because of these traditional values, what people see is only nature's instrumental value and short-term practical value, a situation that has led to the dire consequence of unlimited conquest of nature. Serious environmental crisis has prompted profound changes in human values and forced man to contemplate the significance of nature's intrinsic values for mankind.

In early twentieth century, Albert Schweitzer, through the proposition of "reverence for life," expanded the scope of traditional ethics beyond "human-centrism" that had dominated the history of thoughts. In Schweitzer's opinion, all past ethics were incomplete. In the eyes of traditional ethicists, man and other being were classified, but any attempt to differentiate them, in whichever way, would unavoidably highlight man's superiority over other beings. Therefore, traditional western ethics shaped the morality, law and way of life between man and man, leaving no space for other beings. The interpersonal relation could be restricted by morality or law, but once the relation extended to man and fauna or flora, man's value became the yardstick to gauge other beings. Since man was seen as an advanced species, its trample on and destruction of other species was taken for granted, and its activities harming other lives became a kind of collective unconsciousness. Schweitzer put all living things in the scope of ethics and tried to create an ethical view that pursued harmony between man and other beings, namely reverence for life. According to this principle, man is not the dictator of other beings and is in no position to harm them either. Man cannot gauge other species based on their own standards, but should take the initiative to feel and help other lives. The rules of nature pointed out by Schweitzer indicate that the natural state we are in is a dynamic balance in which all species affect each other. Man is powerful in comparison with other species, which means that it is not only able to seriously damage nature, but also to protect it to the largest extent. If man could fulfill their responsibilities for other beings, all beings could sustain, making it possible for the ecological ethics that man and nature co-exist in harmony to become a reality. The "reverence for life" created by Schweitzer is an important ideological source of modern ecological ethics.

In the mid-twentieth century, the land ethic proposed by famous American ecologist Leopold broke the boundary of traditional ethics and viewed man and nature as a whole ethical world. Leopold's land ethic holds that man, other organic living things than man and inorganic beings on earth jointly constitute a new land community, in which all members are closely related in reciprocal causation. Although man has the ability of reasoning, the interdependence among different species decides that man cannot continue to enjoy the privilege of plundering land resources out of individual interests without performing any obligation. They should change from the role of a land conqueror to a common member of the land community. Keeping the land community complete, stable and beautiful is the basic moral principle of human activities. Leopold tried to inspire people's love and gratitude to land and motivate them to voluntarily take up the moral responsibility for it, thus establishing the value of man and nature in harmony and uniformity. His land ethic kicked off a critical period of the change of environmental ethics as Leopold, from an ethical point of view, extended morality to the whole natural domain for the first time. Even though land ethic, in comparison with rigorous ethics and philosophy, is more like a guideline on environmentalist movement, Leopold is nevertheless known as the "father of ecological ethics."

Rolston in the 1980s carried on and developed Kant's natural teleology and Schweitzer and Leopold's ecological ethics, and systematically elaborated on nature's intrinsic values. He is the founder of modern environmental ethics. According to Rolston, the intrinsic value of nature is the value inherent in some natural scenarios that doesn't need human reference. All living creatures take their own kind as good, which means all living things proactively defend their lives and strive to propagate their species. In the natural ecosystem, the instrumental and intrinsic values of different species transform and blend, and each living creature has an intrinsic purposiveness. The emphasis on nature's intrinsic value is the most basic differentiation of value, and it doesn't negate nature's instrumental value. Rolston expounded on nature's intrinsic values to clarify two things. First, nature's intrinsic value refers to the inherent purposiveness of ecosystem, which is valuable and meaningful for keeping the ecosystem stable, complete and orderly. It can be divided into three levels at least—(1) All existences in the ecosystem, including man, have to observe its rules and norms; (2) Different species in the ecosystem are in a value relation. They are the subjects and objects, and purposes and means of each other, and they satisfy each other's needs while restricting each other as well; (3) There is also a value relation within the same species, such as that between man and man or within a natural species. Second, the value relation between man, the subject, and natural beings, the object, is just one form of value relation. It is not the only form, much less the most important form in the ecosystem. Man is not the only subject in value relation, nor are human rules the ultimate norm for value assessment. Rather, man in a sense has to follow natural rules.

Ecological ethics and values can only affect our living practices after they become "common sense" in life, when they will urge us to rebuild the value coordinates and change our way of life to comply with ecological civilization. To that end, we have to change the current lifestyle. First, we should replace material value with the value of knowledge and wisdom. Instead of marking personal value and success by extracting more material wealth from nature, we should pursue the harmonious development and common evolution, or the "win-win," of man and nature together. Second, we should replace excessive consumption with moderate consumption. The excessive consumption for "flaunting status" or "vanity" largely exceeds the basic needs of survival. It not only deprives other social members of the necessities they live on, but is also a direct source of resource and environmental destruction. Third, we should replace extravagance with simplicity. A simple life is one of moderate consumption centered on improving the quality of life. A simple life aims to meet basic needs rather than pursue luxury and extravagance. A greater diversity of products is produced to meet personalized needs and preferences, convey more knowledge and intelligence and generate less damages to natural environment. It must be noted that a pro-environment lifestyle is richer and more advanced than the traditional one featured by high consumption. It is more consistent with nature, with human needs and with sustainable development of the human society. As long as we can treat nature in a civilized way, we are sure to choose this lifestyle. The whole society should make concerted efforts to establish and implement ecological civilization both in theory and practice—that is the only way for us to build a beautiful China.

The Brilliant Ecological Wisdom

Chengfeng Li

Walking in the villages of Hani ethnic group, I felt like being in a landscape painting of mountains, waters, forests, farmland and lakes. Upon the mountains were lush forests, down in the valley were layers after layers of beautiful terraces. The perennial gurgling stream strung the forests, villages, terraces and rivers perfectly together. It is amid such a beautiful environment that the Hani ethnic group has passed thousands of years. Taking advantage of natural conditions and endowments and fully exerting their intelligence and persistence, the Hani people have written their own culture and history marking the transformation from a nomadic group to a farming one. They have also created the Hani terraces, a magnificent and ecological miracle of agricultural civilization.

1 Rising from Where It Vanished

Migration has been a heavy topic from time immemorial. To survive and sustain, mankind has either voluntarily or been forced to migrate, which involves great hardships and a low survival rate, but almost all successful migrations have worked miracles. In the Shang Dynasty, Pan Geng took the initiative to move the capital to Yin in order to mediate the relation between the aristocracy and the masses, which led to the miracle of the Shang Dynasty lasting more than 300 years and left us a brilliant case of anti-corruption work. In modern times, the CPC-led Chinese Workers' and Peasants' Red Army made strategic withdrawal to maintain revolutionary forces and defeat the KMT's encirclement and attacks, which led to the world-famous miracle

The article was originally published on *China Environment News* on March 6, 2019 and was slightly edited when included in this book.

C. Li (✉)
China Environmental Protection Foundation, Beijing, China

J. Pan et al. (eds.), *Beautiful China: 70 Years Since 1949 and 70 People's Views on Eco-civilization Construction*, https://doi.org/10.1007/978-981-33-6742-5_60

of the "Long March" and took the Chinese revolution from one victory to another. The Hani ethnic group has migrated to where they live now. It originated in the Qiang clan, surnamed Gu, living on the plateau in the juncture of Gansu, Sichuan and Tibet in the upper reaches of the Yangtze River and Yellow River. From the time of Yandi and Huangdi to the period of Spring and Autumn and the Warring States, constant wars forced different ethnic groups into constant confrontation and assimilation, the Qiang clan not being an exception. Part of the Qiang people migrated along the plateau valleys of the Nu River, Lancang River and Jinsha River to the warm south with abundant aquatic plants and dense forests. They passed the Sichuan basin to the Yunnan-Guizhou plateau and moved from Dali to Kunming. To survive and continue their genes, the Hani people buried the hatchet and expressed friendliness and kindness, enduring all kinds of hardships and humiliation to find a place to stay. Having rebuilt their homes and then been driven away for innumerable times, the Hani ethnic group eventually settled down by the Ailao Mountain.

Nobody knows on which day, in which year the Qiang group that walked into the dense forests could no longer be found, and a farming group called "Heyi" began to rise where the Qiang group vanished. Layers after layers of terraces stretched down to the mountain foot, while bundles after bundles of firewood was carried onto the mountain—this is how the Heyi group has lived from generation to generation for thousands of years, with their life closely linked with the terraces. In their own words, the terrace represents the Hani people and vice versa. The farm ridges are like their bones, the red soil and flowing streams their blood and flesh, and the waving rice and swimming crucians their soul.

Migration isn't despair, much less death. It is the resurrection of life and the glory of miracles. That is how Hani people have lived. Based on generations of hard labor and intelligent creations, they have turned a rice-planting civilization in the plain region into a fabulous and splendid miracle of terrace farming culture that has amazed the world for over a thousand years.

2 The Ecological Wisdom of "Iso-Structured Four Elements"

Water is the theme here. The terrace in the morning of late winter and early spring is characterized by fresh and moist air, gurgling water and fast-changing light and shadow. Forests, villages, terraces and rivers all in the same picture create an environment that ecologists call "iso-structured four elements." Water flows from mountains and forests to nourish the villagers, and is then used to irrigate the terraces and diverted along them into rivers; then water in the rivers, due to the subtropical climate, evaporates and condenses and falls again on the forests—the cycle goes on. As a result, water becomes the bond that ties the "four elements" together.

Many years ago, the Hani ancestors were forced to stop in a mountainous area at the elevation of 800–2500 meters because they had no other choice. Even if you drive up the mountain today, the mountain road still has a lot of bends, some of which are steep S-shaped ones. However, the Hani people wouldn't have worked the wonder today without the "choice of no choice" back then. When their ancestors walked deep into the forests to work on nature and fight for survival, they learned how to make use of nature. This reminds me of the story of Yu Gong, the foolish old man who removed the mountains recorded in the *Book of Mountains and Seas*. Were the Hani people inspired by him in any way? Both Yu Gong and Hani people are upbeat and optimistic with the perseverance and philosophical wisdom to overcome difficulties. The difference is that Yu Gong chose to move the obstacle away while Hani people chose to use it. From the modern environmental perspective, a massive project like Yu Gong's moving of mountain can hardly pass the environmental impact assessment, but the "Hani terrace" project, which is equally massive but aimed at the harmonious co-existence between man and nature, would be more acceptable. It seems that the story of Yu Gong is destined to be a legend, but the Hani terrace is a real miracle on earth that has lasted for more than 1000 years, not only rearing generations of Hani people, but also passing down and carrying on a profound spirit of mountainous farming.

Mankind is always forced to take the initiative to fight for survival and the continuation of life. Such potential and wisdom stemming from human instinct has been driving social progress and pushing human civilization forward. Where there is demand, there is struggle; where there is struggle, there is creation; where there is creation, there is progress—that is how the wheels of history keep rolling forward, mostly toward civilization but sometimes toward barbarism. The natural space allows for creation, but some creations are based on adaptation and utilization while others cause pollution and destruction. As we are facing tremendous environmental pressure today, the ecological wisdom of Hani people should enlighten us on what we should and should not do, and what we can and cannot do. But our problem today is that we haven't employed our wisdom to achieve the harmonious co-existence with nature, but have wasted it on struggling and competing with man. The intelligent Chinese nation has created countless miracles in the world, and I expect more in the new era.

3 Beautiful Terrace—A Gift from Nature

The terraces are beautiful because they are natural and, more importantly, because of the children's pure heart. I saw a kindergarten by the Hani terrace, where a female teacher of the Yi ethnic minority was taking a group of Hani children to the field, laughing and romping to their heart's content under the sun. Under the guidance of the teacher, the children were creating paintings by arranging leaves on the paper. Watching them, I couldn't help but think of those children in cities who are "closed up" in kindergartens. They cannot run around or make noises, they have to be very careful and cautious, following the teacher sentence by sentence and drawing on paper what they imagine nature is like, everyone drawing the same, standardized sunshine, air, cars and buildings. I was told that the Hani children play the game of making terrace since they are young. Every boy, with a small hoe in hand, would reclaim terraces on the ground like adults do, following exactly the same procedures even if it is just a game. They dig a ditch first to divert water, then build layers of ridges to channel the water into small terraces. The girls, on the other hand, would each carry a small basket on their back and pretend catching eels and loaches in the completed terraces. Real beauty is never painted. The Hani terraces, as a paradigm of classic farming culture, are hidden among the mountains and forests as an integral part of nature—that is real beauty.

For thousands of years, the Hani people have "polished" and "chiseled" their terraces by hand, creating a peerless piece of land art with its lines, rhythms and pace all impregnated with both cultural and natural beauty. The Hani terraces bask in the shining sun during the day and sleep under the twinkling stars at night. It is a place where the sun, the moon and the stars all seem to be within reach, so beautiful that it is intoxicating and awe-inspiring. Like the "ladder to the sky," the terraces, big or small, high or low, snake up along the mountain layer upon layer and stretch from one mountain to another; along the way the farm ridges meander either into the villages or forests or rivers, but all toward the future. "There are four seasons in the same mountain; the weather changes every ten miles." The terraces in spring are like a green blanket, those in summer like billowing waves, those in autumn like a golden landscape painting, and those in winter like bright mirrors, which reflect the appealing mountains and forests, the agile clouds, and the eternity of nature.

Green farmlands, golden harvests and numerous ponds and lakes—these are just what meet the eyes. The thousand-year-long ecological wisdom is more vividly reflected in how the mountain wetland has nurtured and preserved natural ecology. When the Hani ancestors first arrived here, floods year after year caused serious water and soil losses. To protect their home, they began to build dams and farm ridges on different layers following the local landform to mitigate the impacts of flood and make up for nature's deficiencies. As a result, arable terraces gradually took shape and functioned as a vast artificial wetland on the mountain that could not only fend against floods and droughts, but also moderate the climate. Two years ago, Yunnan was hit by a rare drought that hasn't been seen in a century, but the water-rich Hani terrace wasn't affected at all. In particular, the genes of the red rice growing on the terraces haven't deteriorated in the past over a thousand years, and it is a real green food because it is so sensitive to fertilizers that the application of fertilizers would cause rice blast.

Therefore, the beauty of terraces doesn't stay on the surface, but lies in its ecological wisdom; doesn't stop at harvest, but lies in its harmony with nature. The beautiful terraces have shown us how to let life and nature thrive together into eternity.

We celebrate the Hani terrace because it has followed nature's course and reared generations of Hani people; we celebrate the Hani people because they have worked miracles and contributed the best ecological wisdom to the world.

4 An Eternal Ode

"Like dogs digging holes, you cannot be afraid of slurry in farming; like a cattle plowing the farm, you cannot complain about exhaustion and hardship in working." This is the song that Hani people sing when transplanting rice seedlings and they sing it every year during spring farming. The clear and crisp song flows in the air far and wide, not only expressing the farmers' high spirit and energy, but also conveying their aspiration for a bumper harvest and a happy life. It is their work song.

However, Hani people of the young generation have walked out of their villages into cities over recent years and merged into the modern urban life. They leave the elders and children at home, like in some mountainous villages, and would only come back on festivals, during the Spring Festival or in the farming season. Perhaps they are too accustomed to the beauty of the terraces to be as shocked and amazed as visitors like us. Young Hani people no longer wear dark blue blouses and wide-legged trousers, but prefer trendy shirts and close-fitting jeans. It is the elders that are guarding and inheriting the terraces. Wearing wide pants and holding hoes in hand, they are still mending the farm ridges while aged women, carrying baskets on their back, are either picking loaches or mending the ridges because some ridges are so fragile that a break of them would lead to chain effects, especially in the rain season. The thousand-year-long farming civilization is in danger of having no successors and inheritors.

This reminds me of the speech given by General Secretary Xi Jinping at the central rural work meeting in 2013. "At the end of the day, the key to rural economic and social development lies in people. Without people, without labor force, we cannot achieve food security, modern agriculture or the construction of a new countryside, and our traditional farming culture cannot be well protected and inherited. The farming culture is a valuable fortune in our agriculture and an important component of Chinese culture. It should not be forsaken, but should be carried forward and promoted. If people all leave the villages, who will inherit and carry on the farming culture? I heard that where the Hani rice is planted in Yunnan, the number of people who can sing ancient Hani songs like *Hani Work Song all Year Long* and dance to Hani music is shrinking. We cannot lose what our ancestors have left us in the name of modernization!"

Do not make light of the *Hani Work Song all Year Long*. It relates a complete system of folk culture and knowledge about agricultural production ranging from natural landscape, fauna and flora, production and life to experience and wisdom. The song consists of five parts—introduction, winter, spring, summer and autumn. The "introduction" emphasizes how important the all-year-long work song passed down from previous generations is for the Hani ethnic group, while the other four parts recount in seasonal order the procedures and technical essentials for terrace farming as well as the associated astronomical and calendar knowledge, patterns of natural and climatic changes, festivals and worship ceremonies, and ritual norms. These contents are usually taught through oral instructions and personal examples between master and apprentice, father and son, or mother and daughter, either by the terrace, by the fireplace, on the drinking table or in the dyeing workshop. Year after year, generation after generation, the farming production and life of Hani ethnic group have been passed on for thousands of years through the cycle of seasons, along with the rich knowledge associated with them.

To our relief, the Hani terrace has been listed as a world cultural heritage and a national protected site. A group of young people who used to work in cities have come back to the village to undertake new responsibilities and missions. In the country's tough battle of poverty alleviation, some of them have served as village secretaries or directors, and some of the younger generation are also planning to follow in their steps and come back to their hometown. This reminds me of the kindergarten children that I saw by the terrace. Where will they be when they grow up? Do they have to "get of the mountain" as some media claim? I must say that we urban residents must use our brain when trying to comment on the rural culture and what rural children should do.

"Cuckoo, cuckoo." Spring is here, and a cuckoo is flying over from far away. It is the season to sing the *Hani Work Song all Year Long* again, a song we look forward to lasting forever to continue the wisdom and miracles of the Hani people.

We hail Hani people's ecological wisdom and should learn from it. Guided by the Xi Jinping Thought on Ecological Civilization, let's create wisdom and miracles in all sectors of our big family of all ethnic groups.

Furthering the Reform of the System and the Set of Institutions for Developing an Ecological Civilization

On Xi Jinping's Important Statements on Reform of the System for Developing an Ecological Civilization

Manhong Shen

Xi Jinping's important statements on the reform of the system for developing an ecological civilization form an important part of Xi Jinping Thought on Ecological Civilization, and serve as the fundamental guideline for ecological progress toward building a beautiful China. Therefore, studying the evolution, basic content, main features and historical contributions of Xi Jinping's important statements on the reform of the system for developing an ecological civilization has important practical significance and theoretical value.

1 The Basic Content of Xi Jinping's Important Statements on the Reform of the System for Developing an Ecological Civilization

These statements form an ideological system that mainly includes three parts: the system reform, the mechanism design, and the construction of institutions for promoting ecological progress.

1. **The reform of the system for developing an ecological civilization**

(1) The system approach is adopted to reform administrative departments. The ecological environment is a complex system. It can be divided into such subsystems as the forest ecosystem, wetland ecosystem, lake ecosystem, marine ecosystem, urban ecosystem, and rural ecosystem, but it is not simply the sum of all the subsystems. Therefore, General Secretary Xi Jinping pointed out in the "Explanation" at the Third Plenary Session of the 18th CPC Central Committee that: "Mountains, waters, forests, farmlands and lakes form a living community," and "it is necessary to have one department take charge of all land

M. Shen (✉)
Ningbo University, Ningbo, China

J. Pan et al. (eds.), *Beautiful China: 70 Years Since 1949 and 70 People's Views on Eco-civilization Construction*, https://doi.org/10.1007/978-981-33-6742-5_61

and space uses within the territory, and centralized protection and restoration of mountains, waters, forests, farmlands and lakes." This pointed out the direction for the institutional reform of the State Council. The most eye-catching part of the State Council's latest institutional reform plan involved the Ministry of Natural Resources, the Ministry of Ecology and Environment, and the National Forestry and Grassland Administration. The reformed Ministry of Natural Resources integrated the functions of eight departments, and the Ministry of Ecology and Environment and the National Forestry and Grassland Administration each seven departments. From the perspective of the ecological civilization system, the institutional reform of the State Council has at least resolved to a considerable extent the long-standing problem of decentralization of power for environmental protection.

(2) The accounting system is reformed in accordance with the understanding that "lucid waters and lush mountains are invaluable assets." Promoting economic and social development and strengthening ecological and environmental protection both aim to improve human well-being. Therefore, generally speaking, environmental protection, like economic development, is considered in the performance assessment of government officials. But environmental benefits are slower to come by than economic benefits. That's why government departments are more interested in attracting investment than in protecting the environment. In this case, from a static point of view, the comprehensive benefits, i.e. the sum of ecological and economic benefits, may be increasing while the ecological benefits are decreasing, which means that economic development is at the expense of ecological benefits and thus unsustainable. Therefore, it is necessary to adopt the dynamic perspective and strive to maximize the comprehensive benefits. During his tenure in Zhejiang, Comrade Xi Jinping divided officials' understanding of the relationship between economic development and environmental protection into three stages. First, only economic benefits matter; second, both economic benefits and ecological benefits matter; and third, lucid waters and lush mountains are invaluable assets. The most important is the statement that "lucid waters and lush mountains are invaluable assets." After assuming office as the President of the People's Republic of China and General Secretary of CPC Central Committee, Xi clearly put forward three important conclusions: "We want both ecological and economic benefits, ecological benefits are more important than economic benefits, and lucid waters and lush mountains are invaluable assets." Following Xi's instructions, China has vigorously reformed and explored the resource tax, environmental tax, carbon tax system, ecological compensation, circular subsidies, and low-carbon subsidies, and promoted pilot projects of the paid use and trading of water rights, pollution discharge rights, forest rights and carbon emission rights. The economic value of lucid waters and lush mountains is materializing in more and more sectors.

(3) The responsibility division between departments and localities at different levels is reformed according to the control theory. In ecological and environmental management, there exist prominent conflicts between different departments (such as between the forestry department and the water conservancy department, between the water conservancy department and the environmental protection department), between different localities (such as between upstream and downstream localities, and between localities on the left bank and the right bank of a river), and between departments and localities. In regional comprehensive planning, land planning, and spatial planning, it is common for Area A in the upstream area to plan its water source in its upstream, while the industrial area its downstream, which is followed suit by other areas B and C. In this way, A's industrial area becomes the upstream of B's water source, and B's industrial area becomes the upstream of C's water source, and so on and so forth. This highlights the importance to strengthen the coordination between departments, between localities, and between departments and localities. That's why Xi attaches great importance to space control and advance planning. The report to the 18th CPC National Congress calls for more efforts in ecological conservation, and the first step is to "optimize the pattern of land and space development." The *Decision of the CPC Central Committee and the State Council on Several Major Issues Regarding Comprehensively Deepening Reform* stresses the importance of "establishing a spatial planning system" and "defining the red line for ecological protection." The *Opinions of the CPC Central Committee and the State Council on Accelerating Ecological Progress* specifically elaborates on "strengthening the positioning of main functions and optimizing the development pattern of land and space." *The Master Plan for Reform of the System for Developing an Ecological Civilization* proposes the establishment of eight systems, two of which are about the use of space, namely "a system for the development and protection of land and space" and "a space planning system." It is precisely under the guidance of Xi Jinping's important statements on the reform of the system for developing an ecological civilization that China has gradually established the main function planning system, national park system, ecological red line system, and the system of replacing multiple plans with a master plan.

2. **The mechanism design for promoting ecological progress**

In ecological conservation, there are areas where the government mechanism, market mechanism, and/or social mechanism work, and where they fail, either individually or collectively. For this reason, the mechanism design must take into account the effective range of each mechanism on the one hand, and the checks and balances between different mechanisms on the other. Xi Jinping attaches great importance to the effective range of each mechanism.

(1) The government takes the initiative to take responsibility. At the National Conference on Eco-environmental Protection, Xi Jinping emphasized that the ecological environment is a major political issue concerning the Party's mission

and purpose, and a major social issue concerning people's well-being. The Chinese people are eagerly looking forward to more rapid improvement in the ecological environment. The Party and the government must actively respond to the people's wish, expectation, and concern, vigorously promote ecological progress, provide more high-quality ecological products, and continue to meet the people's growing needs for a beautiful ecological environment. Xi went further to emphasize that local Party and government leaders at all levels are the first responsible persons for ecological and environmental protection in their respective administrative region, and that all relevant departments must earnestly fulfill their respective responsibilities for ecological and environmental protection and work with each other to produce synergy. About 60% of the 47 systems proposed in the *Master Plan for Reform of the System for Developing an Ecological Civilization* are mainly dependent on the government performance. This shows Xi's sense of responsibility and the CPC's mission and spirit.

(2) Full play is given to the role of the market. *The Decision of the CPC Central Committee on Some Major Issues Concerning Comprehensively Deepening the Reform* proposes to "accelerate the price reform of natural resources and their products to fully reflect market supply and demand, resource scarcity, costs and restoration benefits of ecological environment damage… Whoever benefits shall make the compensation. We shall improve the ecological compensation mechanism for key ecological function areas, and promote the establishment of an inter-regional horizontal ecological compensation system. We shall develop the environmental protection market, promote the trading of energy savings, carbon emission rights, pollution discharge rights, and water rights, establish a market-based mechanism that channels social capital to ecological and environmental protection, and promote third-party pollution control." This fully shows that ecological conversation must give full play to not only the role of the green fiscal and taxation mechanism based on Pigou's theory, but also that of the green property rights mechanism based on the Coase's theory. Both approaches are typical of the market economy. In fact, to translate ecological benefits into economic benefits we must rely on the market mechanism, the core of which is the pricing mechanism.

(3) The social governance mechanism shall be constructed. If we compare the general public to the "client" and the environmental protection department the "agent," the agent is naturally subject to the supervision of the client. Therefore, the government must strengthen the disclosure of environmental information, and let the public know the environmental quality to take preventive measures, and wield their environmental rights to supervise the government's environmental protection behavior. The shift from "management" to "governance" of the ecological environment is an important part of the institutional reform for ecological conservation. We must avoid setting the government as the manager against the public as it used to be, and instead build a system of checks and balances in which the government mechanism, the market mechanism, and the social mechanism work together. This is the general trend for environmental

governance. In the report to the 19th CPC National Congress, Xi Jinping clearly proposed to "We will establish an environmental governance system in which government takes the lead, enterprises assume main responsibility, and social organizations and the public also participate." Environmental governance is not only an environmental issue, but also an economic, social, and even political one. Therefore, the construction of an environmental governance system is important for building a national governance system and modernizing national governance.

3. **The construction of institutions for promoting ecological progress**

In addition to system reform and mechanism design, Xi Jinping also attaches great importance to the construction of institutions for promoting ecological progress.

In the section "Enhance system building to promote ecological progress" in the report to the 18th CPC National Congress, it systematically elaborates five systems: the evaluation, reward and punishment system for ecological conservation, the strictest ecological and environmental protection system, the system for paying for resource consumption and compensating for ecological damage, the ecological and environmental supervision and accountability system, and the system for publicity of and education in ecological progress. They lay an important foundation for building a complete set of institutions for promoting ecological progress.

In the *Decision of the CPC Central Committee on Some Major Issues Concerning Comprehensively Deepening the Reform* adopted at the Third Plenary Session of the 18th CPC Central Committee, the Party, for the first time, in the section "Accelerating the system building for ecological conservation," made it clear that "a systematic and complete set of institutions for promoting ecological progress must be established," and in particular explained it in four aspects: improving the property rights system and use control system for natural resources, setting the red line for ecological conservation, enforcing the system for paying for resource consumption and compensating for ecological damage, and reforming the administrative system for ecological and environmental protection.

In the *Opinions of the CPC Central Committee and the State Council on Accelerating Ecological Progress*, it emphasizes "accelerating the establishment of a systematic and complete set of institutions for promoting ecological progress," and proposes ten specific measures: improving laws and regulations and the system of standards, bettering the property rights system and use control system for natural resources, enhancing the ecological and environmental supervision system, strictly respecting the redlines for ecological protection, refining economic policies, promoting market-oriented mechanisms, strengthening ecological compensation mechanism, and improving the performance evaluation system and the accountability system.

The Master Plan for Reform of the System for Developing an Ecological Civilization goes further to emphasize "accelerating the establishment of a systematic and complete set of institutions for promoting ecological progress." Specifically the designed set of institutions includes 47 institutions in eight aspects: the natural resources property rights system (including 5 specific institutions), the land and space development and protection system (4), the spatial planning system (3), the total resource management and the comprehensive conservation system (10), the system for paying for resources consumption and compensating for ecological damage (8), the environmental governance system (6), the environmental governance and ecological protection market system (6), and the performance evaluation and accountability system for ecological conservation (5).

So far, with the strong support from Xi Jinping, China has established a set of institutions for promoting ecological progress, with in place the structure tree for institutions at different levels and the matrix composed of pairs of different institutions by category.

2 The Key Features of Xi Jinping's Important Statements on the Reform of the System for Developing an Ecological Civilization

1. **The reform is rational**

The Master Plan for Reform of the System for Developing an Ecological Civilization advocates six principles while advancing the reform: (1) respecting nature, staying in harmony with nature, and protecting nature; (2) coordinating economic development and ecological conservation; (3) lucid waters and lush mountains are invaluable assets; (4) natural value and natural capital; (5) spatial balance; and (6) mountains, waters, forests, farmlands and lakes form a living community. Among them, the principles that "lucid waters and lush mountains are invaluable assets" and "mountains, waters, forests, farmlands and lakes form a living community" are major innovations in Xi Jinping Thought on Ecological Civilization. The former answers questions such as "why to change" and "what to change" from the perspective of dialectic materialism, and the latter answers the question about "how to change" based on the system theory.

2. **The reform is problem-oriented**

Xi Jinping's important statements on the reform of the system for developing an ecological civilization are distinctly problem-oriented, aimed to effect changes and solve practical problems. First, targeted reform is launched to tackle problems. Wherever there is a problem, efforts are made to ensure it is solved; the more serious the problem is, the higher it is on the priority list. For example, the "10-Chapter Water Pollution Action Plan," "10-Chapter Air Pollution Action Plan" and "10-Chapter Soil Pollution Action Plan" are all introduced to tackle major issues concerning water, air and food quality that are vital to people's well-being. Second, targeted reform is launched to address the root cause of each problem. In response to the decentralized power in the management of water resources and water environment, Xi Jinping vigorously promoted the institutional reform of the State Council in line with the principle that "mountains, waters, forests, farmlands, lakes and grasslands form a living community." The Ministry of Natural Resources, the Ministry of Ecology and Environment, and the National Forestry and Grassland Administration are established to re-define and re-divide responsibilities. Taking water pollution control as an example, the power of administering surface water and groundwater, rivers, lakes and shore water, land water and marine water, is consolidated.

3. **The reform is goal-oriented**

There are broad and specific goals for the system reform. The broad goal is to build a beautiful China, for the system reform is the fundamental guarantee for the building of an ecological civilization. The report to the 18th CPC National Congress for the first time lists the building of a beautiful China as the grand goal of ecological conservation; the report to the 19th CPC National Congress again emphasizes that the Beautiful China Initiative is an important part of the Chinese dream of national rejuvenation. As the title of its ninth section "Speeding up Reform of the System for Developing an Ecological Civilization, and Building a Beautiful China" suggests, the reform is a means toward the goal and purpose of building a beautiful China. The specific goal is to improve the system and mechanism. According to *The Master Plan for Reform of the System for Developing an Ecological Civilization*, the goal of the reform is to "establish by 2020 a systematic, complete set of institutions for developing an ecological civilization that features clearly defined property rights, multi-party participation and equal attention to incentives and constraints, and is composed of eight systems: the property rights system for natural resources, the land and space development and protection system, the space planning system, the total resource management, and comprehensive conservation system, the system for paying for resources consumption and compensating for ecological damage, the environmental governance system, the environmental governance and ecological protection market system, the ecological evaluation and accountability system, promote the modernization of national governance system and capabilities in the field of ecological conservation, and strive to move toward a new era of developing a socialist ecological civilization." It thus can be seen that the specific goal of the reform is to promote the development of the national governance system and the modernization of governance capabilities in the field of ecological conservation.

4. **The system reform for promoting ecological progress is based on top-level design**

The reform of system for developing an ecological civilization was mainly a bottom-up initiative before the 18th CPC National Congress, but since then it has been powerfully pushed ahead from top down and based on top-level design. Xi Jinping has played a decisive role in the key document drafting and decision-making of all the plenary sessions and the Central Committee since the 18th CPC National Congress. According to the *Decision of the CPC Central Committee on Some Major*

Issues Concerning Comprehensively Deepening the Reform, "to develop an ecological civilization, it is necessary to establish a systematic and complete set of institutions for promoting ecological progress, implement the strictest source protection system, system for compensating for ecological damage and the accountability system, improve environmental governance and ecological restoration system, and protect the ecological environment with institutions." Xi Jinping made a comment on it: "We must deepen the reform of the system for developing an ecological civilization, establish the framework of the system and institutionalize and legalize the construction of an ecological civilization as soon as possible." Xi, as head of the Central Leading Group for Deepening Overall Reform, has made forward-looking top-level designs and important decisions such as the *Master Plan for Reform of the System for Developing an Ecological Civilization*, a holistic reform scheme, and specific policies such as "Off-Office Auditing of Natural Resource Assets for Leading Officials." In the course of top-level design of the system for developing an ecological civilization, Xi attached great importance to the system structure, its comprehensive layout, and promoting it as a whole. In accordance with the guiding principles of the Third Plenary Session of the 18th CPC Central Committee, China plans to form a systematic and complete set of institutions for developing an ecological civilization in 2020. In fact, the set of institutions for developing an ecological civilization has basically taken shape. Since the 18th CPC National Congress, China has formed a set of institutions consisting of mandatory institutions, selective institutions for weighing the pros and cons, and guiding institutions for moral education; institutions for source control, process management control, and end processing; legal, administrative, economic, education, and publicity institutions; formal institutions represented by laws and regulations, informal institutions represented by ideology, ethics and morality, and the implementation mechanism represented by information disclosure, rewards and punishments.

3 The Historic Contribution of Xi Jinping's Important Statements on the Reform of the System for Developing an Ecological Civilization

1. **Xi was the first to define the scope of reform of system for developing an ecological civilization**

The ruling CPC not only boldly introduces the "ecological civilization" vision in the most important documents, but also proposes to promote economic, political, cultural, social, and ecological progress at the same time. On this basis, Xi Jinping went further to propose the "reform of system for developing an ecological civilization," "mechanism design for developing an ecological civilization," and "construction of institutions for developing an ecological civilization." The reform of the system mainly focuses on solving the problems with the administrative system, and governance structure and system for developing an ecological civilization at the macro level; the mechanism design on addressing the labor division and mutual checks and balances between the market mechanism with the enterprise as the main player, the government mechanism with the government as the main body, and the social mechanism with social organizations and the public as the main body, at the medium level; the construction of institutions on the design of institutions, the construction of the set of institutions, the optimization, selection and implementation of institutions, at the micro level. By defining the scope of the reform of the system, mechanism design, and construction of institutions, Xi provides a conceptual basis for the deepening of the theory about and opening up new areas for reforming the system, and meanwhile lays a framework foundation for it. For example, within the first-level category of "construction of institutions for developing an ecological civilization," there are second-level categories such as "mandatory regulatory institutions," "selective institutions for weighing pros and cons" and "guiding institutions for moral education." In the secondary category "selective institutions for weighing the pros and cons" there are third-level categories such as fiscal and taxation institutions and property rights institutions for developing an ecological civilization. In the third-level category "property rights institutions for developing an ecological civilization," there are fourth-level categories such as property rights institutions for natural resources, for environmental resources and for climate resources, among others. In this way, the framework is established for the set of institutions for developing an ecological civilization. It can be seen that categorization is an important prerequisite for theoretical innovation and institution design. Xi Jinping's important statements on the reform of system for developing an ecological civilization have laid the theoretical cornerstone of the reform.

2. **Xi's important statements have enriched and developed the methodology of reforming the system for developing an ecological civilization**

The reform of the system is to readjust the interest relationship between the living generation, between the living generation and the future generations, and between man and nature, and requires more than anything else the guidance of science. The theory that "We want both ecological and economic benefits, ecological benefits are more important than economic benefits, and lucid waters and lush mountains are invaluable assets" is a skillful application and evolution of dialectical materialism, and provides the methodological basis for the reform of the system. Dialectical materialism believes in universal connection and eternal development, and upholds three laws: the law of unity of opposites, the law of mutual change of quality and quantity, and the law of negation of the negation. Apply it to the above theory and we can see the internal connection between ecological benefits and economic benefits: that lucid waters and lush mountains are invaluable assets conforms to the law of unity between two opposites, i.e., ecological benefits and economic benefits, and that of mutual change between them, and the law of negation of the negation (from negating the value of lucid waters and lush mountains to protecting them).

To advance the reform of the system for developing an ecological civilization, we must recognize that economic growth and ecological protection are in a unity of opposites. Without a good ecological environment, there will be no natural basis for economic growth. Therefore, ecological protection should be the prerequisite for economic growth. On the other hand, without economic growth, ecological protection won't sustain, so protecting the ecological environment doesn't mean stopping economic growth. We need to balance economic growth and environmental protection to realize green development that benefits both the environment and the economy.

While advancing the reform of the system, we must understand the ecological threshold, environmental threshold and climate threshold. When the ecological damage reaches a certain threshold, it will irreversibly lead to the degradation of the ecosystem. When the environmental pollution reaches a certain threshold, it will irreversibly lead to a decline in environmental quality. When the greenhouse gas emissions reach a certain threshold, it will irreversibly lead to climate warming. Therefore, we must strictly cap water intake, pollutant discharge, and carbon emissions, in order to realize the sustainable utilization of natural resources, environmental resources and climate resources, ecological balance and environmental improvement.

While advancing the reform of the system, we must also realize that "growth—pollution—protective growth" is a negation of negation, and so is "environmental protection—slower growth—better growth." Therefore, we must constantly seek innovation and change the cage for birds. To this end, it is necessary to adopt green innovative technologies and institutions, and use innovation to drive development and transformation.

3. **Promoting the establishment of an overall framework for the reform of the system in China**

Xi Jinping's important statements on the reform of the system for developing an ecological civilization have promoted the establishment of an overall framework for the reform of the system in China.

The reform of the system has put an end to the decentralization of powers and realized coordinated governance, i.e., the coordinated governance of water supply and pollution control, shore and water, surface and underground water, and fresh water and sea water. It has separated ownership from the administrative power, which leads to more efficient allocation of natural resources, environmental resources and climate resources. It has also separated "players" and "referees" and created a mutually restrictive power structure through labor division.

The mechanism design for developing an ecological civilization has transitioned from ecological and environmental management to governance. The government, enterprises, and the public participate in the construction of ecological civilization as equals. The environmental governance system is basically formed, and the capability significantly enhanced. The market mechanism, government mechanism, and social mechanism are no longer independent from each other, but rather collaborating and restraining each other based on clear division of labor. It has also shifted from focus on one or several elements of ecological civilization to comprehensively promoting the coordinated development of ecological culture, ecological industry, ecological consumption, ecological environment, ecological resources, ecological technology, ecological institutions, ecological villages, and ecological cities.

The construction of institutions for developing an ecological civilization has expanded a single institution to a set of institutions, and built a complete set and structure of institutions. China has not only tried institutions imported, but also invented its own institutions; converged the once-separated formal, informal institutions and implementation mechanisms, with informal institutions and implementation mechanisms to support the introduction of formal institutions. The performance of institutions has been much improved. The economic development and ecological protection are mutually dependent and inseparable, leading to a not only safer but also more beautiful ecological environment.

It took China only decades to achieve industrialization which took developed countries centuries. Since the 18th CPC National Congress, China has managed to halt the degeneration of and see much improvement in the environmental quality, a transition that took developed countries at least 30 to 50 years, or more than 100 years at most in the course of industrialization. Xi Jinping pointed out at the National Conference on Eco-environmental Protection: "Since the 18th CPC National Congress, we have carried out a series of fundamental, pioneering and long-term work, accelerated the top-level design and the construction of the set of institutions for promoting ecological progress, strengthened the rule of law, and established and implemented the central supervision and inspection system for environmental protection… spurring historic, defining and across-the-board changes in the ecological and environmental protection drive." This comment is a fair evaluation of what we have achieved, which is there for all to see and is inseparable from Xi Jinping Thought on Ecological Civilization, including his important statements on the reform of the system for developing an ecological civilization.

Progress in and Prospects of the Reform of the System for Developing an Ecological Civilization

Shuzhong Gu

1 Historic Progress in and Outcomes of the Reform of the System for Developing an Ecological Civilization

1. **Historic progress**

The Master Plan for Reform of the System for Developing an Ecological Civilization, introduced in 2015, systematically charts the framework and reform path of China's system for building an ecological civilization, clarifies eight institutional areas and 47 items, and is a fundamental guideline for advancing the reform of the system for developing an ecological civilization. It is followed by a series of special plans for the reform, numbered nearly 50 so far, and covering the main areas of ecological conservation such as general affairs, resources, environment, ecology, and space. These plans are directional and pioneering documents to advance the system reform and construction for building an ecological civilization from different angles and aspects. Especially those about performance evaluation and accountability play an extremely important role in transforming the mindset about development and the assessment methods, and increasing efforts in accountability for environmental protection. At the same time, more and more supporting policies have been rolled out to accelerate ecological progress. While continuing to advance the reform of system for building an ecological civilization, departments and localities have continuously introduced their own supporting policies in the form of notices and decisions for the effective implementation of the system reform.

2. **Historic outcomes**

First, the vision of ecological civilization, in particular the principle that lucid waters and lush mountains are invaluable assets, is deeply rooted in the hearts of the people,

S. Gu (✉)
Research Institute of Resources and Environment Policies, Development Research Center of the State Council, Beijing, China

J. Pan et al. (eds.), *Beautiful China: 70 Years Since 1949 and 70 People's Views on Eco-civilization Construction*, https://doi.org/10.1007/978-981-33-6742-5_62

especially in developed areas and areas plagued by environmental pollution. On the whole, people do not dare, and are increasingly unwilling to pollute or damage the environment. Second, eco-friendly actions are increasingly popular. Concrete actions such as the Blue Sky Action, actions to treat black and odorous water bodies, and actions to improve the living environment have been continuously promoted across the country, and won high recognition and support from the people. Third, the resources reserve and ecological environment have been improved. Compared with 2013, in 2017, in 338 cities at the prefecture level and above, the average concentration of inhalable particulate matters (PMi) decreased by 22.7%; the proportion of water bodies of state-controlled section of surface water at Grade I-III increased to 67.9%, and that of water bodies with quality worse than Grade V decreased to 8.3%. Fourth, ecological conservation has boosted high-quality economic development. Resource consumption, pollutant discharge, and ecological occupancy keep declining, especially energy consumption per unit of GDP, water consumption, and land consumption. Fifth, ecological conservation has improved China's national image and international status. China is evolving from a participant and contributor to an important leader in global green development and ecological civilization construction.

2 New Tasks and Requirements for the Reform of System for Developing an Ecological Civilization

1. New requirements in the new era

The construction of socialism with Chinese characteristics has entered a new era. The people will have an unseen strong demand for a better life and unseen little tolerance for environmental pollution, resource waste and ecological damage; government officials, especially leading ones, will attach unseen importance to ecological conservation; and the international community will closely follow as it never before China's green development and ecological conservation. Ecological conservation has entered a new era.

First, there are at least four signs from an objective point of view. (1) Environmental pollution in China has been effectively contained on the whole, and the environmental (especially air and water) quality of some areas and some regions (especially the Pearl River Delta) has gradually improved. (2) The problem of resource waste is being tackled, and significant progress has been made in the conservation of land, energy and water, and especially the intensity of energy consumption has decreased significantly while the rate of land output has increased significantly. (3) The trend of ecological deterioration has been roughly curbed; the ecological protection pattern, mainly relying on the drawing of redlines, is being accelerated; and the pilot national park system is introduced, and in particular, the forest coverage rate continues to increase. (4) The disorderly land development is being reversed; major functional zones are taking shape; pilot projects to replace multiple plans for local development with one master plan have made important progress; and the spatial planning system has begun to emerge.

Second, there are at least another four signs from a subjective point of view. (1) People are more environment conscious. Government officials at all levels, especially high-ranking ones, are no longer indifferent to, but valuing ecological conservation, and thinking and acting in line with relevant instructions and documents. (2) People are more proactive in ecological conservation. Instead of passive order-taking, people have greater initiative and actively plan and implement the action plan for ecological conservation. (3) People are paying more attention to the effect. People are no longer going through the motions in protecting the environment, but earnestly studying and implementing relevant documents. (4) The ecological conservation efforts are more extensive and systematic, and the vision of ecological civilization has become increasingly popular across different walks of society. Ecological conservation used to be the job of the environmental protection department; now it has engaged almost all departments, and is coordinated with the efforts to drive economic, social, cultural, and political progress.

Third, logically speaking, the core sign is the application of concepts to practice. Ecological conservation involves many aspects such as ideology, institutions, theories and practices. It is undeniable that in the field of ecological conservation, innovation in theories lags behind that in practices, innovation in practices behind institutional innovation, and institutional innovation behind innovation in ideology. In other words, we have the most advanced concepts and the most rigorous institutions, but laggard practices and weak theoretical support. Strictly speaking, this is not a normal logical relationship. On the other hand, it shows top leaders' importance attached to and deliberation on ecological conservation, and China's unprecedented intensity in reforming the system for developing an ecological civilization. In the new era, especially after the 19th CPC National Congress, it is urgent to support new concepts and institutions with innovation in theories and practices, and especially to accelerate the application of advanced ideas to scientific practice.

The new era of ecological conservation raises new and higher requirements for various groups of stakeholders. First, for government officials. It is required that government officials at all levels, especially high-ranking ones, should take the initiative to closely follow the Party Central Committee, put the people at the center, and place ecological conservation in an important position in the Party and government agendas, and even top of the agenda in important ecological function areas and environmentally sensitive areas, and protect the environment while pursuing economic, social, cultural progress and Party building. Government officials at all levels, especially high-ranking ones, must plan ecological conservation in a way responsible to history, and formulate and implement mid-term and long-term ecological conservation plans, and see that a beautiful China is being built. They are required to carefully formulate and fully implement a science-based, rational ecological conservation plan that respects, conforms to, and protects nature. They must refrain from making frequent, unpredictable changes to existing policies, rushing to decisions or statements, and refusing to take responsibilities for any consequences caused. They should advance ecological conservation in a scientific, orderly, and efficient manner.

Second, for enterprises. In the new era of ecological conservation, enterprises are no longer just the main force in economic development, but should also assume their due social responsibilities, especially the primary responsibility for resources conservation, environmental protection and ecological restoration. In addition to good economic credit such as paying due taxes, repaying loans and paying wages on time, enterprises need to have a good ecological and environmental credit which will directly affect corporate survival and development especially after a sound ecological and environmental credit is established for enterprises. And in addition to a good brand image, an enterprise also need a green corporate image, and strives to be a new force and vanguard for resource conservation, environmental friendliness, and ecological conservation.

Third, for citizens. In the new era of ecological conservation, citizens have greater aspiration beyond food, clothing and fun, and a common goal they share is blue sky, green land and clear water. They no longer consider ecological conservation a business of the Party and the government, but increasingly a business of their own; no longer passive, they are more active in ecological conservation. Their initiative will conquer passivity. Citizens discipline themselves no longer out of fear of laws and regulations or other external forces, but more and more out of awe for nature, out of internal force. Their self-discipline will replace heteronomy. Citizens are no longer just after money-making and spending to make others envy or even jealousy, but are more committed to environmental protection and ecological restoration and other good deeds, and they are being praised and even imitated for this. The new model will certainly replace the old one.

2. Fully implementing Xi Jinping Thought on Ecological Civilization

The Xi Jinping Thought on Ecological Civilization provides an important source of thought for the future reform and construction of the system for developing an ecological civilization. First, when dealing with the relationship with nature, we should put into practice the concept that "only when the ecology thrives will civilization thrive" and that "Man and nature should co-exist in harmony." We should more than ever respect, protect and conform to nature, and examine the rationality of the system reform plan from the holistic and long-term perspective. To this end, the reform plan should be adapted to local conditions, follow the laws of nature and science, and must not over-exaggerate the role of subjective initiative. Second, when dealing with the relationship between ecological and environmental protection and economic development, we must put into practice the principle that "lucid waters and lush mountains are invaluable assets," and work harder to protect the ecology and environmental and produce long-term effects. We would rather sacrifice short-term economic development to protect the ecology and environment, and strive to translate the good ecology and environment into economic benefits. This requires the reform plan to focus on the green transition of the economy and fostering new drivers of green development. Third, while establishing the goal and value orientation for the reform of system for developing an ecological civilization, we shall implement Xi Jinping's instruction that "a good ecological environment is the fairest public product, and the most accessible welfare for the people." We shall work harder to generate more ecological dividends for the people, enhance their sense of gain, and make ecological dividends a new green driving force for high-quality development in the new era. This requires that the reform should stress the delivery of ecological dividends and the establishment of a sound incentive mechanism for ecological conservation. Fourth, while promoting ecological conservation, we shall put into practice Xi Jinping's holistic and systematic view that "mountains, waters, forests, farmlands and lakes form a living community" and the strict view to "implement the most rigorous ecological and environmental protection system." It is necessary to strengthen coordination and cooperation between departments and localities under the common goals, and work together to efficiently and sustainably promote ecological progress. This requires the reform to go beyond departmental limits, to be locally universal and coordinated. Meanwhile, it is necessary to promote the law-based reform of the system for developing an ecological civilization, and at the same time make necessary revisions and improvements to existing laws and regulations that do not align with the environmental goals and requirements. Fifth, while dealing with the relationship between domestic ecological conservation and global green or sustainable development, it is necessary to uphold Xi Jinping's national proposal for "jointly building a beautiful China" and win-win global proposal for "jointly seeking the path toward global ecological conservation." We must do our best in domestic ecological conservation and meanwhile, work to fulfill our international obligations for green development, especially responding to climate change, and strive to become an important participant, contributor and pioneer in the world's green development, especially response to climate change. This requires the reform to be universal and inclusive on a global level.

3. **Fully adapting to the new situation where "three phases" co-exists**

At the National Conference on Eco-environmental Protection in May 2018, General Secretary Xi Jinping made it clear that: The drive of ecological conservation has entered a phase with mounting pressure and burden, a phase when more and better ecological products are needed to meet the people's growing demand for a beautiful ecological environment, and a phase when conditions and capabilities are ready for tackling outstanding ecological and environmental problems. The future reform of the system for developing an ecological civilization must adapt to the new situation where these three phases co-exist.

In the past five years, the reform has scored historic achievements, but also exposed some problems which are mainly categorized into two kinds. First, problems with the reform design. The reform design is undoubtedly scientific and effective on the whole, but there are some shortcomings: (1) there are more restraints than incentives—the two should be equally important and we should in particular increase the supply of market-oriented reform proposals. (2) The design should be more targeted and problem oriented, and increase the supply of targeted proposals. (3)

The design should place more emphasis on differentiation, and increase the supply of differentiated assessment proposals. (4) The design should be more systematic and less departmental to prevent fragmentation. (5) The design should be more democratic to fully absorb the exploration experience at the local and even the grassroots level. Second, problems with the reform execution. The reform execution is generally smooth and effective, and has played a fundamental and supporting role in ecological conservation, but many problems have been exposed in investigations and analysis. For example, there exists the phenomenon of implementing documents with documents. In the face of growing special reform plans and supervision requirements, coupled with the weak local work foundation and the lack of working mechanisms and reform motivation, national plans only beget more local plans that exist only on paper. There is also the problem of "zero tolerance for failure." In most pilot projects, "failure" is unacceptable and trial-and-error and fault-tolerant mechanisms are absent; as a result, the lessons and experience that should have been drawn from these pilot projects are often neglected. The third is the problem of a weak foundation, especially in underdeveloped areas where the shortage of funds, technology, equipment, and talents is particularly prominent.

These are generally the problems encountered in reform and development. But left unattended, they will undoubtedly hinder the continuous, smooth and effective progress in the reform of the system for developing an ecological civilization. It is necessary to fully adapt to the new situation of ecological conservation where the three phases co-exist, follow Xi Jinping Thought on Ecological Civilization as the fundamental guideline, and continue to deepen and effectively promote the reform of the system for developing an ecological civilization.

3 Continuing to Deepen and Promote the Reform of the System for Developing an Ecological Civilization in an Efficient Manner

1. Carrying out legislation on ecological conservation when necessary

Legislation on ecological conservation is imperative. First, it is an inevitable requirement that fully reflects the will of the Party and the purpose of the national constitution. The newly-revised *Constitution of the Communist Party of China* lists ecological conservation as an important part of the Five-Sphere Integrated Plan and the Four-Pronged Comprehensive Strategy, and clearly proposes to coordinate economic, political, cultural, social, and ecological progress. The newly-revised *Constitution of the People's Republic of China* clearly calls to "promote the coordinated development in material, political, cultural and ethical, social, and ecological dimensions, and develop China into a great modern socialist country that is prosperous, strong, democratic, culturally advanced, harmonious, and beautiful." Second, it is an inevitable requirement for promoting ecological progress by law. Compared with the actual needs for ecological progress, the urgent need for innovative system for developing an ecological civilization, and especially the urgent need for sustainable, law-based ecological progress, the legislation for ecological progress is lagging behind, leading to the lack of solid legal support for concrete actions in promoting ecological progress in China. Third, it is an inevitable requirement for eliminating inconsistencies, contradictions, and conflicts between relevant laws. In particular, we need a law to clearly define water pollution and its monitoring and treatment responsibilities, and unify the land use standards. Fourth, it is an inevitable requirement for clarifying the basic connotations and attributes of ecological progress, in particular for systematically clarifying the status and role of ecological conservation, the main body and its responsibilities, the direction and goal, basic concepts and principles, and the working systems and mechanisms.

To this end, we may consider advancing the legislation in the following aspects. First, we should clarify the basic purpose of legislation for ecological progress to fully reflect the Party's leadership and role in ecological conservation and the *CPC Constitution*'s provisions on it. The legislation should fully reflect the specific provisions of the newly-revised Constitution of the People's Republic of China on ecological conservation, especially the those on related (legal) responsibilities and obligations; it should highlight the characteristics of the basic law for ecological progress, and strengthen its supremacy over special laws for resources and environment, eliminate the inconsistency among laws, and provides guidance for the future revision of related laws such as the *Water Law*, the *Land Management Law*, the *Mineral Resources Law*, the *Grassland Law* and the *Forest Law*. Second, we should clarify the basic content of legislation for ecological progress. According to the above requirements and the overall plan for the reform of the system for developing an ecological civilization, and in reference to similar laws, we may call the legislation for ecological progress the Ecological Progress Promotion Law, and it should include the following: general principles, the basic connotation and denotation, status and role of ecological progress, the main body and contents of ecological conservation, direction and requirements, concepts and principles, goals and standards; the asset and property rights system for natural resources; land and space development and protection system; resource quantity management and comprehensive conservation system; paid use of resources and ecological compensation system; environmental protection and governance and ecological restoration and compensation system; ecological performance evaluation and accountability system; legal liabilities; and supplementary provisions. Third, we should advance the legislation for ecological progress in an orderly and efficient manner, with a focus on comprehensive coordination of legislative work, to jointly advance the legislation research and drafting, and fully engage functional departments and experts in various fields, to prevent the law from becoming a distinctly departmental law. We should sort out relevant laws, especially those that conflict with each other, to strive to eliminate these conflicts through legislation. We should fully absorb local experience. Fujian Province's *Regulations on the Promotion of Ecological Progress* can be used as a useful reference.

2. **Launching special plans for the reform of the system for developing an ecological civilization in an ongoing, orderly manner**

In the future, the design and formulation of a special plan for the reform of the system for developing an ecological civilization should pay attention to the following five aspects. First, the plan should be corrective and original. In particular, it should be corrective and original enough for the times and the problems it is targeting, to be revolutionary in the strict terms. Second, the plan should be authoritative and effective. Particular attention should be paid to the issuer and the actual effect of the plan, and plans with poor authority and effectiveness should be avoided. Third, the systematicity and alignment between plans, especially the alignment of similar plans in terms of orientation, goals, and measures, to avoid the absence of key plans, overlapping between plans, inconsistent orientations or conflicting goals. Fourth, the logical relationship between and timing sequence of the plans, especially to the cause-and-effect relationship, upstream and downstream relationship, main contradictions and main aspects of contradictions, general and special aspects, and time relevance. Fifth, the structural and hierarchical nature of the plan, especially those determined by the *Master Plan for Reform of the System for Developing an Ecological Civilization* to avoid dislocation and reversion of the hierarchy.

3. **Continuing to increase the effective supply of incentives for ecological conservation**

To promote ecological progress, on the one hand, we must continue to strengthen the "reverse restraint" mechanism, and on the other hand, vigorously strengthen the "positive incentive" mechanism. The key is to establish and promote a mechanism for cashing in the "invaluable assets" of "lucid waters and lush mountains," including an accounting system that highlights the value of natural resource assets, an evaluation system that highlights ecological protection and its importance, a pricing mechanism that is more conscious about resources, environment and ecological costs, and an ecological compensation mechanism that stresses more the value of ecosystem products and services. The priority is to establish a sound compensation mechanism and market mechanism for realizing the value of ecological products and services.

4. **Integrating special plans for the reform of the system for developing an ecological civilization when necessary and appropriately**

Making use of the ongoing institutional reform, to address the lack of coordination and synergy across sectors and departments for the reform of the system for developing an ecological civilization, we should further sort out, consolidate, improve and summarize existing reforms, and give full play to the institutions' guiding and motivating role. It is recommended that ministerial/commission-level policies and measures should be introduced in accordance with relevant timetables and roadmaps for governments in provinces (municipalities, autonomous regions) to strictly follow.

5. **Strengthening coordination of the reform and the construction of the system for developing an ecological civilization**

In some areas of the reform, it is difficult to advance the reform, formulate standards or even implement alone at the provincial level or below. Therefore, it is recommended that a national institutional framework should be in place, with supporting policies and supervision and follow-up guidance, to avoid problems such as delay, repetition, detours, and poor effects of reform.

6. **Timely promoting good practices in the reform and construction of the system for developing an ecological civilization**

It is recommended that the task force for economic system and ecological progress system should take the lead in organizing the annual selection of reform and innovation practices. Then, success cases will be published and their experience and models promoted via government documents, media, and on-site meetings. Appropriate policy freedom may be given to the areas of these cases to speed up the promotion of the reform results.

7. **Strengthening the coordinated advancement of the reform of the system for developing an ecological civilization in key regions and river basins**

For example, in the Yangtze River Economic Belt, we should advance coordinated reform across provinces by "promoting well-coordinated environmental conservation and avoiding excessive development," including establishing a sound Yangtze River ecological and environmental monitoring network system, reforming the comprehensive development and assessment system, improving the ecological compensation mechanism, and strengthening the accountability and audit system. In the Beijing-Tianjin-Hebei region, we should focus on the fight against air and water pollution, and advance coordinated reform in the centralized management of water basins, cross-regional air pollution control, and emission rights trading.

8. **Lending support to the reform of the system for developing an ecological civilization in underdeveloped areas in various aspects**

The central government should offer special support in such areas as cadre training and assignment, financial support, technical support, equipment provision, and advanced experience and models, to ensure that no prefecture, city or league is left behind in the national drive to reform and build the system for developing an ecological civilization system. A beautiful China should be beautiful in every inch of the land.

9. Continuing to reform and improve the mechanism for promoting the reform of the system for developing an ecological civilization

First, we should work hard to strengthen the public engagement mechanism for the reform. While designing the reform, it is necessary to fully solicit and absorb opinions and suggestions from Party committees and governments at all levels, experts and scholars, and all walks of life, and combine top-down and bottom-up measures to improve public engagement in, awareness and recognition of the reform, thereby reducing the resistance to it and improving its efficiency. Second, we should establish a sound scientific evaluation mechanism for the reform. The priority is to scientifically evaluate the basic conditions of resources and ecological environment, predict the development trends, and demonstrate and assess the directions and goals, priorities and measures, and prospects and effects of the reform, so as to make the reform plan more science-based and rational. Although third-party assessment already exists and various departments have each set up their own expert groups, a regular expert demonstration and consultation mechanism is still to be established. Third, we should try to design an upgraded version of the reform plan based on pilot projects. In the past five years, the central government has launched a series of pilot projects in such areas as resources, environment, ecology and space, with varying progress, outcomes, experience, and lessons. We should pay these practices, experience and lessons the attention they deserve, summarize them, and on this basis, make necessary corrections, improvements to existing plans, and produce an upgraded version to meet higher-level requirements of a new stage. Fourth, we should establish a sound trial-and-error and fault-tolerant mechanism for pilot reforms. Although dozens of pilot projects for the reform of the system for developing an ecological civilization have been launched, with progress and outcomes made, the surveyed pilot areas all report that the competent authorities are too strict with the requirements, and allow no individual "correction" to the pilot plan, let alone "error" in the implementation. But in practice, the pilot plan cannot be all-encompassing or perfect from the beginning; "defects" are inevitable and "corrections" to varying degrees are often necessary. A pilot reform is to find a new way and even to "try and err," which is extremely beneficial for the general good, so there must be a trial-and-error and fault-tolerant mechanism.

We are convinced that under the strong leadership of the CPC Central Committee with Comrade Xi Jinping at the core and by following Xi Jinping Thought on Ecological Civilization, through concerted efforts and constant exploration, we will be able to establish a scientific, complete and effective system for developing an ecological civilization, provide solid institutional guarantee for ecological progress, and realize the ambitious goal of building a beautiful China as soon as possible!

Promoting Ecological Progress with Procuratorial Services

Xiangjun Zhang

General Secretary Xi Jinping of the CPC Central Committee once emphasized that, "Ecological and environmental protection must rely on the institutions and the rule of law. Only the most rigorous institutions and rule of law can provide a reliable guarantee for ecological progress." A hallmark of Xi Jinping Thought on Ecological Civilization is its insistent call for legalizing and institutionalizing ecological and environmental protection and using legal weapons to fight pollution, and rule of law to protect the environment. The procuratorial organs are legal supervision organs stipulated in the *Constitution of the People's Republic of China*, represent the public interest, and shoulder major missions and responsibilities in promoting ecological progress in accordance with the law. As an important force in promoting the rule of law and protecting the ecological environment in China, procuratorial organs should strengthen legal supervision in practice to safeguard ecological progress.

1 Earnestly Practicing Xi Jinping Thought on Ecological Civilization, and Continuing to Plan Environment-Related Procuratorial Work and Based on the Five-Sphere Integrated Plan and the Four-Pronged Comprehensive Strategy

Xi Jinping Thought on Ecological Civilization is an important part of Xi Jinping Thought on Socialism with Chinese Characteristics for a new era. It emphasizes that only when the ecology thrives will civilization thrive, that man and nature should co-exist in harmony, that lucid waters and lush mountains are invaluable assets, that a good ecological environment is the most inclusive benefit for the people, and that

X. Zhang (✉)
The Supreme People's Procuratorate of the People's Republic of China, Beijing, China

J. Pan et al. (eds.), *Beautiful China: 70 Years Since 1949 and 70 People's Views on Eco-civilization Construction*, https://doi.org/10.1007/978-981-33-6742-5_63

mountains, waters, forests, farmlands and lakes form a living community. It calls for the most rigorous institutions and rule of law for ecological and environmental protection, for building a beautiful China by all, and for cooperating to build a global ecological civilization. It explains the close relationship between ecological and environmental protection and the survival, rise and fall of a country from the perspectives of politics and history. It explains the close relationship between the quality of ecological environment and the people's well-being in light of the Five-Sphere Integrated Plan, and the quality of ecological environment and state governance in light of the Four-Pronged Comprehensive Strategy. It points out the direction and fundamental guidance for advancing the construction of a beautiful, modern China where man and nature co-exist in harmony. It is also the fundamental guideline for the environment-related procuratorial work in the new era.

To put Xi Jinping Thought on Ecological Civilization into action, procuratorial organs must consciously plan the environment-related procuratorial work by aligning it with Five-Sphere Integrated Plan and the Four-Pronged Comprehensive Strategy for developing socialism with Chinese characteristics, in order to steer in the right direction. In recent years, procuratorial organs have implemented the general requirements of "following the Party's political line, keeping the general picture in mind, seeking development, and prioritizing self-improvement" in procuratorial work. They take the initiative to serve the implementation of major initiatives such as "Beautiful China" and "Rule by Law." In the fight against air pollution, water pollution, soil pollution and comprehensive improvement to the rural environment and other key and difficult issues with environmental governance, they give full play to the role of procuratorial functions in criminal, civil, administrative, and public-interest litigation, work with the central environmental protection inspection team, strengthen cooperation with relevant departments, and launch special actions to increasingly crack down on violations, and provide strong judicial support for pollution control and ecological conservation in accordance with the law. According to statistics, from November to December 2018, in the field of criminal prosecution, 15,095 people were arrested in 9470 criminal cases suspected of damaging the environment and resources and 26,287 in 42,195 cases were prosecuted. In the field of public interest litigation, 59,312 cases related to natural resources, ecology and environment were filed, 53,521 pre-litigation procedural cases were handled, with the rectification rate of 97%, and 1732 related civil public interest actions and civil public interest actions collateral to criminal procedures were filed. Since 2018, the Supreme People's Procuratorate has successively issued the *Opinions on Giving Full Play to the Procuratorial Function to Provide Judicial Support for Fighting the "Three Tough Battles"* and the *Notice on Giving Full Play to the Role of the Procuratorial Function in Pollution Control*. It has also co-issued the *Opinions on Strengthening Cooperation and Coordination in Public-Interest Suits to Fight Pollution in Accordance with the Law* with nine ministries and commissions including the Ministry of Ecology and Environment, striving to play a more active role in promoting ecological progress and improving environmental governance. Especially in the governance of key regional river basins, they follow the national development strategy for such basins, give full play to the procuratorial function, and promote the improvement of environmental governance in key basins in the Beijing-Tianjin-Hebei region, Yangtze River Delta, and Sanjiangyuan Area. For example, aware of the strategic importance of environmental public interest litigation (EPIL) for protecting the ecology of the Yangtze River and promoting the development of the Yangtze River Economic Belt, the Supreme People's Procuratorate held a symposium on the procuratorial work of the Yangtze River Economic Zone, a forum on procuratorial services in the Zone, and explored a working mechanism for public interest litigation (PIL) across the 11 provincial administrative regions along the Yangtze River. It has jointly organized a special action to protect the Mother River with the Ministry of Water Resources, coordinating procuratorial organs and river chief's offices in the nine provincial-level administrative regions along the Yellow River to crack down on illegal occupation, mining, piling and construction along it and use legal weapons to protect the ecological environment of the Yellow River basin. The practice of the procuratorial organs has been affirmed by the central government. In their respective plan for the National Ecological Civilization Pilot Zone, Fujian, Guizhou, Jiangxi, and Hainan all speak positively the environment-related procuratorial work and raise specific requirements.

The environment-related procuratorial work must be closely aligned with the Five-Sphere Integrated Plan and the Four-Pronged Comprehensive Strategy for developing socialism with Chinese characteristics. Special efforts must be made in the following aspects. First, we must combine Xi Jinping Thought on Ecological Civilization with Xi's new concepts, new thoughts, and new strategies for promoting the rule of law in every dimension of state governance, to guide the environment-related procuratorial work. We must follow the new vision of innovative, coordinated, green, and open development that is for everyone, and the concept that "lucid waters and lush mountains are invaluable assets" and the concept of supervision for win-win results, to change our mindset and advance the innovation and development of environment-related procuratorial work in the new era. Second, we must focus on the center while adopting a holistic view, and enhance the acumen and initiative to serve ecological conservation. We shall earnestly implement the *Opinions on Comprehensively Strengthening Ecological and Environmental Protection and Resolutely Fighting Pollution* issued by the CPC Central Committee and the State Council and *Resolution on Comprehensively Strengthening Ecological and Environmental Protection and Promoting Pollution Control in Accordance with the Law* issued by the Standing Committee of National People's Congress, and find out how and where procuratorial powers can serve the high-quality economic development and contribute to ecological and environmental protection, and the governance of key regional river basins. Third, we must truly put people at the center of our work, and allow them to enjoy the benefits of ecological conservation and rule of law. The environment-related procuratorial work should be for the people. We shall employ the rule-of-law thinking and legal means to address the common environmental concerns of the people, and continue to enhance the people's sense of gain, happiness and security. Fourth, we must reform, innovate and respect the pioneering initiative at the community level. Without compromising the integrity and unity of the rule of law for the ecology and environment, we shall encourage and support local procuratorial organs to proceed from local reality and actively explore new regional judicial cooperation models for basin environments that conform to the patterns of environment-related procuratorial work and have local characteristics. We shall continue to promote the innovative practices at the community level and provide useful experience for diversifying the path of specialized procuratorial work for environmental protection.

2 Fully Performing the Four Procuratorial Responsibilities of Criminal, Civil, Administrative and Public Interest Litigation in Accordance with the Law, and Safeguarding Ecological Conservation by Legal Means

General Secretary Xi Jinping pointed out that we must "protect the ecology and environment with the most rigorous institutions and rule of law." "To promote the rule of law in every dimension, the priority is to ensure the strict execution of the law." "The procuratorial, judicial and public-security organs should carry the scale of justice on their shoulders, hold the sword of justice in their hands, and take concrete actions to safeguard social fairness and justice to the true benefits of the people." As the legal-supervision agency, procuratorial organs perform their procuratorial supervision function, and involve in ecological and environmental protection in accordance with the law, to ensure the implementation and compliance of environmental laws and regulations. This is an important task of procuratorial organs to serve ecological conservation. Since December 2018, the Supreme People's Procuratorate has implemented the guiding principles of the Third Plenary Session of the 19th CPC Central Committee, adapted to the people's growing needs for a better life in the new era, and advanced systematic and restructuring reforms of internal institutions. A general framework for legal supervision, with the parallel functions of criminal, civil, administrative and public-interest litigation has taken shape, which is a breakthrough to promote the innovation and development of the entire procuratorial work, including environment-related procuratorial work. Drawing strength from the institutional reform, procuratorial organs should resort to the four measures of punishment, prevention, protection, and education at the same time, and give full play to the role of criminal, civil, administrative, and public-interest litigation prosecution in ecological and environmental conservation.

First, procuratorial organs should temper justice with mercy and resolutely crack down on environmental crimes. On the one hand, environmental pollution must be punished by the toughest law. By approving the arrest and prosecuting suspects, procuratorial organs shall continue to severely crack down on criminal crimes of ecological and environmental damages, and resolutely punish by law illegal logging, illegal mining, illegal occupation of agricultural land, environmental pollution and other common criminal crimes against the ecology and environment. Especially those who have polluted the environment and major rivers in pursuit of personal interests must be strictly held accountable and severely punished by law. In 2018, in response to the resolution issued by the NPC Standing Committee, the Supreme People's Procuratorate actively participated in the fight against pollution, took the lead in formulating regulations for handling criminal cases of environmental pollution, and prosecuted 42,195 people in environmental and resources crimes, registering a year-on-year increase of 21%. It co-launched with the Ministry of Water Resources a special action to protect the Yellow River, coordinating procuratorial organs and river chief's offices in the nine provincial-level administrative regions along the Yellow River to crack down on illegal occupation, mining, piling and construction along it. It established a coordination mechanism for procuratorial organs in 11 provinces and cities along the Yangtze River, and introduced 10 measures to serve the development of the Yangtze River Economic Belt. It guided procuratorial organs in Anhui and Hubei in handling a number of cases of cross-provincial dumping of solid waste, illegal sewage discharge, illegal sand mining, etc., to jointly protect the water basin environment. It continues to exercise supervision on every major case. In 2018, the Supreme People's Procuratorate alone or jointly with the Ministry of Public Security and the Ministry of Ecology and Environment supervised the handling of 56 major environmental pollution cases, including environmental pollution cases in the Yangtze River basin, and with the National Forestry and Grassland Administration, 10 super and major forestry-related criminal cases. It also sent teams to Anhui, Hubei and Sichuan to supervise the handling of cases. On the other hand, procuratorial organs exercise leniency for the confessants and temper justice with mercy in criminal cases. In the case of a minor environmental crime, when the criminal pleads guilty, sincerely repents, and actively repairs and restores the ecology and environment, he/she will be punished with leniency in accordance with the law. Both punishment and education are applied to crack down on existing crimes and prevent future ones. Procuratorial organs shall promptly approve arrests and prosecute the suspects in accordance with the law for serious crimes. In the process of interrogation, reading of the indictment and public prosecution opinions, they should not only emphasize the damages to the environment and resources caused by the crime, but also offer suggestions and methods for improvement. By tempering justice with mercy, we can educate the general public while cracking down on crimes.

Second, procuratorial organs should integrate internal resources and strengthen the supervision of ecological and environmental litigation. Procuratorial organs exercise legal supervision over criminal, civil, and administrative litigation, and play a unique role in promoting strict law enforcement and judicial justice in the field of ecological and environmental conservation. It is necessary to strengthen internal coordination of criminal, civil, administrative and public interest-litigation within the procuratorial organs, and the litigation supervision over cases of destroying the ecological environment. We shall supervise the environmental protection, water conservancy, land and resources, and forestry authorities which are responsible for the regulation of ecological environment and resources to ensure strict law enforcement, and promote fair justice in environmental and resources cases. We shall intensify the supervision over criminal cases. We should urge administrative law enforcement agencies to promptly transfer criminal clues about damaging the ecological environment, and resolutely correct the problems of substituting fines for criminal punishments, and failing to file criminal cases, or failing to hold the guilty accountable. We shall strengthen the review and protest of criminal, civil and administrative judgments in the field of ecological and environment conservation, and correct any mistakes found in a timely manner; exercise whole-process supervision over criminal judgment of penalty or civil judgments of economic compensation and restoration, promptly urge the court's enforcement departments to perform their duties, and make corrections as soon as negligence of duty or inappropriate exercise of power is detected. We shall strengthen the supervision over the execution of administrative non-litigation cases in the field of ecological and environment conservation, and over the courts' acceptance, review and enforcement of non-litigation cases, and over administrative agencies' failure to apply for compulsory execution and the court's failure to execute approved rulings, and safeguard correct administrative decisions and authority in ecological and environmental cases.

Third, we shall vigorously promote EPIL to protect public welfare. General Secretary Xi Jinping once remarked that "Prosecutors, as representatives of public interests, shoulder important responsibilities." China is the first country in the world to allow procuratorial organs to file civil and administrative EPIL. In 2014, Jinsha County Procuratorate sued the Jinsha County Environmental Protection Bureau in Guizhou, making it the first administrative PIL in China since 1949. In 2015, the NPC Standing Committee authorized procuratorial organs to run a two-year PIL pilot project which has delivered remarkable results. In June 2017, the NPC Standing Committee passed a resolution to amend the *Civil Procedure Law* and the *Administrative Procedure Law*, to formally recognize the PIL system. In March 2018, the Supreme People's Procuratorate and the Supreme People's Court jointly issued the *Interpretation on Several Issues Concerning the Application of Laws to Public-Interest Litigation Filed by Procuratorial Organs*, which clarifies the rules for procuratorial organs to file civil and administrative PIL. The system of PIL filed by procuratorial organs is the best path toward public-interest protection under China's national conditions. It is a useful attempt to integrate procuratorial work into promoting the rule of law and deepening reforms in all respects. It has formed a basic structure composed of civil and administrative PIL. The ecological, environmental and resources conservation is the focus of PIL filed by procuratorial organs. In 2018, procuratorial organs across the country filed 59,312 PIL cases for resources and environmental protection, accounting for about half of the total. Among them, 53,521 cases went through pre-litigation procedures, and 97% of them saw corrections made by administrative organs involved; they urged the treatment of 2.11 million *mu* of polluted or damaged farmland, wetland, woodland, and grassland damaged, the removal of 20 million tons of solid waste and other waste, and recovered RMB 3 billion for ecological restoration and environment improvement. In practice, PIL initiated by procuratorial organs is playing an increasingly important role in ecological and environmental protection, displaying a growing institutional advantage for promoting the rule of law in social governance, and has provided a Chinese proposal for effective common-land governance and stronger judicial protection for public welfare. In particular, in pursuit of win-win results, procuratorial organs assist the government in exercising administration in strict accordance with the law, and strive to, ideally, defend public welfare before the litigation. They issue announcements or procuratorial recommendations before the litigation, and urge relevant parties to initiate litigation and administrative organs to perform their duties in accordance with the law, in order to protect public welfare and maximize the social effect with the minimal judicial input. Baijiazhou of Changsha City is located in the first-class protection zone for the drinking water source of Xiangjiang River, but suffers from discharge of untreated sewage. In response, local procuratorial organ issued procuratorial recommendations which were effectively implemented: the islanders were relocated, illegal buildings demolished, and catering ships shut down, restoring the clean water environment. Meanwhile, we should use litigation to educate the public on rule of law. When the procuratorial recommendations are neglected, we can advance with litigation and court trial until the problem is resolved, by which we can set an example for the society at the same time. In Rongchang District of Chongqing, 286 farmers were found to illegally raise livestock and poultry in the prohibited zone along the Laixi River, and directly discharge the manure and sewage into the river. The local procuratorial organ then issued procuratorial recommendations to 16 towns and neighborhoods, filed and won a public interest lawsuit against the Guchang township government for neglecting its environmental regulation duties, thus promoting pollution control in the river section in the prohibited area.

The EPIL system is an important institutional design that uses rule-of-law thinking and methods to tackle environmental pollution, an important measure to modernize the national governance system and capabilities, and also a powerful driving force for the transformation and upgrading of procuratorial work in the new era. In 2019, the NPC Standing Committee will hear and deliberate on the report about the procuratorial work on PIL. We must seize the opportunity to deepen the reform of the system for procuratorial organs to file public interest lawsuits. We shall advocate that administrative PIL is to urge and work with the administrative organ involved to make corrections. With the protection of public interest as our central aim, we should make full use of pre-litigation procedures in accordance with the law. For administrative organs which illegally exercised their power or fail to perform their powers, we can hold talks, round-tables and hearings to stimulate them to rectify their errors, and promote them to actively perform duties, make corrections, and timely repair the damaged ecological environment. We shall continue to promote the establishment of a cross-provincial working mechanism for administrative PIL for ecological, environmental and resources protection, to ensure whole-watershed and cross-regional public interest protection. We shall improve the working mechanism for forensic authentication of ecological and environmental damage, and work to solve the problems of little, expensive and time-consuming authentication.

Fourth, we shall introduce the concept of restorative justice and establish the restorative justice mechanism. The restorative justice theory was originated in North America in the 1960s and 1970s, and was soon put into judicial practice in Anglo-American legal system. Unlike the traditional justice, its core idea is "restoration." In some countries, restorative justice has entered the mainstream of criminal judicatory of environment and resource offenses, and is even regarded by some scholars as a "full-function alternative to the current criminal judicatory of environment and resource offenses." After introduced into China, it was first applied to the study of juvenile crimes and the handling of criminal reconciliation cases. In recent years, in the face of the frequent, common environmental damage and related crimes, restorative justice has been applied to punish environmental crimes and promote "replanting and re-greening and ecological restoration." In 2015, the Supreme People's Procuratorate demanded that procuratorial organs at all levels across the country should actively apply restorative justice to judicial protection of the ecological environment, and strive to realize criminal punishment and ecological and environmental protection at the same time. In practice, some local procuratorial organs prioritize restoration, and urge the defendants to actively restore the environment by ways of replanting and re-greening, stock enhancement and release, restoration within a stipulated time limit, and labor compensation, achieving good results. For example, Fujian has created the model of "specialized legal supervision + practice of restorative justice + socialized comprehensive governance." Jiangxi has introduced the concept of restorative justice into the ecology-related procuratorial work and promoted the establishment of ecological restoration and compensation mechanisms such as replanting and re-greening to realize criminal punishment and ecological and environmental protection at the same time. In Guizhou, the procuratorial organs practice restorative justice, strengthen the education of criminal suspects, and promote replanting and re-greening, and thus the comprehensive improvement in the ecological environment. In Zhejiang, restorative justice has been increasingly applied in cases about ecological and environmental protection. To educate law-violating fishermen to correct their mistakes and actively restore fishery and the water environment, the procuratorial organs work with fishery law enforcement agencies to see to it that fishermen stock fish fry and repay their "environmental debt," thus protecting the ecology and punishing crimes at the same time. As of the end of December 2018, in 30 provincial-level administrative regions, the procuratorial organs had worked with courts, public security organs, and environmental protection departments to establish 2327 procuratorial working mechanisms for ecological and environmental restoration, and 459 ecological and environmental restoration bases. In 2018, 85.91 million trees were replanted, 74.67 million fish fry were proliferated, 84,000 mu of cultivated land was restored, and RMB 360 million for ecological restoration paid.

The principle that "whoever pollutes the environment shall eliminate and control the pollution; whoever damages the environment shall restore it" is the golden rule for ecological and environmental protection and governance. By introducing the concept of restorative justice into the field of ecological and environmental protection, coordinating the applicable criminal, civil, and administrative responsibilities, restoring the ecological environment to the greatest extent, and conditionally granting leniency to those who damaged the environment, we can better implement the relief system with ecological and environmental restoration at the core and solve the dilemma in which enterprises pollute, the people suffer and the government pays." In view of the conflict between restorative justice and traditional criminal justice, insufficient legal basis, and poor supporting institutions, it is necessary to shift the priority of the procuratorial work from punishment to ecological restoration, to make the two equally important and exercise protective justice over ecological restoration while severely cracking down on crimes of ecological and environmental damage. Adapting to the diverse requirements for ecological and environment restoration, we should strengthen legal protection, improve the punishment for environment and resources-related crimes, explore more non-custodial forms of punishment, and gradually remove legal obstacles to restorative justice. We should better the working model for restorative justice in environmental cases, improve the evaluation and supervision mechanism for restorative justice, focus on the application of the ecological restoration results, and establish a sound comprehensive governance mechanism for restorative justice in environmental cases participated by multiple parties. In particular, it's clearly stipulated that for criminals who offer restorative compensation and who strictly perform the agreement, discretionary non-prosecution or discretionary lighter sentencing may be proposed as appropriate.

3 Continuing to Control Pollution from the Source, Strengthening the Effective Alignment Between Procuratorial Services and Social Governance, and Promoting the Establishment of a Long-Term Management Mechanism for Ecological Conservation

General Secretary Xi Jinping once emphasized the need to "actively build an environmental governance system led by the government, with enterprises as the main body, and participated by social organizations and the public." Building a socialist ecological civilization concerns every individual and sector in the society. It requires both a top-down institutional design by the government and mass operations from bottom up. We shall extensively promote the Beautiful China Initiative that is by all and for all, produce a strong synergy, and create an environment in which ecological conservation is valued and practiced by all and at all times. As the legal supervision agency, procuratorial organs play a unique and irreplaceable role in prompting administrative organs to administer according to the law and citizens to abide by law in environmental governance.

First, while handling cases, they should promote environmental governance from the source. The system for developing an ecological civilization, as systematic and complete as it is, covers strict prevention at the source, strict in-process management, and severe punishment for consequences. Procuratorial organs shall better combine case handling with regional governance, promptly issue procuratorial recommendations to relevant departments in the course of case handling, identify and solve common problems in individual cases, and promote the establishment of long-term management mechanisms. We shall urge relevant departments to improve systems, close loopholes, and promote social governance innovation in such areas as ecological conservation and economic management. It is necessary to strengthen investigation and research, analyze problems with ecological conservation, and put forward countermeasures and suggestions for the decision-making of the Party committees and the governments at all levels. We shall promote the construction of a linkage mechanism for law enforcement and justice in environmental cases, and a database about environment-related procuratorial work via information platforms. In particular, we should follow the development trends of the Internet, big data, cloud computing, artificial intelligence, and blockchain technologies, establish an information platform for environment law enforcement and justice, and improve the environmental governance operation mechanism in the Internet era.

Second, we should implement the principle that "whoever enforces the law shall popularize the law" to raise public awareness of rule of law for ecological conservation. Since the 18th CPC National Congress, the public attention to ecological and environmental issues has climbed to an unprecedented level, but people's awareness of environmental rights and participation remains low, and their understanding of the ecological environment limited. Especially when their own interests are not threatened, most people are indifferent or passive toward environmental pollution. To realize the rule of law in ecological conservation, the key is to create a law-abiding society which will promote law enforcement and save judicial resources. While implementing Jinping Thought on Ecological Civilization, we should regard raising public awareness of the rule of law as a fundamental task, uphold the principle that "whoever enforces the law shall popularize the law," effectively strengthen education on laws for ecological conservation, and help enhance the rule-of-law awareness across the society. New ways must be introduced to seek actual results. By interpreting law with individual cases, publishing legal documents online, publishing guiding cases, typical cases and procuratorial announcements, and making public delivery of procuratorial recommendations, procuratorial organs can increase public knowledge about relevant regulations and environmental protection, and guide the public to deeply participate in ecological and environmental protection in an orderly manner. We should provide unimpeded channels for the public to report and expose environment and resources offense, raise public awareness of the rule of law and environmental protection, and encourage and increase public participation in ecological, environmental and resources conservation.

Third, we shall help improve the system for public participation in ecological and environmental protection. Environmental protection cannot rely on the government alone. The abuse of decision-making power and the lack of public participation caused by the "government failure" are the root cause for the legal dilemma facing environmental governance. The people's strong demand for environmental information, knowledge, and culture is unprecedented. In the *Environmental Protection Law* implemented in 2015, there is a chapter "Information Disclosure and Public Engagement" which for the first time stipulates that "citizens, legal persons and other organizations shall be entitled to the right of acquiring the environment information…" The disclosure of environment information concerns the people's right to know, to participate, to express and to supervise. While raising citizens' awareness of the rule of law on the environment, procuratorial organs should, in the course of case handling, actively put forward procuratorial suggestions against the lack of procedural environmental rights granted to citizens, prompt the government to improve the environmental protection supporting system, ensure the effective implementation of relevant policies and measures, promote citizens to exercise the right to know, to participate in environmental affairs, to plead, and to supervise environmental affairs, and thus enhance the public's rational understanding and scientific participation in environmental protection.

Fourth, we should promote the establishment of a sound multi-party settlement mechanism for environmental and resource disputes. The legal supervision power of procuratorial organs is exercised not only throughout environmental criminal litigation, but also in the procuratorial work related to administrative and public-interest litigation: they supervise not only the courts to exercise impartial justice and but also administrative organs to administer according to law, thus playing an important and irreplaceable role in facilitating the settlement of conflicts and disputes over the environment and resources. We must follow and develop the "Fengqiao Model" in the new era, step up efforts to resolve conflicts and disputes over environmental protection, and promote the integration of diverse forces and resources and the linkage between mediation, arbitration, administrative reconsideration, administrative adjudication, administrative litigation, and procuratorial supervision. Organically connect. The goal is to form a multi-party settlement mechanism for environment and resource disputes based on information and resource sharing, complementary advantages, and win-win cooperation, and prevent the escalation of these disputes into criminal cases or mass petitions.

4 Strengthening Team Building and Establishing a Team of Prosecutors with Comprehensive Environmental Knowledge and Legal Expertise

Environmental cases are hard to handle for they often involve knowledge and multiple laws and regulations in such fields as environmental protection, land, forestry, water conservancy, and atmosphere, with high proof standards. To strengthen the environment-related procuratorial work, we must work to set up specialized organs and cultivate specialized talents.

One is to establish a specialized department for legal supervision over environment-related procuratorial work. Specialized procuratorial organs are platforms and carriers to improve the professional level of environment-related procuratorial work. For example, the Huzhou Municipal Procuratorate was the first in Zhejiang Province to open a prosecutor's office in the ecology and environment bureaus at the municipal and county levels, and establish a daily liaison mechanism for case consultation, information sharing, and clue transfer. The Quzhou Municipal Procuratorate of Zhejiang has established a four-in-one working mechanism consisting of "punishment, repair, prevention, and protection," promotes the model of "specialized legal supervision + restorative justice + integrated social management," and prompts the two-level procuratorates across the city to set up specialized organs or task forces tasked with the "approval of arrest, prosecution, litigation supervision, crime prevention, and participation in comprehensive social governance" in environment-related procuratorial work. At the same time, in response to the frequent forest-related cases and increasing environmental-pollution cases in Quzhou, it actively sought government support, opened prosecutor's offices in the two-level forestry public-security organs and the Quzhou Green Industry Cluster Zone, and built new working platforms in different areas. Guizhou has taken the lead in China with 37 specialized procuratorial organs for ecological protection, covering key ecological function areas, key river basins and key forest areas. It has embarked on a "green channel" and "new reform road" for specialized judicial protection for the ecology and environment. Given the current serious environment and resource problems, there are many high-profile environment and resource cases that require in-depth knowledge of areas new to prosecutors. Hence, procuratorial organs are carrying a heavy load to explore the appropriate working mechanisms, and are reforming their internal setup. When conditions permit, procuratorial organs dedicated to environmental cases should be established and dedicated prosecutors assigned to handle criminal cases of environmental damage. An integrated working model of "arrest, prosecution, supervision, and prevention" should be in place to ensure the quality and efficiency of case handling, and crack down on crimes damaging the ecology and environment.

The other is to train specialists. As environmental issues are all-encompassing, complex and diverse, the environment-related procuratorial work is specialized and complex. For environmental cases, procuratorial organs need to deal with almost all environmental regulatory departments. The procuratorial work must be specialized to meet the needs of diverse parties involved. Therefore we should strive to cultivate a group of inter-disciplinary specialists who know how to handle environmental cases. If the prosecutors, though familiar with law, know little about ecology, environment and resources, are unfamiliar with industry standards or regulatory rules, and are not sensitive to data about administrative law enforcement, their judgment will be blinded, which is not good for the fight against crime. At present, the shortage of prosecutors specialized in environmental cases has become an obvious bottleneck preventing procuratorial organs to play their role. That's why talent training is so urgent. We must continue to prioritize capacity building for the environment-related procuratorial work, highlight the importance of field experience, practicality and actual effective of training, and strengthen graded and classified training to cultivate stronger professional skills, capabilities, thinking and professionalism. We shall strengthen on-the-job training and practice, diversify the knowledge structure of prosecutors, and comprehensively improve their ability to apply laws and policies, guard against risks, work with the masses, apply the latest technology, and guide public opinion. To meet the needs of comprehensive environmental protection, we shall work with ecology & environment departments, public security organs, and people's courts to establish a sound joint training mechanism for law enforcers for environmental protection to raise their awareness of community. We shall promote the formation of a law-enforcing and judicial team composed of environmental police, environmental prosecutors, and environmental judges, to provide strong human-resources support for the specialized judicial services for environmental protection.

Research on the Environmental Public Interest Litigation System Under the Guidance of Xi Jinping Thought on Ecological Civilization

Zaicun Wu

The Xi Jinping Thought on Ecological Civilization lays the fundamental guideline for China to construct its environmental public interest litigation (EPIL) system. To follow Xi Jinping Thought on Ecological Civilization is to consciously implement its scientific outlook on nature, on ecological democracy, on green development, and on law-based ecological protection in the construction and improvement of the EPIL system. To this end, we must continue to safeguard the public interest and uphold the democratic concept of co-governance for all, prioritize prevention, safeguard national ecological security, scientifically construct a public interest litigation (PIL) system and improve relevant measures. The ultimate goal is to make EPIL a powerful weapon for building a beautiful China, a key to joint environmental protection for all, and an important means for the people's courts to participate in state governance.

1 The Rich Connotation and Important Value of Xi Jinping Thought on Ecological Civilization

To build an ecological civilization, we need a scientific world view and methodology, the guiding vision of green development, and a path for public engagement and democracy in public interest undertakings. Of course, the most important is the strong support of legal system. Only with a holistic, systematic understanding of the Xi Jinping Thought on Ecological Civilization can we fully implement it in judicial practices for environmental protection.

Although the EPIL system has been discussed academically for a long time, it did not enter the legislation until 2012 when the *Civil Procedure Law* was amended. But the *Civil Procedure Law* provide only general principles and have limited operability,

Z. Wu (✉)
Beijing No 1 Intermediate People's Court, Beijing, China

J. Pan et al. (eds.), *Beautiful China: 70 Years Since 1949 and 70 People's Views on Eco-civilization Construction*, https://doi.org/10.1007/978-981-33-6742-5_64

it is imperative to construct a sound PIL system through judicial practice, for which the Xi Jinping Thought on Ecological Civilization offers valuable guidance.

First, the scientific outlook on nature of the Thought represents the highest goal for constructing the EPIL system. EPIL is a type of PIL for environmental protection, and its system is of course constructed to protect the public interest in environmental protection. The scientific outlook on nature, which advocates respect for nature, conforming to nature, protecting nature, sustainable development and harmonious coexistence of man and nature, inevitably manifests the biggest "public interest" in the system for environmental protection. While defining and improving the qualifications of principal players, the scope, and the compensation system of EPIL, we must follow the scientific outlook on nature to ensure that the EPIL system can truly work.

Second, the concept of green development in Xi Jinping Thought on Ecological Civilization is the ideological basis for constructing a sound EPIL system and effectively balancing the relationship between economic development and environmental protection. Green development requires a balanced relationship between economic development and environmental protection, but PIL is often based on public interest damage, and its institutional value echoes the value in the traditional areas of infringement, with emphasis on the compensation of damage and legal punishment. To uphold the concept of green development, we must place equal emphasis on incentive and punishment, and balance environmental protection and economic development while exploring and designing specific systems. The Supreme People's Court has called for piloting installment payments based on such factors as the subjective fault and operating conditions of the indemnity obligor, and exploring diversified means to assume responsibilities." This is an attempt to balance the relationship between incentive and punishment and in line with the concept of green development.

Third, the concept of ecological democracy in Xi Jinping Thought on Ecological Civilization is a strong guarantee for constructing a sound EPIL system that engages the general public, democratic, and people-centered. As far as environmental protection is concerned, public powers alone produce limited effects. From the evolution and regulatory structure of PIL in various countries, we can see that public engagement has been considered as an important factor since its birth. In other words, the public power of environmental protection is re-allocated to give the public extensive governance authority. To uphold the ecological democracy concept, we must consider as much as possible the procedures and forms of public participation in justice when designing the system, and effectively follow the path of modern environmental governance that is non-for-profit, democratic and participated in by the general public.

Fourth, the concept of law-based ecological protection is the fundamental guarantee for improving the modern environmental governance system by promoting the rule of law. "The most rigorous environmental protection system and law-based environmental protection concept adopted demonstrate the central government's firm determination to promote ecological progress, and at the same time highlights the importance of rule-of-law thinking and methods." Therefore, the people's courts must follow the requirements of Xi Jinping Thought on Ecological Civilization for strengthening institutional and legal support for environmental protection and earnestly implement the basic national policy of saving resources and protecting the environment. They shall continuously reform the PIL system and mechanism, improve the judicial procedures for EPIL, and develop the specialized team of judges. They shall effectively implement the most rigorous systems for source protection, damage compensation and accountability, and continuously improve judicial services for environmental protection in the new era.

2 The Basic Concepts for the People's Courts to Construct and Improve the EPIL System Under the Guidance of Xi Jinping Thought on Ecological Civilization

First, we must safeguard the public interest as soon as possible, in a continuous and complete way. The fundamental purpose of EPIL is to protect the environment, including the restoration of the damaged environment, the cessation of the infringement, and the prevention of possible damage. The filing of EPIL means that the damaged environment has real harm and is in urgent need of protection. Once the judicial procedures are activated for the EPIL, what comes first should be public interest protection, not the punishment and evaluation function of the judgment. Therefore, first, the court should continue to safeguard public interest. From the perspective of protecting the environmental public interest, we should consider not only the insufficient capacity or impure motivation of the social organization in question, but also how to properly exercise judicial power in litigation so that the protection of environmental public interest is prioritized. To this end, in addition to appropriately relaxing the qualification threshold for social organizations to file litigation and boosting their initiative to file PIL, we should improve the efficiency of litigation, and make mediation and withdrawal as easy and convenient as possible; forbid the interruption of requirements and arbitrary withdrawal of the lawsuit and emphasize the continuity of protection; oppose lowering the value of public interest restoration, and emphasize the priority of ecological restoration and the conformity of the trial results to the purpose.

Second, the democratic concept of co-governance for all shall be upheld. Public engagement is the basic principle of environmental law generally followed by countries all over the world. It generally means that "the public has the right to participate in all development decisions related to public environmental rights and interests through certain procedures or channels, and is entitled to corresponding legal protections and remedies to avoid blind decision-making and make decisions in favor of the vital interests and needs of the general public." Under the guidance of Xi Jinping Thought on Ecological Civilization, Article 5 of the *Environmental Protection Law* amended in 2014 stipulates that "Environmental protection adheres to the principles of giving priority to protection, focusing on prevention, conducting comprehensive treatment, engaging the general public, and enforcing accountability for damage." Thus the principle of public engagement in environmental protection is formally written in law. "It is foreseeable that in the long-term future China will firmly, fully and effectively mobilize the public to participate in environmental protection in accordance with the law, which will help to gradually form a sound environmental protection pattern featuring government-enterprise-masses (society) interaction."

Third, the principle of "giving priority to protection and focusing on prevention" should be upheld in protecting environmental public interest. The filing of EPIL means that the damaged environment has real harm and is in urgent need of protection. And "large-scale pollution and environmental damages caused by illegal activities cannot be remedied to a satisfactory extent by financial compensation or environmental restoration and other remedies." Therefore, it is necessary to establish the principle of "giving priority to protection and focusing on prevention." In practice, we must strictly abide by Article 100 and 101 of the *Civil Procedure Law*, and resort to behavior preservation and advance execution as soon as possible to prevent the occurrence and expansion of environmental damage. We shall also give full play to the role of administrative PIL, effectively supervise administrative organs to administer in accordance with the law, and promptly exercise public powers to stop environmental damage that has occurred or is likely to occur.

Fourth, we should strive to safeguard national ecological security. In the National Environmental Protection Outline issued by the State Council in November 2000, the tasks and goals of "maintaining national environmental security" are specified for the first time. At the National Environmental Protection Conference held in May 2018, General Secretary Xi Jinping made a clear-cut proposal to build an "ecologically safe society" in the twenty-first century. It's fair to say that the "ecological security system" is the cornerstone and safeguard of the "national security system." The people's courts should play the role of trials to safeguard ecological security. They should continuously improve the trial system and mechanism for environmental and resource cases and the specialized trial mechanism for such cases, promote the specialized trial of criminal, civil, and administrative litigation related to environment and resources, and properly coordinate the criminal, civil and administrative legal responsibilities that the parties should bear, and promote the integrated environmental protection and restoration. By so doing, they can provide strong judicial support for resolutely fighting pollution, including air pollution, water pollution, soil pollution and agricultural and rural pollution.

3 The Initial Achievements Made by the People's Courts in Constructing the EPIL System Under the Guidance of Xi Jinping Thought on Ecological Civilization

Since the implementation of the PIL system, the people's courts have established the framework for EPIL through theoretical research and practices under the guidance of Xi Jinping's important instructions and especially the Xi Jinping Thought on

Ecological Civilization. The Supreme People's Court has established the basic system for EPIL and has played an important role in the trial of EPIL cases and environmental protection.

First, the courts have properly handled a number of serious environmental pollution problems by trying EPIL cases in accordance with the law. According to statistics, from January 2015 to the end of September 2018, courts across the country accepted 2041 EPIL cases and concluded 1335 of them. Among them, there were 205 civil EPIL filed by social organizations, 98 of which were closed; 1836 filed by procuratorial organs, and 1237 were closed. These include the Taizhou case known for its "astronomical amount of compensation," the serial PIL cases against the pollution of Tengger Desert that have the ears of the central government, the high-profile civil EPIL against Zhenhua Co., Ltd. under Jinghua Group in Dezhou Shandong for air pollution, and the serial cases against the water pollution threatening the high-quality development of the Yangtze River Economic Belt. The courts have sentenced all the responsible persons to undertake environmental restoration or alternative restoration, and even pay for the expensive environmental restoration. By so doing, the courts have protected the environment in a timely and effective manner and solved some long-standing environmental problems.

Second, by hearing EPIL cases in accordance with the law, the courts have continuously improved the judicial capacity for environmental protection and resources conservation. According to the statistics of the Environmental and Resources Tribunal of the Supreme People's Court, since the second half of 2017, specialized environmental-case hearing agencies across the country have continued to grow rapidly in numbers. As of the end of December 2018, the people's courts of 31 provinces, municipalities directly under the central government and autonomous regions had opened a total of 1271 specialized environmental case hearing organs, including 391 trial tribunals, 808 collegiate panels, and 72 circuit tribunals. There were 257 community-level people's courts, 110 intermediate people's courts, and 23 high-level people's courts which had set up specialized trial tribunals for environmental cases. Since the second half of 2017, courts at all levels have strengthened the selection and training of judges, and the building of professional supporting teams, to continuously improve the trial capabilities for environmental cases. Meanwhile, specialists are playing a much bigger role in fact finding, causality determination, determination of restoration plans, and acceptance inspection of restoration results, providing professional basis and intellectual support for judges to properly exercise their discretion in accordance with the law. They either serve as the people's juror to directly participate in the trial of the case; or as an expert assistant applied by one or both parties; or as a member of the expert advisory committee or expert reserve to advise the judges."

Third, by hearing EPIL cases in accordance with the law, the people's courts continuously improve and reform the EPIL system. The serial EPIL cases against the pollution of the Tengger Desert, filed by China Biodiversity Conservation and Green Development Foundation, sparked widespread controversy. In these cases, the Supreme People's Court clarified that whether a social organization is eligible for filing the EPIL should be determined by the following three factors: whether its purpose and scope of business include the maintenance of environmental public interest, whether it is actually engaged in environmental protection activities, and whether the environmental public interest it tries to safeguard is related to its purpose and scope of business. This was the first time the court used specific cases to clarify who can file EPIL at the judicial level, which has promoted the development of the EPIL system, and played an important guiding and exemplary role in the trial of civil EPIL cases.

Fourth, by hearing EPIL cases in accordance with the law, the people's courts continuously enrich and diversify the content and channels of public engagement. In the case of Changzhou Environmental Public Interest Association vs. Chu Weiqing and Changzhou Boshi'er Material Recycling Co., Ltd., the court demanded the assessment company to produce three environmental restoration plans, then published them around the contaminated site, and collect public opinions on them by distributing questionnaires on the spot. The final environmental restoration plan was chosen with the public opinion as an important reference and based on the case situation. In the case of the Lianyungang District People's Procuratorate of Lianyungang City vs. Yin Baoshan et al. for illegal fishing of aquatic products, the court disclosed the ecological restoration plan to the public through local media, the court's official account on Weibo and WeChat, solicited public opinions extensively, and obtained proof of the rationality of the restoration plan issued by relevant functional departments, creating a new mechanism for guiding the public to participate in environmental justice.

Fifth, by hearing EPIL cases in accordance with the law, the people's courts have more effectively prioritized protection. As Professor Lv Zhongmei said, "Environmental restoration requires continuous investment and inspection, which is difficult for the court to follow up, monitor and enforce for a long time." In the Taizhou EPIL case, the court of second instance partially changed the payment method of environmental restoration costs, allowed partial deferral and conditional deduction, guided and encouraged the enterprise concerned to actively upgrade technologies for environmental protection, and thus from the source reduce the possibility of environmental pollution. This practice conformed to the principles proposed in the *Environmental Protection Law* of focusing on prevention and combining prevention with control, and generated judicial, social, and environmental benefits at the same time. It can be regarded as an original way of assuming environmental tort liabilities, and is the fruit of judicial wisdom.

4 The Path for Deeply Implementing Jinping Thought on Ecological Civilization and Improving the EPIL System

1. The theoretical system and conceptual basis for the EPIL improvement

First, we should fully understand the philosophical characteristics embodied in Xi Jinping Thought on Ecological Civilization, and on this basis construct a theoretical basis system for environmental law and EPIL. As mentioned above, the Xi Jinping Thought on Ecological Civilization is rich with theoretical connotations. Among them, the scientific outlook on nature about "harmonious coexistence of man and nature" is powerfully philosophical, reflective and critical. Traditionally Western philosophy emphasizes the human "subjectivity" and discusses the attributes of objects, the relationship between humans and objects, and how people perceive objects from the primary point of view of "humans" as the subject, which is a stance of "anthropocentrism." In contrast, based on Marxism and traditional Chinese philosophy, Xi Jinping put forward the theory of "community of life," which "from the height of the dialectical unity of subject and object, unites mankind and nature in the profound and grand concept of 'community of life,' reveals the essential man-nature relationship and the correct way for humans to understand the world, transcends and bridges the two opposing schools of anthropocentrism and naturalism in western ecological ethics." If we say that the legal system of any country and period is based on the philosophy of the times, then to improve today's environmental law and EPIL system, we must base on the Xi Jinping Thought on Ecological Civilization, a philosophical thinking of the new era. As Professor Lv Zhongmei pointed out, "The object of environmental legal relations has certain subjectivity." But the theory of legal relations based on traditional philosophy does not recognize the independent value of objects. The environmental law is based on environmental philosophy. People-oriented, it regards the natural environment as the object of human understanding and the means for human development, at the same time attaches importance to the value of the natural environment and demands the establishment of harmonious relationship between man, nature and individuals. The object of environmental legal relations under this concept is no longer at the whim of the subject." Therefore, to construct the basic theoretical system for PIL, we need to interpret the dialectical relationship between man and nature with a new philosophical basis and from a new theoretical perspective.

Second, we should straighten out the relationship between the environmental legal system and the traditional legal system. In current judicial practice, EPIL generally invokes the *Tort Liability Law* as the basis of substantive law. But many hard-to-explain theoretical issues will arise by handling environmental torts or EPIL through the tort system. For example, from the perspective of tort liability, which rights or interests of whom are violated by environmental damage? If it is the environmental public interest, then how to define public interest? Is public interest as the direct object of violation too general and too vague in the legal system? The Reform Plan for the Environmental Damage Compensation System is under trial implementation, which gives the provincial governments and below the power to file a lawsuit as the owner of natural resources. What separates PIL from traditional civil litigation is that the plaintiff's qualifications are given by legal techniques, and the plaintiff is not the "private possessor" entity of the public interest; in other words, the plaintiff does not hold the substantive rights. On this basis, when the provincial government or below as the owner of natural resources files a lawsuit claiming environmental damage compensation, some questions will inevitably arise: What is the relationship between the environmental damage compensation lawsuit and the EPIL? Does the filing of environmental damage compensation litigation makes EPIL an ancillary lawsuit other than ownership protection, with the latter's scope greatly squeezed? To answer these questions, we must fully understand the difference between EPIL and traditional law when trying to improve the EPIL system, carefully weigh the similarities and differences between tort liability and environmental damage compensation, and rationally construct a sound EPIL system.

Third, we should continue to improve the compensation system for ecological damage to ensure that the maximum institutional value of EPIL is realized through procedures. *The Interpretation of the Supreme People's Court on Several Issues Concerning the Application of Laws to the Trial of Environmental Civil Public Interest Litigation Cases* stipulates the methods of assuming responsibility for ecological damage. In particular, the court can rule that the defendant should repair or adopt alternative repair plans or pay for the environmental restoration cost. But ecological restoration is a complicated process and covers many tasks, including the investigation and validation of environmental damage, the analysis of the causal relationship between environmental pollution or the behavior of ecological damage and the fact of environmental damage, and the physical quantification of environmental damage at the litigation stage; in the court ruling, the determination of ecological restoration goals and costs and the formulation of restoration plans; and at the execution stage, the implementation of the restoration plan and the evaluation of restoration effects. Therefore, in the next step, we should consider cultivating and expanding specialized appraisal and evaluation agencies, or forming a temporary expert group to strengthen the feasibility of the restoration plan and maximize the restoration effect within the budget. In the repair process, we should strengthen supervision and carefully organize acceptance checks to ensure that the expected effect be achieved. Regarding the management of restoration funds, "we must take a long-term view and continue to promote the establishment of a national management system for environmental restoration funds; meanwhile, based on reality, we can consider transferring the restoration funds to the soil pollution control funds for management or including them into the environment damage compensation funds in accordance with the *Soil Pollution Prevention and Control Law* and the *Reform Plan for the Environmental Damage Compensation System.*"

Fourth, we should continue to form a professional trial pattern based on typical cases. The Supreme People's Court is constantly advancing the construction of the trial system and mechanism for environmental cases. In practice, with the appearance of specialized trial departments and tribunals and the introduction of technical experts, a professional trial model with clear division of labor is taking form. But environmental protection involves different local courts which may disagree in trial thinking and differ in trial techniques, and technical experts may not always be able to appear in court to participate in litigation. Therefore, constructing a sound EPIL guidance case system and giving full play to the guiding value of typical cases for trials is the optimal choice for China to improve the EPIL trial mechanism at this stage. In the next step, we suggest determining the general principles and technically feasible practices for handling cases of similar infringements and damage facts based on analysis of typical cases, so as to provide technical support for other courts and promote specialized trials.

2. **Improving the specific EPIL system**

First, we should strictly review and restrict the exercise of the party's right of disposition to prevent any harm to the public interest. In PIL, the physical disposition involves public interest and must be restricted. One is to determine the exercise of the right of disposition based on the degree of public interest realization. If the plaintiff's claim has indeed been fulfilled through settlement or the restoration is made or danger eliminated by the defendant's actions, the plaintiff may be allowed to withdraw the lawsuit. If the plaintiff's withdrawal of part of the lawsuit does not hinder public interest maintenance, it should also be permitted. Two is to treat the procuratorate and other organizations equally. The withdrawal by the procuratorate must also comply with the principle of public interest maintenance. The procuratorate's exercise of the right to withdraw a PIL lawsuit will defeat the court's authority-based review of public interest maintenance, and does not align with the PIL's ultimate purpose.

Second, we should to continue to improve the procedures for the procuratorate to file PIL. The filing of civil administrative PIL and administrative PIL by the procuratorate should comply with the *Civil Procedure Law* and the *Administrative Procedure Law*, for although the *Constitution* grants the procuratorate the power of legal supervision, their power to file PIL comes from the basic civil and administrative procedural laws. Therefore, the procuratorate should qualify as a plaintiff in the proceedings. There has been a lot of debate about the procuratorate's role in PIL. Though it has been getting clearer since the issuance of the *Interpretation of the Supreme People's Court and the Supreme People's Procuratorate on Several Issues Concerning the Application of Laws in Public Interest Litigation Cases*, there is still room for discussion. For example, if the procuratorate is identified as the plaintiff, should it be referred to as "plaintiff" or "public interest litigant" in the court and in the judgment? What if procuratorates at different levels disagree on whether to appeal or withdraw the case, and on facts and legal issues? Should the opinion of the superior procuratorate prevail, or that of the lower-level procuratorate that filed the lawsuit? And should the court suspend the procedure to wait for the two to resolve the dispute? We need specific answers to these questions. In the administrative PIL filed by the procuratorate, "we must pay attention to the active role of pre-litigation procedures in urging administrative organs to perform their duties, and examine whether the claims of the administrative PIL suit is consistent with the procuratorial recommendations made in the pre-litigation procedure; if not, an explanation should be made to the People's Procuratorate." When reviewing pre-litigation procedures, we must be objective and fair. We must consider not only the actual performance but also the ability of the administrative organ to perform its duties. "We should consider whether the administrative organ has initiated the filing, investigation and other procedures of administrative punishment in a timely manner after receiving the procuratorial recommendations, and whether there are objective obstacles that prevent it from performing its duties within two months. Under special circumstances such as the restoration of vegetation and soil and pollution control, when the respondent administrative agency which is willing to make rectification, due to objective limits such as seasonal climate, construction conditions and period, fails to complete the rectification within the period recommended by the procuratorate, it should not be simply deemed as failure to perform duties according to law."

Third, we should explore the composite litigation model. EPIL involves the entanglement of rights and obligations between multiple parties. A case of environmental damage caused by pollution may involve the private interests of individual rights holders, and cause damage to the ecosystem as a public interest. Hence it may involve infringements, criminal issues, and administrative issues caused by the poor performance of duties by regulatory authorities. In terms of efficiency, it is obviously time-consuming and labor-intensive to handle a case of environmental pollution via separate litigation procedures. Therefore, "an important mission facing current legislation for environmental litigation is to establish the symbiosis between man and nature and realize integrated ecological protection based on substantial interaction and integration of environmental issues." Considering the fact that the damage to public and private interests is based on environmental damage, some scholars suggest the combined trial of environmental civil public- and private-interest litigation cases to resolve the current dilemma faced by separate litigation, which is undoubtedly the most appropriate solution so far, and can accumulate experience for the ultimate legislation and research for environmental (civil) procedures. In addition to the civil PIL and administrative PIL initiated by the procuratorate, Article 20 of the "Interpretation of the Supreme People's Court and the Supreme People's Procuratorate on Several Issues Concerning the Application of Law in Public Interest Litigation Cases Filed by Procuratorates" clearly stipulates the system for procuratorates to file civil PIL collateral to criminal procedures, which has become an important type of EPIL, and whose specific rules are being laid down. The composite litigation model can solve public and private interest issues at the same time and improve the litigation efficiency and realize full, holistic environmental protection at the same time.

Fourth, we should continue to explore the establishment of a punitive compensation system. Punitive compensation refers to the compensation whose amount decided by the court exceeds the actual damages. It is based on the damage suffered by the victim, that is, a special punitive measure based on the loss to be compensated. It is stipulated in the General Principles of the Civil Law, Tort Liability Law, and Consumer Rights Protection Law, and in practice, it can effectively deter offense. In EPIL, there used to be the problem of calculating multiples based on the virtual governance cost, but the multiples here are not punitive compensation but represent the ceiling and floor of the damage amount. Therefore, we may consider applying punitive compensation to actors who intentionally damage or let others damage the environment to further deter future environment damages. Next, we should demonstrate the feasibility of punitive compensation in EPIL, its applicable conditions and how to calculate the penalty amount.

Fifth, we should introduce new ways of engaging the public to truly implement the democratic concept of "co-governance of public interest for all." Public engagement embodies the modern concept of democratic public interest. Its breadth and depth should both be considered. It is necessary to ensure the breadth of public engagement, which can be resolved through extensive public outreach. What's more important is its intensity and depth, which is determined by the professionalism required in EPIL. For one thing, we can effectively apply the expert witness system according to the *Civil Procedure Law*, value the opinions of expert witnesses, and realize checks and balances on appraisal opinions, to restrain the decisive role of appraisal opinions in the court judgment and thus ensure the overall fairness of the judgment. Meanwhile the active use of the expert witness system is itself an effective means to ensure deep public engagement in EPIL. For another, we should continue to expand the coverage of public announcements. In addition to mediation, we can make public the restoration plan, the time limit and means for restoring the original state or eliminating the danger and solicit public opinions, especially expert opinions. We should issue more announcements and notices targeting residents living in damaged areas and specialized agencies. We should purposefully establish contacts with residents and institutions in the affected areas, to solicit their support and suggestions on the prevention of damage, the determination of the restoration plan and the amount of damage compensation, etc., for scientific decision-making. The objection procedure after the announcement is issued should be introduced. For example, a restricted number (such as three or more) of ordinary people may jointly file the objection in writing, with reasons attached; and it should stipulate the legal consequences that may arise when the conditions are not met.

National Ecological Civilization Pilot Zone: From Theory to Practice

Qingzhi Huan

Strictly speaking, the National Ecological Civilization Pilot Zone is just a public policy initiative carried out in a very short period of time. But the winners of this designation have in fact been working in this regard for a long time under other names. More importantly, the *Master Plan for Reform of the System for Developing an Ecological Civilization* and the *Opinions on Establishing National Ecological Civilization Pilot Zones with Uniform Standards* officially give them the national title, which raises our expectation for the orientation and demonstration effect of the practices in the provinces of the Zone. In this paper, I choose Fujian, Hainan, and Guizhou provinces for case study, hoping to promote the academic research on not only the National Ecological Civilization Pilot Zone but also the national ecological civilization pilot demonstration projects in a more general sense.

1 Theoretical Basis for the Construction of National Ecological Civilization Pilot Zone

The direct theoretical or policy basis for the construction of the National Ecological Civilization Pilot Zone is the political report to the 18th National Congress of the Communist Party of China held in November 2012. The report emphasizes that promoting ecological progress is a long-term task of vital importance to the people's wellbeing and China's future, a necessary condition for the lasting and sustainable development of the Chinese nation and global ecological security, and therefore, we must give high priority to making ecological progress and incorporate it into all aspects and the whole process of advancing economic, political, cultural, and social progress, i.e. the so-called Five-Sphere Integrated Plan. Meanwhile, while vigorously promoting ecological progress, we must focus on the four major strategic

Q. Huan (✉)
School of Marxism, Peking University, Beijing, China

J. Pan et al. (eds.), *Beautiful China: 70 Years Since 1949 and 70 People's Views on Eco-civilization Construction*, https://doi.org/10.1007/978-981-33-6742-5_65

deployments of "improving development of China's geographical space," "promoting all-round resource conservation," "intensifying protection of the ecosystem and the environment" and "enhancing system building to promote ecological progress." Particular efforts should be made in institutional construction and innovation for environmental administrative regulation, the use of environmental economic means, the performance evaluation of ecological conservation, rewards and punishments, and ecological conservation publicity and education.

Based on this, the *Decision of the CPC Central Committee on Some Major Issues Concerning Comprehensively Deepening the Reform* adopted at the Third Plenary Session of the 18th CPC Central Committee in November 2013 proposes to accelerate the establishment of a systematic and complete set of institutions for promoting ecological progress by building a beautiful China and deepening the reform of the system for developing an ecological civilization. Under the heading of "Accelerating system building for promoting ecological progress," Article 51 stipulates to "improve the natural resources property rights system and use control system" and Article 53 to "implement the system for paying for resources consumption and compensating for ecological damage." Both broadly fall into the scope of the ecological (environmental) economic system. And "defining ecological protection red line" in Article 52 and "reforming the environmental protection management system" in Article 54 roughly fall into the scope of the environmental regulation and governance system (but the ecological civilization demonstration zone or pilot zone is not mentioned).

The *Opinions on Accelerating Ecological Progress*, reviewed and approved by the Political Bureau of the CPC Central Committee on March 24, 2015, consists of 9 parts and 35 articles, including general requirements (guidelines, basic principles, and main objectives); strengthening the function positioning of main players; promoting technological innovation and structural adjustment; comprehensively promoting conservation, recycling and efficient utilization of resources, and accelerating the fundamental transformation of resources utilization; intensifying the ecosystem and environmental protection and effectively improving the set of institutions for promoting ecological progress; strengthening the statistical monitoring and law enforcement supervision for promoting ecological progress; accelerating the formation of a good social ethos for promoting ecological progress; and effectively strengthening organizational leadership. In the part of "Effectively Strengthening Organizational Leadership—Exploring Effective Models," it clearly states that it is necessary to "formulate the *Master Plan for Reform of the System for Developing an Ecological Civilization* as soon as possible, deepen the construction of ecological civilization pilot zones, and explore effective models for promoting ecological progress in areas at different development stages, with varying resource and environmental endowments, and different function positioning."

On this basis, in September 2015, the CPC Central Committee and the State Council issued the *Master Plan for Reform of the System for Developing an Ecological Civilization*, emphasizing the need to construct eight sound basis systems: the natural resources property rights system, the land and space development and protection system, the spatial planning system, the total resource management and comprehensive conservation system, the system for paying for resources consumption and compensating for ecological damage, the environmental governance system, the environmental governance and ecological protection market system, the performance evaluation and accountability system for ecological conservation. In the section "Supporting the Implementation of the Reform of the System for Developing an Ecological Civilization—Developing Pilot Projects," it clearly proposes to "merge the departmental comprehensive ecological civilization pilot projects to form a national pilot project which each department shall guide and promote in accordance with their respective responsibilities."

It can be seen that although various comprehensive ecological construction pilot projects, including the National Ecological Civilization Pilot Zone project, are more closely related to the strategic deployment and general requirements of the system building for promoting ecological progress, they are not directly part of system construction (such as the natural resources property rights system, the system for paying for resources consumption and compensating for ecological damage, etc.), but an important guarantee for their implementation (just like strengthening the leadership over the system reform for developing an ecological civilization) or efforts with a more comprehensive goal.

It must be noted that before the National Ecological Civilization Pilot Zone, there were already ecological civilization (pilot) demonstration zones in various forms and at different levels across in China, some of which were organized by ministries and commissions under the State Council or by provincial administrative regions. The highest-level among them after the 18th CPC National Congress in 2012 included the National Ecological Civilization Pilot Demonstration Zone with the Ministry of Environmental Protection as the lead organizer, the National Ecological Civilization Pilot Demonstration Zone jointly launched by seven ministries and commissions including the National Development and Reform Commission (NDRC), the National Water Ecological Civilization Pilot City under the leadership of the Ministry of Water Resources, and the National Marine Ecological Civilization Demonstration Zone by the Ministry of Land and Resources.

Among them, the National Ecological Civilization Pilot Demonstration Zone under the leadership of the Ministry of Environmental Protection was the first attempt of its kind, and could be traced back to 1999 when Hainan took the lead in the ecological province (city, county) initiative. Since then, 13 other provinces, municipalities directly under the central government and autonomous regions have followed Hainan in the pilot project of building ecological provinces (cities, counties). In 2008, the Ministry of Environmental Protection issued the *Opinions on Promoting Ecological Progress*, clarifying the guidelines and basic principles for promoting ecological progress, and demanding the establishment of an industrial system, environmental safety, culture, ethics, institutions and mechanisms that meet the requirements of ecological conservation. It also decided to organize a national pilot project for promoting ecological progress. In June 2013, approved by the central government, the Ecological Conservation Demonstration Zone was officially renamed the Ecological Civilization Demonstration Zone. As of October 2013, the Ministry of Environmental Protection had approved 125 national ecological civilization pilot demonstration zones in six batches, including 19 prefecture-level and two cross-administrative regions or river basins, but there was no pilot zone covering an entire province, municipality directly under the central government or autonomous region. And 70% of these zones are in Jiangsu, Zhejiang, Liaoning, Guangdong and Sichuan with a sound ecological environment and a strong economic foundation.

After the 18th CPC National Congress in 2012, ministries and commissions under the State Council obviously shifted more attention to the ecological conservation pilot project, and rolled out their respective plans. In December 2013, the NDRC and five (which became six) other ministries and commissions including jointly proposed the *National Ecological Civilization Pilot Demonstration Zone Plan (Trial)* based on the national function zoning. In June 2014, the NDRC and other five ministries/commissions jointly issued the *Notice on Printing and Distributing the National Ecological Civilization Pilot Demonstration Zone Plan (Trial)*, officially launching the national ecological civilization pilot demonstration zone project. In the end, the project selected 102 administrative regions, river basins and ecological regions, including Miyun District of Beijing, as pilot zones in two batches. In March 2014, the State Council issued the *Several Opinions on Supporting Fujian Province's Further Implementation of the Ecological Province Strategy and Accelerating the Construction of Ecological Civilization Pilot Demonstration Zone*, making Fujian the first ecological civilization pilot demonstration zone (province) directly designated by the State Council. Later, Jiangxi, Guizhou, Yunnan and Qinghai formed the first batch of national ecological civilization pilot demonstration provinces (regions) organized by the NDRC and six other ministries/commissions. After that, in accordance with the *National Ecological Civilization Pilot Zone Plan (Trial)*, *National Ecological Civilization Demonstration Zone Management Regulations (Trial)*, and *Indicators for National Ecological Civilization Demonstration Counties and Cities (Trial)*, these five provinces formulated their respective implementation plans.

In October 2015, the Fifth Plenary Session of the 18th CPC Central Committee decided to establish national ecological civilization pilot zones with uniform standards, with the focus on carrying out comprehensive experiments on system reform for promoting ecological progress, regulating various pilot demonstration projects, and exploring ways and accumulating experience for improving the set of institutions for developing an ecological civilization. On August 22, 2016, the General Office of the CPC Central Committee and the General Office of the State Council issued the *Opinions on the Establishment of National Ecological Civilization Pilot Zones with Uniform Standards* and the *Implementation Plan for National Ecological Civilization Pilot Zone (Fujian)*, making Fujian the country's first National Ecological Civilization Pilot Zone (Province). Fujian was followed by Jiangxi and Guizhou in 2017, and Hainan in 2018, whose respective implementation plans were all subsequently approved. In September 2017, the Ministry of Environmental Protection launched the National Ecological Civilization Demonstration City/County project and the construction of innovation base for practice guided by the theory that lucid waters and lush mountains are invaluable assets. As of the end of 2018, 91 cities/counties including Yanqing District of Beijing had been named demonstration cities/counties, and 29 areas including Anji County in Zhejiang named innovation bases.

2 National Ecological Civilization Pilot Zones: Fujian, Hainan and Guizhou

In addition to east-west geographical balance, an important reason for choosing Fujian, Jiangxi, Guizhou and Hainan among the first batch of national ecological civilization pilot zones is their advantageous natural endowments: they are all located in the south of the Yangtze River, and boast abundant rainfall, high forest coverage, and a large environmental capacity. As a whole, they all adopt the "green development" path or model as I have outlined, although they might vary in the choice of where to start.

1. **Fujian**

Fujian Province is known to have a superior ecological environment, with "80% covered by mountains, 10% by rivers and 10% by farmland," and is an early starter in ecological conservation. In 2000, Fujian put forward the strategic vision of building an "ecological province." At the end of 2004, the *Outline of the Master Plan for Building Fujian into an Ecological Province* was approved by the then State Environmental Protection Administration. In April 2006, the Fujian Provincial Government issued the *Implementation Opinions on the Outline of the Master Plan for Building Fujian into an Ecological Province* to advance the building of an ecological province in all respects. In January 2010, the Fujian Ecological Function Zoning was implemented. In September 2011, the Fujian Provincial Government issued the *Plan for Building Fujian into an Ecological Province in the Twelfth Five-Year Period*. In January 2013, the *Function Zoning Plan of Fujian Province* was promulgated and implemented. For the first time, the province's land territory was clearly divided into four types: areas for optimization, priority areas, restricted areas, and prohibited areas for development. Among them, two-fifths of the province's counties (cities) and 197 areas were listed as restricted and prohibited areas for development. On this basis, in March 2014, the State Council issued the *Several Opinions on Supporting Fujian Province in Deepening the Implementation of the Ecological Province Strategy and Accelerating the Construction of Ecological Civilization Pilot Demonstration Zone*, entering Fujian into the first batch of province-wide ecological civilization pilot demonstration zones directly designated by the State Council, and pushing its Ecological Province drive into a new stage. In August 2016, along with the issuance of the *Implementation Plan for National Ecological Civilization Pilot Zone (Fujian)* by the General Office of the CPC Central Committee and the General Office of the State Council, Fujian officially became the country's first National Ecological Civilization Pilot Zone. Therefore, it's safe to say that Fujian has always been at the forefront of the country in promoting ecological progress.

In a nutshell, since the establishment of the pilot demonstration zone and the national pilot zone in 2014, Fujian has vigorously promoted the system building and institutional innovation for promoting ecological progress, including promoting forestry development, reforming the ways of ecological protection, improving the ecological compensation mechanism for key river basins, promoting the river chief system on all fronts, and vigorously developing green economy according to local conditions.

First, it has vigorously promoted forestry development and the reform of ecological protection measures. By 2015, Fujian had reached the set target of increasing its forest coverage rate to above 65.95% and forest stock volume to 608 million cubic meters, and set new goals to lead the country in terms of forest coverage rate, increase the forest stock volume to 623 million cubic meters, and realize an annual growth rate of over 8% in the total output value of forestry by 2020. To this end, the provincial government issued in 2015 the *Notice on Promoting Forestry Reform and Development and Accelerating the Construction of Ecological Civilization Pilot Demonstration Zone*, and proposed such measures as deepening the administrative reform of forest tenure, improving financial services for forestry, launching commercial forest redemption in key ecological locations, improving the ecological compensation mechanism, scientifically managing and using forest land and wetlands, intensifying the cultivation of forest resources, bettering the assessment and evaluation mechanism for ecological conservation, and continuing to incorporate indicators such as forest coverage, forest stock and forest land retention into the government performance evaluation system.

Second, it has improved the ecological compensation mechanism for key river basins. To further increase ecological compensation in river basins and promote the compensation mechanism to all river basins in the province, the General Office of the Provincial Government issued the Ecological Compensation Measures for Key River Basins (revised edition) in August 2017. It proposed to establish an ecological compensation mechanism covering the entire province and with uniform standards. With support from the provincial government, the funding will be coordinated by cities and counties, to promote the sustainable development of the upper reaches and improve the water environment quality of the entire river basins.

Third, the river chief system is implemented on various fronts. Especially since the General Office of the CPC Central Committee and the General Office of the State Council issued the Opinions on the Comprehensive Implementation of the River Chief System in 2016, Fujian has stayed problem-oriented, relied on innovation as the driving force, and combined six measures. It has formed the "region + basin" system for river management by the river chief, the "patrol inspection + conservation" innovative mechanism for river management and conservation, the "county/district government + society" pattern of holistic river governance, a new model of technology-enabled river governance, the "group + individual" platform for engaging the general public in river conservation, and "rule of law + joint governance" model for river protection. All these efforts have significantly improved the ecology of the river basin, continued to improve the water quality, and significantly enhanced the sense of gain among the general public.

Fourth, the green economy is vigorously developed in accordance with local conditions. Fujian is widely "giving priority to ecological conservation" and promoting "green development." Based on its own ecological endowment and industrial reality, Fujian continues to industrialize ecological conservation and green industrial development at the same time, and speeds up the development of high-tech green economy that can create a lot of jobs and improve environmental quality. It strives to boost high-quality economic development through green development, and catch up with advanced areas while beautifying its environment, so as to benefit the people and realize ecological conservation and economic development at the same time. Its specific measures include vigorously promoting the market-oriented reform of green industries and products, actively greening industries, accelerating the green development of industrial parks, further consolidating and deepening the green utilization of resources, and vigorously creating a favorable environment for green economy.

2. **Hainan**

On February 7, 1999, *Hainan Daily* published the "Decision of Hainan Provincial People's Congress on Building an Ecological Province" on the front page, marking the official start of building Hainan into an "Ecological Province." On March 30 of the same year, the then State Environmental Protection Administration approved Hainan as a national ecological conservation demonstration province. In 2007, the report to the Fifth Party Congress of Hainan clearly put forward the "Ecological Province" strategy. In 2009, as China kicked off its national initiative of building Hainan into an international tourism island, "building a national ecological civilization demonstration zone" became one of the six strategies supporting the initiative. In 2013, General Secretary Xi Jinping paid an inspection visit to Hainan and asked it to "strive to create a good practice of developing socialism with Chinese characteristics and contribute to the building of a beautiful China." Xi also talked about ecological conservation, and especially called for more efforts in "greening" and "protecting the blue sky," to set an example for the national drive of ecological conservation and leave a sustainable "green bank" for future generations. In April 2017, the report to the Seventh Party Congress of Hainan Province pointed out that to build Hainan into a national ecological civilization demonstration zone, it is necessary to protect the ecology from the source to make the mountains lusher, the water more lucid, the sky bluer and the air fresher. Later, the Second Plenary Session of the Seventh Provincial Party Committee adopted the *Decision on Further Strengthening Ecological Conservation and Contributing to the Building of a Beautiful China*, systematically deploying efforts of ecological conservation. The *Decision* has become a programmatic document guiding Hainan's vigorous promotion of ecological progress. In April 2018, on another inspection trip to Hainan, Xi Jinping raised higher requirements for the island's drive to develop an ecological civilization: "Hainan must firmly establish and fully act in line with the concept that lucid waters and lush mountains are invaluable assets. It should pioneer in reforming the system for developing an ecological civilization, and set an example for the rest of the country." He made it clear that "(the central government) supports for Hainan's building of a national ecological civilization pilot zone, encourages Hainan to take a path of harmonious development between man and nature, and accumulate experience for nationwide ecological conservation." Later Hainan was named the fourth National Ecological Civilization Pilot Zone.

In a nutshell, over the years, especially since it became a national pilot zone in 2018, what Hainan has achieved in vigorously promoting ecological progress can be attributed to its steadfast efforts in five aspects under the guidance of Xi Jinping's thought on socialist ecological civilization with Chinese characteristics for a new era.

First, it's steadfast in implementing the "Ecological Province" strategy. Steadfast implementation of the "Ecological Province" strategy is not only an objective requirement for Hainan to apply the new vision for development, but also the basic experience drawn to vigorously promote ecological progress. To implement the "Ecological Province" strategy, we must give priority to ecological conservation when dealing with the relationship between economic development and environmental protection, and try to maximize green economic benefits. In other words, when economic development and environmental protection contradict, environmental protection must prevail; and when economic benefits and ecological benefits contradict, the immediate economic benefits must be discarded.

Second, it's steadfast in safeguarding the "ecological red line." "To contribute to the building of a beautiful China," Hainan must not deviate from its strategic focus of building an "Ecological Province," and a specific measure to do so is strictly safeguarding the "ecological red line." As to the relationship between economic development and environmental protection, if there is no development, then there is no capital for protection, and development at the expense of environment is like "drinking poison to quench thirst." Therefore, to properly handle the relationship between economic development and environmental protection is to strike a balance between the two. Moderation is the key, and the minimum requirement is to safeguard the ecological red line.

Third, it's steadfast in coordinating land and marine development. To promote ecological progress in Hainan, in addition to planning the ecological space of the land territory, we need to establish a land-sea coordinated marine ecological protection and restoration system, and especially strengthen the construction of marine ecological function zones with Sanya and Sansha at the center.

Fourth, it's steadfast in advancing the reform of replacing multiple plans for local development with one master plan. In June 2015, the Central Leading Group for Deepening Overall Reform approved Hainan to launch a province-wide pilot project of replacing multiple plans for local development—including economic and social development plans, urban and rural plans and land use plans—with one master plan. Subsequently, Hainan adopted the *Master Plan of Hainan Province (2015–2030)*, coordinated the whole province in specifying the national function zoning plan, established a uniform spatial planning system, and realized the convergence of various plans. While advancing the reform, Hainan has achieved remarkable results in restraining and promoting county-level economic and social development, urban and rural land use, and environmental protection.

Fifth, it's steadfast in encouraging diversified development paths. There are 19 cities and counties in Hainan, all enjoying good natural conditions for promoting ecological progress. On the basis of their different locations, industrial layouts, and stages of economic and social development, they all strive to balance ecological conservation and economic development, and properly handle urban development by giving priority to environmental protection, and have generated their unique practices and innovations.

3. **Guizhou**

Guizhou is the first province in western China to systematically promote ecological progress. After the Third Plenary Session of the Thirteenth CPC Central Committee, Guizhou ramped up the reform of the system and mechanisms for developing an ecological civilization, and formulated and implemented a series of pilot reforms. In June 2014, it was approved to build a national ecological civilization pilot demonstration zone. In August 2016, it was approved as one of the first three national ecological civilization pilot zones. Subsequently, the Provincial Party Committee and People's Government issued the *Opinions on Promoting Green Development and Building an Ecological Civilization*. In April 2017, Guizhou listed "holistic ecological conservation" as the third strategic action after poverty alleviation and big data. In June, the *Implementation Plan for National Ecological Civilization Pilot Zone (Guizhou)* was adopted at the 36th meeting of the Central Leading Group for Deepening Overall Reform. In November, Guizhou was listed as a pilot province for the ecological product value realization mechanism, driving its reform of the system for developing an ecological civilization into a new era.

In recent years, especially since becoming a national pilot zone in 2018, what Guizhou has done and achieved in vigorously promoting ecological progress can be summarized in five aspects.

First, it has built a strong green barrier, and extended provincial environmental protection inspections to the whole province. In recent years, Guizhou has focused on the full implementation of the *Air Pollution Prevention and Control Action Plan*, *Water Pollution Prevention and Control Action Plan* and *Soil Pollution Prevention and Control Action Plan*. It has introduced action plans and annual implementation plans against air, water and soil pollution and sped up four campaigns aimed to protect the "lush mountains," "lucid water," "blue sky" and "uncontaminated soil." In 2017 alone, it converted 4.774 million acres of farmland into forests, controlled the rocky desertification of 2520 square kilometers, and organized a campaign to fight pollution in ten major pollution sources and drinking water sources, and five actions to comprehensively improve the environment of Caohai. It concluded all the 3478 reports and complaints filed by the people and assigned by the Central Environmental Protection Supervision Group, and provincial environmental protection inspections have covered all cities (prefectures) and Gui'an New Area.

Second, it has improved the green system and is the first to introduce provincial regulations on ecological conservation. Guizhou's work in this area focuses on deepening the reform of key systems for building an ecological civilization, strengthening the legal guarantee for ecological conservation, and strict evaluation and assessment of ecological progress. In May 2014, Guizhou promulgated China's first provincial regulations on ecological conservation: the *Regulations on Promoting Ecological Progress in Guizhou Province*, which was followed by more than 30 supporting regulations. It canceled the GDP assessment in 10 counties located in key ecological function areas, and instead introduced ecological assessment on the Party committees and governments at the city/prefecture level.

Third, it has created 11 national ecological demonstration zones while cultivating a green culture. Via the Eco Forum Global Annual Conference, Guiyang has deepened exchanges and cooperation with the international community in ecological protection and response to climate change. To mark the "Guizhou Ecological Day" on June 18, Guizhou has held a series of activities such as organizing river chiefs to patrol the Mother River as well as patrol inspections to mountains and the city. At the same time, it tries to incorporate ecological education into the national education system and has compiled textbooks for schools and universities and for Party and government officials. Guizhou has created 11 ecological demonstration areas, 2 ecological counties, 56 ecological towns and 14 ecological villages at the national level; 7 ecological counties, 374 ecological towns and 515 ecological villages at the provincial level.

Fourth, it has been vigorously developing the green economy. In 2017, the green industries accounted for 37% of the region's GDP. Thanks to the implementation of green economy doubling plan, the promotion of green transformation and upgrading, the acceleration of the development of the digital economy, and the implementation of poverty alleviation through ecological conservation, Guizhou's green economy has made remarkable progress. In September 2016, the Provincial Party Committee and People's Government issued the *Opinions on Promoting Green Development and Building an Ecological Civilization*, and called for accelerating the development of industries that are based on ecological utilization, recycling and efficient, low-carbon and clean, and contributing to environmental governance, to continuously green the province's economic structure. In terms of poverty alleviation through ecological conservation, Guizhou has relocated nearly 2 million people for poverty alleviation, and provided them with financial and economic support tailored to local conditions, striving to explore a new model of regional poverty alleviation featuring "Internet + ecological conservation + targeted poverty alleviation."

Fifth, it has been building green homes, and has created 157 provincial-level rural demonstration sites. Focusing on strengthening the guiding role of plans, Guizhou is striving to build green towns and beautiful villages. It is speeding up functional zone planning and town planning. And 30% of its counties (districts and cities) have formulated their county-level rural development plan, with the goal of keeping the province's land development intensity below 4.2%. In terms of rural development, it has launched a three-year action for improving the rural living environment, implemented 100 projects for improving the rural environment, and created 157 provincial-level rural demonstration sites and 192 provincial demonstration sites for comprehensive improvement in the rural environment.

3 Theoretical Analysis and Thinking

As mentioned before, the national ecological civilization pilot zones, such as Fujian, Hainan and Guizhou, which are based on previous pilot demonstration projects, especially that organized by the NDRC and other six ministries/commissions, have advantageous natural conditions. Despite their differences, in general, they are still lagging behind in economic and social modernization, or in other words, are less embedded in or dependent on traditional industrial modernization and urbanization model. Therefore, we have reason to assume that it's easier for them to translate the discourse system for building a socialist ecological civilization and its policy assumptions into a green perception of their natural resources and practice of their economic utilization, and to transition from environmental protection to comprehensive transformation covering economic, political, social and cultural aspects. In other words, it is easier to become a radical reform embodying the true meaning of socialist ecological civilization.

The practice of building national ecological civilization pilot zones in Fujian, Hainan and Guizhou has indeed met the above-mentioned innovation goals and produced the desired effects. For example, the ecological civilization and the Five-Sphere Integrated Plan—in the sense of approach and goal—are both accepted as a political and cultural consensus. In other words, the realization of the overall and phased goals of regional ecological progress must rely on a three-dimensional or systematic change in the five spheres, and isolated effort in any sense may lead to local and temporary changes, but will certainly not to the overall goal of building an ecological civilization. Meanwhile, "green development" has become a common approach or model choice for these provinces to build an ecological civilization: while fully ensuring the quality of the existing ecological environment, they are striving for modern development in which man and nature co-exist in harmony, that is, to wisely transform the current "lucid water and lush mountains" before your eyes into "invaluable assets" in your hands. Both practices are innovations with worldwide demonstration significance for philosophical epistemology and the modern development theory. In this regard, their demonstration and guiding role is obvious and huge for national and global ecological conservation.

Specifically, according to the list of key reform tasks identified in the Implementation Plan for National Ecological Civilization Pilot Zone (Fujian), Fujian has launched a pilot reform of the natural resources asset management system, reformed the public policy and governance mechanism centering on ecological cloud platforms and grid management, and developed specialized judicial services for ecological conservation; Hainan is constructing a green industry system led by tourism and modern service industries in accordance with local conditions, and strives to promote the deep integration of Internet, Internet of Things, big data, satellite navigation, and artificial intelligence with the real economy; Guizhou has for a long time been a pioneer in introducing local regulations and system innovation for promoting ecological progress, and publicity and education. All these practices have undoubtedly epoch-making significance or potential for change, and deserve much affirmation and attention.

Of course, judged from the actual realization of the established goals or spillover potentials of these national ecological civilization pilot zones, or from the inevitable requirements of ecological conservation and its theories, Fujian, Hainan and Guizhou are still in an experimental or preliminary stage.

On the theoretical level, we can introduce a three-dimensional theoretical framework consisting of management strategy, spatial dimensions and socialist politics for preliminary analysis. Briefly speaking, we can examine any attempt to build a comprehensive ecological civilization from the five-sphere management philosophy strategy, the spatial dimension of the unity of administrative hierarchy and ecosystem, and the political dimension of whether the socialist orientation is upheld at the same time, and the national ecological civilization pilot zone project is no exception. In other words, a successful national ecological civilization pilot zone must be truly based on or follow the five-sphere management strategy, must strive to maintain the balance or complementarity between administrative management and the ecological space, and must adhere to the political orientation of socialism at the same time. In particular, it needs to be emphasized that the third criterion seems self-evident for most scholars and the public, because the choice of the path of socialism with Chinese characteristics and the leadership of the CPC have made socialism a political correct choice and a dominant ideology in the development of an ecological civilization. But under the hegemony of the international economic and political order and theoretical discourse dominated by capitalism, "socialist" implies a clear and radical political preference and choice of "red and green," which we must always be aware of or not shy from.

From this perspective, in my opinion, the National Ecological Civilization Pilot Area and other forms of pilot demonstration projects, such as the National Ecological Civilization Demonstration City/County project and the innovation bases project under the leadership of the Ministry of Ecology and Environment deserve higher expectation and more attention. We should pay attention to not only the extent and the specific path they take to achieve the specific institutional innovation goals set in the Implementation Plan for National Ecological Civilization Pilot Zone, but also to what extent they have continued to develop and reform an ecological civilization vision and theory that keep pace with the times, which should be ecological and socialist at the same time, and striven to maintain a benign interaction between the two. This is not only important for their direct purpose as pilot zones, but also for China's goal of ecological progress and the grand goal of building a modern, socialist country with Chinese characteristics.

Lawyers' Legal Services for Developing an Ecological Civilization

Yiming Dong

1 The Development of Law-Based Ecological Conservation

The guiding role of the rule-of-law thinking is crucial to promoting ecological conservation. With the ecological civilization drive unfolds, it's imperative to seek institutional and policy innovation to promote institutionalized and law-based practices, strengthen national and local legislation, exercise strict law enforcement, and ramp up judicial support for ecological protection. China's current law-based ecological progress is mainly reflected in such areas as the policy system, legal sources, administration, and judicial remedies, and basically the rule of law in this regard is progressing in both breadth and depth.

1. **The policy system: the top-level design is more and more specific**

According to the *Decision of the CPC Central Committee on Some Major Issues Concerning Comprehensively Deepening the Reform* adopted at the Third Plenary Session of the 18th CPC Central Committee: "To develop an ecological civilization, it is necessary to establish a systematic and complete set of institutions for promoting ecological progress, implement the strictest source protection system, system for compensating for ecological damage and the accountability system, improve environmental governance and ecological restoration system, and protect the ecological environment with institutions." The 19th CPC National Congress proposed that to "speed up the reform of the system for developing an ecological civilization and build a beautiful China," it is necessary to provide more high-quality ecological products to meet the people's growing demand for a beautiful ecological environment, intensify ecosystem protection, and reform the environmental regulation system. At the 2018 National Conference on Eco-environmental Protection, Xi Jinping put forward six principles and five ecological civilization systems for developing an ecological civilization. As people get a clearer and deeper understanding of the connotation

Y. Dong (✉)
Jointide (Beijing) Law Firm, Jinan, China

J. Pan et al. (eds.), *Beautiful China: 70 Years Since 1949 and 70 People's Views on Eco-civilization Construction*, https://doi.org/10.1007/978-981-33-6742-5_66

and content of ecological civilization, the central government has started to deploy the reform of the system for developing an ecological civilization on all fronts, and specify the top-level design of the institutional framework.

(1) The ecological damage compensation system

In 2015, the CPC Central Committee and the State Council successively issued two programmatic documents: the *Opinions of the CPC Central Committee and the State Council on Accelerating Ecological Progress* and *Master Plan for Reform of the System for Developing an Ecological Civilization*. The General Office of the CPC Central Committee and the General Office of the State Council issued the *Pilot Reform Plan for the Environmental Damage Compensation System* and the *Reform Plan for the Environmental Damage Compensation System* in 2015 and 2017, respectively, and gradually established an environmental compensation system nationwide.

(2) The central inspection system for ecological and environmental protection

Starting from the pilot project in Hebei Province in January 2016, the central environmental protection inspection team has completed the first round of inspections on ecological and environmental protection in 31 provinces (autonomous regions, municipalities) and the Xinjiang Production and Construction Corps, and re-inspected 20 provinces (autonomous regions), achieving remarkable results. The General Office of the CPC Central Committee and the General Office of the State Council issued the *Regulations on the Central Inspection on Ecological and Environmental Protection* in June 2019. It is the first Party regulation specifying the framework, procedures, powers and responsibilities of the inspection system, and will play an important role in furthering the inspection work according to law, consolidating political responsibility for developing an ecological civilization and in building a beautiful China.

(3) The environmental public interest litigation (EPIL) system

The EPIL system was first stipulated in the *Civil Procedure Law* revised in 2012. But it was not promoted widely until the 2015 version of *Environmental Protection Law* formally stipulates the qualifications of social organizations that can initiate public interest litigation. The Decision adopted at the Fourth Plenary Session of the 18th CPC Central Committee proposes to "explore and establish a system for procuratorial organs to file public interest litigation." Authorized by the Standing Committee of the National People's Congress, 13 provinces (municipalities directly under the central government) began to pilot public interest litigation filed by procuratorial organs. In June 2017, after the NPC Standing Committee revised the *Civil Procedure Law* and the *Administrative Procedure Law*, the system for procuratorial organs to file public interest litigation was formally established. As an important part of it, the EPIL has since then been enriched and improved, and has become an important part of China's environmental legal system.

2. **Legal sources: a continuously improving legal system**

Since the first meeting of the 13th National People's Congress voted and passed the *Amendments to the Constitution of the People's Republic of China*, which includes the new development vision, the vision of ecological civilization and the requirements for building a beautiful China into the Constitution, China's legal system for ecological conservation has gradually taken shape. Article 9 of the "General Principles" of the *Civil Law* stipulates the "green principle," making environmental protection one of the basic principles of the *Civil Law* and opening a channel for applying the *Civil Law* to environmental and resources protection. Since the 18th CPC National Congress, China has formulated and revised nearly ten laws governing ecological and environmental protection, and countless related judicial interpretations, administrative regulations, departmental rules, local regulations and rules, environmental standards and provisions on environmental legal responsibilities.

3. **Administration: scientific administrative methods**

According to the *Measures for the Accountability of Party and Government Leading Officials for Ecological and Environmental Damage* (Trial) issued in August 2015, a lifetime accountability system for ecological and environmental damage is implemented, under which local Party committees and governments at all levels are responsible for the protection of the ecological environment and resources in their jurisdiction, the main leadership of the Party committee and government concerned are the main responsible persons, and other leading officials concerned take responsibilities according to the scope of their duties. Since the central inspection system for environmental protection was put into practice in January 2016, China has seen unprecedented intensity in inspection and problem solving. While effectively solving a lot of prominent widely-concerned environmental problems, it has also promoted the transformation and upgrading of local industries and improved local mechanisms for environmental protection and ecological conservation. On the whole, China's knowledge about and practice in environmental protection is undergoing historic, defining and holistic changes. In March 2018, the Ministry of Ecology and Environment was established as part of the reform of the Party's and government's organizational structure. It is tasked with the centralized regulation of the discharge of pollutants into the ecological environment and urban and rural areas and administrative law enforcement. By centralizing the once-devolved functions, it will see to it that there are no blind spots in supervision, administration, and governance, and exercise more scientific administration.

4. **Judicial remedies: increasingly diversified channels of judicial remedies**

First of all, judicial organs provide strong judicial services and support for promoting ecological progress and green development by punishing environmental and resources crimes and trying environmental cases such as environmental infringements, environmental civil public interest lawsuits filed by social organizations, environmental public interest lawsuits filed by procuratorial organs, environmental compensation lawsuits filed by provincial and municipal governments, and administrative environmental and resources cases. Secondly, the people's courts continue to build a specialized trial system for environmental and resource cases, by steadily constructing specialized trial organs and promoting the centralized trial model, establishing a centralized jurisdiction system for such cases across administrative divisions, actively promoting regional judicial collaboration, forming a new pattern of judicial protection based on co-governance by multiple parties, and introducing new execution methods. Finally, in recent years, from the Supreme People's Court to local high people's courts, judicial organs have published several judicial interpretations and typical cases that are distinctively pro-environment, covering mining rights disputes, compensation for marine resources and ecological damage, and environmental civil public interest litigation, public interest litigation filed by procuratorial organs, ecological conservation and the development of the Yangtze River Economic Belt, etc.

2 Opportunities for Lawyers to Serve the Development of an Ecological Civilization

As China ramp up its efforts to promote ecological progress, its legal services market for environmental protection is facing golden opportunities. From July 2016 to June 2017, people's courts at all levels accepted a total of 16,373 environmental crime cases, 187,753 environmental civil cases, 39,746 environmental administrative cases of various types, 57 environmental civil public interest litigation cases filed by social organizations, and 791 environmental public interest litigation cases by procuratorial organs. In this context, lawyers need to work hard to provide the most badly-needed legal services, upgrade legal services for environmental protection in both breadth and depth, and provide comprehensive legal services for promoting ecological progress.

1. **Lawyers should assist the government in pollution prevention and control with legal thinking and means**

In the first round of central inspections on ecological and environmental protection, about 27,000 people were held accountable, and 4218 punished, including 686 officials at the departmental/bureau level and above, and 2062 at the division level. It's obvious that local governments and their functional departments take most of the blame for environmental damage and pollution, and they bear huge responsibilities and pressure. Lawyers should actively help the government prevent and control ecological and environmental problems with legal thinking and means. For example, they can (1) provide legal support for the formulation and improvement of policies, regulations and institutions regarding the protection and utilization of the ecological environment and resources; (2) provide legal support for the decision-making and construction of expropriation, demolition, new-type urbanization and other major construction projects that may affect the ecological environment; (3)

provide legality review of major enforcement decisions by ecology & environment departments at all levels and other relevant departments and ensure legality and compliance of administrative penalties; (4) provide provincial and municipal governments with legal services for the compensation of ecological damage. In addition, when the government handles major cases and mass incidents arising from ecological and environmental issues, lawyers should actively help the government resolve social conflicts as soon as possible by legal means and the rule of law.

2. **Lawyers should provide comprehensive green legal services for enterprises**

Amid the ongoing "Environmental Protection Storm" action, enterprises are under tremendous ecological and environmental pressure. From January to April 2019, China issued 36,465 penalty decisions, with a fine of RMB 2.806 billion and an average fine of RMB 77,000 per case. Lawyers should provide enterprises with a full range of green legal services, and especially assist them in guarding against ecological and environmental legal risks at the source. Such services include establishing a mechanism for guarding against environmental legal risks for polluting enterprises; legal remedies for environmental administrative penalties (for hearings, re-considerations, litigation, etc.); emergency response to environmental pollution incidents; services related to the environmental tax.

For example, our green legal service team has developed the "Legal Services for Ensuring Corporate Compliance in Environmental Protection and Building a Risk Control System" product. This integrated solution aimed to help polluting enterprises construct their environmental compliance system and environmental risk control mechanism fully demonstrates our lawyers' approaches, structure, steps and methods for advising enterprises on environmental protection. This product provides such legal services as industry analysis and process search; opening pollutant accounts and conducting discharge analysis; assisting enterprises in understanding environmental regulations and policies; conducting special legal due diligence on enterprises and proposing suggestions on adjustments and solutions accordingly; embedding the environmental compliance and risk control system into corporate management, production and operation; organizing the implementation of work plans and mechanisms.

3. **Lawyers should provide legal services for EPIL**

After several years of rapid development, the EPIL system is increasingly proving its value. Especially the public hearings of major EPIL cases in Changlong, Nanping and Dezhou have won high praise from all walks of life, at home and abroad, for their positive role in supporting public engagement in environmental protection in an orderly, law-based manner, supervising the exercise of powers by administrative agencies, safeguarding the environmental public interest and the people's environmental rights, and serving and safeguarding ecological conservation. Still in its infant stage, the EPIL system in China allows new subjects to participate in environmental governance, and is reshaping the original environmental governance pattern, means and methods. It is also seeking breakthroughs in different fields, trying to solve some deep-seated problems with environmental governance, including transboundary pollution, local governments' assuming responsibilities for environmental quality and the accountability path, and who's responsible for fixing, repairing the long-standing environmental pollution problems, low pollution costs, environmental law enforcement and fair market competition. Environmental lawyers can provide useful proposals and play a pivotal role in tackling these problems.

4. **Lawyers should advise responsible individuals on guarding against environmental legal risks and resolving disputes**

With the continuous advancement of ecological and environmental protection work, the political risks, criminal risks, and civil liability risks have drastically increased for natural persons related to pollution and ecological damage, such as the main responsible persons of local Party committees and governments, the main responsible persons of the ecology & environment authorities and other relevant functional departments, and the legal representatives, actual controllers and direct supervisors of enterprises. Lawyers can provide them with the following legal services to safeguard their lawful rights: guarding against environmental administrative and criminal liability risks for the main leaders and direct supervisors of the Party and government organs; guarding against environmental administrative and criminal liability risks for the main leaders and direct supervisors of the ecological and environmental protection authorities and related departments; guarding against environment criminal liability risks for the legal representatives, actual controllers and direct supervisors of enterprises.

3 Challenges Facing Lawyers in Providing Legal Services for Ecological Conservation

1. **The shortage of professional lawyers seriously restricts business development**

Legal services for ecological conservation require professional knowledge in law, environment, resources, energy, economics, finance, society, etc. Lawyers should be familiar with not only the latest national environmental policies, institutions, laws and regulations, but also the international development trend in such sectors as new energy, low-carbon economy, resource recycling and environmental protection. They must be specialized and inter-disciplinary at the same time, but such lawyers are in severe shortage in reality. The shortage of professional lawyers has severely restricted the development of the green legal service market.

2. **The service scope is limited and hard to expand**

Lawyers' current role in ecological and environmental protection includes representing the client in litigation and advising the client after environmental incidents; more specifically, they participate in environmental cases on behalf of parties involved, provide legal advice to parties, help them mediate with polluting companies, or protect rights through non-litigation channels. But all these are only a small part of the legal services for developing an ecological civilization. It is difficult for the lawyer to develop and provide the right legal services much needed in key areas and links of ecological conservation, and it is even more difficult to form a systematic product of such services. As a result, the lawyer's role in serving the development of an ecological civilization is very limited, and for most lawyers, providing legal services for promoting ecological progress doesn't even count as an independent source of income.

3. **Poor awareness of the rule of law in ecological conservation**

The environmental right is the most basic of human rights and an important part of the right to survival, but it still needs to be interiorized among the people. For one thing, many people have only little awareness of environmental protection which should cover every aspect of life from waste sorting to ecological conservation. For another, it is difficult to exercise one's environmental right. In the face of environmental pollution, many people won't resort to legal means to protect their environmental right, out of fear for or resistance against litigation. The poor awareness of rule of law in building an ecological civilization is also reflected in the production and operation of enterprises and their response to environmental administrative penalties and litigation.

4 Conclusion

To contribute to the grand national blueprint of developing an ecological civilization, lawyers should uphold the concept of environmental protection, work together to make the dream of building a "Beautiful Chinese" come true, jointly develop legal services for building an ecological civilization, including all-round and three-dimensional environmental non-litigation and litigation services for the government, enterprises and citizens, alike, and strive to play our due part in developing an ecological civilization.

Exploring the Road to the Law-Based Green Development of the Yangtze River Economic Belt

Gang Qiao

Ecological civilization is a new form of civilization after the primitive civilization, agricultural civilization, and industrial civilization, and is also known as "green civilization." Since the 18th National Congress of the Communist Party of China, the content of ecological civilization has been continuously enriched and its status has become more prominent. Especially in 2018, ecological civilization was officially written into the Constitution of the People's Republic of China, marking the formal completion of the Five-Sphere Integrated Plan. Green development has become an inevitable requirement for building an ecological civilization, and is an engine for and effective path to ecological progress. Law, as a regulatory means formulated or approved by the state and enforced by the coercive force of the state, plays a decisive role in promoting ecological progress and green development.

The Yangtze River nurtures more than 400 million people living on a quarter of China's land territory, benefiting them with irrigation, navigation and fishing. The Yangtze River Economic Belt is the strategic zone in China's latest round of regional opening-up, an inland river economic belt with global influence, a coordinated development zone for the interaction and cooperation of the east, central and west China, and a pioneering and demonstration zone for ecological conservation. Its plans, policies and institutions should all work to promote green development. Under the guidance of the vision of ecological civilization, we should use legal means to promote the green development of the Yangtze River Economic Belt, and steer it toward a resource-saving and eco-friendly green development path.

G. Qiao (✉)
Western Ecological Law Research Center, Southwest University of Political Science and Law, Chongqing, China

J. Pan et al. (eds.), *Beautiful China: 70 Years Since 1949 and 70 People's Views on Eco-civilization Construction*, https://doi.org/10.1007/978-981-33-6742-5_67

1 The Meaning of the Green Development of the Yangtze River Economic Belt for Developing an Ecological Civilization

Ecological civilization is shaping a form of cultural ethics that pursues harmonious coexistence between man and nature and between people, coordinated development and sustainable prosperity based on a virtuous cycle. It can be said that promoting ecological progress concerns not only the people's well-being, but also the distant future of the nation, and is the necessary foundation for achieving sustainable development. Promoting ecological progress must be guided by the vision of ecological civilization. It is actually the process of implementing the basic vision of ecological civilization. It is generally believed that a concept is a rationalized idea. The vision of ecological civilization reflects a new kind of harmony between man and nature and between human beings, more compliant with the law of nature favorable for the sustainable development of human society. As China moves into a new era in building a socialist ecological civilization, the Xi Jinping Thought on Ecological Civilization has taken shape, whose key message is embodied in the "eight views," including the view that "lucid waters and lush mountains are invaluable assets." The principal contradiction facing Chinese society has evolved. What we now face is the contradiction between unbalanced and inadequate development and the people's ever-growing needs for a better life which must be based on the environment and resources. Green is associated with not only life, but also people's well-being; it is valuable, and the green culture should be promoted. The green development concept advocates respect for nature and the coordinating economic development and environmental protection, so as to realize truly green development. Environmental protection is the red line when we pursue green development. We should work hard to introduce new ideas and transform development methods, rather than blindly pursue economic benefits at the expense of ecological value.

General Secretary Xi Jinping emphasized: "Green development is an inevitable requirement for building a high-quality modern economic system and a fundamental solution to pollution." The ultimate goal of green development is to build an ecological civilization. Following the environmental crisis caused by the traditional extensive development model, green development has become a new type of social development model with the vision to build an ecological civilization. It can be said that the fundamental mission of green development is to promote the development of a socialist ecological civilization and guide the harmonious coexistence of man and nature in China. Likewise, only under the guidance of the vision of ecological civilization can we cultivate a solid foundation for green development.

The 18th CPC National Congress made the strategic decision to "vigorously promote ecological progress." The report of the 19th CPC National Congress clearly proposed to "implement the strictest possible systems for environmental protection, and develop eco-friendly growth models and ways of life." The Yangtze River Economic Belt development strategy is the first national strategy to coordinate river basin management, regional development, industrial transformation, spatial layout optimization, and transformation of development models against the background of promoting ecological progress. It must be based on joint efforts to protect and avoid large-scale development. The Economic Belt should pursue green development by promoting well-coordinated environmental conservation and avoiding excessive development.

2 Dilemmas Facing Law-Based Green Development in the Yangtze River Economic Belt

The Yangtze River is known for its regional prosperity, but much of that is at the expense of the environment. To pursue green development, we must focus on fostering the harmonious relationship between man and nature. Due to the lack of uniform governance standards for the cities, villages, mountains, tributaries, forests, farmlands and lakes along the Yangtze River, fragmented, chaotic administration and absence of supervision across jurisdictions, the ecological environment of the Yangtze River Basin has been deteriorating. By proposing the coordinated regional development strategy of the Yangtze River Economic Zone against the background of promoting ecological progress, the government shows its strong resolve to change the situation with green development. But there lacks strong legal support for promoting green development, mainly due to problems in the legislation and enforcement of rule of law, specifically as listed below.

1. **The dilemma facing the system of law**

We need a sound system of law to support the green development of the Yangtze River Economic Belt. For the integrity of the Yangtze River Economic Belt, the system must consists of connected laws and regulations governing ecological protection, natural resources development, and energy recycling, to better promote and protect green development. But in reality, the system of law is poor and has become the main bottleneck restricting the green development of the Yangtze River Economic Belt. The green development of the Yangtze River Economic Belt involves 11 provincial-level administrative regions in the basin and many different departments such as ecology and environment, natural resources, water conservancy, planning, transport and agriculture, etc. Under the current model of decentralized legislation, the basin lacks a comprehensive, uniform system of law; laws, regulations and rules are poorly aligned with legal departments for resources development and utilization, pollution control, ecological protection and so on, and sometimes there are even conflicts between laws and regulations. For example, the laws governing water resources in the Yangtze River Basin include the *Environmental Protection Law*, the *Water Pollution Prevention and Control Law*, the *Water Law*, the *Soil and Water Conservation Law*, the *Flood Control Law*, and the *Urban and Rural Planning Law*, in addition to a large number of wide-ranging regulations and rules. Judged by their names, it seems that each has its own functions, but together they are not systematic, with some terms too generalized, thus difficult to produce synergy. The Yangtze River Economic Belt is a complete system consisting of cities, villages, mountains, tributaries, forests, farm fields and lakes along the River. Only with a uniform system of law can we meet the actual needs of green development of the region.

2. **The dilemma facing the administrative regulation system**

China adopts a dual management system for water resources based on river basin management and administrative jurisdiction management. In the case of the Yangtze River Economic Belt, the Yangtze River Water Conservancy Commission is the administration for the whole river basin, and in in-basin provinces and cities, the protection and utilization of water resources are the responsibility of local water resources departments and ecology and environment departments. But in fact, there lacks division of responsibilities at the basin level and the local level, and even at the local level, water management responsibilities are not clearly divided between the provincial, municipal, and county levels. What's more, the situation varies in different administrative regions, and the water affairs involve multiple functional departments. As a result, problems such as multi-channel management, overlapping functions and absence of supervision occur in practice, and even the co-existence of excessive regulation by multiple departments and the absence of supervision. According to statistics, the power to manage the Yangtze River Basin is divided between 15 ministries or commissions with 76 functions at the national level, and between 19 provincial-level governments with more than 100 functions at the local level. The separation of interests and the institutional barriers for administration and regulation have resulted into the lack of overall consideration for the ecological environment of the river basin, which seriously hinders the green development of the Yangtze River Economic Belt.

3. **The dilemma facing the judicial cooperation mechanism**

The judicial system is the last line of defense for social fairness and justice. The use of judicial means to protect the environment and natural resources, promote green development along the Yangtze River Economic Belt, and contribute to ecological progress is the right way for the judicial system to serve national development. We should base on judicial functions support the green development of the Yangtze River Economic Belt by giving priority to ecological conservation. Considering the flowing waters of the Yangtze River and the spillover effect of environmental problems, the Yangtze River Economic Belt is most prone to cross-regional environmental disputes, which cannot be resolved by a single judicial agency. That's why judicial collaboration is called forth. In the narrow sense, judicial cooperation refers to the cooperation between courts in different areas to perform certain functions, including the service of documents, investigation and execution. As judicial activities diversify, the concept of judicial cooperation is expanding. It is no longer limited to the courts, but extended to judicial agencies including public security organs, procuratorates and courts, as well as administrative law enforcement agencies in cooperation with judicial agencies. And their cooperation has also expanded from investigation, service and execution to case information sharing, optimization of resource allocation, case linkage, judicial services, and uniform judgment standards. There are many problems with the judicial cooperation mechanism of the Yangtze River Economic Belt, such as insufficient legislative support, jurisdictional disputes in cross-regional environmental cases, poor information exchange, inconsistent judgment standards for regional environmental cases, and obstacles to the collection of evidence in cross-regional cases. Without a specialized and effective judicial coordination mechanism, it is difficult to effectively guide and regulate the green development of the Yangtze River Economic Belt.

4. **The dilemma facing the environmental governance system**

Governance refers to the sum of the many ways the public and private sectors manage their common affairs. The governance of the Yangtze River Economic Belt used to rely on direct government control: the powers were allocated and environmental issues tackled by administrative means, resulting in the little participation of social organizations, enterprises and the public. But the status of the ecological environment is directly related to the public's own interests, and blind reliance on government control only makes it difficult to maximize the effect of environmental governance. The government is responsible for not only environmental governance, but also economic development and for improving the people's living standards. Thus it is inevitable to weigh the loss and gains of governance, and sometimes environmental benefits are sacrificed for economic benefits. In contrast, under the green development model, in pursuit of development we must safeguard the red line of ecological protection, and economic benefits must come at the expense of ecological value. This raises higher requirements for the environmental governance system.

3 Measures for Promoting Law-Based Green Development in the Yangtze River Economic Belt

General Secretary Xi Jinping once emphasized that "to promote the development of the Yangtze River Economic Belt, we must consider the long-term interests of the nation, take the path of green development by prioritizing ecological conservation, and keep our Mother River running clean forever." In recent years, the central government has rolled out the top-level design based on the "Outline of the Yangtze River Economic Belt Development Plan" and supported by the "Environmental Protection Plan of the Yangtze River Economic Belt" and other special plans. But promoting the green development of the Yangtze River Economic Belt should be a systematic project. It requires not only the change in mindset of the government, enterprises, and society, but also specific institutional reforms and relevant supporting measures. The Xi Jinping Thought on Ecological Civilization has made it clearly to "apply the strictest system to protect the ecological environment." The rule of law is an important means to accelerate and effectively guarantee the modernization of the government's governance system and capabilities, and also an important guarantee for long-term stability of the country. In better promoting and protecting the green development of the Yangtze River Economic Belt under the guidance of the vision of ecological civilization, the rule of law which is authoritative, stable and compulsory will play a fundamental role. It can be said that green development would not be possible without the guidance, protection and regulation of law. It should be written in law that the Yangtze River Economic Belt is for green development, to achieve "good governance with a good law," and see to it that the Economic Belt truly prioritizes ecological conservation and embarks on the green development path.

1. **Improving the system of law**

Under the guidance of the vision of ecological civilization, the law is an indispensable booster for the green development of Yangtze River Economic Belt. Good governance requires good laws. Considering the Yangtze River Basin as a whole and its functions, we need a systematic set of laws and regulations to govern the river basin. Therefore, the legislation should follow the principle of "centralized planning, supervision and pollution control, and uniform standards" to construct a uniform legal system governing environmental protection in the Economic Belt. Recent years have heard loud cries for a comprehensive law on the protection of Yangtze River. The advocates argue that: "To prioritize ecological conservation and promote green development in the Yangtze River Economic Belt, we need a comprehensive law that covers and coordinates all elements across the entire river basin." The focus of such law should be to realize the green development of the Yangtze River Economic Belt, eliminate the inconsistency between previous departmental or local regulations, comprehensively adjust the protection, development, and utilization of the Economic Belt, establish relevant legal liability systems, and thus reflect and meet the special development characteristics and practical needs of the Yangtze River Basin. In addition, it is necessary to strengthen the legislative coordination between departments, between local governments, and between departments and localities on promoting and protecting the green development of the Yangtze River Economic Belt. In accordance with the reality of different localities and departments, we shall improve legislation under the guidance of the *Law on the Protection of the Yangtze River* and existing *Environmental Protection Law* and the *Water Law*. Meanwhile, we will strengthen legislation coordination between different localities through mechanisms such as joint provincial- and municipal-level meetings and contracts, and use legal means to jointly promote the green development of the Yangtze River Economic Belt.

2. **Improving the administrative regulation mechanism**

In essence, the integration of the Yangtze River Economic Belt is not only about centralized governance of water resources utilization and protection and shipping, but also the coordinated governance of industries, transport, ecology, culture and social undertakings across the entire Economic Belt. As China keeps modernizing its national governance system and capabilities, to achieve the integrated green development of the Yangtze River Economic Belt, we must remove the barriers of local interests and institutional barriers, and use law to improve the administrative regulation mechanism. First, a unified river basin management agency should be established. We may reform the Yangtze River Water Conservancy Commission and the Changjiang River Administration of Navigational Affairs and set up a dedicated administration for Yangtze River Economic Belt, which shall directly report to the State Council, not any ministry or commission under the State Council. The administration will consist of administrative heads in provinces and cities along the Yangtze River to realize centralized leadership, coordinate departments and exert direct control. Second, we need to rationally allocate administrative powers, strengthen cooperation between localities and departments, and establish a regular mechanism for consultation and cooperation.

3. **Improving the judicial cooperation mechanism**

We should give full play to the role of judicial organs in safeguarding green development and embody the vision of ecological civilization in judicial work. Based on green development and in-basin coordination, we should explore the model of cooperation between in-basin courts and procuratorial organs, between public security organs, procuratorial organs and courts of the same locality, and between courts and ecology & environment authorities in the Yangtze River Economic Belt. We should introduce new legal institutional designs for judicial cooperation, promote the law-based, regular model of judicial cooperation, and form a basic framework of "top-level design + independent exploration" and a judicial collaboration model for cases about the protection and development of ecological environment and natural resources. The goal is to realize judicial cooperation in environmental cases, including joint development of judicial standards through consultation, standardizing case filing and litigation services, and realizing cross-regional centralized jurisdiction. Specific mechanisms involved include the information sharing mechanism for major environmental cases, judicial execution coordination mechanism, judicial service coordination mechanism, judicial big data application mechanism, the uniform interaction and communication mechanism for law application, and the coordination mechanism for cross-regional environmental public interest litigation and environmental damage compensation.

4. **Improving the environmental governance system**

Environmental crisis is a common challenge for mankind. The building of an ecological civilization requires the concerted efforts of the government, the society, enterprises and individuals. The report to the 19th CPC National Congress clearly states that: "We will step up institution building in social governance and improve the law-based social governance model under which Party committees exercise leadership, government assumes responsibility, non-governmental actors provide assistance, and the public get involved." While vigorously promoting the green development of the Yangtze River Economic Belt, we need to change the traditional environmental governance model and encourage co-governance by multiple parties. We should provide social organizations, enterprises and the public channels to supervise and manage affairs related to the green development of the Yangtze River Economic Belt, and build diversified public participation mechanisms including the information disclosure and feedback mechanism, seminars and the demonstration mechanism. At the same time, we should introduce a more flexible market-based governance model, implement the water rights trading system, ecological compensation system, and emission rights trading system to combine administrative regulation with market regulation, motivate co-governance parties and replace direct government control with co-governance by multiple parties over the green development and environmental governance in the Yangtze River Economic Belt.

4 Conclusion

Along with the increasingly heated environmental problems and increasingly severe environmental governance situation, social and economic development has entered a critical period of transformation. People are increasingly aware that only by promoting ecological progress and green development can we create a hopeful future for mankind. As we reform to promote the rule of law on all fronts and vigorously advocate the vision of ecological civilization, we must employ legal means to effectively resolve the complex and diverse conflicts of interest in the Yangtze River Basin, and blaze a path for law-based green development of the Yangtze River Economic Belt.

Building a Global Ecological Civilization and an Ecological Security System

Global Transformation of China's Ecological Civilization Paradigm as Part of the *2030 Agenda for Sustainable Development*

Jiahua Pan

Increasingly recognized by the international community, the Chinese paradigm for the building and development of ecological civilization provides a direction and approach for global transition from industrial civilization to ecological civilization. China has done a great job in low-carbon development and climate change mitigation, and has made outstanding contributions to achieving the Millennium Development Goals set by the United Nations. At the heart of the 2030 Agenda for Sustainable Development are five dimensions—people, planet, prosperity, peace and partnership, collectively known as the 5P's—for achieving harmony between man and nature. We can say that the action plan proposed in the Agenda goes beyond the three pillars of "economy, society, and environment" under industrial civilization, and it involves China's great contributions to establishing ecological civilization. If industrial civilization could be regarded as a revolutionary innovation in human development by the West, then China's building of ecological civilization would be a fundamental contribution to global sustainable development by adding Eastern wisdom to the cause.

The original text, published in the 3rd issue of *Dongyue Discussions* in 2018, has the title "Strategic Recognition, Development Paradigm and Strategic Measures for the Building of Ecological Civilization in a New Era." It was abridged to be included in this book.

J. Pan (✉)
Chinese Academy of Social Sciences, Beijing, China

J. Pan et al. (eds.), *Beautiful China: 70 Years Since 1949 and 70 People's Views on Eco-civilization Construction*, https://doi.org/10.1007/978-981-33-6742-5_68

1 Ethical Values and Development Goals of Ecological Civilization

Industrial civilization was based on utilitarian ethical values emphasizing the importance of utility of human beings—or man's value and well-being—that brings progress. When dealing with the relationship between man and nature, industrial civilization advocated that human beings are a master over nature, and 'all that are meaningful to humanity' have 'value.' As such, humanity could satisfy all their needs by exploiting and destroying nature. Also, from a generational perspective, the ethical values of industrial civilization prioritized the utility of contemporary people, while ignoring the well-being of future generations and society. Such values lead to the fact that workers in industrial civilization were deemed economically rational machinery, with the relationship among people reduced to one based on barter and money, and that between man and nature becoming one in which the former are entitled to exploiting the latter for free. "Survival of the fittest" became the norm of the relationship among people in a dog-eat-dog world, and "anthropocentrism," the belief that "each man is the maker of his own fate," prompted people to relentlessly conquer nature. In industrial civilization, development goals were to maximize profits and accumulate wealth as much as possible without considering the outside world or the unwanted costs caused to others or society. Under the values of industrial civilization, carrying capacity, or the environment's ability to accommodate a particular activity without unacceptable impact, was seen as something that could increase with technological innovation, and there was no need to establish regulations in terms of tapping natural resources.

Ecological civilization advocates the ethical values of harmony between man and nature, and emphasizes that people should respect, conform to and preserve nature. It inherits ancient and traditional Eastern ecological wisdom, draws on the essence of the Chinese civilization—where "unification of nature and man" is accentuated, the three realms of Heaven-Earth-Man is established, and the Taoist idea to follow the laws of nature is adopted—and emphasizes the integration of man and nature, rather than transforming and conquering nature or the concept that man's determination will triumph over nature. Recognizing and respecting the value of nature is the ethical values of ecological civilization that is based on ecological and social justice, namely, respect for human rights and fair sharing of natural resources.

Since the 18th National Congress of the CPC, Comrade Xi Jinping's series of important speeches, special expositions and important instructions on the building of ecological civilization have enhanced our understanding that ecological civilization is different from industrial civilization. For example, he proposed the "two mountains" theory that human civilization evolves around the idea of "harmony between man and nature, and harmony between the economy and society," explaining the dialectical relationship between economic development and ecological environment protection in the process of comprehensively deepening reform in China. He also emphasized "ecology priority" that the ecosystem is entitled to. Contrast to economic priority, this stance provides a criterion for coordinating economic, social, and environmental conflicts. The principle of ecological priority is against the traditional development thinking with economic benefits at the core, and focuses on coordinated development of economy, environment and society, involving ecological law priority, ecological capital priority, and ecological benefit priority.

We should pass down ecological civilization and develop the innovative genes of industrial civilization—something that's different from industrial civilization. Innovation under ecological civilization is not only to maximize benefits, but to achieve sustainable economic benefits based on the premise of ensuring and improving ecological efficiency. In other words, as pointed out by the 19th National Congress, the value system of ecological civilization in a new era is, in essence, to pursue a model of sustainable development featuring increased production, higher living standards, and healthy ecosystems. It is a path of ecological prosperity, social happiness, and harmonious co-existence between man and nature. The social relationship based on this value system is not a web of intrigue, vicious competition, and a zero-sum game; instead, it's about mutual benefits, a win-win situation, and harmonious coexistence. We should not exhaust natural resources as the natural ecosystem has only limited carrying capacity and threshold constraints.

2 Cornerstone of the Natural Values and Theoretical Economics of Ecological Civilization

The value theory of industrial civilization is based on the labor value theory of classical economics. Physical labor is turned into average social labor, and people make products for exchange purposes, and create value in the process. As such, value is measured by labor, with labor results distributed based on the number of labor hours, and the value of nature is ignored or even negated. In Western Europe, where classical economics was born, people used to think that natural resources—for example, relatively constant water supply, the atmosphere with a relatively stable chemical structure, and continued solar radiation—were unlimited without scarcity. They were not for exchange purposes and are valueless as a result. The system design under industrial civilization focused on protecting the accumulated wealth through labor, while ignoring laborers with limited abilities to create value as well as nature deemed as something without market value.

Ecological civilization first recognizes the value of nature and believes that there is no substitute for nature. Human labor does not contribute to the output and services that nature provides, but the products and services created and provided by nature have features like regeneration (reproduction), cultivation (self-restoration), and interaction (interdependence), and they are systematic (as a whole). And nature also offers the inorganic environment, materials and media as well. The preservation and appreciation of natural assets come from the value of nature's labor. Comrade Xi Jinping offered a famous scientific thesis that "green mountains and clear waters are as good as mountains of gold and silver," which fully embodies the concept of respecting nature, accentuating natural resources, and seeking harmony between man and nature. He pointed out, "We should establish the concept of natural value and natural capital because natural ecology is valuable, and protecting nature is the process of increasing natural value and natural capital." This has laid the foundation for the theory of natural value implied by the above "clear waters and green mountains" phrase. It fully embodies the concept of respecting nature, accentuating natural resources, and seeking harmony between man and nature—a theoretical innovation of Marxist political economy and also the new theoretical economics for building an ecological civilization in a new era that provides a theoretical basis for the transformation and enhancement of industrial civilization, the transition to ecological civilization, and the approach to ecological civilization. It shows that under the ecological civilization system, value is measured not only with the value increment through human labor, but also by that created by nature's labor. Since labor results must be shared among all the parties involved, nature itself should also share a certain proportion of the output by nature and social labor, so that man and nature can at least work together to achieve simple reproduction, and the system at large and its elements can be sustainable. Therefore, the system designed under the value system of ecological civilization is not for capital itself, but for the people and nature that create capital.

3 Industrial Theory and Growth Momentum of Ecological Civilization

In the industrial civilization, people believed that capital, labor, and land were the basic factors of production, and the growth of material capital was achieved through investment and production. However, natural factors of production—the land, for example—were regarded as something that didn't create value, and in the process of distribution, invested capital is rewarded with interest, labor was paid with wages, and land got rent. But land rent went to the owner, instead of the land itself, with no thought of the land's recuperation or increase of its productivity. As such, industrial civilization achieved growth through environmental consumption, and its growth inevitably faced the "triple ceiling effect," that is, "consumption and demand saturation," "resource constraints," and "asset stock saturation," all of which in turn limited the development of industrial civilization. As western countries are entering the post-industrialization stage, they are losing growth momentum, and the gap between the rich and the poor is also widening. Moreover, many developing countries failing to avoid the development path of "pollution before governance" became "pollution refuges" for developed countries—a situation unwittingly resulting in increased regional and global environmental pollution and ecological degradation. The growth under industrial civilization has long been criticized by western academic circles. British economist and philosopher John Moore took the lead in proposing "Static Economy," a concept that promotes the idea that the population, economic aggregate and scale, and natural environment should basically remain stable. In the 1960s, resource depletion and environmental pollution called widespread attention to the limits of industrialization and economic growth. In view of this, American economist Kenneth Boulding proposed the "Spaceship Economy"; Donella Meadows and others promoted a zero-growth economy in *The Limits to Growth*; Herman Edward Daly, an American ecological economist, argued in favor of a "steady state economy" that entails stabilized population as well as energy and material consumption with limited fluctuations. That being said, these theories were either too biased or divorced from reality, or involve methodological dilemmas, which is why they couldn't be realized. And it's more difficult to use them to guide practice. The fundamental contradictions and problems of western industrial civilization are yet to be resolved, with its theories, methods and practices still causing puzzles and confronting difficulties.

Ecological civilization pursues ecologically neutral economic development, emphasizing the overall growth of labor value and natural value, so that natural assets can keep increasing, with less loss, coupled with repair and expansion. The conversion of natural assets, per capita material consumption, stock of fixed assets, and technical efficiency have become potential growth factors—among which technical efficiency offers the most important growth momentum in an ecologically neutral environment. For a long time, we have simply equated (economic) development with GDP growth, focusing on the production of labor products without realizing that ecological products are also one of the necessities for human survival and advancement. We also haven't formed the concept of nature's value and capital in the ecological environment. As such, while utilizing the natural environment and resources, we've failed to properly handle the relationship between man and nature as well as restrain our destructive behavior. Comrade Xi Jinping pointed out, "Throughout the world history, to protect the environment is to protect productivity, and to improve the environment is to boost productivity."—a stance that, to a large extent, gives a brand new dimension to "productivity," or to be more specific, modern green productivity, that ecological civilization will reply on for achieving growth. Besides understanding, changing and utilizing nature, this green productivity must gradually create within itself capabilities of preserving nature, including ecological balance and restoration, protection of the original habitats, environmental monitoring, and pollution prevention and control, in order to have "green waters and green mountains surround gold mountains and silver mountains." Currently, technologies and intellectual resources characterized by ecological technology, loop utilization technology, system management science and complex systems engineering, clean energy and environmental protection industry technology are becoming the internal driving factors of productivity enhancement and economic growth, thus resulting in the rise of ecological production methods. Moreover, the modern green shift in the industrial structure has greatly demonstrated, from a practical viewpoint, that Comrade Xi Jinping's ecological productivity theory is scientific, accurate and forward-looking.

4 Strategic Measures for Building an Ecological Civilization in a New Era

1. Ecology as a priority to achieve green development

Putting ecology as a priority means regarding the preservation of the ecological environment as a primary task, and seeing respect for, conformity to, and protection of nature as the prerequisites and criteria for development: ecological advantages are development advantages, and achieving ecological advantages means achieving development advantages. Basically, it is to seek harmony between humanity and nature, adapting production activities to the carrying capacity of the natural resources and environment, with the aim of maintaining ecological balance while keeping stable functions of the ecosystem for promoting green development, which integrates the criteria "ecology as a priority" into production, life, culture and political governance in all aspects to advance environmental protection. The goal of green development lies in development, with "green" indicating the direction of development, in which low-carbon, clean, and high-efficiency development should be employed as the desired methods and approach, while "development" is a comprehensive reform involving production, life, culture, and ecology, striving to adapt social and economic activities to the carrying capacity of the natural resources and environment through the transformation of the development model, while paying attention to the quality and benefits of development, and realizing all-round human development. Ecology as a priority and green development creatively combine to achieve both protection of nature and economic development.

Ecology as a priority and green development are a major theoretical breakthrough in terms of China's building of an ecological civilization. Since the two concepts are correlated, realizing the mode of ecology as a priority and green development requires mechanism innovation in the four aspects of technology, path, culture and system before turning ecological advantages into development advantages.

The first is to use technological innovation to create new economic formats and provide new governance methods. We should innovate production technologies, tap resources, give birth to new forms of ecological economy like ecological agriculture, industry and tourism, and promote clean production processes and diversify the industrial structure. Meanwhile, we should strengthen energy-saving and resource recycling technologies, advance the development and utilization of new energy sources and materials, improve resource utilization efficiency, innovate emission reduction and pollution control methods, and enhance ecological environment governance capabilities.

The second is to innovate development path (approach) to move towards new industrialization and urbanization. European and American countries used to follow the industrial development path of "treatment after pollution" and the urbanization path of "destruction before transformation," a fact that lead to waste of resources, environmental pollution, serious ecological damage, causing the so-called "urban diseases," which China are now suffering. Therefore, we should never repeat the same mistakes in our five-pronged socialist modernization in a new era. We must take "greening" as the forerunner before blazing a new trail to industrialization featuring "high scientific and technological content, good economic returns, low resources consumption, little environmental pollution." What's more, "greening" should be adopted to lead China's urbanization to usher in "green industries, green transportation, and green buildings." In short, we should stick to path innovation to protect the ecosystem from damage that would later requires restoration, and to reduce development costs as much as possible.

The third is to rely on cultural innovation to bring about a new trend of ecological civilization. Chinese traditional culture has the genes of ecological civilization—for example, the "ternary of Heaven-Earth-Man" in the *I Ching*, or *Book of Changes*, the Taoist idea to follow the laws of nature, and "unification of nature and man," a well-known Confucian concept—which promote harmonious coexistence between man and nature. Emphasizing respect for and conformation to nature, as well as ecology as a priority and green development, draws on our traditional ecological concept and has been promoted in contemporary China. It is necessary to cultivate ecological culture, promote socialist ecological civilization concept throughout society, instill ecological preservation into people and encourage them to carry it out, and promote diligence and thrift in society to enhance China's cultural soft power before creating the image of a "civilized, ecological and beautiful China."

The fourth is to adopt system innovation to improve the function of resource and wealth compensation and the transformation to social costs. The series of institutional innovations—the natural resource asset property rights transaction system, the ecological environment compensation system, the green GDP and the green GEP (i.e., the ecological system gross product value), and the red line system of ecological protection, for instance—will call attention to the value of resources and the environment, and clarify the standards for environmental protection and economic development. Resources and wealth will be used for environmental compensation, to make up for the depletion of resources, and improve the compensation for resource damage. Meanwhile, natural resources will be guided into social capital and invested in social construction so that the benefits reaped from using natural resources will be shared among all the people, old and young, throughout society.

2. **Adopting a holistic approach to conserving our mountains, rivers, forests, farmlands, lakes, and grasslands, and promoting building an ecological civilization from an ecosystem view**

Since the 18th National Congress, the CPC Central Committee with Comrade Xi Jinping at its core have been deepening the reform of the ecological civilization system, striving to promote the modernization of the national governance capacity and governance system for building an ecological civilization, a systematic project in which our "mountains, rivers, forests, farmlands, lakes and grasslands" were sloppily managed—for example, the property rights of natural resources and assets were not clearly documented, ownership was unidentified, and the rights and interests were not in place, all of which lead to common, chaotic management, infamously described with the term "nine dragons controlling water" and the like, to refer to the involvement, not always well coordinated, of multiple government units in the management of natural resources. Management was fragmented and distributed across government departments, with considerable overlap, in terms of central or local financial support or departmental interests; however, when it comes to supervision or administrative accountability, there was often a regulatory vacuum where "everyone was in charge yet no one was responsible."

To solve this problem, we must advance ecological civilization based on the system engineering thinking. In the report of the 19th National Congress, Comrade Xi Jinping extended the scientific thesis that "mountains, rivers, forests, farmlands, lakes, and grasslands form a community with a shared future" to "adopting a holistic approach to conserving our mountains, rivers, forests, farmlands, lakes, and grasslands." Biological communities form an organic whole with the environment in which they live through the exchange of energy and matter, and they are interconnected, interacting, and inseparable. A complete ecosystem includes elements such as mountains, rivers, forests, farmlands, lakes and grasslands. Only when these elements reach a dynamic and systematic balance can the sustainable development of human society be guaranteed. In view of this, to advance building an ecological civilization with systematic engineering and to coordinate the management of mountains, forests, farmlands, lakes, and grasslands, we must start with top-level design, break stereotypical thinking, and establish a strict ecological environment protection and supervision system. Moreover, we must strengthen the natural resource property rights and the mechanism of offering rights for all natural ecosystems such as grasslands, forests, wetlands, oceans, and rivers. We must integrate all protected areas like natural reserves, forest parks, and geological parks to form overall and comprehensive management that offers nature with time and space for recuperation. Finally, we must protect forests, lakes, and wetlands to expand green ecological space, and define the ceiling for using natural resources, the bottom line for exploiting the environment, and the red line for utilizing the ecological system in an all-round and system-wide manner.

3. Completing building an ecological civilization institutional system

Building an ecological civilization is a revolutionary transformation and fundamental change, which must rely on a systematic and complete system. In terms of China's environment protection, imperfect systems and mechanisms, along with incomplete rule of law, have caused serious problems that haven't been solved for a long time. Comrade Xi Jinping pointed out, "Only by implementing the strictest possible system for environmental protection and rule of law can we provide a reliable guarantee for building an ecological civilization."

Since the 18th National Congress, the CPC Central Committee with Comrade Xi Jinping at its core have successively proposed a series of target systems, assessment methods, rewards and punishment mechanisms that meet the requirements of ecological civilization. What's more, mechanisms like the natural resource property rights, use control, the ecological red line, pay for use, ecological compensation, and the management system, have been continuously improved, strengthened and highlighted. The "multiple pillars" reform of the ecological civilization system have been established with emerging beneficial results. Moreover, the report of the 19th National Congress emphasized, once again, the importance of establishing an ecological civilization system from many aspects, including "stepping up efforts to establish a legal and policy framework that promotes green production and consumption," "improving environment management systems," and "enforcing stricter pollutants discharge standards and seeing to it that polluters are held accountable. We will improve our systems for credibility assessment based on environmental protection performance, for mandatory release of environmental information, and for imposing severe punishment for environmental violations."—all of which have put forward new requirements for building an ecological civilization system in a new era.

In general, building an ecological civilization requires efforts in the following aspects: The first is to accelerate the improvement of the property rights system for natural resources and assets and the use control system, delineate the red line for ecological protection, and promote the "natural balance sheet," in order to provide a fundamental legal basis and system guarantee for instilling the "concept of nature's value and capital"; The second is to clarify the spatial planning system by "combing regulations into one" with integrated planning for land use, urban construction, industrial layout, pollution control, and ecological protection; The third is to accelerate the establishment of an ecological compensation mechanism and an ecological compensation fiscal policy system, strengthen the overall conservation system, improve the function of resource wealth compensation and the transformation to social costs, and lay the foundation for creating a legal system for green production and consumption; The fourth is to speed up the establishment of the target evaluation and accountability system for ecological civilization construction, supervise the party and government leaders at all levels and guide them to actively advance ecological civilization to reverse their idea of "pursuing GDP only" so that they can gain a new concept about political performance in the ecological civilization governance system; The fifth is to speed up the establishment of an environmental damage assessment system, deepen the reform of the national environmental damage compensation system, and promote the concept that "the environment has value, and those who damage it will be subject to punishment" to support ecological restoration and the like.

4. **Contributing to global ecological security**

The report of the 19th National Congress clearly required that we should have a strong commitment to socialist ecological civilization and work to develop a new model of modernization with humans developing in harmony with nature. This was not only a signal, but also a promise to the world, and a Chinese approach to building a community with a shared future for mankind.

China has taken responsibility for handling global ecological security issues by advancing the negotiation, conclusion, effectiveness and implementation of *the Paris*

Agreement, and by actively assuming responsibilities and obligations in the implementation process of the United Nations' *2030 Agenda for Sustainable Development*. President Trump's announcement of his intention to withdraw from the *Paris Agreement* to circumvent the US's environmental responsibilities, along with his "de-globalization," has created obstacles to the global response to climate change. However, mitigating climate change is still a global consensus and strong political will. China should unswervingly join the cause and take the lead to make greater contributions to handling global ecological security issues such as climate change, biodiversity protection, and marine pollution management. Promoting global sustainability goes hand in hand with building China's ecological civilization.

First, China's current contributions to the building of global ecological civilization involve improving its own environment and protecting global ecology at the same time. Therefore, we need to put in more efforts than developed countries and other developing nations. From the perspective of system theory, China is part of the world, and without the security of the world's members, there would be no security for all countries—which explains why governing one's environment is a contribution to global ecological security. As the world's largest developing country and industrial power, China needs to first solve its own arduous environmental problem, adhere to governance by all the people, adopt source prevention, and mitigate air pollution to keep it clean for all beings; prevent and reduce water pollution to clean up polluted waters; control and mitigate soil pollution to eliminate the source of poisoning in the food chain; strengthen solid waste management and garbage disposal to promote resource regeneration and increase the carrying capacity of the land. Second, China should follow the trend of international sustainable development, continuously deepening the understanding of sustainable development, exploring correct development ideas based on its specific national conditions. At home, we should promote economic transformation, strengthen building an ecological civilization, and actively realize the 2030 sustainable development goals; internationally, we should create a dialogue mechanism, promote cooperation and exchanges, pursue green development, eliminate absolute poverty, build a global ecological governance system, and promote global low-carbon transformation. From another perspective, we can say that leading the world towards ecological civilization while maintaining global ecological security will be the most powerful counterattack to the fallacy of "China's threat to environmental security" in the international arena. This will be the most fundamental contributions made by China, a great socialist nation in the East, to sustainable development for mankind.

Guiding Global Climate Governance and Domestic Response Actions with Xi Jinping Thought on Ecological Civilization

Jiankun He

Promoting the building of ecological civilization while taking the path of sustainable development in a green and low-carbon circle is China's basic strategy to reduce the ever-increasing resource and environmental constraints and build a socialist modern power in a new era. It is also a global response to ecological crisis, among which climate change is the most serious threat, to achieve harmony between man and nature as well as the sustainable development of human society. China should follow Xi Jinping Thought on Ecological Civilization to thoroughly improve its ecological environment and build a beautiful China, while contributing its wisdom and approach to help preserve the earth's ecology, and taking the lead in building global ecological civilization.

1 Guiding the Global Climate Governance and Cooperation Process with Xi Jinping Thought on Ecological Civilization and the Concept of a Community with a Shared Future

The report of the 19th National Congress listed climate change as an unconventional threat to a global security and a common challenge facing mankind. The core of tackling climate change is to reduce carbon dioxide and other greenhouse gas emissions from human activities, control the concentration of greenhouse gases in the atmosphere, and slow down climate warming. The *Paris Agreement* adopted by the United Nations Climate Conference at the end of 2015 includes commitments from all the parties through a "bottom-top" mechanism, aiming to cut global greenhouse gas emissions to limit the global temperature increase in this century to 2 °C,

J. He (✉)
Tsinghua University, Beijing, China

J. Pan et al. (eds.), *Beautiful China: 70 Years Since 1949 and 70 People's Views on Eco-civilization Construction*, https://doi.org/10.1007/978-981-33-6742-5_69

while pursuing means to limit the increase to 1.5 °C. Under the accord, the world is committed to achieving net zero carbon emissions or "carbon neutrality" by the second half of the twenty-first century. The major means is to achieve a low-carbon energy system and turn economic development into a green and low-carbon mode, with the aim of ensuring sustainable economic and social progress while creating a win-win situation for economic development and addressing climate change by controlling and reducing greenhouse gas emissions to achieving balance between economic development, social progress, and ecological preservation.

Taking Xi Jinping Thought on Ecological Civilization to advance the building of global ecological civilization and achieving harmony between man and nature is the fundamental and the only way to deal with climate change under the framework of global sustainable development. Comrade Xi Jinping pointed out, "Climate change is vital for people's lives and the future of mankind." He emphasized that China should "work with others to build global ecological civilization, join global environmental governance, propose solutions for global environmental protection and sustainable development, and take the lead in the world's response and cooperation to address climate change." Xi Jinping Thought on Ecological Civilization advocates the establishment of a green, low-carbon, and sustainable economic system, the construction of a clean, low-carbon, safe and efficient energy system, the development of green finance, the establishment of a legal system and policies for green production and consumption, in order to form a joint effort to tackle the earth's ecological crisis with climate change as the most urgent issue, achieve harmony between man and nature, and contribute Chinese wisdom and approach to help build a beautiful world, while effectively advancing global environmental governance and building global ecological civilization.

The current global ecological crisis is mainly caused by the uncontrolled consumption of the earth's resources and the discharge of waste into the environment since the industrial revolution—a development mode that has far exceeded the earth's carrying capacity given its limited resources and fragile environment. Over the past century, the emissions of carbon dioxide and other greenhouse gases from extensive fossil energy consumption and land use have been rapidly causing global warming, posing a serious threat to the world's ecology and the survival and development of human society. During their industrialization and modernization, developing countries can no longer follow developed countries' traditional development modes, characterized by high resource consumption and high pollutant emissions. And they should instead embark on a resource-saving and environment-friendly green development path. Comrade Xi Jinping proposed ideas that offer guidance for all the countries, particularly developing ones, including: Ensuring harmony between man and nature; lucid waters and lush mountains are invaluable assets; a good ecological environment is the fairest public products and the most accessible welfare for people; mountains, rivers, forests, farmlands, lakes and grasslands form a community with a shared future; and the strictest possible systems and rule of law should be adopted for environmental protection. His purpose is to create an ecological cultural system based on ecological values, coordinate the relationship between economic development, social progress and environmental protection, promote low-carbon energy consumption and economic progress, and embark on a climate-suitable low-carbon economy development path—all of which are of universal guiding significance. Xi Jinping also put forward the idea of building a global ecological civilization, stressing that "it is necessary to build an ecological system that respects nature and promotes green development. It is necessary to resolve the contradictions brought about by industrial civilization, achieve harmony between man and nature, and achieve sustainable development of the world and all-round development of mankind." It will become an important guiding ideology for global cooperation in responding to the ecological crisis, promoting the transformation of human society from industrial civilization to ecological civilization, leading the overhaul of the world's economic and social development methods, and achieving harmony between man and nature.

China is "taking a driving seat in international cooperation to respond to climate change, China has become an important participant, contributor, and torchbearer in the global endeavor for ecological civilization," and has played a positive role in promoting the conclusion and entry into force of the *Paris Agreement*. While implementing the *Paris Agreement*, we adhere to Xi Jinping Thought on Ecological Civilization as a guide at home, firmly following the path of increasing production, achieving affluence and maintaining good ecology by carrying out the strictest ecological and environmental protection system to build a beautiful China. We adhere to the sustainable development concept of a green and low-carbon circle, which is consistent with the development path of a climate-friendly, low-carbon economy advocated by the *Paris Agreement*. China has achieved great results in energy reform and carbon dioxide emission reduction. By combining the response to climate change with domestic sustainable development, we have created a win-win situation for the economy, people's livelihood, energy, the environment, and carbon dioxide emission reduction, making China a vital contributor and leader in the world's energy transformation and transition to a low-carbon economy. In short, China's energy utilization and economic transformation, new urbanization construction, industrial transformation and upgrading, environmental governance, and other successful experiences and cases, as well as its energy-saving and carbon-reduction policy systems and ecological civilization system construction, can all be imitated and referenced by other developing countries. Moreover, China's remarkable achievements and successful experience in building an ecological civilization and environmental protection, along with its ideas and practices of creating an economic system, institutional system, and ecological safety system based on ecological civilization, will also play an active leading role in building a global ecological civilization and promoting global sustainable development.

The report of the 19th National Congress emphasized that China would follow a path of peaceful development, work to build a community with a shared future for mankind, uphold the global governance concept of extensive consultation, joint construction, and shared benefits, actively participate in the reform and construction of the global governance system, and keep contributing Chinese wisdom and approach. The report also stressed that we should be good friends to the environment, cooperate to tackle climate change, protect our planet for the sake of human survival, and enhance global ecological security. Xi Jinping Thought on Ecological Civilization and the concept of building a community with a shared future for mankind combine to serve as Chinese wisdom and solutions for global environmental governance. We advocate a new global governance concept of mutual respect, fairness, justice, and win-win cooperation, and regard climate change as an opportunity for all countries to pursue sustainable development, and promote mutually beneficial cooperation and development, which is conducive to expanding voluntary cooperation between countries, expanding the convergence of interests with other countries, and turning climate change negotiations from a "zero-sum game" into win-win cooperation. In terms of climate governance concepts and cooperation methods, China has shown a new type of leadership different from that of the United States and Europe—which is being increasingly recognized worldwide. Cooperation to address climate change truly meets the interests of all countries, with more converging points of interests and room for win-win cooperation than other global risks and regional hotspot issues in political, economic, social and other fields, and thus has brighter prospects for cooperation. Tackling climate change provides an international stage for China to demonstrate its leadership in the reform and construction of global governance, making it possible for it to serve as a pioneer in promoting mutual respect, fairness, justice, and win-win cooperation in a new era to build a community with a shared future for mankind.

Addressing climate change is related to the common interests of mankind, and thus requires global cooperation. Comrade Xi Jinping emphasized that "we must implement a national strategy to actively address climate change, promote and guide the establishment of a fair, reasonable, cooperative and win-win global governance system, and follow the vision of a community with a shared future for mankind, live up to the responsibility as a major country." He also proposed other ideas, such as "working with other countries to build a beautiful homeland with good ecology for mankind," that have become an important guiding ideology leading the reform and construction of environmental governance systems worldwide.

The global climate governance mechanism and international negotiation rules laid out in the *United Nations Framework Convention on Climate Change* adhere to the principles of multilateral participation and consensus of all parties, and offer provisions for conscious cooperation to promote ecological security of the earth and the common interests of mankind. However, there are contradictions and differences between the parties in the sharing of responsibilities and obligations, unwittingly resulting in complex multi-party games—a situation that requires influential powers to intervene and take the lead. To play a leading role in climate change, China must show its ability to coordinate the positions and interests of all parties, guide the formulation of global climate governance mechanisms and action plans that reflect fairness, reasonableness, and win-win cooperation, and promote political consensus and willingness of all parties to showing influence and charisma in seeking the common goal, considering the positions of all parties, and the balance of the interests of all parties, thereby promoting the joint efforts and strengthening actions. China should adopt Xi Jinping's idea of building a global ecological civilization and the concept of building a community with a shared future for mankind to actively lead global cooperation and actions in tackling climate change, so that China can take the international moral vantage point to enhance its national image and leadership, maintain and gain more interests, and demonstrate its soft power.

2 Changing Values and Development Methods Based on Xi Jinping Thought on Ecological Civilization

It is very urgent to curb greenhouse gas emissions to limit the global temperature increase to 2 °C in this century. The key to achieving net zero carbon emissions by the second half of the twenty-first century depends on the transformation of the energy system and development patterns in the next 20–30 years. Failure to do so would result in global heating of 3–4 °C—a possibility that would have an irreversibly catastrophic and devastating impact on the earth's ecological security. Therefore, it is urgent for all countries to keep up their response.

To respond to climate change under the global framework of sustainable development and to achieve both sustainable economic and social development of all countries and the earth's ecological security, we must change the development concepts, values, codes of conduct, and production and consumption methods under industrial civilization to move towards ecological civilization. Comrade Xi Jinping emphasized the need to "establish and improve a cultural system based on the values of ecological civilization," and transform economic and social development methods by changing ideological concepts. China must use Xi Jinping Thought on Ecological Civilization as a guide to promote domestic and even global awareness of ecological risk management, transform ideas and concepts, prompt countries, cities, enterprises, and the public to voluntarily reduce greenhouse gas emissions, and get people involved in joint efforts. Xi Jinping pointed out that a good ecological environment is the fairest public product that is the most accessible welfare for the people. He insisted on ecological civilization that will benefit the people and meet their growing needs for a beautiful ecological environment. Moreover, it of great significance to promote China's traditional culture of "unification between nature and man" and "the Taoist idea to follow the laws of nature," promote a simple and modest way of life, help transform values and consumption patterns worldwide, and enhance the influence and soft power of Chinese traditional culture and socialist values on the international community.

While guiding the transformation of social development concepts and values with the idea of ecological civilization, China has been strengthening its ecological civilization and governance system, incorporating it into the legal system to coordinate all aspects involving the economy, society, and resources and the environment to effectively prevent and control ecological and environmental risks. Xi Jinping emphasized the need to use the strictest system and rule of law to protect the ecological environment, promote system innovation, accelerate building an ecological civilization management system, establish a target responsibility system with improving the quality of the ecological environment as the core, and build a governance system and modernization of governance capabilities to protect the ecological civilization system, and create an ecological security system focusing on the benign the ecosystem as well as effective prevention and control of environmental risks—a series of thoughts and measures that will play an active role in leading and demonstrating the building of an ecological civilization around the world.

To achieve the dual goals of sustained human economic and social development and control of global temperature rise and protection of the earth's ecological security, we should focus on promoting the revolutionary transformation of the energy system and realizing the low-carbon transformation of economic and social development. This is also the fundamental approach and key to building global ecological civilization. The ever-increasing consumption of fossil energy since the Industrial Revolution has caused the emissions of greenhouse gases, along with other common pollutants like sulfur dioxide, nitrogen oxides and smoke, which have been damaging the world's ecology. Whether it is to protect and improve the regional ecological environment or to address climate change, it is necessary to control and reduce fossil energy consumption before a new energy system with zero pollution and zero carbon emissions emerge to replace the current system based on fossil energy with high pollution and high carbon emissions. Therefore, the revolutionary transformation of the energy system has become a new trend, a strategic choice for countries to respond to global climate change and promote domestic sustainable development, as well as an important field and focus of global ecological civilization construction. Making revolutionary changes in the energy system to promote economic and social development onto a sustainable development path characterized by a green and low-carbon circle is the fundamental way to turn industrial civilization into ecological civilization.

China has incorporated the building of an ecological civilization into the overall layout of its five-pronged socialist modernization, and put it in a prominent position. Under the guidance of Xi Jinping Thought on Ecological Civilization, we have achieved world-renowned results: for example, we are revolutionizing energy production and consumption, turning our development mode into a green one, strengthening the building of the ecological civilization system, speeding up the establishment of legal systems and policies for green production and consumption, creating and improving a green, low-carbon economy, advancing the revolution of energy production and consumption, building a clean, low-carbon, safe and efficient energy system, and advocating a simple and moderate, green and low-carbon lifestyle. With successful practices in transforming its development mode through our guiding ideology, strategic thinking, implementation approach, and policy guarantee measures, we will offer our experience and contribute to the transformation of the global development model, while influencing and leading global energy and development model transformation.

China has been improving environmental quality and responding to climate change at home and abroad, strengthening the building of the ecological civilization system, and incorporating energy conservation and carbon reduction into the national economic and social development plan. During the 11th Five-Year Plan period, we formulated a binding indicator for reducing energy intensity per unit of GDP; during the 12th Five-Year Plan period, we added the carbon dioxide intensity reduction indicator for GDP; during the 13th Five-Year Plan period, we further incorporated the total energy consumption control target. We broke down the indicators and distributed them to provinces and cities to strengthen the target accountability system of governments at all levels. While establishing and improving a series of fiscal, taxation and financial policy systems, we were building our national carbon market to combine government regulatory measures and market means as a strategy for response to climate change.

Currently, China adheres to the new development concept. While economic development is shifting from high-speed to high-quality growth, we should focus on changing the development mode, optimizing the economic structure, transforming the growth momentum, deepening supply-side reforms, and increasing productivity, all of which will help turn the extensive growth mode, driven by increased inputs of production factors at the expense of resources and the environment, into a green and low-carbon development path driven by innovation. By strengthening energy conservation, improving energy and output efficiency, developing new and renewable energy sources, and accelerating the low-carbon energy structure, we have effectively curbed rapid growth in energy consumption and carbon dioxide emissions, reduced the energy intensity per unit of GDP and sped up the downward trend of carbon dioxide intensity. From 2005 to 2018, China's intensity of carbon dioxide per unit of GDP dropped by about 48%, achieving ahead of schedule the voluntary emission reduction target of 40–45% that it promised at the Copenhagen Climate Conference in 2009, while laying the foundation for realizing the voluntary emission reduction target that it committed itself in the *Paris Agreement*. With its aggressive emission reduction targets and effective actions and practices, advanced low-carbon energy technology innovation and industrialization development, and the construction of high-efficiency energy and low-carbon infrastructure in the green Belt and Road Initiative, China has been, and will be, taking the lead as a catalyst for advancing global energy source reform and low-carbon economic transformation.

3 Establishing and Implementing a Medium- and Long-Term Low-Carbon Development Strategy that Is Consistent with the Goals and Strategies of Socialist Modernization in a New Era, and Is Compatible with the Global Deep Decarbonization Goal and Emission Reduction Approach

The 19th National Congress put forward the goal and basic strategy of building China into a modern socialist country by 2050. It also listed climate change as an unconventional threat to national security, promoting the reform and construction of the global environmental governance system to contribute to global ecological security. China's long-term low-carbon emission strategy must be consistent with its two-stage development goal of modernization by 2050, and guided by the goal of following the pathway to deep decarbonization—limiting the global temperature increase to 2 °C—under the *Paris Agreement*. We should promote the low-carbon reform of the energy system, create a sustainable economic development model with a green and low-carbon cycle, study the technological innovation roadmap for achieving net zero emissions as soon as possible after the mid-twenty-first century, make the revolution goal of deep decarbonization, and follow a low-carbon economic development path of limiting the global temperature increase to 2 °C, in order to make contributions to the world's ecological security and the common interests of humanity that are in parallel to our rising overall national strength and international influence.

In the first stage from 2020 to 2035, China, with 15 years of hard work, will see that a socialist modernization is basically achieved, and the goal of building a beautiful China is basically attained. This is consistent with the timetable in which China is committed to achieving the carbon dioxide emission reduction target, with a synergistic effect on adopted measures, which is conducive to reducing carbon dioxide emissions. The goal is to thoroughly improve the ecological environment, forming an overall spatial arrangement, production method, and lifestyle that save resources and preserve the environment. The urban environment will become beautiful, harmonious, and livable to meet people's demand for beautiful environment and ecological products. And we will effectively promote production and consumption revolution to quickly form a clean, low-carbon, safe, and efficient new energy and economic system characterized by a green and low-carbon circle that will effectively reduce carbon dioxide emissions. Moreover, we will achieve the targets of reducing the carbon dioxide emission intensity per unit GDP in 2030 by 60–65% compared with 2005, non-fossil energy accounting for 20% of the energy consumption mix, and carbon dioxide emissions peaking by around 2030—all the targets are also conducive to fundamentally controlling and reducing conventional pollutants to help improve environmental quality and build a beautiful China. Therefore, it is necessary to realize the implementation and action plan of the Intended Nationally Determined Contributions (INDC), and make a specific timetable for the peaking of carbon dioxide emissions ASAP and the peak emission control target. The 19th National Congress also proposed the goal and measures for strengthening actions and deepening emission reduction by 2035, in order to get connected with the emission reduction goals and measures for deep decarbonization in the second stage. Meanwhile, we must adapt to the situation where countries are required to strengthen and update their intended contribution targets under the *Paris Agreement*.

In the second stage from 2035 to the middle of the twenty-first century, after a further 15 years of hard work, China will be developed into a great modern socialist country featuring harmonious coexistence between man and nature, green mode of development and way of life. At the same time, an energy transition and low-carbon development strategy compatible with the deep decarbonization pathway under the *Paris Agreement* to limit global temperature rise of no more than 2 °C is established and implemented. As China's ecological environment has fundamentally improved at this stage, the strategy for addressing climate change must go beyond the internal needs of domestic resource conservation, environmental protection and sustainable development, with more consideration directed to reducing carbon dioxide emissions to ensure the planet's ecological security—starting from the historical mission of making new and greater contributions to mankind, to making the goal and measures to significantly reduce carbon dioxide emissions, compared with the peak year, by 2050 and achieve zero emissions. We should take the goal of actively responding to climate change as an important part of the overall goal of building China as a modern and powerful socialist country, and make contributions to the global response to climate change that are in line with our national strength and international influence at that time, in order to assume our responsibility and take the lead to pursue global ecological civilization development and the common interests of mankind and to build a modern socialist China and ensure global ecological security at the same time.

At present, China is at the intersection of the 13th Five-Year Plan period for building a moderately prosperous society in all respects, and the 14th Five-Year Plan, where our country has embarked on a new journey of socialist modernization. We should follow the proposal of the 19th National Congress to accelerate the reform of the ecological civilization system, promote a sound economic structure that facilitates green, low-carbon, and circular development, win the war on pollution prevention and control, build a beautiful China, and contribute to global ecological security. During the 13th Five-Year Plan and 14th Five-Year Plan periods, it is necessary to use Xi Jinping Thought on Ecological Civilization as a guide to give play to the synergistic effect of carbon dioxide emission reduction and preservation of the environment, and to coordinate deployment and strengthen actions. Currently, it is necessary to combine the strategic deployment of building a moderately prosperous society in all respects, and while implementing a series of policy measures such as promoting the building of an ecological civilization and fighting the war on pollution, we should improve the ecological environment and reduce carbon dioxide emissions at the same time; moreover, while pursuing the short-term target of dealing with and preventing regional environmental pollution, we should strive to achieve the long-term goal of low-carbon development and carbon dioxide emission reduction. We should coordinate governance of the economy, energy, and the environment with response to climate change to create a win-win situation for all.

Besides achieving urgent emission reduction targets and following emission reduction paths, we should encourage breakthroughs in revolutionary advanced technologies, such as large-scale energy storage technology and smart grid technology with a large proportion of renewable electricity grids in place; Bio-energy with carbon capture and storage (BECCS), a process of extracting bio-energy from biomass and capturing and storing the carbon, thereby achieving negative carbon dioxide emissions; the preparation, storage and utilization technology of hydrogen, considered as a clean zero-carbon secondary source of energy; and zero-carbon production technology for raw materials such as chemicals, steel, and cement etc. It is necessary to strengthen R&D and demonstration projects of these revolutionary technologies, and increase investment to make breakthroughs as soon as possible to quickly industrialize them before turning them into mature and affordable products for realizing zero carbon dioxide emissions. China must strengthen its R&D and demonstration of advanced technologies, and carry out forward-looking deployments to obtain the core competitiveness of advanced technologies. The transition to low-carbon energy use and economy in the global response to climate change will trigger a fundamental change in the economic and social development mode, and will also reshape the political, economic, and technological competition across the globe. The development of advanced energy technologies and low-carbon industries will represent a country's core competitiveness, which will in turn help expand its international market share and influence, form new economic growth points and create new employment opportunities, and lead the world's trends in technological innovation and economic transformation.

Strategic Thinking and Theoretical Practice of Marine Environmental Protection in a New Era

Daoming Guan

Since the 18th National Congress of the CPC, the Central Committee has been attaching great importance to the building of ecological civilization with unprecedented theoretical vigor and promotion efforts as well as in-depth reform. China is a large country bordering the ocean that has a significant strategic role in ensuring national security, promoting economic and social development, building an ecological civilization, and advancing cultural prosperity. Enhancing marine habitat protection is an important part of building an ecological civilization and building China as a maritime power, and it is also of great significance to creating a beautiful China. Moreover, marine ecological and environmental issues have formed a bottleneck and shortcomings that delay the economic and social development of China's coastal areas. To atone for the above shortcomings and solve the marine environmental problems that the people are deeply concerned about, we should follow General Secretary Xi Jinping Thought on Ecological Civilization and important expositions on building China as a maritime power as a guide, respecting, conforming to and preserving the ocean to protect and improve the marine ecological environment—a responsibility which we should take to enhance China to a maritime power and build a beautiful China, and which is also related to the welfare and future of the people.

1 History of China's Marine Environmental Protection

General Secretary Xi Jinping pointed out, "The global economic development has shown an obvious trajectory from the inland to the ocean, and then from the ocean to the world in terms of a country's path towards prosperity." The rapid development of China's coastal areas has been a process of continuous development towards the sea. Given the increasingly important role the ocean plays in spurring the national

D. Guan (✉)
National Marine Environmental Monitoring Center, Dalian, China

J. Pan et al. (eds.), *Beautiful China: 70 Years Since 1949 and 70 People's Views on Eco-civilization Construction*, https://doi.org/10.1007/978-981-33-6742-5_70

economy, China's marine industry has been flourishing, with its management going through four stages of industrial management, navy management, administrative management, and comprehensive management. Moreover, the preservation of the marine ecological environment started with establishing a marine environment monitoring network to keep an eye on the ocean, and then gradually moved towards a legal and professional track; after the institutional reform, the responsibility of marine environmental protection was integrated into the newly formed Ministry of Ecology and Environment—an approach combining the management of the land and ocean. As such, marine environmental protection tasks have gradually shifted from "fragmentation" and "five dragons governing the sea"—a phrase referring to the involvement, usually not well-coordinated, of multiple units in charge—to land-sea coordination and integrated management, with the tasks gradually moving from "focusing on prevention and control of pollution in coastal waters" to "equal emphasis on prevention of marine pollution, ecological protection, risk prevention, and global governance."

(1) Oceanic monitoring. As basic marine management and an important means of preserving the marine environment, oceanic monitoring was initiated in the 1970s, with everything starting from scratch—from the construction of the Bohai Sea and the Yellow Sea marine environmental monitoring network to the completion of the national marine environmental monitoring network. What's more, the marine environment monitoring items have been expanded from water environment monitoring only to multi-medium and multi-regional monitoring, such as monitoring the sedimentary environment, biological quality, ecology, marine functional areas, pollution sources, etc., with the monitoring scope first covering the waters under China's jurisdiction before being extended to the Western Pacific, an area that is closely related to China's ecological security.
(2) The protected marine area. Since our first marine reserve was created in 1963, China has undergone sporadic and stagnant development, then recovery and rapid development, and has now entered a stage of high-speed development with the natural reserve system in place. Up to now, there are 271 protected marine areas of all types at various levels, which account for 4.1% of the sea area under China's jurisdiction, across 11 coastal provinces (municipalities and autonomous regions), and cover a number of typical marine ecosystems and rare, endangered species, thus achieving the protection of diverse marine life forms.
(3) Oceanic restoration. The recent years have seen the implementation of air, water, and soil pollution prevention plans and the war on pollution, which have helped improve the ecological environment and marine habitats across or along China. The strict reclamation management and the "Blue Bay" ecological environment improvement and restoration project have helped clean up more than 600 illegal (or unreasonable) sewage outlets into the sea, with more than 260 km of restored shorelines, over 1200 hectares of restored beaches, and

more than 4000 hectares of restored wetlands, thus curbing the ecological degradation of coastlines, coastal wetlands and key bays.

Although the overall quality of China's marine environment has stabilized and been improved, and our local ecosystems restored, we are still at the peak of pollution emissions and environmental risks as well as the period of ecological degradation superimposed onto frequent disasters, all of which present serious threats to our marine ecological environment. Therefore, we must pay heed to marine ecological conservation, curb marine environmental pollution, and preserve marine biodiversity to keep the ocean clean and beautiful.

2 Overall Thinking and Strategic Arrangement of Marine Environmental Protection in a New Era

1. New requirements of the Party Central Committee for marine environmental protection

Since the 18th National Congress, China has incorporated the building of an ecological civilization into the overall layout of its Five-sphere Integrated Plan for realizing socialist modernization with Chinese characteristics. The Party Central Committee has promulgated a series of important documents, such as the *Opinions on Accelerating the Construction of Ecological Civilization*, the *Overall Plan for the Reform of the Ecological Civilization System*, and the *Measures for Evaluation and Assessment of the Goals Achieved under the Construction of Ecological Civilization*. The Committee has also made major decisions and deployments to enhance ecological environment protection, improve ecological civilization, and build a beautiful China, and has incorporated the requirements for ecological civilization and building a beautiful China into the Constitution, thus laying a firm foundation for Xi Jinping Thought on Ecological Civilization, expanding and deepening the construction of ecological civilization, offering guidance for continued ecological and environmental protection, and providing guidelines for conserving marine habitats in a new era.

General Secretary Xi Jinping has clearly pointed out that "how the ocean industry develops and whether the ocean issues are handled well are related to the survival and development of our people, as well as the prosperity of our country," and made a series of important expositions on building China as a maritime power while protecting our marine ecological environment. The 18th and 19th National Congresses of the CPC successively made major strategic deployments of "building China into a strong maritime country" and "pursuing coordinated land and marine development and stepping up efforts to build China into a strong maritime country" The two meetings also arranged important tasks for protecting the marine habitats by issuing important policy documents, including *Water Pollution Prevention and Control Action Plan, Promoting China's Ecological Civilization Construction to a New Level, Opinions on Comprehensively Strengthening Ecological and Environmental Protection and Resolutely Winning the War on Pollution*, and *Action Plan for the Comprehensive Management of the Bohai Sea*. Xi Jinping Thought on Ecological Civilization and his important expositions on building China as a strong maritime country have pointed out the direction and goal for our marine industry, further clarifying marine environment conservation efforts in a new era.

2. **Overall thinking and strategic arrangement of marine environmental protection in a new era**

a. **The General Thinking**

China is now in a decisive period of building a moderately prosperous society in all respects, a critical period for achieving its first centenary goal, and a vital period when it is seeing economic growth shifts, structural adjustment, and transformation of economic momentum. In an open framework of linked management of the land and sea, and East-West mutual aid and opening up, the ocean has been a key strategic location for high-quality development. Therefore, marine habitat conservation is vital to expanding the blue economic space and achieving high-quality progress. The reform of our national institutions has resulted in major changes to oceanic protection in terms of our work patterns, objectives, methods, and areas. Therefore, it is necessary to fulfill the new requirements of the Party Central Committee for marine conservation, adopt the national development strategy in a problem-oriented, demand-driven, and natural-law-based manner, further clarify the overall thinking of marine environmental protection, and enhance targeted, accurate, and scientific efforts in protecting the oceanic habitats in a new era.

Based on the characteristics of the marine ecosystem and the current marine ecological environment, the overall idea is initially proposed as follows: Using General Secretary Xi Jinping Thought on Ecological Civilization and important expositions on building China as a maritime power as a guide to thoroughly follow the spirit of the relevant instructions of the Party Central Committee and the State Council, and following the principle of "coordinated management of the land and sea as well as land planning based on oceanic conservation," with green development and renovation as the approach, and system building and capacity improvement as the guarantee, and technological innovation and international cooperation as the support. Efforts will be made to solve serious problems facing the marine ecological environment, improve oceanic quality, make the marine ecological system healthy and stable, continue to enhance the supply capacity of high-quality marine ecological products, realize the modernization of the marine ecological environment governance system and governance capabilities, create "rich, beautiful marine habitats in harmony with humanity" that the people desire, and offer strong support for high-quality economic progress.

b. **Strategic Arrangements**

The 19th National Congress made the two centenary goals and two-stage strategic arrangements for building socialism with Chinese characteristics in a new era. It also made "building a beautiful China" a vital part of our journey of building a strong socialist country, incorporating "the provision of more high-quality ecological goods to meet people's ever-growing demand for a beautiful environment" into our strategy for enhancing people's livelihood, and corresponding to the overall layout of the Five-sphere Integrated Plan – all of which were major strategic tasks in our new journey of fully building a modern socialist China. Given the fact that the ecological and environmental problems are not accumulated in a day, coupled with oceanic mobility, complexity, and hysteresis, it is impossible to improve the quality of the marine ecological environment overnight. Therefore, we need make strategic arrangements for oceanic conservation in a pragmatic and far-sighted way, and continue to refine, deepen, and optimize the timetable, roadmap and project charter for preserving marine ecological environment.

Now, it is necessary to make short-term plans to focus on the comprehensive management of the Bohai Sea, while making long-term efforts to help realize the vision of preserving the beautiful ocean; it is also necessary to lay out plans for near-shore management, and coordinate land and sea management and to expand efforts to include the deep sea and the polar regions. To implement the "two centenary goals," given the characteristics and requirements of marine environmental protection, our oceanic conservation was initially divided into three periods, namely the key breakthrough period (2019–2020), the comprehensive improvement period (2020–2035), and the consolidation and enhancement period (2035–2050).

(1) The key breakthrough period: We should focus on making breakthroughs to win the battle in the comprehensive management of the Bohai Sea. We must ensure the effective improvement of the sea's ecological environment to realize the goal of the *Action Plan for the Comprehensive Management of the Bohai Sea*. Meanwhile, we should put in efforts to create a pioneering area for pollution control and environmental improvement across China, keep continued improvement of the oceanic habitats, solve the serious problems plaguing the marine ecological environment, curb marine ecological degradation, continue to improve the quality of marine ecological environment, and keep the ocean clean and healthy.

(2) The comprehensive improvement period: After achieving solid results and accumulating experience in key areas, we should expand and deepen our efforts to prevent pollution and help restore natural habitats, improve the service functions of the marine ecosystem, effectively maintain the biological resources of the ocean, effectively prevent and control environmental risks, greatly improve the marine ecological environment, make significant progress in building the marine environmental governance system and governance capacity in a bid to keep the ocean healthy.

(3) The consolidation and enhancement period: We should consolidate the results of marine environmental protection and governance, thoroughly improve the marine ecological environment, make the marine ecological system healthy and stable, continue to improve our ability to provide marine ecological goods, make the ocean more accessible to the public, raise people's awareness of

marine environment protection and encourage them to conserve oceanic habitats, modernize the marine ecological environment governance system and governance capabilities, and ensure the security of the marine ecological environment to keep the ocean beautiful.

In terms of the direction and content, the first is to pursue green development and achieve prevention at the source. Ecological and environmental issues, in essence, are linked to the way people develop and live. The fundamental solution to curbing pollution is to speed up green development. Only by drastically reducing pollutant emissions from the source can the ecological environment be significantly improved. Since rivers go to the sea, all the pollutants from human activities—on the land, rivers, or the ocean—would directly or indirectly impact the marine ecological environment. Therefore, tracing, preventing and controlling the source of the pollutants are crucial to improving the marine habitats.

The second is to protect and restore the ecological environment. We should follow the principle of giving priority to ecological protection to involve the ocean into the community of "mountains, rivers, forests, farmlands, lakes and grasslands"—a natural unit with a shared future, with the aim of comprehensively protecting and restoring the marine ecosystem in an all-round manner covering all the regions, while putting more efforts into the conservation of marine biological resources and helping marine ecology restore itself. What's more, we should adhere to the principle of natural restoration as the main approach with artificial restoration as the supplement, systematically implement a wide range of policies to link the ecological habitats of the land with that of the ocean, restore the ecological and hydrodynamic regions like beaches, coastal wetlands, and bays, enhance the service functions of marine ecosystems, improve our ability to offer high-quality marine ecological goods, and expand marine ecological capacity.

The third is to make innovations and enhance capabilities. Institutional construction, capacity enhancement and technological innovation are vital to the conservation and improvement of the marine ecological environment. We should create a strict and rigorous legal system to ensure that the decision-making and deployment of marine environmental protection take root and take effect; it is necessary to improve monitoring, supervision and emergency response, and to actively promote intelligent monitoring, modernization of supervision, and normalization of emergency response; finally, we should give full play to scientific and technological innovation, incubate the domestic environmental protection and the technology industry, protect marine ecology, and build a beautiful China.

The fourth is to seek international cooperation and global governance. Protecting the marine ecological environment is a common challenge for humanity and also a responsibility the world should take. We should regard the ocean as part of the natural community with a shared future, and create a modern marine environmental governance system based on our national conditions—a system led by the government, with enterprises as the main body, and participation by social organizations and the public. Furthermore, we should coordinate domestic governance and international cooperation, build an efficient and fair global marine environmental governance system, get deeply involved in the handling of emerging marine environmental issues like marine debris, micro plastics and unwanted deoxygenated marine environment, and contribute Chinese wisdom and approach.

3 Practical Approach and Measures for Improving the Marine Ecological Environment

Since the land and the sea are closely linked, improving the marine ecological environment is inseparable from emission reduction and prevention of land-based pollution, as well as the management and control of human development activities. Therefore, in this new era, we should seek, from the perspective of coordinated management of the land and sea, a new solution to protecting the marine ecological environment. It is necessary to strengthen the overall planning and organic connection between the land and sea, enhance the coordination of land-sea pollution prevention, and pursue ecological environment protection to thoroughly improve the marine habitats.

1. **Coordinating land and sea management and planning land use based on oceanic conservation and strengthening comprehensive control of land and sea pollution**

The marine ecological environment problems are manifested in the sea, but the root of these problems comes from the land. The management of estuaries is the key to improving the coastal marine environment. In the new era, we should take advantage of institutional reforms to comprehensively resolve land-sea environmental pollution, covering areas "from the mountain tops to the ocean," and establish a model where rivers and the ocean are managed simultaneously, with river pollution, instead of oceanic pollution, being handled first. We should assess pollutant load into the sea, find out the total amount of nitrogen and phosphorus entering the sea from the rivers, and identify the key areas and major industries for pollution control. We should follow quality control in coastal waters to identify the total (and reduced) amount of river pollutants entering the sea, carry out refined pollution source management to control the nitrogen and phosphorus emissions from agriculture, animal husbandry and urban domestic sewage from the source, and effectively reduce the amount of nitrogen and phosphorus entering the ocean from the key river basins to help enhance the environment of major estuaries and bays.

Moreover, we should create a pollution management reversal mechanism of "land planning based on oceanic conservation" and a joint prevention and control mechanism for oceanic areas and river basins to give full play to the "river chief" system and effectively combine it with the "bay chief" system, making the local governments along coastal areas and rivers responsible for pollution prevention and control, and implementing regional pollution reduction and environmental remediation tasks in different regions.

2. **Respecting the integral and systematic ecosystem and ensuring ecological security by linking land with sea**

Ecology is a unified natural system, an organic chain of interdependence and close connection. We must adhere to the principle of ecological priority and green development, along with the idea of natural restoration as the main approach with artificial restoration as the supplement, to establish a land-sea ecosystem protection and restoration mechanism, and carry out integrated ecological protection and restoration projects of mountains, rivers, forests, lakes, grasslands, and the sea for protection of important ecosystems and restoration of damaged ecosystems. Through marine ecological restoration projects—for example, the "Blue Bay," the "South Mangrove, North Tamarisk" and the "Ecological Island"—we should restore important ecological corridors and nodes such as coastal wetlands, natural shorelines, key estuaries and bays, special islands, and biological migration channels, in order to enhance the functions of the coasts, oceanic areas and islands, maintain biodiversity, link the ecological boundaries of land and sea, and build a land-sea ecological safety network that is interconnected and mutually beneficial.

From the perspective of improving the integral ecological network, we should further strengthen the selection and planning of protected areas, effectively conserve the ecologically sensitive areas such as typical wetland ecosystems, rare species habitats and migration routes, economic species bait and breeding areas, and rescue protected marine areas with an important ecological status to incorporate the ocean into the ecological network.

3. **Constructing coastal retreat lines and buffer zones, and continuously making the ocean more accessible to the public**

Currently, there exists inconsistency in terms of the management boundary of the land and sea, resulting in a large number of overlapping and conflicting areas in land development and oceanic management. We should conduct coastline surveys to clarify the boundaries of coastline management, and carry out the most stringent coastline management control system, strengthen the red line and reclamation systems for marine ecological protection, rationally arrange production, living and ecological boundaries to maximize ecological coastlines, expand waterfronts and beaches accessible to the public, create spaces for exposure to nature in bays to make people better understand the benefits of improved marine ecological environment and thus feel happy about it.

Moreover, we should strengthen the management of the land adjacent to coastlines, and encourage industries along the coast to optimize and improve their structural adjustment and transformation for the purpose of protecting the marine ecological environment. We should also formulate and implement a coastal construction retreat line system, strictly observing the 200-meter retreat line and the retention rate of the natural coastline, while demolishing unreasonable buildings on beaches, and creating a buffer zone of "coastal inland zone–tidal transition zone–coastal water zone." Finally, we should achieve "naturalization, ecological civilization, and green planting" and create gentle slopes into the sea, maintain the integrity of important coastal forest ecosystems such as coral reefs, mangroves, beach dunes, in order to recreate the ecological landscape of green coasts, golden beaches, and blue bays.

4. **Strengthening the building of monitoring, emergency response, and supervision systems to enhance conservation of the marine ecological environment**

We should advance effective integration and innovation of the "Internet +" and "big data" with monitoring and supervision of emergency response to the marine ecological environment, and improve the national marine ecological environment monitoring network, supervision system, and emergency response capacity building. We should also plan the monitoring layout of shore-based, nearshore, and offshore waters, as well as islands and reefs in the South China Sea and oceanic and polar regions, with overall planning and stations offering multiple functions, in addition to strengthening coordinated tasks like monitoring, pollution discharge supervision, development supervision, industrial regulation, emergency response, risk warning, etc., all of which combine to realize the transition from monitoring of the quality of the marine environment to comprehensive and accurate monitoring of the marine ecological environment. We should incorporate the risks of the marine ecological environment into general management, and strengthen the monitoring and supervision of the exploitation, storage and transportation of oil and gas, of the production, storage and transportation of hazardous chemicals, and radioactive pollution, and establish a multi-level ecological environment risk prevention system for strict prevention beforehand, strict management during the process, and disposal afterwards. Finally, we should promote the supervision and administration of marine environmental protection, strengthen the law enforcement of marine resources and habitats, and enhance the supervision of marine engineering and dumping during and after the process, in order to form a new supervision pattern with a sounder system, stronger supervision and stricter protection.

5. **Strengthening rule of law and policy support, and improving the marine environmental protection system**

We should speed up legislation and enhance policy support to help protect the marine ecological environment by making marine laws and regulations based on our national conditions, improving the top-level legal design of our national marine policies and marine system, actively stepping up efforts to the building of the system and policies for the conservation of the marine ecological environment, and implementing the system of "Three Lines and One List" in the marine field. Also, we should speed up the revision of the marine environmental protection laws and its supporting regulations, revise related laws, regulations and normative documents such as the *Marine Dumping Management Regulations*, the *Management Regulations on the Prevention of Marine Engineering Pollution and Damage to the Marine Environment*, in order to establish a marine legal system and supporting laws and regulations that comply with the new responsibilities, positioning and institutions, and are suitable for the protection of the marine ecological environment. Finally, we should create an inter-regional coordination and cooperation mechanism for tackling cross-region and cross-boundary marine pollution, and strengthen vertical guidance and horizontal linkage; what's more, we should reconstruct the "river basin-estuary-sea" environmental quality standard system to effectively combine standards and data from different regions and varied media.

6. **Actively getting involved in global oceanic governance and building a new pattern of marine environmental governance**

We should actively participate in global oceanic governance, jointly promote the blue economy and maritime cooperation under the Belt and Road Initiative, focus on solving serious issues like ocean pollution, marine debris and micro plastics, ocean acidification, deoxygenation, and reduced marine biodiversity, and create a marine environmental governance system that gets everyone involved—a system led by the government, with enterprises as the main body, and participation by social organizations and the public. Furthermore, we should keep guiding international response to climate change, strengthen capacity building in the marine sector to address climate change, vigorously promote international maritime exchange and cooperation, further exert China's leading role in international bilateral and multilateral maritime governance, and actively enhance multidisciplinary collaborative research on nature's influence on society and the economy across the world to marshal joint response to global marine environmental issues.

China is Capable of Leading Global Climate Governance

Guiyang Zhuang

1 China's Contribution to Global Climate Governance

With experience in international climate negotiations, China has gained an increasingly mature concept of climate governance. From the initial emphasis on "impossibility" to undertaking emission reduction obligations, then the belief that mandatory emission reductions are "inappropriate," to setting voluntary quantitative emission reduction targets and the 2030 carbon emissions peak, China has shifted from a passive follower to an active leader in global climate governance, thus showing its willingness to take responsibility for emission reduction. This is because China—now the number one carbon emitter and the second largest economy—is facing external pressure and has internal motivation to voluntarily change its role. From 2013 to 2016, China's average contribution to world economic growth reached about 30%, surpassing the combined total of the United States, the Eurozone and Japan, and ranking first in the world. As Klaus Schwab, Executive Chairman of the World Economic Forum, said, "China now has a strong voice." China is accumulating strength from the three levels of implements, institutions, and concepts, consolidating leading advantages while gradually taking the lead in global climate governance to advance sustainable development worldwide.

1. **The implementation level: supply of material public products**

Financial and technical issues are the focus of climate negotiations. Developed countries have responsibility for greenhouse gas emissions and are more capable of handling climate governance. They should take the lead in solving global climate problems and assume more obligations, while supporting developing countries financially and technically. However, developed countries have been slumped in economic recession since 2008, with their economic strength greatly reduced by increased

G. Zhuang (✉)
Research Insitute for Eco-civilization, Chinese Academy of Social Sciences, Beijing, China

J. Pan et al. (eds.), *Beautiful China: 70 Years Since 1949 and 70 People's Views on Eco-civilization Construction*, https://doi.org/10.1007/978-981-33-6742-5_71

economic uncertainties. Faced with a huge capital and technology gap, they have stagnated and even withdrawn.

In contrast, with years of steady and high-speed economic development, China has greatly improved its overall national strength while actively exploring the path of transformation to tackle its economic difficulties. Based on its national conditions, China not only firmly defends the basic development rights of developing countries in climate change conferences, but also provides assistance to developing countries in green technology transfer, financial support, and education. In terms of funding, at the United Nations Sustainable Development Summit, General Secretary Xi Jinping announced the establishment of "South-South Cooperation Assistance Fund" to financially support developing countries along with other channels such as the Silk Road Fund, the BRICS Fund and the Asian Infrastructure Investment Bank. Meanwhile, China will continue to increase its investment in the least developed countries and cancel the intergovernmental interest-free loan debts of the least developed countries, landlocked developing countries and developing small-island nations that were not paid off at the end of 2015. In terms of technological support, China will set up an international development knowledge centre to discuss the construction of a global energy network to promote clean and green ways to meet global electricity demand. In 2016, it initiated the "Ten-Hundred-Thousand" project of South-South cooperation on climate change—establishing 10 low-carbon demonstration zones, launching 100 mitigation and adaptation projects, and initiating 1,000 cooperation projects for climate change training in developing countries. In 2017, China started carbon emission trading pilot schemes nationwide to see how to cut emissions through market mechanisms—measures fully showing its courage to assume responsibilities of global climate governance as a major and responsible country.

2. **The institutional level: supply of institutional public goods**

In the face of climate change, now a global public health issue, setting the agenda and shaping issues are the basis for achieving consensus among all parties, and it requires both teamwork and contributions from responsible countries willing to take the lead. At critical moments of international climate negotiations when difficult issues arise and make negotiations reaching deadlock, countries with a leading status should strive to balance the interests of all parties while adhering to the bottom line, and use flexible solutions to reach consensus. China has always been an active participant in international climate negotiations, with presence in every international climate convention. China adheres to the principle of "common but differentiated responsibilities" and adopts a consensus decision-making mechanism under the framework of the United Nations, thus making outstanding contributions to the conclusion, entry into force and implementation of the *Paris Agreement*.

At the beginning of the negotiation for the *Paris Agreement*, in view of the large differences between the North and South camps, China started bilateral negotiations with major countries, making progress step by step through mediation and seeking the most basic consensus. It held talks on climate change with India, Brazil, the European

Union, the United States, and France before releasing a series of documents such as the *China–US Joint Statement on Climate Change*, the *China-EU Joint Statement on Climate Change*, and the *Joint Statement by the Heads of China and France*—statements that laid the foundation for the conclusion of the *Paris Agreement*. At the Paris Climate Conference, General Secretary Xi Jinping pointed out that whether an international agreement is successful or not lies in its ability to solve current contradictions and prepare for the future. As such, he put forward four expectations for the *Paris Agreement*: conducive to achieving the objectives of the Convention and leading green development; conducive to rallying global efforts and encouraging extensive participation; conducive to increasing investments and prompting action; conducive to taking the national conditions of all countries into account, while being pragmatic and effective. The four expectations laid the tone for the negotiations of all parties in terms of adhering to the principles of the Framework Convention, determining the institutional arrangements, pointing out the direction of green and low-carbon development, and providing financial and technical support. They also voiced on behalf of the majority of developing countries, which were finally realized in the *Paris Agreement*. The UN Climate Conference took place in Bonn, Germany, in November 2017, and it reached a deadlock. At the time, China adopted a "bridging" approach, that is, when big differences arose in the negotiations, it would take the lead to extract the most opposing views and try to find the "greatest common divisor" (consensus)—an approach that fully demonstrated China's political influence and Chinese wisdom recognized by all parties. China's contributions to the international climate negotiation at the institutional level have been widely recognized by the international community, a fact that has in turn enhanced China's voice in global climate governance.

In addition to the UN climate negotiation mechanism, China has proposed the integration of green development into all aspects of the Belt and Road Initiative to seek a regional cooperation development model based on sustainable development. For example, at the 2014 APEC High-level Roundtable on Green Development, China initiated a global green supply chain and value chain cooperation initiative to promote industrial upgrading and a transition to green development; at the 2016 G20 Hangzhou Summit, China made, for the first time, the green finance initiative an important topic of the summit; at the 2017 BRICS summit, China proposed to create a green technology cooperation and exchange platform and promoted partnership for sustainable urban development, thus achieving important results. China will appear on more international stages, and work with others to come up with practical and effective institutional plans.

3. **The spiritual level: supply of conceptual public goods**

Since the 18th National Congress of the CPC, China has successively proposed green development, the Belt and Road Initiative, and the idea of a community with a shared future for mankind, all of which have been recognized by the international community in terms of theory and content, and therefore reflect China's innovative thinking increasingly recognized by the world. Moreover, the 19th National Congress called on the people of all countries to work together to build a community with a shared future for mankind to build an open, inclusive, clean, and beautiful world that enjoys lasting peace, universal security, and common prosperity. These brilliant ideas will shine in the process of global climate governance.

China is striving to build a community with a shared future for mankind and work with others to pursue global ecological civilization. General Secretary Xi Jinping repeatedly stated on diplomatic occasions that the international community is uniting as a "world of shared destiny." Faced with complex economy and global issues, no country can survive on its own. China not only regards response to climate change as its due international obligation, presents sincerity, determination and Chinese wisdom in climate change negotiations and climate governance actions, but also provides its experience in building domestic ecological civilization and embarking on the path of green transformation as Chinese wisdom for global climate governance, which helps to achieve national and global development at the same time, and therefore projecting a commendable image on the international stage.

China adheres to a correct view of justice and benefit, seeks the "greatest common divisor" in terms of the interests of all parties, and promotes international cooperation in climate governance. At the Paris Climate Conference, China proposed a win–win global climate governance concept, advocating the global governance concepts like "creating a future of win–win cooperation, with each country making contribution to the best of its ability everyone doing their best, and working together for win–win results," "creating a future of the rule of law, fairness and justice," and "creating a future of inclusiveness, mutual learning and common development," in a way that allows all countries to find the most suitable measures based on their respective national conditions—an approach that is in line with the Chinese traditional culture of "seeking harmony without uniformity," and "If big rivers have water, small ones are full; if small rivers have water, big ones are full." Climate governance is not a "zero-sum game." Since addressing climate change is a common cause of mankind, developed countries should take the initiative to cut emissions of greenhouse gases whereas developing countries should not repeat the old-fashioned path of industrial civilization and high-carbon emission development.

China has always adhered to the principle of "common but differentiated responsibilities" in promoting global climate governance. As such, General Secretary Xi Jinping pointed out, "Developed and developing countries have different historical responsibilities for climate change, and different development needs and capabilities. Just like in a car race: it would be neither reasonable nor fair to apply the same speed requirements to cars which have run far ahead and those which have only just left the starting point. Developed countries should do more and lead the way in addressing climate change. This is in keeping with the important principles laid down in the UNFCCC, such as 'common but differentiated responsibilities,' equity and respective capabilities. This is also the hope of all developing countries." Meanwhile, China will not shirk its responsibilities and will keep making contributions by urging developed countries to shoulder their responsibilities and fulfill their emission reduction commitments, and help developing countries mitigate emissions and adapt to climate change.

2 China's Strategies of Becoming a Leader in Global Climate Governance

It's necessary to make it clear that becoming a "leader" means that we should have a leading edge, play an exemplary role, assume due responsibilities, and be committed to promoting public interests. In this regard, the report of the 19th National Congress pointed out the direction for China's future strategic choices in global climate governance: "The whole Party must be completely clear about this fundamental dimension of our national context, and must base our work on this most important reality—the primary stage of socialism" and "get actively involved in global environmental governance and fulfill our commitments on emissions reduction." On the one hand, we should make domestic emission reduction efforts and projects that help developing countries overseas according to our national conditions, and we should do everything in our capacity, and follow objective development laws; on the other hand, taking the lead in addressing climate change means greater responsibilities for cutting emissions. At present, China's primary task is to keep promoting green and low-carbon development at home, accelerate ecological civilization, lay the foundation for domestic economic development, and accumulate advantages in mitigating and adaptation to climate change, in order to help realize global emission reduction targets.

1. **China's proposal to be a leader is a strategic response to international expectations**

Currently, the global climate governance system has seen a new situation. On the one hand, the world economy is slowly recovering, with an improving global green development trend; on the other hand, the center of global climate governance is gradually moving eastward. The US's withdrawal from the *Paris Agreement* has shaken the confidence of the international community in achieving low-carbon growth and emission reduction targets. Therefore, the United States has since lost its leadership in global climate governance. What's more, the EU is now losing strength, having internal differences, and is slow in action. By contrast, China is playing a constructive role in reforming and adjusting the global environmental governance system.

Over the past few years, China has been more and more active in advising and providing Chinese approach at the organizations of the *United Nations Framework Convention on Climate Change* and within the multilateral mechanism framework, and has become an important force in promoting global governance. During this period, using the "Xi-style diplomacy," China has proactively participated in global governance. Some western countries and international organizations—including leaders of the United States, France, the United Nations, and the International Monetary Fund—even advocated on multilateral and bilateral occasions that China should play a leading role in global climate governance. China has now achieved initial results in its building of ecological civilization, and its contributions to the world have received recognition from the international community. As such, the world is beginning to join China's initiatives. Broad-minded, China is marshalling global efforts for development and laying the foundation for its leadership across the globe.

Assuming international responsibilities and obligations is in the fundamental and long-term interests of our country. Given that China is still a developing country and faces arduous tasks, we will take as many international responsibilities and obligations as we can—an attitude that has won China the attention of the world and has also paved the way for China to lead global climate change governance.

In the face of the global climate change crisis, China has always held high the banner of peace, development, cooperation, and win–win results, calling on all countries to build a community with a shared future for mankind, guiding international cooperation in tackling climate change, and playing a "leading role" in building global ecological civilization. This is China's strategic response to international expectations, an approach that is totally different from arbitrary "hegemonism," and also a deviation from "leadership" characterized by control and obedience as well as master and servant. China's leadership emphasizes its conformity to the objective trend of the evolving global climate governance pattern, and its willingness to create a sense of responsibility, take the lead in addressing climate change, and maintain a global relationship where all countries live in peace, and man coexist with nature. As a major and responsible country, China will address the global climate change crisis, fulfill its emission reduction commitments by setting an example, contribute Chinese solutions to the world with its effective and practical experience in building an ecological civilization, and build a new international climate governance order featuring win–win cooperation and shared benefits.

2. **Holding high the banner of building an ecological civilization and providing Chinese approach to global climate governance**

The 18th National Congress officially incorporated the building of an ecological civilization into the overall layout of our Five-sphere Integrated Plan for realizing socialist modernization with Chinese characteristics. Since then, building a beautiful China has been a vital part of the great Chinese dream, thus opening up the road to building an ecological civilization of green, low-carbon growth and a green ecology. With the construction of ecological civilization, China's carbon emissions continue to decline while its GDP is maintaining medium-to-high speed growth—a fact indicating that China is gradually getting rid of its dependence on carbon emissions for economic growth, and the people are happier with the improved ecology, which has in turn enhanced understanding and support of enterprises and the public for building an ecological civilization across the country.

Building an ecological civilization is to comply with the general trend and strategy of the transformation and development of world civilization, and to comply with the long-term goal of sustainable development. Since issues concerning climate change are complex, extensive, and irreversible, addressing this global crisis is at the core of building an ecological civilization around the world. Leading global climate governance will inevitably be the "entry" to building a global ecological civilization; moreover, covering economic, political, cultural and ecological fields, building an ecological civilization is the only way to solve the global crisis of climate change. China is working hard to build ecological civilization and share the benefits with other countries through specific measures such as forming strategic ideas for ecological civilization construction, expanding green transformation and development models, and creating an ecological civilization system.

In May 2016, the United Nations Environment Programme released a report titled "Lucid Waters and Lush Mountains Are Invaluable Assets: Strategies and Actions for China's Ecological Civilization," which introduces the guiding principles, basic concepts, policies and measures for building of an ecological civilization in China, particularly the integration of ecological civilization into its national development plans—an approach showing that China is determined to embark on the path of development through a green and low-carbon circle, avoid the predicament of the development paradigm of industrial civilization, and provide Chinese wisdom and solutions for achieving global ecological security and sustainable development.

While building an ecological civilization at home, China has also actively participated in global environmental governance, implemented international environmental conventions, and achieved remarkable results. With the help of the green Belt and Road Initiative, and South-South cooperation projects, China has been sharing its experience in green development and environmental protection, in an effort to promote China's green technology and standards to a larger audience. Moreover, China has advocated building a community with a shared future for mankind, seeking a new climate governance plan for mutual benefits and win–win results that provide new opportunities and development models for resolving the global environmental crisis. In this process, China's brilliant ecological ideas are constructive to the establishment of global ecological civilization, the formation of theoretical framework, and policy arrangements.

Finally, China will hold high the banner of building an ecological civilization to participate in global climate and environmental governance, adjust its role and position, guide the global efforts in building a community with a shared future for mankind, advocate shared responsibility for climate change, and lead developing countries to tackle climate change, in order to share its benefits reaped from building an ecological civilization and transition to green development with other developing countries, while making contributions to global climate security and sustainable development.

3. Adhering to multilateralism and leading the new process of global climate governance

Although the political structure of a multipolar world has been formed, the power of speech of Western countries such as Europe and the United States in global governance has yet to be fundamentally changed. With the change in the model of the US's development as the priority and the decline of the European Union as a whole, the international community has expressed a different voice, saying that China is actively participating in global governance and thus should lead global climate governance. On January 17, 2017, President Xi Jinping emphasized at the Davos Summit in

Switzerland that the *Paris Agreement* is in line with the general direction of global development and the results are hard to come by, and should be held together and cannot be given up. He also stressed the need to raise the consciousness of building a community with a shared future for mankind, and promoted shared responsibility and mutual accountability to advance development across the world. This was a Chinese leader's high-profile response to the problems of global governance and international order that the international community is concerned about, including climate change, and it clearly expressed China's willingness to adhere to multilateralism and the *Paris Agreement*.

Since the Copenhagen Climate Change Conference, China has been expressing more and more voices in climate governance, and it has played a pivotal role in shaping new mechanisms for global climate governance. However, China must be aware that many aspects are yet to be improved, with tons of problems to be resolved, and therefore it still lacks the power to lead global climate governance alone. Moreover, it is still in the primary stage of socialism and has to enhance its strength. China has to strengthen its ecological civilization, deepen the concept of green development, and improve its independent innovation capabilities to ensure leadership in a desired manner. In addition, the United States, the European Union, Japan, India and other major power groups may not be willing to be led. Therefore, before achieving the carbon emission peak target in 2030, China must increase international coordination and cooperation with the BASIC countries—a bloc of four large newly industrialized countries including Brazil, South Africa, India and China—the European Union, and the United States, in order to explore flexible and effective climate negotiation models, jointly lead the climate negotiation process, implement the negotiation results, and set an example for the world.

The entry into force of the *Paris Agreement* marks the compliance stage of global climate governance after 2020. The top-level design of global climate governance is facing reconstruction, and China will assume more international responsibilities. On the one hand, China needs to make a successful transition as soon as possible, set an example for countries around the world, and accumulate leadership capabilities; on the other hand, China should make long-term choices at the strategic level for its participation in global climate governance, offer assistance to developing countries based on its own capabilities, and help underdeveloped countries enhance their insufficient capacities in terms of capital, technology, institutions and policies. From the perspective of the complexity of climate governance and the trend of multipolar world development, China's leadership will not be exclusive, and will reflect inclusiveness and co-construction instead.

In the short to medium term (2020–2035), China must respond to changes in climate governance, deepen the concept of green development, lay the foundation for ecological civilization, continue to leverage its advantages and participate in global climate governance, and implement the goal of voluntary contributions of the *Paris Agreement* to showcase its achievements of low-carbon development, while enhancing its international influence through South-South climate cooperation. Moreover, China should strengthen its cooperation with the United States, the Central Europe, and the BASIC countries, interact with the international community according to the principles of "inclusiveness, cooperation, mutual trust, and win–win results," and explore open, inclusive and mutually beneficial leadership with Chinese characteristics.

From a long-term perspective (2035–2050), China should focus on top-level global governance design that goes beyond climate issues. In the new pattern of global governance, China will lead and coordinate in more areas instead of being a condescending leader in the traditional sense. Only in this way can China get all parties actively involved, and marshal forces to solve the collective dilemma.

3 Conclusion

Generally speaking, when China initially participated in international negotiations, it had relatively less power of speech, and mainly focused on asserting its own rights, called for Western developed countries to fulfill their obligations to global climate governance, and provided financial and technical support to developing countries, with the international order being mainly dominated by developed countries. However, due to the continued increase of its carbon dioxide emissions, China was facing unprecedented pressure from the North and South camps, and also from the need for transformation at home. With its increasing national strength and deepening participation in international affairs, China should transit from a "participant" to a "leader" in climate governance—an inevitable choice based on China's current development stage and in keeping up with the times. Therefore, China should seize the opportunities to lead global climate governance, strive to create a clean and stable climate environment, create opportunities for green economic development, strengthen its soft power, and improve its status in international cooperation.

Interpreting Ecological Civilization with the Latest Results of International Research on Sustainable Development and the Universal Language

Dajian Zhu

On June 7, 2019, President Xi Jinping delivered a speech titled "Insisting on Sustainable Development and Creating a Prosperous World" at the 23rd plenary session of the St. Petersburg International Economic Forum. He stated that sustainable development meets the best interests of all parties and offers a chance for cooperation as the "golden key" to solve the problems the world now faces. As Xi Jinping's first speech on sustainable development, it contains high-profile contents that boost the confidence of sustainable development researchers and practitioners.

There are not many research texts that interpret ecological civilization from the perspective of sustainable development. It is very important to promote the dialogue between sustainable development and ecological civilization. The first is to use the universal language of international research on sustainable development to interpret the thoughts, policies and practices of and tell the story about China's ecological civilization to advance the 2030 Sustainable Development Goals (SDGs) adopted by the United Nations. The second is to integrate the latest achievements and ideas of international sustainable development research into China's ecological civilization, so that we can deepen the theories, strategies and policies of ecological civilization while building a beautiful China and realizing socialist modernization with Chinese characteristics.

1 The Latest Results of Research on Sustainable Development and Their Theoretical Insights

Over the past decade, especially since the United Nations Conference on Sustainable Development, or Rio+20, held in 2012, a series of important new results and ideas have emerged in international research on sustainable development. These latest

D. Zhu (✉)
School of Economics and Management, Tongji University, Shanghai, China

J. Pan et al. (eds.), *Beautiful China: 70 Years Since 1949 and 70 People's Views on Eco-civilization Construction*, https://doi.org/10.1007/978-981-33-6742-5_72

achievements and ideas can be summarized in three dimensions: object, process, and subject. Interpreting ecological civilization with them can dispel fallacies of ecological civilization construction, provide new theoretical insights for advancing ecological civilization, and enhance China's power of speech in disseminating its theory of ecological civilization internationally.

(1) The object dimension: Sustainable development requires the decoupling of human progress from environmental pressures. In 2009, scientists published a paper on *Nature* that proposed planetary boundaries and proved that the scale of economic and social development is constrained by the earth. This finding suggested that the essence of sustainable development is the pursuit of social and economic prosperity within the constraining boundaries of natural capital, and the goal is to achieve absolute decoupling of economic and social development from resource and environmental pressures. Such decoupling can be divided into two aspects: one is the decoupling of economic growth from environmental pressure, namely improving resource productivity; the other is the decoupling of social welfare from economic growth, namely improving the efficiency of economic services. Over the past four decades, China has made considerable progress in dealing with resource productivity between the economy and environment. We need to make corresponding progress in the efficiency of services between the economy and society.

From the object perspective, ecological civilization includes two aspects: the environment and development, as well as ecology and civilization. Neither the resource and environmental protection without improved economic and social civilization nor the economic and social development without resource and environmental protection can be regarded as ecological civilization. The common myth is that people often see ecological civilization as the fusion of resource and environmental protection and economic and social progress, thus leading to the understanding or practice of seeking civilization without conserving ecology or vice versa—the former is represented by GDP-oriented economic growth, whereas the latter by the idea that ecological civilization is environmental protection. For example, regions with relatively underdeveloped economies often claim to be or are considered as the areas of good ecological civilization. In fact, when evaluating developing regions from the perspective of ecological civilization, we should examine whether their ecological environment promotes economic and social development; by the same token, when measuring developed regions, we should see that whether their economic and social development comes with a good environment with bountiful resources.

(2) The process dimension: Sustainable development requires the change of the whole causal chain process. Although it stems from resource and environmental problems, the solution should go beyond resource and environmental problems. From the perspective of causality, there are four different ways of interpreting human understanding and the evolution of the relationship among the economy, society and environment. First, the growth model. The environment is deemed as an insignificant subsystem of the economy and society, which was

a mainstream concept in international development thinking before the United Nations Conference on the Human Environment in 1972. Second, the parallel model. The economy, society, and environment are in a parallel relationship in sustainable development. Environmental issues received attention but end-of-pipe control was emphasized. This is a mainstream concept from 1972 to 1992. Third, the intersection model. The interface of the economy, society, and environment was improved, yet with a contradiction between the improvement of material efficiency and the expansion of material scale—a mainstream concept from 1992 to 2012. Fourth, the inclusion model. Absolutely decoupling of economic and social development from resource and environmental consumption is accentuated. The Doughnut economics, a visual framework for sustainable development proposed by Oxford economist Kate Raworth in 2012, is the latest achievement of the inclusion model that shares the same values and methodology with China's ecological civilization emphasizing the ecological red line. As such, it can provide ideological insights for improving the theories and policies of ecological civilization.

From the process perspective, ecological civilization is a new environmentalism or deep green revolution that transcends end-of-pipe pollution control. The pressure-state-response management, namely the PSR model that ecological civilization emphasizes turns post treatment into active prevention. The common myth is that people often emphasize end-of-pipe pollution control without changing the growth model, instead of emphasizing source-oriented and life-cycle material and energy flow control.

(3) The subject dimension: Three levels of cooperative governance to promote sustainable development. Since 2012, people have increasingly emphasized that sustainable development needs to be expanded into a four-in-one system involving the economy, society, environment, and governance, with cooperative governance as a vital part. In 2012, Cavagnaro proposed that sustainable development requires three levels of cooperative governance. The first level is macro management and social management; the second level is organizational cooperation and public–private cooperation; and the third level is public participation and individual management. Incorporating cooperative governance into the theoretical system for sustainable development is to give full play to various organizations and public–private cooperation, thereby achieving sustainable development through their respective actions under a common goal, namely "thinking globally, and acting locally."

From the subject perspective, the development of ecological civilization involves the interaction and extensive participation of stakeholders such as the government, enterprises, society, and citizens, rather than the willingness and mobilization of the government alone, or even the top-to-bottom willingness and mobilization of the central government. The common myth is that people often see promoting ecological civilization as a task of the resource and environmental protection departments, rather than collaboration among varied units and network governance across society. The traditional government system often has conflicts in terms of the goals and methods of different departments. Ecological civilization requires shifting government management from fragmentation to integration: The first one is the mutual enhancement of goals. Departments with different goals, such as development departments and environmental protection ones, need to seek a common ground and balance between conserving ecology and promoting civilization. The second is the mutual enhancement of methods. Government management approach usually includes three methods: regulation, the market and public participation. Different methods should support each other.

2 Using Latest Research Results to Analyze Three Scenarios of China's Ecological Civilization

The report at the 19th National Congress of the Communist Party of China put forward the goal of ecological civilization construction from 2020 to 2050 compatible with China's modernization, which offers specific indicators and is divided into two stages, each with 15 years. Based on the updated concept that ecological civilization means the decoupling of economic and social development from resource and environmental consumption, I think that if we use a two-dimensional matrix of human development and ecological footprint and the research results of the above-mentioned doughnut economics to discuss China's ecological civilization, there are three different ways, or scenarios—the model A, model B, and model C. And China should follow the path of the mode C in its pursuit of ecological civilization and green development.

(1) The model A: pursuing growth before going green. A country following this model first goes from low human development with a low ecological footprint to high human development with a high ecological footprint, and then reduces its ecological footprint to pursue the target of high human development with a low ecological footprint. This is the "brown first, green later" development model adopted by developed countries which are now in the outer layer of the doughnut, with the per capita GDP generally reaching about USD 30,000, and the human development index being above 0.8. However, their per capita carbon dioxide emission is about 10 tons, and the per capita ecological footprint generally exceeds the ecological supply capacity by twice or more (over 3.5 global hectares). The rapid economic growth in China over the past 40 years has significantly impacted the earth's resources and environment, and its per capita ecological footprint has exceeded the ecological carrying capacity of the earth. According to a WWF research, under the business-as-usual scenario, China would exceed the global per capita level in per capita ecological footprint, reaching 2.9 global hectares in 2030. Therefore, it should build ecological civilization to reverse such a trend of high consumption, high pollution and high growth.

(2) The model B: leaping forward to pursue the target. It is hoped that developing countries immediately go from low human development with a low ecological footprint to high human development with a low ecological footprint required for sustainable development, which means reducing the ecological footprint while increasing the human development level. This line of thinking, though seems reasonable, doesn't work. The reason is that the model B often requires a high degree of ideological, institutional and technological innovation, but developing countries who are generally restricted by outmoded concepts, old-fashioned systems, and old technologies, always incline to follow what has been set before them. Since the reform and opening up, China has repeatedly proposed to avoid the western path of "pollution before treatment"; however, its per capita ecological footprint has now reached 2.2 global hectares, an amount still lower than the global ones of 2.6 hectares, but higher than the 1.7 hectares, or the earth's per capita ecological carrying capacity. As such, it is no longer possible for China to follow the model B, which would negatively affect its modernization process without satisfying the people's basic needs, and also fail to meet the inherent requirements of sustainable development and ecological civilization.

(3) The model C: pursuing green competitiveness. The significance of this model for the development of China is to increase its per capita GDP to an amount between USD 20,000 and 30,000, and the human development index to more than 0.8. China's per capita ecological footprint may rise to an amount that would exceed the per capita carrying capacity of the earth, but it would be strictly kept under the average global per capita ecological footprint level, or 2.6 global hectares, far less than 3.5 hectares of developed countries. This is a grand goal of green development. If China could achieve it, it would reach the modernization level of developed countries with a lower per capita ecological footprint. Therefore, China's goal of achieving ecological civilization by 2035 needs to have upper and lower limits in terms of its ecological footprint goal—the per capita ecological footprint close to the ecological carrying capacity at that time makes the lower limit, and the amount never exceeding that of developed countries makes the upper one. By 2020, China will begin to enter the middle tier of the doughnut, with the per capita GDP close to USD 10,000 and the human development index standing at about 0.8. If it can meet the basic requirements of ecological civilization through smart growth that is green, inclusive, and innovative, it is possible for China to stay in the middle tier where it can achieve high economic and social development through acceptable consumption of the earth's natural capital. At that time, it must strive to realize three goals marked by 20%—namely, a population less than 20% of the world, a total economic volume exceeding 20% of the world, and a reduced ecological footprint impact smaller than 20% of the world.

(4) The theoretical significance of the mode C for China's development. For a long time to come, the main task of China will still be to achieve industrialization, urbanization, and modernization. Compared with the ecological civilization of western post-industrial societies, the ecological civilization with Chinese

characteristics is to combine industrial civilization with ecological civilization, or to use the principles of ecological civilization to transform traditional industrial civilization. In essence, Chinese ecological civilization is actually new industrial civilization. Generally speaking, in the years to come, China should neither follow the traditional industrial civilization nor enter the post-industrial ecological civilization in advance; instead, it should seek ecological industrial civilization with its own characteristics. The mainstream ecological civilization in the next 20 to 30 years should emphasize three integrations: the first is to integrate into new urbanization, emphasizing that urban development should shift from spatial sprawl and material waste to compact space and material intensiveness; the second is to integrate into the transformation of industrialization, emphasizing that industrial development should shift from a linear, high-carbon economy to a circular and low-carbon one; the third is to integrate into modern life, emphasizing that the life mode should shift from ownership-orientation to share-orientation.

3 Strengthening Cooperative Governance for Ecological Civilization for the Post-2020 Five-Year Plans

Starting this year, China will study and formulate the 14th Five-Year Plan for national economic and social development, the first top-level policy blueprint for realizing China's modernization by 2035. Traditionally, the five-year plan is deemed a development plan without much attention to cooperative governance. In fact, the goal of ecological civilization is about development at a superficial level and about governance at a deeper level. Therefore, it is necessary to accentuate cooperative governance for building ecological civilization in the five-year plans conceived for periods after 2020.

(1) The post-2020 five-year plans need to develop the inclusion model. China's five-year plans over the past four decades have evolved, from a simple to a more complicated form, in terms of ecological civilization and green development. A brief dip into history reveals that the guiding ideology of the "6th Five-Year Plan," the "7th Five-Year Plan" and the "8th Five-Year Plan" was the growth model, where social progress received the same amount of attention as economic growth, yet the resource and environment sectors were not specially listed. The guiding ideology of the "9th Five-Year Plan" and the "10th Five-Year Plan" was the parallel model, in which the resource and environment sectors were paid equal attention as economic growth and social development, thanks to the introduction of the concept of sustainable development, but they were limited to the end-of-pie pollution control. The guiding ideology of the "11th Five-Year Plan" and the "12th Five-Year Plan" was the intersection model, where the concepts of low-carbon economy, cycle economy, energy intensity, and resource productivity were emphasized, with green development shifting from the end to the beginning of the economic and social process. The guiding

ideology of the "13th Five-Year Plan" began to reveal trace of the inclusion model, and stressed concepts such as the ecological red line and ecological functional zoning, using the ecological red line to force the transformation of the growth model. In fact, China will achieve a per capita GDP of USD 10,000 in 2020, moving from the inner to middle tiers of the doughnut economics. At that time, it will enter a new stage of development and need to solve two major issues and strengthen the macro management of ecological civilization. While conceiving the post-2020 five-year plans, China have to deal with the relationship among the economy, society, environment, and governance from the perspective of full coverage and penetration of ecological civilization.

(2) Macro management as well as cooperation and integration among the governments. Combing cooperative governance with the building of ecological civilization first requires the cooperation and integration among the governments that are conducive to ecological civilization, including planning, institutional, policy, and indicator integration. In terms of planning integration, the post-2020 five-year plans should go beyond the traditional resource and environmental sectors and make these fields the mainstream of economic and social development—for example, emphasizing red line constraints and ecological thresholds in the resource and environment sectors; stressing content enhancement and efficiency improvement in terms of economic growth; accentuating ecological justice and green consumption in social development; and underlining adaptive and mitigation management in cooperative governance. Regarding to system integration, there are two aspects that need to be improved in the post-2020 five-year plans: one is that departments like the development and reform commission should do a better job in conducting top-level design and coordinating the building of ecological civilization, instead of repeating what the resource and environmental departments used to do. The other is that all development departments should integrate ecological civilization into their professional fields to promote the ecological and green processes in all areas of the economy and society. And the traditional resource and environmental departments should better strengthen the ecological red lines, environmental bottom lines, and resource ceilings, in a bid to force the transformation of the growth model in various fields, rather than putting all efforts into end-of-pie governance. As for indicator integration, while conceiving the post-2020 five-year plans, we should strengthen the compound indicators combing the environment and development—for example, resource productivity indexes, such as the economic output per unit of land, economic output per unit of energy consumption, economic output per unit of water consumption, and economic output per unit of waste, can be used to measure the development level of the green economy; in green development, human development per unit of ecological footprint can be used to measure the development and type of local ecological civilization; the relative and absolute decoupling of development and environment can be used to illustrate the progress of ecological civilization. In terms of policy integration, the institutional construction and

policy design for ecological civilization must include three links of "determination of scale, allocation of property rights, and market transactions." Since the ecological scale and fair distribution are determined by political and management mechanisms beyond the market, it is a cooperative governance process that integrates governmental, market and social mechanisms.

(3) Organizational cooperation and interface management between the public and private sectors. In the process of ecological civilization, strategic innovation and collaborative innovation require organizational changes for ecological civilization and sustainable development and also enhanced interface management between different organizations, as well as cooperation between the public and private sectors—all of which are conducive to ecological civilization. In terms of organizational cooperation, while conceiving the post-2020 five-year plans, we should emphasize that the construction of ecological civilization requires both the top-down efforts of the government and the extensive bottom-up participation of the market and social organizations. We need to coordinate and integrate all organizations, with each taking responsibilities related to their business. In terms of the enhanced vision, the construction of ecological civilization requires all types of organizations, regardless of their interest preferences, to seek their special interests by considering the big picture involving economic, social, and environmental benefits based on the premise that other interests are not reduced but even increase. In terms of interface management, while designing the post-2020 five-year plans, we should emphasize that the interface management between organizations targeting ecological civilization needs to focus on the intersection of internal and external areas to find priorities that meet the requirements of ecological civilization and sustainable development. On such a basis, we come up with a strategy that is conducive to promoting ecological civilization before entering the management process of operational planning, execution, and evaluation.

(4) Enhancing the people's capacities in building ecological civilization. While conceiving the post-2020 five-year plans, we should combine the ecological civilization with cultural construction, foster education, publicity, research, and international promotion activities with Chinese characteristics, and cultivate new ethics and character of the public participation in ecological civilization. Specifically, we should enhance environmental education to ecological civilization education, promote lifelong education of ecological civilization for citizenship, and strengthen knowledge creation and international cooperation related to new ethics and character. What's more, we should not only disseminate and apply new ethics and character, but also create and innovate knowledge in this area. It is necessary to facilitate in-depth academic research in this area—particularly the study of the relationship between the new ideas of ecological civilization and that of sustainable development—by institutions of higher learning as well as the humanities and social sciences. For example, how do the four requirements of the new ethics of ecological civilization connect with the four capitals of sustainable development? How should we connect them with Maslow's theory in psychology comprising a five-tier model of human

needs? And how should we explain the inner connection between China's Five-Sphere Integrated Plan and the new ethics and character of ecological civilization? Finally, we should work with international organizations to tell how China has managed to combine the construction of ecological civilization with cultural construction in a proper way, so as to instill Chinese thinking and practice of building ecological civilization into international education of sustainable development.

Paradigm Shift in International Climate Governance: From Cost Game to Joint Action

Mou Wang

It has been 27 years since international climate governance was established in 1992 when the *United Nations Framework Convention on Climate Change* (UNFCCC, hereafter referred to as the Convention) was signed following the enhanced climate and environmental awareness of the international community in the 1980s. During this period, adjustments in the international economy, emissions of greenhouse gases and political patterns led to dynamic changes in the structure, functions, and mechanisms of international climate governance. Teasing out and analyzing the process and characteristics of dynamically evolving international climate governance allows us to better understand the trend of climate governance and to participate in international climate governance more efficiently.

1 The Basic Course of International Climate Governance

With the increased number of extreme climate events and deeper scientific research on climate change issues, the international community has become increasingly aware that greenhouse gas emissions from human activities have threatened the safety and development of human society. Emissions of greenhouse gases from any region could have negative consequences for the whole world. In the 1970s, to effectively mitigate climate change, the international community was trying to address this global crisis through international collaboration. The Convention was finally adopted at the United Nations Conference on Environment and Development in 1992. It was signed by 154 participating countries and the heads of state or high-level representatives of

M. Wang (✉)
The Research Centre for Sustainable Development, The Chinese Academy of Social Sciences, Beijing, China

J. Pan et al. (eds.), *Beautiful China: 70 Years Since 1949 and 70 People's Views on Eco-civilization Construction*, https://doi.org/10.1007/978-981-33-6742-5_73

the European Community, and came into force in March 1994, thus laying an international institutional foundation for all countries to work closely to tackle climate change.

The objective of the Convention is to "stabilize greenhouse gas concentrations at a level that would prevent dangerous anthropogenic (human induced) interference with the climate system." And it clearly stipulates that developed and developing countries have "common but differentiated responsibilities." In other words, all parties are obliged to take action to deal with climate change, but developed countries have historical and realistic responsibilities for climate change and should assume more obligations, whereas developing countries whose primary task is to develop domestic economy and eradicate poverty should also take measures to reduce greenhouse gas emissions and pursue low-carbon development.

Since the Convention only sets general, non-legal binding targets of the greenhouse gas emission reduction, it is not possible to achieve its ultimate goal. The first Conference of the Parties (COP) in 1995 decided to reach a legally binding protocol through negotiation, and in the third COP, which took place in 1997 in Kyoto, Japan, all parties adopted the Kyoto Protocol, a landmark treaty which extends the Convention. The Kyoto Protocol sets a legally binding quantitative emission reduction target for Annex I countries (developed countries and economies in transition), and introduces three "flexibility mechanisms": Emissions Trading (ET), Joint Implementation (JI), and Clean Development Mechanism (CDM). In 2007, the 13th COP was held in Bali, Indonesia, where the parties reached the Bali Action Plan that outlined the roadmap and basic framework for the establishment of an international climate system after 2012, and brought the United States, which was staying away from international cooperation at the time, back to the negotiation table. In 2011, the 17th COP was held in Durban, where all the parties reached the Durban mandate, which opened the negotiation process of the international climate system after 2020, and discussed how to put in more effort into reducing emissions before 2020. The 18th COP, held in Doha in 2012, clearly launched the second commitment period of the Kyoto Protocol. All parties, including the United States, agreed on emission reduction targets, adaptation mechanisms, financial mechanisms and technical cooperation up to 2020, in addition to producing a package of documents on the long-term cooperative action working group. At the 2015 Paris Conference (the 21st COP), the Paris Agreement, adopted through much effort from all parties including the United States and China, basically clarified the institutional arrangements and cooperation model for international climate governance from 2020 to 2030.

2 The Basic Pattern and Evolution of International Climate Governance

From the UN's launch of global climate change negotiations in the mid and late 1980s to the 2007 Bali Climate Change Conference, the international climate pattern basically involved two camps: the developed countries, represented by the EU and the United States, and the developing countries, represented by the Group of 77 and China. Although the interests and goals of the two camps were quite different, their overall structure was basically fixed. Specifically, the United States was the rule maker first, and then the rule-breaker after the Kyoto Protocol, whereas the EU upheld international morality, always taking the lead in addressing global climate change and abiding by the Convention and the Kyoto Protocol. Although the interests and goals of the United States and the EU were not exactly the same, belonging to the developed country group, they basically assumed the same stance towards developing countries. By contrast, due to differences in development stages and needs, developing countries or small groups of them had varied interests; however, they adhered to the principle of "common but differentiated responsibilities" and took nearly the same stance on requiring developed countries to reduce emissions and take financial and technological obligations.

After the Durban Platform that launched a new round of negotiations was adopted in 2011, the United States managed to reduce its total carbon emissions, thanks to the financial crisis and the "Shale Revolution," the combination of hydraulic fracturing and horizontal drilling enabling the United States to greatly increase its production of oil and natural gas. Therefore, the Obama administration endeavoured to be the leader of global climate change and the maker of a new climate agreement. The EU was relatively weak in the international climate pattern and difficult to take the lead; however, it still hoped to actively work with others to advance the international governance. By contrast, emerging economies represented by China had enhanced national strength, with GDP per capita generally reaching or exceeding the level of the middle and high income, and some of them even close to the EU in terms of per capita carbon emissions—a situation that showed the "duality" of them and the developed countries, as well as a widened gap between their stance and that of the least developed and small island countries regarding emission reduction and capital contribution. As such, a new group of developing countries with a similar stance was emerging.

The basic pattern of today's global climate change can be summarized as: north–south interweaving, edging closer to the north in the south, division within the north, and north–south spectrum—which can be roughly described as "two camps," "three major sectors," and five categories of economies. The two camps of the "north and south" that traditionally divide developing and developed countries faintly exist. The three major sectors of developed, emerging and low-income countries are roughly identifiable—developed economies can be classified into two types: the countries with a rapid population growth, represented by the United States, and the ones with a stable or declining population, represented by the EU and Japan; emerging economies can also be classified into two types: the countries with a stable population growth, represented by China and the ones with a rapid population growth, such as India; low-income economies are mainly low-income countries. These countries may divide and reorganize in the future, but as a whole, they will exist for a long time.

3 The Structure and Mechanism of International Climate Governance

Response to climate change and control of greenhouse gas emissions may, to some extent, limit a country's development and negatively affect its economic and political interests, but also may become an important area of international cooperation. The world must rationally mitigate climate change through international institutional arrangements, clarify the responsibilities that all countries should take, and promote international cooperation to achieve a win–win situation between advancing human society and protecting global climate. From 1979 when the World Meteorological Organization (WMO) convened the first World Climate Conference calling for protecting the global climate to 1990 when international climate negotiations were kicked off, global response to climate change entered an institutional and legal track. International cooperation mechanisms to address climate change now mainly include two categories: the climate convention mechanisms and non-climate convention mechanisms that are regular, irregular, international, regional, industry, and professional. All of these mechanisms have played different roles in international cooperation on climate change because of their different positioning and functions.

The climate convention mechanisms. The Convention was passed and open for signature in Rio, Brazil in 1992. On March 21, 1994, the Convention went into effect and became an important legal basis for international climate governance. Its ultimate objective is to "stabilize greenhouse gas concentrations at a level that would prevent dangerous anthropogenic (human induced) interference with the climate system." And it states that "such a level should be achieved within a time-frame sufficient to allow ecosystems to adapt naturally to climate change, to ensure that food production is not threatened, and to enable economic development to proceed in a sustainable manner." The 1997 Kyoto Protocol, the 2009 Copenhagen Accord, the 2012 Doha Resolution, and the 2015 Paris Agreement made important achievements, which clarified the goals of cooperation in global climate governance in different periods as well as the international institutional framework and pointed out the direction for global cooperative action. While carried out by the Convention parties with varied views who were trying to arrive at some compromise, the negotiation process of the international climate system moved forward with difficulty, eventually becoming the core platform for advancing international climate governance and forming the legal basis for international cooperation. Response to climate change requires the joint efforts of all countries, and it is necessary to uphold and improve the cooperation platform of the Convention, so that the international climate governance plays a leading role in achieving global emission reduction targets and ensuring climate security.

The non-climate convention mechanisms. To promote the negotiation of the Convention, the parties also carried out a variety of activities and practices outside the convention system, or cooperative mechanisms supplementing the convention mechanism, which played a positive role in enhancing mutual understanding and reaching a consensus. In terms of their nature, these mechanisms can be divided into three types: political, technical and economically incentive. First, the political mechanisms include the United Nations Climate Summit, the Forum on Millennium Development Goals, the Energy and Climate Forum of Economic Powers, the G20, the G8, APEC meetings, etc. The common feature of these mechanisms is that heads of government or high-level officials participate in consultation and reach political consensus on some major issues, but generally do not discuss specific technical details. Political non-convention mechanisms like the United Nations Climate Summit usually play an important role in overall, long-term, and political issues. Because of the high-level participation, at such conferences, especially summits of the heads of states, leaders often solve some long-term problems the technical groups face under the Convention, thereby advancing the negotiation process of the Convention. Second, technical non-convention mechanisms mainly include cooperation mechanisms such as the International Civil Aviation Organization, the International Maritime Organization, and the UN Secretary General's High-Level Advisory Group on Climate Change. These mechanisms carry out special research and discussion on specific issues in the negotiation of the Convention, and feed back the results and suggestions, so as to promote the negotiation process of related issues under the Convention. Technical mechanisms have limitations: first, climate change is not the main business of these institutions or mechanisms, and their focus and purposes may be different from those of the Convention; next, varied mechanisms or institutions also have their own rules of procedure and guiding principles, which may be different from those of the Convention—a fact that may lead to mismatches in understanding. Finally, economically incentive/binding mechanisms, including climate change-related trade mechanisms, standard-setting mechanisms related to production activities and domestic and foreign market expansion, and other non-convention consultation mechanisms. Economically incentive measures are a supplementary negotiation topic in the negotiation of the Convention. Most of the time, the negotiation is not about the core concern of the Convention, but the negotiated issues are closely relevant to the operations of the real economy and the development interests of related industries and fields. These mechanisms, such as trade mechanisms and standard-setting mechanisms, have been developed for a long time and existed before the climate change issue became an international governance mechanism. However, after the emergence of the climate change governance mechanism, the boundaries between these various mechanisms are blurred. As such, discussion and consultation of climate change issues through these mechanisms includes not only technical issues, but also political and principle ones.

We should build an efficient and fair international cooperation mechanism, the purpose of which is to promote cooperation among countries around the world to coordinate the governance of climate change issues. A fair and efficient international cooperation mechanism is the basis for improving international cooperation governance and achieving the goal of international cooperation governance as well. From the perspective of the comprehensive influence of various mechanisms in international climate governance in terms of the role, function, binding force, and degree of participation, the Convention should undoubtedly play a leading role in the process of international cooperation in climate governance, while the non-convention cooperative mechanisms should serve as a supplement that helps advance the negotiation process. Such a governance mechanism can not only reflect the principle of fairness in international cooperation (the maximum degree of participation), but ensure the efficiency of international cooperation, thanks to the focus and legal binding power of the Convention.

4 Evolution of the International Climate Governance Responsibility System

A dip into the evolving international climate governance pattern reveals that the 1992 Convention described two major camps: Annex I and non-Annex I, or the north and south camps. The 1997 Kyoto Protocol further divided Annex I countries into developed countries and economies in transition, thus giving rise to three camps, namely developed countries, developing countries and economies in transition. Parties to the Kyoto Protocol agreed on a second commitment period in 2007 and the Bali Road Map committing to a "dual track" approach under the Convention. The 2009 Copenhagen Accord no longer distinguished between Annex I and non-Annex I countries, and due to the eastward expansion of the EU, the definition of economies in transition was basically cancelled. The 2015 Paris Agreement emphasized the "nationally determined contributions" consistent in legal expressions and regardless of where a country is, and the self-positioning differences among countries can only be seen through their varied contributions. So far, the global climate governance pattern basically blurs the division between the north and south camps, showing a continuous, spectral evolution.

Basically, the evolution of the Convention from 1992 to 2015 was a process of adding responsibilities and obligations to developing countries. At first, developing countries gave priority to economic and social development, with virtually no responsibility or obligation to mitigate climate change. After the adoption of the Copenhagen Accord, the subsequent Cancun Agreements and a series of Doha resolutions, many developing countries proposed control or reduction targets for greenhouse gas emissions, based on certain amount of global cooperation, including financial and technical support from developed countries and corresponding supportive measures. Nevertheless, the targets of developed and developing countries are listed into two documents with different methods of supervision and assessment. The 2015 Paris Agreement required or encouraged developing countries to propose mitigation targets or even donate money to the Convention's financial mechanism. In fact, many developing countries put forward contribution targets (including mitigation, adaptation, finance, and technology goals). The goals of all countries, including developed and developing countries, are included in the same table with generally the same methods of supervision and assessment. Regarding the commitments, we can say that developing countries basically converge with developed countries in responsibilities and obligations.

This trend came with an economic reason. From the year 2000 onwards, with the rapid economic progress of developing countries, especially emerging economies, the international economic pattern was undergoing significant changes. The share of developed countries (ECD countries) in the world economy was declining year by year, from over 70% of global GDP in 2000 to 62% in 2017; their share of export trade in the world was also declining year by year, dropping from around 75% in 1998 to 58% in 2017. From the perspective of the global emission pattern, the greenhouse gas emissions by developing countries were rising rapidly, with the proportion of developed countries dropping from 66% in 1990 to 41.4% in 2012. Changes in the world's economic, trade and emission patterns will likely affect a country's willingness to participate in global governance, including international climate governance. Developed countries may become more conservative in terms of their willingness to make investment, propose cooperation methods, take emission reduction actions, and pursue trade protection, and also demand actions from developing countries who may be more active in their response according to their economic progress.

The per capita cumulative emission is an indicator that better reflects a country's historical responsibility. According to the World Resources Institute's CAIT database, the per capita cumulative emissions from developed countries are generally very high, with the United States, the United Kingdom, and Germany all exceeding 1,000 tons of carbon dioxide emissions per capita, while developing countries generally do not exceed 100 tons. China's value of 104 tons is in the middle among developing countries, and India stands only at 29 tons. Climate change is caused by greenhouse gas emissions over the course of history. Judging from the per capita cumulative emissions of each country, we can identify how much responsibility a country should take in international cooperation to address climate change and how much room left for its emissions in the future. Therefore, although the will ingness and ability of developing countries to act have been strengthened compared with when the Convention was reached, developed countries still keep the same responsibilities and obligations to intensify their efforts to reduce greenhouse gas emissions in international climate governance.

5 The Establishment of a New Paradigm for International Climate Governance

The Paris Conference is undoubtedly another landmark meeting in international climate governance after the Copenhagen Conference. The heads of state of major parties such as China and the United States gathered in Paris and issued a political declaration of active cooperation and action—an extremely important and efficient global political mobilization. Compared with the Copenhagen meeting, the heads of state expressed more active and independent actions, rather than just willingness to take cooperative actions, and their stance on climate governance also shifted from previous concerns about the cost of climate governance to actively seeking opportunities of economic growth and driving forces from international climate governance. This is a cognitive shift that also laid a more solid foundation for subsequent negotiations and global response.

The new paradigm of international climate governance: active commitment and joint action. Based on their economic and social development and the inclusive mechanism of nationally determined contributions, some developing countries, including China, have shown their willingness to cooperate by putting forward a more positive nationally determined contribution goal than they did in previous climate agreements. In the document of nationally appropriate mitigation actions of the Copenhagen Accord (developing countries' nationally determined emission reduction action targets before 2020), developing countries proposed their emission reduction targets based on the premise that they could obtain funding, technologies, and capacity building support. However, in the nationally determined contribution target system under the Paris Agreement, more developing countries showed a positive attitude in taking action on their own initiative, and they were very flexible in the negotiation of the financial mechanism, transparency, and the inventory mechanism, showing their willingness and commitment to make joint efforts.

Under the Paris Agreement, 188 parties submitted nationally determined contribution targets to be implemented mostly from 2021 to 2030, with some from 2021 to 2025. Each country repeatedly discussed and adjusted their targets at home before proposing. With the ratification process of the Paris Agreement, the nationally determined contribution targets put forward by all countries will be further established as domestic legally binding action goals that will in turn ensure the realization of the Paris Agreement goals. The subsequent negotiations will also focus on the implementation of the Paris Agreement, and prompt all parties to implement the Agreement by passing the legally binding and supportive rules covering transparency, global inventory, and financial mechanisms. From the Copenhagen Accord to the Paris Agreement, the cooperation model of all parties under international climate governance gradually shifted from the model where developed countries led and developing countries followed to a new paradigm of independent contributions, active commitments, and joint actions by all parties. Regarding the relationship between climate change and economic development, there was also a significant change in understanding and concepts, from the previous view of climate change as an incremental cost of economic development to a new idea of climate change as a catalyst for economic growth as well as green transformation and development. Such positive changes have established a new paradigm for international climate governance and prompted a leap forward in achieving global climate security through international cooperation.

6 Prospects of International Climate Governance

The Paris Agreement and the 2030 Sustainable Development Goals have become a new driving force for global green development. Research by the International Energy Agency shows that in 2015 global energy-related carbon dioxide emissions remained the same as that in 2013, yet the world saw a global economic growth of more than 3%. It indicates that global economic growth is being decoupled from the increase in carbon emissions—a situation directly related to the rapid development of renewable energy and the continuous shrinking of the coal industry. In 2017, 70% of the global net added power generation came from renewable energy resources. Global investment in renewable energy power generation is more than twice the total investment in fossil fuel and nuclear power generation. Global climate governance has achieved initial results. The year 2015 witnessed two important international achievements in global green development, namely the Paris Agreement and the 2030 Sustainable Development Goals. The results of the Paris Agreement, authorized by the Rio Conference, automatically became a part of the 2030 Sustainable Development Goals. Therefore, the two processes are highly interrelated and can also work together and promote each other to serve as a new force driving the green development around the world.

International climate governance has become a resilient trend. The Paris Agreement, a milestone achievement made by more than 100 heads of state, including the heads of China and the United States, is a treaty in which all countries expressed their willingness to act more independently instead of just taking cooperative actions, thus forging a new paradigm involving joint actions for international climate governance. The United States withdrew from the Paris Agreement just over half a year after the Agreement came into effect, which is an unwanted action with a negative impact on the global joint effort to mitigate climate change. However, the international community has been adapting pretty well because it has seen similarly subversive situation caused by a shift in political power in the United States. Against the background of a decline in the global influence of America's politics and economy, and the rapid rise in the influence of emerging economies concerned about environmental and development issues, the United States Republican administration's announcement, once again, of its plan to withdraw from international climate governance process has far less impact than the Bush administration's withdrawal from the Kyoto Protocol. The United States, a developed country with the greatest historical emission responsibility, as well as the strongest economic strength and capacity for taking action to mitigate climate change, ignores the common willingness and efforts of the international community to cope with climate change and insists on withdrawing from the Paris Agreement. It has been, and will be, condemned by the international community. To offset the negative consequences of its withdrawal, other parties (including China), international organizations and the private sector, who are concerned about economic progress and environmental security, will be more pragmatic and engage in close cooperation to fully realize the goals of the Paris Agreement and to ensure global climate security.

The new paradigm of independent contributions, active commitments, and joint actions established by the Paris Agreement will last, and international climate governance will surely become a new driving force for global green transformation and development. Global cooperative response to climate change has shifted from a model driven by the government to a model based on the market and capital. International climate governance has basically established the resilience against policy risks, and future actions will keep expanding its breadth and depth to ensure climate security and sustainable economic and social development.

Preventing Environmental Risks from the Perspective of Territorial Space Planning

He Xu

China is a country that has a vast territory, with great differences in various regions in terms of economic and social development, natural resources, and the diversity and fragility of the ecosystem. Therefore, the types and degrees of environmental security issues faced by regions across China are different. In general, accidental and cumulative environmental risks exist side by side in China, where its ecosystems of air, water and soil are facing complex and considerable risks. The 13th Five-Year Plan is vital for China's economic restructuring, social transformation, and the building of a moderately prosperous society in all respects. From the period of the 13th Five-Year Plan onwards, China will still face formidable environmental risks that impose serious constraints on the sustainable advance of its economy and society. The development of ecological civilization inevitably comes with various environmental risks arising from the intensified contradictions between human and nature. In a sense, the concept of ecological civilization is the goal of preventing environmental risks. In other words, ecological civilization aims to realize recycling of material and energy on the premise that environmental risks can be effectively prevented and eliminated.

At the National Conference on Eco-environmental Protection held in May 2018, General Secretary Xi Jinping pointed out that we must effectively forestall environmental risks. Environmental security is a key component of national security and an important safeguard for the continuation of healthy socioeconomic development. We must make management of environmental risks a regular practice, carefully constructing a multi-leveled environmental risk prevention system that monitors these risks at every stage. We must speed up the structural reforms in the ecological domain, focus on implementing reform measures that have already been launched, and formulate new reform plans in a timely manner. Prevention of environmental risks has become a major issue for China, and environmental risk issues—latent and continuous in nature—are closely linked to the spatial distribution of China's

H. Xu (✉)
Institute of Ecological Civilization, Nankai University, Tianjin, China

J. Pan et al. (eds.), *Beautiful China: 70 Years Since 1949 and 70 People's Views on Eco-civilization Construction*, https://doi.org/10.1007/978-981-33-6742-5_74

population, economy, and environmental elements. Only by using the thinking and tools of spatial governance and giving an insight into its spatial differentiation and temporal dynamics can we effectively forestall environmental risks. Territorial space planning in a new era will gradually morph into a complete, unified planning system, or a comprehensive management system based on a spatial perspective that plays a vital role in solving problems faced by the ecological environment and preventing environmental risks. As a part of the ecological civilization system, it is of great significance to advance China's ecological civilization.

1 The Current Situation of Environmental Risk Prevention

With rapid economic development and accelerated urbanization process, China is facing formidable environmental risks, with a large number of environmental emergencies and risks related to public health and ecological safety, which are posing a major threat to the smooth progress of its national economy, healthy development of society, and environmental protection. For one thing, environmental emergencies frequently occur in China. Statistics show that, from 2010 to 2017, a total of 3,627 environmental emergencies happened—for instance, the incident of heavy metal pollution from Zijin Mining Group in Fujian; the blood lead incident in Taizhou, Zhejiang; the cadmium pollution incident along the Longjiang River in Guangxi; and the tap water contamination incident in Lanzhou. Moreover, the explosion of a hazardous chemical warehouse in Binhai New Area, Tianjin in 2015 is another proof that China still faces frightful risks of frequent, sudden pollution incidents. For another, regional and cumulative environmental risks are looming. In recent years, thanks to increased public awareness of environmental protection and enhanced enforcement and supervision of relevant laws, many pollution incidents have been disclosed, such as the ones that took place in the Tengger Desert, the Qilihai wetland, Tianjin, and the Qilian Mountains. These are a living proof that China is facing tremendous risks when it comes to environmental protection, and can no longer ignore regional and cumulative environmental problems resulted from blind development. From the 13th Five-Year Plan period onwards, China will remain in a shift stage in economic growth rate, a painful period of structural adjustment, and a phase of industrialization, urbanization, continued utilization of natural resources, and social transformation—a time of stalemate between social and economic development and conservation of the increasingly complicated ecological environment.

Over the past decade, China has been improving its environmental risk prevention and management system, and has made great efforts to assess environmental risks, manage environmental emergency plans, and conduct environmental risk inspection and classification in key industries. At the national level, environmental risk prevention has received particularly high attention, but it is generally managed in an event-driven model with a big gap between the actual "need" of forestalling risks and the effective "supply" of related policies and measures. It is very hard to use the current environmental management system and measures to fully reduce the structural and regional risks faced by China's ecological environment. Since the environmental risk management system in China is still in its infancy, there are management problems like "stressing emergency handling while neglecting preventive measures" and "tackling sudden pollution incidents while ignoring chronic risks to public health." Moreover, the environmental management model has yet to morph into a risk control-oriented one. Compared with other aspects of environmental protection, management of environmental risks still requires suitable systems, policies and technologies in place, and such risks are posing a serious challenge to the development of ecological civilization in China.

In his speech at the National Conference on Eco-environmental Protection in 2018, General Secretary Xi Jinping said that "our efforts to build an ecological civilization are now in a crucial phase in which we must carry forward despite heavy strain and immense pressure, a decisive stage in which we will supply more high-quality ecological goods to meet the growing demands of the people for a pristine environment, and also a period of opportunity in which we have the conditions and abilities necessary to resolve prominent environmental issues." That being said, China has yet to form a decision-making system based on environmental risks, and fully incorporate environmental risk assessment into the formulation of major strategies and plans. What's more, the fragmented environmental management has failed to eliminate a wide variety of environmental risks. Compared with economic construction, environmental risk prevention is still relatively insufficient. We touch on how to avoid environmental risks while implementing plans and strategies, but have not yet decided whether we should, or how to, make strategies and plans through specific methods like environmental risk assessment.

2 Territorial Space Planning in a New Era

1. The international spatial planning system

According to their political, economic and social development as well as status quo, countries around the world adopt different approaches to create, manage and implement their spatial planning systems and gain different results. In the national governance system, "spatial planning" refers to the management of spatial development and/or land use in a specific country and region, and in Euro-English, it is a collective term for "the entire system of the government's management of spatial development." The EU-level spatial policies and the spatial planning of EU countries (domestic urban and rural planning management system) are closely connected with frequent interactions, and they have been a key issue in the European planning community. In the context of European integration, the term "spatial planning" is widely used in the planning field, with flexible usages emerging, and each country has a different form and term for it. For example, "regional spatial planning" appears in the planning literature of the United Kingdom, which refers to the transition from a top-down, goal-driven approach in the past to a more localized one. The

United Kingdom traditionally calls it "town and country planning," Germany and Austria call it "spatial planning (Raumplanung)," and France calls it "urban planning (Urbanisme)" or "territorial improvement (Amenagement du Territoire)."

If we tease out spatial planning systems of developed countries which have been relatively improved for a long time, we can tell that they have different types but still have something in common to be learnt for reshaping China's spatial planning system. The first is arranging and implementing territorial space planning based on sustainable development. Currently, developed countries opt for spatial planning that no longer gives priority to economic development but to their ecological environments and sustainable use of land or other natural and energy resources, while paying attention to fair public services like providing housing and creating infrastructure in society—a fact particularly prominent in the planning systems of the United Kingdom, Germany and Japan. In this new era, we should follow these developed countries in territorial space planning and focus on the ecological environment, resources and energy, while adopting high-quality development methods that emphasize both development and protection, as well as efficiency and fairness. The second is macroeconomic regulation and control, often abbreviated macro-control. For most countries, national spatial planning is an indispensable way of improving their market systems, enhancing competitiveness, and conducting macro-control. It is also the central government's standpoint to prevent and correct market failures under a completely free economic system—a government intervention. This is exactly the basic principle of reforming China's spatial planning system to improve the national governance system and modernize the governance capabilities.

2. China's spatial planning system

China's spatial planning took shape during the planned economy period, which began in the mid-1950s, yet without legislative guarantee for a long time. In February 1979, the *Forest Law of the People's Republic of China (Trial)* was issued, marking the first requirement for legally binding planning. With the promulgation of subsequent laws such as the Land Management Law, Urban Planning Law, Environmental Protection Law, and Water Law, spatial planning kept evolving, with an increasing number of the types of arrangements that are mostly compatible with laws and management, and respond to the management requirements of different industries with a strong departmental flavor.

China's old spatial planning system is complex and imperfect. With varied purposes, problems and needs, many outmoded spatial plans are self-contained. Laws that clearly define a complete spatial planning system are yet to be made, but a consensus has been reached: China's spatial planning system should mainly consist of planning regarding to the national economic and social development, overall land use, urban and rural development, and ecological protection, in addition to special plans for other fields like transportation development planning. Spatial planning represents a "vertical and horizontal" network. Vertically, different plans are connected at all administrative levels, with departments adopting "top-down" management; horizontally, "multiple plans are adopted at the same time"—a situation where these plans are intertwined horizontally and vertically to form a complex planning system. With continued economic and social progress, the spatial planning system in China is seeing more and more problems, and its scientificity and solemnity have been challenged by the public, thus giving rise to contradictions and unwanted issues that impose constraints on the healthy and sustainable development of the national economy. At the lower administrative level with smaller room for maneuver, due to the lack of connection and coordination between departmental plans, planning of the same specific space often results in conflicts of control and overlapped controlled places, as well as reduced urban management efficiency, waste of land resources, and environmental destruction.

The national governance system and governance capabilities are a mutually complementary organic whole. Only with a good national governance system can governance capabilities be improved, and only by improving national governance capabilities can the effectiveness of the national governance system be brought into full play. Space governance directly or indirectly affects the structure and process of governmental, market and social governance through a series of institutional arrangements that control and guide the elements of territorial space, so that it can fully reflect China's optimized space development pattern while advancing its strategic intentions and values of sustainable economic and social development. On November 12, 2013, the *Decision of the CPC Central Committee on Several Important Issues of Comprehensively Deepening Reform* adopted at the Third Plenary Session of the 18th CPC Central Committee took "advancing the modernization of China's system and capability for governance" as the overall goal of comprehensively deepening reform, and went all out to achieve this lofty goal. This signifies that China's spatial planning system will be further integrated, spatial contradictions and conflicts between different plans effectively resolved, and the spatial scope more coordinated and unified. What's more, the collaborative governance that entails planning in various fields will enhance the role of spatial planning in space development and protection, and provide, for national development goals, guarantee for the implementation of spatial planning, as well as ecological protection, environmental management and control, and measures for preventing risks.

3 Spatial Planning and Prevention of Environmental Risks

With the former Ministry of Land and Resources as the backbone, the 2018 institutional reform integrated the planning, resource management and other functions of relevant departments to establish the Ministry of Natural Resources, which exercises the responsibilities of supervising spatial planning and managing the use of all territorial space, including centralized management of natural resources such as mountains, rivers, forests, farmlands, lakes and grasslands. As the basis of spatial development, territorial space planning in a new era is an important way for China to promote ecological civilization, fully coordinate economic and social development, rationally and efficiently allocate resources, and coordinate economic and social progress with environmental protection. Spatial planning will determine the management of the use of territorial space, which serves as an important basis and starting point for forestalling environmental risks. Risk sources and environmental risk receptors are two critical factors of environmental risks, as well as two necessary conditions for sudden and cumulative environmental events. In the context of limited land resources, the spatial relationship between risk sources and sensitive receptors in the ecological environment determines the likelihood of environmental risks.

At present, China's spatial planning is mainly to implement the national security strategies and main functional strategies, integrate multiple goals, and delineate the red line for urban, agricultural, ecological space as well as the boundaries for ecological protection, permanent basic farmlands, and urban development (collectively referred to as the "three zones and three lines"), optimize the urbanization, agricultural production and ecological protection patterns, manage development intensity, implement major control lines, coordinate various types of spatial planning, so as to form a unified big picture of territorial space development, conservation, utilization, and restoration. Sudden and cumulative environmental (pollution) events with layout-related problems can be prevented or totally avoided through spatial planning to delineate the "three zones and three lines" or to manage the use of different spaces. The key to integrating the requirements of preventing environmental risks in spatial planning is to pursue binding by environment protection beforehand. By optimizing the distribution of risk sources and environmental receptors, we can reduce sudden environmental events, or by focusing on ecological safety and resource consumption, with improvement of environmental quality at the core, scientifically and effectively control the use of space to reduce or avoid cumulative environmental events. Meanwhile, spatial planning should merge with ecological protection goals, environmental quality control lines, and resource consumption "ceilings," besides setting strict requirements for access to the ecological environment as vital constraints on implementation of spatial planning, so as to strictly control regional spaces for reduced environmental risks. Moreover, we should take the carrying capacity of natural resources and environmental capacity as the basis for strategic decision-making and preconditions for planning, and take resource and environmental conditions into consideration while putting constraints on spatial planning to force transformation of economic development, thereby achieving harmony between human and nature. We should link territorial space delimitation and zoning with the division of environmental management and control units, in order to coordinate the use of territorial space with access to the ecological environment for more rational and coordinated spatial planning. We should combine policies that offer access to the ecological environment with spatial planning, taking the development goals and positioning of each region as the guide, to propose practical control requirements that coordinate environmental protection with regional development, and help implement measures for managing the environment and forestalling environmental risks.

4 Suggestions

Establishing a spatial planning system is a basic, all-inclusive, and arduous task that is problem-oriented, goal-oriented and based on current and future conditions and overall planning. By the same token, it is an equally arduous task to gradually build a multi-level risk prevention system characterized by strict prevention beforehand, strict management during the process, and disposal afterwards in a systematical manner. Environmental risks span a wide range of fields, which means we should base the building of a comprehensive environmental risk prevention system on specific and actual conditions, and from the perspectives of laws, institutional functions and policies, to optimize implementation methods, management measures, and guarantee conditions for environmental risk prevention.

To improve the mechanism of spatial planning and environmental risk prevention, and enhance the capacities of territorial space governance, we propose the following suggestions.

The first is to improve the laws and regulations related to spatial planning, with the aim of providing a strong legal basis for planning preparation, approval, supervision and implementation, and ensuring smooth spatial planning. Environmental protection must rely on the rule of law. For one thing, the top-level design should be strengthened. We recommend that the general provisions of the basic law of territorial space planning clarify the principles of environmental priority and ecological safety, regard the prevention of environmental risks as the basis, and clarify the status of forestalling such risks in laws related to territorial space planning. For another, since the laws of territorial space planning are mainly used as a guide in planning, they should be relatively easy to adopt. Therefore, we recommend that the prevention of environmental risks be deemed the purpose or one of the main contents of managing territorial space planning at all levels, and that prevention of environmental risks be included into relevant legal provisions guiding the formulation and implementation of territorial space planning.

The second is to establish an overall and coordinated departmental management mechanism. A cross-departmental, collaborative management mechanism should be created to break down administrative barriers. With the Ministry of Natural Resources and the Ministry of Ecology and Environment taking the lead, followed by the Ministry of Agriculture and Rural Affairs, the National Forestry Administration and other relevant ministries and commissions, a unified management system with a clear division of labor should be established to give full play to the system aimed at preventing environmental risks.

The third is to establish a well-arranged supervision mechanism. The enhanced transparency of environmental risk-related information—a vital part of public participation—can not only prompt the government to disseminate risk-related information to the public, but also help the people more objectively understand environmental risks, thereby contributing to a more stable society thanks to reduced risks. We should clarify the importance of supervision and feedback to territorial space planning, build a safeguard mechanism for avoiding environmental risks based on space management, get the public involved, create online and offline platforms for public supervision, establish efficient supervision and feedback mechanisms, and improve the effectiveness of public supervision as vital measures for ensuring management tasks to maintain the orderly progress of related work.

The fourth is to establish an environmental assessment mechanism that is combined with the mechanism for preventing environmental risks, and include ecological and environmental indicators into the system for assessing the performance of officials involved. The fewer environmental incidents, the better the ecological environment will become, and the higher the evaluation will score. In this way, local governments can be prompted and mobilized, with enhanced efficiency of them and related units, to promote economic and social development in a manner that protects the ecological environment.

The fifth is to clarify the relationship between laws and regulations, clarify the relationship among the laws, regulations, technical guidelines related to territorial space planning, strengthen the connection between them, and delimit the "three zones and three lines" to establish a mechanism for forestalling environmental risks based on the spatial planning system.

Building Solid Ecological Barriers and Promoting the Development of Ecological Environment

Yulong Sun

1 Getting a Proper Insight into Gansu's Ecological Status and Role

Gansu is located at the intersection of the Loess Plateau, the Qinghai-Tibet Plateau, and the Inner Mongolian Plateau, and within the catchments of the Yangtze River, the Yellow River and the inland river. It is an important water conservation and replenishment region for the Yangtze and Yellow Rivers, and also a blocking area where the Tengger, Badain Jaran and Kumtag Deserts converge and turn southwards. Moreover, Gansu is a strategic channel for the Eurasian Land Bridge, and a transportation hub linking southwest and northwest China. Therefore, it is strategically important in China's ecological construction. Building an important ecological security barrier for the northwest and even the whole country is one of the strategic definitions for Gansu's development—a goal that is related to the sustainable development of Gansu's economy and society, and that will also affect ecological security across China. All in all, Gansu's ecological status and role are mainly manifested in following aspects:

First, Gansu serves as a reservoir for the Yellow and Yangtze Rivers. Gannan and Longnan have large wetlands, grasslands and forests, and more than 120 main and tributaries of the Yellow River, Tao River, Daxia River, and Bailong River, all of which combine to provide irreplaceable ecological adjustment functions of conserving water and soil. To protect the area's ecological environment, we should reduce the sediment and pollutants flowing from the upper reaches to the middle and lower reaches of the Yellow and Yangtze Rivers.

Second, Gansu is a vital area for water conservation and an ecological barrier. Its glaciers, forests, grasslands and wetlands provide a solid foundation for water conservation at the northern foot of the Qilian Mountains, conserving and nurturing 57 rivers, large and small, which collectively constitute the four river systems—Ruo

Y. Sun (✉)
Department of Ecology and Environment of Gansu Province, Lanzhou, China

J. Pan et al. (eds.), *Beautiful China: 70 Years Since 1949 and 70 People's Views on Eco-civilization Construction*, https://doi.org/10.1007/978-981-33-6742-5_75

Shui, the Shiyang River, the Shule River and the Haerteng River—in western China, with numerous oases distributed across the region that block the convergence of Tengger and Badain Jaran Deserts. The deserts and semi-desert grasslands along the northern Qilian Mountains also help protect the oases along the Hexi Corridor.

Third, Gansu is a region that curbs sandstorms. The province gives birth to sandstorms and is on the path through which the sandstorms generally pass while moving eastwards. As such, it plays an important role in preventing desertification in northwest China, protecting the forests that help conserve water in the upper reaches of the Yellow and Yangtze Rivers, reducing soil erosion in the Loess Plateau and the Qinghai-Tibet Plateau, and restoring and protecting the ecological environment within the catchments of the inland river, thus reducing the sand and dust blowing from northwest to southeast China.

2 Resolutely Implementing Xi Jinping Thought on Ecological Civilization, and Getting the Significance of the Qilian Mountains' Ecological Environment Protection and Development from a Strategic Perspective

As a "lifeline" for the Hexi Corridor in Gansu, the Qilian Mountains, designated a national nature reserve in 1988, has diverse ecosystems offering a great ecological service function with an important role for water conservation. With a unique geographical location, this region helps stop the Tengger, Badain Jaran and Kumtag Deserts from sprawling southwards, maintain the oases along the Hexi Corridor, and conserve the run-offs to the Yellow River, thus forming a vital ecological barrier in northwest China.

General Secretary Xi Jinping always pays great attention to the development of ecological civilization and environmental protection, and calls for speeding up the construction of an ecological security system for creating a benign loop of the ecosystem as well as effective prevention and control of environmental risks. On January 16, 2017, Focus Report featured a report on the damaged ecological environment in the Qilian Mountains national nature reserve, and from February 12 to March 3, the central inspection team carried out a special inspection of this issue. Then, the meeting of the Standing Committee of the Political Bureau of the Central Committee and the Party Group Meeting of the State Council listened to the report on the inspection. Later, the former conducted an in-depth analysis of ecological destruction in the reserve, and notified the issue nationwide. Moreover, the State Council meeting studied the supervision, protection and restoration of the ecological environment in the reserve; On June 1, the Office of the Central Government and the Office of the State Council issued "A Notification of the Supervision and Treatment of Ecological and Environmental Issues in Gansu's Qilian Mountains National Nature Reserve and the Lessons that Have Been Learned," a report showing, at a deeper level, that the Central Committee with comrade Xi Jinping at its core were stressing the development of ecological civilization and ecological protection, and also reflecting that the Party Central Committee was determined to do a good job in building ecological civilization and was confident in adhering to green development.

1. **The profile of the Qilian Mountains in Gansu**

The Qilian Mountains is a huge mountain system located in the northeastern part of the Qinghai-Tibet Plateau, to northeast Qinghai, and on the western border of Gansu. It is made up of several parallel mountains and wide valleys running from northwest to southeast. The part of the Qilian Mountains in Gansu mainly includes five cities, namely Wuwei, Jinchang, Zhangye, Jiuquan, Jiayuguan, and a part of Lanzhou—a populated region serving as Gansu's most important industrial hub and agricultural production base. The Qilian Mountains has a wide variety of glaciers, snow-capped mountains, forests, wetlands and grasslands, and its 2,859 glaciers, large and small, collectively have an ice reserve of 81.12 billion cubic meters. Its forest ecosystems are a vital base for preserving biodiversity of China and even the world—the most important natural barrier that ensures ecological safety in northwest China. The forests help store precipitation in the mountainous areas, forming 56 inland rivers in the Shiyang River, Ruo Shui, and the Shule River systems. Over the years, the average self-produced surface water resources have exceeded 5.6 billion cubic meters, and the groundwater resources, calculated without double counting, have reached 470 million cubic meters—all of which are used to irrigate 0.7 million hectares of farmlands, 1.1 million hectares of woodlands, and 8 million hectares of grasslands in the Hexi Corridor and Ejina, a banner in the far west of Inner Mongolia, and to ensure the supply of water to thousands of industrial and mining enterprises, such as Jinchuan Group and Jiuquan Iron and Steel (Group), and the Jiuquan Satellite Launch Center. All in all, it's very important to protect the glaciers and ecological environment in the Qilian Mountains, improve the forests' capabilities of conserving water, and conserve the Shiyang River, Ruo Shui, and the Shule River, along with the Yellow River. It's also vital to curb the merging of the Tengger, Badain Jaran and Kumtag Deserts, protect biodiversity, optimize the regional ecological environment, and promote social harmony along the Hexi Corridor. All of these can ensure ecological security and smooth land transportation across China, keep China's energy strategic channel unblocked, achieve national unity, prosperity and development, and stabilize China's borders.

2. **The overall issues faced by ecosystems of the Qilian Mountains**

Due to global warming and excessive human activities, the Qilian Mountains is becoming arid, with some areas seeing retreated glaciers and snow lines, vegetation degradation, increased desertification, less run-offs from the mountainous region, and declined biodiversity—environmental degradation that needs to be reversed. The ecological environments of the Qilian Mountains and the Hexi Corridor are now in a state characterized by partial improvement yet overall deterioration, with restoration and conservation far behind destruction and degradation, thus revealing an apparent conflict between economic development and ecological protection.

First, the climate is becoming warmer, snow lines retreating, and glaciers shrinking. With global warming, the Qilian Mountains has seen greatly retreated snow lines and glaciers, grassland and forest degradation, and reduced rare, precious wildlife. According to data from the Gansu Meteorological Bureau, over the past five decades, the average annual temperature in the Qilian Mountains has been rising at rate above the national level. From 1956 to 2013, glaciers and ice reserves in the Hexi Inner Flow Region decreased by 12.6% and 11.5% respectively. Moreover, glacier shrinkage, from the mid-1980s onwards, has been the worst since 1956. What's worse, while the glaciers are decreasing, they are becoming thinner, with an average thinning of 5–20 meters, and the fluctuation of the snow line reaches 100–140 m as well. Since 1999, the amount of snow and ice melting in June and July has increased sharply, and the snow line keeps moving upwards. In recent years, the snow line in the Qilian Mountains has been rising at an average annual rate of 2–6 meters, and the snow line in some areas has been rising even by 12.5–22.5 m per year.

Second, the edge of forests is moving up, grasslands are degrading, and soil erosion is becoming more and more serious. Many people live in the Qilian Mountains, or even at the heart of the reserve. In the national reserve alone, there are 35 villages and towns with 142,000 people, and more than 10,000 households of over 39,000 people living in the core area and buffer zones all year round. The problem of "forestry right and grassland right certificates" issued for one place at the same time is pretty common, with apparent conflicts between agriculture and animal husbandry, as well as forestry and animal husbandry. Moreover, deforestation, land reclamation, and overgrazing are banned but still present. In recent years, hydropower and mineral resources have been rapidly tapped into in the Qilian Mountains, resulting in indiscriminate mining and digging; tourism and road construction have increased sharply, making it impossible to stop deforestation and grassland destruction. The conflict between tapping into resources and conserving the ecological environment is the main culprit for the deterioration of the Qilian Mountains. Meanwhile, the forests and grasslands are extremely fragile and susceptible and thus are very difficult to restore. Years of efforts have been made to curb ecological degradation, but the trend of "partial improvement yet overall deterioration" is yet to be reversed. Restoration is obviously way behind degradation. The "three challenges"—grassland degradation, desertization and salinization—are intensifying, resulting in increased bare lands, deteriorating and accelerating soil erosion, and decreased capacities of water conservation.

Third, the increasing population and social progress have caused water shortages. Since the reform and opening up, the Hexi Corridor has seen substantial economic and social development contributing to a greatly increased economic volume and population. With a decline in water conservation capacities of the Qilian Mountains and the continued decrease in the total amount of water, water shortages are becoming more and more serious. The per capita water resources along the Hexi Corridor is 1,270 cubic meters, representing one-half of the national average, and the per *mu* of arable land is 475 cubic meters, or one-third of the national average. There exists an apparent conflict between supply and demand of water, and the way water is used is poorly arranged with low efficiency. There is intense competition of water resources between upstream and downstream areas, the left and right banks of rivers, cities and rural areas, industry and agriculture, and industry/agriculture and ecology. What's more, over-exploitation of groundwater has severely depleted water resources—for example, water utilization in Ruo Shui and the Shiyang River reaches as high as 106% and 142% respectively. Over-exploitation has also depleted springs in the plains in front of the mountains, made lakes and wetlands smaller, caused declining groundwater levels, and resulted in deteriorating water quality as well as vegetation degradation, which led to a decrease in groundwater that aggravated vegetation degradation, creating a vicious circle. Eventually, desertification, sandification, and salinization are speeding up in the northern part of the Corridor, one of the main areas that give birth to sandstorms in China.

Fourth, a non-smooth mechanism and insufficient investment make it difficult to conserve the environment. The ecological protection and construction of the Qilian Mountains mainly rely on various nature reserves, where there exist contradictions between the management agencies and local governments, industrial and mining enterprises and farmers and herdsmen in terms of how to seek balance between environmental protection and economic development. Take the Qilian Mountains national nature reserve as an example. The reserve's ecological protection and construction task involves more than 10 ethnic groups of 630,000 people in 71 towns and villages, in 3 cities and 8 counties along the Hexi Corridor within Gansu and Qinghai provinces. The reserve management bureau spends a lot of time and energy on coordination, which has significantly reduced their time for doing routine work, because its subordinate protection stations and forest police stations share the same staff with local state-owned forest farms—a poor-designed, dual management mode in that the same people are required to carry out measures from the reserve management bureau and local governments at the same time. What's worse, the expenditure of the 1,500-plus management and protection personnel are covered by local governments, nearly half of which are monitoring expenditure against income. These employees have a poor working condition with a low salary. Without a decent job, they tend to live in hardship and cannot keep their mind on work. Therefore, the ecological protection and construction of the Qilian Mountains is in an undesirable situation of a serious shortage of investment, an understaffed workforce for extremely difficult tasks, outmoded infrastructure, and slow progress in vegetation restoration.

Fifth, unclear responsibilities and unbalanced interests make ecological compensation difficult to implement. When it comes to the ecological protection and construction of the Qilian Mountains, various nature reserves, upstream local governments and the residents have made great contributions, even paying a high price for losing opportunities for development—a situation characterized by unclear responsibilities and unbalanced interests, namely "management by a few, yet benefits for others; upstream areas bearing the burden, but downstream ones benefiting." The *Regulations on the Management of the Qilian Mountains National Nature Reserve in Gansu Province* stipulated that "3% of the total water resource fee collected from the areas benefiting from the Qilian Mountains' forests conserving water shall be used to protect and nurture the forests conserving water in the protected areas" and the policy that in areas with slopes of over 25 degrees, farmlands shall be returned to forests and grazing lands to grasslands has not been implemented. Especially in the catchments of Ruo Shui, an unfair system that transfers a large amount of water from the Ruo Shui administration bureau to the lower reaches of Ejina has severely limited water available for industrial development and agricultural production in the middle and upper reaches, and affected the economic and social progress of the Zhangye oasis. Zhangye City is implementing the most stringent water-saving system that ensures discharging water into downstream areas by limiting its use of water for developing agriculture, industry and ecological conservation. However, precious water resources are being wasted in Ejina, a banner with a lot of lakes and rivers. The ecological compensation system promoted by the central government is yet to be implemented. With this kind of water resources management mechanism and ecological protection mechanism at the expense of the Hexi Corridor, the residents living in upstream areas are very unwilling to protect the ecological environment.

3 Ensuring Green Development by Clarifying Thinking, Planning Scientifically, and Coordinating Construction Efforts

Gansu's ecological construction and environmental protection are vital to building an important ecological barrier in China, and are also the basis for reducing the gap between Gansu and the rest of the country in terms of pursuing a moderately prosperous society while ensuring its green and sustainable development. It is necessary to increase the government's investment and adopt reform and innovation to explore new mechanisms for building ecological civilization. Moreover, we should work hard to build a coordinated linkage mechanism for ecological protection, economic development and improvement of people's livelihood, a long-term ecological compensation mechanism and a multi-investment and financing mechanism. We should also give full play to the market, get social entities involved in ecological protection and construction, and create a "triple win" situation where ecological protection, economic progress, and improved livelihood for the people are achieved simultaneously. In addition, it is necessary to establish a proper guidance and evaluation mechanism for promoting scientific development, implement the most stringent source protection system, strictly observe the red line of ecological protection, and build a red line for environmental protection.

1. **Gansu should pay attention to the following four points while formulating strategies**

The first is to strive for development while giving priority to environmental protection. Environmental capacity should be regarded as an important basis for regional arrangements, the environmental standards deemed a vital condition for market access, environmental costs seen as an important factor in the price formation mechanism, environmental management regarded as an important way of adjusting regional economic structure and optimizing industrial upgrading, and environmental safety seen as an important guarantee for maintaining social stability, achieving harmony in society, and advancing green development.

The second is to work hard to develop a circular economy and green industries. It is necessary to construct 7 circular economical bases, nurture 16 circular economic and industrial chains, and strive to form circular economic and industrial clusters. What's more, we should promote a loop-type production method and accelerate the formation of a loop-type system entailing industry, agriculture, and service industry. Finally, we should closely follow the Belt and Road Initiative, develop green industries based on clean energy, characteristic agriculture, ecological and cultural red tourism, modern logistics and service industries, in line with local resource advantages and actual conditions.

The third is to implement a strategy focusing on environmental protection and resource conservation. For one thing, it is necessary to identify the key points of ecological construction and environmental protection based on the climatic characteristics and ecological conditions of different regions in Gansu, and the main functional area planning. The west focuses on controlling wind and sand hazards and speeding up ecological restoration, the central region on comprehensive management and increasing vegetation coverage, the east on curbing soil erosion and developing ecological industries, and the south on strengthening ecological restoration and developing industries with local characteristics. For another, it is necessary to adhere to the marketization of resource development, the industrialization of resource application, the maximization of resource benefits, and the continuity of resource utilization, abandon sloppily expansion, and strengthen concentrated mining.

The fourth is to develop industries around water conservation. It is necessary to adopt the most stringent system of managing water resources, and rationally allocate water resources to agriculture, industries, and ecological conservation according to "uniform discharging, quota management, paid use, and market regulation" for surface water and groundwater to ensure the basic water volume for maintaining the ecological environment of regions and river basins.

2. **Gansu should focus on the following four aspects in promoting ecological construction and protection**

The first is to work harder to conserve water in the Qilian Mountains. Gansu should also speed up its ecological conservation plan to protect its forests, grasslands, wetlands, deserts and other ecosystems, as well as wildlife and foliage to help protect the nature reserve.

The second is to accelerate the comprehensive management of the catchments of the Shiyang River, Ruo Shui, and the Shule River. It is necessary to adopt a project for controlling sand and restoring ecology in areas along the Shiyang River, strengthen the construction of the Ruo Shui wetland nature reserve, and rationally use water resources, and advance ecological protection in Dunhuang. Gansu should actively facilitate a project for controlling desertization and salinization in the three river systems, and accelerate the establishment of sand-controlling shelterbelts at the north of the Hexi Corridor.

The third is to restore and protect the ecological environment that conserves water in Gannan. It is necessary to speed up ecological protection and construction projects for the important water supply ecological function area of the Yellow River within Gannan, strengthen the protection of wetlands in the area, restore water conservation, and do a good job in controlling soil erosion and geological disaster prevention in the basin of the Bailong River.

The fourth is to advance the comprehensive management of soil erosion in the Loess Plateau and the mountains in southern Gansu. It is necessary to steadily implement the comprehensive management plan for the Loess Plateau, actively carry out key ecological restoration projects for the middle and upper reaches of the Yellow River, and comprehensively manage small and medium-sized rivers like the Wei, Jing and Tao Rivers, in order to conserve the sources of the rivers.

3. **Gansu should fulfil the following four tasks to ensure green development**

The first is to ensure environmental protection. It is necessary to open up green channels for projects that meet the requirements of industrial policies and environmental protection—in particular, projects benefiting the people's livelihood, infrastructure and poverty alleviation projects, the "3341" engineering projects, and the construction projects for the Chinese Civilization Inheritance & Innovation Zone and the National Ecological Security Barrier Comprehensive Experimental Area. However, Gansu must strictly restrict and prohibit projects for industries of high energy consumption, high emissions, and overcapacity.

The second is to vigorously promote energy conservation and emission reduction. It is necessary to coordinate the comprehensive prevention and control of pollutants from key industries, pollution factors, river basins, and regions, and resolutely curb the excessive growth of high energy-consuming and high-polluting industries, focus on energy conservation in industries, construction and transportation fields, and public institutions by developing wind energy, solar energy, photovoltaic power generation, hydropower, and other clean, new energy resources.

The third is to improve comprehensive environmental management. It is necessary to incorporate green development into the overall arrangement of economic and social development, fully adopt action plans for managing air, water, and soil pollution, win the three major battles to build a moderately prosperous society in all respects, take environmental protection supervision responsibilities, implement a pollution discharge permit system, achieve the normalization of environmental protection in businesses, strengthen the five basic measures of environmental law enforcement and disclosure of environmental information, and effectively strengthen the comprehensive environmental management.

The fourth is to promote the development of environmental protection industries. It is necessary to actively promote the industrialization of new materials such as non-ferrous metal alloy materials, rare-earth materials, new chemical materials, and battery materials, and accelerate the development of new pillar industries like biomedicine, processing of Chinese medicinal materials, and production of characteristic Tibetan medicine. Moreover, it is necessary to develop the energy-saving and environmental protection equipment manufacturing industry, promote high-efficiency and energy-saving products, and enhance the competitiveness and technological innovation capabilities of the energy-saving and environmental protection equipment industry.

Afterword

At the 70th anniversary of the founding of the People's Republic of China, the compilation of *Beautiful China 1949–2019: 70 People's Views on Ecological Civilization*—a collection of theories and facts co-sponsored by the Think tank for Eco-civilization Studies of Chinese Academy of Social Sciences (CASS), the Research Institute of Resources and Environment Policies of the Development Research Center of the State Council, China Ecological Civilization Research and Promotion Association, Chinese Academy of Environmental Planning, and China Environment Publishing Group—has come to an end. It is an impressive corpus of theories celebrating the 70th anniversary of the founding of the People's Republic of China, a heavy responsibility assumed by think tanks on ecological civilization and those committed to building ecological civilization, who have worked together to provide innovative input highlighting this historic achievement.

This book is based on the history of building China's ecological civilization over the past seven decades and guided by Xi Jinping Thought on Ecological Civilization, covering a variety of themes on the basic theories of ecological civilization and the ecological cultural system, green development and the ecological industry, deepening the reform of the ecological civilization system, and the construction of global ecological civilization and ecological security system. With concentrated, comprehensive and systematic topics, this pastiche offers an extensive and in-depth analysis of China's development of ecological civilization over the past 70 years. It teases out the twists and turns towards the modernization of China's national governance system and governance capabilities, and discusses the theories, implementation, experience, insights, challenges and prospects of the scientific thesis that "lucid water and lush mountains are invaluable assets" in a new era of socialist ecological civilization. Moreover, it highlights and demonstrates the all-round and groundbreaking historical achievements and fundamental changes that have taken place in China's development of ecological civilization under the guidance of Xi Jinping Thought on Ecological Civilization since the 18th National Congress of the CPC. It inspires a lot of workers for building ecological civilization to unite more closely around the Party Central Committee with Comrade Xi Jinping at the core in a broader time and perspective, and also encourages them to remain true to their

J. Pan et al. (eds.), *Beautiful China: 70 Years Since 1949 and 70 People's Views on Eco-civilization Construction*, https://doi.org/10.1007/978-981-33-6742-5

original aspiration and keep their mission firmly in mind, turn their patriotic spirit into practical actions, and in the new historical journey, strive for a decisive victory in constructing a moderately prosperous society in all respects, winning the battle against pollution, and building China into a great modern socialist country that is prosperous, strong, democratic, culturally advanced, harmonious and beautiful.

As the Chinese saying goes, the last leg of a journey marks the halfway point. With joint efforts by all the organizers, the event and publication of the book completed on schedule. However, at the 70th anniversary of the founding of the People's Republic of China, a "historic moment of great significance," it is difficult to form an "impressive literary topography" that roughly reflects China's development of ecological civilization over the last seven decades, as well as its great achievements and valuable experiences. In short, it leaves a lot to be desired for two reasons. First, to be honest, given the fact that the Party's, and China's, development of ecological civilization is facing severe challenges, whether the selected authors' contributions can be carried out or truly reflect the actual situations is still up in the air and needs to stand the test of time. This book merely represents the voice, opinions and standpoints of the theoretical workers and builders of ecological civilization. By contrast, in a new era of socialist ecological civilization, Xi Jinping Thought on Ecological Civilization has become widely known, and building ecological civilization is highly complicated, interdisciplinary, and systematic, covering all aspects of economic, political, cultural and social construction. Many theoretical and practical "experts" from all walks of life who care about, support, and pursue the development of ecological civilization are coming out, with the business and industry sectors as the main catalyst for advancing China's ecological civilization. Unfortunately, due to reasons beyond our control, they were not extensively mobilized—a fact that is regrettable for the editor-in-chief team. Second, since the selected articles are generally too long for the limited space available, we simply cannot keep all of their explanatory pictures and statistical tables, and would appreciate their understanding on the matter.

Below is a list of the book's editors (or authors) based on the order of their participation: Pan Jiahua, member of the Chinese Academy of Social Sciences, Executive Vice Chairman of the CASS Think tank for Eco-civilization Studies; Gao Shiji, Director of the Research Institute of Resources and Environment Policies of the Development Research Center of the State Council; Li Qingrui, Executive Vice President of China Ecological Civilization Research and Promotion Association; Wang Jinnan, academician of the Chinese Academy of Engineering and Dean of Chinese Academy of Environmental Planning; Wu Dekai, Secretary of the Party Committee of China Environment Publishing Group; and Huang Chengliang, Director of the Theory Department of the CASS Think tank for Eco-civilization Studies.

Xu Yuhong, the president of the Fourth Branch of the China Environment Publishing Group, and editor Zhao Yan spend much time editing the book, and young researchers and graduate students helped a lot. We would like to thank them for their dedication throughout the project.

Editors
October 1, 2019

Bibliography

https://news.xinhuanet.com/newscenter/2007-10/24/content_6938568.htm.
https://www.clubofrome.org/report/the-limits-to-growth/.
https://www.gov.cn/2011lh/content_1825838.htm.
https://www.scio.gov.cn/xwfbh/xwbfbh/yg/2/Document/1436286/1436286.htm.
https://www.un.org/en/ga/search/view_doc.asp?symbol=A/RES/70/1.
https://www.xj.xinhuanet.com/2012-11/19/c_113722546.htm.

Baker, M., & Wong, C. (2013). The delusion of strategic spatial planning: What's left after the labour government's english regional experiment? [J]. *Planning Practice and Research, 28*(1), 83–103.

Bao, S. (2014). Coordinated development of the new model of urbanization and ecological civilization development. *Qiu Shi*, 859–863.

Can the Hu Line be crossed? [EB/OL]. http://www.cssn.cn/shx/shx_bjtj/201501/t20150116_1481742.shtml.

Cao, M. (2002). Theoretical basis for ecology law. *Chinese Journal of Law, 9*, 98–107.

Cao, G., Yu, F., Qin, C., et al. (2019). Research on China's environmental security and governance strategies [J]. *Environmental Protection, 47*(8), 13–15.

Chen, K. (2011). *From direct control to democratic consultation—A study on the legislation coordination against water pollution and the construction of legal environment in the Yangtze River Basin* (p. 66). Shanghai: Fudan University Press.

Chen, J, & Ma, J. (2017). *Collection of national policies on environmental protection* (2nd ed.). Beijing: National Academy of Governance Press.

Chen, W. (2019, January 1). Zhou Yuan. An extraordinary 2018 for Hainan. *Hainan Daily*.

Cheng, X. *Vigorously promoting the development of the national ecological civilization pilot zone and creating a "Guizhou Model" of beautiful China [OL].* http://news.gzw.net/2018/0705/1290860.shtml.

"China has to cross the 'Environmental Protection Gate' during its rise, environmental incidents come at a high price." Published online at https://news.sohu.com/20090828/n266295203.shtml [cited August 28, 2009].

Chinese Society for Environmental Sciences. (2012). *Collection of academic papers on ecological civilization*. Beijing: China Environmental Science Press.

Chinese Academy of Social Sciences & Central Archives. (1991). *Economic materials of the People's Republic of China (1949–1952): Agriculture* (p. 741). Beijing: Social Sciences Literature Press.

Chinese Academy of Social Sciences & Central Archives. (1998). *Economic materials of the People's Republic of China (1953–1957): Agriculture* (pp. 921, 924). Beijing: China Price Press.

Chongqing Environmental Protection Bureau. (1997). *Environmental protection history of Chongqing (for internal use)* (p. 3).

J. Pan et al. (eds.), *Beautiful China: 70 Years Since 1949 and 70 People's Views on Eco-civilization Construction*, https://doi.org/10.1007/978-981-33-6742-5

Chu, C., Xu, S., Jiang, G. et al. (2014). Study of natural ecosystem protection in Chinese cities. *Ecological Economy, 30* (9), 162–171.

Collected Works of Marx and Engels. (1995). *(Vol. 1)* (p. 66). Beijing: People's Publishing House.

Commoner, B. (1997). *The closing circle: Nature, man and technology* (p. 37). Changchun: Jilin People's Press.

Complete Works of Marx and Engels. (2001). (Vol. 44, p. 10). Beijing: People's Publishing House.

"Contemporary China" Series Editorial Board. (1985). *Forestry in contemporary China* (p. 84). Beijing: China Social Sciences Press.

Dai, Y., Kang, Y., & Xiong, X. (2017). *China's energy and carbon emission scenarios and energy transition and low-carbon development roadmap.* Beijing: China Environment Press.

Department of Treaties and International Law of the Ministry of Foreign Affairs. (2019a). Voice of China: Legal viewpoints on "towards a global pact for the environment." *Frontiers of International Law in China.*

Department of Treaties and International Law of the Ministry of Foreign Affairs. (2019b). Discussion of the international community about environmental governance: A review of the Fourth United Nations Conference on the Human Environment. *Frontiers of International Law in China.*

Ding, Y. (2018). Hainan in its thirties: Upgrade itself for developing an ecological civilization. *Environment Economy, 8,* 12–13.

Documents of the 19th National Congress of the Communist Party of China. Beijing: People's Publishing House, 2017.

Dong, J., Libin, W., Jing, G., et al. (2017, August 3). Creating a new situation of ecological civilization—a record of CPC central committee with Xi Jinping at the core leading ecological progress since the 18th CPC national congress. *The Economic Daily.*

Drucker, P., F.,. (1993). *Post-capitalist society.* Butterworth-Heinemann Ltd. 2–3, 5, 166.

Du, Y., & Bao, C. (2018, March 27). Advancing the building of a beautiful countryside. *Jiefang Daily.*

During the 23rd collective study session of the Politburo of the CPC Central Committee, Xi Jinping emphasizes the importance of promoting the concept of safety-centered development and ensuring personal and property safety of citizens. (2015, May 31). *People's Daily,* (01).

Editorial board of *Qiu Shi.* (2019). Walking to a New Realm of Ecological Progress in the New Era under the Guidance of Xi Jinping Thought on Ecological Civilization. *Qiu Shi* (3).

Engels. (1971). *Dialectics of nature* [M].// Central Compilation & Translation Bureau of Works by Marx, Engels, Lenin and Stalin. *Complete works of marx and engels (Vol. 20)* [M] (p. 519). Beijing: People's Publishing House.

Environmental Protection Bureau of Qiqihar City. (1989). *Qiqihar city's environmental protection history* (pp. 96, 99). Harbin: Heilongjiang Science and Technology Press.

Environmental and Economic Policy Research Center of the Ministry of Ecology and Environment. (2018). Transition of role from participant to contributor to leader—Review of environmental protection actions in China. *Ziguangge, 11,* 49.

"Environmental inspection by central authorities covered all 31 Provinces in 2016 and 2017." Published online at https://www.xinhuanet.com/2017-11/07/c_1121916536.htm [cited November 7, 2017].

"Environmental protection industry may register annual growth rate of more than 20% during the 13th five-year period, total investment to reach RMB 17 trillion." Published online at https://finance.sina.com.cn/china/20151102/222723655477.shtml [cited November 2, 2015].

"Environmental Protection History of Hubei Province" Editorial Board. (1989). *Environmental protection history of Hubei Province* (pp. 76–77). Beijing: China Environmental Science Press.

"Environmental Protection History of Shanghai" Editorial Board. (1998). *Environmental protection history of Shanghai* (p. 26). Shanghai: Shanghai Academy of Social Sciences Press.

Establish the framework of the system for developing an ecological civilization as soon as possible. *China Youth Daily* (2016, December 3).

Fan, N., Zhang, Y., Lv, Y. et al. (2014). Status Quo of ecosystem protection and evaluation of protection level—A case study of Jiangxi Province. *Acta Ecologica Sinica, 34* (12), 3341–3349.

Fang, S. (2016). *Ecological civilization, innovation economy* (pp. 220–227). Beijing: China Environment Publishing House.
Fang, S., et al. Study of Innovative Economic Development Path in Socialist Ecological Civilization (for internal use): 219.
Fetscher, I. (1978). Conditions for the survival of humanity: On the dialectics of progress. *Universitas, 20*(3), 161–172.
Floridi, L. (2014). *The fourth revolution: How the infosphere is reshaping human reality*. Oxford University Press.
Floridi, L. (2015). The Onlife manifesto, Being human in a hyperconnected era. Springer Open.
Friends of Nature. (2001). *Environmental alarm bells in the 20th century*. Beijing: Huaxia Publishing House.
Fu, H. (2002). *An inquiry into ecological ethics*. Beijing: Huaxia Publishing House.
Fujian: Report on the reform of the national ecological civilization pilot zone [EB/OL]. http://politics.rmlt.com.cn/2018/1010/529956.shtml.
Gare, A. (2017). *The philosophical foundations of ecological civilization: A manifesto for the future*. London and New York: Routledge.
General Secretary (Sep 27, 2015) Xi Jinping attended the United Nations development summit and delivered an important speech: Seeking common and sustainable development for win–win cooperation [EB/OL]. Xinhuanet. https://www.xinhuanet.com.
General Secretary (Sep 28, 2015) Xi Jinping attended the 70th United Nations general assembly at the United Nations headquarter in New York and delivered an important speech at the general debate: Working together for win–win cooperation and build a community with a shared future for mankind [EB/OL]. Xinhuanet. https://www.xinhuanet.com.
General Office of the Central Committee of the Communist Party of China. (1971, April 1). *Minutes of the 1971 national planning meeting*. Archives of Hebei Province, file number 940-011-0017-001.
Gillies, M., Perkins, D. H., Roemer, M., et al. (1989). *Economics of development*. Beijing: Economic Science Press.
Gu, M. (1998). Premier Zhou is the founder of environmental movement in China. In Q. Li (Ed.), *The days with Zhou Enlai—Memories of the staff of the Xihua Hall* (p. 332). Beijing: Central Party Literature Press.
Guiding Committee for the Compilation and Review of Materials for Training of Government Officials Nationwide. *Promoting Ecological Civilization, Building a Beautiful China*. Beijing: People's Publishing House, Party Building Books Publishing House, 2019.
Guizhou's Eight Institutional Innovations and Experiments for Developing an Ecological Civilization: Green is Gold. *China Economic Herald*, 2018, July 16.
Haken, H. (2018). *Secrets of nature: Synergetics*. Translated by Ling Fuhua. Shanghai: Shanghai Translation Publishing House.
Han, D., Liu, J., & Jin, S. (2019). Retrospect and prospect of China's environmental protection policies for agriculture and rural areas over the last 40 years. *Environment and Sustainable Development, 44*(2), 16–21.
Han, Q., et al. (2018). *Understanding the New Era* [M] (pp. 173, 183–184). Beijing: China Fangzheng Press.
Han, M., & Yang, P. (2018). The application and perfection of preventive liability in environmental public interest litigation. *Economic Research Guide, 34*.
Hawken, P. (1983). *The next economy*. New York: Holt, Rinehart and Winston 78.
He, J. (2014). The revolution of the new energy system is the only way to ecological civilization. *Journal of China University of Geosciences, 2*, 1–10. (Social Science Edition).
He, J. (2018). New thinking on long-term strategies for climate change and low-carbon development in a new era. *Wuhan University Journal, 4*, 13–21. (Social Sciences Edition).
Heilmann, S., Melton, O., & Shi, L. (2013). Planning: The core mechanism for China's policy-making. *Open Times, 2013*(6), 8–31.

Herman, E. D. & John B., C., Jr. (2015). *For the common good: Redirecting the economy toward community, the environment, and a sustainable future*. Beijing: Central Compilation & Translation Press.

Hong, D. (2014). Growth and limitations of public environmental awareness. *Green Leaf* (4).

Hong, D. (2016). The chinese path of composite environmental governance. *Journal of the Party School of the Central Committee of the C.P.C* (3).

Hu, A. (2011). China's green development and its 12th five-year plan. *Farm Economic Management, 2011*(4), 10–21.

Hu, J. (2012). *Firmly march on the path of socialism with chinese characteristics and strive to complete the building of a moderately prosperous society in all respects—Report to the eighteenth national congress of the communist party of china (offprint)*. Beijing: People's Publishing House.

Hu, A. (2012). Important approaches to green development in China. *China Environment News, 2012*(5), 2.

Hu, A., & Honghua, M. (2005). Green development and green rising: On China's development path. *Journal of the Party School of Tianjin Committee of the CPC, 2005*(1), 19–30.

Hu, A., & Zhou, S. (2014). Green development: Functional definition, mechanism analysis and development strategy. *China Population, Resources and Environment, 24*(1), 14–20.

Huang, S., & Lin, S. (1986). *Healthcare in contemporary China (part 1)* (p. 108). Beijing: China Social Sciences Press.

Huang, D., Wang, Y., & Kang, X. (1997). *Man and nature series—An introduction to the relationship between man and nature*. Wuhan: Hubei Science & Technology Press.

Huang, Z., Wang, Y., Zhang, C., et al. (2017). A study of the comprehensive reform strategy for environmental protection and economic development in the upper reaches of the Yangtze River. *Journal of Lake Sciences, 29*(2), 257–265.

Huang, C. L., & Yu, M.-C. (2010). *Ecological civilization: A comprehensive transformation of human society*. Beijing: Central Party School Press.

Huang, Y. Promote the deep integration of industrial development and ecological resources [EB/OL]. http://news.cssn.cn/zx/bwyc/201811/t20181120_4778530_1.shtml.

Huang, X. (2018, August 24). Colorful Guizhou: flowers blooming in the experimental field. *People's Daily*.

Huo, X. (2017). What Can We Learn from the Principle of Lucid Waters and Lush Mountains Being Invaluable Assets Proposed by Xi Jinping? [EB/OL]. (2017-08-09). [2015-08-06]. https://news.xinhuanet.com/politics/2015-08/06/c_1116159476.htm.

Huo, T., et al. (2012, June 12). Attendees talking about the 1992 United Nations Conference on Environment and Development. *China Environment News*.

Intergovernmental Panel on Climate Change (IPCC). (2018, May 18). *The synthesis report of the fifth assessment report of the IPCC*. Intergovernmental Panel on Climate Change. http://ar5-syr.ipcc.ch/ipcc/ipcc/resources/pdf/IPCC_SynthesisReport.pdf.

Interview with Hu Weilie, Director of the Eighth Procuratorate Department of the Supreme People's Procuratorate: Summarize experience and better perform public interest litigation duties. (2019, February 28). *Jiancha Ribao Newspaper*.

Janicke, M., Monch, H., Ranneberg, T., & Simonis, U. E. (1988). Economics structure and environmental impacts. *Environmental Monitoring and Assessment*.

Jiang, W., Qin, C., Wang, Q., et al. (2015). Why should we pay attention to both quantity and quality in fine management: International experience in environmental quality management. *Environmental Economics, 2015*(8), 16–17.

Jiang, B. (2019). The development in practice and institutional improvement of environmental public interest litigation in China. *Journal of Law Application, 1*.

Jiao, J. (2006). Ecological civilization: Basic patterns of modern civilization. *Jiangsu Social Sciences, 1*, 74–78.

Jin, R. (2002). *Introduction to environmental law* (p. 11). Beijing: Contemporary World Press.

Jinping, Xi. (2014). *The Governance of China* (p. 211). Beijing: Foreign Languages Press.

Jintao, H. (2017, October 25). Hold high the great banner of socialism with Chinese characteristics and strive for new victories in building a moderately prosperous society in all respects—Speech by Hu Jintao to the 17th National Congress of the CPC. *People's Daily*.
Karlsson, C., Hjerpe, M., & Parker, C. (2012). The legitimacy of leadership in international climate change negotiations [J]. *Ambio, 41*, 46–55.
Le, X., Luan, S., & Wan, J. (2003). On the innovation of China's policies for rural environmental protection. *Chinese Journal of Environmental Management, 3*, 1–4.
Lenin, V. (1990). *Collected works of Lenin*. Beijing: People's Publishing House.
Leopold, A. (1987). *A sand county almanac: And sketches here and there*. Oxford University Press.
Li, H. (2015). *Marxist Ecological View and Contemporary Chinese Practice [M]*. Beijing: People's Publishing House.
Li, Q. (2016). Green lifestyle: Interpretation from the perspective of political science [J]. *Journal of Nanjing Normal University, 5*, 45.
Li, G. (2017). Strengthen environment protection and build a "beautiful China." *Environment, 12*, 14–16.
Li, S. (2017). *Interpretation of the general principles of the civil law of the People's Republic of China*. Beijing: Law Press.
Li, A., & Zhou, S. (2018). Development of environmental law in China: Review and outlook. *Environmental Protection, 20*, 28.
Li, S. (1990). On ecological awareness and ecological civilization. *Journal of Southwest Minzu University (Philosophy and Social Sciences)*, (2).
Li, H. The only solution for the problems facing China economic and social development [N]. Guangming Daily. May 18, 2016.
Li, G. (2019, June 18) Furthering the development of the central inspection system on ecological and environmental protection in accordance with law. *People's Daily*.
Literature Research Office of the Central Committee of the Communist Party of China. (2007). *Biography of Zhou Enlai (1949–1976) (part 2)* (p. 375). Beijing: Central Party Literature Press.
Literature Research Office of the CPC Central Committee. (2017). *Excerpts from Xi Jinping's Statement on the Building of Socialist Ecological Civilization [M]*. Beijing: Central Party Literature Press.
Liu, P. (1996). Sustainable development theory and China's Agenda 21. *Frontiers of Earth Science, 1*, 1–9.
Liu, X. (1999). *On ecological civilization* (p. 30). Changsha: Hunan Education Publishing House.
Liu, S., et al. (2002). *Sustainable development of corporate economy* (p. 170). Beijing: China Environmental Science Press.
Liu, S. (2011). The green development: A perspective of scientific outlook on development. *Contemporary Economic Research, 2011*(5), 65–70.
Liu, G., & Yue, W. (2013). Education's basic role in building ecological civilization. *Educational Research, 407*(12), 10–17.
Liu, Z. (2016). China's contribution to global climate governance [J]. *Seeking Truth*, (7).
Liu, Y. Reinterpreting the declaration of the united nations conference on the human environment: A valuable effort to construct china's environmental discourse system [N]. China Environment News. 14 January 2015.
Liu, C. (2010). Rethinking the logic behind environmental courts—A case study of Guiyang environmental tribunal and Qingzhen environmental tribunal. *Law Review, 1*.
Liu, J. (2018, December 16). *The closing ceremony of the United Nations Climate Change Conference and the proposal put forward by the Chinese delegation for the Paris Agreement* [EB/OL] (The Paper).
"Local Chronicle of Beijing" Editorial Board. (2004). *Local chronicle, municipal affairs chronicle, and environmental protection chronicle of Beijing* (pp. 121, 184, 255, 262, 272). Beijing: Beijing Publishing Group.
"Local Chronicle of Fujian Province" Editorial Board. (2008). *Chronicle of Fujian Province: Environmental protection* (p. 132). Fuzhou: Fujian People's Publishing House.

"Local Chronicle of Harbin" Editorial Board. (1998). *Chronicle of Harbin: Environmental protection supervision* (p. 98). Harbin: Heilongjiang People's Publishing House.
Lu, S. (2006). *Space for ecological criticism* (p. 73). Shanghai: East China Normal University Press.
Luo, J., & Wang, Y. (2013). PM 2.5 pollution control pathway and green and low-carbon development. In J. Xue & Z. Zhao (Eds.), *Report of low-carbon economic development in China* (pp. 34–52). Beijing: Social Sciences Literature Press.
Lv, S. (2008). Environmental ethics of ecological civilization. *Journal ofNanjing Forestry University (Humanities and Social Sciences Edition), 9,* 139–144.
Lv, Z., et al. (2017). *Judicial specialization in environmental law: Status quo investigation and system restructuring.* Beijing: Law Press.
Lv, Z. (2018). Looking for the new jurisprudence of legislation for the Yangtze River Basin—From the perspective of methodology. *Chinese Journal of Law, 6,* 68.
Lv, Z., & Chen, H. (2016). On legislation for the Yangtze River. *Environmental Protection, 18,* 33.
Lv, Z. (2014a). On legal mechanism for decision-making of matters concerning the development of ecological civilization. *China Legal Science, 3.*
Lv, Z. (2014b). The past and present of environmental protection law. *Journal of Political Science and Law, 5.*
Lv, Z. (2015, October 8). Establish a public participation system that is consistent on substantive and procedural fronts. *China Environment News.*
Lv, Z. (2017, October 19). The "green" footprint of environmental rule of law. *Newspaper of People's Political Consultative Conference.*
Lv, Z., & Liu, C. (2019). Two-year observation of judicial specialization in environmental justice. *Applied Jurisprudence in China, 2.*
Lv, Z., & Zhang, Z. (2017). Status quo of judicial specialization in environmental law and typification of environmental cases. *Applied Jurisprudence in China, 6.*
Lv, Z. (2008). Where is the path for the return of environmental law? Rethinking the relationship between environmental law and traditional departmental law. *Tsinghua University Law Journal, 5.*
Lv, Z. (2016). Environmental judicial rationality must go beyond astronamical amount of compensation: An analysis of Taizhou environmental public interest litigation case. *China Legal Science, 3.*
Lv, Z. (2018). Reflections on the research of environmental law in the new era. *Journal of China University of Political Science and Law, 4.*
Lv, Z., & Liu, C. (2019). Innovative development of specialized and professional environmental justice: 2017–2018 observations. *China Review of Administration of Justice, 2.*
Lv Zhongmei's interpretation of soil pollution prevention and control law: Solving the pollution problem by legal means [EB/OL] (2018, January 9). (The Paper).
Ma, K. (2005). Implementing scientific outlook on development, strongly promoting the development of circular economy—Speech at the national circular economy work meeting. Published online at https://www.sdpc.gov.cn/hjbh/fzxhjj/t20050914_45796.htm [cited December 20, 2012].
Marx, E., L. & Stalin. (1995). Central compilation & translation bureau of works. In *Collected works of marx and engels (Vol. 4) (2nd Edition)* [M] (p. 224, 705). Beijing: People's Publishing House.
Marx, K. (1972a). *Marx and Engels collected works* (Vol. II). Beijing: People's Publishing House.
Marx, K. (1972b). *Marx and Engels collected works* (Vol. III). Beijing: People's Publishing House.
Marx, K. (1972c). *Marx and Engels collected works* (Vol. IV). Beijing: People's Publishing House.
Marx, K. (1995). *Marx and Engels collected works* (Vol. I). Beijing: People's Publishing House.
Marx. (2009). *A critique of political economy (Manuscripts of 1857–1858).//* Marx and Engels. *Collected works of marx and engels (Vol. 8).* Compiled and translated by Central Compilation and Translation Bureau. Beijing: People's Publishing House.
Marx. (1979). *Economic & philosophic manuscripts of 1844* (p. 49). Translated by Liu Pikun. Beijing: People's Publishing House.

Marx and Engels. (1956). *The German ideology* [M].// Central Compilation & Translation Bureau of Works by Marx, Engels, Lenin and Stalin. *Complete Works of Marx and Engels (Vol. 3, pp. 23–24) (1st Edition)* [M]. Beijing: People's Publishing House.
Mei, F. (2015). Ecological civilization: A turning point in human civilization. *Ecological Economy., 31*(11), 176–179.
Merchant, C. (1999). *The death of nature* (p. 110). Changchun: Jilin People's Press.
Miao, D., Liao, F., & Qi, sX. (2011). Issues concerning urbanization in China in the context of ecological civilization. *Journal of Fujian Normal University (Philosophy and Social Sciences Edition), 1*, 24.
Miao, C. (2008). Marx and Engels' thoughts on ecological civilization thought and their contemporary meaning. In *A study of the theories of Mao Zhedong and Deng Xiaoping* (Vol. 5, pp. 52–56, 85).
Military Management Commission of the Ministry of Health. (1971, April 27). *Notice on investigation of air and water pollution caused by industrial waste.* Archives of Hubei Province, file number SZ67-04-0157-009.
Military Management Committee of the Ministry of Health. (1971, February 7). *The 4th five-year plan for National Health Service (1971–1975) (draft).* Hunan Provincial Archives, file No. 212-2-21-2.
Military Management Committee of the Ministry of Health. (1971, February 12). *Main points of health work plan in 1971 (draft).* Hunan Provincial Archives, file number 212-2-21-1.
Ministry of Ecology and Environment. (2018). *2017 Bulletin of Environmental Situation in China.* Published online at https://ww.zhb.gov.cn/hjzl/zghjzkgb/lnzghjzkgb/ [cited May 22, 2018].
Ministry of Education. *Catalog of degree granting and talent training programs (updated in April 2018)* [EB/OL]. https://www.moe.gov.cn/s78/A22/xwb_left/moe_833/201804/t20180419_333655.html.
Ministry of Environmental Protection. (2008). Exploration of environmental protection practices with Chinese characteristics—Environmental protection in China in the past of 30 years. *Environmental Protection, 15*, 24.
Monks, F. 2003. (2002). China human development report 2002: Making green development a choice. *China Quarterly, 19*(100), 539–541.
Morin, E. (2002). *Methods: The nature of nature* (p. 205). Beijing: Peking University Press.
Morrison, R. (1995). *Ecological democracy*, (p. 3, 8, 10, 11). Boston: South End Press.
Nadin, V., & Shaw, D. (2010). Transnational collaboration in the Atlantic Region [M]. In S. Duehr, C. Colomb, & V. Nadin (Eds.), *European spatial planning and territorial cooperation* (pp. 22–38). London: Routledge.
Naess, A. (1989). *Ecology, community and lifestyle: Outline of an ecosophy.* Cambridge University Press.
National Conference on Industrial Wastewater Treatment and Comprehensive Utilization (December 1959). Archives of Hubei Province, file number: SZ112-2-64-2.
National People's Congress. *Civil Procedure Law.*
Niu, W. (2013). Theoretical connotations and measurement model of ecological civilization. *Bulletin of Chinese Academy of Sciences, 38*(2), 163–172.
Notice of the State Council on Printing and Distributing the National Plan for Functional Zones [EB/OL]. http://www.gov.cn/zwgk/2011-06/08/content_1879180.htm.
Odum, E. (1981). *Fundamentals of ecology* (p. 261). Beijing: People's Education Press.
Opinions of the CPC Central Committee and the State Council on Unifying the Planning System to Improve the Strategic Guidance Role of National Development Plans (Zhongfa [2018] No. 44).
Opinions of the Central Committee of the Communist Party of China and the State Council on comprehensively strengthening environmental protection and pollution prevention and control. (2018, June 25). *People's Daily*, (01).
Opinions of the State Council of the CPC Central Committee on Comprehensively Strengthening Eco-environmental Protection and Resolutely Implementing Pollution Prevention and Control (June 16, 2018).

Opinions of the central committee of the communist party of china and the state council on accelerating ecological progress. *Guangming Daily*, (02) (2015, May 06).
Ouyang, Z. (2017). Demographic dividend or brain dividend? *Journal of Nanjing Forestry University (Humanities and Social Sciences Edition), 1,* 1–10.
Ouyang, Z. (2019). On three natures and ecological civilization. *Journal of Ningxia Communist Party Institute, 2,* 5–11.
Pan, Y. (2006). On socialist ecological civilization. *Green Leaves, 10,* 10–18.
Pan, J., & Chen, Z. (2016). *The transformation agenda for sustainable development in 2030: Global perspectives and China's experience.* Beijing: Social Sciences Academic Press.
Pan, J., & Li, M. (2018). *Guizhou's experience in developing the national ecological civilization pilot zone* (p. 5). Beijing: Social Sciences Academic Press.
Pan, J., Wang, M., Chao, Q., et al. (2016a). A new paradigm for addressing climate change in the post-paris era: shared responsibility and active action [M]. In W. Wang & G. Zheng (Eds.), *Report on Climate Change* (pp. 1–17). Beijing: Social Sciences Academic Press.
Pan, J., Wang, M., Chao, Q., et al. (2016b). A path to Paris: Change and constancy in the international responsibility system [M]. In W. Wang & G. Zheng (Eds.), *Report on Climate Change* (pp. 1–23). Beijing: Social Sciences Academic Press.
Pan, J. (2017, November 23). *Contributing China's approach to global ecological security. Chinese Social Sciences Today, 01.*
Party Literature Research Center of the CPC Central Committee. (2017). *Compiled statements by Xi Jinping on building socialist ecological civilization.* Beijing: Central Party Literature Press.
Piketty, T. (2014). *Capital in the twenty-first century.* Beijing: China CITIC Press.
Pu, X. (2016, February 23). Give teeth to environmental protection law (revision). *Legal Daily.*
Publicity Department of the CPC Central Committee. (2014a). *Collection of important remarks by General secretary Xi Jinping.* Beijing: Xuexi Publishing House, People's Publishing House.
Publicity Department of the CPC Central Committee. (2014b). *Collection of important remarks by general secretary Xi Jinping.* Beijing: Xuexi Publishing House, People's Publishing House, 121–122.
Qian, S. (2007). *The past, current situation and future of technological revolutions.* Guangzhou: Guangdong Higher Education Press.
Qin, C., Su, J., Wang, Q., et al. (2018a). A study of theories, practices and policies concerning "Green is Gold". *Research of Environmental Sciences, 31*(6), 985–990.
Qin, C., Wan, J., & Wang, Q. (2018b, September 18). Develop a strategic roadmap for the beautiful China campaign. *China Environment News.*
Qingzhi, X. (2016a). Research on ecological civilization demonstration zones from the perspective of triple theories. *Journal of Beijing Administrative College, 1*(1), 17–25.
Qingzhi, Y. (2016b). The path of ecological modernization for developing an ecological civilization. *Yuejiang Academic Journal, 6,* 23–35.
Qingzhi, X. (2017a). Development of ecological civilization demonstration zones from a three-dimensional perspective: Evaluation and prospects. *Journal of China University of Geosciences, 5,* 54–63. (Social Science Edition).
Qingzhi, X. (2017b). The green development path toward developing an ecological civilization: Taking Jiangxi as an example. *Journal of Poyang Lake, 1,* 29–41.
Qingzhi, X. (2019). *Philosophical research on the practice of ecological civilization pilot demonstration zones.* Beijing: China Forestry Publishing House.
Qiu, B. (2015). How to develop a successful top-level design. *Tsinghua Management Review, 4,* 8–13.
Qu, G. (1988). Outline of China's environmental protection cause. *Environmental Protection, 3,* 2–5.
Qu, G. (1989). An environmental protection model with Chinese characteristics—Qu Geping's work report at the Third National Environmental Protection Conference (Abstract). *World Environment, 3,* 13–14.

Qu, G. (2000). *Dreams and expectations: The past and future of environmentalism in China* (p. 50). Beijing: China Environmental Science Press.
Qu, G. (2013). Review of environmental protection in China in the past 40 years—A speech at the "environmental protection in China in the past 40 years" academic forum at the Chinese University of Hong Kong. *Journal of Environmental Management College of China, 23*(4), 1–6.
Qu, G., & Peng, J. (2010). *Awakening of environmental awareness—UN Conference on Human Environment and China's First National Environmental Protection Conference* (pp. 463–470). Beijing: China Environmental Science Press.
Reform Office of the Ministry of Environmental Protection. (2015). *Collection of documents on the reform of the system for developing an ecological civilization, 283*, 202–203.
Ren, Y, & Qiu, S. (2018). *Accelerating the Building of Ecological Civilization System* (13)
Ren, Y. (2018). Seizing the Key of Ecological Structural Reform. People's Daily, 2018-04-12.
Ren, Y. (2018). Theoretical Reflection on Xi Jinping Thought on Ecological Civilization. China Environment News, 2018-05-29.
Report at the 19th National Congress of the CPC. http://www.gov.cn/ zhuanti/2017-10/27/content_5234876. htm. Accessed October 18, 2017.
Response to global climate change is related to the interests and well-being of mankind, and it is a "battle" that all countries must work together to win [EB/OL]. China Youth Network. Nov 29, 2015. https://www.youth.cn.
Rifkin, J. (2009). *The third industrial revolution* (Z. Tiwei & Y. Sun, Trans.). Beijing: CITIC Press.
Rolston, H. (2000). *Philosophy gone wild,* (p. 18). Translated by Liu Er and Ye Ping. Changchun: Jilin People's Press.
Rolston. (1994). *Environmental Ethics: Values in and duties to the natural world.// Foreign philosophies of natural science*. Beijing: China Social Sciences Press, 2907.
Rolston. (2000). *Environmental Ethics: Values in and duties to the natural world* (p. 454). Beijing: China Social Sciences Press.
Roy, M. (2006). *Ecological democracy*. Beijing: China Environmental Science Press.
Schweitzer, A. (1996). *Reverence for life* (p. 91932). Shanghai: Shanghai Academy of Social Sciences Press.
She, Z. (2008). Ecological culture: Public qualities necessary for developing ecological civilization. *Journal of Nanjing Forestry University (Humanities and Social Sciences Edition), 8*(3), 150–158.
Shen, X. (1993). Toward ecological civilization. *Solar Energy*, (3).
Shi, S. (2015). The level evaluation and improvement measures for the development of ecological civilization pilot and demonstration zones: A case study of Fujian Province. *Southeast Academic Research, 5*, 67–73.
Shi, C. (2017). On the ecological restoration in environmental civil public interest litigation: On the rationality of the supreme people's court's judicial interpretation. *Journal of Zhengzhou University, 2* (Philosophy and Social Science Edition).
Shouqiu, C. (2017). *Laws and systems for developing an ecological civilization* (p. 1). Beijing: China Legal Publishing House.
Simon, J. L. (1984). *The economics of population growth*. Beijing: Peking University Press.
Singer, P. (1999). *Animal liberation*. Translated by Meng Xiangsen and Qian Yongxiang. Beijing: Guangming Daily Press.
Song, L. (2007). Theories and practices of ecological civilization. *Social Sciences in Nanjing, 12*, 3–9.
Song, Y. (2010). On community-level "self-organization" of environmental protection in China's countryside. *China Population, Resources and Environment, 1*, 81–86.
State Council. (1972, February 18). *Minutes of the 1972 National Planning Meeting (draft)*. Archives of Hebei Province, file number 924-8-14-1.
"State council information office introduces the 2016 annual report on China's policies and actions for addressing climate change." Published online at https://www.gov.cn/xinwen/2016-11/01/content_5127079.htm.

Sun, Y. (2015). On the establishment of legal system for ecological civilization under the rule of law. *Journal of Southwest Minzu University (Humanities and Social Science*, (5), 85–88.

Sun, N. (Feb 1, 2016). *Global Climate Governance: Can China Be a Leader?* [EB/OL]. New Meteorological Press. https://www.zgqxb.com.cn/zhuant/20160201/2016020101/201602/t20160201_59389.html.

Supreme People's Court. *China Environmental Justice Development Report 2017–2018.*

Supreme People's Court. *Trial of Environmental Cases in China 2017–2018.*

Supreme People's Court. *White Paper on the Trial of Environmental Cases in China 2016–2017.*

Supreme People's Court. (2016). *Opinions on giving full play to the role of adjudication and providing judicial services and support for promoting ecological progress and green development.*

Supreme People's Court. (2018). Opinions on the in-depth study and implementation of Xi Jinping thought on ecological civilization and on providing judicial services and support for environmental protection in the new era.

Ten Years in Promoting China's Green Transformation. Beijing: China Environmental Publishing Group, 2017.

The decision of the CPC central committee on some major issues concerning comprehensively deepening the reform. Beijing: People's Publishing House, 52 (2013).

The editorial committee. (2015). *Eco-culture development in Beijing: Theories and practices* (p. 33). Beijing: China Forestry Press.

The report shows that global renewable energy power generation increased significantly in 2017 [OL]. https://www.gov.cn/xinwen/2018-06/05/content5296397.htm.

The World Commission on Environment and Development (WCED). (1987). *Our common future* (p. 383). Oxford: Oxford University Press.

The Editorial Board. (1994). *Environmental governance in China in the past 20 years* (pp. 4, 16). Beijing: China Environmental Science Press.

The Editorial Board. (1996). *Environmental governance in Nanjing* (pp. 218, 311). Beijing: China Environmental Science Press.

The Literature Research Office of the Central Committee of the Communist Party of China. (1999a). *Collection of important documents since the 14th National Congress of the Communist Party of China (Part 2)* (p. 1995). Beijing: People's Publishing House.

The Literature Research Office of the Central Committee of the Communist Party of China. (1999b). *Collection of important documents since the 16th National Congress of the Communist Party of China (Part 1)* (p. 1995). Beijing: Central Party Literature Press.

The Literature Research Office of the Central Committee of the Communist Party of China. (1999c). *Collected works of Mao Zedong* (Vol. 6, p. 191). Beijing: People's Publishing House.

The Literature Research Office of the Central Committee of the Communist Party of China. (2003). *Excerpts of monographs of Mao Zedong (Part 2)* (p. 1657). Beijing: Central Party Literature Press.

The Literature Research Office of the Central Committee of the Communist Party of China and the State Forestry Administration. (1999). *Zhou Enlai's thoughts on forestry* (pp. 43, 55). Beijing: Central Party Literature Press.

The Roadmap for Green Growth during the "13thFive-Year Plan" Period: Strengthen the Reverse Coercion Mechanism for Environmental Governance [EB/OL]. http://news.bjx.com.cn/html/20151116/681475-3.shtml.

The Propaganda Department of the CPC Central Committee. (2018). *Thirty Lectures on Xi Jinping Thought on Socialism with Chinese Characteristics for a New Era [M].* Beijing: Xuexi Publishing House.

The National Bureau of Statistics. (2019). *China's statistics abstract in 2019.* Beijing: China Statistics Press.

The Paris Agreement, an agreement within the United Nations Framework Convention on Climate Change (UNFCCC). (2015, December 12). https://unfccc.int/resource/docs/2015/cop21/chi/l09c.pdf.

The Sustainable Development Strategy Research Group of the Chinese Academy of Sciences. (2008). *China's sustainable development strategy report 2008—Policy review and prospects*. Beijing: Science Press.

The Sustainable Development Strategy Research Group of the Chinese Academy of Sciences. (2014). *China's sustainable development strategy report 2014—Build an ecological civilization system*. Beijing: Science Press.

The Sustainable Development Strategy Research Group of the Chinese Academy of Sciences. (2015). *China's sustainable development strategy report 2015—Reshape the environmental governance system*. Beijing: Science Press.

The Chinese Government has approved the Kyoto Protocol. (2002). *China Population, Resources and Environment, 4*.

The United Nations Environment Program Released a Report Entitled "Green is Gold: The Strategy and Actions of China's Ecological Civilization". (2016, May 28). *People's Daily*.

The decision of the CPC central committee on some major issues concerning comprehensively deepening the reform. *People's Daily* (2013, November 12).

The master plan for reform of the system for developing an ecological civilization. *People's Daily* (2015, September 21).

Toffler, A. & Toffler, H. (1995). *Creating a new civilization: The politics of the third wave*. Atlanta: Turner Publishing, Inc., 19, 21, 27, 31, 36, 42.

Toynbee, A. (2001). *Mankind and mother earth: A Narrative history of the world* (p. 523). Translated by Xu Bo, et al. Shanghai: Shanghai People's Publishing House.

Tu, F. (2019). Research on the approach for preventing environmental risks from the perspective of ecological civilization [J]. *Social Sciences Review, 34*(3), 60–63.

UNEP. (2016). Green is gold: The strategy and action of China's ecological civilization [R]. https://reliefweb.int/report/china/green-gold-strategy-and-actions-chinas-ecological-civilization.

UNFCCC. (1992). https://unfccc.int/files/essential_background/background_publications_htmlpdf/application/pdf.

Wan, J. (2010). A new perspective on the five-year plan for environmental protection. *World Environment, 6*, 16–17.

Wang, C. (2010). Entities in public interest environmental lawsuits in China and disputes. *Journal of National Prosecutors College, 18*(3), 3–6.

Wang, J. (2011). *Environmental law* (Second ed., pp. 106–107). Beijing: Peking University Press.

Wang, S. (2013). Scientific development of thse new model of urbanization—Connotations, goals and path. *Theory Monthly, 4*, 166.

Wang, C. (2014). Establishment of legal system for building ecological civilization. *China Legal Science, 3*, 34–53.

Wang, J. (2014). The concept of ecological compensation—Based on the legislative explanation for the draft ecological compensation ordinance. *Journal of China University of Geosciences (Social Sciences Edition), 14*(1), 1–8.

Wang, C. (2014). On the construction of the legal system for developing an ecological civilization. *China Legal Science, 3*, 37.

Wang, X., & Bai, N. (1986). *Poverty of abundance*. Chengdu: Sichuan People's Publishing House.

Wang, J., Cao, G., Cao, D., et al. (2013). National environmental risk prevention and control as well as framework construction for management system [J]. *China Environmental Science, 33*(1), 186–191.

Wang, H., & Gao, S. (2016). The emergence, early period and policies of green development in China: Study into the phased characteristics [J]. *Reform, 3*, 6–26.

Wang, Y., & Huang, B. (2019). China's national park system reform: Review and prospect. *Biodiversity, 27*(2), 117–122.

Wang, J., Liu, N., & Jiang, H. (2014). Environmental planning system innovation under the new environmental protection law. *Environmental Protection, 42*(13), 10–13.

Wang, Y., Sun, H., & Zhao, J. (2012). Policy review and outlook on China's sustainable development since 1992. *Chinese Geographical Science, 22*(1), 1–9.

Wang, J., Wan, J., Wang, Q., et al. (2018). 40 years of reform and opening-up and China's environmental planning. *Chinese Journal of Environmental Management, 10*(6), 5–18.

Wang, M. (2017). Implementing the paris agreement to promote green development [J]. *Special Issue of the Academy of Social Sciences*, (416).

Wang, Y. (2012). Going green is the first step of becoming a great power. In *Caijing annual edition: Forecasts & strategies* (pp. 290–293).

Wang, J., Jiang, H., He, J., et al. (2017). The strategy and tasks of the building of the socialist ecological civilization with Chinese characteristics in a new era. *China Environmental Management, 4*.

Wang, M. (2018, May 23). Development, achievements and experience of Hainan's drive to develop an ecological civilization. *Hainan Daily*.

Wang, T., & Wu, Y. (2016). Research on the integrated development of new-type urbanization and ecological civilization pilot zone in Fujian Province. *Fujian Forum, 10*, 207–213 (Humanities and Social Sciences Edition).

Wang, L. (2000). Research on punitive compensation. *Social Sciences in China, 4*.

Wei, S. (2019). On the value guidance of the concept of green development to the building of an ecological civilization. *Science of Law, 2*, 25.

Wei, H., & Lu, L. (2019). The connotation, characteristics and value of Xi Jinping thought on ecological civilization for the times. *Journal of Xi'an Jiaotong University, 39*(3).

Wei, J., Zhou, L., & Xi, J. (2013). *I would rather have lucid waters and lush mountains than mountains of gold and silver*. [EB/OL]. http://www.youth.cn.

Weimin, Y. (2003). Tasks and direction of China's planning system reform. *Macroeconomic Management, 2003*(4), 4–8.

Weimin, Y. (2010). *The Theory and Practice of Development Planning*. Beijing: Tsinghua University Press.

Wenyuan, N. (2012). Theory and practice of China's sustainable development. *Bulletin of Chinese Academy of Sciences, 27*(3), 280–290.

World Wildlife Fund. (2016). *Living planet report 2016*.

World Bank. https://data.worldbank.org.cn/.

World Bank. (1997). *Five years after Rio: Innovations in environmental policy*. Washington, DC: World Bank.

World Bank. (2003). *World development report 2003*. Beijing: China Financial and Economic Publishing House.

Wu, X. (2012). Ecological Progress and culture and education. *Forward Position, 305*(3), 143–144.

Wu, S., Wang, Y., Yue, Liu, et al. (2018). Firmly establish and fully implement the concept that "lucid waters and lush mountains are invaluable assets" [J]. *Environment and Sustainable Development, 4*, 10–11.

Wu, Y. (2013). On applying the rule-of-law thinking to promote ecological progress. *Theory and Contemporary, 12*.

Wuhan Environmental Protection Bureau. (1991). *Environmental governance in Wuhan* (p. 149). Beijing: China Environmental Science Press.

Xi, J. (2014). *The governance of China* (p. 211). Beijing: Foreign Languages Press.

Xi, J. (2017). *The Governance of China II [M]*. Beijing: Foreign Language Press.

Xi, J. (2017). *Secure a decisive victory in building a moderately prosperous society in all respects and strive for the great success of socialism with Chinese characteristics for a new Era—Report at the 19th national congress of the communist party of China*. Beijing: People's Publishing House.

Xi, J. (2018, May 9). Resolutely win the battle against pollution and elevate ecological progress to a new level. *People's Daily*.

Xi, J. (2019a). Stresses maintaining the strategic resolve of building an ecological civilization, and further fortifying the green ecological barrier in Inner Mongolia when participating in deliberations of the Inner Mongolian delegation. *Guangming Daily*, March 6, 2019.

Xi, J. (2019b). Speech at the National Conference on Eco-environmental Protection. *Qiu Shi*, 2019 (3).

Xi, J. (2019). Deeply understand the new development concepts. *Qiu Shi* (10).
Xi, J. (2006, April 24). Green is gold. *People's Daily* (01).
Xi, J. (2012). Xi Jinping's speech during inspection in Guangdong. *Bulletin of the General Office of the CPC Central Committee* (26).
Xi, J. (2013a, July 21). Congratulations to the 2013 annual meeting of the Guiyang international forum on ecological civilization. *People's Daily* (01).
Xi, J. (2013b, September 7). *Speech at Nazarbayev University in Kazakhstan* [EB/OL]. Xinhua News Agency.
Xi, J. (2013c, April 10). *Speech at the end of the inspection in Hainan* (Document of the CPC Committee of Hainan Province [2013] No. 6).
Xi, J. (2013d, May 24). *Speech at the sixth collective study session of the Politburo of the Chinese Communist Party* [EB/OL]. Xinhua News Agency.
Xi, J. (2013e). Speech on the economic situation in the first quarter at the meeting of the Politburo Standing Committee. *Bulletin of the General Office of the CPC Central Committee* (11).
Xi, J. (2019a). In-depth understanding of new development concepts. *Qiu Shi*, (10).
Xi, J. (2019b). Pushing China's development of an ecological civilization to a new stage. *Qiu Shi*, (3).
Xi, J. (2019). Promoting China's Ecological Civilization Development to a New Level. *Qiu Shi*, (3).
Xi, J. (2015). Zhejiang: China: A New Vision for Development. Hangzhou: Zhejiang People's Publishing House, pp. 125–257.
Xi, J., & Qui, S. (2019). *Promoting the Building of Ecological Civilization to a New Level*, (3).
Xi, J. (2019). Pushing China's development of an ecological civilization to a new stage. *Qiu Shi* (3).
Xi, J. (2016). *Selection of important documents since the 18th national congress of the communist party of China* (Vol. 2, pp. 826). Beijing: Central Party Literature Press.
Xi, J. (2006). *Doing the practical work and walking in the forefront: Thinking on and practice of advancing Zhejiang's new development* (Vol. 192, pp. 194–195). Beijing: CPC Central Party School Press.
Xi, J. (2007). *Zhijiang Xinyu* (Vol. 13, pp. 30). Hangzhou: Zhejiang Publishing United Group, Zhejiang People's Publishing House.
Xi, J. (2013, November 12). Explanatory notes for the "Decision of the CPC central committee on some major issues concerning comprehensively deepening the reform." *People's Daily*.
Xi Jinping's speech at the opening ceremony of the Paris conference on climate change: Working together to build a win–win, fair and reasonable climate change governance mechanism [EB/OL]. Xinhuanet. (Dec 1, 2015). https://www.xinhuanet.com.
Xi Jinping's unforgettable experience in Yan'an, [EB/OL]. Cnwest.com, 2012-11-15. https://news.cnwest.com/content/2012-II/15/content_767695_3.htm.
Xi Jinping on Socialist Ecological Civilization. Beijing: Central Party Literature Press, 2017.
Xiao, X. (2006). *Environment and society: Human perspectives on environmental issues*. Beijing: Higher Education Press.
Xie, Z., et al. (2017). *General report on China's low-carbon development macro-strategy research*. Beijing: People's Publishing House.
Xie, Z., & Wang, Y. (2017). Building a socialist ecological civilization governance system with Chinese characteristics. *China Institutional Reform and Management, 10*, 10–14.
Xie, X., Wu, Y., Feng, Z., et al. (2015). Survey of Chinese resource enterprises' green behaviors. *China Population Resources and Environment* (25), 6.
Xie, Z. (2019). Timeline of environment protection in China in the past 40 years—From waste control to the development of ecological civilization. *Chinese Journal of Environmental Management, 11*(4).
Xing, Y. (2017). A Civilization may thrive if its natural surroundings thrive, and will suffer if its natural surroundings suffer – Overview of CPC Central Committee with Xi Jinping at the core promoting ecological progress since the 18th CPC National Congress. Published online at https://www.qstheory.cn/zoology/2017-06/19/c_1121167567.htm [cited June 19, 2017].

Xinhua News Agency. (2013, May 24). *Xi Jinping's speech at the sixth collective study of the 18th Central Political Bureau.*

Xinhua News Agency. Xi Jinping Attends the National Conference on Eco-environmental Protection and Delivers an Important Speech [EB/OL]. May 19, 2018. http://www.gov.cn/xinwen/2018-05/19/content_5292116.htm.

Xu, H. (2018). A community of shared future for mankind and international law. *Chinese Review of International Law, 5.*

Xu, Z. The Harmony and Integration of Ecological Civilization and the Civilization of the Rule of Law. *Tsinghua Journal of Rule of Law, 22.*

Yang, K. (1989). *On regional development in China.* Beijing: China Ocean Press.

Yang, K. (1993). *Toward spatial integration—Market economy and regional development in China.* Chengdu: Sichuan People's Publishing House.

Yang, Y. (2012). *Development Planning* (pp. 24–25). Beijing: Tsinghua University Press.

Yang, J. (2016). On the layout of law for China's green development. *Law Review, 4,* 160.

Yang, J. (2019, June 20). *China's contribution to the Paris Agreement* [EB/OL]. Xinhuanet. https://www.xinhuanet.com/world/bldh/index.htm.

Yang, J. (Nov 23, 2015). Actively assuming international responsibilities and obligations [N]. *People's Daily*, (6).

Yang, W. (2013, November 23). Establish a comprehensive ecological civilization system. *Guangming Daily* (02)

Yang, H., & Zhang, Y. Preventing environmental risks and advancing the development of ecological civilization [N]. *Chinese Social Sciences Today*, 2019-06-06 (001).

Ye, B. (2017). The legal mechanism for the coordination of national economic and social development planning in the Yangtze River economic zone. *Journal of China University of Political Science and Law, 4,* 14.

Ye, Q. (1992). *Collected works of Ye Qianji [M]*, (pp. 80–81). Beijing: Social Sciences Academic Press.

Ye Q. (2014). Collected works of Ye Qianji [M]. Beijing: Social Sciences Academic Press.

Yi, J. Introduction to promoting the modernization of national governance system and governance capability [EB/OL]. 23 June 2014. http://www.cssn.cn/zt/zt_xkzt/zt_zzxzt/tjgjzltxhzlnlxdh/tjgjzltxdy/201406/t20140623_1222111.shtml.

Yu, M. (1991). *The Philosophy of ecology.* Kunming: Yunnan People's Press.

Yu, M. (2010). *Environmental philosophy: The theoretical basis for ecological civilization.* Beijing: China Environmental Science Press.

Yu, K. (2014). Promoting the modernization of the national governance system and capabilities. *Frontline, 1,* 5–8.

Yu, K. (2015). Scientific outlook on development and ecological civilization. *Marxism & Reality, 4,* 4–5.

Yu, G., & Yu, X. (2013). Network of ecosystem studies and natural ecosystem protection in China. *Bulletin of Chinese Academy of Sciences, 28*(2), 275–283.

Yu, M., -C. (1986). Issues concerning ecological culture. (abstract of speech). *Xue Tan* (12).

Yu, M., -C. (1988). *Ecology and society.* Beijing: Science Press.

Yu, M., -C. (1989). Issues concerning ecological culture. *Studies in Dialectics of Nature* (4).

Yu, M., -C. (1996). *New century of culture—Theoretical explanations of ecological culture.* Beijing: Beijing Forestry University Press.

Yu, M., -C. (2010). *Ecological civilization is a new human civilization.* Beijing: Central Compilation & Translation Press, September, 2010.

Yunnan Provincial Environmental Protection Committee. (1994). *Chronicle of Yunnan: Environmental protection* (Vol. 6, pp. 43–44). Kunming: Yunnan People's Publishing House.

Zeng, J. (2004). *Thoughts of nature: On western ideas of ecological ethics.* Beijing: China Social Sciences Press.

Zhang, X. (2013). *The cycle of civilization: New Era of ecological civilization and rejuvenation of Chinese civilization.* Beijing: China Social Publishing House.

Zhang, Y., & Zhang, Y. (2015). New features and virtual form of information civilization. *Information Technology, 7,* 101.
Zhang, B. *National spatial governance and spatial planning* [EB/OL]. March 19, 2019. http = ://www.cssn.cn/zhcspd/zhcspd_tt/201903/t20190319_4849876.html.
Zhang, X. (2009). The "new economic revolution" behind the financial crisis and China's counter strategy. *Journal of Chinese Academy of Governance,* (5).
Zhang, H. (1992). *Productive practices and ecological civilization: Philosophical thinking of environmental issues.* Beijing: China Agriculture Press. preface, 4.
Zhang, P., Yang, Z. (2015). A brief analysis of the characteristics of environmental mass events in China in the past decade. *Journal of China University of Geosciences (Social Sciences Edition)* (2).
Zhang, W., & Huang, C. (2010, September 10). The ecological civilization theory is derived from the scientific development theory. *China Environment News,* (02).
Zhang, X. (2018). The dilemma of and way out for parallel hearing of environmental civil public and private interest litigation cases. *China Legal Science, 5.*
Zhe, X. (2004). Making Ecological Culture to Take Root in Society. Zhejiang Daily, 2004-05-08 (01).
Zheng, T. (2018). China is leading the world in developing ecological civilization: An Interview with John B. Cobb, Jr., a member of the American Academy of Arts and Sciences. *Green China,* (8A), 34–37.
Zhong, Z. (2019, March 7). Fujian promotes the development of China's first national ecological civilization pilot zone. *People's Daily.*
Zhong, Z. (2018, August 13). Ecological Fujian: A green response. *People's Daily.*
Zhou, H. (2013). On several relations of ecological civilization. *China Reform, 3,* 90–99.
Zhou, H. (2019). Making overall arrangements, avoiding mistakes in ecological progress. *The Economic Daily,* April 15, 2019.
Zhou, Q. (2016, December 18). Continuing to prioritizing ecological conservation and green development, and serving and safeguarding the development of the Yangtze River economic belt with judicial functions. *People's Court Daily,* 2016-12-18.
Zhu, X. (2016). *Research on the comprehensive prevention of and legal remedy mechanism for ecological damage* (p. 85). Beijing: Law Press.
Zhu, D. (2013, December 21). In-depth analysis of the development of an ecological civilization system. *Xinmin Evening News* (A04).
Zhu, L., Sun, Q., & Ju, C. (2014). Flaws on management policies for improving living environments in rural areas and countermeasures. *Journal of Ecology and Rural Environment, 30*(6), 811–815.